Next-Generation Cognitive Radar Systems

Other volumes in this series:

Volume 1 **Optimised Radar Processors** A. Farina (Editor)
Volume 3 **Weibull Radar Clutter** M. Sekine and Y. Mao
Volume 4 **Advanced Radar Techniques and Systems** G. Galati (Editor)
Volume 7 **Ultra-Wideband Radar Measurements: Analysis and processing** L. Yu. Astanin and A.A. Kostylev
Volume 8 **Aviation Weather Surveillance Systems: Advanced radar and surface sensors for flight safety and air traffic management** P.R. Mahapatra
Volume 10 **Radar Techniques Using Array Antennas** W. Wirth
Volume 11 **Air and Spaceborne Radar Systems: An introduction** P. Lacomme (Editor)
Volume 13 **Introduction to RF Stealth** D. Lynch
Volume 14 **Applications of Space-Time Adaptive Processing** R. Klemm (Editor)
Volume 15 **Ground Penetrating Radar, 2nd Edition** D. Daniels
Volume 16 **Target Detection by Marine Radar** J. Briggs
Volume 17 **Strapdown Inertial Navigation Technology, 2nd Edition** D. Titterton and J. Weston
Volume 18 **Introduction to Radar Target Recognition** P. Tait
Volume 19 **Radar Imaging and Holography** A. Pasmurov and S. Zinovjev
Volume 20 **Sea Clutter: Scattering, the K distribution and radar performance** K. Ward, R. Tough and S. Watts
Volume 21 **Principles of Space-Time Adaptive Processing, 3rd Edition** R. Klemm
Volume 22 **Waveform Design and Diversity for Advanced Radar Systems** F. Gini, A. De Maio and L.K. Patton
Volume 23 **Tracking Filter Engineering: The Gauss–Newton and Polynomial Filters** N. Morrison
Volume 25 **Sea Clutter: Scattering, the K distribution and radar performance, 2nd Edition** K. Ward, R. Tough and S. Watts
Volume 33 **Radar Automatic Target Recognition (ATR) and Non-Cooperative Target Recognition** D. Blacknell and H. Griffiths (Editors)
Volume 26 **Radar Techniques Using Array Antennas, 2nd Edition** W. Wirth
Volume 101 **Introduction to Airborne Radar, 2nd Edition** G.W. Stimson
Volume 530 **Radar Sea Clutter: Modelling and target detection** Luke Rosenburg and Simon Watts
Volume 534 **New Methodologies for Understanding Radar Data** Amit Kumar Mishra and Stefan Brüggenwirth
Volume 537 **Ocean Remote Sensing Technologies: High frequency, marine and GNSS-based radar** Weimin Huang and Eric W. Gill (Editors)
Volume 550 **Fundamentals of Inertial Navigation Systems and Aiding** M. Braasch
Volume 551 **Theory and Methods for Distributed Data Fusion Applications** F. Govaers
Volume 553 **Modern Radar for Automotive Applications** Z. Peng, C. Li and F. Uysal (Editors)

Next-Generation Cognitive Radar Systems

Edited by
Kumar Vijay Mishra, Bhavani Shankar M.R. and
Muralidhar Rangaswamy

The Institution of Engineering and Technology

Published by SciTech Publishing, an imprint of The Institution of Engineering and Technology, London, United Kingdom

The Institution of Engineering and Technology is registered as a Charity in England & Wales (no. 211014) and Scotland (no. SC038698).

First published 2023

The Institution of Engineering and Technology
Futures Place
Six Hills Way, Stevenage
Hertfordshire, SG1 2AU, United Kingdom

www.theiet.org

British Library Cataloguing in Publication Data
A catalogue record for this product is available from the British Library

ISBN 978-1-83953-474-4 (hardback)
ISBN 978-1-83953-475-1 (PDF)

Typeset in India by MPS Limited

Credit for cover Image: Airborne Interception Radar based on Active Electronically Scanned Array (AESA) technology designed and developed by Electronics and Radar Development Establishment (LRDE) of Defence Research and Development Organisation (DRDO) under Ministry of Defence, Govt. of India.
Photo Credit: LRDE
Supplied by editors

"Dedicated, with supreme reverence and humility, to Shri Ganesh – the remover of all obstacles – and Shri Hanuman – the eradicator of all troubles.
To my Mom Shraddha Mishra and the memory of my Dad Shyam Bihari Mishra.
To my brothers Kumar Digvijay Mishra and Kumar Jay Mishra."

– K.V.M.

"To my wife and kids for their support and to my parents for their blessings."

– M.R.B.S.

"Dedicated to my loving family for their outstanding support and to my beloved parents for their blessings."

– M.R.

Contents

About the editors

Kumar Vijay Mishra is a senior fellow with the United States DEVCOM Army Research Laboratory (ARL), Adelphi, USA. His research interests include radar systems, signal processing, remote sensing, and electromagnetics. He is the recipient of several prestigious fellowships and awards including the US National Academies ARL Harry Diamond Distinguished Fellowship, Viterbi Fellowship, and IET Premium Award. He is chair (2023–2026) of the International Union of Radio Science (URSI) Commission C.

Bhavani Shankar M.R. is currently assistant professor at the Interdisciplinary Centre for Security, Reliability and Trust at The University of Luxembourg. His research interests include design and optimization of MIMO communication systems, automotive radar and array processing, polynomial signal processing, and satellite communication systems. He was a co-recipient of the 2014 Distinguished Contributions to Satellite Communications Award, from the Satellite and Space Communications Technical Committee of the IEEE Communications Society.

Muralidhar Rangaswamy is the technical lead for radar sensing at the Sensors Directorate of the Air Force Research Laboratory, USA. His research interests include radar signal processing and statistical communication theory. He has co-authored more than 180 refereed journal and conference papers. Additionally, he is a contributor to eight books and is a co-inventor on three US patents. He has received numerous IEEE, Air Force, and NATO awards.

List of editors

Kumar Vijay Mishra, United States DEVCOM Army Research Laboratory, Adelphi, MD, USA
Bhavani Shankar M.R., Interdisciplinary Centre for Security, Reliability and Trust (SnT), University of Luxembourg, Luxembourg
Muralidhar Rangaswamy, United States Air Force Research Laboratory, Wright-Patterson Air Force Base, OH, USA

List of contributors

Ravi Adve, University of Toronto
Khaled AlHujaili, Taibah University
Bhashyam Balaji, Defence Research and Development, Canada
Kristine Bell, Metron, Inc.
Jameson S. Bergin, Information Systems Laboratories, Inc.
Arindam Bose, University of Illinois at Chicago
Aaron Brandewie, The Ohio State University
Stefan Bruggenwirth, Fraunhofer Institute for High Frequency Physics and Radar Techniques
Alexander Charlish, Fraunhofer FKIE
Jonas Myhre Christiansen, Norwegian Defence Research Establishment
Alfonso Farina, Rome, Italy
Marco Frasca, MBDA Italia S.p.A.
Xiangyu Gao, University of Washington at Seattle
Baocheng Geng, University of Alabama at Birmingham
Sandeep Gogineni, Information Systems Laboratories, Inc.
Hugh Griffiths, University College London
Joseph R. Guerci, Information Systems Laboratories, Inc.
Ali C. Gurbuz, Mississippi State University
Sevgi Z. Gurbuz, The University of Alabama at Tuscaloosa
Folker Hoffmann, Fraunhofer FKIE
Sian Jin, University of Washington at Seattle
Joel T. Johnson, The Ohio State University
Bosung Kang, University of Dayton Research Institute
Christopher Kreucher, Centauri, Ann Arbor
Vikram Krishnamurthy, Cornell University
Jian Li, University of Florida
Kumar Vijay Mishra, United States DEVCOM Army Research Laboratory
Vishal Monga, Pennsylvania State University
Hoan K. Nguyen, Information Systems Laboratories, Inc.
Roland Oechslin, Armasuisse, Switzerland
Björn Ottersten, University of Luxembourg
Daniel P. Palomar, Hong Kong University of Science and Technology
Kunal Pattanayak, Cornell University
Muralidhar Rangaswamy, United States Air Force Research Laboratory
Taylor Reininger, Johns Hopkins University

Matthew Ritchie, University College London
Sumit Roy, University of Washington at Seattle
Mathini Sellathurai, Herriot Watt University
Pawan Setlur, United States Air Force, Wright-Patterson Air Force Base
Mahdi Shaghaghi, University of Toronto
Bhavani Shankar M.R., University of Luxembourg
George Shehata, University of Toronto
Chenguang Shi, Nanjing University of Aeronautics and Astronautics
Graeme Smith, Johns Hopkins University
Mojtaba Soltanalian, University of Illinois at Chicago
Pramod K. Varshney, Syracuse University
Fei Wang, Nanjing University of Aeronautics and Astronautics
Linlong Wu, University of Luxembourg
Jianjiang Zhou, Nanjing University of Aeronautics and Astronautics

List of reviewers

Ravi Adve, University of Toronto
Kristine Bell, Metron, Inc.
Alexander Charlish, Fraunhofer FKIE
Sandeep Gogineni, Information Systems Laboratories, Inc.
Hugh Griffiths, University College London
Joseph R. Guerci, Information Systems Laboratories, Inc.
Sevgi Z. Gurbuz, The University of Alabama, Tuscaloosa
Bosung Kang, University of Dayton Research Institute
Vikram Krishnamurthy, Cornell University
Kumar Vijay Mishra, United States DEVCOM Army Research Laboratory
Vishal Monga, Pennsylvania State University
Roland Oechslin, Armasuisse, Switzerland
Sumit Roy, University of Washington, Seattle
Mathini Sellathurai, Herriot Watt University
Pawan Setlur, United States Air Force, Wright-Patterson Air Force Base
Bhavani Shankar M.R., University of Luxembourg
Mojtaba Soltanalian, University of Illinois, Chicago
Pramod K. Varshney, Syracuse University
Linlong Wu, University of Luxembourg

Preface

The title of this book was initially "Beyond Cognitive Radar." But, on advice and further discussions, we changed it to "Next-Generation Cognitive Radar Systems." This title, in essence, captures the ongoing frenetic research on various theoretical questions and enabling technologies for cognitive radars. During the past two decades, introducing cognition in radar has heralded a new era of radar system design and engineering. These systems offer advanced sensing capabilities by simultaneously optimizing both transmit and receive processing in response to the changes in the target environment. Research on cognitive radars have revealed unique opportunities in a variety of civilian and defence applications by affording greater control of transmitters and higher adaptability of receivers than their non-cognitive counterparts.

At the heart of cognitive radar lies the key question: if a radar is embodied with a reasonable cognitive model, would it interact with targets and other entities with cognition like that of a human. This requires understanding the cognitive model of humans themselves, which is an active neurobiological research area of its own. In general, Benjamin Bloom's taxonomy proposed in 1950s and its variants are considered benchmarks for classifying various human cognitive abilities. Cognition in wireless systems itself falls under the broad umbrella if cognitive cyberphysical systems, wherein a machine is equipped with a human-like cognitive capabilities that provides the basis for human–machine interactions. While it is difficult to trace the origin of the term "cognitive systems," its closest counterpart "cybernetics" was coined by Norbert Weiner in 1948. Hollnagel and Woods were the first to describe cognitive systems in detail in their 1983 paper "*Cognitive Systems Engineering: New Wine in New Bottles*" published in the journal *International Journal of Man-Machine Studies*. The term quickly gained currency eventually finding its way to *cognitive radio* in the 1999 paper "*Cognitive Radio: Making Software Radios More Personal*" by Mitola and Maguire that appeared in IEEE personal communications journal. Thereafter, initial developments in cognitive radars followed the ideas from cognitive radio literature of early 2000s.

However, the obviously different application-specific requirements and design options led to the concept of cognitive radars as an independent and prolific research topic in mid-2000s. Simon Haykin's landmark 2006 paper "*Cognitive Radar: A Way of the Future*" published in the *IEEE Signal Processing Magazine*, along with Joseph Guerci's 2010 book "*Cognitive Radar: The Knowledge-aided Fully Adaptive Approach*" (second edition published in 2020), laid the conceptual groundwork for many preliminary applications. Today, with the proliferation of sensing to perform various tasks in many novel applications such as automotive radar, the target environments are no longer as benign as those considered in the early cognitive radars. Further, conventional cognitive radars face several challenges in not only making intelligent abstraction of received data in real-time but also adapting sensing techniques to a highly dynamic and complex environment. To address challenges beyond the conventional cognitive radars, there has been a surge of interest to enhance, enable, and engineer novel processing methods to achieve more complex levels of cognition.

To equip radar practitioners with state-of-the-art tools for the next generation of highly sophisticated cognitive radar systems, we are delighted to edit this book published by the Institution of Engineering and Technology (IET) falls under the prestigious SciTech/IET series on Radar, Electromagnetics & Signal Processing Technologies. This book is aimed at bringing together contributions from leading well-qualified researchers who are engaged in the forefront of research and development of next-generation cognitive abilities in radar engineering.

There already exist several excellent books on cognitive radars (e.g., by Simon Haykin and Joseph Guerci), which provide insights on realizing optimized and adaptive behavior within the realm of classical cognition in radars. Given the significant developments and applications that have emerged in this area during the last 5 years—including the use of new tools/theories such as deep learning, sparse reconstruction, non-convex optimization, game theory, stochastic control, and quantum theory—the concept of cognition itself has been refined such that it now transcends Bloom's classical cognition levels applied to radars so far. Hence, there is a need to put together the most significant and successful cognitive radars concepts in a tutorial fashion from the experts themselves who developed those results. The book goes beyond a high-level understanding of new concepts by including detailed empathetical treatment of each new cognitive method. This will aid researchers to develop a deep understanding of novel cognitive radar concepts and support relevant graduate courses. However, we do not foresee that the existing cognitive radar processing will disappear altogether. Rather, this book provides the mathematical machinery with applications where the conventional processing is inappropriate, unreliable, or inaccurate, and where we indeed need to look beyond the existing cognitive radar frameworks.

The book complements the existing cognitive radar literature by assembling in-depth theory in a single reference, which also covers latest efforts in hardware prototyping. Some key concepts included beyond classical cognition are metacognition, inverse cognition, meta-level and adversarial tracking, and quantum sensing. Some chapters dwell upon the applications of emerging processing paradigms such as deep learning, game theory, stochastic control, sparse reconstruction, and mathematical optimization. Finally, the book also highlights several future research directions.

Our intention is that either chapter can be read independently, but that they also complement each other by examining emerging challenges for cognitive radar systems. Further, each chapter features recent advances in the theory and applications of advanced cognitive radar tools to address these challenges. While the chapters are sequenced to achieve these goals in a lucid mathematical manner, we impose no requirement that the chapters are read in a specific order. Yet if the reader finds it suitable to read the chapters in the order they appear, we will feel the book has achieved its purpose.

We thank all contributing authors for submitting their high-quality contributions. We sincerely acknowledge the support and help from all the reviewers for their timely and comprehensive evaluations of the manuscripts that improved the quality of this book.

Finally, we are grateful to the IET Press Editorial Board and the staff members Nicki Dennis, Olivia Wilkins, and Sarah Lynch for their support, feedback, and guidance.

Kumar Vijay Mishra
Adelphi, MD, USA

Bhavani Shankar M.R.
University of Luxembourg

Muralidhar Rangaswamy
Wright-Patterson Air Force Base, OH, USA

Acknowledgments

K.V.M. acknowledges support from the National Academies of Sciences, Engineering, and Medicine via Army Research Laboratory Harry Diamond Distinguished Fellowship.

M.R.B.S. acknowledges support from the ERC AGNOSTIC under Grant EC/H2020/ERC2016ADG/742648 and in part by FNR CORE SPRINGER under grant C18/IS/12734677.

M.R. was supported by the Air Force Office of Scientific Research under Project 20RYCOR051 and under Project 20RYCOR052.

Part I

Fundamentals

Chapter 1
Beyond cognitive radar

Kumar Vijay Mishra[1] Bhavani Shankar M.R.[2] and Muralidhar Rangaswamy[3]

In this chapter, we describe the essential features and concepts of emerging and futuristic cognitive radar systems. Since the introduction of cognitive radars in 2000s, the signal processing landscape has undergone major transformations. We describe cognition beyond its classical definition and focus on the developments to enable these new features during the past few years. We then describe the structure of the book.

Cognitive radar gained significant attention in the last decade because of its ability to adapt both the transmitter and the receiver to changes in the environment and provide flexibility for different scenarios as compared to conventional radar systems [1–3]. Applications considered for radar cognition included waveform design [4–7], target detection and tracking [8,9], and spectrum sensing/sharing [10–13]. Cognitive radar design requires reconfigurable circuitry for many subsystems such as power amplifiers, waveform generator, and antenna arrays [14].

Radar cognition was first introduced by Simon Haykin [1] and later expanded by Joseph Guerci [2]. Here, a cognitive radar was presented as a dynamic closed loop system employing three key steps: Sense, Learn, and Adapt. These three stages form a cognitive cycle or a perception–action cycle, a key feature in any cognitive system (Figure 1.1). Based on the obtained awareness, operational parameters of transmitters and receivers in each subsystem are adjusted to optimize their performance. The concept of cognitive radar has its origins in neurobiological systems. In general, studies devoted to cognitive radars involve: (a) performance indicators of cognitive state, (b) computational cognitive modeling, (c) automated knowledge capture, (d) augmented cognition, and (e) product applications of cognitive radars.

1.1 Aspects of cognition

After nearly two decades of niche applications and theory, cognitive radar technology is realistically close to field deployment more than before. Research in cognitive

[1]United States DEVCOM Army Research Laboratory, Adelphi, MD, USA
[2]Interdisciplinary Centre for Security, Reliability and Trust, University of Luxembourg, Luxembourg
[3]Air Force Research Laboratory, Wright Patterson Air Force Base, Dayton, OH, USA

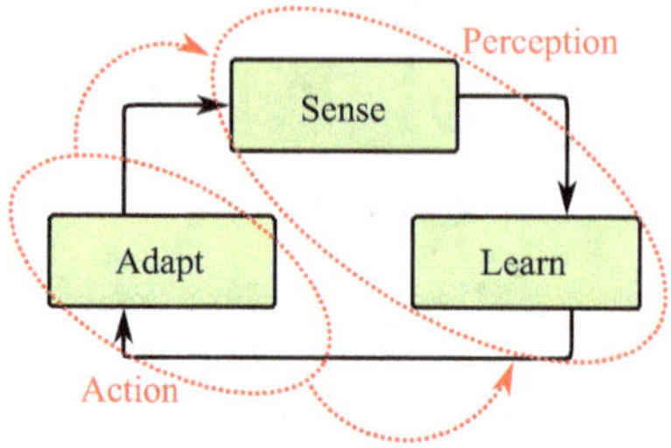

Figure 1.1 Classical cognitive radar cycle

Table 1.1 Adapting Bloom's taxonomy [15] to cognition radar [16]

Problem scenario	**Architecture/strategy**
Level 1	Prior measurements or databases in radar
Level 2	Learning-based algorithms for adaptive radar
Level 3	Knowledge-aided processing for a dynamic scene
Levels 5 and 6	Higher-order cognition to interconnect, decide, and synthesize

radars has become very popular during the past 10 years. This timeframe witnessed a dramatic progress in understanding of cognitive radar beyond its classical cognitive cycle of sense–learn–adapt. In general, cognitive abilities of radar are benchmarked analogously to human cognition. In 1956, Bloom *et al.* [15] described six levels of cognitive abilities. This was recently re-stated in the context of cognitive radars in [16]. Table 1.1 summarizes these features.

In recent years, there have been attempts to ascribe additional features of neurobiological cognition to cognitive radars. For instance, the concept of *metacognition* from neurobiology denotes the process of knowing and controlling cognition. A metacognitive radar is at the confluence of cognition and learning driven by the challenges in implementing advanced cognitive features (cf. Chapter 19). In general, metacognitive radars coordinate between multiple cognitive cycles to ensure they do not unbalance—*hyper-cognition*—the cognitive operations [17] and also to select the best possible cognitive solution [18].

Recent cognition literature [19,20] envisages situations when the target itself may become cognitive. In this *inverse cognition* scenario, a target may be equipped with cognitive abilities that predict the actions of a cognitive radar trying to detect the target and guard against it (cf. Chapter 2). When the target is equipped with such cognitive abilities, the cognitive radar must simultaneously consider two actions: mask its cognitive abilities and continue to estimate the target parameters. The former objective has been considered metacognition while the latter may be more appropriately termed *inverse–inverse cognition*. Finally, *super-cognition* is ascribed to enabling cognition in legacy radars. Within Bloom's taxonomy, often *ultra-cognition* is a term used for building cognitive strategy databases.

1.2 Key technology enablers

Cognitive radars reveal unique opportunities in remote sensing by encouraging greater control of transmitters and higher adaptability of receivers than their non-cognitive counterparts. The last few years have also witnessed the growing use of various algorithms to accomplish different stages of sense–learn–adapt cycle of cognitive radars such as building awareness, waveform optimization at transmitter, receive filter optimization, interference management, resource allocation, transmitter/receiver selection, target detection using deep learning, and waveform classification. Here, we list some of these broad approaches that are expected to witness growth in cognitive radar domain in the near future.

1.2.1 Convex and non-convex optimization

Since a cognitive radar employs both transmit and receive functions to enhance channel/target estimation and the radar optimizes a spatio-temporal transmit and receive strategy. This optimization process involves solving mathematical optimization problems, which yield optimum transmit and receive functions that maximize a performance metric such as output signal-to-interference ratio (SINR). In general, additional waveform or transceiver constraints are also imposed in such formulations. Earlier approaches focused on leveraging developments in convex optimization (cf. Chapter 5) but recent approaches have also focused on non-convex objective functions and approaches, including the use of sparse representation (cf. Chapter 8) and low-bit sensing (cf. Chapter 13).

1.2.2 Control-theoretic tools

A wide variety of stochastic optimization problems in cognitive radar involve the use of control (cf. Chapter 10) and game-theoretic tools (cf. Chapter 11). These stochastic optimization communities have conducted research covering techniques and applications such as decision trees, stochastic search, optimal stopping, optimal control, (partially observable) Markov decision processes (MDPs/POMDPs), approximate dynamic programming, reinforcement learning, model predictive control, stochastic programming, ranking and selection, and multiarmed bandit problems. Similarly, inverse cognition [20] often involves the development of appropriate stochastic filters for a target to recognize the cognition on the radar [21,22].

1.2.3 Learning techniques

Some radar applications, such as space-time adaptive processing, are characterized by big data. Imparting cognition to these systems requires software-defined intelligent decision-making in high-noise scenarios based on feature extraction from received data. Several cognitive radar functions, such as antenna, spectrum or beam selection, typically also involve high-latency combinatorial optimization—a task machine/deep learning networks are known to accomplish with low latency [23]. In particular, reinforcement learning allows the radar to learn optimal policies, agile spectrum use or efficient use of degrees-of-freedom (DoFs). Radar performance such as successful detection, high-quality estimation, tracking or maximizing SINR could be written as a reward that is maximized by using a reinforcement learning, see, e.g., [24] and references therein.

Learning techniques are also useful for the cognitive recognition of unknown waveforms in electronic warfare scenarios, achieving sparse representation for various radar data sources, and design of optimal waveforms for cognitive target tracking and detection (cf. Chapter 12). Certain deep learning techniques interpret data by exploiting temporal correlation that radar-received data often exhibits. There are opportunities to apply transfer learning when sufficient training data is unavailable for the newly deployed cognitive radar [25].

1.2.4 Operationalization

Despite significant interest in cognitive radar theory and applications, substantial challenges remain in designing operational systems. Although cognitive radar implementations are guided by their specific applications, such as spectrum sharing, cognitive tracking, enhanced target localization and efficient scene classification, most of them usually comprise on-the-fly reconfigurability of hardware and intelligent software-defined functions. Due to the centrality of transmitter adaptability in cognitive radar, agile circuitry is required to shift through carefully designed and highly parameterized cognitive radar waveforms. In some cases, it is essential that the individual radio-frequency chains leading up to the antenna array elements be available for selective use. Some current prototypes employ novel low-complexity and dynamic feedback mechanisms between the transmitter and the receiver to facilitate cognition, especially in applications such as detection of dynamic interference. At the signal processing and algorithmic level, cognitive radar systems have been recently developed to exploit techniques such as learning networks (cf. Chapter 12) and low-complexity algorithms. A major implementation challenge remaining is the determination of cognitive radar performance and design criteria such as dynamic range, sampling rates, array designs, bandwidth, latency, and tuning speed that would optimally trade-off with signal-to-noise-ratio and subsequent probability of detection. Some current state-of-the-art implementations [26,27] demonstrate the use of reconfigurable circuitry in cognitive radars at a testbed level, see Chapters 16 and 17 for more details.

1.3 Organization of the book

The book is organized as follows.

- **Part I: Fundamentals.** Chapters 1–5 lay out the fundamentals for the book. These include emerging new paradigms of cognitive radar such as inverse cognition and metacognition; human–machine collaboration paradigm for cognitive sensing, cognitive radar channel estimation, and mathematical concepts to deal with the optimization of cognitive radars.

- **Part II: Design methodologies**. Chapters 6–12 deal with various design methodologies. These chapters present the audience with a selection of topics from waveform optimization, enabling transmit adaptivity, sparse techniques, resource allocation strategies for tracking and classification, stochastic control paradigms, game theoretic approaches, and neural-network-based system concepts.
- **Part III: Beyond cognition—from theory to practice**. Here, the book examines interesting applications, and current strides towards implementation and offers a peek into the emerging trends. Chapters 13–19 deal with concepts of one-bit enabled cognitive radar; spectrum sharing methodologies; use of cognitive radars in the automotive sector; architectures and experiments; the futuristic cognitive quantum radar; and developments in radar metacognition.

We now summarize the contributions of each chapter:

Chapter 2: Adversarial radar inference: inverse tracking, identifying cognition, and designing smart interference: This chapter introduces inverse cognition, which is an adversarial signal processing problem where a target attempts to infer actions of a hostile cognitive radar through observation and probing of the channel state. The goal is to avoid tracking and detection through advanced inference techniques.

Chapter 3: Information integration from human and sensing data for cognitive radar: This chapter discusses mathematical aspects of human–machine networks under cognitive biases. Since the human behavior in decision making is quite complex and uncertain, the cognitive radar sense–learn–adapt cycle is affected by the involvement of human input. The analysis of human–machine collaborative decision making is an important cognitive radar research area, where the goal is to optimize the system performance based on appropriate modeling of the human behavior.

Chapter 4: Channel estimation for cognitive fully adaptive radar: Future radar will be endowed with large volumes of environmental data. In many radio waves transmission scenarios, good physical models for wave propagation in the environment may exist including estimates of reflection coefficients of materials or accurate topographic maps of the environment. In fact, various levels of knowledge of this information can be used to estimate radar and radio channels using methods including ray-tracing. This chapter introduces recent research in establishing new physics-infused (model-based) data-driven approaches to optimize and adapt cognitive radar channel estimation.

Chapter 5: Convex optimization for cognitive radar: There has been significant developments in convex and non-convex optimization techniques during the past decade. The cognitive radar community has benefitted from these advances in the development of efficient processing algorithms and cognitive strategies. This chapter reviews the relevant optimization concepts and illustrates their applications in advanced cognitive radar processing.

Chapter 6: Cognition-enabled waveform design for ambiguity function shaping: Transmit waveform forms a key ingredient in a radar that affects significantly on the quality of the backscatter echoes, from which the environmental parameters are inferred by estimation and learning techniques. Further, the waveform design based on the extracted information will further strengthen the radar performance in the next illumination. This chapter focuses on the latter aspect to illustrate how to design waveforms using elaborate optimization techniques under specific circumstances by exploiting the prior knowledge obtained by a cognitive radar.

Chapter 7: Training-based adaptive transmit–receive beamforming for MIMO radars: Transmit adaptivity, along with that of the receiver, can complement each other to truly maximize the SINR. Central to this adaptivity is the availability of information and this chapter investigates how a cognitive radar might acquire the information needed to implement transmit adaptivity. It develops a training model to obtain the needed second-order statistics and illustrate how transmit adaptivity differs from that at the receiver during implementation.

Chapter 8: Random projections and sparse techniques in radar: In the last decade, new approaches to radar signal processing have been introduced that allow the radar to perform signal detection and parameter estimation from much fewer measurements than that required by Nyquist sampling, be it temporal, spectral or spatial domains. These systems exploit the fact that very few targets occupy the radar environment, hence facilitating the use of sparse reconstruction methods in signal recovery. This chapter investigates the applicability, shortcomings, and future directions of these techniques for target detection and estimation in cognitive radars.

Chapter 9: Fully adaptive radar resource allocation for tracking and classification: Once the target has been detected, tracking its movement is of immense interest. Although early cognitive radar works identified tracking as a common application, its theoretical development in the context of the use of prior information (Bayesian tracking) that results in more efficient transmit resource allocation is a recent idea. This chapter details the modeling and resource allocation in a cognitive target tracking problem.

Chapter 10: Stochastic control for cognitive radar: While control theory lies at the heart of many radar tracking algorithms, recent developments in stochastic control have been imported into cognitive radar design and processing, such as knowledge exploitation, perception, action, memory, intelligence, and attention. This chapter describes the cognitive characteristic of anticipation in enhancing the cognitive radar performance.

Chapter 11: Applications of game theory in cognitive radar: The increasing demand for scarce spectrum among multiple entities, particularly radar and wireless communications, motivates spectrum sharing paradigms. Several tools are used for enabling sharing and this chapter addresses the problem of power allocation for the cognitive multistatic radar system in a spectral coexistence scenario through the game theoretic formulation.

Chapter 12: The role of neural networks in cognitive radar: Cognition in big data radar systems requires software-defined intelligent decision-making in dynamic high-noise scenarios based on feature extraction from raw and processed target echoes. This chapter focuses on cognitive radar target classification, which is a task machine/deep learning networks are known to accomplish with low latency.

Chapter 13: One-bit cognitive radar: High-resolution sampling with conventional analog-to-digital-converters (ADCs) can be very costly and energy-consuming for many modern applications. This is further accentuated by increased demands in sensing and radar signal processing. This chapter explores the low-resolution 1-bit paradigm for radar functionalities and evaluates its performance in comparison to existing methods.

Chapter 14: Cognitive radar and spectrum sharing: Modern radar and communications systems are increasingly characterized by their ability to jointly access, operate, and manage a common spectrum. This spectrum-sharing paradigm offers efficient use of limited spectrum, low cost, compact size, safe operations, and improved performance. The management of dynamic spectrum requirements is essential for a smooth operation cognitive radar in interference-ridden frequencies. The chapter summarizes various strategies to achieve the same.

Chapter 15: Cognition in automotive radars: Automotive industry is an emerging market where the use of radar is becoming commonplace. Further, the sensing problems encountered there in motivate the need for cognition. This chapter brings out the need for greater operational intelligence under dense vehicular radar cases. Emerging themes involving intelligent signal processing and machine intelligence are introduced, and their impact on radar imaging and object recognition is explored.

Chapter 16: A canonical cognitive radar architecture: Various cognitive radar definitions or architectures have been described in recent years. The impetuses for cognitive radar also vary from advanced military applications in contested environments, civilian applications in highly congested electromagnetic spectrum operations, to advanced autonomous vehicle applications. This chapter provides a generalized canonical cognitive radar architecture that can accommodate all the known cognitive radar elements currently described, and how it can be implemented using existing and emerging embedded computing architectures.

Chapter 17: Advances in cognitive radar experiments: Although cognitive radar implementations are guided by their specific applications, most of them usually comprise on-the-fly reconfigurability of hardware and intelligent software-defined functions. Due to the centrality of transmitter adaptability in cognitive radar, agile circuitry is required to shift through carefully designed and highly parameterized cognitive radar waveforms. This chapter surveys emerging requirements, design recommendations, and recent hardware prototypes toward realizing the promise of cognitive radars.

Chapter 18: Quantum radar and cognition: looking for a potential cross fertilization: Recent years have witnessed several efforts toward the realization of a reasonably close equivalent of quantum radars. The purpose of this chapter has been to capture the relations between the concepts of cognition and quantum physics, leading to potential hybridization between cognitive radar and quantum radar.

Chapter 19: Metacognitive radar: Similar to the origin of the neurobiological concept of cognition, metacognition also originates from neurobiological research on problem-solving and learning. Broadly defined as the process of learning to learn, metacognition improves the application of knowledge in domains beyond the immediate context in which it was learned. This supplement describes basic features of a metacognitive radar and then illustrates its application with some examples such as antenna selection and resource sharing between radar and communications.

Epilogue: The book concludes with a discussion on the future outlook of the subject.

References

[1] Haykin S. Cognitive radar: A way of the future. *IEEE Signal Processing Magazine*. 2006;23(1):30–40.
[2] Guerci JR. Cognitive radar: A knowledge-aided fully adaptive approach. In: *IEEE Radar Conference*; 2010. p. 1365–1370.
[3] Smith GE, Cammenga Z, Mitchell A, *et al*. Experiments with cognitive radar. *IEEE Aerospace and Electronic Systems Magazine*. 2016;31(12):34–46.
[4] Chen P and Wu L. Waveform design for multiple extended targets in temporally correlated cognitive radar system. *IET Radar, Sonar & Navigation*. 2016;10(2):398–410.
[5] Kilani MB, Nijsure Y, Gagnon G, *et al*. Cognitive waveform and receiver selection mechanism for multistatic radar. *IET Radar, Sonar & Navigation*. 2016;10(2):417–425.
[6] Mishra KV and Eldar YC. Performance of time delay estimation in a cognitive radar. In: *IEEE International Conference on Acoustics, Speech and Signal Processing*; 2017. p. 3141–3145.
[7] Mishra KV, Eldar YC, Shoshan E, *et al*. A Cognitive Sub-Nyquist MIMO Radar Prototype. arXiv preprint arXiv:180709126. 2018.
[8] Bell KL, Baker CJ, Smith GE, *et al*. Cognitive radar framework for target detection and tracking. *IEEE Journal of Selected Topics in Signal Processing*. 2015;9(8):1427–1439.
[9] Goodman NA, Venkata PR, and Neifeld MA. Adaptive waveform design and sequential hypothesis testing for target recognition with active sensors. *IEEE Journal of Selected Topics in Signal Processing*. 2007;1(1):105–113.
[10] Stinco P, Greco MS, and Gini F. Spectrum sensing and sharing for cognitive radars. *IET Radar, Sonar & Navigation*. 2016;10(3):595–602.
[11] Cohen D, Mishra KV, and Eldar YC. Spectrum sharing radar: Coexistence via Xampling. *IEEE Transactions on Aerospace and Electronic Systems*. 2018 3;29:1279–1296.
[12] Mishra KV and Eldar YC. Sub-Nyquist Radar: Principles and Prototypes. arXiv preprint arXiv:180301819. 2018.
[13] Na S, Mishra KV, Liu Y, *et al*. TenDSuR: Tensor-based 3D sub-Nyquist radar. *IEEE Signal Processing Letters*. 2019;26(2):237–241.
[14] Baylis C, Fellows M, Cohen L, *et al*. Solving the spectrum crisis: Intelligent, reconfigurable microwave transmitter amplifiers for cognitive radar. *IEEE Microwave Magazine*. 2014;15(5):94–107.
[15] Bloom BS, Englehart M, Furst E, *et al*. Taxonomy of educational objectives: The classification of educational goals. In *Handbook 1: Cognitive Domain*. Longmans; 1956.

[16] Gurbuz SZ, Griffiths HD, Charlish A, *et al.* An overview of cognitive radar: Past, present, and future. *IEEE Aerospace and Electronic Systems Magazine*. 2019;34(12):6–18.

[17] Greenspan M. Potential pitfalls of cognitive radars. In: *2014 IEEE Radar Conference*. IEEE; 2014. p. 1288–1290.

[18] Mishra KV, Shankar MB, and Ottersten B. Toward metacognitive radars: Concept and applications. In: *2020 IEEE International Radar Conference (RADAR)*. IEEE; 2020. p. 77–82.

[19] Krishnamurthy V, Angley D, Evans R, *et al.* Identifying cognitive radars – Inverse reinforcement learning using revealed preferences. *IEEE Transactions on Signal Processing*. 2020;68:4529–4542.

[20] Krishnamurthy V and Rangaswamy M. How to calibrate your adversary's capabilities? Inverse filtering for counter-autonomous systems. *IEEE Transactions on Signal Processing*. 2019;67(24):6511–6525.

[21] Singh H, Chattopadhyay A, and Mishra KV. Inverse extended Kalman filter—Part I: Fundamentals. *IEEE Transactions on Signal Processing*. 2023;71: 2936–2951.

[22] Singh H, Chattopadhyay A, and Mishra KV. Inverse extended Kalman filter—Part II: Highly non-linear and uncertain systems. *IEEE Transactions on Signal Processing*. 2023;71:2936–2951.

[23] Elbir AM, Mishra KV, and Eldar YC. Cognitive radar antenna selection via deep learning. *IET Radar, Sonar & Navigation*. 2019;13(6):871–880.

[24] Elbir AM, Mishra KV, Vorobyov SA, *et al.* Twenty-five years of advances in beamforming: From convex and nonconvex optimization to learning techniques. *IEEE Signal Processing Magazine*. 2023;40(4):118–131.

[25] Elbir AM and Mishra KV. Sparse array selection across arbitrary sensor geometries with deep transfer learning. *IEEE Transactions on Cognitive Communications and Networking*. 2020;7(1):255–264.

[26] Egbert A, Goad A, Baylis C, *et al.* Continuous real-time circuit reconfiguration to maximize average output power in cognitive radar transmitters. *IEEE Transactions on Aerospace and Electronic Systems*. 2022;58(3):1514–1527.

[27] Christiansen JM and Smith GE. Development and calibration of a low-cost radar testbed based on the universal software radio peripheral. *IEEE Aerospace and Electronic Systems Magazine*. 2019;34(12):50–60.

Chapter 2

Adversarial radar inference: inverse tracking, identifying cognition, and designing smart interference

Vikram Krishnamurthy[1] Kunal Pattanayak[1] Sandeep Gogineni[2] Bosung Kang[3] and Muralidhar Rangaswamy[4]

This chapter considers three inter-related adversarial inference problems involving cognitive radars. We first discuss inverse tracking of the radar to estimate the adversary's estimate of "us" based on the radar's actions and calibrate the radar's sensing accuracy. Second, using revealed preference from microeconomics, we formulate a non-parametric test to identify if the cognitive radar is a constrained utility maximizer with signal processing constraints. We consider two radar functionalities, namely, beam allocation and waveform design, with respect to which the cognitive radar is assumed to maximize its utility and construct a set-valued estimator for the radar's utility function. Finally, we discuss how to engineer interference at the physical layer level to confuse the radar which forces it to change its transmit waveform.

2.1 Introduction

Cognitive sensors are reconfigurable sensors that optimize their sensing mechanism and transmit functionalities. The concept of cognitive radar [1–4] has evolved over the last two decades and a common aspect is the sense–learn–adapt paradigm. A cognitive fully adaptive radar enables the joint optimization of the adaptive transmit and receive functions by sensing (estimating) the radar channel that includes clutter and other interfering signals [5,6]; see also [7] for a stochastic control-based discussion of cognition in radars. The results in this chapter build on the recent paper [8] for adversarial radar inference and develop adversarial inference algorithms for multiple layers of abstraction: inference design based on Wiener filters at the pulse/waveform

[1]School of Electrical & Computer Engineering, Cornell University, Ithaca, NY, USA
[2]Information Systems Laboratories, Inc., San Diego, CA, USA
[3]University of Dayton Research Institute, Dayton, OH, USA
[4]Air Force Research Laboratory, Wright Patterson Air Force Base, Dayton, OH, USA

level, inverse Kalman filters at the Bayesian tracking level, and revealed preference techniques for estimating the adversary's utility function at the systems level.

2.1.1 Objectives

This chapter achieves the following adversarial inference objectives as shown schematically in Figure 2.1. The framework in this chapter involves an adversarial signal processing problem comprising "us" and an "adversary". "Us" refers to an asset such as a drone/UAV or electromagnetic signal that probes an "adversary" cognitive radar. Figure 2.2 shows the schematic setup. A cognitive sensor observes our kinematic state x_k in noise as the observation y_k. It then uses a Bayesian tracker to update its posterior distribution π_k of our state x_k and chooses an action u_k based on this posterior. We observe the sensor's action in noise as a_k. Given knowledge of "our" state sequence $\{x_k\}$ and the observed actions $\{a_k\}$ taken by the adversary's sensor, we focus on the following inter-related aspects:

1. Inverse tracking and estimating the adversary's sensor gain: Suppose the adversary radar observes our state in noise; updates its posterior distribution π_k of our state x_k using a Bayesian tracker, and then chooses an action u_k based on this posterior. Given knowledge of "our" state and sequence of noisy measurements $\{a_k\}$ of the adversary's actions $\{u_k\}$, how can the adversary radar's posterior distribution (random measure) be estimated? We will develop an inverse Bayesian filter for tracking the radar's posterior belief of our state and present an example involving the Kalman filter where the inverse filtering problem admits a finite-dimensional characterization.

A related question is: How to remotely estimate the adversary radar's sensor observation likelihood when it is estimating us? This is important because it tells us how accurate the adversary's sensor is; in the context of Figure 2.2 it tells us, how accurately the adversary tracks our drone. The data we have access to is our state (probe signal) sequence $\{x_k\}$ and measurements of the adversary's radar actions $\{a_k\}$. Estimating the adversary's sensor accuracy is non-trivial with several challenges. First, even though we know our state and state dynamics model (transition law), the adversary does not. The adversary needs to estimate our state and state transition law based on our trajectory; and we need to estimate the adversary's estimate of our state transition law. Second, computing the MLE of the adversary's sensor gain also requires inverse filtering.

2. Revealed preferences and identifying cognitive radars: Suppose the cognitive radar is a constrained utility maximizer that optimizes its actions a_k subject to physical level (Bayesian filter) constraints. How can we detect this utility maximization behavior? The actions a_k can be viewed as resources the radar adaptively allocates to maximize its utility. We consider two such resource allocation problems, namely,

- *Beam allocation*: The radar adaptively switches its beam while tracking multiple targets.
- *Waveform design*: The radar adaptively designs its waveform while ensuring the signal-to-interference-plus-noise ratio (SINR) exceeds a pre-defined threshold.

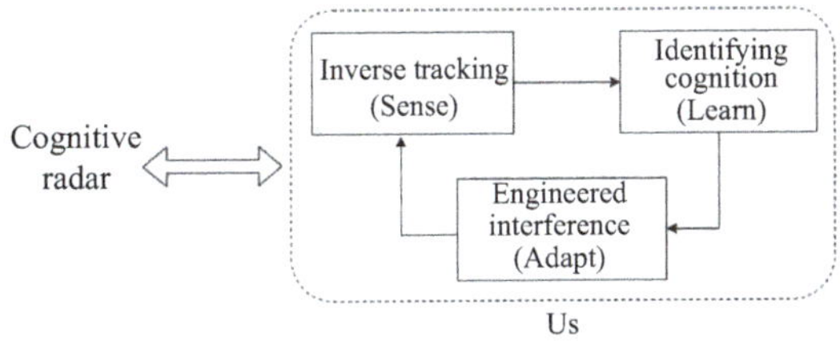

Figure 2.1 *Schematic illustrating the main ideas in the chapter. The three components on the right are inter-related and constitute the sense–learn–adapt paradigm of the observer ("us") reacting to a reactive system such as the cognitive radar.*

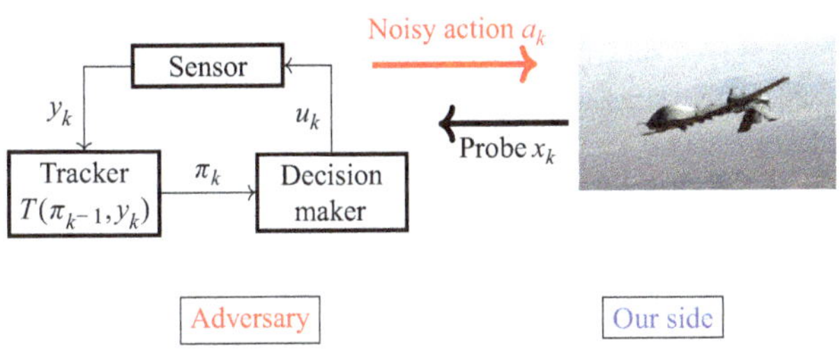

Figure 2.2 *Schematic of adversarial inference problem. Our side is a drone/UAV or electromagnetic signal that probes the adversary's cognitive radar system.*

Nonparametric detection of utility maximization behavior is the central theme of *revealed preference* in microeconomics. A remarkable result is *Afriat's theorem*: it provides a necessary and sufficient condition for a finite dataset to have originated from a utility maximizer. We will develop constrained set-valued utility estimation methods that account for signal processing constraints introduced by the Bayesian tracker for performing adaptive beam allocation and waveform design, respectively.

3. Smart signal dependent interference: We next consider the adversary radar choosing its transmit waveform for target tracking by implementing a Wiener filter to maximize its signal-to-clutter-plus-noise ratio (SCNR*). By observing the optimal waveform chosen by the radar, the aim is to develop a smart strategy to estimate the adversary cognitive radar channels followed by signal-dependent interference generation mechanism to confuse the adversary radar.

2.1.2 Perspective

The adversarial dynamics considered in this chapter fit naturally within the so-called Dynamic Data and Information Processing (DDIP) paradigm. The adversary's radar senses, adapts, and learns from us. In turn, we adapt, sense, and learn from the

*The terms SCNR and SINR are used interchangeably in the chapter.

adversary. So in simple terms, we are modeling and analyzing the interaction of two DDIP systems. In this context, this chapter has three major themes as shown schematically in Figure 2.1: inverse filtering which is a Bayesian framework for interacting DDIP systems, inverse cognitive sensing which is a non-parametric approach for utility estimation for interacting DDIP systems, and interference design to confuse the adversarial DDIP system.

This work is also motivated by the design of counter-autonomous systems: given measurements of the actions of an autonomous adversary, how can our counter-autonomous system estimate the underlying belief of the adversary, identify if the adversary is cognitive (constrained utility maximizer), and design appropriate probing signals to confuse the adversary. This chapter generalizes and contextualizes recent works in adversarial signal processing [9,10] which only deal with specific radar functionalities. Instead, this chapter views the cognitive radar as a holistic system operating at three stages of sophistication unifies the three inter-related aspects of adversarial signal processing, namely, inverse tracking, identifying cognition, and designing interference. The three components complement one another and constitute this chapter's adversarial signal processing sense–learn–adapt (SLA) paradigm of Figure 2.1.

2.1.3 Organization

We conclude this section with a brief outline of the key results of the following sections, and their relevance to the sense, learn, and adapt elements of the SLA paradigm of Figure 2.1.

Sense: In Section 2.2, we discuss inverse tracking techniques to estimate the sensor accuracy of an adversary radar. We mainly focus on the inverse Kalman filter and illustrate in carefully chosen examples how the adversary sensor's accuracy can be estimated. This constitutes the "sensing" aspect of the SLA paradigm.

Learn: In Section 2.3, we abstractly view the adversarial radar as a cognitive decision-maker that maximizes a utility function subject to physical resource constraints. Specifically, we show that if the cognitive radar optimizes its waveform to maintain its SINR above a threshold, then we can identify (and hence, "learn") the utility function of the radar. The utility function provides deeper knowledge of the radar's behavior and constitutes the "learn" element of the SLA paradigm.

Adapt: In Section 2.4, we consider a slightly modified setup where the radar chooses its waveform to maximize its SCNR. We show that by intelligently probing the radar with interference signals and observing the changes in the radar's waveform, we can confuse the adversary's radar by decreasing its SCNR. This adaptive signal processing algorithm is justified only if the "sense" and "learn" aspect of the SLA paradigm functions properly, that is, the counter-adversarial system knows how the radar will react to changes in its environment.

Finally, we emphasize that the three main aspects of inverse tracking (sensing the estimate of the adversary), identifying utility maximization (learning the adversary's utility function), and adaptive interference (adapting our response) are instances of the general paradigm of SLA in counter-adversarial systems. As mentioned earlier, our formulation deals with the interaction of two such SLA systems.

2.2 Inverse tracking and estimating adversary's sensor

This section discusses inverse tracking in an adversarial system as illustrated schematically in Figure 2.2. Our main ideas involve estimating the adversary's estimate of us and estimating the adversary's sensor observation likelihood.

2.2.1 Background and preliminary work

We start by formulating the problem which involves two entities; "us" and "adversary". With $k = 1, 2,...$ denoting discrete time, the model has the following dynamics:

$$\begin{aligned} x_k &\sim P_{x_{k-1},x} = p(x|x_{k-1}), \quad x_0 \sim \pi_0 \\ y_k &\sim B_{x_k,y} = p(y|x_k) \\ \pi_k &= T(\pi_{k-1}, y_k) = p(x_k|y_{1:k}) \\ a_k &\sim G_{\pi_k,a} = p(a|\pi_k) \end{aligned} \tag{2.1}$$

Let us explain the notation in (2.1):

- $x_k \in \mathscr{X}$ is our Markovian state with transition kernel $P_{x_{k-1},x}$, prior π_0 and state space $\mathscr{X}$.
- y_k is the adversary's noisy observation of our state x_k; with observation likelihood (the likelihood of the observation given our Markovian state) $B_{x_k,y}$.
- π_k is the adversary's belief (posterior) of our state x_k where $y_{1:k}$ denotes the observation sequence $y_1, \ldots, y_k$. The operator T in (2.1) is the classical Bayesian optimal filter that computes the posterior belief of the state given observation y and current belief π:

$$T(\pi, y) = \text{vec}\left(\frac{B_{x,y} \int_{\mathscr{X}} P_{\zeta,x}\, \pi(\zeta)\, d\zeta}{\int_{\mathscr{X}} B_{x,y} \int_{\mathscr{X}} P_{\zeta,x}\, \pi(\zeta)\, d\zeta dx}, x \in \mathscr{X} \right) \tag{2.2}$$

  Let Π denote the space of all such beliefs. When the state space $\mathscr{X}$ is finite, then Π is the unit $X - 1$ dimensional simplex of X-dimensional probability mass functions.
- a_k denotes our measurement of the adversary's action based on its current belief π_k. The adversary chooses an action u_k as a (possibly) stochastic function of π_k and we obtain a noisy measurement of u_k as a_k. We encode this as G_{π_k,a_k}, the conditional probability of observing action a_k given the adversary's belief π_k. Although not explicitly shown, G abstracts two stochastic maps: (1) the map from the adversary's belief π_k to its action u_k, and (2) the map from the adversary's action u_k to our noisy measurement a_k of this action.

Figure 2.2 displays a schematic and graphical representation of the model (2.1). The schematic model shows "us" and the adversary's variables.

Aim: Referring to model (2.1) and Figure 2.2, we address the following questions in this section:

1. How to estimate the adversary's belief given measurements of its actions (which are based on its filtered estimate of our state)? In other words, assuming probability distributions P, B, G are known,† we aim to estimate the adversary's belief π_k at each time k, by computing posterior $p(\pi_k \mid \pi_0, x_{0:k}, a_{1:k})$.
2. How to estimate the adversary's observation kernel B, i.e., its sensor gain? This tells us how accurate the adversary's sensor is.

From a practical point of view, estimating the adversary's belief and sensor parameters allows us to calibrate its accuracy and predict (in a Bayesian sense) future actions of the adversary.

Related Works. In recent works [11–13], the mapping from belief π to adversary's action u was assumed deterministic. In comparison, our proposed research here assumes a probabilistic map between π and a and we develop Bayesian filtering algorithms for estimating the posterior along with maximum-likelihood estimation (MLE) algorithms for estimating the underlying model. Estimating/reconstructing the posterior given decisions based on the posterior is studied in microeconomics under the area of social learning [14] and game-theoretic generalizations [15]. There are strong parallels between inverse filtering and Bayesian social learning [14,16–18]; the key difference is that social learning aims to estimate the underlying state given noisy posteriors, whereas our aim is to estimate the posterior given noisy measurements of the posterior and the underlying state. Recently, the authors of [19] used cascaded Kalman filters for LQG control over communication channels. This work motivates the design of the function ϕ in (2.8) below that maps the adversary's belief to its action; see also footnote below. Our inverse Kalman filtering results [9] have been recently extended to non-linear processes by [20]. In [21], the authors investigate the inverse problem of trajectory identification based on target measurements, where the target is assumed to follow a constant velocity model. Finally, in the field of inverse problems, the authors of [22] propose an ensemble Kalman filter approach to estimate the true (fixed) state of a system given a noisy observation of the system's response to the true state. The authors of [22] assume the forward operator mapping the state to the response. In comparison, our inverse Kalman filter in Section 2.2.2 generalizes the results of [22] to the case where the ground truth is time-varying and can be modeled as a linear Gaussian system.

2.2.2 *Inverse tracking algorithms*

How to estimate the adversary's posterior distribution of us?

Here we discuss inverse tracking for the model (2.1). Define the posterior distribution $\rho_k(\pi_k) = p(\pi_k | a_{1:k}, x_{0:k})$ of the adversary's posterior distribution given our state sequence $x_{0:k}$ and actions $a_{1:k}$. Note that the posterior $\rho_k(\cdot)$ is a *random measure*

†As mentioned in the footnote on page 21, this assumption simplifies the setup; otherwise, we need to estimate the adversary's estimate of us, which makes our task substantially complex.

since it is a posterior distribution of the adversary's posterior distribution (belief) π_k. By using a discrete-time version of Girsanov's theorem and appropriate change of measure[‡] [24] (or a careful application of Bayes rule), we can derive the following functional recursion for ρ_k (see [9])

$$\rho_{k+1}(\pi) = \frac{G_{\pi,a_{k+1}} \int_{\Pi} B_{x_{k+1}y_{\pi_k,\pi}} \; \rho_k(\pi_k) d\pi_k}{\int_{\Pi} G_{\pi,a_{k+1}} \int_{\Pi} B_{x_{k+1}y_{\pi_k,\pi}} \; \rho_k(\pi_k) d\pi_k \, d\pi} \tag{2.3}$$

Here $y_{\pi_k,\pi}$ is the observation such that $\pi = T(\pi_k, y)$ where T is the adversary's filter (2.2). We call (2.3) the *optimal inverse filter* since it yields the Bayesian posterior of the adversary's belief given our state and noisy measurements of the adversary's actions.

Example: inverse Kalman filter

We consider a special case of (2.3) where the inverse filtering problem admits a finite-dimensional characterization in terms of the Kalman filter. Consider a linear Gaussian state-space model

$$\begin{aligned} x_{k+1} &= A x_k + w_k, \quad x_0 \sim \pi_0 \\ y_k &= C x_k + v_k \end{aligned} \tag{2.4}$$

where $x_k \in \mathscr{X} = \mathbb{R}^X$ is "our" state with initial density $\pi_0 \sim \mathbf{N}(\hat{x}_0, \Sigma_0)$, $y_k \in \mathbf{Y} = \mathbb{R}^Y$ denotes the adversary's observations, $w_k \sim \mathbf{N}(0, Q_k)$, $v_k \sim \mathbf{N}(0, R_k)$ and $\{w_k\}$, $\{v_k\}$ are mutually independent i.i.d. processes. Here, $\mathbf{N}(\mu, C)$ denotes the normal distribution with mean μ and covariance matrix C.

Based on observations $y_{1:k}$, the adversary computes the belief $\pi_k = \mathbf{N}(\hat{x}_k, \Sigma_k)$ where $\hat{x}_k$ is the conditional mean state estimate and Σ_k is the covariance; these are computed via the classical Kalman filter equations[§]:

$$\begin{aligned} \Sigma_{k+1|k} &= A \Sigma_k A' + Q_k \\ S_{k+1} &= C \Sigma_{k+1|k} C' + R_k \\ \hat{x}_{k+1} &= A \hat{x}_k + \Sigma_{k+1|k} C' S_{k+1}^{-1} (y_{k+1} - C A \hat{x}_k) \\ \Sigma_{k+1} &= \Sigma_{k+1|k} - \Sigma_{k+1|k} C' S_{k+1}^{-1} C \Sigma_{k+1|k} \end{aligned} \tag{2.5}$$

[‡]This chapter deals with discrete time. Although we will not pursue it here, the recent paper [23] uses a similar continuous-time formulation. This yields interesting results involving Malliavin derivatives and stochastic calculus.

[§]For localization problems, we will use the information filter form:

$$\Sigma_{k+1}^{-1} = \Sigma_{k+1|k}^{-1} + C' R^{-1} C, \quad \Psi_{k+1} = \Sigma_{k+1} C' R^{-1} \tag{2.6}$$

Similarly, the inverse Kalman filter in information form reads

$$\bar{\Sigma}_{k+1}^{-1} = \bar{\Sigma}_{k+1|k}^{-1} + \bar{C}'_{k+1} \bar{R}^{-1} \bar{C}_{k+1}, \; \bar{\Psi}_{k+1} = \bar{\Sigma}_{k+1} \bar{C}_{k+1}^{\pi} \bar{R}^{-1}. \tag{2.7}$$

The adversary then chooses its action as $\bar{a}_k = \phi(\Sigma_k)\hat{x}_k$ for some pre-specified function[‖] ϕ. We measure the adversary's action as

$$a_k = \phi(\Sigma_k)\hat{x}_k + \varepsilon_k, \quad \varepsilon_k \sim \text{iid } \mathbf{N}(0, \sigma_\varepsilon^2) \tag{2.8}$$

The Kalman covariance Σ_k is deterministic and fully determined by the model parameters. Hence, we only need to estimate $\hat{x}_k$ at each time k given $a_{1:k}, x_{0:k}$ to estimate the belief $\pi_k = \mathbf{N}(\hat{x}_k, \Sigma_k)$. Substituting (2.4) for y_{k+1} in (2.5), we see that (2.5) and (2.8) constitute a linear Gaussian system with unobserved state $\hat{x}_k$, observations a_k, and known exogenous input x_k:

$$\begin{aligned}
\hat{x}_{k+1} &= (\boldsymbol{I} - \psi_{k+1}C)A\hat{x}_k + \psi_{k+1}v_{k+1} + \psi_{k+1}Cx_{k+1} \\
a_k &= \phi(\Sigma_k)\hat{x}_k + \varepsilon_k, \quad \varepsilon_k \sim \text{iid } \mathbf{N}(0, \sigma_\varepsilon^2), \\
&\text{where } \psi_{k+1} = \Sigma_{k+1|k}C'S_{k+1}^{-1}.
\end{aligned} \tag{2.9}$$

ψ_{k+1} is called the Kalman gain and $\boldsymbol{I}$ is the identity matrix.

To summarize, our filtered estimate of the adversary's filtered estimate $\hat{x}_k$ given measurements $a_{1:k}, x_{0:k}$ is achieved by running "our" Kalman filter on the linear Gaussian state-space model (2.9), where $\hat{x}_k$, ψ_k, Σ_k in (2.9) are generated by the adversary's Kalman filter. Therefore, our Kalman filter uses the parameters

$$\begin{aligned}
\bar{A}_k &= (\boldsymbol{I} - \psi_{k+1}C)A, \; \bar{F}_k = \psi_{k+1}C, \; \bar{C}_k = \phi(\Sigma_k), \\
\bar{Q}_k &= \psi_{k+1}\,\psi_{k+1}', \; \bar{R}_k = \sigma
\end{aligned} \tag{2.10}$$

The equations of our inverse Kalman filter for estimating the adversary's estimate of our state are:

$$\begin{aligned}
\bar{\Sigma}_{k+1|k} &= \bar{A}_k\bar{\Sigma}_k\bar{A}_k' + \bar{Q}_k \\
\bar{S}_{k+1} &= \bar{C}_{k+1}\bar{\Sigma}_{k+1|k}\bar{C}_{k+1}' + \bar{R}_k \\
\hat{\hat{x}}_{k+1} &= \bar{A}_k\hat{\hat{x}}_k + \bar{\Sigma}_{k+1|k}\bar{C}_{k+1}'\bar{S}_{k+1}^{-1} \\
&\quad \times \left[a_{k+1} - \bar{C}_{k+1}\left(\bar{A}_k\hat{x}_k + \bar{F}_kx_{k+1}\right)\right] \\
\bar{\Sigma}_{k+1} &= \bar{\Sigma}_{k+1|k} - \bar{\Sigma}_{k+1|k}\bar{C}_{k+1}'\bar{S}_{k+1}^{-1}\bar{C}_{k+1}\bar{\Sigma}_{k+1|k}
\end{aligned} \tag{2.11}$$

Note $\hat{\hat{x}}_k$ and $\bar{\Sigma}_k$ denote our conditional mean estimate and covariance of the adversary's conditional mean $\hat{x}_k$. The computational cost of the inverse Kalman filter is identical to the classical Kalman filter, namely $O(X^2)$ computations at each time step.

[‖] In general, the action a_k is a function of the state estimate and covariance matrix. Choosing the action a_k as a linear function of the state estimate is for convenience and motivates the inverse Kalman filter discussed later. Moreover, it mimics linear quadratic Gaussian (LQG) control where the feedback is a linear function of the state estimate. In LQG control, the feedback gain is obtained from the backward Riccati equation. Here we weigh the feedback by a nonlinear function of the Kalman covariance matrix (forward Riccati equation) to allow for incorporating uncertainty of the estimate into the choice of the action a_k.

Remarks:

1. As discussed in [9], inverse Hidden Markov model (HMM) filters and inverse particle filters can also be derived to solve the inverse tracking problem. For example, the inverse HMM filter deals with the case when π_k is computed via an HMM filter and the estimates of the HMM filter are observed in noise. In this case, the inverse filter has a computational cost that grows exponentially with the size of the observation alphabet.
2. A general approximate solution for (2.3) involves sequential Markov chain Monte-Carlo (particle filtering). In particle filtering, cases where it is possible to sample from the so-called optimal importance function are of significant interest [25,26]. In inverse filtering, [9] shows that the optimal importance function can be determined explicitly due to the structure of the inverse filtering problem. Specifically, in our case, the "optimal" importance density is $\pi^* = p(\pi_k, y_k|\pi_{k-1}, y_{k-1}, x_k, a_k)$. Note that in our case

$$\begin{aligned}\pi^* &= p(\pi_k|\pi_{k-1}, y_k)p(y_k|x_k, a_k) \\ &= \delta\left(\pi_k - T(\pi_{k-1}, y_k)\right) p(y_k|x_k) \end{aligned} \tag{2.12}$$

   is straightforward to sample form. There has been a substantial amount of recent research in finite sample concentration bounds for the particle filter [27,28]. In future work, such results can be used to evaluate the sample complexity of the inverse particle filter.
3. Equation (2.11) implicitly requires knowledge of the adversary's sensor gain C. If C is unknown, Examples 2.2.3 and 2.2.4 provide algorithms for estimating C. A practical implementation of the inverse Kalman filter would involve interleaving the algorithms in Examples 2.2.3 and 2.2.4 for estimating C, and the inverse filtering equations of (2.11) to estimate the adversary's posterior distribution of our state using the estimated C.

2.2.3 Estimating the adversary's sensor gain

In this section, we discuss how to estimate the adversary's sensor observation kernel B in (2.1) which quantifies the accuracy of the adversary's sensors. We assume that B is parameterized by an M-dimensional vector $\theta \in \Theta$ where Θ is a compact subset of $\mathbb{R}^M$. Denote the parameterized observation kernel as B^θ. Assume that both us and the adversary know¶ P (state transition kernel) and G (probabilistic map from adversary's belief to its action). As mentioned earlier, the stochastic kernel G in (2.1) is a composition of two stochastic kernels: (1) the map from the adversary's belief π_k to its action u_k, and (2) the map from the adversary's action u_k to our measurement a_k of this action.

¶Otherwise the adversary estimates P as $\hat{P}$ and we need to estimate the adversary's estimate of us, namely $\hat{\hat{P}}$. This makes the estimation task substantially more complex. In future work, we will examine conditions under which the MLE in this setup is identifiable and consistent.

Then, given our state sequence $x_{0:N}$ and adversary's action sequence $u_{1:N}$, our aim is to compute the MLE of θ. That is, with $L_N(\theta)$ denoting the log-likelihood, the aim is to compute

$$\theta^* = \underset{\theta\in\Theta}{\operatorname{argmax}}\, L_N(\theta), \; L_N(\theta) = \log p(x_{0:N}, a_{1:N}|\theta). \tag{2.13}$$

The likelihood can be evaluated from the un-normalized inverse filtering recursion (2.3)

$$L_N(\theta) = \log \int_\Pi q_N^\theta(\pi) d\pi,$$

$$q_{k+1}^\theta(\pi) = G_{\pi,a_{k+1}} \int_\Pi B^\theta_{x_{k+1} y^\theta_{\pi_k,\pi}} q_k^\theta(\pi_k) d\pi_k, \tag{2.14}$$

initialized by setting $q_0^\theta(\pi_0) = \pi_0$. Here $y^\theta_{\pi_k,\pi}$ is the observation such that $\pi = T(\pi_k, y)$ where T is the adversary's filter (2.2) with variable B parametrized by θ. Given (2.14), a local stationary point of the likelihood can be computed using a general-purpose numerical optimization algorithm.

2.2.4 Example. Estimating adversary's gain in linear Gaussian case

The aim of this section is to provide insight into the nature of estimating the adversary's sensor gain via numerical examples. Consider the setup in Section 2.2.2 where our dynamics are linear Gaussian and the adversary observes our state linearly in Gaussian noise (2.4). The adversary estimates our state using a Kalman filter, and we estimate the adversary's estimate using the inverse Kalman filter (2.9). Using (2.9) and (2.10), the log-likelihood for the adversary's observation gain matrix $\theta = C$ based on our measurements is**

$$L_N(\theta) = \text{const} - \frac{1}{2}\sum_{k=1}^{N} \log|\bar{S}_k^\theta| - \frac{1}{2}\sum_{k=1}^{N} \iota_k' (\bar{S}_k^\theta)^{-1} \iota_k$$

$$\iota_k = a_k - \bar{C}_k^\theta \bar{A}_{k-1}^\theta \hat{\hat{x}}_{k-1} - \bar{F}_{k-1}^\theta x_{k-1} \tag{2.15}$$

where ι_k are the innovations of the inverse Kalman filter (2.11). In (2.15), our state x_{k-1} is known to us and, therefore, is a known exogenous input. Also note from (2.10) that $\bar{A}_k, \bar{F}_k$ are explicit functions of C, while $\bar{C}_k$ and $\bar{Q}_k$ depend on C via the adversary's Kalman filter.

The log-likelihood for the adversary's observation gain matrix $\theta = C$ can be evaluated using (2.15). To provide insight, Figure 2.3 displays the log-likelihood versus adversary's gain matrix C in the scalar case for 1,000 equally spaced data points over the interval $C = (0, 10]$. The four sub-figures correspond to true values $C^o = 2.5, 3.5$ of C, respectively.

Each sub-figure in Figure 2.3 has two plots. The plot in red is the log-likelihood of $\hat{C} \in (0, 10]$ evaluated based on the adversary's observations using the standard

**The variable θ is introduced only for notational clarity.

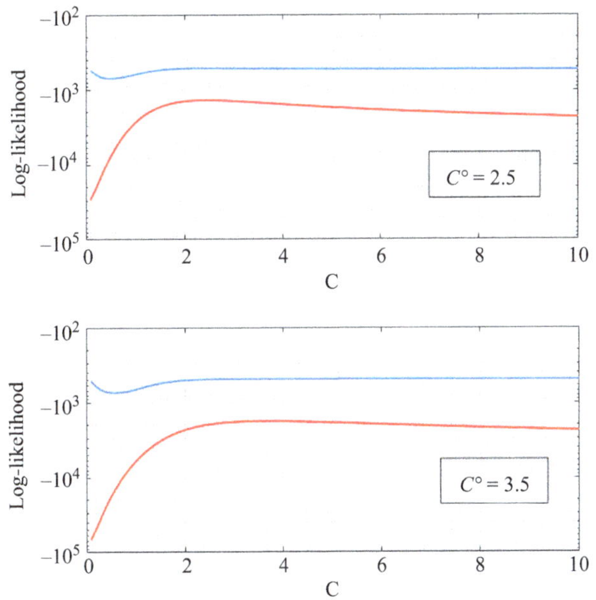

Figure 2.3 *Log-likelihood as a function of adversary's gain* $C \in (0, 10]$ *when true value is* C^o. *The red curves denote the log-likelihood of* C *given the adversary's measurements of our state. The blue curves denote the log-likelihood of* C *using the inverse Kalman filter given our observations of the adversary's action* u_k. *The plots show that it is more difficult to compute the MLE* (2.13) *for the inverse filtering problem due to the almost flat likelihood (blue curves) compared to red curves.*

Kalman filter. (This is the classical log-likelihood of the observation gain of a Gaussian state-space model.) The plot in blue is the log-likelihood of $C \in (0, 10]$ computed using our measurements of the adversary's action using the inverse Kalman filter (where the adversary first estimates our state using a Kalman filter)—we call this the inverse case.

Figure 2.3 shows that the log-likelihood in the inverse case (blue plots) has a less pronounced maximum compared to the standard case (red plots). Therefore, numerical algorithms for computing the MLE of the adversary's gain C^o using our observations of the adversary's actions (via the inverse Kalman filter) will converge much more slowly than the classical MLE (based on the adversary's observations). This is intuitive since our estimate of the adversary's parameter is based on the adversary's estimate of our state and so has more noise.

Sensitivity of MLE. It is important to evaluate the sensitivity of the MLE of C wrt covariance matrices Q_k, R_k in the state-space model (2.4). For example, the sensitivity wrt Q_k reveals how sensitive the MLE is wrt our maneuver covariance since from (2.4), Q_k determines our maneuvers. Our sensitivity analysis evaluates the variation of the second derivative of the log-likelihood of C computed at the true gain C^o to small changes in Q_k and R_k. Table 2.1 displays our sensitivity results wrt the scalar setup of Figure 2.3. Table 2.1 comprises two sensitivity values,

$$\eta_Q = \frac{\partial}{\partial Q_k}\left(\frac{\partial^2 L_N(\theta)}{\partial \theta^2}\right)\Big|_{\theta=C^o} \text{ and}$$
$$\eta_R = \frac{\partial}{\partial R_k}\left(\frac{\partial^2 L_N(\theta)}{\partial \theta^2}\right)\Big|_{\theta=C^o}, \tag{2.16}$$

evaluated for both the inverse case (that uses the inverse Kalman filter (2.15)) and the classic case where the adversary's observations are known. $\eta_{(\cdot)}$ measures the change in the sharpness of the log-likelihood plot around the true sensor gain wrt change in the noise covariance. Note that the experimental setup of Figure 2.3 assumes that the covariances Q_k, R_k are constant over time index k, hence, we drop the subscript in the LHS of (2.16).

Table 2.1 shows that the second derivative of the log-likelihood is more sensitive (in magnitude) to the adversary's observation covariance R_k than the maneuver covariance Q_k. Also, it is observed that the sensitivity of the log-likelihood is higher for lower sensor gain C^o. This observation is consistent with intuition since a larger gain C implies a larger SNR (signal-to-noise ratio) of the observation y_k which intuitively suggests the estimate of C is more robust to changes in maneuver covariance and observation noise covariance.

Cramér-Rao (CR) bounds. It is instructive to compare the CR bounds for MLE of C for the classic model versus that of the inverse Kalman filter model. Table 2.2 displays the CR bounds (reciprocal of Fisher information) for the four examples considered above evaluated using via the algorithm in [29]. It shows that the covariance lower bound for the inverse case is substantially higher than that for the classic case. This is consistent with the intuition that estimating the adversary's parameter based on its actions (which is based on its estimate of us) is more difficult than directly estimating C in a classical state-space model based on the adversary's observations of our state that determines its actions.

Table 2.1 Comparison of sensitivity values (2.16) *for log-likelihood of C wrt noise covariances* Q_k, R_k (2.4)*—classical model versus inverse Kalman filter model*

	C^o	Classic	Inverse
η_Q	2.5	−43.45	−6.46
	3.5	−25.16	−2.77
η_R	2.5	−189.39	−50.04
	3.5	−65.27	−30.55

Table 2.2 Comparison of Cramér-Rao bounds for C—classical model vs inverse Kalman filter model

C^o	Classic	Inverse
0.5	0.24×10^{-3}	5.3×10^{-3}
1.5	1.2×10^{-3}	37×10^{-3}
2	2.1×10^{-3}	70×10^{-3}
3	4.6×10^{-3}	336×10^{-3}

Consistency of MLE. The above example (Figure 2.3) shows that the likelihood surface of $L_N(\theta) = \log p(x_{0:N}, a_{1:N}|\theta)$ is flat and, hence, computing the MLE numerically can be difficult. Even in the case when we observe the adversary's actions perfectly, [11] shows that non-trivial observability conditions need to be imposed on the system parameters.

For the linear Gaussian case where we observe the adversary's Kalman filter in noise, strong consistency of the MLE for the adversary's gain matrix C can be established fairly straightforwardly. Specifically, if we assume that state matrix A is stable, and the state-space model is an identifiable minimal realization, then the adversary's Kalman filter variables converge to steady-state values geometrically fast in k [30] implying that asymptotically the inverse Kalman filter system is stable linear time invariant. Then, the MLE θ^* for the adversary's observation matrix C is unique and strongly consistent [31].

2.3 Identifying utility maximization in a cognitive radar

The previous section was concerned with estimating the adversary's posterior belief and sensor accuracy. This section discusses detecting utility maximization behavior and estimating the adversary's utility function in the context of cognitive radars. As described in the introduction, inverse tracking, identifying utility maximization, and designing interference to confuse the radar constitute our adversarial setting.

Cognitive radars [32] use the perception–action cycle of cognition to sense the environment and learn from it relevant information about the target and the environment. The cognitive radars then tune the radar sensor to optimally satisfy their mission objectives. Based on its tracked estimates, the cognitive radar adaptively optimizes its waveform, aperture, dwell time, and revisit rate. In other words, a cognitive radar is a constrained utility maximizer.

This section is motivated by the next logical step, namely, *identifying a cognitive radar* from the actions of the radar. The adversary cognitive radar observes our state in noise; it uses a Bayesian estimator (target tracking algorithm) to update its posterior distribution of our state and then chooses an action based on this posterior.

From the intercepted emissions of an adversary's radar, we address the following question: Are the adversary sensor's actions consistent with optimizing a monotone utility function (i.e., is the cognitive sensor behavior rational in an economics sense)? If so how can we estimate a utility function of the adversary's cognitive sensor that is consistent with its actions? The main synthesis/analysis framework we will use is that of revealed preferences [33–35] from microeconomics which aims to determine preferences by observing choices. The results presented below are developed in detail in the recent work [10]; however, the SINR constraint formulation in Section 2.3.3 for detecting waveform optimization is new. Related work that develops adversarial inference strategies at a higher level of abstraction than the tracking level includes [36]. The author of [36] places counter unmanned autonomous systems at a level of abstraction above the physical sensors/actuators/weapons and datalink layers; and below the human controller layer.

2.3.1 Background. Revealed preferences and Afriat's theorem

Non-parametric detection of utility maximization behavior is studied in the area of revealed preferences in microeconomics. A key result is the following:

Definition 1 ([37,38]). *A system is a* utility maximizer *if for every probe $\alpha_n \in \mathbb{R}^m_+$, the response $\beta_n \in \mathbb{R}^m$ satisfies*

$$\beta_n \in \underset{\alpha_n'\beta \leq 1}{\operatorname{argmax}}\, U(\beta) \tag{2.17}$$

where $U(\beta)$ is a monotone *utility function.*

In economics, α_n is the price vector and β_n the consumption vector. Then $\alpha_n'\beta \leq 1$ is a natural budget constraint[††] for a consumer with 1 dollar. Given a dataset of price and consumption vectors, the aim is to determine if the consumer is a utility maximizer (rational) in the sense of (2.17).

The key result is the following theorem due to Afriat [33,35,37–39].

Theorem 1 (Afriat's theorem [37]). *Given a data set $\mathscr{D} = \{(\alpha_n, \beta_n), n \in \{1, 2, \ldots, N\}\}$, the following statements are equivalent:*

1. *The system is a utility maximizer and there exists a monotonically increasing, continuous, and concave utility function that satisfies (2.17).*
2. *There exist positive reals $u_t, \lambda_t > 0,\ t = 1, 2, \ldots, N$, such that the following inequalities hold.*

$$u_s - u_t - \lambda_t \alpha_t'(\beta_s - \beta_t) \leq 0\ \forall t, s \in \{1, 2, \ldots, N\}. \tag{2.18}$$

[††]The budget constraint $\alpha_n'\beta \leq 1$ is without loss of generality, and can be replaced by $\alpha_n'\beta \leq c$ for any positive constant c. A more general nonlinear budget incorporating spectral constraints will be discussed later.

The monotone, concave utility function[‡‡] *given by*

$$U(\beta) = \min_{t\in\{1,2,\ldots,N\}} \{u_t + \lambda_t \alpha_t'(\beta - \beta_t)\} \tag{2.19}$$

constructed using u_t *and* λ_t *defined in* (2.18) *rationalizes the dataset by satisfying* (2.17).

3. *The data set* $\mathscr{D}$ *satisfies the Generalized Axiom of Revealed Preference (GARP) also called cyclic consistency, namely for any* $t \leq N$, $\alpha_t'\beta_t \geq \alpha_t'\beta_{t+1}\ \forall t \leq k - 1 \implies \alpha_k'\beta_k \leq \alpha_k'\beta_1$.

Afriat's theorem tests for economics-based rationality; its remarkable property is that it gives a *necessary and sufficient condition* for a system to be a utility maximizer based on the system's input–output response.[§§] Although GARP in statement 3 in Theorem 1 is not critical to the developments in this chapter, it is of high significance in microeconomic theory and is stated here for completeness. The feasibility of the set of inequalities (2.18) can be checked using a linear programming solver; alternatively GARP can be checked using Warshall's algorithm with $O(N^3)$ computations [40,43].

The recovered utility using (2.19) is not unique; indeed any positive monotone increasing transformation of (2.19) also satisfies Afriat's theorem; that is, the utility function constructed is ordinal. This is the reason why the budget constraint $\alpha_n'\beta \leq 1$ is without generality; it can be scaled by an arbitrary positive constant and Theorem 1 still holds. In signal processing terminology, Afriat's theorem can be viewed as a set-valued system identification of an *argmax* system; set-valued since (2.19) yields a set of utility functions that rationalize the finite dataset $\mathscr{D}$.

2.3.2 Beam allocation: revealed preference test

This section constructs a test to identify a cognitive radar that switches its beam adaptively between targets. This example is based on [10] and is presented here for completeness; see also [44,45] for a POMDP-based formulation of adaptive beam allocation in radars. The setup is schematically shown in Figure 2.4. We view each component i of the probe signal $\alpha_n(i)$ as the trace of the precision matrix (inverse covariance) of target i. We use the trace of the precision matrix of each target in our probe signal—this allows us to consider multiple targets. Since the adversarial radar is assumed to be stationary, the target covariance used to define the probe for the radar is indeed the maneuver covariance.

The setup in Figure 2.4 differs significantly from the setup in Figure 2.2 considered in the previous section. First, the adversary in the current setup is an economically

‡‡As pointed out in [40], a remarkable feature of Afriat's theorem is that if the dataset can be rationalized by a monotone utility function, then it can be rationalized by a continuous, concave, monotonic utility function. Put another way, continuity and concavity cannot be refuted with a finite dataset.

§§In complete analogy to Afriat's theorem and revealed preference, there has been extensive progress in the related area of Bayesian revealed preference [41]. Bayesian revealed preference has been used in the recent paper [42] to quantify user engagement behavior in social multimedia platforms like YouTube. Although not explored here, it is instructive to use Bayesian revealed preference results for refinement of the results presented here.

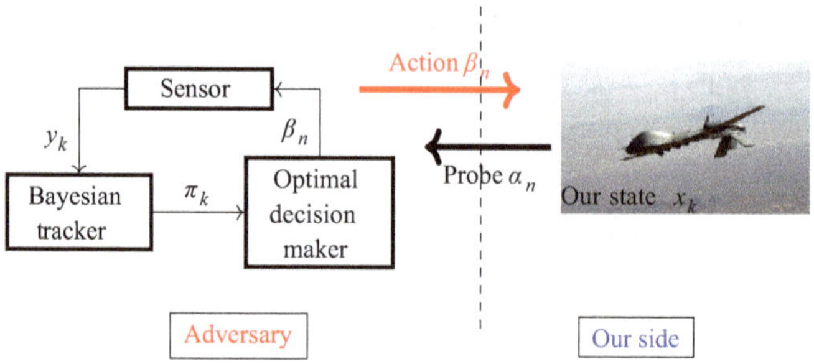

Figure 2.4 Schematic of adversarial inference problem. Our side is a drone/UAV or electromagnetic signal that probes the adversary's cognitive radar system. k denotes a fast time scale and n denotes a slow time scale. Our state x_k, parameterized by α_n (purposeful acceleration maneuvers), probes the adversary radar. Based on the noisy observation y_k of our state, the adversary radar responds with action β_n. Our aim is to determine if the adversary radar is economic rational, i.e., is its response β_n generated by constrained optimizing a utility function?

rational agent. In Figure 2.2, the adversary is only specified at a lower level of abstraction as using a Bayesian filter to track our maneuvers. Second, this section abstracts adversary's actions at the fast time scale indexed by k by an appropriately defined response at the slow time scale indexed by n. The previous section's analysis was confined to the actions generated only at the fast time scale k. Lastly, Figure 2.4 assumes the abstracted response β_k of the adversary is measured accurately by us as opposed to a noisy measurement a_k of the adversary's action u_k in Figure 2.2.

Suppose a radar adaptively switches its beam between m targets where these m targets are controlled by us. As in (2.4), on the fast time scale indexed by k, each target i has linear Gaussian dynamics and the adversary radar obtains linear Gaussian measurements:

$$\begin{aligned} x_{k+1}^i &= A x_k^i + w_k^i, \quad x_0 \sim \pi_0 \\ y_k^i &= C x_k^i + v_k^i, \quad i = 1, 2, \ldots, m \end{aligned} \tag{2.20}$$

Here $w_k^i \sim \mathbf{N}(0, Q_n(i))$, $v_k^i \sim \mathbf{N}(0, R_n(i))$. Recall from Figure 2.4 that n indexes the epoch (slow time scale) and k indexes the fast time scale within the epoch. We assume that both $Q_n(i)$ and $R_n(i)$ are known to us and the adversary.

The adversary's radar tracks our m targets using Kalman filter trackers. The fraction of time the radar allocates to each target i in epoch n is $\beta_n(i)$. The price the radar pays for each target i at the beginning of epoch n is the trace of the predicted *accuracy* of target i. Recall that this is the trace of the inverse of the predicted covariance at epoch n using the Kalman predictor

$$\alpha_n(i) = \mathrm{Tr}(\Sigma_{n|n-1}^{-1}(i)), \quad i = 1, \ldots, m \tag{2.21}$$

The predicted covariance $\Sigma_{n|n-1}(i)$ is a deterministic function of the maneuver covariance $Q_n(i)$ of target i. So the probe[‖‖] $\alpha_n(i)$ is a signal that we can choose, since it is a deterministic function of the maneuver covariance $Q_n(i)$ of target i. We abstract the target's covariance by its trace denoted by $\alpha_n(i)$. Note also that the observation noise covariance $R_n(i)$ depends on the adversary's radar response $\beta_n(i)$, i.e., the fraction of time allocated to target i. We assume that each target i can estimate the fraction of time $\beta_n(i)$ the adversary's radar allocates to it using a radar detector.

Given the time series $\alpha_n, \beta_n, n = 1, \ldots, N$, our aim is to detect if the adversary's radar is cognitive. We assume that a cognitive radar optimizes its beam allocation as the following constrained optimization:

$$\beta_n = \operatorname*{argmax}_{\beta} \; U(\beta) \tag{2.22}$$

$$\text{s.t.} \quad \beta' \alpha_n \le p_*, \tag{2.23}$$

where $U(\cdot)$ is the adversary radar's utility function (unknown to us) and $p_* \in \mathbb{R}_+$ is a pre-specified average accuracy of all m targets.

The economics-based rationale for the budget constraint is natural: for targets that are cheaper (lower accuracy $\alpha_n(i)$), the radar has incentive to devote more time $\beta_n(i)$. However, given its resource constraints, the radar can achieve at most an average accuracy of p_* over all targets.

The setup in (2.23) is directly amenable to Afriat's Theorem 1. Thus, (2.18) can be used to test if the radar satisfies utility maximization in its beam scheduling (2.23) and also estimate the set of utility functions (2.19). Furthermore (as in Afriat's theorem), since the utility is ordinal, p_* can be chosen as 1 without loss of generality (and therefore does not need to be known by us).

2.3.3 *Waveform adaptation: revealed preference test for non-linear budgets*

In the previous subsection, we tested for cognitivity of a radar by viewing it as an abstract system that switches its beam adaptively between targets. Here, we discuss cognitivity with respect to waveform design. Specifically, we construct a test to identify cognitive behavior of an adversary radar that optimizes its waveform based on the SINR of the target measurement. By using a generalization of Afriat's theorem (Theorem 1) to non-linear budgets, our main aim is to detect if a radar intelligently chooses its waveform to maximize an underlying utility subject to signal processing constraints. Our setup below differs from [10] since we introduce the SINR as a nonlinear budget constraint; in comparison [10] uses a spectral budget constraint.

We start by briefly outlining the generalized utility maximization setup.

[‖‖] In comparison to (2.4), the velocity and acceleration elements of x_k^i in (2.20) must be multiplied by normalization factors Δt and $(\Delta t)^2$, respectively, for (2.21) to be dimensionally correct, where Δt is the time duration between two discrete time instants on the fast time scale.

Definition 2 ([34]). *A system is a* generalized utility maximizer *if for every probe* $\alpha_n \in \mathbb{R}^m_+$, *the response* $\beta_n \in \mathbb{R}^m$ *satisfies*

$$\beta_n \in \underset{g_n(\beta)\leq 0}{\operatorname{argmax}}\, U(\beta) \tag{2.24}$$

where $U(\beta)$ *is a* monotone *utility function and* $g_n(\cdot)$ *is monotonically increasing in* β.

The above utility maximization model generalizes Definition 1 since the budget constraint $g_n(\beta) \leq 0$ can accommodate non-linear budgets and includes the linear budget constraint of Definition 1 as a special case. The result below provides an explicit test for a system that maximizes utility in the sense of Definition 2 and constructs a set of utility functions that rationalize the decisions β_n of the utility maximizer.

Theorem 2 (Test for rationality with nonlinear budget [34]). *Let* $B_n = \{\beta \in \mathbb{R}^m_+ | g_n(\beta) \leq 0\}$ *with* $g_n : \mathbb{R}^m \rightarrow \mathbb{R}$ *an increasing, continuous function and* $g_n(\beta_n) = 0$ *for* $n = 1, \dots N$. *Then the following conditions are equivalent:*

1. *There exists a monotone continuous utility function* U *that rationalizes the data set* $\{\beta_n, B_n\}, n = 1, \dots N$. *That is*

$$\beta_n = \underset{\beta}{\operatorname{argmax}}\, U(\beta), \quad g_n(\beta) \leq 0$$

2. *There exist positive reals* $u_t, \lambda_t > 0,\ t = 1, 2, \dots, N$, *such that the following inequalities hold:*

$$u_s - u_t - \lambda_t g_t(\beta_s) \leq 0 \quad \forall t, s \in \{1, 2, \dots N\} \tag{2.25}$$

The monotone, concave utility function given by

$$U(\beta) = \min_{t\in\{1,\dots N\}} \{u_t + \lambda_t g_t(\beta)\} \tag{2.26}$$

constructed using u_t *and* λ_t *defined in (2.25) rationalizes the data set by satisfying (2.24).*

3. *The data set* $\{\beta_n, B_n\}, n = 1, \dots, N$ *satisfies GARP:*

$$g_t(\beta_j) \leq g_t(\beta_t) \implies g_j(\beta_t) \geq 0 \tag{2.27}$$

Like Afriat's theorem, the above result provides a *necessary and sufficient condition* for a system to be a utility maximizer based on the system's input–output response. In spite of a non-linear budget constraint, it can be easily verified that the constructed utility function $U(\beta)$ (2.26) is ordinal since any positive monotone increasing transformation of (2.26) satisfies the GARP inequalities (2.27).

We now justify the non-linear budget constraint in (2.24) in the context of the cognitive radar by formulating an optimization problem the radar solves equivalent to Definition 2. Suppose we observe the radar over $n = 1, 2, \dots, N$ time epochs (slow varying time scale). At the nth epoch, we probe the radar with an interference vector $\alpha_n \in \mathbb{R}^M$. The radar responds with waveform $\beta_n \in \mathbb{R}^M_+$. We assume that the chosen waveform β_n maximizes the radar's underlying utility function while ensuring the radar's SINR exceeds a particular threshold $\delta > 0$, where the SINR of the radar given probe α and response β is defined as

$$\mathrm{SINR}\,(\alpha, \beta) = \frac{\beta' Q \beta}{\beta' P(\alpha) \beta + \gamma}. \tag{2.28}$$

In (2.28), the radar's signal power (numerator) and interference power (first term in denominator) are assumed to be quadratic forms of $Q, P(\alpha)$, respectively, where $Q, P(\alpha) \in \mathbb{R}^{M \times M}$ are positive definite matrices known to us. The term $\gamma > 0$ is the noise power. The SINR definition in (2.28) is a more general formulation of the SCNR (2.34) of a cognitive radar derived in Section 2.4 using clutter response models [46]. The matrices $Q, P(\alpha)$ are analogous to the covariance of the channel impulse response matrices $H_t(\cdot)$ and $H_p(\cdot)$ corresponding to the target and clutter (external interference) channels, respectively (see Section 2.4.1 for a discussion).

Having defined the SINR above in (2.28), we now formalize the radar's response β_n given probe α_n, $n = 1, 2, \ldots$ as the solution of the following constrained optimization problem:

$$\begin{aligned} \beta_n \; &\in \operatorname*{argmax}_{\beta} U(\beta) \\ \text{s.t.} \quad & \mathrm{SINR}\,(\alpha_n, \beta) \geq \delta \end{aligned} \tag{2.29}$$

Clearly, the above setup falls under the non-linear utility maximization setup in Definition 2 by defining the non-linear budget $g_n(\,\cdot\,)$ as $g_n(\beta) = \delta - SIR(\alpha_n, \beta)$ where $SIR(\,\cdot\,)$ is defined in (2.28). It only remains to show that this definition of $g_n(\beta)$ is monotonically increasing in β. Theorem 3 stated below establishes two conditions that are sufficient for $g_n(\beta)$ to be monotonically increasing in β.

Theorem 3. *Suppose that the adversary radar uses the SINR constraint* (2.29). *Then* $g_n(\beta) = \delta - \mathrm{SIR}\,(\alpha_n, \beta)$ *is monotonically increasing in* β *if the following two conditions hold.*

1. *The matrix* Q *is a diagonal matrix with off-diagonal elements equal to zero.*
2. *The matrix* $\left(\frac{c_{P(\alpha_n)}}{d_Q} P(\alpha_n) - Q\right)$ *is component-wise less than* 0 *for all* $n \in \{1, 2, \ldots, N\}$, *where* $c_{P(\alpha_n)} > 0$ *and* $d_Q > 0$ *denote the smallest and largest eigenvalues of* $P(\alpha_n)$ *and* Q, *respectively.*

The proof of Theorem 3 follows from elementary calculus and is omitted for brevity. Hence, assuming the two conditions hold in Theorem 3, we can use the results from Theorem 2 to test if the radar satisfies utility maximization in its waveform design (2.29) and also estimate the set of feasible utility functions $U(\cdot)$ (2.29) that rationalizes the radar's responses $\{\beta_n\}$.

2.4 Designing smart interference to confuse cognitive radar

This section discusses how we can engineer external interference (a probing signal) at the physical layer level to confuse a cognitive radar. By abstracting the probing signal to a channel in the frequency domain, our objective is to minimize the signal power of the interference generated by us while ensuring that the SCNR of the radar does not exceed a pre-defined threshold. The setup is schematically shown in Figure 2.5. Note that the level of abstraction used in this section is at the Wiener filter pulse/waveform level; whereas the previous two sections were at the system level (which uses the utility maximization framework) and tracker level (which uses the Kalman filter formalism), respectively. This is consistent with the design theme of sense globally (high level of abstraction) and act locally (lower level of abstraction). Design of smart interference can also be used in situations when the adversary is trying to eavesdrop 'Us' and our aim is to confuse the eavesdropping adversary [47].

As can be seen in the SCNR expression in (2.34), the interference signal power manifests as additional clutter perceived by the radar in the denominator thus forcing the SCNR to go down. The radar then re-designs its waveform to maximize its SCNR given our interference signal. We observe the radar's chosen waveform in noise. Our task can thus be re-formulated as choosing the interference signal with minimal power while ensuring that with probability at least $1 - \varepsilon$, the optimized SCNR lies below a threshold level Δ (ε, Δ are user-defined quantities). This approach closely follows the formulation in Section 2.3.3 where the cognitive radar chooses the optimal waveform while ensuring the SINR exceeds a threshold value. Further, the SCNR of the adversary's radar defined in (2.35) below can be interpreted as a monotone function of the radar's utility function in the abstracted setup of Section 2.3.3, since in complete analogy to the utility maximization model of Section 2.3.3, this section assumes that the radar maximizes its SCNR in the presence of smart interference signals (probes).

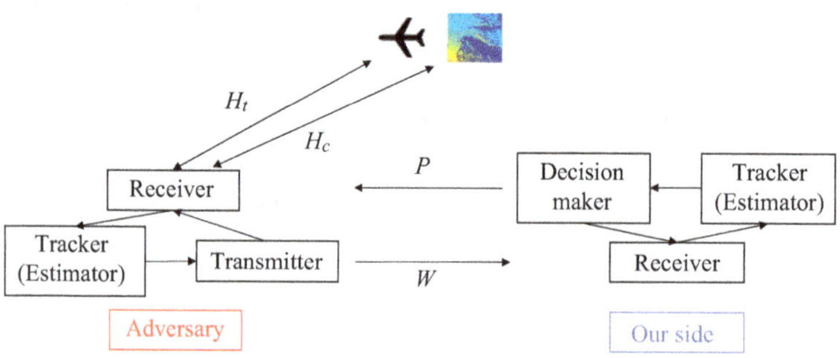

Figure 2.5 Schematic of transmit channel H_t, clutter channel H_c and interference channel P involving an adversarial cognitive radar and us. We observe the radar's waveform W in noise. The aim is to engineer the interference channel P to confuse the cognitive radar.

2.4.1 Interference signal model

We first characterize how a cognitive radar optimally chooses its waveform based on its perceived interference. The radar's objective is to choose the optimal waveform that maximizes its signal-to-interference-plus-noise (SINR) ratio.

Suppose we observe the radar over $l = 1, 2 \ldots L$ pulses, where each pulse comprises $n = 1, 2, \ldots N$ discrete time steps. A single-input single-output (SISO) radar system has two channel impulse responses, one for the target and the other for clutter. Let $w(n)$ denote the radar transmit waveform and $h_t(n)$, $h_c(n)$ denote the target and clutter channel impulse responses, respectively. Then, the radar measurements corresponding to the lth pulse can be expressed as

$$x(n,l) = h_t(n,l) \circledast w(n,l) + h_c(n,l) \circledast w(n,l) + e_r(n,l) \tag{2.30}$$

where $\circledast$ represents a convolution operator and $e_r(n,l)$ is the radar measurement noise modeled as an i.i.d. random variable with zero mean and known variance σ_r^2. We model the radar's measurement using the stochastic Green's function impulse response model presented in [46], where the radar's electromagnetic channel is modeled using a physics-based impulse response.

Since convolution in the time domain can be expressed as multiplication in the frequency domain (with notation in upper case), we can express the measurements in the frequency domain as follows:

$$X(k,l) = H_t(k,l)W(k,l) + H_c(k,l)W(k,l) + E_r(k,l) \tag{2.31}$$

where $k \in \mathcal{K} = \{1, \ldots, K\}$ is the frequency bin index. Equation (2.31) can be extended to an $I \times J$ MIMO radar and the received signal at the jth receiver is given by

$$X_j(k,l) = \sum_{i=1}^{I} H_{t_{ij}}(k,l)W_i(k,l) + H_{c_{ij}}(k,l)W_i(k,l) + E_{r,j}(k,l), \tag{2.32}$$

$\forall k \in \{1, \ldots K\}$. Using matrices and vectors obtained by stacking and concatenating (2.32) for all i, j, and k, the MIMO radar measurement model at the lth pulse in vector–matrix form can be expressed as

$$\boldsymbol{X}(l) = \boldsymbol{H}_t(l)\boldsymbol{W}(l) + \boldsymbol{H}_c(l)\boldsymbol{W}(l) + \boldsymbol{E}_r(l) \tag{2.33}$$

where $\boldsymbol{X}(l) \in \mathbb{C}^{(J\times K)\times 1}$ is the received signal vector, $\boldsymbol{H}_c(l), \boldsymbol{H}_t(l) \in \mathbb{C}^{(J\times K)\times(I\times J\times K)}$ are the effective transmit and clutter channel impulse response matrices, respectively, and $\boldsymbol{W}(l) \in \mathbb{C}^{(I\times J\times K)\times 1}$ is the radar's effective waveform vector. $\boldsymbol{E}_r(l) \in \mathbb{C}^{(J\times K)\times 1}$ is the effective additive noise vector modeled as a zero mean i.i.d. random variable (independent over pulses) with covariance matrix $C_r \in \mathbb{R}^{(J\times K)\times(J\times K)}$, $C = (\sigma_r^2/K)\boldsymbol{I} = \tilde{\sigma}_r^2\boldsymbol{I}$. The block diagram in Figure 2.5 shows the entire procedure for this model.

2.4.2 Smart interference for confusing the radar

The aim of this section is to design optimal interference signals (to confuse the adversary cognitive radar) by solving a probabilistically constrained optimization problem.

At the beginning of the lth pulse, the adversary radar transmits a pilot signal to estimate the transmit and clutter channel impulse responses $\boldsymbol{H}_t(l)$ and $\boldsymbol{H}_c(l)$, respectively. Assuming that it has a perfect estimate of $\boldsymbol{H}_t(l)$ and $\boldsymbol{H}_c(l)$, the radar then chooses the optimal waveform $\boldsymbol{W}^*(l)$ such that SCNR defined below in (2.34) is maximized. The radar's waveform $\boldsymbol{W}^*(l)$ is the solution to the following optimization problem

$$\boldsymbol{W}^*(l) = \underset{\boldsymbol{W}(l):\|\boldsymbol{W}(l)_2=1\|}{\text{argmax}} \ \text{SCNR}(\boldsymbol{H}_t(l), \boldsymbol{H}_c(l), \boldsymbol{W}(l)), \tag{2.34}$$

where the SCNR is defined as

$$\text{SCNR}(\boldsymbol{H}_t, \boldsymbol{H}_c, \boldsymbol{W}) = \frac{\|\boldsymbol{H}_t \boldsymbol{W}\|_2^2}{\mathbb{E}\{\|\boldsymbol{H}_c \boldsymbol{W} + \boldsymbol{E}_r\|_2^2\}}. \tag{2.35}$$

Denote the maximum SCNR achieved in (2.34) as

$$\text{SCNR}_{\max}(\boldsymbol{H}_t(l), \boldsymbol{H}_c(l), \sigma_r^2) = \text{SCNR}(\boldsymbol{H}_t(l), \boldsymbol{H}_c(l), \boldsymbol{W}^*(l)). \tag{2.36}$$

Given $\boldsymbol{H}_t(l), \boldsymbol{H}_c(l)$ and the radar's measurement noise power σ_r^2, the radar generates an optimal waveform at the lth pulse using (2.34) as the solution to the following eigenvector problem [5]:

$$\begin{aligned} \boldsymbol{A}\,\boldsymbol{W}^\star(l) &= \lambda_l \boldsymbol{W}^\star(l) \\ \boldsymbol{A} &= \left((\boldsymbol{H}_c(l)'\boldsymbol{H}_c(l) + \tilde{\sigma}_r^2 \boldsymbol{I})^{-1} \boldsymbol{H}_t(l)'\boldsymbol{H}_t(l)\right), \end{aligned}$$

Here $(\cdot)'$ denotes the Hermitian transpose operator.

As an external observer, we send a sequence of probe signals $P = \{\boldsymbol{H}_p(l), l \in \{1, 2, \dots L\}\}$ over L pulses to confuse the adversary radar and degrade its performance. The interference signal $\boldsymbol{H}_p(l-1)$ at the $(l-1)$th affects only radar's clutter channel impulse response $\boldsymbol{H}_c(l)$ at the lth pulse which subsequently results in the change of optimal waveform (2.34) chosen by the radar $\boldsymbol{W}^*(l)$. We measure the optimal waveform at the lth pulse in noise as $\boldsymbol{Y}(l)$. We assume constant transmit and clutter channel impulse responses $\boldsymbol{H}_t, \boldsymbol{H}_c$ in the absence of the probe signals P. The dynamics of our interaction with the adversary radar due to probe P are as follows and shown schematically in Figure 2.6:

$$\boldsymbol{H}_c(l) = \boldsymbol{H}_c + \boldsymbol{H}_p(l-1) \tag{2.37}$$

$$\boldsymbol{H}_t(l) = \boldsymbol{H}_t \tag{2.38}$$

$$\begin{aligned} \big((\boldsymbol{H}_c(l)' \quad & \boldsymbol{H}_c(l) + \tilde{\sigma}_r^2 \boldsymbol{I})^{-1} \boldsymbol{H}_t(l)'\boldsymbol{H}_t(l)\big)\boldsymbol{W}^*(l) \\ &= \lambda_l \boldsymbol{W}^*(l) \end{aligned} \tag{2.39}$$

$$\boldsymbol{Y}(l) = \boldsymbol{W}^*(l) + \boldsymbol{E}_o(l). \tag{2.40}$$

In (2.40), $\boldsymbol{E}_o(l)$ is our measurement noise modeled as a zero mean i.i.d. random variable (independent over pulses) sampled from a known pdf f_o with zero mean and covariance $C_o = (\sigma_o^2/K)\boldsymbol{I} = \tilde{\sigma}_o^2 \boldsymbol{I}$.

Our objective is to optimally design the probe signals $P^* = \{\boldsymbol{H}_p^*(l), l \in \{1, \dots L\}\}$ that minimizes the interference signal power such that for a pre-defined $\Delta > 0$, there

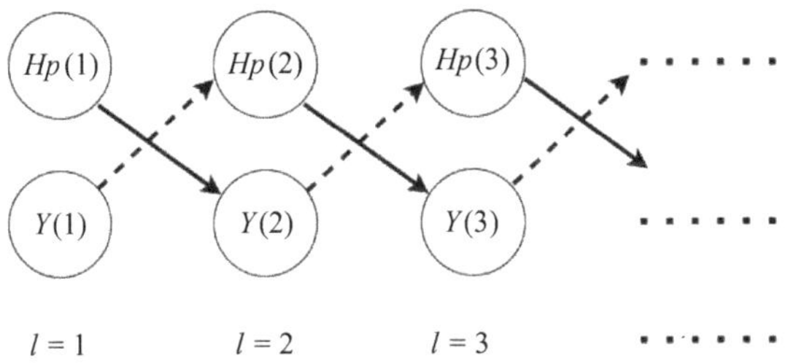

Figure 2.6 Schematic of smart interference design to confuse the cognitive radar. The interference signal at the lth pulse affects the waveform choice of the radar in the $(l+1)$th pulse. We record the noisy waveform measurement $\boldsymbol{Y}(l+1)$ and generate the interference signal for the $(l+2)$th pulse.

exists $\varepsilon \in [0, 1)$ such that the probability that the SCNR of the radar lies below Δ exceeds $(1-\varepsilon)$, for all $l = 1, 2, \dots L$:

$$\begin{aligned} &\min_{\{\boldsymbol{H}_p(l), l \in \{1,2,\dots L\}\}} \sum_{l=1}^{L} \boldsymbol{H}_p(l)' \boldsymbol{H}_p(l) \\ &\text{s.t. } \mathbb{P}_{f_o}(\,\text{SCNR}\,(\boldsymbol{H}_t(l), \boldsymbol{H}_c(l), \boldsymbol{Y}(l)) \leq \Delta) \geq 1-\varepsilon, \\ &\forall l \in \{1, 2, 3, \dots L\}. \end{aligned} \tag{2.41}$$

Here, $\mathbb{P}_f(\cdot)$ denotes the probability wrt pdf f. The design parameter Δ is the SCNR (performance) upper bound of the cognitive radar. To confuse the radar, our task is to ensure that the SCNR of the radar is less than Δ with probability at least $1-\varepsilon$. Hence, ε is the maximum probability of failure to confuse the radar with our smart interference signals. Although not shown explicitly, the $\text{SCNR}_{\max}$ expression in (2.41) depends on our interference signal $\boldsymbol{H}_p$ as depicted in (2.37).

Solving the non-convex optimization problem (2.41) is challenging except for trivial cases. It involves two inter-related components: (i) estimating the transmit and clutter channel impulse responses $\boldsymbol{H}_t, \boldsymbol{H}_c$ from observation $\boldsymbol{Y}(l)$ and (ii) using the estimated value of channel impulse responses to generate the interference signal $\boldsymbol{H}_p(l)$. Moreover, solving for $\boldsymbol{H}_c$ and $\boldsymbol{H}_t$ from recursive equations (2.37) through (2.40) for $l = 1, \dots, L$ is a challenging problem since it does not have an analytical closed form solution.

With the above formulation, we can now discuss the construction of smart interference to confuse the radar. The cognitive radar maximizes its energy in the direction of its target impulse response and transfer function. As soon as we have an accurate estimate of the target channel transfer function from the L pulses, we can immediately generate signal dependent interference that nulls the target returns. Even if the clutter channel impulse response changes after we perform our estimation, since the target channel is stationary for longer durations, the signal-dependent interference will work successfully for several pulses after we conclude the estimate. The main take away from this approach is that we are exploiting the fact that the cognitive radar provides information about its channel by optimizing the waveform with respect to its environment.

2.4.3 Numerical example illustrating design of smart interference

We conclude this section with a numerical example that illustrates the smart interference framework developed above. The simulation setup is as follows:

- $L = 2$ pulses (optimization horizon in (2.41)).
- Impulse response matrices for transmit channel $\boldsymbol{H}_t = [7\ 7]$, clutter channel $\boldsymbol{H}_c = [1\ 1]$, and adversary radar noise covariance $\tilde{\sigma}_r^2 = 1$ (2.33).
- *Design parameters*: SCNR upper bound $\Delta = \{2.8, 3, 3.2\}$, minimum probability of success
  $\varepsilon = 0.2, 0.3$ (2.41).
- Probe signals for pulse index:

$$l = 1,\ \boldsymbol{H}_p(1) = [0.2r\ 0.5r], \tag{2.42}$$

$$l = 2,\ \boldsymbol{H}_p(2) = [0.4r\ 0.4r]. \tag{2.43}$$

  The smart interference parameter $r > 0$ parametrizes the magnitude of the probe signals. The aim is to find the optimal probe signals $\boldsymbol{H}_p(l),\ l = 1, 2$ parametrized by r in (2.43) that solves (2.41). The corresponding value of r is our optimal smart interference parameter.
- Our measurement noise covariance is $\tilde{\sigma}_o^2 = 0.1$ (2.40).

Figure 2.7 displays the performance of the cognitive radar as our smart interference parameter r is varied. It shows that increasing r leads to increased confusion (worse SCNR performance) of the cognitive radar. Specifically, we plot the LHS of (2.41), namely, $\mathbb{P}_{f_o}(\text{SCNR}(\boldsymbol{H}_t(l), \boldsymbol{H}_c(l), \boldsymbol{Y}(l)) \leq \Delta)$, for SCNR upper bound $\Delta \in \{2.8, 3, 3.2\}$. Recall that this is the probability with which the maximum SCNR of the radar (2.36) lies below Δ.

To glean insight from Figure 2.7, let $r^*(\Delta, \varepsilon)$ denote the optimal smart interference parameter such that solves (2.41) for design parameters Δ and ε. Figure 2.7 shows that $r^*(\Delta, \varepsilon)$ decreases with both design parameters Δ and ε. This can be justified as follows. For a fixed value of failure probability ε, increasing the upper bound Δ implies that the constraint (2.41) is satisfied for smaller r, hence, the optimal interference parameter $r^*(\Delta, \varepsilon)$ decreases with Δ. Recall ε upper bounds the probability with which the maximum SCNR of the radar exceeds Δ. Increasing ε (or equivalently, relaxing the maximum probability of failure) allows us to decrease the magnitude of the probe signals without violating the constraint in (2.41) for a fixed Δ. Hence, $r^*(\Delta, \varepsilon)$ decreases with both Δ and ε.

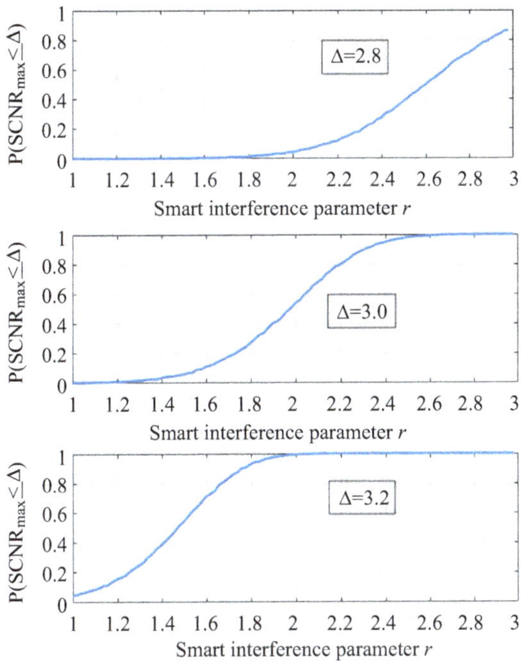

Figure 2.7 The figure illustrates the performance of the cognitive radar as our smart interference parameter r in (2.43) *is varied. The plots display the LHS of* (2.41)*, namely, probability that the radar's maximum SCNR (which depends on r* (2.43)*) is smaller than threshold* Δ*. The probability curves are plotted for* $\Delta = 2.8, 3, 3.2$ *and signify the extent of SCNR degradation as a function of the magnitude of the probe signal.*

2.5 Stochastic gradient-based iterative smart interference

Section 2.4 involved designing an optimal smart interference scheme by "Us" to confuse an adversarial cognitive radar. Recall that by "Us," we mean the counter-adversarial system trying to confuse the adversarial cognitive radar. Recall from Section 2.4.2 that the solution to the optimization problem (2.46) generates the optimal probe signal sequence that sufficiently confuses the adversarial cognitive radar with minimum power. In this section, we generalize the optimization problem (2.41) and assume that the time-varying clutter channel response $\boldsymbol{H}_c(l)$ in (2.41) is measured in noise by us. We then propose a stochastic gradient-based iterative algorithm to solve this generalized optimization problem.

This section is organized as follows. In Section 2.5.1, we state the generalized optimization problem for smart interference with measurement noise and the key assumptions needed for the convergence of the stochastic gradient-based iterative algorithms presented in Section 2.5.2 to a locally optimal solution. In Section 2.5.2, we present two iterative algorithms for evaluating a locally optimal solution to the optimization problem stated in Section 2.5.1. The first algorithm is an augmented Lagrangian-based primal dual algorithm which assumes known gradients of the objective function and the constraints of the optimization problem. The second algorithm is a stochastic approximation extension of the first algorithm, which uses approximations of the gradients numerically evaluated via simulation. The reader is encouraged to refer to [48] for the proofs of convergence of the proposed algorithms.

2.5.1 Smart interference with measurement noise

In this section, we state the optimization problem of (2.41) generalized to noisy measurements of the clutter channel $\boldsymbol{H}_c$ using compact notation. First, we define $\bar{\boldsymbol{H}}_p(l) = \text{vec}(\boldsymbol{H}_p(l))$ and $\boldsymbol{H} = [\bar{\boldsymbol{H}}_p(l)', l \in \{1, 2, \dots L\}]$ (thus, $\boldsymbol{H}$ is an $(m \times n \times L)$ dimensional real-valued vector). The objective function and constraints, respectively, in (2.41) can be compactly expressed as

$$C(\boldsymbol{H}) \triangleq \|\boldsymbol{H}\|_2^2,$$

$$B_l(\boldsymbol{H}) \triangleq \mathbb{P}_{f_0}\{SCNR(\boldsymbol{H}_t, \hat{\boldsymbol{H}}_c(l) + \boldsymbol{H}_p(l), \boldsymbol{Y}(l)) \geq 1 - \delta\} - \varepsilon, \; l = 1, 2, \dots, L \quad (2.44)$$

$$\hat{\boldsymbol{H}}_c(l) = \boldsymbol{H}_c + \xi_l, \text{vec}(\xi_l) \sim f_l, \quad (2.45)$$

where f_l is the pdf for r.v. ξ_l. We make two observations here. First, note that the constraint $\boldsymbol{B}_l(\cdot)$ specializes to the constraint in (2.41) by setting $\xi_l = 0$ for all l. Second, the expression within the expectation operator in (2.45) contains two noise sources: (1) measurement of the optimal waveform $\boldsymbol{W}^*(l)$ and (2) measurement of the clutter channel of the cognitive adversarial radar. Comparatively, (2.41) in Section 2.4 included only noise source (1). Given the above compact notation, our generalization of the optimization problem (2.41) can be stated as

$$\boldsymbol{H}^* \triangleq \{\boldsymbol{H}_p^*(l), l = 1, 2, \dots, L\} = \underset{\boldsymbol{H} \in \mathbb{R}^{mnL \times 1}}{\text{argmin}} \; C(\boldsymbol{H}),$$

$$\text{s.t. } B_l(\boldsymbol{H}) \leq 0, \; l = 1, 2, \dots L, \quad (2.46)$$

where the objective $C(\cdot)$ and constraints $B_l(\cdot)$ are defined in (2.45). The solution of the optimization problem (2.46), $\boldsymbol{H}^*$, is the optimal smart interference signal sequence for the generalized interference design problem.

2.5.2 Algorithms for solving constrained optimization problem (2.41)

In this section, we propose two algorithms to solve the constrained optimization problem (2.46): (1) a deterministic iterative algorithm [(2.51) and (2.52)] which assumes the gradients $\nabla_{\boldsymbol{H}} B_l(\boldsymbol{H})$ can be computed in closed form, and (2) a stochastic approximation version of the algorithm in (1), Algorithm 1, which uses a simultaneous perturbation method to approximate the gradients $\nabla_{\boldsymbol{H}} B_l(\boldsymbol{H})$ when the gradients cannot be computed.

We start by stating the key assumptions about the counter adversarial system, "Us" and the optimization problem so that the iterative algorithms outlined below converge to a locally optimal solution of the optimization problem (2.46).

2.5.2.1 Assumptions

We make the following assumptions about "Us" that solves the optimization problem (2.41).

1. "Us," the counter-autonomous system has access to $\widehat{\boldsymbol{H}}_t, \widehat{\boldsymbol{H}_c}$, which are unbiased estimates of the true transmit and clutter channel impulse response vectors $\boldsymbol{H}_t, \boldsymbol{H}_c$.
2. The functions $C(\boldsymbol{H})$, $B_l(\boldsymbol{H}) \in C^2$ (twice continuously differentiable).
3. The minima $\boldsymbol{H}^*$ of (2.41) is regular, i.e., $\nabla_{\boldsymbol{H}} B_l(\boldsymbol{H}^*)$, $l = 1, 2, \ldots L$ are linearly independent.

If the above assumptions hold, then $\boldsymbol{H}^*$ belongs to the set of Kuhn–Tucker points

$$\begin{aligned} \mathrm{KT} = \{&\boldsymbol{H} \in \mathbb{R}^{(mnL)\times 1} : \exists \lambda_l \geq 0,\ l = 1, 2, \ldots L \text{ s.t. } \nabla_{\boldsymbol{H}} C(\boldsymbol{H}) \\ &+ \sum_{l=1}^{L} \lambda_l \nabla_{\boldsymbol{H}} B_l(\boldsymbol{H}) = 0,\ \lambda_l B_l(\boldsymbol{H}) = 0\} \end{aligned} \tag{2.47}$$

Moreover, $\boldsymbol{H}^*$ satisfies the second-order sufficiency condition $\nabla^2_{\boldsymbol{H}} C(\boldsymbol{H}^*) + \sum_{l=1}^{L} \lambda_l \nabla^2_{\boldsymbol{H}} B_l(\boldsymbol{H}^*) > 0$ (positive definite) for any $\lambda_1, \ldots \lambda_L > 0$ s.t. $\sum_{l=1}^{L} \lambda_l \nabla_{\boldsymbol{H}} B_l (\boldsymbol{H}^*) = 0$ and $\nabla_{\boldsymbol{H}} B_l(\boldsymbol{H}^*) = 0$.

2.5.2.2 Deterministic algorithm for optimization problem (2.46)

Consider the optimization problem (2.46). Suppose the gradients $\nabla_{\boldsymbol{H}} B_l(\boldsymbol{H})$ can be computed for any $\boldsymbol{H} \in \mathbb{R}^{mnL}$. In the context of (2.46), this assumption means that we know the noise pdfs f_l, $l = 1, 2, \ldots L$ in (2.45). A widely used deterministic optimization method for handling constraints is based on the method of Lagrange multipliers and uses a first-order primal dual algorithm [49, p. 446] described later; the authors of [50] provide a stochastic approximation extension to the deterministic methods discussed here. First, convert the inequality constraints in (2.46) to equality constraints by introducing the variables $z = (z_1, z_2, \ldots z_L)$ so that $B_l(\boldsymbol{H}) + z_l^2 = 0$. Define $\psi \triangleq (\boldsymbol{H}, z)$, $B_l(\psi) \triangleq B_l(\boldsymbol{H}) + z_l^2$. Define the Lagrangian

$$\mathscr{L}(\psi, \lambda) \triangleq C(\boldsymbol{H}) + \sum_{l=1}^{L} \lambda_l B_l(\psi) \tag{2.48}$$

In order to converge, a primal dual algorithm operating on the Lagrangian requires the Lagrangian to be locally convex at the optimum, i.e., Hessian to be positive definite at the optimum (which is much more restrictive that the second-order sufficiency condition of Assumption 2 in Section 2.5.2.1).

We can "convexify" the problem by adding a penalty term to the objective function (2.46). The resulting problem is:

$$\min_{\psi \in \Psi} C(\boldsymbol{H}) + \rho/2 \sum_{l=1}^{L} (B_l(\psi))^2, \text{ s.t. } B_l(\boldsymbol{H}) \leq 0,\ l = 1, 2, \ldots L, \tag{2.49}$$

where ρ is a large positive constant. As shown in [51, p. 429], the optimum of the above problem is identical to that of 2.46. Define the augmented Lagrangian,

$$\mathscr{L}_\rho(\psi, \lambda) \triangleq C(\boldsymbol{H}) + \sum_{l=1}^{L} \lambda_l B_l(\psi) + \rho/2 \sum_{l=1}^{L} (B_l(\psi))^2 \tag{2.50}$$

Note that although the original Lagrangian may not be convex near the solution (and, hence, the primal dual algorithm does not work), for sufficiently large ρ, the last term in $\mathscr{L}_\rho$ "convexifies" the Lagrangian. For sufficiently large ρ, [49] shows that the augmented Lagrangian is locally convex.

Primal dual algorithm. We now present the following primal dual algorithm operating on the augmented Lagrangian $\mathscr{L}_\rho(\psi, \lambda)$ [49, p. 396]:

$$\begin{aligned} \boldsymbol{H}^\varepsilon(n+1) &= \boldsymbol{H}^\varepsilon(n+1) - \varepsilon\,(\nabla_{\boldsymbol{H}}(C(\boldsymbol{H})) + \nabla_{\boldsymbol{H}} B(\boldsymbol{H}^\varepsilon(n)) \\ &\quad [\lambda^\varepsilon(n) + \rho B(\boldsymbol{H}^\varepsilon(n))]) & (2.51) \\ \lambda^\varepsilon(n+1) &= \max\,[0, \lambda^\varepsilon(n) + B(\boldsymbol{H}^\varepsilon(n))], & (2.52) \end{aligned}$$

where $\varepsilon > 0$ denotes the step size and $B(\boldsymbol{H}) = [B_1(\boldsymbol{H})\ B_2(\boldsymbol{H})\ \ldots B_L(\boldsymbol{H})]'$. In [52], it is shown that there exists a step size $\bar{\varepsilon} > 0$ such that the iterative procedure (2.51) and (2.52) converges to a local Kuhn–Tucker pair $(\boldsymbol{H}^*, \lambda^*)$ for sufficiently large ρ.

Remark: For optimization problem (2.46), the gradient of the objective function $\nabla_{\boldsymbol{H}} C(\boldsymbol{H}) = 2\boldsymbol{H}$. In practical scenarios, the noise pdfs $f_l, l = 1, 2, \ldots L$ (2.45) are not known, hence, computing the gradients $\nabla_{\boldsymbol{H}} B_l(\boldsymbol{H})$ of the constraints (2.45) by "Us" is a non-trivial task. To tackle this problem, we present below a stochastic approximation extension of the above iterative procedure where the gradients $\nabla_{\boldsymbol{H}} B_l(\boldsymbol{H})$ are replaced with numerically evaluated approximations via simulation.

2.5.2.3 Stochastic approximation extension for primal dual algorithm using SPSA

In this section, we present a stochastic approximation-based algorithm to solve the constrained optimization problem (2.46) when the gradients $\nabla_{\boldsymbol{H}} B_l(\boldsymbol{H})$ can only be estimated via simulation. This algorithm follows the iterative procedure of the deterministic primal dual algorithm (2.51), (2.52) by replacing $\nabla_{\boldsymbol{H}} B_l$ in (2.51), (2.52) with the approximated gradients $\widehat{\nabla_{\boldsymbol{H}} B_l}$. Existing works in literature [52,53] propose non-parametric measure valued (weak derivative) gradient estimation methods for online optimization of Markov Decision Processes (MDP). However, here we use the simultaneous perturbation stochastic approximation (SPSA) algorithm [54,55] for gradient approximation.

The SPSA algorithm is a gradient-based stochastic optimization algorithm where the gradient $\nabla_{\boldsymbol{H}} B_l(\boldsymbol{H})$ is estimated numerically by random perturbation. With respect to (2.46), the SPSA algorithm approximates the gradient $\widehat{\nabla_{\boldsymbol{H}} B_l}$ by using only two measurements. The SPSA approach has all components of the vector $\boldsymbol{H}$ randomly perturbed simultaneously with a random vector Δ. Two measurements of $B_l(\boldsymbol{H})$ are obtained for $\boldsymbol{H} \pm \Delta$ via simulation and the gradient is approximated as

$$\widehat{\nabla_H B_l} = \frac{\hat{B}_l(\boldsymbol{H} + \omega\Delta) - \hat{B}_l(\boldsymbol{H} - \omega\Delta)}{2\omega}\Delta.$$

In the above gradient approximation, Δ_i, the ith component of Δ is an i.i.d. Bernoulli random variable with $p(+1) = p(-1) = 0.5$ and ω is an appropriately chosen gradient step size. The nice property of the SPSA algorithm is that estimating the gradient $\widehat{\nabla_H B_l}$ $(l = 1, 2, \ldots L)$ requires only two measurements of $B_l(\boldsymbol{H})$ corrupted by noise (Δ) per iteration in contrast to finite difference stochastic approximation methods (e.g., Kiefer Wolfowitz algorithm), which performs $2d$ function evaluations per iteration to approximate the gradient, where $d = mnL$ is the dimension of $\boldsymbol{H}$.

The complete SPSA algorithm that can be used by "Us" for solving the constrained optimization problem (2.46) is outlined in Algorithm 1. For decreasing step size $\eta_k = 1/k$ $(i \in \{1, 2\})$ in Algorithm 1, the SPSA algorithm in Algorithm 1 converges to a Kuhn–Tucker point in the set KT with probability 1. For constant step size $\eta_k = \eta > 0$, it converges weakly in probability (see [55] for a detailed exposition).

Extensions

The results in this chapter lead to several interesting future extensions. There is strong motivation to determine analytic performance bounds for inverse tracking/filtering and estimation of the adversary's sensor gain. Another aspect (not considered here) is when the adversary does not know the transition kernel of our dynamics; the adversary then needs to estimate this transition kernel, and we need to estimate the estimate of this transition kernel. In future work, we will design the smart interference problem (2.41) as a stochastic control problem; since dynamic programming is intractable we will explore limited look-ahead policies and open-loop feedback control.

Regarding identifying cognitive radars, it is worthwhile developing statistical tests for utility maximization when the response of the utility maximizing adversarial radar is observed in noise; see Varian's work [43] on noisy revealed preference. Ongoing research in developing a dynamic revealed that preference framework will be used to extend the beam allocation problem of Section 2.3.2 to a multi-horizon setup where we analyze batches of adversary responses over multiple slow time scale epochs. Another natural extension is to a Bayesian context, namely, identifying a radar that is a Bayesian utility maximizer. We refer to [41] for seminal work in this area stemming from behavioral economics.

Finally, in the design of controlled interference, it is worthwhile considering a game-theoretic setting where the cognitive radar (adversary) and us interact dynamically. Also, in future work, it is worthwhile to develop a stochastic gradient algorithm for estimating the optimal probe signal.

Algorithm 1: Optimal interference using SPSA

Given noisy cognitive radar measurements $\hat{\boldsymbol{H}}_c,\ \boldsymbol{Y}$, compute optimal probe sequence $\boldsymbol{H}^*$ that solves (2.46).
Step 1. Choose initial values $\boldsymbol{H}_0 \in \mathbb{R}^{mnL}$ for the probe signal and $\lambda_l(0) > 0,\ l = 1, 2, \ldots L$ for the Lagrange multipliers. Also, choose a sufficiently large penalty parameter $\rho > 0$.
Step 2. For iterations $k \geq 0$, obtain the set of measurements $\hat{\boldsymbol{H}}_c(r, l), \boldsymbol{Y}(r, l)$ by probing the cognitive radar with probe signals $\vec{\boldsymbol{H}_k} = (\boldsymbol{H}_k(1), \ldots, \boldsymbol{H}_k(L))$ for R successive time horizons (the probe signals are kept identical over the R iterations) of L time steps each.¶¶ Here, r indexes the fast time scale and l indexes the slow time scale. Estimate $B_l(\boldsymbol{H}_k)$ as

$$\hat{B}_l(\boldsymbol{H}_k) = \frac{1}{R}\sum_{r=1}^{R} SCNR(\boldsymbol{H}_t, \hat{\boldsymbol{H}}_c(r, l) + \boldsymbol{H}_k(l), \boldsymbol{Y}(r, l)) \geq 1 - \delta\} - \varepsilon \tag{2.53}$$

The parameter R controls the accuracy of the empirical estimate $\hat{B}_l(\boldsymbol{H}_k)$.
Step 3. Compute the gradient estimates $\widehat{\nabla_{\boldsymbol{H}} B_l}(\boldsymbol{H}_k)$ for all $l = 1, 2, \ldots L$ for updating $\boldsymbol{H}_k$

$$\widehat{\nabla_{\boldsymbol{H}} B_l}(\boldsymbol{H}_k) = \frac{\hat{B}_l(\boldsymbol{H}_k + \omega_k \Delta_k) - \hat{B}_l(\boldsymbol{H}_k - \omega_k \Delta_k)}{2\omega_k}\Delta_k. \tag{2.54}$$

In the above gradient estimate, gradient step size $\omega_k = \frac{\omega}{(k+1)^\gamma}$ with $\gamma \in [0.5, 1]$ and $\omega > 0$, and perturbations $\Delta_k(i) = \begin{cases} +1 & \text{with prob. } 0.5, \\ -1 & \text{with prob. } 0.5. \end{cases}$
Step 4. Update $\boldsymbol{H}_k$ with step size $\eta_k = \frac{\eta}{(k+1+s)^\xi},\ \xi \in [0.5, 1],\ \eta > 0$ as

$$\boldsymbol{H}_{k+1} = \boldsymbol{H}_k - \eta_k\left(2\boldsymbol{H}_k + \sum_{l=1}^{L}(\lambda_l(k) + \rho\hat{B}_l(\boldsymbol{H}_k))\widehat{\nabla_{\boldsymbol{H}}\hat{B}_l}(\boldsymbol{H}_k)\right),$$

$$\lambda_l(k+1) = \max(0, \lambda_l(k) + \rho\hat{B}_l(\boldsymbol{H}_k)).$$

Set $k \rightarrow k + 1$, go to step 2.

Acknowledgment

This research was partially supported by the Army Research Office Grant W911NF-21-1-0093 and the Air Force Office of Scientific Research Grant FA9550-22-1-0016.

References

[1] E. K. P. Chong, C. Kreucher, and A. Hero. Partially observable Markov decision process approximations for adaptive sensing. *Discrete Event Dynamic Systems*, 19(3):377–422, 2009.

[2] V. Krishnamurthy and D. Djonin. Structured threshold policies for dynamic sensor scheduling—a partially observed Markov decision process approach. *IEEE Transactions on Signal Processing*, 55(10):4938–4957, 2007.

[3] V. Krishnamurthy and D. Djonin. Optimal threshold policies for multivariate POMDPs in radar resource management. *IEEE Transactions on Signal Processing*, 57(10), 2009.

[4] S. Haykin. Cognitive dynamic systems: radar, control, and radio [point of view]. *Proceedings of the IEEE*, 100(7):2095–2103, 2012.

[5] J. S. Bergin, J. R., R. M. Guerci, and M. Rangaswamy. MIMO Clutter Discrete Probing for Cognitive Radar. In *IEEE International Radar Conference*, April 2015, pp. 1666–1670.

[6] J. R. Guerci, J. S. Bergin, R. J. Guerci, M. Khanin, and M. Rangaswamy. A new MIMO clutter model for cognitive radar. In *IEEE Radar Conference*, May 2016.

[7] A. Charlish and F. Hoffmann. Anticipation in cognitive radar using stochastic control. In *2015 IEEE Radar Conference (RadarCon)*, IEEE, 2015, pp. 1692–1697.

[8] V. Krishnamurthy, K. Pattanayak, S. Gogineni, B. Kang, and M. Rangaswamy. Adversarial radar inference: Inverse tracking, identifying cognition and designing smart interference, 2020. *arXiv preprint arXiv:2008.01559*.

[9] V. Krishnamurthy and M. Rangaswamy. How to calibrate your adversary's capabilities? Inverse filtering for counter-autonomous systems. *IEEE Transactions on Signal Processing*, 67(24):6511–6525, 2019.

[10] V. Krishnamurthy, D. Angley, R. Evans, and W. Moran. Identifying cognitive radars – inverse reinforcement learning using revealed preferences. *IEEE Transactions on Signal Processing*, 2019 (in press; also available on arxiv: https://arxiv.org/abs/1912.00331).

[11] R. Mattila, C. Rojas, V. Krishnamurthy, and B. Wahlberg. Inverse filtering for hidden Markov models. In *Advances in Neural Information Processing Systems*, 2017, pp. 4204–4213.

[12] R. Mattila, C. Rojas, V. Krishnamurthy, and B. Wahlberg. Inverse filtering for linear Gaussian state-space models. In *Proceedings of IEEE Conference on Decision and Control*, Miami, FL, USA, pp. 5556–5561, 2018.

[13] R. Mattila, I. Lourenço, C. R. Rojas, V. Krishnamurthy, and B. Wahlberg. Estimating private beliefs of Bayesian agents based on observed decisions. *IEEE Control Systems Letters*, 3:523–528, 2019.

[14] C. Chamley. *Rational Herds: Economic Models of Social Learning*. Cambridge: Cambridge University Press, 2004.

[15] G. Angeletos, C. Hellwig, and A. Pavan. Dynamic global games of regime change: Learning, multiplicity, and the timing of attacks. *Econometrica*, 75(3):711–756, 2007.

[16] V. Krishnamurthy. *Partially Observed Markov Decision Processes. From Filtering to Controlled Sensing*. Cambridge: Cambridge University Press, 2016.

[17] V. Krishnamurthy. Quickest detection POMDPs with social learning: Interaction of local and global decision makers. *IEEE Transactions on Information Theory*, 58(8):5563–5587, 2012.

[18] V. Krishnamurthy. Bayesian sequential detection with phase-distributed change time and nonlinear penalty – a lattice programming POMDP approach. *IEEE Transactions on Information Theory*, 57(3):7096–7124, Oct. 2011.

[19] C.-C. Huang, B. Amini, and R. R. Bitmead. Predictive coding and control. *IEEE Transactions on Control of Network Systems*, 6(2):906–918, 2018.

[20] H. Singh, A. Chattopadhyay, and K. V. Mishra. Inverse extended Kalman filter, 2022. *arXiv preprint arXiv:2201.01539*.

[21] D. Ciuonzo, P. K. Willett, and Y. Bar-Shalom. Tracking the tracker from its passive sonar ML-PDA estimates. *IEEE Transactions on Aerospace and Electronic Systems*, 50(1):573–590, 2014.

[22] M. A. Iglesias, K. J. Law, and A. M. Stuart. Ensemble Kalman methods for inverse problems. *Inverse Problems*, 29(4):045001, 2013.

[23] V. Krishnamurthy, E. Leoff, and J. Sass. Filter-based stochastic volatility in continuous-time hidden Markov models. *Econometrics and Statistics*, 6:1–21, 2018.

[24] R. J. Elliott, L. Aggoun, and J. B. Moore. *Hidden Markov Models – Estimation and Control*. New York, NY: Springer-Verlag, 1995.

[25] B. Ristic, S. Arulampalam, and N. Gordon. *Beyond the Kalman Filter: Particle Filters for Tracking Applications*. Artech, 2004.

[26] O. Cappe, E. Moulines, and T. Ryden. *Inference in Hidden Markov Models*. New York, NY: Springer-Verlag, 2005.

[27] P. Del Moral and E. Rio. Concentration inequalities for mean field particle models. *Annals of Applied Probability*, 21(3):1017–1052, 2011.

[28] J. Marion. *Finite Sample Bounds and Path Selection for Sequential Monte Carlo*. PhD thesis, Duke University, 2018.

[29] J. Cavanaugh and R. Shumway. On computing the expected fisher information matrix for state space model parameters. *Statistics & Probability Letters*, 26:347–355, 1996.

[30] B. D. O. Anderson and J. B. Moore. *Optimal Filtering*. Englewood Cliffs, NJ: Prentice Hall, 1979.

[31] P. Caines. *Linear Stochastic Systems*. New York, NY: Wiley, 1988.

[32] S. Haykin. Cognitive radar. *IEEE Signal Processing Magazine*, pages 30–40, 2006.

[33] H. Varian. Revealed preference and its applications. *The Economic Journal*, 122(560):332–338, 2012.

[34] F. Forges and E. Minelli. Afriat's theorem for general budget sets. *Journal of Economic Theory*, 144(1):135–145, 2009.

[35] W. Diewert. Afriat's theorem and some extensions to choice under uncertainty. *The Economic Journal*, 122(560):305–331, 2012.

[36] A. Kuptel. Counter unmanned autonomous systems (cuaxs): Priorities. policy. future capabilities. *Multinational Capability Development Campaign (MCDC)*, 2017.
[37] S. Afriat. The construction of utility functions from expenditure data. *International economic review*, 8(1):67–77, 1967.
[38] S. Afriat. *Logic of Choice and Economic Theory*. Oxford: Clarendon Press, 1987.
[39] H. Varian. Non-parametric tests of consumer behaviour. *The Review of Economic Studies*, 50(1):99–110, 1983.
[40] H. Varian. The nonparametric approach to demand analysis. *Econometrica*, 50(1):945–973, 1982.
[41] A. Caplin and M. Dean. Revealed preference, rational inattention, and costly information acquisition. *The American Economic Review*, 105(7):2183–2203, 2015.
[42] W. Hoiles, V. Krishnamurthy, and K. Pattanayak. Rationally inattentive inverse reinforcement learning explains YouTube commenting behavior. *Journal of Machine Learning Research*, 21(170):1–39, 2020.
[43] H. Varian. Revealed preference. In *Samuelsonian Economics and the Twenty-First Century*, Oxford: Oxford University Press, 2006, pp. 99–115.
[44] J. Seo, Y. Sung, G. Lee, and D. Kim. Training beam sequence design for millimeter-wave MIMO systems: a POMDP framework. *IEEE Transactions on Signal Processing*, 64(5):1228–1242, 2015.
[45] D. Zhang, A. Li, H. Chen, N. Wei, M. Ding, Y. Li, and B. Vucetic. Beam allocation for millimeter-wave MIMO tracking systems. *IEEE Transactions on Vehicular Technology*, 69(2):1595–1611, 2019.
[46] J. Guerci, J. Bergin, R. Guerci, M. Khanin, and M. Rangaswamy. A new MIMO clutter model for cognitive radar. In *2016 IEEE Radar Conference (RadarConf)*. IEEE, 2016, pp. 1 6.
[47] N. Su, F. Liu, Z. Wei, Y.-F. Liu, and C. Masouros. Secure dual-functional radar-communication transmission: exploiting interference for resilience against target eavesdropping, 2021. *arXiv preprint arXiv:2107. 04747*.
[48] F. V. Abad and V. Krishnamurthy. Constrained stochastic approximation algorithms for adaptive control of constrained Markov decision processes. In *42nd IEEE Conference on Decision and Control*, 2003, pp. 2823–2828.
[49] D. P. Bertsekas. Nonlinear programming. *Journal of the Operational Research Society*, 48(3):334–334, 1997.
[50] H. Kushner and G. G. Yin. *Stochastic Approximation and Recursive Algorithms and Applications*, vol. 35. New York, NY: Springer Science & Business Media, 2003.
[51] D. G. Luenberger and Y. Ye. *Linear and Nonlinear Programming*, vol. 2. New York, NY: Springer, 1984.
[52] F. J. V. Abad and V. Krishnamurthy. Self learning control of constrained Markov decision processes—a gradient approach. *Les Cahiers du GERAD ISSN*, 711:2440, 2003.

[53] V. Krishnamurthy and F. V. Abad. Real-time reinforcement learning of constrained Markov decision processes with weak derivatives, 2011. *arXiv preprint arXiv:1110.4946*.

[54] J. C. Spall. An overview of the simultaneous perturbation method for efficient optimization. *Johns Hopkins APL Technical Digest*, 19(4):482–492, 1998.

[55] J. C. Spall. Multivariate stochastic approximation using a simultaneous perturbation gradient approximation. *IEEE Transactions on Automatic Control*, 37(3):332–341, 1992.

Chapter 3
Information integration from human and sensing data for cognitive radar

Baocheng Geng[1] Pramod K. Varshney[2] and Muralidhar Rangaswamy[3]

Cognitive radar, according to IEEE Standard Radar Definitions 686 [1], is a "radar system that in some sense displays intelligence, adapting its operation and its processing in response to a changing environment and target scene." In particular, both the active and passive sensors embedded in a cognitive radar allow it to perceive/learn the dynamically changing environments, e.g., targets, clutter, RF interference, and terrain map. To attain optimized performance for tasks such as detection, tracking and classification, the controller in a cognitive radar adapts the radar architecture and adjusts the resource allocation policy in real time [2–4]. For a wide range of applications, different techniques and methods of adaptation have been proposed, e.g., adaptive revisit time scheduling, waveform selection, antenna beam pattern, and spectrum sharing, to advance the mathematical foundations, assessment and evaluation in the context of cognitive radar [5–10]. Cognitive radar systems and their applications have also been studied from the game-theoretic, learning-theoretic and control-theoretic point of view in different contexts [11–13].

While cognitive approaches and techniques have led to great progress in improving the radar performance in a number of areas, one key challenge of cognitive radar design and implementation is its interaction with the end users, i.e., how to bring humans in the loop for decision making and control. In critical situations such as national security and natural disaster forecasting, incorporating human cognitive strengths and expertise is imperative to improve decision quality and enhance situational awareness (SA). For instance, in electronic warfare (EW) systems, the detection of an adversary radar is required before designing appropriate countermeasures. In such scenarios where the course and success of the campaign depends on a small detail being observed or missed, automatic sensor-only decision making may not be sufficient and it is necessary to incorporate human(s) in the loop of decision making, command and control.

[1]Department of Computer Science, University of Alabama at Birmingham, Birmingham, AL, USA
[2]Department of Electrical Engineering and Computer Science, Syracuse University, Syracuse, NY, USA
[3]Air Force Research Laboratory, Wright Patterson Air Force Base, Dayton, OH, USA

In many applications, humans serve as sensors as well, e.g., scouts monitoring a phenomenon of interest (PoI) to gather intelligence. In next generation cognitive radar systems, it is desirable to establish a framework to capture attributes suggested by human-based sources of information so that information from both the physical sensors and humans can be employed for inference. However, unlike traditional physical sensors/machines* that take objective measurements, humans are subjective in expressing their opinions or decisions. Modeling and analysis of human decision making need to take several factors into account including cognitive biases of humans, mechanisms to handle uncertainties and noise as well as unpredictability of humans, in contrast to decision-making processes consisting of only machine agents.

There have been research efforts that exploit the theory of signal processing and information fusion to analyze and incorporate human-specific factors in decision-making. The authors in [14] employed quantization of prior probabilities to model the fact that humans make categorical perceptions instead of continuous observations for collaborative decision making in a Bayesian framework. In [15,16], the authors have studied the group decision making performance when the human agents are assumed to use random thresholds to make threshold-based binary decisions. Considering that humans are affected by the starting point beliefs, the impact of selecting, ordering, and presentation of data on human decision-making performance has been studied in [17]. In the collaborative human decision-making paradigm, different schemes and fusion rules have been developed to ameliorate the unreliability and uncertainty of human crowd workers [18,19]. Moreover, in [20,21], the authors have included prospect theory (PT) to characterize human cognitive biases such as risk aversion and have studied human decision-making behavior in realistic environments. Information fusion of human-based and machine-based sources of information has also been explored in [22,23] for different scenarios. In [22], the authors show that human cognitive strengths can utilize multimedia data for better interpretation of data. A user refinement stage has been exploited together with the Joint Director of Labs (JDL) fusion model to incorporate human behavioral factors and judgment in decision-making [23].

Battlefield of the future will require seamless integration of the human and the machine expertise where they simultaneously work within the same environment model to understand and solve problems. According to [24], humans surpass machines in their ability to improvise and use flexible procedures, exercise judgment, and reason inductively. Moreover, machines outperform humans in responding quickly, storing a large amount of information, performing routine tasks, and reason deductively (including computational ability). Advanced cognition in future radar systems seeks to build an augmented human–machine symbiosis and merge the best of the human with the best of the machine [25].

We depict a typical human–sensor collaboration network in Figure 3.1, where the subjective opinions of humans and the objective measurements of sensors are aggregated for decision fusion. On the one hand, the measurements taken by the physical

*The terms “sensors” and “machines” will be used interchangeably throughout this chapter.

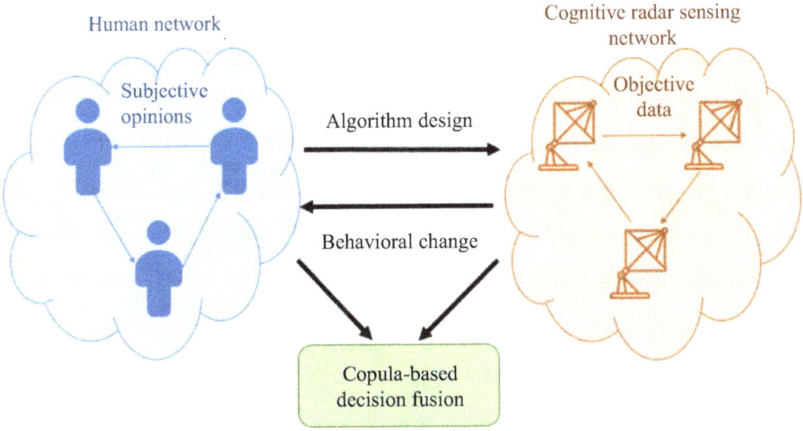

Figure 3.1 Human–sensor collaboration framework for cognitive sensing

sensors affect the behavior, actions, and decisions of the humans. On the other hand, the behavior of humans also determines the optimal decision-making algorithm design in the human–sensor networks. To maximize system performance, efficient implementations of the human–sensor network should be designed in a holistic manner based on the appropriate modeling of human behavior. In this chapter, we provide an overview of these challenges and focus on three specific problems: (i) integration of human decisions with decisions from physical sensors for decision-making, (ii) usage of the behavioral economics concept PT to model human decision-making under cognitive biases, and (iii) human–sensor collaboration for semi-autonomous binary decision-making and Copula-based decision fusion.

The rest of this chapter is organized as follows. In Section 3.1, we present a work that shows how the presence of human sensors can be incorporated into the statistical signal processing framework. We also derive the asymptotic performance of such a human–machine integrated system when the humans possess auxiliary/side information that is not available to the machine. We employ the behavioral economics concept of PT to model human cognitive biases and study the behavior of human decision-making under the binary hypothesis testing framework in Section 3.2. A novel human–machine collaboration paradigm is discussed in Section 3.3 to solve the binary hypothesis testing problem, where the dependency of the human's knowledge and the machine's observation is characterized using Copula theory. Finally, we provide a summary of the current challenges and some research directions related to this problem domain in Section 3.4 before concluding in Section 3.5.

Throughout this chapter, we use an arrow on top of a lowercase letter to denote vectors, e.g., $\vec{x}$. We represent the set of real numbers by $\mathbb{R}$. We denote the transpose by $(\cdot)^T$ and we use $Pr(\cdot)$ to denote probability. $N(\mu, \sigma^2)$ denotes the Gaussian distribution with mean μ and variance σ^2. The notations we use in this chapter are summarized in Table 3.1.

Table 3.1 Glossary of notations

Section 3.1	
x_i	The observation of the ith sensor/human
d_i	The ith sensor's local decision based on x_i
t_i	The ith human's local decision based on x_i
τ_i	The ith sensor's decision threshold
ς_i	The ith human's decision threshold
w_i	The ith human's side information
v_i	The ith human's local decision based on x_i and w_i
C	Chernoff information
Section 3.2	
$\mathrm{v}(\cdot)$	Value function in prospect theory
$\mathrm{w}(\cdot)$	Probability weighting function in prospect theory
α	Probability distortion coefficient in $\mathrm{w}(\cdot)$
β	Loss aversion parameter in $\mathrm{v}(\cdot)$
λ	Diminishing marginal utility parameter in $\mathrm{v}(\cdot)$
Section 3.3	
r	Machine's observation
s	Human's side information
β	Accuracy of the human's side information
C	Copula function
ρ	Correlation parameter

3.1 Integration of human decisions with physical sensors in binary hypothesis testing

In this section, we investigate the problem of decision-making using data collected by physical sensor/device and sources of information from humans. Consider that we have $L = L_1 + L_2$ agents (L_1 physical sensors and L_2 human sensors) that collaboratively perform a binary hypothesis testing task, where the hypotheses are denoted by H_0 and H_1, respectively. We assume that the conditional probability density functions (PDFs) of the observation x under H_j are denoted as $f_j(x)$ for $j = 0, 1$. The observations at the ith agent are represented by x_i, $i = 1, \ldots, L$, which are assumed to be independent and identically distributed (iid) under H_0 and H_1.

3.1.1 Decision fusion for physical sensors and human sensors

In this subsection, we analyze the decision fusion performance when the human agents and physical sensors use different thresholds to make local decisions regarding a given PoI. First, we provide some background on the fusion of binary decisions made by physical sensors only [26]. We consider that the ith sensor's binary decision d_i is made by comparing its observation x_i with a decision threshold

$$d_i = \begin{cases} 1, & \text{if } x_i \geq \tau_i \\ 0, & \text{if } x_i < \tau_i \end{cases} \tag{3.1}$$

for $i = 1, \ldots, L_1$, where τ_i is the decision threshold used by the ith sensor. The fusion center (FC) makes the final decision regarding the true hypothesis based on the decision vector $\vec{d} = [d_1, \ldots, d_{L_1}]$. Let $p_{d,i}$ and $p_{f,i}$ represent the probability of detection and probability of false alarm of the ith sensor when providing its decision d_i. To determine the true hypothesis at the FC based on some observed evidence, it was shown in [26] that the likelihood ratio test (LRT) is optimal as it has the minimum average probability of error:

$$\frac{Pr(\vec{d}|H_1)}{Pr(\vec{d}|H_0)} = \frac{\prod_{i=1}^{L_1} p_{d,i}^{d_i}(1-p_{d,i})^{1-d_i}}{\prod_{i=1}^{L_1} p_{f,i}^{d_i}(1-p_{f,i})^{1-d_i}} \underset{H_0}{\overset{H_1}{\gtrless}} \eta_0 \tag{3.2}$$

where η_0 is an appropriate threshold. In such a distributed detection setup, the authors in [27] proved that it is asymptotically optimal for the physical sensors to use the same decision threshold in order to achieve the best detection accuracy. As a result, we assume that $\tau_i = \tau$, $p_{f,i} = p_f$ and $p_{d,i} = p_d$ for $i = 1, \ldots, L_1$. In this case, the log-likelihood ratio (LLR) can be computed as $G_1 \sum_i d_i + G_0$ with $G_1 = \log \frac{p_d(1-p_f)}{p_f(1-p_d)}$ and $G_0 = L_1 \log \frac{1-p_d}{1-p_f}$. The decision rule (3.2) for final decision making becomes

$$d = \begin{cases} 1, & \text{if } \sum_i d_i \geq \eta_1 \\ 0, & \text{otherwise} \end{cases} \tag{3.3}$$

where $\eta_1 = \frac{\log \eta_0 - G_0}{G_1}$. As $\Lambda_1(\vec{d}) = \sum_i d_i$ follows binomial distribution under both hypotheses, the probabilities of false alarm and detection for the decision rule (3.3) can be computed as [28]

$$P_{f,1} = Pr(\Lambda_1 \geq \eta_1 | H_0) = 1 - \sum_{k=0}^{\lfloor \eta_1 \rfloor} \binom{L_1}{k} p_f^k (1-p_f)^{L_1-k} \tag{3.4}$$

$$P_{d,1} = Pr(\Lambda_1 \geq \eta_1 | H_1) = 1 - \sum_{k=0}^{\lfloor \eta_1 \rfloor} \binom{L_1}{k} p_d^k (1-p_d)^{L_1-k} \tag{3.5}$$

respectively, where the second subscript "1" in $P_{f,1}$ and $P_{d,1}$ refers to case 1 that is composed of L_1 physical sensors. We consider that L_1 is large enough so that $\Lambda_1(\vec{d})$ can be approximated by a Gaussian random variable and the probabilities of false alarm and detection can be computed as

$$P_{f,1} \approx Q\left(\frac{\eta_1 - L_1 p_f}{\sqrt{L_1 p_f(1-p_f)}}\right), \; P_{d,1} \approx Q\left(\frac{\eta_1 - L_1 p_d}{\sqrt{L_1 p_d(1-p_d)}}\right) \tag{3.6}$$

respectively, where $Q(x)$ is the complement of the cumulative distribution function (CDF) of the standard normal distribution with $Q(x) = \frac{1}{\sqrt{2\pi}} \int_x^\infty e^{-u^2/2} du$.

Under the Neyman–Pearson criterion, the aim is to maximize the probability of detection under a false-alarm constraint. If the FC chooses the decision threshold that satisfies $P_{f,1} \leq \omega_f$, the corresponding $P_{d,1}$ can be approximated as [28]

$$P_{d,1} \approx Q\left(\frac{\sqrt{p_f(1-p_f)}Q^{-1}(\omega_f) + \sqrt{L_1}(p_f - p_d)}{\sqrt{p_d(1-p_d)}}\right) \tag{3.7}$$

where $Q^{-1}(\cdot)$ is the inverse function of $Q(x)$. As both p_f and p_d in (3.7) are determined by τ such that $p_f = \int_\tau^\infty f_0(x)dx$ and $p_d = \int_\tau^\infty f_1(x)dx$, the optimal τ that maximizes $P_{d,1}$ is expressed as

$$\arg\max_\tau Q\left(\frac{\sqrt{p_f(\tau)(1-p_f(\tau))}Q^{-1}(\omega_f)+\sqrt{L_1}(p_f(\tau)-p_d(\tau))}{\sqrt{p_d(\tau)(1-p_d(\tau))}}\right) \tag{3.8}$$

In case that the physical sensors are not able to collect sufficiently informative observations, or when the system demands a higher level of detection accuracy such as adversarial target detection and medical diagnosis, incorporation of human sensors-based sources of information is desirable. To characterize the cognitive biases and uncertainties of each human in decision-making, we consider that human sensors employ random thresholds to make binary decisions as discussed in [15].

Consider that L_2 human agents have the same iid observations as the physical sensors x_i for $i = L_1 + 1, \ldots, L_2$, and the ith human's decision rule is to compare the observation of a random threshold ς_i

$$t_i = \begin{cases} 1, & \text{if } x_i \geq \varsigma_i \\ 0, & \text{if } x_i < \varsigma_i \end{cases} \tag{3.9}$$

for $i = L_1 + 1, \ldots, L_1 + L_2$. Unlike physical sensors that can be programmed to use fixed thresholds, we consider that humans use random decision thresholds because of their individual cognitive biases and uncertainties, where the PDF of ς_i is given by, $f_{\varsigma_i}(\varsigma_i)$, which can be estimated by collecting experimental data. For the ith human that employs (3.9), the expected values of probabilities of detection and false alarm can be computed as $\theta_{t,i} = Pr(\tilde{x}_i \geq 0|H_0)$ and $\gamma_{t,i} = Pr(\tilde{x}_i \geq 0|H_1)$, respectively, with $\tilde{x}_i = x_i - \varsigma_i$. If the statistical distribution of ς_i is available, the PDF of $\tilde{x}_i$ under both H_0 and H_1 can be expressed as $f_{\tilde{x}_i}(\tilde{x}_i|H_j) = \int_{-\infty}^\infty f_i(\tilde{x}_i + \varsigma_i) f_{\varsigma_i}(\varsigma_i) d\varsigma_i$ for $j = 0, 1$. Hence, we have $\theta_{t,i} = \int_0^\infty f_{\tilde{x}_i}(\tilde{x}_i|H_0)d\tilde{x}_i$ and $\gamma_{t,i} = \int_0^\infty f_{\tilde{x}_i}(\tilde{x}_i|H_1)d\tilde{x}_i$. Define $\vec{t} = [t_{L_1+1}, \ldots, t_{L_1+L+2}]$ and the LLR at the FC is given by

$$\Lambda_2(\vec{d}, \vec{t}) = G_1 \sum_{i=1}^{L_1} d_i + G_0 + \sum_{i=L_1+1}^{L_1+L_2} t_i \tilde{G}_{1,i} + \tilde{G}_0 \tag{3.10}$$

where $\tilde{G}_{1,i} = \log \frac{\gamma_{t,i}(1-\theta_{t,i})}{\theta_{t,i}(1-\gamma_{t,i})}$ and $\tilde{G}_0 = \sum_{i=L_1+1}^{L_1+L_2} \log \frac{1-\gamma_{t,i}}{1-\theta_{t,i}}$. When the final decision d is made by comparing (3.10) to a suitably designed threshold η_2, the probabilities of detection and false alarm are given by $P_{d,2} = Pr(\Lambda_2(\vec{d},\vec{t}) \geq \eta_2|H_1)$ and $P_{f,2} = Pr(\Lambda_2(\vec{d},\vec{t}) \geq \eta_2|H_0)$, respectively, where the second subscript "2" in $P_{f,2}$ and $P_{d,2}$ refers to case 2 that is composed of L_1 physical sensors and L_2 human sensors.

By utilizing the statistical information of the human sensors' threshold (i.e., the PDFs of ς_i), it is possible to design the optimal threshold τ for the physical sensors so that the decision accuracy of the entire system can be maximized. With a total of $L = L_1 + L_2$ sensors, the LLR at the FC is $\Lambda_2(\vec{d},\vec{t}) = G_1 \sum_{i=1}^{L_1} d_i + G_0 + \sum_{i=L_1+1}^{L_1+L_2} t_i\tilde{G}_{1,i} + \tilde{G}_0$. Let $\tilde{\Lambda}_2(\vec{d},\vec{t}) = G_1 \sum_{i=1}^{L_1} d_i + \sum_{i=L_1+1}^{L_1+L_2} t_i\tilde{G}_{1,i}$ and note that $\sum_i d_i$ can be approximated by Gaussian random variables under both hypotheses. Define $\tilde{\Lambda}_{2,1}(\vec{d}) = G_1 \sum_i d_i$ and $\tilde{\Lambda}_{2,2}(\vec{t}) = \sum_{i=L_1+1}^{L_1+L_2} t_i\tilde{G}_{1,i}$. For the ease of presentation, we assume that the PDF of threshold ς_i employed by all the human agents are the same, then, we have $\tilde{G}_{1,i} = \tilde{G}_1$, $\theta_{t,i} = \theta_t$, and $\gamma_{t,i} = \gamma_t$ for $i = L_1+1,\ldots,L_1+L_2$. As the locals' decisions of the human sensors are independent, $\tilde{\Lambda}_{2,2}/\tilde{G}_1$ represents a sum of iid Bernoulli random variables. Exploiting Gaussian approximation, we get $\tilde{\Lambda}_{2,2}(\vec{t})|H_1 \sim N(L_2\tilde{G}_1\gamma_t, L_2\tilde{G}_1^2\gamma_t(1-\gamma_t))$ and $\tilde{\Lambda}_{2,2}(\vec{t})|H_0 \sim N(L_2\tilde{G}_1\theta_t, L_2\tilde{G}_1^2\theta_t(1-\theta_t))$. Hence, $\tilde{\Lambda}_2$ approximately follows the Gaussian distributions with $\tilde{\Lambda}_2|H_j \sim N(\mu_j(\tau), \sigma_j^2(\tau))$ for $j = 0,1$ where $\mu_0(\tau) = L_1G_1(\tau)p_f(\tau) + L_2\tilde{G}_1\theta_t$, $\mu_1(\tau) = L_1G_1p_d(\tau) + L_2\tilde{G}_1\gamma_t$, $\sigma_0^2(\tau) = L_1G_1^2(\tau)p_f(\tau)(1-p_f(\tau)) + L_2\tilde{G}_1^2\theta_t(1-\theta_t)$ and $\sigma_1^2(\tau) = L_1G_1^2(\tau)p_d(\tau)(1-p_d(\tau)) + L_2\tilde{G}_1^2\gamma_t(1-\gamma_t)$. Moreover, define $\tilde{\eta}_2 = \eta_2 - G_0 - \tilde{G}_0$ and we get

$$P_{d,2}(\tau) \approx Q\left(\frac{\tilde{\eta}_2 - \mu_1(\tau)}{\sigma_1(\tau)}\right), P_{f,2}(\tau) \approx Q\left(\frac{\tilde{\eta}_2 - \mu_0(\tau)}{\sigma_0(\tau)}\right) \tag{3.11}$$

The optimal τ^* for optimized system performance can be obtained as

$$\tau^* = \arg\max_{\tau} P_{d,2} \text{ given that } P_{f,2} \leq \omega_f,$$

where ω_f is the constraint on the probability of false alarm.

3.1.2 Asymptotic system performance when humans possess side information

We analyze the asymptotic performance of the likelihood ratio (LR)-based collaborative human–sensor decision-making systems in this subsection. In the previous analysis, the evaluation of system performance is obtained by the Gaussian approximation under certain regularity conditions on τ as $N \to \infty$. To avoid such an approximation, we evaluate the asymptotic performance of this system via Chernoff information, which could be used to approximate the probability of error for a LR-based decision fusion rule.

The Chernoff distance (Chernoff information) between two PDFs $f(z|H_0)$ and $f(z|H_1)$ is defined as

$$C \triangleq -\min_{0\leq\lambda\leq1} \log \int_z \left(\frac{f(z|H_1)}{f(z|H_0)}\right)^{\lambda} f(z|H_0)\mathrm{d}z \tag{3.12}$$

where $f(z|H_0)$ and $f(z|H_1)$ are the conditional joint PDFs of z under hypotheses H_0 and H_1, respectively. The probability of error at the FC can be expressed as $p_e \approx 2^{-nC}$, where n represents the number of data samples. Intuitively, p_e decreases exponentially as the number of data samples n increases, and C represents the decay rate. It is desired to have a larger Chernoff information in order to minimize p_e.

We consider that the PDFs of the humans' decision thresholds follow the same distribution $f_\varsigma(\varsigma)$. We denote θ_t and γ_t as the expected values of probabilities of false alarm and detection when t_i is made based on (3.9). In this situation, the normalized Chernoff information of the L_2 human sensors' decision $\vec{t}$ is given by [29]

$$\frac{1}{L_2}C_h = -\min_{0\leq\lambda\leq1} log\left(\left(\frac{\gamma_t}{\theta_t}\right)^{\lambda}\theta_t + \left(\frac{1-\gamma_t}{1-\theta_t}\right)^{\lambda}(1-\theta_t)\right) \tag{3.13}$$

However, in tasks requiring human situation awareness, individuals may have access to additional information about the PoI beyond the shared attributes x observed by both the physical sensors and human sensors. For instance, humans are able to observe some phenomena such as the activities of animals, which cannot be easily observed by physical sensors, before a natural disaster. We name this kind of information as human sensors' side information. In particular, assume that in addition to the common observation x_i, humans sensors possess side information w_i related to the PoI for $i = L_1 + 1, \ldots, L_2$. To model the error behavior exhibited in side information, w_i is modeled as an independent Bernoulli random variable with $Pr(w_i = 1|H_0) = \theta_{w,i}$ and $Pr(w_i = 1|H_1) = \gamma_{w,i}$. We further assume that $\theta_{w,i} = \theta_w$ and $\gamma_{w,i} = \gamma_w$ for $i = L_1 + 1, \ldots, L_2$.

The overall decision v_i of the ith human sensor is made based on t_i in (3.9) and the side information w_i. In the following, we consider two operations, i.e., OR rule and AND rule, that humans may use to incorporate t_i and w_i into the overall decision v_i.

OR operation [28]

The decision rule when employing the OR operation to incorporate side information is given as follows:

$$v_i = \begin{cases} 1 & x_i \geq \tau_i \quad \text{or} \quad w_i = 1 \\ 0 & \text{otherwise} \end{cases} \tag{3.14}$$

for $i = L_1 + 1, \ldots, L_2$. When the FC performs LR-based fusion with the humans' local decisions $\vec{v} = [v_{L_1+1}, \ldots, v_{L_2}]^T$, the PDF of $\vec{v}$ under H_1 and H_0 needs to be computed:

$$f(\vec{v}|H_1) = \prod_{i=L_1+1}^{L_2} \gamma_w v_i + (1-\gamma_w)(\gamma_t{}^{v_i}(1-\gamma_t)) \tag{3.15}$$

$$f(\vec{v}|H_0) = \prod_{i=L_1+1}^{L_2} \theta_w v_i + (1-\theta_w)(\theta_t{}^{v_i}(1-\theta_t)) \tag{3.16}$$

The asymptotic performance of the L_2 human sensors can be computed via Chernoff information. When employing the OR operation to incorporate side information, the normalized Chernoff information of the L_2 human decisions can be written as

$$\frac{1}{L_2}C_h^{OR} = -\min_{0\leq\lambda\leq 1} \log\left(T_1^{OR}(1-\theta_w)(1-\theta_t) + T_2^{OR}(\theta_w + (1-\theta_w)\theta_t)\right) \quad (3.17)$$

where $T_1^{OR} = \left(\frac{(1-\gamma_w)(1-\gamma_t)}{(1-\theta_w)(1-\theta_t)}\right)^\lambda$ and $T_2^{OR} = \left(\frac{\gamma_w+(1-\gamma_w)\gamma_t}{\theta_w+(1-\theta_w)\theta_t}\right)^\lambda$.

AND operation [30]

The decision rule when utilizing the AND operation to incorporate human decision t_i and side information w_i is given by

$$v_i = \begin{cases} 1 & x_i \geq \tau_i \quad \text{and} \quad w_i = 1 \\ 0 & \text{otherwise} \end{cases} \quad (3.18)$$

for $i = L_1+1, \ldots, L_2$. The PDF of $\vec{v} = [v_{L_1+1}, \ldots, v_{L_2}]^T$ under H_1 and H_0 can be expressed as

$$f(\vec{v}|H_1) = \prod_{i=L_1+1}^{L_2} \gamma_w \gamma_t^{v_i}(1-\gamma_t)^{1-v_i} + (1-\gamma_w)(1-v_i) \quad (3.19)$$

$$f(\vec{v}|H_0) = \prod_{i=L_1+1}^{L_2} \theta_w \theta_t^{v_n}(1-\theta_t)^{1-v_i} + (1-\theta_w)(1-v_i) \quad (3.20)$$

Similar to the results for the OR operation, the normalized Chernoff information of the L_2 human decisions for the AND operation is given by

$$\frac{1}{L_2}C_h^{AND} = -\min_{0\leq\lambda\leq 1} \log\left(T_1^{AND}(\theta_w(1-\theta_t) + 1 - \theta_w) + T_2^{AND}\theta_w\theta_t\right) \quad (3.21)$$

where $T_1^{AND} = \left(\frac{\gamma_w(1-\gamma_t)+1-\gamma_w}{\theta_w(1-\theta_t)+1-\theta_w}\right)^\lambda$ and $T_2^{AND} = \left(\frac{\gamma_w\gamma_t}{\theta_w\theta_t}\right)^\lambda$.

In the above two cases, we can derive the optimal λ^* by setting $\frac{\partial C_h^{OR}}{\partial\lambda} = 0$ and $\frac{\partial C_h^{AND}}{\partial\lambda} = 0$, respectively. Given the PDF $f_\varsigma(\varsigma)$ of the human thresholds, we can find the conditions under which incorporating the side information improves the detection performance, i.e., when $\gamma_w \in \{\gamma_w | C_h^{OR}(\gamma_w) \geq C_h\}$ for the OR operation or $\gamma_w \in \{\gamma_w | C_h^{AND}(\gamma_w) \geq C_h\}$ for the AND operation, the side information helps improve the quality of human sensors' decisions.

Recall that when the physical sensors use the same decision threshold τ in (3.1), the likelihood function of $\vec{d} = [d_1, \ldots, d_{L_1}]$ under H_1 and H_0 is expressed as

$$f(\vec{d}|H_1) = \prod_{i=1}^{L_1} p_d^{d_i}(1-p_d)^{1-d_i}, \quad f(\vec{d}|H_0) = \prod_{i=1}^{L_1} p_f^{d_i}(1-p_f)^{1-d_i} \quad (3.22)$$

The normalized Chernoff information of the L_1 physical sensors' decisions is given by

$$\frac{1}{L_1}C_p = -\min_{0\leq\lambda\leq 1} \log\left(\left(\frac{p_d}{p_f}\right)^\lambda p_f + \left(\frac{1-p_d}{1-p_f}\right)^\lambda (1-p_f)\right) \quad (3.23)$$

Finally, the overall Chernoff information of the integrated system composed of L_1 physical sensors and L_2 human sensors can be approximately computed as $C_o^{OR} \approx C_p + C_h^{OR}$ when humans adopt the OR operation rule, and $C_o^{AND} \approx C_p + C_h^{AND}$ when humans adopt the AND operation rule.

We present some simulation results for illustration. It is assumed that $z_i|H_j \sim exp(x_j)$ for $i = 1, 2, \ldots, L_1 + L_2$ and $j = 0, 1$, where $x \sim exp(z)$ means that x follows an exponential PDF $f(x) = ze^{-zx}$ for $x \geq 0$ and the probability is 0 otherwise. In Figures 3.2 and 3.3, we plot the performance of the system composed of only human sensors and without the participation of physical sensors to show the impact of side information on humans' decisions more clearly. The thresholds of the human sensors are assumed to follow the Gaussian distribution with mean and standard deviation (μ_τ, σ_τ). The impact of OR and AND operations on the asymptotic performance of this system is illustrated in Figure 3.2. It shows how the normalized Chernoff information changes with respect to the parameter of the PDF of threshold μ_τ and the accuracy of side information γ_w. It is observed that the system performs even worse than the system with no side information if $\gamma_w = 0.6$. This indicates that the accuracy of side information is not in the set $\{\gamma_w | Ch_h^{OR}(\gamma_w) \geq C_h\}$ or $\{\gamma_w | Ch_h^{AND}(\gamma_w) \geq C_h\}$, and, therefore, the side information will not improve the quality of humans' decisions. When γ_w increases beyond a specific threshold, the side information helps improve the human's performance. Furthermore, when μ_τ is in the region A as shown in Figure 3.2,

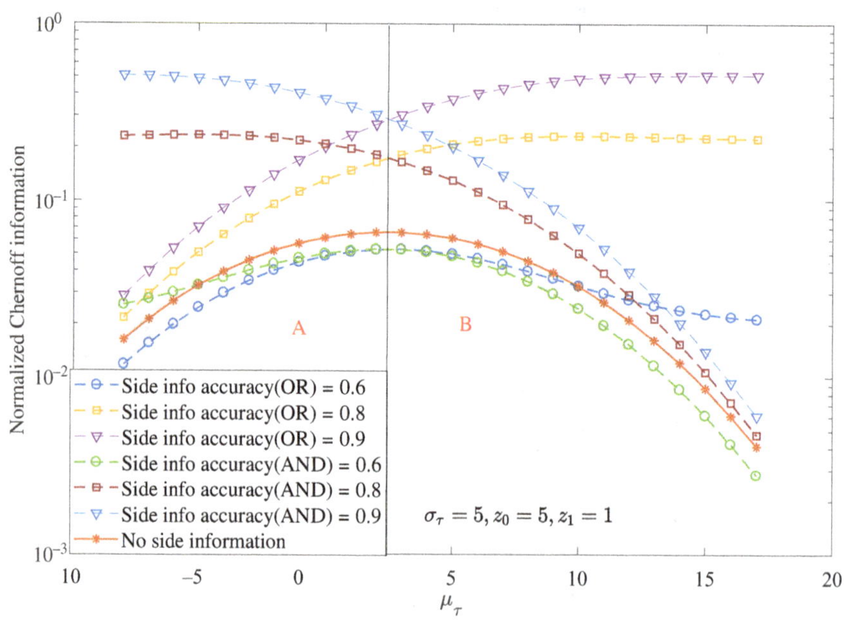

Figure 3.2 Normalized Chernoff information of collaborative human decision making system as a function of μ_τ

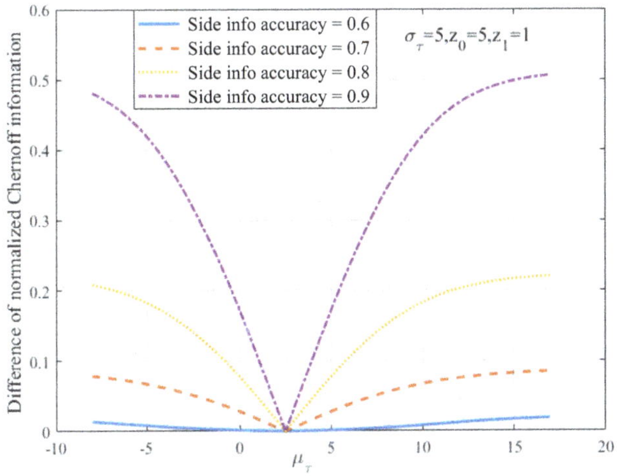

Figure 3.3 Comparison of the Chernoff information when the humans employ AND operation and OR operation for different values of μ_τ

the OR operation is a better choice to incorporate side information. Otherwise, the AND operation performs better.

To further identify the regions *A* and *B* more clearly, the difference of the Chernoff information when the humans employ the OR operation and the AND operation $|Ch_h^{OR} - Ch_h^{AND}|$ is plotted in Figure 3.3. It is observed that the difference decreases to 0 at a certain fixed μ_τ irrespective of how the accuracy of side information changes. Hence, the point that delineates the regions corresponding to the two rules, namely OR and AND rules, is a fixed point that depends on the statistical information of the humans' thresholds and has nothing to do with the accuracy of side information.

In Figure 3.4, we compute the performance of the integrated system composed of $L_1 = 20$ physical sensors and $L_2 = 20$ human sensors. The observations x_i for $i = 1, \ldots, 40$ are assumed to follow exponential distributions. Out of the 20 human sensors, we consider that the thresholds of 10 human sensors follow the Gaussian distribution $N(\mu_\tau^1, \sigma_\tau)$, and the thresholds of the other 10 human sensors follow $N(\mu_\tau^2, \sigma_\tau)$. It can be observed that the system incorporating side information significantly outperforms the system with no side information when the accuracy of side information is 0.8.

3.2 Prospect theoretic utility-based human decision making in multi-agent systems

In this section, we study how human cognitive biases may cause the random thresholds of humans to differ from one person to another in binary decision making. The Nobel-prize-winning PT is utilized to characterize the human cognitive biases in decision-making. PT provides a psychologically accurate framework to describe the

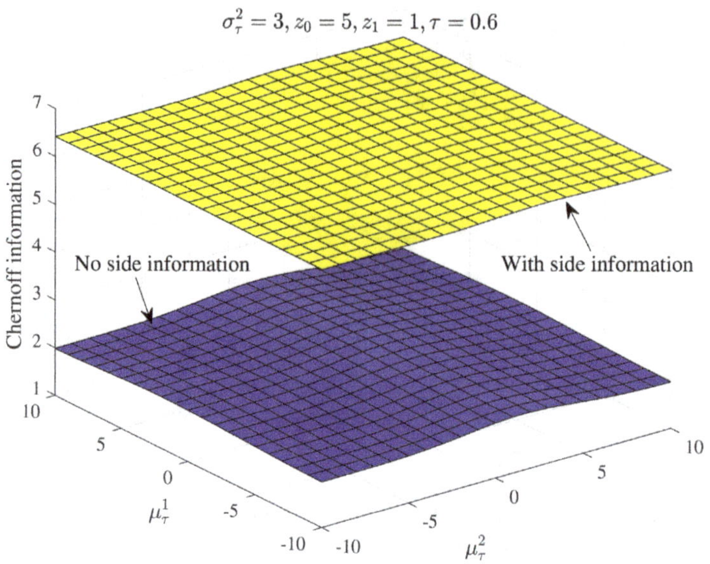

Figure 3.4 Comparison of the system performance when humans make decisions with side information and without side information

way people choose between probabilistic alternatives that involve risk. There are two main properties of PT: (1) there is a value function that suggests humans have asymmetric valuations towards gains and losses, as one strongly prefers avoiding losses than achieving gains. (2) Humans are risk averse to gains and risk seeking for losses in the sense that they over-emphasize low probability events and under emphasize high probability events. This is reflected in the probability weighting function.

According to the PT [31], quantitative outcomes x are represented through the lens of a monotonically increasing *value function* $\mathrm{v}(x)$, illustrated in Figure 3.5(a), that is convex below a reference point ϵ for which $\mathrm{v}(\epsilon) = 0$, and concave for gains above it. In turn, probabilities are represented by an inverse-S-shaped weighting function $\mathrm{w}(p)$, depicted in Figure 3.5(b), where the horizontal axis is real probability and the vertical axis denotes subjective probability. It can be seen that a human usually *overweights* small probabilities (e.g., $p < .2$), is *insensitive* toward moderate probabilities (e.g., $.2 \le p < .7$), and *underweights* large probabilities (e.g., $p > .7$).

Following Tversky and Kahneman (1992), we assume that the value function is:

$$\mathrm{v}(x) = \begin{cases} x^\lambda, & \text{for } x \ge 0, \\ -\beta|x|^\lambda, & \text{for } x < 0, \end{cases} \tag{3.24}$$

where β is the loss aversion parameter and λ characterizes the phenomenon of diminishing marginal utility, which says that as the total number of units of gain (or loss) increases, the utility of an additional unit of gain (or loss) to a person decreases.

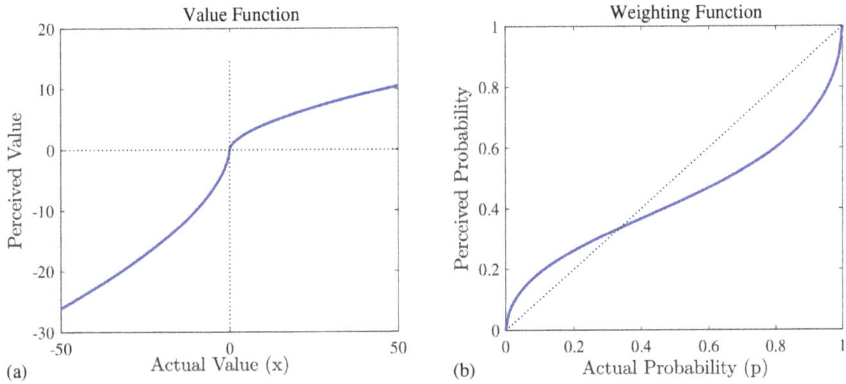

Figure 3.5 (a) Value function and (b) weighting function in prospect theory

The probability weighting function is given by

$$w(p) = \frac{p^{\alpha}}{(p^{\alpha} + (1-p)^{\alpha})^{\frac{1}{\alpha}}}, \tag{3.25}$$

where α characterizes the degree of distortion. Both the value function and the weighting function are used to determine the subjective utility U of a choice option with probabilistic outcomes, with the option maximizing U being preferred.

Some research efforts have been made to investigate how PT affects the behavior of humans in decision making. The authors in [32] studied a simplified form of PT to analyze the behavioral difference of pessimistic and optimistic human decision-makers. It has been shown that the LRT may or may not be optimal in PT-based binary hypothesis testing [33]. In these works, a human decision-maker uses the following rule to decide which hypothesis is true out of H_0 and H_1

$$d = \begin{cases} 1; & \text{if } r \in X_1 \\ 0; & \text{otherwise} \end{cases} \tag{3.26}$$

where r is the observation regarding the PoI and R_1 denotes the acceptance region of H_1. The expected behavioral risk under Bayesian formulation is computed by applying the value and weighting functions from PT in the following:

$$b(R_1) = \sum_{i=0}^{1}\sum_{j=0}^{1} w\left(Pr(\text{Declare } H_i|H_j \text{ is true})\right) \cdot v(c_{ij}) \tag{3.27}$$

where c_{ij} represents the cost of deciding H_i when the true hypothesis is H_j for $i,j = 0, 1$. When the human is subject to cognitive biases, the optimal $R_1{}^*$ is designed to minimize the human's behavioral Bayesian risk $R_1{}^* = \arg\min_{R_1 \in X} b(R_1)$.

In the above Bayesian formulation, the human is assumed to design the decision rule (i.e., the acceptance region of H_1) beforehand and employs the same decision rule no matter what the observation is. However, this method is not reasonable from a psychology point of view. In decision-making, psychologists show that instead of averaging over all possible observations and constructing a fixed decision rule, decision-makers first make some observations and based on these observations, they choose the action that yields the highest expected utility [34–36]. In the following, we proceed with the utility-based approaches to analyze the behavior of human decision-making under cognitive biases modeled by PT.

3.2.1 Subjective utility-based hypothesis testing

We first provide some background on the utility-based method for binary hypothesis testing problems under expected utility theory, where the decision-makers are considered to be rational. Rather than minimizing the Bayesian risk in (3.27), the goal for the decision-maker is to choose the hypothesis that yields the largest expected utility. Let U_{ij} denote the utility of declaring H_i when the true hypothesis is H_j, for $i, j \in \{0, 1\}$. Here, U_{00} and U_{11} denote the utilities of correct decisions and their values are usually positive. On the other hand, U_{10} and U_{01} denote the utilities of wrong decisions and their values are usually negative. When the observation is r, the expected utility of a rational decision-maker in declaring H_0 and H_1 is given by:

$$\begin{aligned} \mathbf{EU}(\text{Declare } H_0) &= Pr(H_0|r)U_{00} + Pr(H_1|r)U_{01} \\ \mathbf{EU}(\text{Declare } H_1) &= Pr(H_0|r)U_{10} + Pr(H_1|r)U_{11}, \end{aligned} \tag{3.28}$$

where $Pr(H_i|r)$ represents the probability that H_i is true if the observation is r,

$$Pr(H_i|r) = \frac{f(r|H_i)\pi_i}{f(r)} = \frac{f_i(r)\pi_i}{f(r)} \tag{3.29}$$

for $i = 0, 1$, respectively, where $f(\cdot)$ and $f_i(\cdot)$ are the appropriate PDFs and π_i denotes the prior probability of H_i. When the observation is r, the decision-maker decides hypothesis H_0 or H_1 whichever has a higher expected utility

$$\mathbf{EU}(\text{Declare } H_1) \underset{H_0}{\overset{H_1}{\gtrless}} \mathbf{EU}(\text{Declare } H_0). \tag{3.30}$$

Substitute the expression of $Pr(H_i|r)$ given in (3.29) into (3.28), and we obtain

$$\begin{aligned} \mathbf{EU}(\text{Declare } H_0) &= \frac{f_0(r)\pi_0}{f(r)}U_{00} + \frac{f_1(r)\pi_1}{f(r)}U_{01} \\ \mathbf{EU}(\text{Declare } H_1) &= \frac{f_0(r)\pi_0}{f(r)}U_{10} + \frac{f_1(r)\pi_1}{f(r)}U_{11} \end{aligned}$$

By substituting the expressions of $\mathbf{EU}(\text{Declare } H_0)$ and $\mathbf{EU}(\text{Declare } H_1)$ into (3.30), the utility-based decision rule becomes the classical LRT:

$$\frac{f_1(r)}{f_0(r)} \underset{H_0}{\overset{H_1}{\gtrless}} \frac{\pi_0(U_{00} - U_{10})}{\pi_1(U_{11} - U_{01})} \triangleq \eta. \tag{3.31}$$

which is known to be optimal that minimizes the Bayesian cost.

In statistical signal detection literature, the decision-making agent is always considered to be rational and the goal is to maximize some expected utility. Under expected utility theory, decision-makers are rational in the sense that they are capable of calculating the expected utility of each outcome without biases. For instance, a typical property of rational decision-makers is that they are indifferent between several actions if their expected utilities are the same. However, because of human cognitive biases in perceiving the utilities and the probabilities, a human decision-maker usually prefers a deterministic gain over a probabilistic gain even if the two alternatives have the same expected utility. In many scenarios where the decisions are made by humans, cognitive biases may cause the results to deviate from the outcomes predicted under expected utility theory. In contrast to rational decision-makers who select the hypothesis that maximizes the expected utilities, human decision-makers act to maximize their subjective utilities, which is usually distorted because of cognitive biases. When computing the subjective utility of declaring H_0 and H_1, we exploit PT by applying the value function $\mathrm{v}(\cdot)$ given in (3.24) on the utilities and applying the probability weighting function $\mathrm{w}(\cdot)$ given in (3.25) on the probabilities. In this case, when the observation is r, the subjective utilities of deciding H_0 and H_1 are:

$$\begin{aligned} \mathbf{SU}(\text{Declare } H_0) &= \mathrm{w}\left(Pr(H_0|r)\right)\mathrm{v}(U_{00}) + \mathrm{w}\left(Pr(H_1|r)\right)\mathrm{v}(U_{01}) \\ \mathbf{SU}(\text{Declare } H_1) &= \mathrm{w}\left(Pr(H_0|r)\right)\mathrm{v}(U_{10}) + \mathrm{w}\left(Pr(H_1|r)\right)\mathrm{v}(U_{11}). \end{aligned} \tag{3.32}$$

It is known that humans select the alternative which has a higher subjective utility given observation r:

$$\mathbf{SU}(\text{Declare } H_1) \underset{H_0}{\overset{H_1}{\gtrless}} \mathbf{SU}(\text{Declare } H_0). \tag{3.33}$$

Exploiting (3.32) and (3.33), the subjective utility-based decision rule of human decision makers becomes:

$$\frac{\mathrm{w}\left(Pr(H_1|r)\right)}{\mathrm{w}\left(Pr(H_0|r)\right)} \underset{H_0}{\overset{H_1}{\gtrless}} \frac{\mathrm{v}(U_{00}) - \mathrm{v}(U_{10})}{\mathrm{v}(U_{11}) - \mathrm{v}(U_{01})} \triangleq \frac{V_{00} - V_{10}}{V_{11} - V_{01}}, \tag{3.34}$$

where $V_{00}, V_{01}, V_{10}, V_{11}$ are the subjective utilities by applying the value function (3.24) on $U_{00}, U_{01}, U_{10}, U_{11}$, respectively. We consider that the correct decisions' utilities V_{00} and V_{11} are positive, while wrong decision's utilities V_{01} and V_{10} are negative. Substituting the expression of the weighting function given in (3.25) and the expression of $Pr(H_i|r)$ given in (3.29), and using that $Pr(H_1|r) = 1 - Pr(H_0|r)$, we obtain $\frac{\mathrm{w}(Pr(H_1|r))}{\mathrm{w}(Pr(H_0|r))} = \frac{Pr(H_1|r)^\alpha}{Pr(H_0|r)^\alpha}$. Consequently, the decision rule given in (3.34) becomes

$$\frac{f_1(r)}{f_0(r)} \underset{H_0}{\overset{H_1}{\gtrless}} \left(\frac{V_{00} - V_{10}}{V_{11} - V_{01}}\right)^{\frac{1}{\alpha}} \frac{\pi_0}{\pi_1} \triangleq \eta_p. \tag{3.35}$$

Hence, the human's decision rule is in the form of a LRT with threshold η_p [21].

Theorem 3.1. *Under prospect theoretic framework, the optimal subjective utility-based decision rule reduces to an LRT. The threshold of the LRT, η_p, is a monotonous function of parameters α and β.*

In many application scenarios, the LR $\Lambda(r) = \frac{f_1(r)}{f_0(r)}$ is strictly increasing or decreasing with respect to r. For instance, this is true when $f_1(r)$ and $f_0(r)$ are Gaussian PDFs with different means and the same variance. Gaussian distributions are widely used as they represent the nature of a large number of problems in signal processing. In the rest of this section, we study human decision-making for the binary hypothesis testing problem, and, in particular, the observations under both hypotheses are assumed to follow Gaussian distributions given by:

$$H_0 : r \sim N(m_0, \sigma_s^2), \quad H_1 : r \sim N(m_1, \sigma_s^2) \tag{3.36}$$

where m_0 and m_1 represent the means of the signal under H_0 and H_1, respectively, and σ_s^2 represents the variance of the signal under both hypotheses. We assume that $m_0 < m_1$ and the diminishing marginal utility parameter λ in (3.24) is set to be a fixed value 0.88. We focus on studying how behavioral parameters α and β in PT affect the human decision-making performance. When the observations follow Gaussian distributions (3.36), the LRT (3.35) becomes

$$\frac{f_1(r)}{f_0(r)} = e^{\frac{2(m_1-m_0)r-(m_1^2-m_0^2)}{2\sigma_s^2}} \underset{H_0}{\overset{H_1}{\gtrless}} \eta_p \tag{3.37}$$

which is equivalent to $r \underset{H_0}{\overset{H_1}{\gtrless}} t = F(\alpha, \beta) = \frac{2\ln \eta_p \sigma_s^2 + (m_1^2 - m_0^2)}{2(m_1 - m_0)}$, where the cognitively biased threshold t is a monotone function F of PT parameters α and β (because η_p is an inherent function of α and β).

In contrast to physical sensors whose decision thresholds can be set to be fixed values, humans have uncertainties in decision-making due to complex factors such as time constraint, emotion, and environment. Individual level uncertainty is a prominent feature in human behavior. Variability exists in human perception and decision-making even when the external conditions, such as the sensory signals and the task environment, are kept the same [37]. It is also known as trial-to-trial variability in psychology literature, i.e., differences of responses are prominent when the same experiment is conducted multiple times using the same human subject. From the psychological perspective, the reasons that cause variability are: (a) the initial state of the neural circuitry is unlikely to be the same at the beginning of each trial, and (b) noises penetrate in each part of the nervous system, from the perception of outside observations to the process of decision-making. These two reasons cause inevitable uncertainties in human decision-making, where the uncertainties depend on factors such as time constraints, emotion, outside, and environment [38].

Inspired by the above discussion, we model the humans' decision thresholds as random variables [15,28,39]. In particular, the human's behaviorally biased decision threshold is modeled as $\tau = F(\alpha, \beta) + v$, where $v \sim N(0, \sigma_\tau^2)$. We use σ_τ^2 to denote the variance associated with a humans while making a decision because of uncertainty. Note that τ is considered to be a Gaussian random variable, whose mean is determined by the average values of human behavioral parameters and the variance σ_τ^2 is impacted by the decision uncertainties. A larger value of σ_τ^2 represents higher uncertainty of a person in decision making. To quantify the individual uncertainty associated with the human decision threshold in real applications, one may perform the experiments as in [31] on the same human under different conditions, e.g., time constraints, emotions, and locations. In each trial of the experiment, the values of behavioral parameters α, β, and λ of the human can be computed using regression. Since the variability of α, β and λ causes the human to change his/her decision threshold, the variance of the decision threshold σ_τ^2 can be derived by analyzing the statistics.

For the hypothesis testing problem (3.36), if a human uses a random decision threshold $\tau \sim N(m_\tau, \sigma_\tau^2)$, the probabilities of false alarm and detection can be computed as [21]

$$P_F = Q\left(\frac{m_\tau - m_0}{\sqrt{\sigma_s^2 + \sigma_\tau^2}}\right), \quad P_D = Q\left(\frac{m_\tau - m_1}{\sqrt{\sigma_s^2 + \sigma_\tau^2}}\right), \tag{3.38}$$

where $Q(x) = \frac{1}{\sqrt{2\pi}} \int_x^\infty \exp\left(-\frac{u^2}{2}\right) du$.

Next, we investigate the impact of decision uncertainty quantified in terms of σ_τ^2 on the quality of human decisions. For a human decision-maker who employs a random decision threshold $\tau \sim N(m_\tau, \sigma_\tau^2)$ to make a decision in the binary hypothesis testing framework (3.36), the following theorem quantifies the relationship between the human decision-making performance and the decision-making uncertainty [21].

Theorem 3.2. *There is a pair of values $\{\underline{m}_\tau, \overline{m}_\tau\}$ where $\underline{m}_\tau < \overline{m}_\tau$ and both $\underline{m}_\tau$ and $\overline{m}_\tau$ can be obtained by solving m_τ in the following equation:*

$$e^{\frac{2(m_1 - m_0)m_\tau - (m_1^2 - m_0^2)}{2\sigma_s^2}} \times \left(\frac{m_\tau - m_1}{m_\tau - m_0}\right) = \eta.$$

The pair $\{\underline{m}_\tau, \overline{m}_\tau\}$ has the following properties: (a) For humans with $\underline{m}_\tau \le m_\tau \le \overline{m}_\tau$, the expected utility while making a decision monotonically decreases as σ_τ^2 becomes larger, i.e., the expected utility while making a decision is maximized for decision uncertainty $\sigma_\tau^{2*} = 0$. (b) For humans with $m_\tau > \overline{m}_\tau$ and $m_\tau < \underline{m}_\tau$, the expected utility is unimodal, i.e., first increases then decreases, as σ_τ^2 becomes larger. The optimal decision uncertainty σ_τ^{2*} is greater than 0 and satisfies:

$$e^{\frac{2(m_1 - m_0)m_\tau - (m_1^2 - m_0^2)}{2(\sigma_s^2 + \sigma_\tau^{2*})}} \times \left(\frac{m_\tau - m_1}{m_\tau - m_0}\right) = \eta.$$

Definition 3.1. *In the hypothesis testing framework analyzed above, if a human's expected utility in decision-making strictly decreases as σ_τ^2 becomes larger, i.e., $\underline{m}_\tau \le m_\tau \le \overline{m}_\tau$ and the best decision-making performance is achieved for $\sigma_\tau^{2*} = 0$, the human is called* ***reasonable****. If the best decision-making performance in terms of expected utility is achieved for decision uncertainty $\sigma_\tau^{2*} > 0$, i.e., $m_\tau > \overline{m}_\tau$ or $m_\tau < \underline{m}_\tau$, the person is called* ***extremely biased****.*

We provide some simulation results to illustrate the performance when a human employs the random decision threshold that follows $N(m_\tau, \sigma_\tau^2)$ in the hypothesis testing problem (3.36) with the following parameters: $\pi_0 = 0.7, \pi_1 = 0.3, U_{11} = U_{00} = 20, U_{01} = -80, U_{10} = -20, m_0 = 0, m_1 = 5$, and $\sigma_s^2 = 2.25$. With this setup, it can be computed that $\underline{m}_\tau = -0.025$ and $\overline{m}_\tau = 5.015$. Hence, the left-side extremely biased

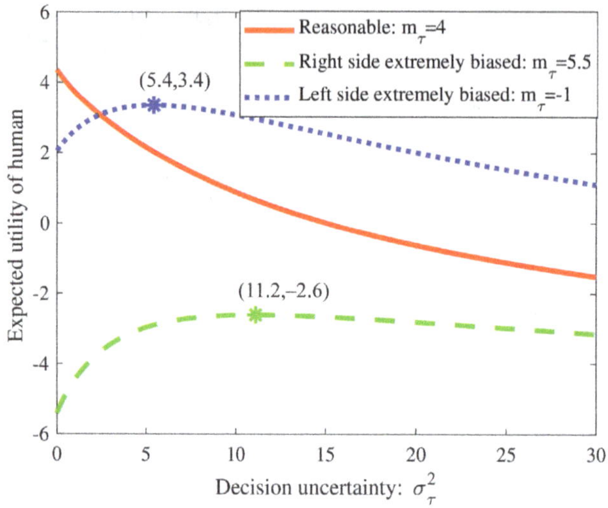

Figure 3.6 Expected utility of a human agent as decision uncertainty σ_τ^2 increases

interval, the reasonable interval and the right-side extremely biased interval of the human in terms of m_τ are $(-\infty, -0.025)$, $[-0.025, 5.015]$, and $(5.015, \infty)$, respectively. Figure 3.6 shows the expected utility of decision-making as a function of the uncertainty of decision threshold σ_τ^2 for three different values of m_τ. We can see that the expected utility of a reasonable human is monotonically decreasing with respect to σ_τ^2. For extremely biased human decision-makers, the optimal value of decision uncertainty to achieve the maximum expected utility σ_τ^{2*} is greater than 0. Note that in this particular example, left-side extremely biased humans whose decision threshold $m_\tau < \underline{m}_\tau$ have higher expected utilities than right-side extremely biased human agents whose decision threshold $m_\tau > \overline{m}_\tau$. The reason for this phenomenon in this example is that the penalty of missed detection ($U_{01} = -80$) is more significant than the penalty of false alarm ($U_{10} = -20$). Right-side extremely biased humans with larger values of decision thresholds are more likely to be penalized by missed detections causing their performance to be worse. Furthermore, it can be seen in Figure 3.6 that a left-side extremely biased human performs better than a reasonable human when σ_τ^2 is greater than a certain value. It is because that as σ_τ^2 becomes larger, a reasonable human decision-maker is more likely to employ larger values of decision thresholds than a left-side extremely biased human, which causes the performance of the rational human to degrade due to the larger penalty associated with missed detections.

Remark 3.1. *The decision-making performance of extremely biased humans is improved due to the presence of decision uncertainty. Before the decision uncertainty reaches a certain value, the decision-making performance increases as the decision*

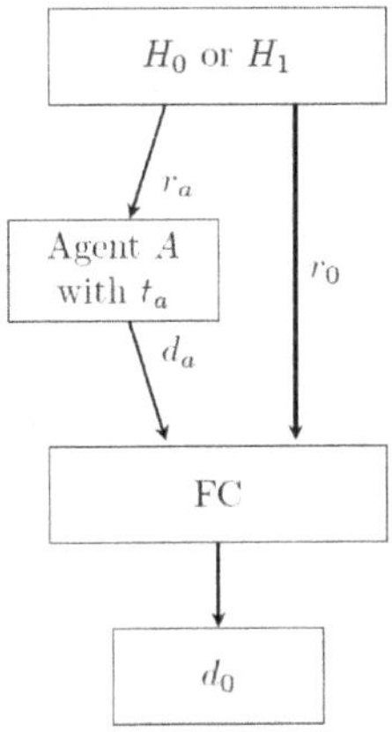

Figure 3.7 Human participating in decision making as an information source

uncertain becomes larger, and after that point is reached, the decision-making performance starts to decrease as the decision uncertainty continues to increase. This phenomenon is analogous to noise-enhanced signal detection [40] where the quality of a suboptimal detector can be improved by adding noise under certain conditions. It is also known as stochastic resonance in the signal processing literature [40,41].

It should be noted that PT is a static concept to characterize human decision-making under cognitive biases. More sophisticated models of social learning (group think) in behavioral economics need to be incorporated to model the dynamics in human networks, where one human may influence other humans in team decision-making. In these scenarios, nonstandard information structures arise and can result in counter-intuitive behavior such as information cascades [42].

3.2.2 Decision fusion involving human participation

In this subsection, we analyze the impact of an individual's behavioral biases (cognitive biases and uncertainties) on the performance of two decision-making systems that involve human participation.

Human participates in decision making as an information source, FC is a rational machine

As shown in Figure 3.7, we first consider the scenario where a human agent acts as an information source to support the FC in making the final decision with the FC being rational (unbiased).

We consider that the FC observes r_0 and the human agent A observes r_a through independent observation channels. Identical and independently distributed additive Gaussian noises are assumed to exist in the observation channels of both the FC and the human agent A. The observations made by the FC and the agent A are denoted by r_0 and r_a to emphasize that the observations are received over two independent channels. Specifically, the human agent A makes a local decision on which hypothesis is present by comparing r_a to a threshold t_a:

$$d_a = \begin{cases} 1 & \text{if } r_a \geq t_a \\ 0 & \text{if } r_a < t_a \end{cases}$$

For the ease of analysis, we first assume t_a to be a fixed decision threshold determined by the human's behavioral parameters α_a and β_a, such that $t_a = F(\alpha_a, \beta_a)$. Decision-making uncertainty of human agent A will be considered later in this subsection. When the FC receives the decision of agent A, $d_a = j \in \{0, 1\}$, it makes the final decision d_0 by fusing d_a and its own observation r_0. Given d_a and r_0, the FC's expected utilities to declare H_0 and H_1 are given by:

$$\begin{aligned} \mathbf{EU}(\text{Declare } H_0) &= Pr(H_0|r_0, d_a = j)U_{00} + Pr(H_1|r_0, d_a = j)U_{01} \\ \mathbf{EU}(\text{Declare } H_1) &= Pr(H_0|r_0, d_a = j)U_{10} + Pr(H_1|r_0, d_a = j)U_{11}, \end{aligned}$$

respectively. Selecting the hypothesis that yields the larger expected utility gives the following decision rule:

$$\frac{Pr(H_1|r_0, d_a = j)}{Pr(H_0|r_0, d_a = j)} \underset{H_0}{\overset{H_1}{\gtrless}} \frac{U_{10} - U_{00}}{U_{01} - U_{11}}, \tag{3.39}$$

where $Pr(H_i|r_0, d_a = j)$ represents the probability that H_i is true given r_0 and $d_a = j$. Note that

$$Pr(H_i|r_0, d_a = j) = \frac{\pi_i Pr(d_a = j|H_i) f(r_0|H_i)}{f(r_0, d_a = j)}$$

for $i,j \in \{0, 1\}$. Denote the probabilities of false alarm and detection of agent A as $Pr(d_a = 1|H_0) = P_F^a$ and $Pr(d_a = 1|H_1) = P_D^a$, respectively. After substituting the expressions for $Pr(H_i|r_0, d_a = j)$, the decision rule (3.39) becomes:

$$\frac{f_1(r_0)}{f_0(r_0)} \underset{H_0}{\overset{H_1}{\gtrless}} \frac{1 - P_F^a}{1 - P_D^a} \frac{\pi_0(U_{10} - U_{00})}{\pi_1(U_{01} - U_{11})} = \frac{1 - P_F^a}{1 - P_D^a}\eta, \quad \text{if} \quad d_a = 0, \tag{3.40}$$

$$\frac{f_1(r_0)}{f_0(r_0)} \underset{H_0}{\overset{H_1}{\gtrless}} \frac{P_F^a}{P_D^a} \frac{\pi_0(U_{10} - U_{00})}{\pi_1(U_{01} - U_{11})} = \frac{P_F^a}{P_D^a}\eta, \quad \text{if} \quad d_a = 1. \tag{3.41}$$

By solving for r_0 in $\frac{f_1(r_0)}{f_0(r_0)} = \frac{1-P_F^a}{1-P_D^a}\eta$ for $d_a = 0$, and $\frac{f_1(r_0)}{f_0(r_0)} = \frac{P_F^a}{P_D^a}\eta$ for $d_a = 1$, the FC's decision thresholds applicable to observation r_0 for final decision-making can be derived, and we denote them as t_0 and t_1, respectively. When observations under H_0 and H_1 follow Gaussian distributions (3.36), the probabilities of false alarm and detection are given by $P_F^a = Q(\frac{t_a - m_0}{\sigma_s})$ and $P_D^a = Q(\frac{t_a - m_1}{\sigma_s})$. The overall performance of the FC in terms of probabilities of false alarm and detection can be written as:

$$\begin{aligned} p_f &= \sum_{j=0}^{1} Pr(d_0 = 1 | d_a = j, H_0) Pr(d_a = j | H_0) \\ &= P_F^a Q\left(\frac{t_1 - m_0}{\sigma_s}\right) + (1 - P_F^a) Q\left(\frac{t_0 - m_0}{\sigma_s}\right), \end{aligned}$$

and

$$\begin{aligned} p_d &= \sum_{j=0}^{1} Pr(d_0 = 1 | d_a = j, H_1) Pr(d_a = j | H_1) \\ &= P_D^a Q\left(\frac{t_1 - m_1}{\sigma_s}\right) + (1 - P_D^a) Q\left(\frac{t_0 - m_1}{\sigma_s}\right), \end{aligned}$$

respectively. Hence, the FC's expected utility for decision-making is:

$$U = \pi_0(1 - p_f)U_{00} + \pi_0 p_f U_{10} + \pi_1(1 - p_d)U_{01} + \pi_1 p_d U_{11}. \tag{3.42}$$

Some simulations are performed for the same hypothesis testing problem as discussed in Section 3.2.1. In Figure 3.8, when human agent A's decision threshold t_a changes, i.e., the cognitive bias of the human varies, we show the expected utilities in decision-making of agent A and that of the FC. It can be seen that the threshold that yields the maximum decision-making performance for agent A by itself is different from the threshold that yields the best decision-making performance for the FC. In other words, a rationally behaving human who acts to maximize his/her expected utility (with decision threshold 3.28 indicated by the red dot) does not necessarily provide the maximum expected utility for the FC. In this particular example, a human who has some bias (with decision threshold equal to 3.41 indicated by the blue dot) results in a higher expected utility for the FC. The strategy of choosing the properly biased person differs when the specific setup of the hypothesis testing problem changes. By computing the optimal value of the decision threshold of agent A that helps the FC achieve the best decision making performance, we can determine the values of α and β corresponding to the most suitable cognitively biased person, to perform the task in different scenarios.

Next, we consider the uncertainties in human decision-making and model the decision threshold employed by agent A as a Gaussian random variable $\tau_a \sim N(m_{\tau_a}, \sigma_{\tau_a}^2)$. In this scenario, P_F^a and P_D^a can be computed using (3.38), and the optimal decision rule at the FC can be derived analogous to the previous analysis. The FC's expected utility in decision-making can also be computed. In the following, we illustrate the FC's decision-making performance for different values of agent A's decision-making uncertainty.

Continuing with the previous parameters for the hypothesis testing problem, Figure 3.9 plots the FC's expected utility in decision-making as a function of the mean decision threshold of agent A. In the red, green, and blue curves, agent A's decision-making variances are $\sigma_{\tau_a}^2 = 0$, $\sigma_{\tau_a}^2 = 1$ and $\sigma_{\tau_a}^2 = 4$, respectively. In the

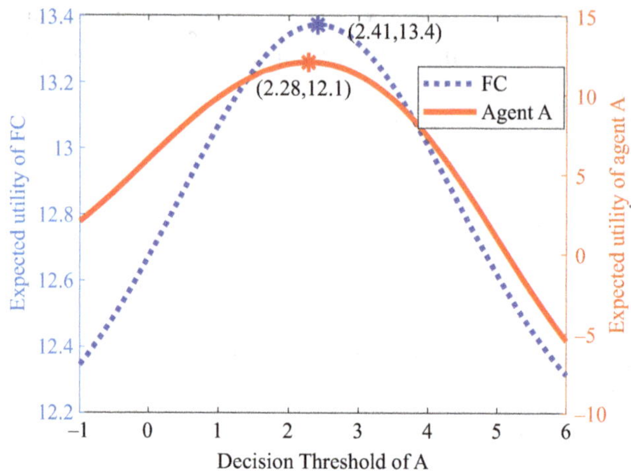

Figure 3.8 Expected utility as a function of threshold t_a used by agent A

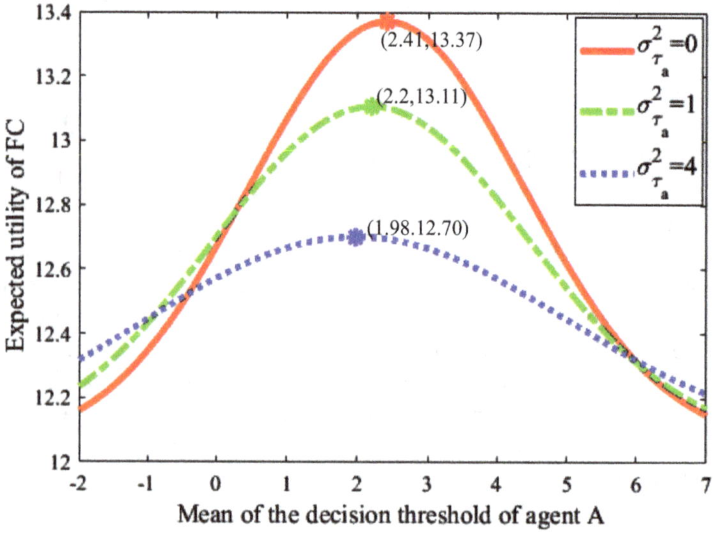

Figure 3.9 Expected utility of the FC as a function of the mean threshold of agent A

middle range of m_{τ_a} where the human decision-maker is reasonable, the red curve with smallest decision making variance performs better than the other two curves. In practical applications, it is preferable to have human agents who are reasonable in that their decision-making is expected to be of higher quality compared to extremely biased humans and, in addition, their performance is more predictable in the presence of decision-making uncertainty. On the extreme left or extreme right of the curve, i.e., when the decision threshold is significantly biased, a higher variance surprisingly yields better decision performance for the FC. Intuitively, for extremely biased humans whose decision thresholds significantly deviate from being rational, a higher variance is more probable to "rectify" the biased thresholds to be close to be optimal. However, for reasonable humans whose behavioral thresholds are already close to the optimal, a large variance is more likely to push their behavioral thresholds away from the optimal values. As a result, a large value of variance helps improve the FC's utility if the human is extremely biased, while it degrades the performance if the human is behaving rationally. This observation is consistent with our previous results about the impact of decision uncertainty on the a single human's decision-making performance as discussed in Figure 3.6.

Physical sensor acts as the information source and human is the decision maker at the FC

Looking at the system shown in Figure 3.7 from a different perspective, we study the scenario where A is a physical sensor that employs a fixed decision threshold t_a. On the other hand, a behaviorally biased human with PT parameters α, β and decision-making uncertainty σ_{FC}^2 acts as the FC to make the final decision. Here, the physical sensor A sends its binary decisions $d_a = j \in \{0, 1\}$ to help the FC in making the decision d_0. Since the FC is biased, we exploit $\mathrm{v}(\cdot)$ and $\mathrm{w}(\cdot)$ when computing the FC's subjective utility of deciding either H_0 or H_1 being true. If agent A sends its decision $d_a = j$, the subjective utilities of deciding H_0 and H_1 are given by

$$\begin{aligned}&\mathbf{SU}(\text{Declare } H_0)\\ &= \mathrm{w}\left(Pr(H_0|r_0, d_a = j)\right) V_{00} + \mathrm{w}\left(Pr(H_1|r_0, d_a = j)\right) V_{01}\end{aligned} \tag{3.43}$$

and

$$\begin{aligned}&\mathbf{SU}(\text{Declare } H_1))\\ &= \mathrm{w}\left(Pr(H_0|r_0, d_a = j)\right) V_{10} + \mathrm{w}\left(Pr(H_1|r_0, d_a = j)\right) V_{11}.\end{aligned} \tag{3.44}$$

The human decision-maker at the FC determines d_0 by choosing the hypothesis that results in a larger value of subjective utility. By assuming that the FC's observation r_0 and agent A's decision d_a are independent, the LR at the FC is increasing as a function of r_0. In this case, the FC uses a threshold-based decision rule and the mean of the decision threshold m_{FC}^j can be obtained by setting (3.43) equal to (3.44) for $j = 0, 1$, and the variance of the decision threshold is assumed to be σ_{FC}^2. Upon receiving $d_a = j$, the FC compares r_0 to a decision threshold τ_j^f to make the final decision, where $\tau_j^f \sim N(m_{FC}^j, \sigma_{FC}^2)$ for $j = 0, 1$. In particular, the decision rule at the behaviorally biased FC is expressed as

$$\frac{f_1(r_0)}{f_0(r_0)} \underset{H_0}{\overset{H_1}{\gtrless}} \frac{1 - P_F^a}{1 - P_D^a} \left(\frac{V_{00} - V_{10}}{V_{11} - V_{01}}\right)^{\frac{1}{\alpha}} \frac{\pi_0}{\pi_1} = \frac{1 - P_F^a}{1 - P_D^a} \eta_p, \quad \text{if} \quad d_a = 0, \tag{3.45}$$

$$\frac{f_1(r_0)}{f_0(r_0)} \underset{H_0}{\overset{H_1}{\gtrless}} \frac{P_F^a}{P_D^a}\left(\frac{V_{00}-V_{10}}{V_{11}-V_{01}}\right)^{\frac{1}{\alpha}}\frac{\pi_0}{\pi_1} = \frac{P_F^a}{P_D^a}\eta_p, \quad \text{if} \quad d_a = 1. \tag{3.46}$$

The probabilities of false alarm and detection of the human decision maker at the FC are given by:

$$\begin{aligned} p_f &= \sum_{j=0}^{1} Pr(d_0 = 1|d_a = j, H_0)Pr(d_a = j|H_0) \\ &= Q\left(\frac{t_a - m_0}{\sigma_s}\right)Q\left(\frac{m_{FC}^1 - m_0}{\sqrt{\sigma_s^2 + \sigma_{FC}^2}}\right) + \left(1 - Q\left(\frac{t_a - m_0}{\sigma_s}\right)\right)Q\left(\frac{m_{FC}^0 - m_0}{\sqrt{\sigma_s^2 + \sigma_{FC}^2}}\right) \end{aligned}$$

and

$$\begin{aligned} p_d &= \sum_{j=0}^{1} Pr(d_0 = 1|d_a = j, H_1)Pr(d_a = j|H_1) \\ &= Q\left(\frac{t_a - m_1}{\sigma_s}\right)Q\left(\frac{m_{FC}^1 - m_1}{\sqrt{\sigma_s^2 + \sigma_{FC}^2}}\right) + \left(1 - Q(\frac{t_a - m_1}{\sigma_s})\right)Q\left(\frac{m_{FC}^0 - m_1}{\sqrt{\sigma_s^2 + \sigma_{FC}^2}}\right), \end{aligned}$$

where $Pr(d_0 = 1|d_a = j, H_i) = Q\left(\frac{m_{FC}^j - m_i}{\sqrt{\sigma_s^2 + \sigma_{FC}^2}}\right)$ for $i,j = \{0, 1\}$ can be computed using (3.38). Again, the expected utility of the FC in decision-making can be obtained using (3.42).

In Figure 3.10, we plot the expected utility of the FC as a function of the decision threshold t_a of the physical sensor A. In Figure 3.10(a), the red curve denotes the baseline scenario where the FC is rational, and, in the green and blue curves, the FC is behaviorally biased with $\beta = 1.5$ and $\beta = 2$, respectively. When the FC is biased, we assume the probability distortion parameter to be $\alpha = 0.72$. It can be seen that the FC achieves higher expected utility when it acts rationally. On the other hand, the peak points on these curves (denoted by the red, green, and blue dots) indicate that if the FC has different behavioral properties, the optimal decision threshold of A while assisting the FC to achieve the best performance is different. In the system considered here, we can adjust the threshold of the physical sensor A in order to help the FC/human achieve the best decision quality.

To find the optimal threshold of the physical sensor t_a^*, let us denote the thresholds of the LRT in (3.45) and (3.46) as $\eta_{p0} = \frac{1-P_F^a}{1-P_D^a}\eta_p$ and $\eta_{p1} = \frac{P_F^a}{P_D^a}\eta_p$, respectively. Let $g(x) = \frac{2\sigma_s^2 \log x + (m_1^2 - m_0^2)}{2(m_1 - m_0)}$, which is the inverse function of the LR $\frac{f_1(r)}{f_0(r)}$. In a manner analogous to that in [43], we can show that the decision threshold t_a^* employed by the physical sensor that minimizes the FC's expected cost satisfies the following condition:

$$G(t_a^*) = \frac{Q\left(\frac{g(\eta_{p1}) - m_0}{\sigma_s + \sigma_{FC}}\right) - Q\left(\frac{g(\eta_{p0}) - m_0}{\sigma_s + \sigma_{FC}}\right)}{Q\left(\frac{g(\eta_{p1}) - m_1}{\sigma_s + \sigma_{FC}}\right) - Q\left(\frac{g(\eta_{p0}) - m_1}{\sigma_s + \sigma_{FC}}\right)}\eta. \tag{3.47}$$

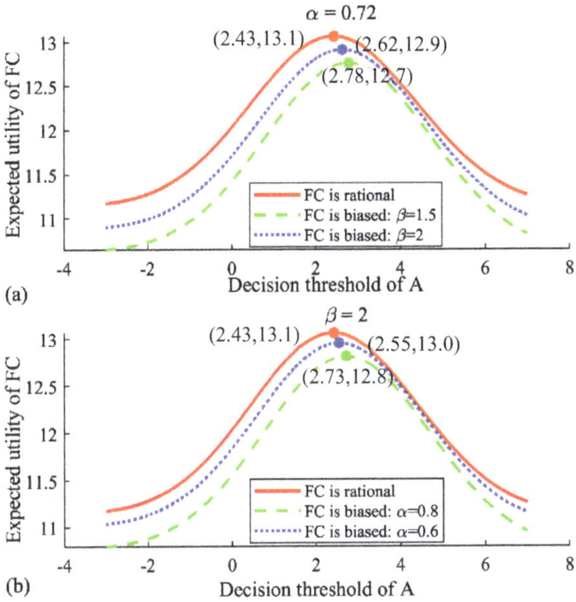

Figure 3.10 Expected utility of FC as a function of the decision threshold of agent A, when FC has behavioral biases

A brief sketch of the proof is shown in the following.

Proof. Exploiting the independence assumption, the Bayesian cost of human while making a decision is:

$$\sum_{d_0,d_a,H_k} \int \pi_k Pr(d_0|d_a, r_0)Pr(d_a|r_a)Pr(r_a|H_k)Pr(r_0|H_k)c_{d_0k}dr_0dr_a$$

for $d_0, d_a, k \in \{0, 1\}$. By summing d_a over $\{0, 1\}$, ignoring the constant factors and using the fact $Pr(d_a = 1|r_a) = 1 - Pr(d_a = 0|r_a)$, we have

$$\int_{r_a} Pr(d_a = 0|r_a) \sum_{d_0,H_k} \int_{r_0} \pi_k Pr(r_0|H_k)Pr(r_a|H_k)c_{d_0k} \times [Pr(d_0|d_a = 0, r_0) - Pr(d_0|d_a = 1, r_0)]\, dr_0dr_a, \tag{3.48}$$

which is minimized by setting $Pr(d_a = 0|r_a) = 0$ if

$$\sum_{d_0,H_k} \int_{r_0} \pi_k Pr(r_0|H_k)Pr(r_a|H_k)c_{d_0k} \times [Pr(d_0|d_a = 0, r_0) - Pr(d_0|d_a = 1, r_0)]\, dr_0 \geq 0 \tag{3.49}$$

and setting $Pr(d_a = 1|r_a) = 0$ if (3.49) does not hold.

Note that $\int Pr(r_0|H_k)Pr(d_0|d_a, r_0)dr_0 = Pr(d_0|d_a, H_k)$ and $Pr(d_0 = 0|d_a = 0, r_0) \geq Pr(d_0 = 0|d_a = 1, r_0)$. By setting (3.49) equal to 0 and summing over H_k for $k = \{0, 1\}$, we obtain the condition that must be satisfied by the optimal decision threshold t_a^* of the physical sensor:

$$G(t_a^*) = \frac{\sum_{d_0=0}^{1} \pi_0 c_{d_0 0} \left[Pr(d_0|d_a = 1, H_0) - Pr(d_0|d_a = 0, H_0)\right]}{\sum_{d_0=0}^{1} \pi_1 c_{d_0 1} \left[Pr(d_0|d_a = 0, H_1) - Pr(d_0|d_a = 1, H_1)\right]},$$

Lastly, substituting $Pr(d_0 = 1|d_a = j, H_k) = Q(\frac{g(\eta_{pj}) - m_k}{\sigma_s + \sigma_{FC}})$ and $Pr(d_0 = 0|d_a = j, H_k) = 1 - Q(\frac{g(\eta_{pj}) - m_k}{\sigma_s + \sigma_{FC}})$ for $j, k = \{0, 1\}$ and after simplification, the condition in (3.47) follows. □

Another interesting phenomenon in Figure 3.10(a) is that in this decision making system, a more loss averse FC (indicated by the blue curve that has $\beta = 2$) performs better than a less loss averse FC (indicated by the green curve that has $\beta = 1.5$) for the entire interval of t_a. It is because of the fact that behavioral parameters α, β and λ jointly impact the threshold used by a biased human. In our scenario, a larger value of β cancels out the effect of α and λ, making the threshold used by this human closer to that of a rational decision maker. In Figure 3.10(b), we set the human's loss aversion parameter to be $\beta = 2$ and show the expected utility in decision making with respect to t_a as α takes two values of $0.8, 0.6$. Similar to the observations in Figure 3.10(a), a more biased probability distortion parameter $\alpha = 0.6$ helps the FC achieve better decision-making performance than $\alpha = 0.8$ when $\beta = 2$. In general, decision-making quality of a human does not depend on one single behavioral parameter, instead all the parameters should be considered together in a holistic manner.

3.3 Human–machine collaboration for binary decision-making under correlated observations

In this section, we present a novel framework where the human and the machine collaboratively perform signal detection tasks [44]. In particular, we consider that the machine observes a continuous signal regarding the PoI, while the human possesses a categorical judgment through experience regarding the PoI through other information sources, inductive reasoning, experience, or other sources of information that are not available to the machine. The human decision and the machine observation are assumed to be statistically dependent. We attempt to use copulas to model the dependence between a continuous and a discrete random variable in the context of sensor fusion. Copulas have been used in the context of distributed detection for modeling the dependence between continuous data from different modalities [45,46]. The use of copulas is very attractive for sensor fusion applications due to their ability to model dependent observations with arbitrary marginals and complex dependence structures. A comprehensive study of copulas is presented in [47].

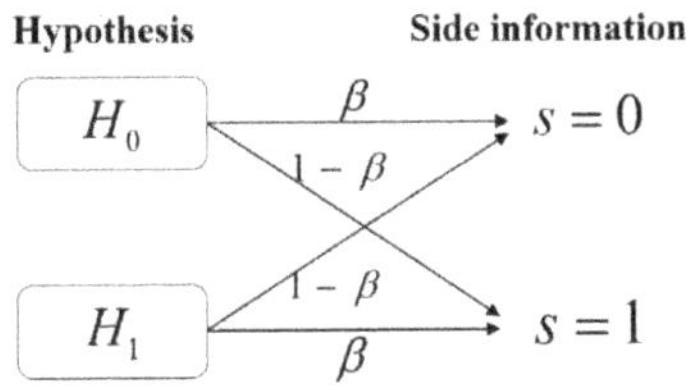

Figure 3.11 Human side information modeled as a BSC

We evaluate the performance of such a system and derive expressions for the probability of detection P_D and the probability of false P_{FA} using Copula density functions. It is shown that when the machine's observation falls into a certain region, there is no need to ask for human decisions as they do not improve the detection accuracy. Hence, we may save human's participation when the machine's observation falls into that region while maintaining the same system performance.

3.3.1 Human–machine collaboration model

Consider that a human and a machine work together to solve hypothesis-testing problems where the two hypotheses are denoted as H_0 and H_1. We consider that the machine acquires an objective measurement r regarding the PoI, where r is assumed to be a continuous random variable. The conditional PDFs of r under the two hypotheses are denoted by $f_0(r)$ and $f_1(r)$, respectively. On the other hand, we consider that the human does not observe r, but possesses an additional *side information* s regarding the PoI through his/her experience or other sources of information that are not available to the machine. For example, (a) in weather forecasting, the machine may measure the objective temperature data and the human may observe the activities of animal migration, which are fused together to predict whether it is going to be a cold winter. (b) In airplane/submarine navigation systems, the machine or physical sensors collect objective data such as velocity and altitude, which are combined with the human's judgment on the topography to determine the maneuver strategy. Beyond these two illustrative examples, this formulation is suitable to model numerous semi-autonomous systems for SA and command and control, both in military and civilian domains, that involve human participation.

While the machine's observation r is a continuous random variable, we consider that the human has limited information processing ability and only makes categorical perceptions [48]. In this section, we assume that the side information s of the human is binary, and human's error behavior of s can be modeled via a binary symmetric channel (BSC) shown in Figure 3.11. In particular, $Pr(s = i|H_i \text{ is true}) = \beta$ for $i = 0, 1$, where in this subsection, we use β to represent the accuracy of the human and we assume that $\beta \geq 0.5$ (the smallest possible value of accuracy $1/2$ represents the response s is a random guess).

When the machine's observation r and the human side information s are fused to make the final decision, we know that the FC's optimal decision rule that achieves minimum Bayesian cost is the LRT [26]

$$L_f(r) = \frac{Pr(r,s|H_1)}{Pr(r,s|H_0)} \underset{H_0}{\overset{H_1}{\gtrless}} \frac{\pi_0(c_{10}-c_{00})}{\pi_1(c_{01}-c_{11})} \triangleq \eta, \tag{3.50}$$

where π_i represents the priors of H_i for $i = \{0, 1\}$. We denote c_{ij} as the *cost* of deciding H_i when the true hypothesis is H_j for $i,j = \{0, 1\}$. Under the assumption that r and s are independent of each other, the decision rule (3.50) becomes $L_f(r) = \frac{f_1(r)(1-\beta)}{f_0(r)\beta}$ if $s = 0$ and $L_f(r) = \frac{f_1(r)\beta}{f_0(r)(1-\beta)}$ if $s = 1$. However, the dependence of the machine observation and the human side information cannot be ignored in practice. It is important to account for the dependencies in order to more accurately evaluate the system performance. Next, we consider that the human decision and the machine observation are dependent and use Copula theory to model these dependencies.

3.3.2 Copula-based decision fusion at the FC

In the following, we investigate the decision fusion of the human–machine collaborative sensing structure and show that there is a region where human decision is required to improve the detection performance. Copula theory characterizes the dependence between the machine observation and the human decision. A copula is a joint distribution function on uniform marginals that can model the dependence between random variables with arbitrary marginal densities. A well-studied result in the copula literature is Sklar's Theorem which lays the foundation of copula theory.

Sklar's theorem: For random variables $X_1, \ldots, X_d$, the joint CDF can be modeled as:

$$F(x_1, \ldots, x_d) = C(F_{X_1}(x_1), \ldots, F_{X_d}(x_d)) \tag{3.51}$$

Further, the copula C uniquely models $F(x_1, \ldots, x_d)$ if $X_1, \ldots, X_d$ are continuous. When the random variables are not all continuous, the copula C is unique on *ran* $F_1(\cdot) \times \cdots \times$ *ran* $F_d(\cdot)$ where "*ran*" refers to the range of the CDF.

Sklar's theorem allows for the use of copulas to model the joint distribution of a discrete random variable and a continuous random variable. However, there are an infinite number of copulas that can describe the same joint distribution. While the non-uniqueness of the copula does not pose a problem for modeling dependent data, it does not allow for non-parametric inference of the copula from the data, in that the copula cannot be learned from the data.

We assume that the dependence between the machine observation r and the human decision s exists only under hypothesis H_1.[†] From the decision rule given in (3.50),

[†]For example, we consider that under H_0, the signal from the PoI is absent. Hence, r is sampled from the additive white Gaussian noise (AWGN) process, which is independent of the human side information s. The model can be easily generalized to the situations where r and s are dependent under both H_0 and H_1.

we incorporate dependencies between r and s, and obtain the copula-based LRT at the FC

$$L_c(r, s_i) \triangleq \frac{f_1(r)\left[\frac{\partial C(u, F_{h,1}(s_i))}{\partial u} - \frac{\partial C(u, F_{h,1}(s_{i-1}))}{\partial u}\right]}{f_0(r)p_{h,0}(s_i)} \mathop{\gtreqless}_{H_0}^{H_1} \eta \tag{3.52}$$

where $p_{h,0}(s_i) = \beta$ if $s_i = 0$, $p_{h,0}(s_i) = 1 - \beta$ if $s_i = 1$, $u = F_1(r)$ with $F_1(\cdot)$ representing the CDF of the machine's observation under H_1 and $F_{h,1}(\cdot)$ representing the CDF of the human decision under H_1. The procedures of deriving the joint density function in the numerator of (3.52) are described in the following:

$$Pr\{S \leq s, R \leq r\} = C(F_1(r), F_{h,1}(s)) \tag{3.53}$$

$$Pr\{S \leq s | R = r\} f_1(r) \overset{(a)}{=} \frac{\partial C(F_1(r), F_{h,1}(s))}{\partial r} \tag{3.54}$$

$$Pr\{S = s_i | R = r\} f_1(r) \overset{(b)}{=} \frac{\partial C(F_1(r), F_{h,1}(s_i))}{\partial r} - \frac{\partial C(F_1(r), F_{h,1}(s_{i-1}))}{\partial r} \tag{3.55}$$

$$Pr\{S = s_i, R = r\} f_1(r) \overset{(c)}{=} f_1(r)\left[\frac{\partial C(F_1(r), F_{h,1}(s_i))}{\partial r} - \frac{\partial C(F_1(r), F_{h,1}(s_{i-1}))}{\partial r}\right] \tag{3.56}$$

$$f_{R,S}(r, s) \overset{(d)}{=} f_1(r)\left[\frac{\partial C(u, F_{h,1}(s_i))}{\partial u} - \frac{\partial C(u, F_{h,1}(s_{i-1}))}{\partial u}\right] \tag{3.57}$$

where s_i and s_{i-1} are in the support $\tilde{S}$ of S and $s_{i-1} < s < s_i$. The support of S takes on discrete values s_k. Here, s_i is the value determined by the discrete variable and $s_{i-1} = \sup\{s : s < s_i, s \in \tilde{S}\}$, i.e., the preceding point in the support $\tilde{S}$. Also, $u = F_1(r)$, which is the CDF of the machine observation under H_1. We set $s_0 = -\infty$, where s_0 is a point in $\tilde{S}$ so that $F_{h,1}(s_0) = 0$. Since $Pr\{S \leq s, R \leq r\} = \int_{-\infty}^{r} \sum_{v:\{v \in \tilde{S}, v \leq s\}} f_{R,S}(u, v)du = \int_{-\infty}^{r} \sum_{v:\{v \in \tilde{S}, v \leq s\}} f_{S|R}(v|u)f_R(u)du$, step (a) is derived by finding the partial derivative with respect to r using Leibniz's integral rule; step (b) comes from the definition of CDF for a discrete variable; step (c) is obtained from the definition of the conditional CDF; and step (d) follows from the chain rule for derivatives.

Example: We compute the joint PDF for the machine observation $R \sim N(\alpha, \sigma^2)$ and $S \sim$ Bernoulli(β) using a bivariate Gaussian copula $C_G(u, v|\rho)$. The Gaussian copula has a correlation parameter ρ which represents the amount of dependency between the variables R and S. From (3.57), we have:

$$f_{R,S}(r, s) = f_1(r)\left[\frac{\partial C_G(u, F_{h,1}(s_i)|\rho)}{\partial u} - \frac{\partial C_G(u, F_{h,1}(s_{i-1})|\rho)}{\partial u}\right] \tag{3.58}$$

For a Gaussian copula, $\frac{\partial C_G(u,v|\rho)}{\partial u} = \Phi\left[\frac{\Phi^{-1}(v)-\rho\Phi^{-1}(u)}{\sqrt{(1-\rho^2)}}\right]$ where $\Phi(x) = \int_{-\infty}^{x} \frac{1}{\sqrt{2\pi}} e^{\frac{-t^2}{2}} dt$ and $u = F_R(r)$. The expression for the derivative is given in [49]. Then, the joint PDF of r and s is expressed as:

$$f_{R,S}(r,s) = \frac{1}{\sqrt{2\pi\sigma^2}} e^{\frac{-(r-\alpha)^2}{2\sigma^2}} \left\{\Phi\left[\frac{\Phi^{-1}(F_S(s_i)) - \rho\Phi^{-1}(u)}{\sqrt{(1-\rho^2)}}\right] - \Phi\left[\frac{\Phi^{-1}(F_S(s_{i-1})) - \rho\Phi^{-1}(u)}{\sqrt{(1-\rho^2)}}\right]\right\} \tag{3.59}$$

We use the LRT to determine when it is necessary to request side information from the human. More specifically, we determine when the human's decision can augment the machine's observation to yield improved performance, so that the overall probability of error decreases. It can be seen from (3.52) that the LRT when the human decisions are $s_i = 1$ and $s_i = 0$ simplifies to

$$\frac{f_1(r)}{f_0(r)}\left[1 - \frac{\partial C(u, 1-\beta)}{\partial u}\right] \underset{H_0}{\overset{H_1}{\gtrless}} \eta(1-\beta) \tag{3.60}$$

$$\frac{f_1(r)}{f_0(r)}\left[\frac{\partial C(u, 1-\beta)}{\partial u}\right] \underset{H_0}{\overset{H_1}{\gtrless}} \eta(\beta) \tag{3.61}$$

respectively. Given a fixed value of β, the terms on the left side of (3.60) and (3.61) are functions of r. If the value of r is such that either,

$$\left\{\frac{f_1(r)}{f_0(r)}\left[\frac{\partial C(u, 1-\beta)}{\partial u}\right], \frac{f_1(r)}{f_0(r)}\left[1 - \frac{\partial C(u, 1-\beta)}{\partial u}\right]\right\} > \eta\beta \quad \text{or} \tag{3.62}$$

$$\left\{\frac{f_1(r)}{f_0(r)}\left[\frac{\partial C(u, 1-\beta)}{\partial u}\right], \frac{f_1(r)}{f_0(r)}\left[1 - \frac{\partial C(u, 1-\beta)}{\partial u}\right]\right\} < \eta(1-\beta) \tag{3.63}$$

the FC decides H_1 and H_0, respectively, to be true, regardless of the value of s. Hence, in these two regions indicated by (3.62) and (3.63), the side information of the human does not improve the global decision solely based on the machine's observation.

Next, we show that when the marginal densities under the two hypotheses H_1 and H_0 have Gaussian distributions with shifted-means, i.e., $f_1(r|\alpha_1) \sim N(\alpha_1, \sigma^2)$, $f_0(r|\alpha_0) \sim N(\alpha_0, \sigma^2)$, and the signal-to-noise ratio (SNR) $\frac{\alpha_1-\alpha_0}{\sigma^2}$ is large enough, the human decision is not required when the machine observation r is outside a region $A = \left[d', d''\right]$ for any value of η. We derive the SNR conditions in the following paragraph.

The LR $L_m(r) = \frac{f_1(r)}{f_0(r)}$ of the machine's observation is a strictly increasing function of r given that the distributions $f_1(\cdot|\alpha_1)$ and $f_0(\cdot|\alpha_0)$ are from the same family and $\alpha_1 > \alpha_0$. Let $\phi(r|\beta) = \frac{\partial C(u,1-\beta)}{\partial u}$, then it is easy to see that the functions $\phi_0(r|\beta) = L_m(r)\phi(r|\beta)$ and $\phi_1(r|\beta) = L_m(r)(1-\phi(r|\beta))$ are strictly increasing if

$$\frac{d}{dr}(L_m(r)\phi(r|\beta)) > 0 \tag{3.64}$$

and

$$\frac{d}{dr}(L_m(r)(1-\phi(r|\beta))) > 0 \tag{3.65}$$

After simplifying the derivatives in the two equations, we have the conditions

$$\frac{\alpha_1-\alpha_0}{\sigma^2} > \max\left\{\sup_{r\in\mathbb{R}}\left(\frac{-\phi'(r|\beta)}{\phi(r|\beta)}\right), \sup_{r\in\mathbb{R}}\left(\frac{\phi'(r|\beta)}{1-\phi(r|\beta)}\right)\right\} \tag{3.66}$$

If the SNR satisfies (3.66), $\phi_1(r|\beta)$ and $\phi_0(r|\beta)$ are injective functions and invertible. Then, the region $A = [d', d'']$ that requires the human's decision to improve the system performance is given by:

$$d' = \max\{\phi_1^{-1}(\eta\beta), \phi_0^{-1}(\eta\beta)\} \tag{3.67}$$

and

$$d'' = \min\{\phi_1^{-1}(\eta(1-\beta)), \phi_0^{-1}(\eta(1-\beta))\} \tag{3.68}$$

The derivatives of the Gaussian copula and the Student's-t copula with respect to the variable u are needed in order to compute the LR. The procedures of deriving these derivatives can be found in [49] and are summarized in Table 3.2.

It can be shown that for a Gaussian copula, the derivative of $\phi(r|\beta)$, $\phi'(r|\beta) = \frac{-\rho}{\sqrt{1-\rho^2}}f_1(r)\frac{f\left(\frac{\phi^{-1}(v)-\rho\phi^{-1}(u)}{\sqrt{1-\rho^2}}\right)}{f\left(\frac{\phi^{-1}(u)}{2}\right)}$, where $f_1(\cdot)$ is the marginal PDF under H_1 and $f(\cdot)$ is the PDF of a standard normal distribution. Also, $\phi(r|\beta) \in [0, 1]$ and when either $\phi(r|\beta) = 0$ or $1-\phi(r|\beta) = 0$, it is required that the SNR $\frac{\alpha_1-\alpha_0}{\sigma^2}$ must be infinitely large for both $\rho > 0$ and $\rho < 0$ so that $\phi_1(r|\beta)$ and $\phi_0(r|\beta)$ are strictly increasing for all values of r and for the region A to be of the form $[d', d'']$. When $\rho = 0$, we have the product copula Π which corresponds to the scenario where the human decision and the machine observation are independent and, in this case, any SNR > 0 is sufficient.

Table 3.2 Derivatives of the Gaussian and Student's-t copulas

Copula derivatives	
Copula family	**Partial derivative** $\frac{\partial C(u,v)}{\partial u}$
Gaussian	$\Phi\left\{\frac{\Phi^{-1}(F_S(s_i))-\rho\Phi^{-1}(u)}{\sqrt{(1-\rho^2)}}\right\}$
Student's-t	$t_{v+1}\left\{\frac{t_v^{-1}(F_s(s_i))-\rho t_v^{-1}(u)}{\sqrt{\frac{\left(v+(t_v^{-1}(u))^2\right)\left(1-\rho^2\right)}{(v+1)}}}\right\}$

3.3.3 Performance evaluation

Next, we derive the expression for the probability of error P_e of the system under the Bayesian criterion. We first compute the expressions for the probability of detection of P_D and the probability of false alarm P_{FA}. With the definition of copulas, we get:

$$Pr\{R > r|S = s_i, H_1\} = 1 - C(F_1(r), F_{h,1}(s_i)) + C(F_1(r), F_{h,1}(s_{i-1})) \tag{3.69}$$

where $F_1(r)$ is the CDF of the machine observation r. It should be noted that the LR simplifies to $\frac{\phi_1(r|\beta)}{1-\beta}$ when $s_i = 1$ and $\frac{\phi_0(r|\beta)}{\beta}$ when $s_i = 0$. The functions $\phi_1(r|\beta)$ and $\phi_0(r|\beta)$ are monotonically increasing if the SNR satisfies the conditions in (3.66) and, hence, are injective and invertible. The inverse functions of $\phi_1(\cdot|\beta)$ and $\phi_0(\cdot|\beta)$ are denoted as $\phi_1^{-1}(\cdot)$ and $\phi_0^{-1}(\cdot)$, respectively. We denote $c_1 = \phi_1^{-1}(\eta(1-\beta))$ and $c_0 = \phi_0^{-1}(\eta\beta)$. The LRT conditioned on the values of s_i is expressed as:

$$r \underset{H_0}{\overset{H_1}{\gtrless}} c_1 \qquad \text{when } s_i = 1 \tag{3.70}$$

$$r \underset{H_0}{\overset{H_1}{\gtrless}} c_0 \qquad \text{when } s_i = 0 \tag{3.71}$$

The probability of detection P_D is then expressed as:

$$\begin{aligned} P_D &= Pr\{R > c_1|S = 1, H_1\}Pr\{S = 1|H_1\} \\ &\quad + Pr\{R > c_0|S = 0, H_1\}Pr\{S = 0|H_1\} \end{aligned} \tag{3.72}$$

Using equations (3.69), (3.70), and (3.71), P_D can be given by:

$$\begin{aligned} P_D &= 1 - \beta F_1(c_1) - C(F_1(c_1), 1-\beta) \\ &\quad + \beta\,[C(F_1(c_1), 1-\beta) - C(F_1(c_0), 1-\beta)] \end{aligned} \tag{3.73}$$

where $Pr(R > c_1|S = 1, H_1) = 1 - C(F_1(c_1), 1) + C(F_1(c_1), F_{h,1}(0)) = 1 - F_1(c_1) + C(F_1(c_1), 1-\beta)$, due to the fact that $C(u, 1) = u$ for any copula C [47]. Moreover, we obtain $Pr(R > c_0|S = 1, H_1) = 1 - C(F_1(c_1), F_{h,1}(0)) + C(F_1(c_1), F_{h,1}(-\infty)) = 1 - C(F_1(c_1), 1-\beta)$ due to the fact that $C(u, 0) = 0$ [47].

We further derive the expression for the probability of false alarm P_{FA}

$$\begin{aligned} P_{FA} &= Pr\{R > c_1|S = 1, H_0\}Pr\{S = 1|H_0\} \\ &\quad + Pr\{R > c_0|S = 0, H_0\}Pr\{S = 0|H_0\} \end{aligned} \tag{3.74}$$

which can be written as

$$\begin{aligned} P_{FA} &= \beta(1 - F_0(c_1)) + (1-\beta)(1 - F_0(c_0)) \\ &= \beta(F_0(c_0) - F_1(c_1)) + 1 - F_0(c_0) \end{aligned} \tag{3.75}$$

In the Bayesian setting, the probability of error P_e is in the form of

$$P_e = \pi_1(1 - P_D) + \pi_0(P_{FA}) \tag{3.76}$$

Some experiments are conducted for the illustration of the performance. First, we show the receiver operator characteristics (ROC) for the Gaussian copula and the Student's-t copula families. We set the SNR of the machine's observation to be $\frac{\alpha_1 - \alpha_0}{\sigma^2} = 3$ and the human's accuracy to be $\beta = 0.8$. The ROC is plotted for different values of the copula parameter ρ (common to both families).

In Figures 3.12 and 3.13, we plot the ROC for a Gaussian copula and a Student's-t copula for different values of the dependence parameter ρ. The degree of freedom in the Student's-t copula is set equal to 3 in all the curves. It can be seen that for both the Gaussian copula and the Student's-t copula, when the human decision and the machine observation are positively correlated, there is an overlap of information that will reduce the overall information that can be used for decision-making. On the other hand, when the data is negatively correlated, the human decision and the machine observation could act as critics of each other and prevent false alarms or missed detection, thus enhancing the overall performance of the system.

In Figure 3.14, we set $\eta = 0.4$, $\beta = 0.8$. For a Gaussian copula with $\rho = 0.3090$, we show the range of the region where the human's decision is not required. We plot the functions $\phi_1(r)$ and $\phi_0(r)$ for two different values of SNR: 0.75 and 0.8. For the conditions given in (3.62) and (3.63), the region that does not require human assistance is larger when the machine's observation has a higher value of SNR. It should be noted that in such a case, the regions that require human decisions are not continuous.

In Figure 3.15, we illustrate how the performance of the detection system degrades if we neglect the dependence between the human decision and the machine observations. The SNR of the marginals is set to 3 and the parameter $\beta = 0.8$. It can be seen that for both $\rho = 0.7071$ and $\rho = -0.7071$ with a Gaussian copula, if dependencies are ignored, the probability of detection of P_D for different values of probability of false alarm P_{FA} is considerably smaller.

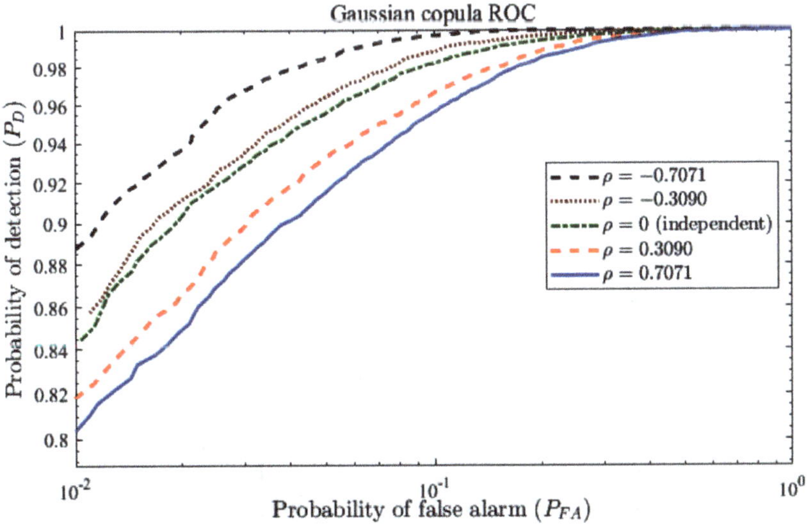

Figure 3.12 ROC for a Gaussian copula for different ρ

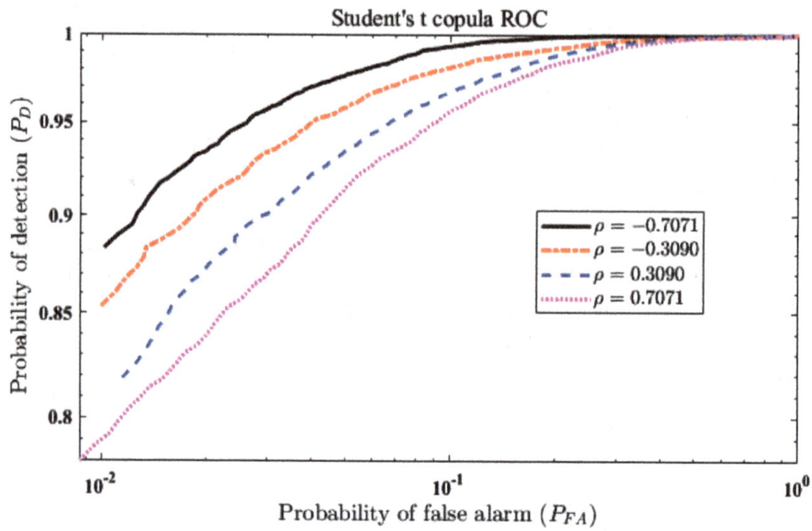

Figure 3.13 ROC for a Student's-t copula for different ρ and v = 3

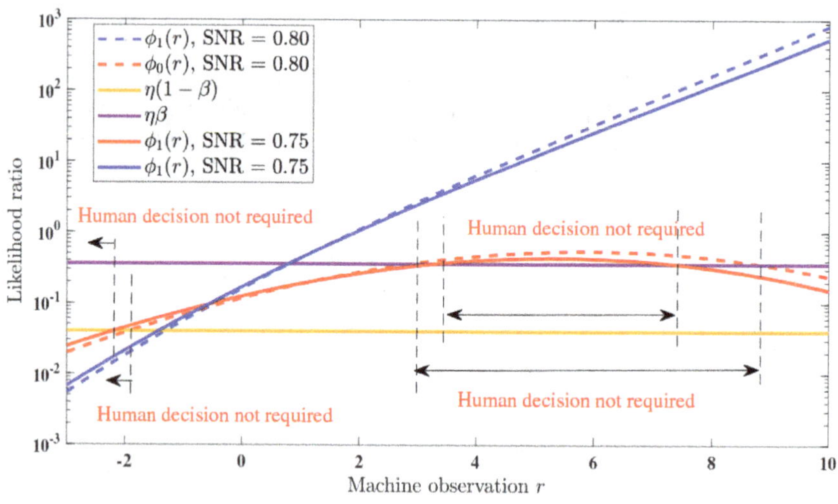

Figure 3.14 Effect of low SNRs on the human decision region

3.4 Current challenges in human–machine teaming

Human behavior and decision-making are complex processes that represent the intricate interplay between the psychological activity within humans and the influence of outside environment. A fully cognitive human–machine (radar) collaborative sensing architecture requires the radar to be able to understand, anticipate, and augment the performance of the human; and the human to have the ability to support, supervise,

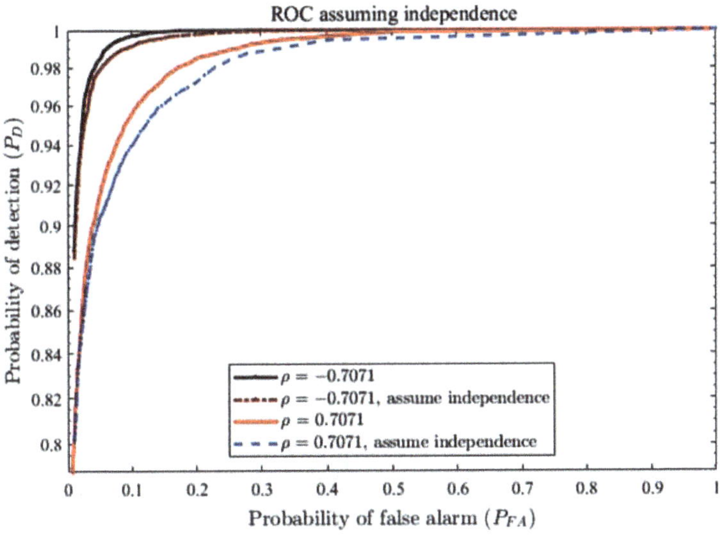

Figure 3.15 Loss in performance when ignoring dependence between human decision and machine observation

and enhance the automation conducted by the radar [25]. To advance the objective of designing an interactive symbiosis where humans and radars are tightly coupled together in cognitive sensing, several research directions include, but not limited to the following.

1. **Behavioral informatics**: For the machine to better understand human behavior in different applications, it is necessary to explore research findings in psychology that characterizes how human behavior is impacted by time constraints, memory limitations, emotion state as well as stimulus from the outside environment. To achieve "that the order of magnitude increases in available, net thinking power resulting from linked human-machine dyads" [50], it becomes imperative to perform human cognitive state sensing for designing efficient communication interfaces between the human and the machine. Another interesting topic is the real-time prediction of human cognitive workload based on sensor-based brain signals such as electroencephalogram (EEG), as well as the design of system augmentations such as offloading tasks or assisting users with modality-specific support.
2. **Trust in autonomous systems**: In human–machine collaboration, the authors in [51] have described trust as "the willingness of a party to be vulnerable to the actions of another party based on the expectation that the other will perform a particular action important to the trustor, irrespective of the ability to monitor or control that other party." In particular, many autonomous systems employed in high stake applications are black boxes that do not explain how the decision are made. The challenge is to develop a quantitative definition of trust and

establish clear guidelines to construct human–machine transparency and enhance calibrated trust between the human and the machine.

3. **Situational awareness**: SA refers to the user's familiarity of the task environment, the perception of the task status, and the anticipation of future states. If humans are not appropriately incorporated in the loop, it is very likely that the human is not aware of or not familiar with the machine's task execution. In such a situation where there is over-reliance on machine automation, the human's understanding of the work environment, i.e., SA, is jeopardized. The loss of SA (also referred to as complacency or automation-induced decision biases in different works) compromises the human's level of expertise and the ability to perform the automated tasks manually in the case of unpredictable automation failure and it may cause severe breakdown in critical applications like autopilot and submarine navigation systems. Hence, the concerns of SA must be addressed in the design of human–machine symbiosis to prevent irreparable damage.
4. **Herding, nudging, and incentives**: Humans are also known to be subject to herding and nudging phenomena. To elicit desirable outputs from humans, future research work can proceed with some explorations along these lines. (a) The optimum design and task allocation of collaborative human–machine networks. This will include changes in the strategies of individual nodes, e.g., adapting the threshold of some or all the nodes or shaping the input to selected nodes during the inference process. (b) The suitable distribution of the tasks and workload to be performed by humans and machines leading to semi-autonomous systems. (c) Another important consideration will be the incentivization measures of humans to actively engage in the inference process, which can be posed in a reinforcement learning-based framework.

3.5 Summary

In almost all cognitive radar systems developed so far, the objective function is based solely on objective performance measures and is devoid of any human perception considerations. The incorporation of human in the loop will generalize cognitive systems by allowing humans and machines being tightly coupled in the same working environment for advanced interaction. In this chapter, we reviewed the state-of-the-art in how human behavior and decision-making can be modeled in the statistical signal processing framework, with emphasis on three topics. First, the decision-making problem in human–machine networks was considered where the humans employ random thresholds and the machines employ fixed thresholds to make binary local decisions. The asymptotic performance of the integrated system was derived by considering humans might possess side information regarding the PoI. Next, the behavioral economics concept of PT was exploited to study the behavior of human binary decision-making under cognitive biases, where the thresholds employed by humans were modeled as a Gaussian random variable. Two decision-making systems involving humans' participation were discussed, and we showed the impact of human cognitive biases on the decision-making performance. Finally, we discussed a collaborative environment to solve binary hypothesis testing problems where Copula theory was used to model the dependency of the machine's observation and the human's side information. The conditions under which the human's side information will yield improved detection performance were derived.

While the topics and research directions discussed in this chapter might serve as starting points to advance the next-generation cognitive radar systems that involve human participations, novel theoretical frameworks for collaborative human–machine decision-making in complex environments require inputs from different disciplines such as statistical signal processing, artificial intelligence, machine learning, economics, experimental psychology, and neuroscience. The ultimate goal is to merge the best of humans with the best of machines in task environments so that humans and machines can interact and complement each other. Application scenarios may include: decisions are made autonomously by machines or the decisions are made by a human or a semi-autonomous system where humans and machines collaborate in making the final decision. Developments in this area are envisaged to result in a significant revolution in the design of many autonomous and semi-autonomous systems for SA and command and control, both in military and civilian applications, that involve human–machine collaboration.

References

[1] IEEE Standard for Radar Definitions. IEEE Std 686-2017 (Revision of IEEE Std 686-2008). 2017, pp. 1–54.

[2] Haykin S. Cognitive radar: a way of the future. *IEEE Signal Processing Magazine*. 2006;23(1):30–40.

[3] Haykin S, Xue Y, and Setoodeh P. Cognitive radar: step toward bridging the gap between neuroscience and engineering. *Proceedings of the IEEE*. 2012;100(11):3102–3130.

[4] Gurbuz SZ, Griffiths HD, Charlish A, *et al.* An overview of cognitive radar: past, present, and future. *IEEE Aerospace and Electronic Systems Magazine*. 2019;34(12):6–18.

[5] Bell KL, Baker CJ, Smith GE, *et al.* Fully adaptive radar for target tracking. Part I: single target tracking. In: *2014 IEEE Radar Conference*. Piscataway, NJ: IEEE, 2014, pp. 0303–0308.

[6] Mitchell AE, Smith GE, Bell KL, *et al.* Hierarchical fully adaptive radar. *IET Radar, Sonar & Navigation*. 2018;12(12):1371–1379.

[7] Mitchell AE, Smith GE, Bell KL, *et al.* Fully adaptive radar cost function design. In: *2018 IEEE Radar Conference (RadarConf18)*. Piscataway, NJ: IEEE, 2018, pp. 1301–1306.

[8] Blunt SD and Mokole EL. Overview of radar waveform diversity. *IEEE Aerospace and Electronic Systems Magazine*. 2016;31(11):2–42.

[9] Guerci J, Guerci R, Ranagaswamy M, *et al.* CoFAR: cognitive fully adaptive radar. In: *2014 IEEE Radar Conference*. Piscataway, NJ: IEEE, 2014, pp. 0984–0989.

[10] Charlish A, Hoffmann F, Klemm R, *et al*. Cognitive radar management. In *Novel Radar Techniques and Applications: Waveform Diversity and Cognitive Radar and Target Tracking and Data Fusion*, vol. 2. Stevenage: SciTech Publishing Inc., 2017, pp. 157–193.

[11] Shi C, Wang F, Sellathurai M, *et al*. Non-cooperative game theoretic power allocation strategy for distributed multiple-radar architecture in a spectrum sharing environment. *IEEE Access*. 2018;6:17787–17800.

[12] Smith GE, Gurbuz SZ, Brüggenwirth S, *et al*. Neural networks amp; machine learning in cognitive radar. In: *2020 IEEE Radar Conference (RadarConf20)*, 2020, pp. 1–6.

[13] Charlish A and Hoffmann F. Anticipation in cognitive radar using stochastic control. In: *2015 IEEE Radar Conference (RadarCon)*, 2015, pp. 1692–1697.

[14] Rhim JB, Varshney LR, and Goyal VK. Quantization of prior probabilities for collaborative distributed hypothesis testing. *IEEE Transactions on Signal Processing*. 2012;60(9):4537–4550.

[15] Wimalajeewa T and Varshney PK. Collaborative human decision making with random local thresholds. *IEEE Transactions on Signal Processing*. 2013;61(11):2975–2989.

[16] Geng B and Varshney PK. On decision making in human–machine networks. In: *2019 IEEE 16th International Conference on Mobile Ad Hoc and Sensor Systems (MASS)*. Piscataway, NJ: IEEE, 2019, pp. 37–45.

[17] Mourad S and Tewfik A. Real-time data selection and ordering for cognitive bias mitigation. In: *2016 IEEE International Conference on Acoustics, Speech and Signal Processing (ICASSP)*. Piscataway, NJ: IEEE, 2016, pp. 4393–4397.

[18] Sánchez-Charles D, Nin J, Solé M, *et al*. Worker ranking determination in crowdsourcing platforms using aggregation functions. In: *2014 IEEE International Conference on Fuzzy Systems (FUZZ-IEEE)*. Piscataway, NJ: IEEE, 2014, pp. 1801–1808.

[19] Geng B, Li Q, and Varshney PK. Decision tree design for classification in crowdsourcing systems. In: *2018 52nd Asilomar Conference on Signals, Systems, and Computers*. Piscataway, NJ: IEEE, 2018, pp. 859–863.

[20] Geng B, Li Q, and Varshney PK. Prospect theory based crowdsourcing for classification in the presence of spammers. *IEEE Transactions on Signal Processing*. 2020;68:4083–4093.

[21] Geng B, Brahma S, Wimalajeewa T, *et al*. Prospect theoretic utility based human decision making in multi-agent systems. *IEEE Transactions on Signal Processing*. 2020;68:1091–1104.

[22] Blasch EP, Rogers SK, Holloway H, *et al*. QuEST for information fusion in multimedia reports. *International Journal of Monitoring and Surveillance Technologies Research (IJMSTR)*. 2014;2(3):1–30.

[23] Blasch EP and Plano S. Level 5: user refinement to aid the fusion process. In: *Multisensor, Multisource Information Fusion: Architectures, Algorithms, and Applications 2003*. vol. 5099. International Society for Optics and Photonics, 2003, pp. 288–298.

[24] Hoffman RR, Feltovich PJ, Ford KM, *et al.* A rose by any other name ... would probably be given an acronym [cognitive systems engineering]. *IEEE Intelligent Systems*. 2002;17(4):72–80.

[25] Grigsby SS. Artificial intelligence for advanced human-machine symbiosis. In: *International Conference on Augmented Cognition*. New York, NY: Springer, 2018. pp. 255–266.

[26] Varshney PK. *Distributed Detection and Data Fusion*. New York, NY: Springer Science & Business Media, 2012.

[27] Tsitsiklis JN. Decentralized detection by a large number of sensors. *Mathematics of Control, Signals and Systems*. 1988;1(2):167–182.

[28] Wimalajeewa T, Varshney PK, and Rangaswamy M. On integrating human decisions with physical sensors for binary decision making. In: *2018 21st International Conference on Information Fusion (FUSION)*. Piscataway, NJ: IEEE, 2018, pp. 1–5.

[29] Wimalajeewa T and Varshney PK. Asymptotic performance of categorical decision making with random thresholds. *IEEE Signal Processing Letters*. 2014;21(8):994–997.

[30] Quan C, Geng B, and Varshney PK. Asymptotic performance in heterogeneous human-machine inference networks. In: *2020 54th Asilomar Conference on Signals, Systems, and Computers*. Piscataway, NJ: IEEE, 2020.

[31] Tversky A and Kahneman D. Advances in prospect theory: cumulative representation of uncertainty. *Journal of Risk and Uncertainty*. 1992;5(4): 297–323.

[32] Nadendla VSS, Brahma S, and Varshney PK. Towards the design of prospect-theory based human decision rules for hypothesis testing. In: *2016 54th Annual Allerton Conference on Communication, Control, and Computing (Allerton)*. Piscataway, NJ: IEEE, 2016, pp. 766–773.

[33] Gezici S and Varshney PK. On the optimality of likelihood ratio test for prospect theory-based binary hypothesis testing. *IEEE Signal Processing Letters*. 2018;25(12):1845–1849.

[34] Plous S. *The Psychology of Judgment and Decision Making*. New York: McGraw-Hill Book Company, 1993.

[35] Edwards W. The theory of decision making. *Psychological Bulletin*. 1954;51(4):380.

[36] Poletiek FH. *Hypothesis-Testing Behaviour*. London: Psychology Press, 2013.

[37] Chaudhuri R and Fiete I. Computational principles of memory. *Nature Neuroscience*. 2016;19(3):394.

[38] Faisal AA, Selen LP, and Wolpert DM. Noise in the nervous system. *Nature Reviews Neuroscience*. 2008;9(4):292–303.

[39] Sorkin RD, West R, and Robinson DE. Group performance depends on the majority rule. *Psychological Science*. 1998;9(6):456–463.

[40] Chen H, Varshney LR, and Varshney PK. Noise-enhanced information systems. *Proceedings of the IEEE*. 2014;102(10):1607–1621.

[41] Kay S. Can detectability be improved by adding noise? *IEEE Signal Processing Letters*. 2000;7(1):8–10.

[42] Bhatt S and Krishnamurthy V. Controlled sequential information fusion with social sensors. *IEEE Transactions on Automatic Control*. 2020;66:5893–5908.
[43] Geng B, Varshney PK, and Rangaswamy M. On amelioration of human cognitive biases in binary decision making. In: *2019 IEEE Global Conference on Signal and Information Processing (GlobalSIP)*. Piscataway, NJ: IEEE, 2019, pp. 1–5.
[44] Sriranga N, Geng B, and Varshney PK. On human assisted decision making for machines using correlated observations. In: *2020 54th Asilomar Conference on Signals, Systems, and Computers*. Piscataway, NJ: IEEE, 2020.
[45] Iyengar SG, Varshney PK, and Damarla T. A parametric copula-based framework for hypothesis testing using heterogeneous data. *IEEE Transactions on Signal Processing*. 2011;59(5):2308–2319.
[46] Zhang S, Geng B, Varshney PK, *et al.* Fusion of deep neural networks for activity recognition: a regular vine copula based approach. In: *2019 22th International Conference on Information Fusion (FUSION)*. Piscataway, NJ: IEEE, 2019, pp. 1–7.
[47] Nelsen RB. *An Introduction to Copulas*. Berlin: Springer Science & Business Media, 2007.
[48] Miller GA. The magical number seven, plus or minus two: some limits on our capacity for processing information. *Psychological Review*. 1956;63(2):81.
[49] Aas K, Czado C, Frigessi A, *et al.* Pair-copula constructions of multiple dependence. *Insurance: Mathematics and Economics*. 2009;44(2):182–198.
[50] Schmorrow DD and Kruse A. Augmented cognition. *Berkshire Encyclopedia of Human–Computer Interaction*. 2004;1:54–59.
[51] Mayer RC, Davis JH, and Schoorman FD. An integrative model of organizational trust. *Academy of Management Review*. 1995;20(3):709–734.

Chapter 4
Channel estimation for cognitive fully adaptive radar

Sandeep Gogineni[1], Bosung Kang[2], Muralidhar Rangaswamy[3], Jameson S. Bergin[1] and Joseph R. Guerci[1]

In this chapter, we present an overview of state-of-the-art radio frequency (RF) clutter modeling and simulation (M&S) techniques. Traditional statistical approximation-based methods will be reviewed followed by more accurate physics-based stochastic transfer function clutter models that facilitate site-specific simulations anywhere on earth. The various factors that go into the computation of these transfer functions will be presented, followed by a formulation of the cognitive radar framework under the stochastic transfer function model. The usability of cognitive radar algorithms and techniques is highly reliant on having accurate knowledge of the channel transfer functions. We present different algorithms to estimate these transfer functions. Finally, we introduce a radar challenge dataset that can enable testing and benchmarking of all cognitive radar algorithms and techniques.

4.1 Introduction

Radio-frequency (RF) signals are used in a multitude of defense, commercial, and civilian applications that are critical to the safety and security of mankind. Most of the RF applications like radar include target detection, localization, and tracking in the presence of intentional and unintentional interference. In this chapter, although we focus on radar applications, the techniques presented herein are relevant to all RF applications. In a radar system, an RF transmitter sends out signals to illuminate a scene of surveillance to infer about the scene and the targets present based on the echo signal measured at the receiver. In an ideal world without any interfering signals, accomplishing these tasks is fairly trivial. However, in a practical setting, the RF signals at the receiver are almost always corrupted by interfering signals. A major

[1]Information Systems Laboratories, Inc., San Diego, CA, USA
[2]University of Dayton Research Institute, Dayton, OH, USA
[3]Air Force Research Laboratory, Wright Patterson Air Force Base, Dayton, OH, USA

source of interference is reflections of ground clutter which are highly dependent on the terrain present in the illuminated scene.

Targets of interest can be obscured by these ground clutter reflections and this interference is even more prominent when the radar system is flying in the air, looking at ground targets. Therefore, the development of any new radar technique is heavily dependent on accurately modeling these ground clutter reflections. The models are critical not only in the development stage but also in the testing and evaluation phase. There is a scarcity of publicly available measured data for RF applications. The measured data is expensive to collect and limited to very specific scenarios. Even when collected, the data is sensitive in nature and not readily available to test new algorithms and techniques. Therefore, most of the radar research, development, and testing rely upon accurately modeling and simulating the data.

The traditional approach to clutter modeling treats the clutter returns to be random vectors with unknown covariance matrices [1–5]. Initially, the covariance matrices were assumed to be constant for any given scenario since traditional radar systems always transmitted fixed waveforms. With the advent of cognitive radar systems that are capable of adapting transmit waveforms, these models have been modified to treat the covariance matrices as a function of the transmit waveform. However, even with this change, these traditional models that have been used for several decades are still a statistical approximation as they essentially treat the clutter signals to be fully random in nature.

In reality, the clutter signals measured at any scene always include a deterministic component that is dependent on the physical features of the scene that has been illuminated. For example, the mountains, rivers, lakes, etc. within a scene do not move and, hence, if we collect radar data over the same scene on multiple days, we will have a common deterministic component to these measurements. There will be a random component as well, owing to other variations such as the swaying of trees and waves on the water surface. In the absence of any information about the fixed features present in the scene, the random models described in the previous paragraph can be used. However, when we have access to real-world environmental databases, an M&S tool must be able to faithfully replicate these site-specific features. Inspired by this, in [6], an alternate approach to clutter modeling using a "stochastic transfer function" (Green's function impulse response in the time domain) approach has been presented for this problem. This results in a fundamental physics-based scattering model that can be used to accurately simulate RF data.

In this chapter, the traditional covariance-based model will be described in more detail, followed by the stochastic transfer function model. All the realistic components that go into the computation of the stochastic transfer functions will be described (see also [7]). A formulation of the cognitive radar framework under the stochastic transfer function model will be presented. The usability of cognitive radar algorithms and techniques is highly reliant on having accurate knowledge of the channel transfer functions. We present different algorithms to estimate these channel transfer functions. Finally, we introduce a radar challenge dataset that can enable testing and benchmarking of all cognitive radar algorithms and techniques.

4.2 Traditional covariance-based statistical model

Traditional space–time adaptive processing (STAP) literature treats radar returns from the ground clutter as completely random with a pre-described probability distribution [8,9]. Considerable history underlies the exposition of [9]. Early work in this direction involved the collection and analysis of experimental data [10,11], which attempted to fit two-parameter families of distributions to describe the heavy-tailed behavior of clutter returns corresponding to high-resolution radar for false-alarm regulation. In an attempt to account for the pulse-to-pulse correlation as well as the first-order probability density function, endogenous and exogenous clutter models were developed [12–17]. The corresponding problem for coherent processing in Gaussian clutter received much attention from the 1950s [18–22]. Extensions of these treatises to account for CFAR behavior of the underlying adaptive processor were undertaken in [23–26]. All of these treatises use statistical approximations for clutter as described herein.

In this section, we provide a brief overview of these traditional approaches to clutter modeling. These models essentially treat clutter as an additive "colored noise" process with various approximate probability distribution models [27]. Figure 4.1 depicts the basic clutter physics model under consideration for a generally monostatic airborne moving target indicator radar (for both airborne- and ground-based targets, AMTI, and/or GMTI)—although the approach developed can be easily generalized to other configurations such as bi-/multi-statics.

As can be seen from Figure 4.1, the clutter returns corresponding to a particular range bin of interest can be expressed as a weighted summation of the returns from individual clutter patches present in that ring. Let there be N_c clutter patches in a

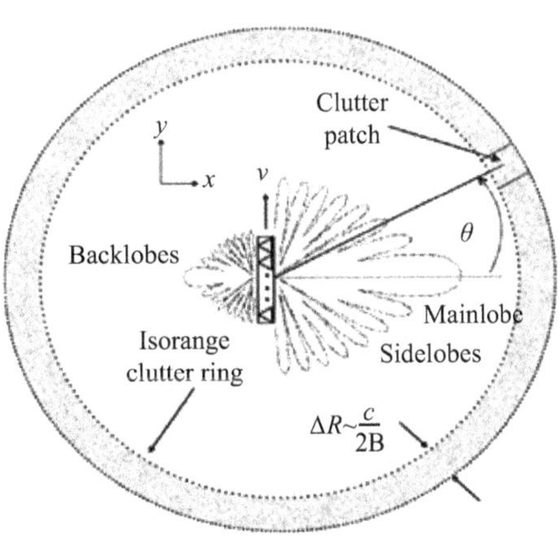

Figure 4.1 Traditional covariance-based clutter model: illustration of the monostatic iso-clutter range cell observed from a stand-off airborne radar

given range bin of interest. Then the clutter response corresponding to that range bin can be expressed as

$$\boldsymbol{x}_{\mathrm{c}} = \sum_{i=1}^{N_{\mathrm{c}}} \gamma_i \boldsymbol{v}_i, \tag{4.1}$$

where $\boldsymbol{x}_{\mathrm{c}}$ is the complex-valued NM-dimensional space–time total clutter return for a given range bin associated with N spatial and M temporal receive degrees-of-freedom (DoFs). $\boldsymbol{v}_i$ is the space–time steering vector to the ith clutter patch and is a Kronecker product of the temporal and spatial steering vectors.

While the steering vectors are deterministic, the traditional clutter models treat the complex scalar reflectivity corresponding to each patch to be zero-mean random variables. These variables γ_i denote the amplitude corresponding to the ith clutter patch and they are a function of the intrinsic clutter reflectivity and the transmit–receive antenna patterns. Given this model, the associated space–time clutter covariance matrix can be expressed as

$$\mathrm{E}\left\{\boldsymbol{x}_c \boldsymbol{x}_c^H\right\} = \sum_{i=1}^{N_{\mathrm{c}}} \sum_{j=1}^{N_{\mathrm{c}}} \mathrm{E}\left\{\gamma_i \gamma_j^*\right\} \boldsymbol{v}_i \boldsymbol{v}_j^H, \tag{4.2}$$

where E {.} denotes expectation operation. Under the assumption that these coefficients corresponding to different clutter patches are independent, we can express the clutter covariance matrix as

$$\mathrm{E}\left\{\boldsymbol{x}_c \boldsymbol{x}_c^H\right\} = \sum_{i=1}^{N_{\mathrm{c}}} G_i \boldsymbol{v}_i \boldsymbol{v}_j^H. \tag{4.3}$$

While this traditional approach has been used for the past several decades, it is essentially a statistical approximation and has not been derived from physics like the model described in the next section. Additionally, all the transmit DoFs have been collapsed into a single complex reflectivity random variable and, hence, appear non-linearly and indirectly in the above equations. Also, under this traditional clutter model, as we can see the clutter returns are independent of the transmitted radar waveform. While this assumption was acceptable for conventional radar systems that transmit a fixed waveform, it is highly unrealistic to assume this model for more modern cognitive radar systems that continuously adapt the transmit waveform to match the operating environment. An important implication of bringing to bear the transmit DoF is the generation of signal-dependent interference. In classical space–time adaptive radar processing, the problem is one of designing a finite impulse response (FIR) filter to adapt to an unknown interference covariance matrix. However, in a given adaptation window, the covariance matrix albeit unknown remains fixed. This fact makes it possible to collect replicas of training data sharing the same covariance structure to form an estimate of the covariance matrix. However, when the transmit DoF are brought to bear, the observed covariance matrix on receive is a non-linear function of the transmit signal. As a consequence, each realization of training data now corresponds to a different covariance matrix. Therefore using such training data for covariance matrix estimation yields an inaccurate estimate of the covariance matrix at best and a singular estimate of the covariance matrix at worst, thereby seriously degrading the performance/implementation of the adaptive processor. Therefore, an advanced clutter modeling approach that can capture the signal-dependent nature of ground clutter returns is required. We shall describe one such modeling approach in the next section.

4.3 Stochastic transfer function model

Contrary to the covariance-based model, the stochastic transfer function model treats the radar measurements according to the block diagram described in Figure 4.3. This is an accurate representation of the signals since the radar electromagnetic signal travels through the channel interacting with the different components present in the channel in a linear fashion as described by Maxwell's equations. Due to the linear nature of these interactions, the overall channel impact can be represented using an impulse response (Green's functions impulse response) in the time-domain or the corresponding stochastic transfer function in the frequency domain. The stochastic aspect of the transfer function comes from the random components present in the scene such as intrinsic clutter motion. Note that this new approach to clutter modeling in Figure 4.3 separates the radar data into target and clutter channels. The main focus of this chapter is the clutter channel.

Let $\boldsymbol{s(n)}$ denotes the transmit waveform and $\boldsymbol{h_c(n)}$, $\boldsymbol{h_t(n)}$ denotes Green's function impulse response for the clutter and target channels, respectively. Additionally, let $\boldsymbol{n(n)}$ represents the additive thermal noise. Then, the measurements at the radar receiver for time instant n can be represented as

$$\boldsymbol{y(n)} = \boldsymbol{h_c(n)} \circledast \boldsymbol{s(n)} + \boldsymbol{h_t(n)} \circledast \boldsymbol{s(n)} + \boldsymbol{n(n)}, \tag{4.4}$$

where $\circledast$ denotes the convolution operation. Convolution in the time domain can be represented using multiplication in the frequency domain. Therefore, the measurement model at frequency bin k can be represented as

$$\boldsymbol{Y(k)} = \boldsymbol{H_c(k)S(k)} + \boldsymbol{H_t(k)S(k)} + \boldsymbol{N(k)}, \tag{4.5}$$

where $\boldsymbol{H_c(k)}$ and $\boldsymbol{H_t(k)}$ denote the clutter and target stochastic transfer functions, respectively.

Having zeroed-in on the physics-based linear model from the above equation, the natural next question would be what goes into the computation of these impulse responses/transfer functions. For the above model to be accurate, the transfer functions must capture the interaction of a transmitted ideal delta function with every component present in the scene. For example, for the clutter channel, the scene (which is typically several square kilometers in size) has to be broken down into extremely small patches and the impact of each individual patch on the received data has to be modeled. The overall clutter returns are a summation of the returns from each individual clutter patch. In other words, the reflectivity of each individual patch along with the propagation attenuation have to be accurately captured to have a realistic model.

For any given patch, whether there is a line-of-sight (LOS) component based on the transmitter and receiver locations needs to be determined first. If an LOS does indeed exist, then the reflectivity of that patch depends on the tilt angle of the patch, operating frequency band, type of material present in the patch, etc. A sophisticated M&S tool incorporates all these factors while computing the transfer function. For example, Figure 4.2 demonstrates the monostatic scattering polynomials as a function of the grazing angle for different types of terrain at X-band. For other frequency bands, the scattering polynomials will be quite different. These scattering polynomials have been extensively studied in the literature [27–30]. In addition to the scattering polynomials demonstrated in Figure 4.2, developing scattering models for ocean surfaces involves additional challenges as a result of moving ship-effects. The Physics-Based Ocean Surface and Scattering (PBOSS) model described in [29,30] incorporates both environmental conditions (i.e., atmospheric and oceanographic) and moving ship effects (i.e., Kelvin and near-field/narrow-V wakes) to generate realizations of the ocean surface and the spatially varying scattering properties of the ocean. The PBOSS model has been incorporated into the high-fidelity M&S tool RFView® [31] to provide RF phenomenology characterization of ocean environments. We can leverage this existing modeling capability, incorporating the effects of the surrounding ocean surface and its scattering properties, to adaptively and optimally design waveforms for the detection of submarines and ships. An example rendering of the ocean surface model output using a ray tracer [32,33] is shown in Figure 4.5. Clouds, sky, and fog are included in the rendering for realism.

Additionally, the impact of the propagation medium needs to be implemented along with the scattering model. The channel is highly dependent on the characteristics of the propagation path between the radar and the targets and clutter patches

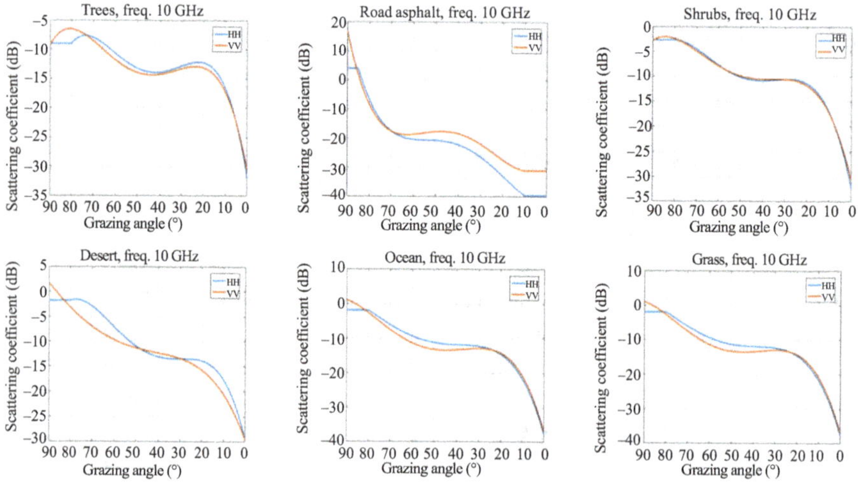

Figure 4.2 The polarimetric scattering coefficient as a function of grazing angle for different landcover types at X-band

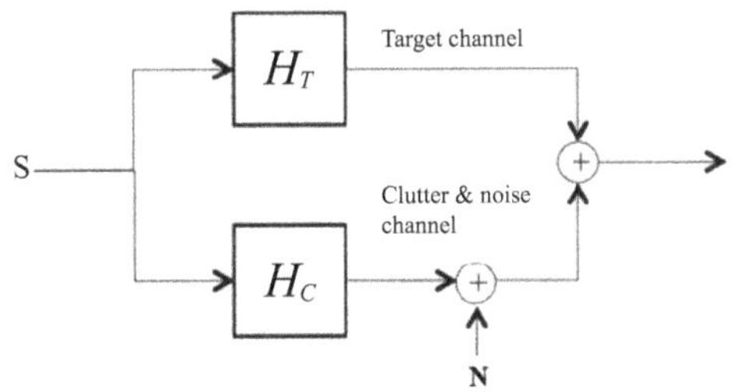

Figure 4.3 Illustration of the stochastic transfer function model

in scene. At higher frequencies such as X-band, the propagation is often dominated by LOS and can be approximated by simple models that identify blockages along the propagation path caused by terrain and buildings in the scene and simply apply a larger attenuation to "shadow" regions. When higher fidelity is required or when simulating systems operating a lower frequencies, it may be necessary to include more advanced propagation modes such as multipath, diffraction, and ducting. Modeling these modes typically involves analysis of the terrain profile between the radar and the target (or clutter patch) to determine the most appropriate mode or combination of modes of propagation to employ for predicting the propagation loss. A good example of this type of model is the SEKE [34] model developed by the MIT Lincoln Laboratory. This model includes multiple knife-edge diffraction, spherical earth diffraction, and multipath to predict the site-specific propagation loss along a specified terrain profile typically extracted from a terrain database such as DTED.

While these types of models are somewhat ad hoc, they are relatively computationally efficient and can provide very realistic results. Effects such as ducting which can dominate propagation in environments with more complex atmospheric conditions typically require more sophisticated and generally more computationally intensive methods such as parabolic wave equation solutions [35]. The advanced propagation model (APM) [36] developed by the SPAWAR Systems Center is an example of a propagation code that includes this type of mode. This APM code allows the atmosphere to be specified with an index of refraction that varies both vertically and horizontally within the plane of propagation between a radar and target/clutter. This allows for accurate simulations of the well-known ducting phenomenon often encountered in maritime and littoral environments.

In addition to high-fidelity environmental modeling, precision modeling of all RF subsystems and components is crucial to again capture many real-world effects. For example, Figure 4.4 shows the difference (antenna pattern) between a standard aperture model using approximations and idealizations, and one that includes a variety of real-world RF component imperfections. This degree of realism is essential if the simulated data is to serve as a testbed for radar algorithms. After incorporating all these real-world hardware effects, scattering, propagation models, and environmental interactions with ground clutter, an advanced M&S tool needs to compute the raw

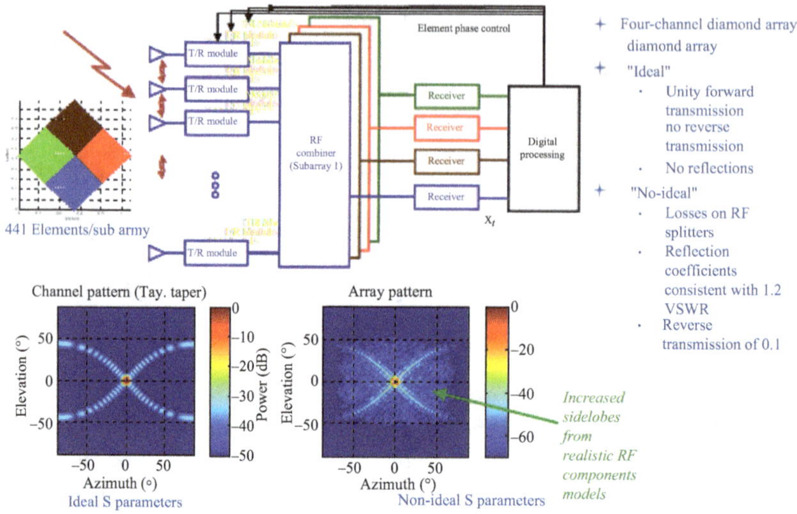

Figure 4.4 Example of high-fidelity active electronically scanned array (AESA) that captures many real-world RF imperfections

I&Q measurements at the radar receivers along with the true EM propagation channel impulse responses. Having described the stochastic transfer function-based radar clutter model, we will now present the cognitive radar framework in the next section.

4.4 Cognitive radar framework

All the examples presented in this chapter have been generated using high-fidelity RF M&S tool RFView [31] which generates the data using the stochastic transfer function model presented in the previous section. Before we present the cognitive radar framework, we start with a simple monostatic GMTI radar example with fixed transmit waveform. Monostatic radar systems have colocated transmitter and receiver. We consider an airborne X-band (10 GHz) radar flying along the coast of southern California looking at a ground-moving target (see Figure 4.6). The radar is moving at an altitude of 1,000 m with a speed of 125 m/s. The simulation spans a range swath of 20 km with a linear frequency modulated (LFM) waveform of bandwidth 5 MHz and 65 pulses with a pulse repetition frequency (PRF) of 2,100 Hz. We now look at the different layers that go into the calculation of the impulse responses. First, for each patch in the scene, the presence or absence of an LOS component needs to be computed.

Figure 4.7 shows the terrain map of the simulated scene. We can clearly notice that the scene has mountainous terrain and one huge mountain peak that is denoted by the

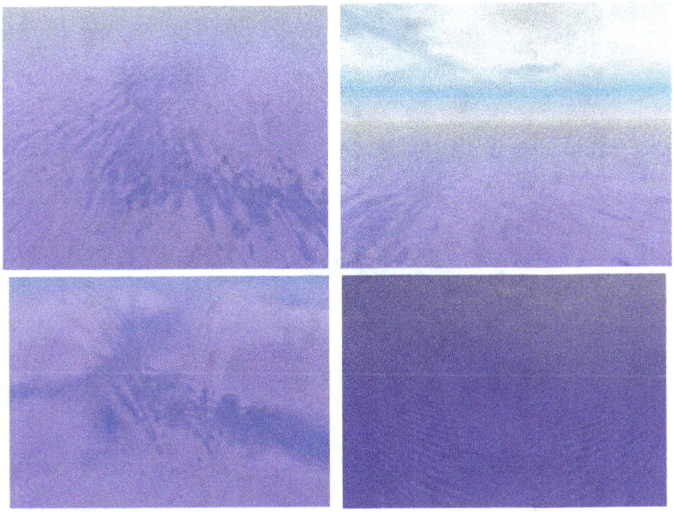

Figure 4.5 Renderings of ocean surface realizations from PBOSS model. Bottom: ocean surfaces including Kelvin wakes generated from a moving ship.

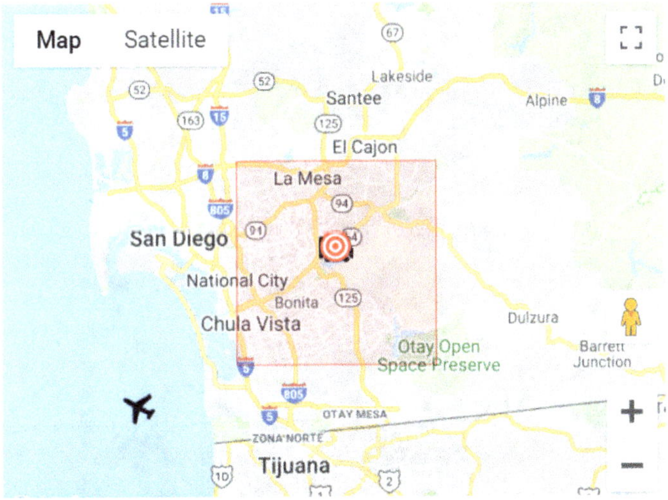

Figure 4.6 Google maps illustration of the simulated monostatic scene with airborne radar and ground target

red region in the terrain map. This information on the terrain is obtained from publicly available terrain databases and they span the entire earth. Given this terrain map, for each patch, the LOS map is demonstrated in Figure 4.8. We can observe that the region behind the mountain peak is the shadow region that cannot be penetrated/illuminated by the radar signal. Hence, the shadow regions do not contribute to the clutter returns.

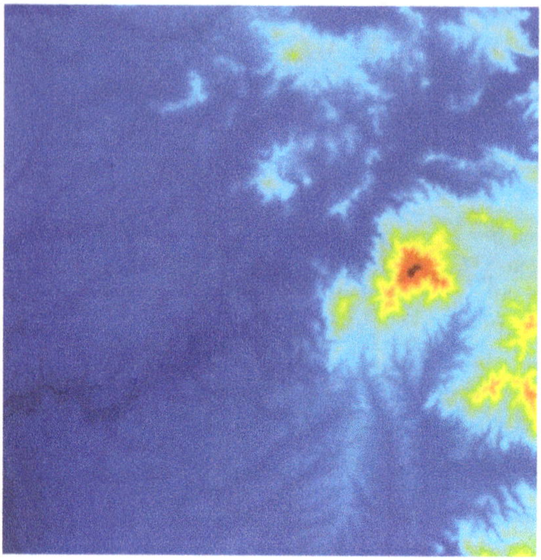

Figure 4.7 Terrain map of the simulated scene

Figure 4.8 LOS map of the simulated scene

Next, for the patches which indeed have an LOS component, the reflectivity has to be computed using the scattering polynomials described in the previous section. As mentioned earlier, these scattering polynomials vary for different types of terrain. Hence, it is important to use environmental databases that describe the type of terrain present in each clutter patch. Figure 4.9 demonstrates the different types

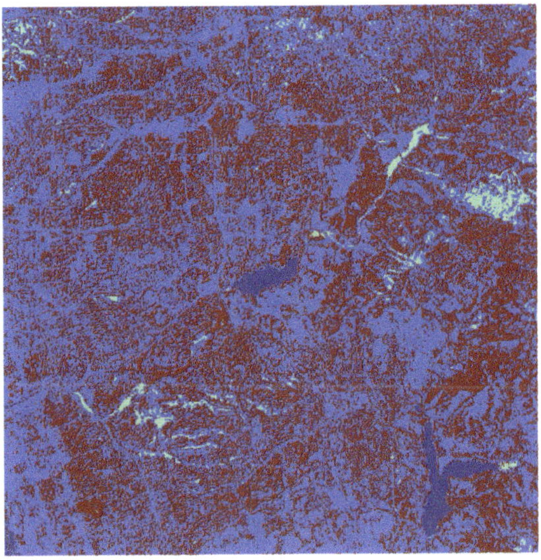

Figure 4.9 Land cover map of the simulated scene

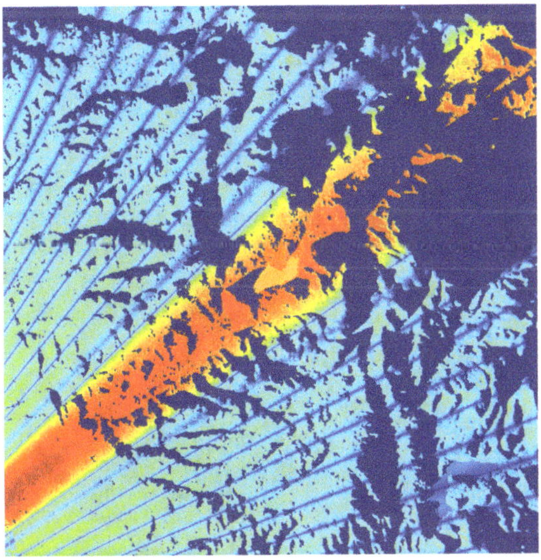

Figure 4.10 Final clutter map of the simulated scene

of terrain present in each patch. Each terrain type leads to a unique clutter response and leads to the overall clutter map with reflectivity from each patch demonstrated in Figure 4.10. Note that this clutter map also shows the effect of the radar main beam and side lobes and the reflectivity from each patch also depends on the incident energy from the radar beam.

Having demonstrated the different components that go into the computation of the RF clutter map for this monostatic GMTI example, we now calculate Green's function impulse response for the ground clutter and target channels. For the clutter channel, the impulse response is computed as a the summation of the responses from each individual clutter patch in the clutter map. Note that along with the reflectivity, each patch also induces a different delay and Doppler component on the incident RF signal. Typically any given scene can contain hundreds of thousands or even millions of patches. However, due to the inherently parallel nature of the computations, the impulse responses can be computed in near real time using high-performance-computing clusters or GPUs. Recent advances in accelerated computing making it feasible to use these advanced methods for realistic RF clutter simulation. For this example, Figure 4.11 demonstrates the impulse response that is summation of both the clutter and target channels. As we can clearly see, the one big peak corresponds to the target and it shows up at the appropriate time instance based on the location of the target. The rest of the impulse response is specific to the clutter scene that has been simulated in this example. Note that the response at any range-bin in the impulse response can be a cumulative effect of responses from multiple clutter patches.

Given the impulse response displayed in Figure 4.11, we can calculate the raw IQ data at the radar receiver as a convolution of the radar transmit waveform and the impulse response with additive noise. Note that the impulse response is site specific and accurately captures all the local features present in the simulated scene. Therefore, the IQ data generated using this approach is very realistic compared to approximate statistical methods that have been used for several decades. Simple beamforming

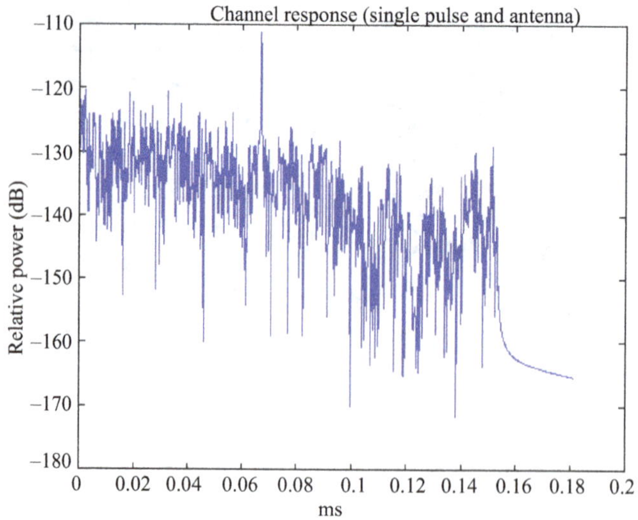

Figure 4.11 Green's function impulse response of the clutter+target channel for the simulated monostatic GMTI example

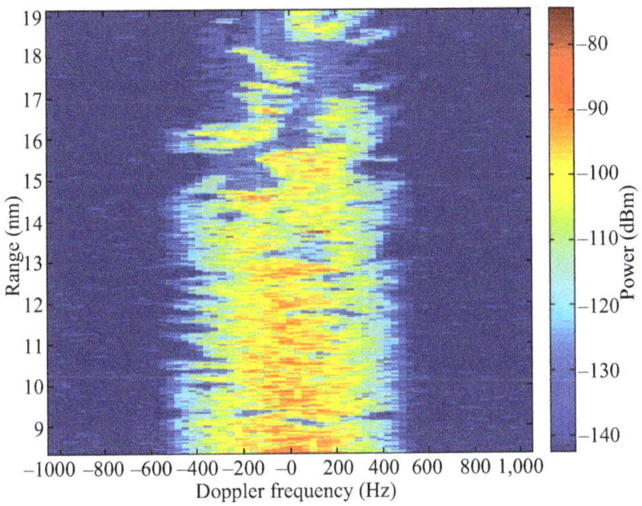

Figure 4.12 Range–Doppler plot after processing the raw IQ data generated by the simulator

and matched filtering of the data generated using the simulator produces the range–Doppler plot as demonstrated in Figure 4.12. The patterns of clutter that show up in this plot are again site specific and if we were to repeat this example at a different location, the generated plots would match the operating environment instead of just using average statistics for several scenes.

Next, we present a bistatic GMTI example. In a bistatic scenario, the radar transmitter and the receiver are present in different physical locations. As a result of this, the underlying computations for the scattering coefficients of each patch present in the scene is completely different compared to the monostatic case. Even the LOS computations need to take into account the present of a direct path from both the transmitter and the receiver to the simulated patch. Therefore, the shadow regions will also be different. We consider the same scenario and radar parameters as we used for the monostatic example above. However, now the transmitter is moved to a different location as shown with a blue aircraft symbol in Figure 4.13. The transmitter is flying at same altitude of 1, 000 m as the receiver.

We plot the channel impulse response for this bistatic GMTI example in Figure 4.14 and clearly notice the difference compared to the impulse response for the monostatic example in Figure 4.11. Similarly, after processing the receiver IQ data, we obtain the range–Doppler plot in Figure 4.15. As expected, this range–Doppler plot captures the effects of the bistatic geometry of the simulation. The delays and Doppler frequencies are now a function of the locations of both the airborne transmitter and the receiver.

The previous two examples represent a traditional monostatic and bistatic radar system with fixed transmit functions. We now simulate a more advanced radar system

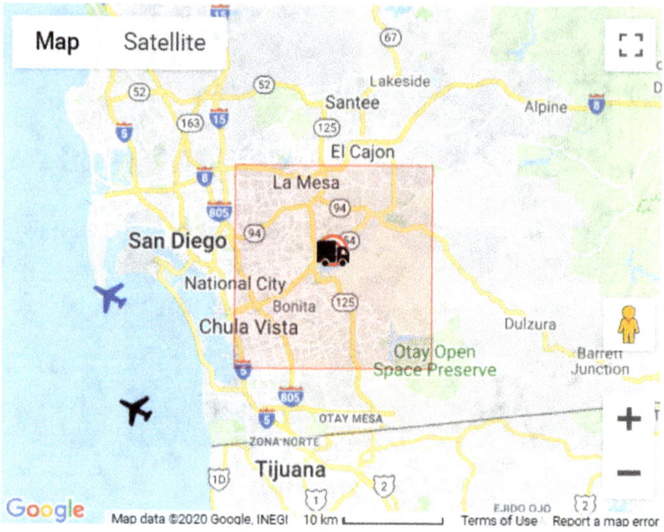

Figure 4.13 Google maps illustration of the simulated bistatic scene with airborne radar transmitter (blue), receiver (black), and ground target

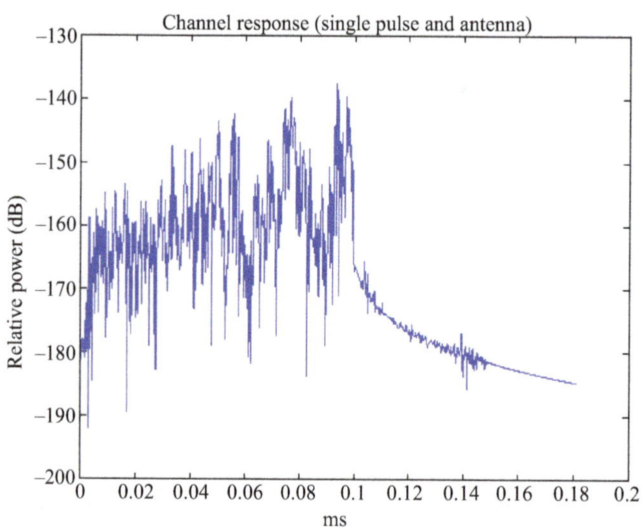

Figure 4.14 Channel impulse response for the simulated bistatic GMTI example

called CoFAR or CR in short. It is very important for an M&S tool to simulate emerging technologies and systems along with the traditional ones. In fact, emerging technologies and algorithms need the most data for testing and evaluation. Cognitive radar (CR) has emerged as key-enabling technology to meet the demands of ever

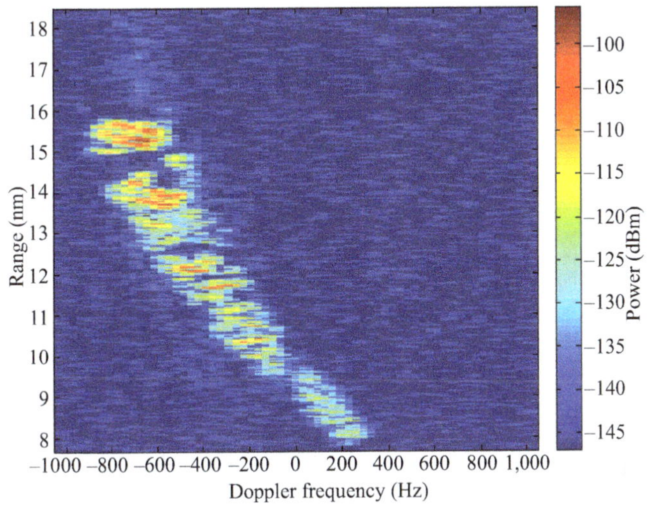

Figure 4.15 Range–Doppler plot after processing the raw IQ data generated by the simulator for bistatic example

increasingly complex, congested, and contested RF operating environments [1]. While a number of CR architectures have been proposed in recent years, a common thread is the ability to adapt to complex interference/target environments in a manner not possible using traditional adaptive methods.

For example, in conventional STAP, it is assumed that a sufficient set of wide sense stationary (WSS) training data is available to allow for convergence of the adaptive weights [8]. Though a variety of "robust" or reduced-rank training methods have been proposed over the past 25 years [8], there are still many real-world scenarios where even these methods are insufficient. These environments are routinely encountered, for example, in dense urban and/or highly mountainous terrain, and/or in highly contested environments. In contrast, CR uses a plurality of advanced knowledge-aided (KA) and artificial intelligence (AI) methods to adapt in a far more sophisticated and effective manner.

For example, in urban terrain, a KA CR would have access to a detailed terrain/building map and real-time ray-tracing tools in order to adapt with extreme precision to targets anywhere in the scene, even those behind buildings (see [37] and [38] for recent work in this area). In cognitive fully adaptive radar (CoFAR) [1], the presence of a fully adaptive transmitter allows for active multichannel probing to support advanced signal-dependent channel estimation. All CR architectures have some form of a sense–learn–adapt (SLA) decision process. The latest CR architectures differ mainly in the ways in which each of these steps is performed.

Figure 4.16 (see also [39]) shows, at a high level, some of the major elements of a CoFAR system. Key-processing elements include:

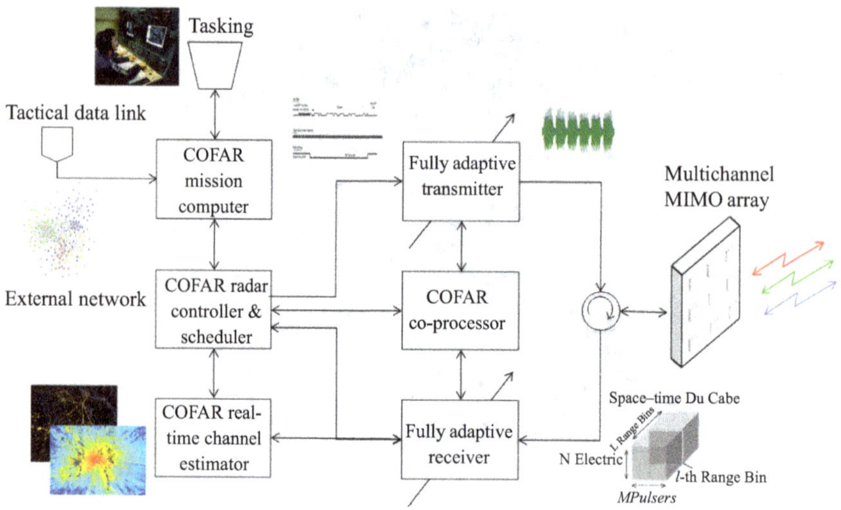

Figure 4.16 Major elements of a CoFAR

- CoFAR controller and scheduler: Performs optimal real-time resource allocation and radar scheduling. It receives mission objectives and has access to all requisite knowledge-bases and compute resources to effectively enable optimal decisioning.
- CoFAR real-time channel estimator: Performs advanced multidimensional channel estimation using a plurality of methods including KA processing, real-time ray-tracing, active MIMO probing, and/or machine-learning techniques.
- CoFAR co-processor: Performs extremely low-latency adaption (potentially intra-pulse). Mostly applicable in advanced electronic warfare applications.
- Fully adaptive receiver: Features the usual adaptive receiver capabilities such as adaptive beamforming and pulse compression.
- Fully adaptive transmitter: A relatively new feature to radar front-end. Extremely useful for pro-active channel probing and support of advanced adaptive waveforms.

In many respects, a key goal of all the above is channel estimation. "As goes channel knowledge, so goes performance." In the CoFAR context, the channel consists of clutter (terrain, unwanted background targets), targets, atmospheric, meteorological effects, and intentional, and/or unintentional RFI. To capture real-world environmental effects, and to present the CR with a meaningfully challenging simulation environment, clutter often presents the greatest challenge. Green's function impulse response method that we described in the previous section exactly addresses this issue and provides an accurate site-specific testbed to evaluate these advanced CoFAR techniques. In this chapter, we present one such example that involved an advance CoFAR system that optimally adapts its transmit waveform to match the operating environment to achieve optimal radar performance. This is in contrast to traditional radar systems that transmit a fixed waveform.

Given the measurement model in the previous section, the goal of CR waveform design is to find the optimal waveform $\boldsymbol{S}$ (stacked into a vector) such that it maximizes the signal-to-clutter-plus-noise-ratio (SCNR) subject to energy constraint $\boldsymbol{S}^H\boldsymbol{S} = 1$

$$\mathrm{SCNR} = \frac{E\left\{\|\boldsymbol{H}_t\boldsymbol{S}\|^2\right\}}{E\left\{\|\boldsymbol{H}_c\boldsymbol{S} + \boldsymbol{N}\|^2\right\}}, \tag{4.6}$$

where $\boldsymbol{H}_t$ and $\boldsymbol{H}_c$ denote the target and clutter channel stochastic transfer functions and $\boldsymbol{N}$ denotes the additive noise. Also, $E\{.\}$ denotes the expectation operator. Note that the transfer functions still have a random component that can be induced by intrinsic clutter motion and other factors, which is why we use the expectation operator. The solution to this optimization problem can be easily shown to satisfy the following generalized eigenvector equation

$$\lambda\left(E\left\{\boldsymbol{H}_c^H\boldsymbol{H}_c\right\} + \sigma^2\boldsymbol{I}\right)\boldsymbol{S} = E\left\{\boldsymbol{H}_t^H\boldsymbol{H}_t\right\}\boldsymbol{S}, \tag{4.7}$$

where σ^2 denotes the additive noise variance. Since the matrix $E\left\{\boldsymbol{H}_c^H\boldsymbol{H}_c\right\} + \sigma^2\boldsymbol{I}$ is always invertible, we can further write down the optimal waveform as the eigenvector of the following matrix:

$$\left(E\left\{\boldsymbol{H}_c^H\boldsymbol{H}_c\right\} + \sigma^2\boldsymbol{I}\right)^{-1} E\left\{\boldsymbol{H}_t^H\boldsymbol{H}_t\right\}\boldsymbol{S} = \lambda\boldsymbol{S}. \tag{4.8}$$

As we can see from the measurement model described in Figure 4.3, the radar clutter and target channel impulse responses (or transfer functions) are independent of the transmit waveform itself even though the IQ data at the receiver is signal-dependent clutter. This approach to modeling makes it feasible to generate simulated data for any arbitrary waveform that has been designed by CR. The new choice of waveform (one example of optimal waveform design is described earlier) will be convolved with the appropriate channel impulse responses. This ability to simulate realistic radar data for rapidly adapting radar waveforms makes this approach to M&S a perfect match to test different CR algorithms without having to resort to expensive measurement campaigns which are further limited by the number of algorithms that can be tested in a single data collect mission. The channel impulse responses corresponding to different locations on earth can be simulated to test the generality of the CR techniques. Further, multiple CPIs of data can be generated to simulate the changing dynamics of these channel impulse responses and their impact on the CR performance.

Figure 4.17 shows an airborne X-band radar surveillance scenario of a north-bound offshore aircraft flying off the coast of southern California. The region has significant heterogenous terrain features and thus presents an interesting real-world clutter challenge. In the presence of flat terrain with no clutter discretes, the spectrum will be flat and as a result there would not be any advantage as a result of adapting the transmit waveform. The heterogenous terrain in this example ensures strong CoFAR performance potential. Shown in Figure 4.18 is the theoretical performance gain (tight bound) using the optimal pulse shape prescribed above (max gain = ratio of max to min eigenvalue in dB). As expected, the maximum potential gain is achieved in

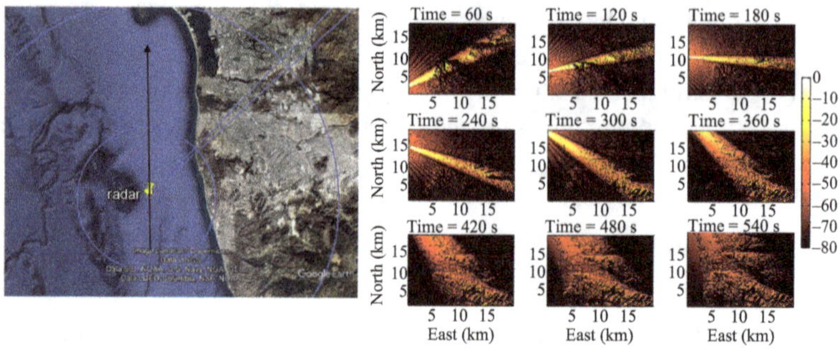

Figure 4.17 X-band site-specific airborne GMTI radar scenario off the coast of southern California. Left: scenario location and geometry. Right: radar beam pointing positions at different portions of the flight.

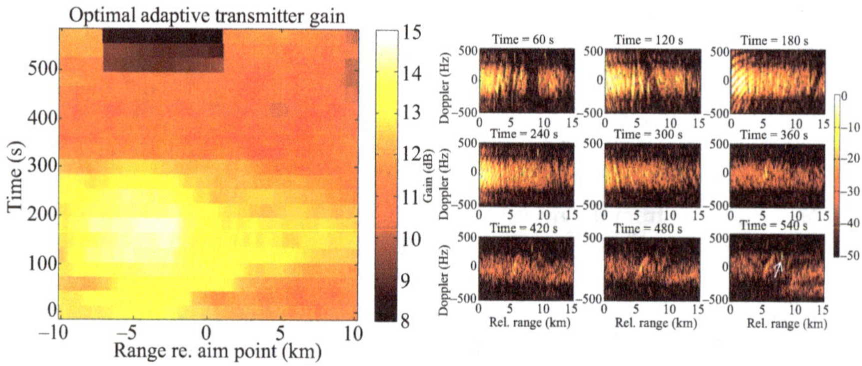

Figure 4.18 Left: Optimal maximum gain (dB) using adaptive waveforms as a function of time and range of interest. Note in general maximum gain is achieved in regions with the strongest heterogenous clutter (as expected). Right: Corresponding range–Doppler plots. Note that there is no gain when the clutter is weak in the presence of a single discrete (again to be expected).

those regions with the strongest heterogenous clutter since this produces significant eigenvalue spread. Also as expected, there is no gain in areas of weak clutter where only a single large discrete (impulse) is present since this yields a flat eigenspectrum. Note, however, that these results assume that the optimizer has access to the true channel transfer functions. However, in reality, these channel impulse responses and their corresponding transfer functions are not known ahead of time and they need to be estimated from the measured data [40,41]. The rest of this chapter is focused on multiple algorithms to estimate the channel transfer functions from measured data.

4.5 Unconstrained channel estimation algorithms

The cognitive radar results from the previous section can be achieved only when we have accurate channel state information. As goes channel knowledge, so goes the performance of your cognitive radar system. In this section, we present some basic channel estimation algorithms using the stochastic transfer function model presented earlier in the chapter.

4.5.1 SISO/SIMO channel estimation

Let the unknown impulse response corresponding to the channel and pulse indices of interest be denoted by $h_c(n)$ and the probing waveform be denoted by $w(n)$. Then, the probing measurements can be expressed as:

$$x_c(n) = h_c(n) \circledast w(n) + n(n).$$

We will look at this measurement model in the frequency domain. Convolution in the time domain can be expressed as multiplication in the frequency domain. Therefore, we can express the measurements in the frequency domain (we use upper case) as below:

$$X_c(k) = H_c(k) \circledast W(k) + N(k), \tag{4.9}$$

where $k = 1, \ldots, N$ is the frequency bin index. Now, the unknown parameters to be estimated are the entries of the transfer function $H_c(k)$. Therefore, a simple channel estimator can be expressed as

$$\widehat{\boldsymbol{H}_c(k)} = \frac{X_c(k)}{W(k)}. \tag{4.10}$$

Note that before performing the above operation, we may need to pad zeros to ensure that the vectors are of the same length. Finally, we estimate the unknown channel transfer function parameters as

$$\widehat{\boldsymbol{h}_c(n)} = \mathscr{F}^{-1}\left\{\widehat{\boldsymbol{H}_c(k)}\right\}, \tag{4.11}$$

where $\mathscr{F}^{-1}$ denotes the inverse Fourier transform. The performance of this estimator has been demonstrated in [40].

4.5.2 MIMO channel estimation

In this subsection (see also [42]), we consider a MIMO radar setup to extend the SISO channel estimator from the previous sub-section. This is a more challenging problem right away because of the necessity to estimate Green's function clutter channel impulse responses corresponding to all the individual bistatic pairs within the MIMO radar configuration of interest. We need multiple probing waveforms to estimate all the channel coefficients. We shall start with the basic 2×2 MIMO model to derive some results on the MIMO channel probing techniques. These results can be extended to the case involving an arbitrary number of transmitters without loss of generality.

In the time domain, let us define the first set of sampled transmit probing waveforms from the two transmitters as $w_1(n)$ and $w_2(n)$. Then, the measured signal at the first receiver can be expressed as

$$x_{c1}(n) = h_{c11}(n) \circledast w_1(n) + h_{c21}(n) \circledast w_2(n) + n_1(n),$$

where $h_{c11}(n)$ denotes the clutter channel impulse response between the first transmitter and the first receiver; $h_{c21}(n)$ denotes the clutter channel impulse response between the second transmitter and the first receiver; $n_1(n)$ denotes the additive thermal noise.

Similarly, the measured signal at the second receiver can be expressed as

$$x_{c2}(n) = h_{c12}(n) \circledast w_1(n) + h_{c22}(n) \circledast w_2(n) + n_2(n).$$

Note that so far we have not imposed any constraints on the relationship between the probing waveforms $w_1(n)$ and $w_2(n)$. Also, these are just the first set of probing waveforms and we need multiple of them to estimate the MIMO channel. Convolution in the time domain can be expressed as a simple multiplication in the frequency domain. Therefore, we will convert the measurement model above into the frequency domain:

$$X_{c1}(k) = H_{c11}(k)W_1(k) + H_{c21}(k)W_2(k) + N_1(k),$$

$$X_{c2}(k) = H_{c12}(k)W_1(k) + H_{c22}(k)W_2(k) + N_2(k),$$

where $k = 1, \ldots, N$ is the frequency bin index. Note that we need to pad zeros to the time domain impulse response and/or the transmit waveform to ensure that they are of the same length while applying the FFT to transform them to the frequency domain and also because convolution operation results in the output vector being longer than the constituent input vectors. Therefore, we need the input vectors to be padded with zeros to ensure that inverse Fourier transforms of the above equations lead to time domain vectors of appropriate lengths. Vectorizing the above equations, for a fixed frequency bin k, we have

$$\boldsymbol{X}_c(\boldsymbol{k}) = \boldsymbol{H}_c(\boldsymbol{k})\boldsymbol{W}(\boldsymbol{k}) + \boldsymbol{N}(\boldsymbol{k}),$$

where

$$\boldsymbol{X}_c(\boldsymbol{k}) = \begin{bmatrix} X_{c1}(k) \\ X_{c2}(k) \end{bmatrix},$$

$$\boldsymbol{H}_c(\boldsymbol{k}) = \begin{bmatrix} H_{c11}(k) & H_{c21}(k) \\ H_{c12}(k) & H_{c22}(k) \end{bmatrix},$$

and the noise and waveform vectors are given by

$$\boldsymbol{W}(\boldsymbol{k}) = \begin{bmatrix} W_1(k) \\ W_2(k) \end{bmatrix}, \boldsymbol{N}(\boldsymbol{k}) = \begin{bmatrix} N_1(k) \\ N_2(k) \end{bmatrix}.$$

$\boldsymbol{W}(\boldsymbol{k})$ is just a single probing vector at the kth frequency and, therefore, not sufficient to estimate the entire channel matrix $\boldsymbol{H}_c(\boldsymbol{k})$. Note that we use bold font to represent vectors and matrices while using a normal font to represent scalar parameters.

Let us assume that we have P probing vectors at the frequency k with the pth vector denoted by $W_p(k)$, then arranging all the probes into a matrix, we have

$$\boldsymbol{W}_{\mathrm{AP}}(\boldsymbol{k}) = [\boldsymbol{W}_{\boldsymbol{1}}(\boldsymbol{k}) \ldots \boldsymbol{W}_{\boldsymbol{P}}(\boldsymbol{k})],$$

where the subscript AP denotes All Probes. Similarly, arranging the measurements from all the probes into a matrix, we get

$$\boldsymbol{X}_{\mathrm{cAP}}(\boldsymbol{k}) = [\boldsymbol{X}_{\boldsymbol{c1}}(\boldsymbol{k}) \ldots \boldsymbol{X}_{\boldsymbol{cP}}(\boldsymbol{k})].$$

Therefore, the final measurement model for channel estimation in the frequency domain using all the probing vectors can be expressed as

$$\boldsymbol{X}_{\mathrm{cAP}}(\boldsymbol{k}) = \boldsymbol{H}_{\mathrm{c}}(\boldsymbol{k})\boldsymbol{W}_{\mathrm{AP}}(\boldsymbol{k}) + \boldsymbol{N}_{\mathrm{AP}}(\boldsymbol{k}),$$

where $\boldsymbol{W}_{\mathrm{AP}}(\boldsymbol{k})$ is a known matrix because it contains all the probing vectors that we transmitted. Therefore, the goal is to estimate $\boldsymbol{H}_{\mathrm{c}}(\boldsymbol{k})$ from $\boldsymbol{X}_{\mathrm{cAP}}(\boldsymbol{k})$ based on our knowledge of $\boldsymbol{W}_{\mathrm{AP}}(\boldsymbol{k})$. The simple least squares solution to this estimation problem can be expressed as

$$\widehat{\boldsymbol{H}_{\mathrm{c}}(\boldsymbol{k})} = \boldsymbol{X}_{\mathrm{cAP}}(\boldsymbol{k})\boldsymbol{W}_{\mathrm{AP}}^{\boldsymbol{H}}(\boldsymbol{k})\left(\boldsymbol{W}_{\mathrm{AP}}(\boldsymbol{k})\boldsymbol{W}_{\mathrm{AP}}^{\boldsymbol{H}}(\boldsymbol{k})\right)^{-1}.$$

However, this solution can be computed only when the matrix $\boldsymbol{W}_{\mathrm{AP}}(\boldsymbol{k})\boldsymbol{W}_{\mathrm{AP}}^{\boldsymbol{H}}(\boldsymbol{k})$ is invertible. In the case of a 2×2 MIMO system, the matrix is invertible only when $\boldsymbol{W}_{\mathrm{AP}}(\boldsymbol{k})$ is a full rank 2 matrix. Therefore, we need to transmit a minimum of 2 linearly independent vectors to estimate the 2×2 MIMO channel matrix.

4.5.3 Minimal probing strategies

In general, based on the result from the previous section, if we have M transmitters, then we need a minimum of M linearly independent vectors at each of the N transmit frequency bins to estimate the MIMO channel matrix. Hence, we need a minimum total of MN probing vectors if the goal is to estimate the entire impulse response of each of the constituent bistatic channels within the MIMO system. However, in most practical applications, the dominant clutter returns are spread across only a limited number of range bins or fast-time samples. For example, these could be from buildings or other dominant localized scatterers on ground which are called clutter discretes. When the goal is to only estimate that component of the impulse response corresponding to the clutter discretes, we can make use of prior knowledge about the location of the clutter discretes to estimate the clutter channel with fewer probing vectors than the MN vectors that are needed to estimate the entire impulse response as described earlier.

In the formulation that we develop here, multiple clutter discretes are allowed to be present in the channel impulse responses (see Figure 4.19). Estimating the responses from these discretes is essential to obtain an accurate estimate of the radar channel which in turn is critical to implement the optimal CoFAR waveform and filter solutions. The most important question to be answered here is whether we can quantify the reduction in the number of required probing vectors provided that we know that the strong clutter discretes span only R out of the total possible N range bins in the fast-time sample domain.

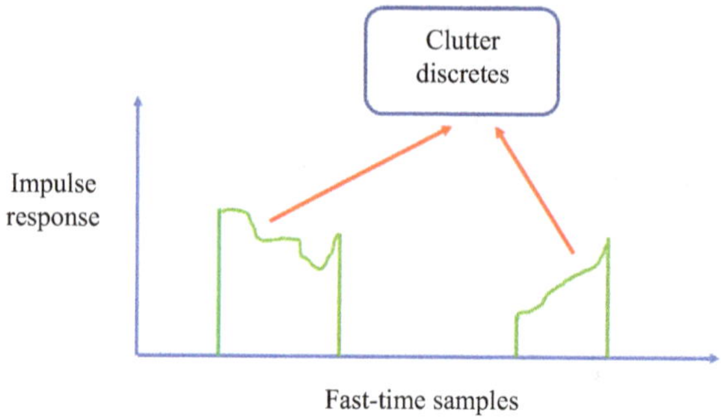

Figure 4.19 Clutter discretes only occupy a few samples in the fast-time sample domain

4.5.3.1 Single receiver

In this scenario, for each transmitter, there is only one unknown channel impulse response that needs to be estimated. Note that a sharp impulse signal in the time domain is spread across all frequencies in the frequency domain. So, while the clutter discretes described earlier are present across only R range bins or fast time samples, they are present across all N frequency domain bins. Let us stack the following for the channel impulse response of interest:

$$\boldsymbol{h}_{\mathrm{c}} = [h_{\mathrm{c}}(1), \ldots, h_{\mathrm{c}}(N)]^{\mathrm{T}},$$

$$\boldsymbol{H}_{\mathrm{c}} = [H_{\mathrm{c}}(1), \ldots, H_{\mathrm{c}}(N)]^{\mathrm{T}}.$$

Therefore, we have:

$$\boldsymbol{H}_{\mathrm{c}} = \boldsymbol{D}\boldsymbol{h}_{\mathrm{c}},$$

where $\boldsymbol{D}$ denotes the $N \times N$ DFT matrix. It is important to note that in the above equation, since only R range bins in $\boldsymbol{h}_{\mathrm{c}}$ are populated, only R columns of the DFT matrix $\boldsymbol{D}$ contribute to the above equation. Let $\mathscr{R}$ denote the set which contains those R indices corresponding to the clutter discretes:

$$\mathscr{R} = \{r_1, r_2, \ldots, r_R\}.$$

In other words, H_{c} is a weighted linear combination of those columns of the DFT matrix that are present in the set $\mathscr{R}$. Hence, we can express the transfer function vector as

$$\boldsymbol{H}_{\mathrm{c}} = \boldsymbol{D}_{\mathscr{R}}\boldsymbol{h}_{\mathrm{c}\mathscr{R}},$$

where $\boldsymbol{D}_{\mathscr{R}}$ denotes the $N \times R$ select-column DFT matrix and $\boldsymbol{h}_{c\mathscr{R}}$ denotes an $R \times 1$ vector that contains the unknown coefficients corresponding to the clutter discretes. We can express the select column DFT matrix as below:

$$\boldsymbol{D}_{\mathscr{R}} = \begin{bmatrix} 1 & 1 & \cdots & 1 \\ \omega^{(r_1-1)} & \omega^{(r_2-1)} & \cdots & \omega^{(r_R-1)} \\ \omega^{2(r_1-1)} & \omega^{2(r_2-1)} & \cdots & \omega^{2(r_R-1)} \\ \vdots & \vdots & \ddots & \vdots \\ \omega^{(N-1)(r_1-1)} & \omega^{(N-1)(r_2-1)} & \cdots & \omega^{(N-1)(r_R-1)}, \end{bmatrix}$$

where $\omega = e^{\frac{-j2\pi}{N}}$.

Further, it is important to note that $\boldsymbol{D}_{\mathscr{R}}$ is a full rank matrix (rank R). So, we can always find R linearly independent rows of $\boldsymbol{D}_{\mathscr{R}}$. If we choose only those frequencies in $\boldsymbol{H}_c$ corresponding to these linearly independent rows, we get the following vector:

$$\boldsymbol{H}_{c\mathrm{TRUN}} = \boldsymbol{D}_{\mathscr{R}\mathrm{TRUN}} \boldsymbol{h}_{c\mathscr{R}},$$

where $\boldsymbol{D}_{\mathscr{R}\mathrm{TRUN}}$ is an invertible $R \times R$ matrix. Therefore,

$$\boldsymbol{h}_{c\mathscr{R}} = \boldsymbol{D}^{-1}_{\mathscr{R}\mathrm{TRUN}} \boldsymbol{H}_{c\mathrm{TRUN}}.$$

So, the impulse response corresponding to all the clutter discretes can be estimated by probing just R frequencies. The only requirement is for those R frequencies to correspond to linearly independent rows in $\boldsymbol{D}_{\mathscr{R}}$. The other $N - R$ frequencies can be used for the primary radar objective of target detection, estimation, or tracking. Let a set of R row indices corresponding to linearly independent rows of $\boldsymbol{D}_{\mathscr{R}}$ be denoted as

$$\mathscr{P} = \{p_1, p_2, \ldots, p_R\}.$$

Note that the choice of $\mathscr{P}$ is not unique. Several combinations of R probing frequencies can lead to linearly independent rows of $\boldsymbol{D}_{\mathscr{R}}$. The goal here is to find at least one set of R probing frequency indices in $\mathscr{P}$ that always lead to linearly independent rows. For the general indices in $\mathscr{P}$, we can express the $R \times R$ matrix

$$\boldsymbol{D}_{\mathscr{R}\mathrm{TRUN}} = \begin{bmatrix} \omega^{(p_1-1)(r_1-1)} & \omega^{(p_1-1)(r_2-1)} & \cdots & \omega^{(p_1-1)(r_R-1)} \\ \omega^{(p_2-1)(r_1-1)} & \omega^{(p_2-1)(r_2-1)} & \cdots & \omega^{(p_2-1)(r_R-1)} \\ \omega^{(p_3-1)(r_1-1)} & \omega^{(p_3-1)(r_2-1)} & \cdots & \omega^{(p_3-1)(r_R-1)} \\ \vdots & \vdots & \ddots & \vdots \\ \omega^{(p_R-1)(r_1-1)} & \omega^{(p_R-1)(r_2-1)} & \cdots & \omega^{(p_R-1)(r_R-1)} \end{bmatrix}.$$

Unlike a full DFT matrix, the above matrix is not a Vandermonde matrix and, hence, it is not straightforward to derive conditions on the invertibility of this matrix. It depends on the choices of the sets $\mathscr{R}$ and $\mathscr{P}$. Now, we will investigate if the choice of $\mathscr{P}$ such that all the R frequencies chosen are uniformly spaced will lead to an invertible $D_{\mathscr{R}\mathbf{TRUN}}$. Without loss of generality, for the ease of notation, let us assume that N is a multiple of R: $N = R \times P$. Then

$$\mathscr{P} = \{1, P+1, \ldots, (R-1)P+1\}.$$

Then, the inner product between the ith and jth columns of $D_{\mathscr{R}_{\text{TRUN}}}$ can be expressed as

$$\boldsymbol{D}'_{\mathscr{R}_{\text{TRUN}}}(:,i)\boldsymbol{D}_{\mathscr{R}_{\text{TRUN}}}(:,j) = \sum_{k=1}^{R} \omega^{((k-1)P+1)(r_j-r_i)},$$

$$= \omega^{(r_j-r_i)} \sum_{k=1}^{R} \omega^{(k-1)P(r_j-r_i)} = \omega^{(r_j-r_i)} \sum_{k=0}^{R-1} e^{\frac{-j2\pi kP(r_j-r_i)}{N}},$$

$$= \omega^{(r_j-r_i)} \sum_{k=0}^{R-1} e^{\frac{-j2\pi k(r_j-r_i)}{R}} = 0, \forall i \neq j.$$

Since $\boldsymbol{D}_{\mathscr{R}_{\text{TRUN}}}$ is a square matrix, from the above result, it is guaranteed that $\boldsymbol{D}_{\mathscr{R}_{\text{TRUN}}}$ will always have orthogonal rows and columns. Therefore, by probing only R equally spaced frequencies, we can estimate the channel responses corresponding to clutter discretes that are distributed across R fast-time samples or range bins. This is a significant reduction compared to the number of probing vectors required when the goal is to estimate the entire impulse response. The only prior knowledge that we have used here is the locations of the range bins where the clutter discretes of interest are present.

4.5.3.2 Multiple receivers

In the earlier subsection, we have proved that uniformly spaced frequency probing is sufficient to estimate the impulse responses corresponding to clutter discretes observed at a receiver. In the presence of multiple receivers, for each transmitter, we have multiple channel impulse responses to estimate, each corresponding to one receiver. The range bins that the clutter discretes are present are different from receiver to receiver. However, the result on uniform frequency sample probing that we proved above is independent of the entries of the set $\mathscr{R}$. Therefore, even though $\mathscr{R}$ is different for each receiver, the exact same frequencies can be probed to estimate all the different clutter discrete impulse responses.

4.6 Constrained channel estimation algorithm

In the previous section, we have looked at channel estimation algorithms in an unconstrained setting. In this section, we will further improve upon the channel estimation accuracy by imposing a realistic constraint. As a refresher, based on the signal model described in this chapter, the channel estimation problem from one realization with no constraints can be expressed by a least squares problem in time domain as follows:

$$\hat{\boldsymbol{h}}_{\text{LS}} = \arg\min_{\boldsymbol{h}} ||\boldsymbol{y} - \boldsymbol{h} * \boldsymbol{x}||^2 \tag{4.12}$$

$$= \arg\min_{\boldsymbol{h}} ||\boldsymbol{y} - \boldsymbol{X}\boldsymbol{h}||^2 \tag{4.13}$$

where $\boldsymbol{y}$, $\boldsymbol{x}$, $\boldsymbol{h}$ denote the measurement, waveform, and channel (to be estimated) vectors, respectively. Also, $\boldsymbol{X}$ is a convolution matrix corresponding to $\boldsymbol{x}$ so that the convolution can be expressed by a matrix–vector multiplication. It is well known that the unconstrained least squares (4.13) has an analytical closed form solution which is given by

$$\hat{\boldsymbol{h}}_{\mathrm{LS}} = (\boldsymbol{X}^H \boldsymbol{X})^{-1} \boldsymbol{X}^H \boldsymbol{y} \tag{4.14}$$

Now, we introduce an additional constraint for the unconstrained least squares problem (4.13) in the form of the cosine similarity constraint and solve the resulting optimization problems in an efficient way. The cosine similarity constraint enforces the magnitude of complex cosine similarity measurement between the previous channel impulse response and the estimated channel impulse response for the current pulse to be greater than a given threshold. This is true in general because the impulse responses corresponding to adjacent pulses are highly correlated. We incorporate this cosine similarity constraint into the least squares problem [43,44].

4.6.1 Cosine similarity measurement

The cosine similarity measures the similarity between two non-zero vectors of an inner product space by measuring the cosine of the angle between them. For two non-zero complex vectors $\boldsymbol{x}$ and $\boldsymbol{y}$, it is defined by

$$\mathrm{cs}\,(\boldsymbol{x}, \boldsymbol{y}) = \frac{\boldsymbol{x}^H \boldsymbol{y}}{\|\boldsymbol{x}\|_2 \|\boldsymbol{y}\|_2} \tag{4.15}$$

It can be interpreted as an inner product normalized by the modulus of each vector. The magnitude of the cosine similarity is always between 0 and 1. A value of 0 means that the two vectors are at 90° to each other and have no match. The closer the cosine value to 1, the smaller the angle and the greater the similarity between the vectors. We first investigate the cosine similarity measurement of adjacent channel impulse responses to see how close the adjacent channel transfer functions are to each other. Figure 4.20 shows the absolute value of the cosine similarity of true channel transfer functions generated using high-fidelity RF M&S tool RFView. We plot four cosine similarity values $|\,\mathrm{cs}\,(\boldsymbol{h}_{i-1}, \boldsymbol{h}_i)|$, $|\,\mathrm{cs}\,(\boldsymbol{h}_{i-2}, \boldsymbol{h}_i)|$, $|\,\mathrm{cs}\,(\boldsymbol{h}_{i-3}, \boldsymbol{h}_i)|$, and $|\,\mathrm{cs}\,(\boldsymbol{h}_{i-4}, \boldsymbol{h}_i)|$ for 64 consecutive pulses in one channel. It is shown that the absolute value of the cosine similarity of channel transfer functions of nearby pulses is greater in general, i.e., $|\,\mathrm{cs}\,(\boldsymbol{h}_{i-1}, \boldsymbol{h}_i)| > |\,\mathrm{cs}\,(\boldsymbol{h}_{i-2}, \boldsymbol{h}_i)| > |\,\mathrm{cs}\,(\boldsymbol{h}_{i-3}, \boldsymbol{h}_i)| > |\,\mathrm{cs}\,(\boldsymbol{h}_{i-4}, \boldsymbol{h}_i)|$.

Figure 4.21 plots the absolute values of the cosine similarity between the unconstrained least squares solution (4.13), $\hat{\boldsymbol{h}}_{\mathrm{LS}}$, and the previous channel transfer function for three different SNR values of 0 dB, 10 dB, and 20 dB. We observe that the least squares solution at the lower SNR values achieves lower cosine similarity measurements and, hence, results in inaccurate estimates of the channel impulse response, which yields high mean squared error (MSE). Our goal is to "pull up" the cosine similarity measurement of the estimated channel impulse response and the previous channel impulse response so that cosine similarity of the estimated channel impulse response is close to the true cosine similarity measurement by adding the cosine similarity constraint in the least squares problem, which in turn generates a more accurate estimate of the channel transfer function even under low SNR values.

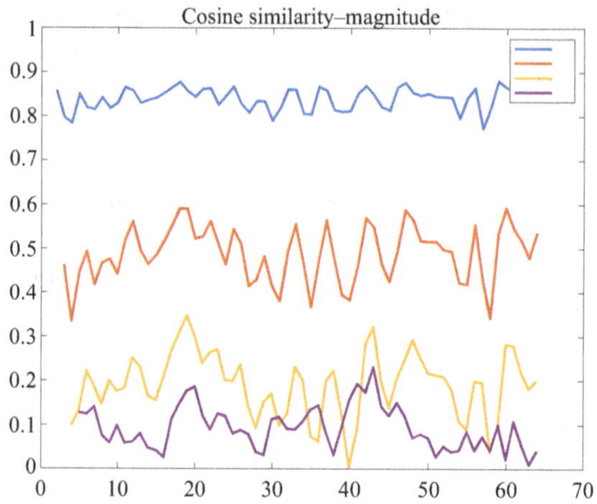

Figure 4.20 Magnitude of the cosine similarity between true adjacent channel impulses responses

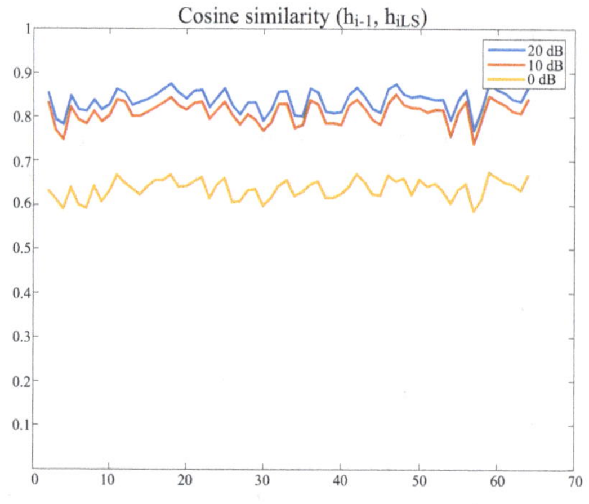

Figure 4.21 Magnitude of the cosine similarity for the unconstrained least squares solution (4.13) *for three different SNR values*

4.6.2 Channel estimation under the cosine similarity constraint: non-convex QCQP

We enforce the estimated channel impulse response to have the cosine similarity measurement greater than or equal to the true cosine similarity value in the least squares problem. Then the optimization problem can be written as

$$\begin{cases} \min\limits_{\boldsymbol{h}} & \|\boldsymbol{y} - \boldsymbol{X}\boldsymbol{h}\|_2^2 \\ \text{s.t} & |\,\mathrm{cs}\,(\boldsymbol{h}_1, \boldsymbol{h})| \geq \tau \end{cases} \tag{4.16}$$

where $\boldsymbol{h}_1$ is a channel transfer function for the previous pulse and τ is a desired level of the cosine similarity measurement, which is assumed to be given.*

Since the constraint in (4.16) is a non-convex hard constraint, we take a square on both sides of the constraint and obtain an equivalent optimization problem

$$\begin{cases} \min\limits_{\boldsymbol{h}} & \|\boldsymbol{y} - \boldsymbol{X}\boldsymbol{h}\|_2^2 \\ \text{s.t} & |\,\mathrm{cs}\,(\boldsymbol{h}_1, \boldsymbol{h})|^2 \geq \tau^2 \end{cases} \tag{4.17}$$

From the definition of the cosine similarity measurement (4.15), the optimization problem (4.17) can be rewritten as

$$\begin{cases} \min\limits_{\boldsymbol{h}} & \|\boldsymbol{y} - \boldsymbol{X}\boldsymbol{h}\|_2^2 \\ \text{s.t} & \dfrac{|\boldsymbol{h}_1^H \boldsymbol{h}|^2}{\|\boldsymbol{h}_1\|_2^2 \|\boldsymbol{h}\|_2^2} \geq \tau^2 \end{cases} \tag{4.18}$$

The constraint can be rewritten and simplified as

$$\frac{|\boldsymbol{h}_1^H \boldsymbol{h}|^2}{\|\boldsymbol{h}_1\|_2^2 \|\boldsymbol{h}\|_2^2} \geq \tau^2 \tag{4.19}$$

$$\Leftrightarrow \quad \boldsymbol{h}^H \boldsymbol{h}_1 \boldsymbol{h}_1^H \boldsymbol{h} \geq \tau^2 \|\boldsymbol{h}_1\|_2^2 \boldsymbol{h}^H \boldsymbol{h} \tag{4.20}$$

$$\Leftrightarrow \boldsymbol{h}^H (\boldsymbol{h}_1 \boldsymbol{h}_1^H - \tau^2 \|\boldsymbol{h}_1\|_2^2 \boldsymbol{I}) \boldsymbol{h} \geq 0 \tag{4.21}$$

Now the optimization problem (4.18) becomes

$$\begin{cases} \min\limits_{\boldsymbol{h}} & \|\boldsymbol{y} - \boldsymbol{X}\boldsymbol{h}\|_2^2 \\ \text{s.t} & \boldsymbol{h}^H \tilde{\boldsymbol{H}}_1 \boldsymbol{h} \geq 0 \end{cases} \tag{4.22}$$

where $\tilde{\boldsymbol{H}}_1 = \boldsymbol{h}_1 \boldsymbol{h}_1^H - \tau^2 \|\boldsymbol{h}_1\|_2^2 \boldsymbol{I}$. Since $\boldsymbol{h}_1 \boldsymbol{h}_1^H$ is a rank-one matrix with one positive singular value $\|\boldsymbol{h}_1\|_2^2$ and $0 < \tau^2 < 1$, the matrix $\hat{\boldsymbol{H}}_1$ is neither positive (semi-)definite nor negative (semi-)definite. Therefore, the constraint is a non-convex constraint and the optimization problem is a non-convex optimization problem. However, since both the objective function and the constraint are in a quadratic form of $\boldsymbol{h}$, it is a non-convex quadratically constrained quadratic programming (QCQP). Though the optimization problem (4.22) is not a convex problem and a non-convex QCQP is in general NP-hard, it is shown that a non-convex QCQP with one constraint is solvable in polynomial time [45,46] since strong duality holds and the Lagrangian relaxation produces the optimal value of (4.22). We solve the problem by the semidefinite relaxation (SDR) approach as follows.

*Note that the exact value of τ is determined by the true channel impulse responses and is not known in practice. However, it also depends on the environment which the radar is flying over and assuming a value of τ does not require knowledge of the true channel impulse responses. For example, it is acceptable to set a value of τ approximately to 0.8 for all the pulses in Figure 4.20.

We first introduce a rank-one matrix $H = hh^H$. Then the objective function and the constraint become

$$\begin{aligned} \|\boldsymbol{y} - \boldsymbol{X}\boldsymbol{h}\|_2^2 &= \boldsymbol{h}^H\boldsymbol{X}^H\boldsymbol{X}\boldsymbol{h} - \boldsymbol{h}^H\boldsymbol{X}^H\boldsymbol{y} - \boldsymbol{y}^H\boldsymbol{X}\boldsymbol{h} + \boldsymbol{y}^H\boldsymbol{y} \\ &= \operatorname{tr}\{\boldsymbol{X}^H\boldsymbol{X}\boldsymbol{H}\} - 2\Re\mathfrak{e}\{\boldsymbol{y}^H\boldsymbol{X}\boldsymbol{h}\} + \boldsymbol{y}^H\boldsymbol{y} \end{aligned} \tag{4.23}$$

and

$$\boldsymbol{h}^H\tilde{\boldsymbol{H}}_1\boldsymbol{h} = \operatorname{tr}\{\tilde{\boldsymbol{H}}_1\boldsymbol{H}\} \tag{4.24}$$

Now we have an equivalent optimization problem given by

$$\begin{cases} \min\limits_{h,H} \ \operatorname{tr}\{\boldsymbol{X}^H\boldsymbol{X}\boldsymbol{H}\} - 2\Re\mathfrak{e}\{\boldsymbol{y}^H\boldsymbol{X}\boldsymbol{h}\} + \boldsymbol{y}^H\boldsymbol{y} \\ \text{s.t} \qquad \operatorname{tr}\{\tilde{\boldsymbol{H}}_1\boldsymbol{H}\} \geq 0 \\ \qquad\qquad \boldsymbol{H} = \boldsymbol{h}\boldsymbol{h}^H \end{cases} \tag{4.25}$$

We have a linear objective function, one linear inequality constraint, and a nonlinear equality constraint in the optimization problem (4.25). Since the optimization problem (4.25) is still not a convex problem due to the nonlinear equality constraint, we relax the nonlinear constraint to an inequality $\boldsymbol{H} \succeq \boldsymbol{h}\boldsymbol{h}^H$. Then we obtain

$$\begin{cases} \min\limits_{h,H} \ \operatorname{tr}\{\boldsymbol{X}^H\boldsymbol{X}\boldsymbol{H}\} - 2\Re\mathfrak{e}\{\boldsymbol{y}^H\boldsymbol{X}\boldsymbol{h}\} + \boldsymbol{y}^H\boldsymbol{y} \\ \text{s.t} \qquad \operatorname{tr}\{\tilde{\boldsymbol{H}}_1\boldsymbol{H}\} \geq 0 \\ \qquad\qquad \boldsymbol{H} \succeq \boldsymbol{h}\boldsymbol{h}^H \end{cases} \tag{4.26}$$

Lastly the inequality constraint $\boldsymbol{H} \succeq \boldsymbol{h}\boldsymbol{h}^H$ can be expressed as a linear matrix inequality by using a Schur complement, which gives

$$\begin{cases} \min\limits_{h,H} \ \operatorname{tr}\{\boldsymbol{X}^H\boldsymbol{X}\boldsymbol{H}\} - 2\Re\mathfrak{e}\{\boldsymbol{y}^H\boldsymbol{X}\boldsymbol{h}\} + \boldsymbol{y}^H\boldsymbol{y} \\ \text{s.t} \qquad \operatorname{tr}\{\tilde{\boldsymbol{H}}_1\boldsymbol{H}\} \geq 0 \\ \qquad\qquad \begin{bmatrix} \boldsymbol{H} & \boldsymbol{h} \\ \boldsymbol{h}^H & 1 \end{bmatrix} \succeq \boldsymbol{0} \end{cases} \tag{4.27}$$

This is semidefinite programming (SDP) [47,48]. The optimization problem (4.27) is not equivalent to the original problem (4.22) due to the relaxation of the nonlinear equality constraint. However, since we minimize the same objective function over a larger set, it is obvious that the solution to (4.27) is less than or equal to the optimal value of (4.22) and, therefore, if $\boldsymbol{H} = \boldsymbol{h}\boldsymbol{h}^H$ at the optimum of (4.27), then h will also be a solution of (4.22). Since strong duality holds between (4.22) and (4.27), provided (4.22) is strictly feasible, we can solve the SDP (4.27) which is convex instead of solving the non-convex problem (4.22). We can now solve the optimization problem (4.27) with CVX SDP solver and it is always guaranteed to obtain the optimal solution $\boldsymbol{H}^\star$ and $\boldsymbol{h}^\star$ such that $\boldsymbol{H}^\star$ is a rank-one matrix and $\boldsymbol{H}^\star = \boldsymbol{h}^{\star H}\boldsymbol{h}^\star$ for the SDP as relaxation of the non-convex QCQP with one constraint.

4.6.3 *Performance comparison using numerical simulation*

In this subsection, we provide simulation results for the proposed estimation methods using realistic data obtained from high-fidelity M&S software RFView. As described earlier, RFView uses publicly available terrain data and land cover types to accurately model ground clutter returns for RF systems by dividing the entire clutter region into individual clutter patches. For this example, we have a monostatic radar platform flying along the coast of Southern California with a speed of 100 m/s. Note that the speed is very important for interpreting the stationarity of the channel impulse responses with respect to the pulse indices. The parameters of this simulation are described in the table below.

Figure 4.22 shows the normalized MSE and the achieved cosine similarity measurement for the channel impulse responses estimated from the unconstrained LS solution and the NQCQP solution for multiple independent realizations of data. We

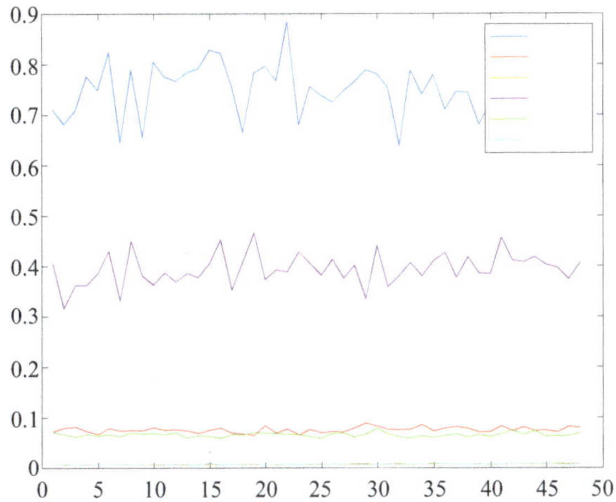

Figure 4.22 Normalized MSE for Channel 1 Pulse 2 for multiple independent realizations of data

Table 4.1 Parameters corresponding to the example simulated using RFView

Parameter	**Value**
Latitude position of radar platform	32.4275–32.4277
Longitude position of radar platform	242.8007
Height of the radar platform	1,000.1 m
Speed of radar platform	100 m/s
Number of pulses	64
Transmit center frequency	1.0000e+10 Hz
Number of transmit channels	1
Number of receive channels	16
Bandwidth	5,000,000 Hz
Transmit waveform	LFM waveform

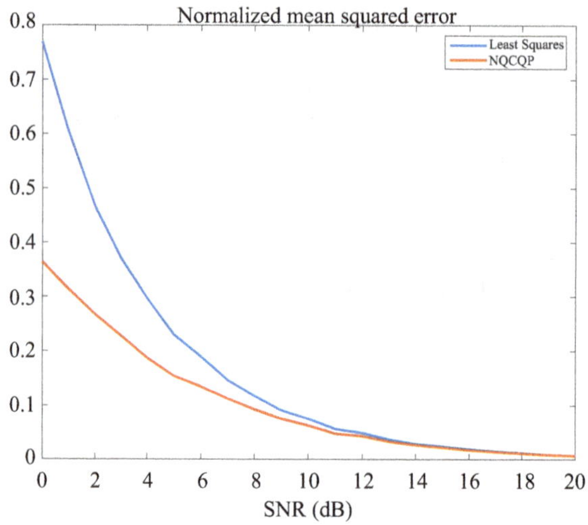

Figure 4.23 Normalized MSE versus SNR for Channel 1 Pulse 2

clearly observe from the plots that the NQCQP solution offers better performance in terms of providing lower normalized MSE for three different values of SNR. This is achieved by increasing the cosine similarity with the adjacent impulse response corresponding to the previous pulse index. Similarly, Figure 4.23 plots the same as a function of SNR. We observe that the performance improvement is significant in the lower SNR regime. More advanced constraints along with knowledge-aided channel estimation are topics of ongoing research.

4.7 Cognitive fully adaptive radar challenge dataset

In this chapter, we have presented a stochastic transfer function-based model to cognitive radar. To conclude this chapter, we present a CoFAR challenge dataset that was generated using this new modeling approach. High-fidelity, physics-based, site-specific modeling, and simulation software RFView was used to generate Green's functions and the corresponding simulated measurements for this dataset. The main purpose of the dataset described here is to provide radar researchers with a common dataset to benchmark their results and compare with existing algorithms. Along with ground clutter induced by the terrain, we have also included a few clutter discretes in the form of buildings. This dataset can be used to test radar detection and estimation algorithms along with CoFAR concepts for radar waveform design. The data was generated using the stochastic transfer function signal model presented in this chapter.

Given this signal model, CoFAR research can be broadly classified into two tasks:

- Using known channel impulse responses, designing optimal radar transmit waveforms to maximize target detection and estimation performance.

- Estimating the channel impulse responses from measured data.

Both these tasks are equally important for practical CoFAR systems. Ultimately, in a CoFAR system, allocating resources between these two tasks is a tradeoff. One would like to allocate resources to channel estimation to obtain accurate estimates of the channel impulse responses while not compromising too much on the primary radar objective of target detection and estimation. Some basic ideas for optimal waveform design using impulse-response-based modeling as well as channel estimation algorithms have been discussed in this chapter. It is our hope that the CoFAR community can develop more advanced waveform design algorithms, further incorporating practical constraints on the waveforms. Our dataset can be used to test the performance of these algorithms for pulse-to-pulse waveform design as well as CPI-to-CPI waveform design. The dataset presented here can be used to perform both the waveform design and channel impulse response estimation tasks. This dataset contains data cubes as well as the corresponding channel impulse responses for two scenarios, each of which is described in this section.

This challenge dataset was initially distributed at the inaugural High Fidelity RF Modeling and Simulation Workshop in August 2020 [49]. The dataset can be accessed/downloaded by all readers by creating a free trial account at [31].

4.7.1 Scenario 1

The first scenario is supposed to be a beginner dataset with few targets, ground clutter, and couple of clutter discretes (like buildings). This scenario involves an airborne monostatic radar flying over the Pacific Ocean near the coast of San Diego looking down for ground moving targets. The data spans several coherent processing intervals as the platform is moving with constant velocity along the coastline.

Along with the simulated data for this scenario, we have provided the true channel impulse responses for clutter and targets. This data spans 30 CPIs, 32 spatial channels, 64 pulses, and $2,334$ range bins. Basic beamforming and delay-Doppler processing of the data cube gives 30 range–Doppler plots, one for each CPI. Along with the data, a reference video containing 30 frames is also provided. Note that we used standard delay-Doppler processing and also assumed that a fixed waveform was transmitted. The goal is for the readers to test their own algorithms and optimally designed waveforms to improve target detection and estimation performance and obtain better results than the plots we demonstrate here with basic signal processing. The details of all the parameters chosen for this simulation are described in a user guide provided along with the challenge dataset.

For example, in the 6th CPI, we obtain the plot in Figure 4.25. Three targets and two clutter discretes can be identified from this plot. The other target which is much weaker and farther away from the other three targets is not visible in this plot. Also, from this plot, we can clearly observe the littoral nature of the simulated environment as we can see regions of water within the ground clutter as water has weaker reflectivity compared to land. As the radar platform moves along and drags its beam along, we present the range–Doppler plot for the 26th CPI in Figure 4.26. Now, returns from the weaker target are also picked up by the radar. The other targets are still visible

Figure 4.24 Scenario 1 *of the challenge dataset with* 4 *targets and* 2 *clutter discretes*

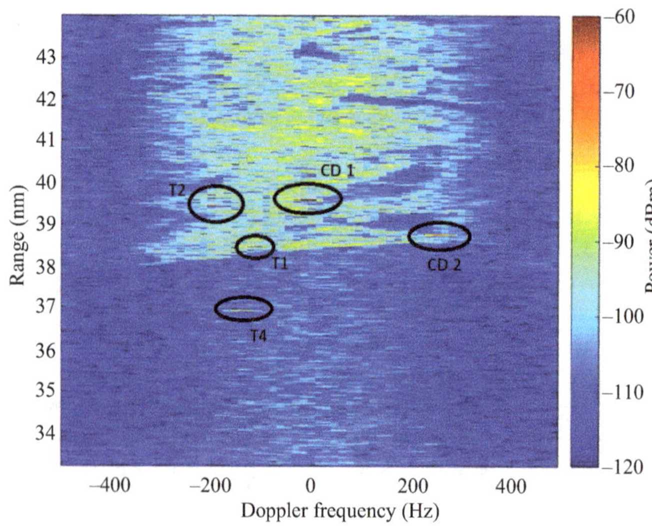

Figure 4.25 Range–Doppler plot from the 6*th CPI in scenario* 1

in the range–Doppler plot, but they have moved around from their positions in the previous plot (Figure 4.25). Since the dataset also includes the true channel impulse responses corresponding to multiple pulses and CPIs, this dataset can be used to test fully adaptive radar optimal waveform design algorithms where waveforms are

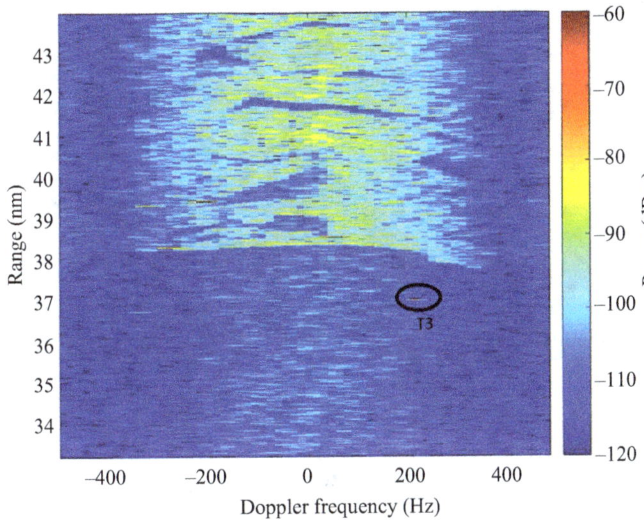

Figure 4.26 Range–Doppler plot from the 26th CPI in scenario 1

changed from pulse to pulse or from CPI to CPI. From the provided true channel impulse responses, data cubes can be generated for any transmit waveform using the model described in this chapter.

4.7.2 Scenario 2

In scenario 1, while we had ground clutter and couple of strong clutter discretes, they were spaced relatively far from the targets of interest. Hence, the detection and the estimation of targets are not so difficult. Now, we move along to a more challenging dataset. In addition to the targets and strong clutter discretes present in scenario 1, this scenario contains several (150) clutter discretes in the form of small buildings (30 m × 30 m × 6 m) arranged in a cluster very close to Target No. 1. This makes this scenario more challenging than scenario 1. The target locations and radar parameters remain the same as scenario 1. 150 buildings were added next to Target No. 1 arranged in a 50 × 3 grid. This is indeed a region populated by buildings as can be seen in satellite images (see Figure 4.27). While all the buildings in this simulation were approximated to be of the same size, in more advanced datasets, each building can be modeled to be of the exact shape and size as in reality.

In Figure 4.28, we have the range–Doppler plot for the 26th CPI in scenario 2. We observe that the cluster of buildings is much stronger compared to Target 1 and the target is not clearly distinguishable from the cluster of clutter discretes. Note that we have added this cluster of buildings all along the road on which this target was moving. This is a challenging scenario where CoFAR techniques for waveform design are needed to suppress the dominating clutter discretes. It is our hope that users of this dataset will devise advanced CoFAR techniques that can mitigate the effects of these

Figure 4.27 Cluster of buildings included as part of scenario 2

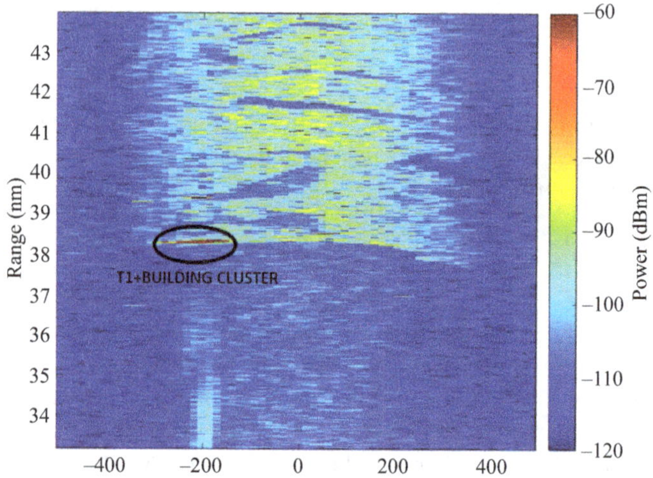

Figure 4.28 Range–Doppler plot from the 26th CPI in scenario 2

clutter discretes using adaptive signal processing techniques and optimal waveform design.

4.8 Concluding remarks

In this chapter, we have reviewed advanced M&S techniques for modeling ground clutter using a stochastic transfer function approach. This approach is contrary to the traditional covariance-based techniques and it lends itself well to accurately simulate various radar scenarios and applications including cognitive radar. After formulating the cognitive radar waveform optimization problem using this modeling approach, we have presented multiple channel estimation algorithms. Channel knowledge is critical to the performance of cognitive radar systems. Lastly, we describe a new CoFAR challenge dataset that users can download, test, and benchmark state-of-the-art cognitive radar algorithms and techniques. It is our endeavor to generate and provide more advanced datasets involving realistic buildings and other cultural features.

Information-theoretic approach to channel estimation is presented in Chapter 10. Also, the prospect theory in decision making is described in Chapter 11. Additionally, incorporating available knowledge databases into the channel estimation process can achieve improved estimation accuracy. This topic of knowledge-aided channel estimation is also a topic our ongoing research.

References

[1] J. R. Guerci, *Cognitive Radar: The Knowledge-Aided Fully Adaptive Approach*. Norwood, MA: Artech House, 2010.

[2] J. R. Guerci, "Optimal and adaptive MIMO waveform design," In W. L. Melvin and J. A. Scheer, eds., *Principles of Modern Radar: Advanced Techniques*. Stevenage: SciTech Publishing, 2013.

[3] J. R. Guerci, "Optimal radar waveform design," In R. Chellappa and S. Theodoridis, eds., *Academic Press Library in Signal Processing, Communications and Radar Signal Processing*, vol. 2, pp. 729–758, 2014.

[4] S. Kay, "Optimal signal design for detection of Gaussian point targets in stationary gaussian clutter/reverberation," *IEEE Journal of Selected Topics in Signal Processing*, vol. 1, pp. 31–41, 2007.

[5] S. U. Pillai, H. S. Oh, D. C. Youla, and J. R. Guerci, "Optimal transmit–receiver design in the presence of signal-dependent interference and channel noise," *IEEE Transactions on Information Theory*, vol. 46, pp. 577–584, 2000.

[6] J. R. Guerci, J. S. Bergin, R. J. Guerci, M. Khanin, and M. Rangaswamy, "A new MIMO clutter model for cognitive radar," In *IEEE Radar Conference*, Philadelphia, PA, May 2016.

[7] S. Gogineni, J. R. Guerci, H. K. Nguyen, *et al.*, "High fidelity RF clutter modeling and simulation," *IEEE Aerospace and Electronic Systems Magazine*, vol. 37, no. 11, pp. 24–43, 2022.

[8] J. R. Guerci, *Space–Time Adaptive Processing for Radar*. Norwood, MA: Artech House, 2014.

[9] J. Ward, "Space–time adaptive processing for airborne radar," In *1995 International Conference on Acoustics, Speech, and Signal Processing*, 1994.

[10] S. F. George, "The detection of nonfluctuating targets in log-normal clutter," *NRL Report*, vol. 6796, Oct. 1968.

[11] G. V. Trunk, "Ocean surveillance statistical considerations," *NRL Report*, Nov. 1968.

[12] A. Farina, A. Russo, and F. Scannapieco, "Radar detection in coherent Weibull clutter," *IEEE Transactions on Acoustics, Speech, and Signal Processing*, vol. 35, pp. 893–895, 1987.
[13] G. Li and K.-B. Yu, "Modelling and simulation of coherent Weibull clutter," *IEE Proceedings F (Radar and Signal Processing)*, vol. 136, pp. 2–12, 1989.
[14] E. Conte and M. Longo, "Characterisation of radar clutter as a spherically invariant random process," *IEE Proceedings F (Communications, Radar and Signal Processing)*, vol. 134, pp. 191–197, 1987.
[15] M. Rangaswamy, D. D. Weiner, and A. Ozturk, "Non-Gaussian random vector identification using spherically invariant random processes," *IEEE Transactions on Aerospace and Electronic Systems*, vol. 29, pp. 111–124, 1993.
[16] M. Rangaswamy, D. D. Weiner, and A. Ozturk, "Computer generation of correlated non-Gaussian radar clutter," *IEEE Transactions on Aerospace and Electronic Systems*, vol. 31, pp. 106–116, 1995.
[17] K. J. Sangston and K. R. Gerlach, "Coherent detection of radar targets in a non-gaussian background," *IEEE Transactions on Aerospace and Electronic Systems*, vol. 30, pp. 330–340, 1994.
[18] I. Reed, "On the use of Laguerre polynomials in treating the envelope and phase components of narrow-band Gaussian noise," *IRE Transactions on Information Theory*, vol. 5, pp. 102–105, 1959.
[19] R. J. Howell and J. W. Stuntz, "Radar system for discriminating against area targets," U.S. Patent US2 879 504A, 1959. https://patents.google.com/patent/US2879504A/en.
[20] S. P. Applebaum, "Adaptive arrays," *Technical Report, Syracuse University Research Corporation*, 1966.
[21] B. Widrow, P. E. Mantey, L. J. Griffiths, and B. B. Goode, "Adaptive antenna systems," *Proceedings of the IEEE*, vol. 55, pp. 2143–2159, 1967.
[22] I. Reed, J. Mallett, and L. Brennan, "Rapid convergence rate in adaptive arrays," *IEEE Transactions on Aerospace and Electronic Systems*, vol. 10, pp. 853–863, 1974.
[23] E. J. Kelly, "An adaptive detection algorithm," *IEEE Transactions on Aerospace and Electronic Systems*, vol. 22, pp. 115–127, 1986.
[24] E. J. Kelly, "Performance of an adaptive detection algorithm; rejection of unwanted signals," *IEEE Transactions on Aerospace and Electronic Systems*, vol. 25, pp. 122–133, 1989.
[25] F. C. Robey, D. R. Fuhrmann, E. J. Kelly, and R. Nitzberg, "A CFAR adaptive matched filter detector," *IEEE Transactions on Aerospace and Electronic Systems*, vol. 28, pp. 208–216, 1992.
[26] S. Kraut and L. L. Scharf, "The CFAR adaptive subspace detector is a scale-invariant GLRT," *IEEE Transactions on Signal Processing*, vol. 47, pp. 2538–2541, 1999.
[27] B. Billingsley, *Low-Angle Radar Land Clutter: Measurements and Empirical Models*. Stevenage: SciTech Publishing, 2002.

[28] F. T. Ulaby and M. C. Dobson, *Handbook of Radar Scattering Statistics for Terrain*. Norwood, MA: Artech House, 1989.

[29] B. C. Watson and J. Bergin, "Ocean scattering model," *Information Systems Laboratories Technical Note ISL-SCRO-TN-09-005*, Apr. 2009.

[30] B. C. Watson and J. Bergin, "Ocean surface model," *Information Systems Laboratories Technical Note ISL-SCRO-TN-09-006*, Apr. 2009.

[31] https://rfview.islinc.com/RFView/login.jsp.

[32] J. C. Gonzato and B. L. Saec, "On modelling and rendering ocean scenes," *The Journal of Visualization and Computer Animation*, vol. 11, pp. 27–37, 2000.

[33] H. Qu, F. Qiu, N. Zhang, A. Kaufman, and M. Wan, "Ray tracing height fields," In *Proceedings in Computer Graphics International*, pp. 202–207, 2003.

[34] S. Ayasli, "SEKE: a computer model for low altitude radar propagation over irregular terrain," *IEEE Transactions on Antennas and Propagation*, vol. 34, pp. 1013–1023, 1986/1988.

[35] G. D. Dockery, "Modeling electromagnetic wave propagation in the troposphere using the parabolic equation," *IEEE Transactions on Antennas and Propagation*, vol. 36, pp. 1464–1470, 1988.

[36] A. E. Barrios, W. L. Patterson, and R. A. Sprague, "Advanced propagation model (APM) version 2.1.04 computer software configuration item (CSCI) documents," *SPAWAR TECHNICAL DOCUMENT 3214*, Feb. 2007. https://apps.dtic.mil/dtic/tr/fulltext/u2/a464098.pdf.

[37] L. B. Fertig and J. M. B. J. R. Guerci, "Knowledge-aided processing for multipath exploitation radar (mer)," *IEEE Aerospace and Electronic Systems Magazine*, vol. 32, pp. 24–36, 2017.

[38] B. C. Watson and J. R. Guerci, *Non Line of Sight Radar*. Norwood, MA: Artech House, 2019.

[39] S. Gogineni, J. R. Guerci, H. K. Nguyen, J. S. Bergin, B. C. Watson, and M. Rangaswamy, "Modeling and simulation of cognitive radar," In *IEEE Radar Conference*, Florence, Italy, Sep. 2020.

[40] S. Gogineni, M. Rangaswamy, J. R. Guerci, J. S. Bergin, and D. R. Kirk, "Estimation of radar channel state information," In *Proceedings of the IEEE Radar Conference*, Boston, MA, Apr. 2019.

[41] J. R. Guerci, J. S. Bergin, S. Gogineni, and M. Rangaswamy, "Non-orthogonal radar probing for MIMO channel estimation," In *Proceedings of the IEEE Radar Conference*, Boston, MA, Apr. 2019.

[42] S. Gogineni, M. Rangaswamy, J. R. Guerci, J. S. Bergin, and D. R. Kirk, "Impulse response estimation for wideband multi channel radar systems," in *Proceedings of the IEEE International Radar Conference*, Washington, DC, Apr. 2020.

[43] B. Kang, S. Gogineni, M. Rangaswamy, and J. R. Guerci, "Constrained maximum likelihood channel estimation for massive MIMO radar," In *Proceedings of the 54th Asilomar Conference on Signals, Systems and Computer*, Pacific Grove, CA, Nov. 2020.

[44] B. Kang, S. Gogineni, M. Rangaswamy, J. R. Guerci, and E. Blasch, "Adaptive channel estimation for cognitive fully adaptive radar," In *IET Radar, Sonar & Navigation*, Dec. 2021.

[45] S. Boyd and L. Vandenberghe, *Convex Optimization*, 2nd ed. Cambridge: Cambridge University Press, 2004.

[46] E. Feron, "Nonconvex quadratic programming, semidefinite relaxations and randomization algorithms in information and decision systems," In T. E. Djaferis and I. C. Schick, eds., *System Theory: Modeling Analysis and Control*. Berlin: Springer, 2000.

[47] L. Vandenberghe and S. Boyd, "Semidefinite Programming," *SIAM Review*, vol. 38, no. 1, pp. 49–95, 1996.

[48] S. Boyd and L. Vandenberghe, "Semidefinite programming relaxations of non-convex problems in control and combinatorial optimization," In A. Paulraj, V. Roychowdhury, and C. D. Schaper, eds., *Communications, Computation, Control and Signal Processing: A Tribute to Thomas Kailath*. Berlin: Springer, 1997, pp. 279–288.

[49] https://events.vtools.ieee.org/m/236218.

Chapter 5

Convex optimization for cognitive radar

Bosung Kang[1], Khaled AlHujaili[2], Muralidhar Rangaswamy[3] and Vishal Monga[4]

5.1 Introduction

A confluence of factors continues to increase the complexity and challenges of modern high-performance radars [1]. Cognitive radar was described and introduced as an advanced form of radar system for the first time by Haykin [2] to meet the challenges of increasingly complex operating environments. On the contrary to a conventional radar, a cognitive radar includes an adaptive transmitter in addition to an adaptive receiver, which entails a number of new adaptation and knowledge-aided methods. A fundamental goal of a cognitive radar is to sense, learn, and adapt (SLA) to a complex environment [1]. The cognitive radar continuously learns about the environment and updates the receiver with relevant information on the environment. Then the transmitter continually adjusts its signal in intelligent manner based on the sensed environment such as the size and range of the targets and clutter. This closed-loop dynamic system is the key aspect of the cognitive radar. The basic structure of cognitive radar is shown in Figure 5.1. The learning process starts when the receiver collects the returns from the target and scatterers. From these returns, the cognitive radar system acquires the required information about its external environment. The transmitter uses the obtained information to alter its transmission and, hence, compensates for the changes in the environment that are captured through the receiver's previous interactions.

For a cognitive radar, both transmit and receive functions are utilized in new ways to enhance channel estimation and the radar optimizes a spatio-temporal transmit and receive strategy. This optimization process involves solving mathematical optimization problems. For example, it has been shown that the optimum transmit and receive functions maximize output signal-to-interference-plus-noise ratio (SINR), which means that it is required to solve an SINR maximization problem to obtain the optimum transmit and receive function. Such mathematical optimization problems may be convex and the solution can be efficiently obtained by numerical approaches

[1]University of Dayton Research Institute, University of Dayton, Dayton, OH, USA
[2]Electrical Engineering Department, Taibah University, Saudi Arabia
[3]Air Force Research Laboratory, Wright Patterson Air Force Base, Dayton, OH, USA
[4]The Department of Electrical Engineering, The Pennsylvania State University, University Park, PA, USA

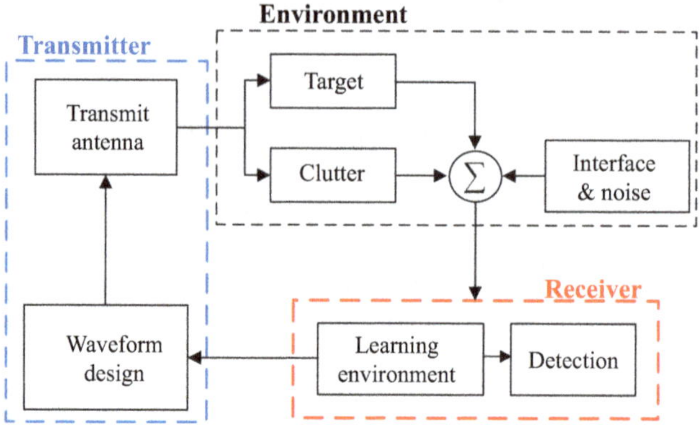

Figure 5.1 Basic structure of cognitive radar system

or closed form solutions in some cases. However, the optimization problems are in general non-convex and it is hard to obtain a solution particularly when practical constraints are enforced in the optimization problems to ensure performance of the radar and satisfy the hardware requirements under complex environment.

In this chapter, we introduce several kinds of optimization problems that are widely and actively studied in cognitive radar applications and how the problems can be solved using mathematical techniques and principles of convex optimization. We first highlight the importance and purposes of waveform design problems in cognitive radar and practical challenges in the waveform design problems, which is followed by the principles and fundamentals of convex optimization. Typical approaches to solve convex optimization problems and non-convex optimization problems are also introduced. Lastly, successful examples of waveform design algorithms that solve hard non-convex optimization problems using the principals of convex optimization are discussed.

5.1.1 Waveform design problems in cognitive radar

The basic principle in radar is to illuminate a certain region of interest by transmitting a radio-frequency (RF) electromagnetic (EM) signal and receive its echo caused by an object of interest known as a *target* and other not-of-interest objects. Utilizing this echo along with the knowledge about the transmitted signal, the radar performs various functions such as detection, tracking, imaging, and classification. In addition to the target return, the received echo contains, as shown in Figure 5.2, signal-dependent returns from objects not of interest known as *clutter*, EM returns from other radiators known as *interference*, and *noise* [3].

5.1.1.1 Waveform design: background and motivation

In the presence of those unwanted contributions, the ability to extract the target return or suppress the unwanted returns is encouraged to enhance the functionality of the

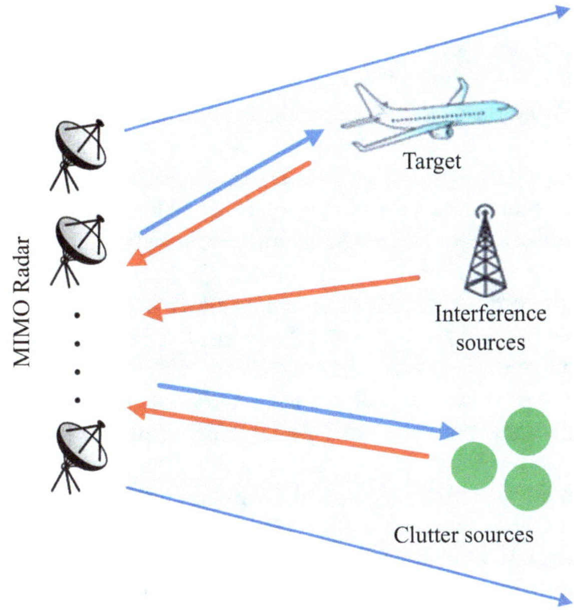

Figure 5.2 Crowded environment

radar systems. Towards this goal, during the last decades, many techniques have been proposed to adaptively suppress those aforementioned unwanted returns at the receiver side [4]. Examples include constant false alarm rate (CFAR) detectors, space–time adaptive processing (STAP), and rejecting range-ambiguous scatterer returns. On the other hand, the transmitter could be also incorporated into this task due to the dependence of the target and clutter returns on the transmit waveform. Accordingly, adjusting this waveform helps in reducing the effect of the unwanted returns and, hence, leads to better extracting of the target return. This concept of adaptively adjusting the transmitted signal was proposed by H. Van Trees in 1965 [5] and is known in the literature of the radar signal processing as *the waveform design*. The concept of the waveform design cannot be applied in a radar system without exploiting the knowledge about its surrounding *environment*. The surrounding environment includes all objects that contribute to the received signal and affect the performance of radar [4]. The knowledge about the environment is obtainable under the framework of the cognitive radar mentioned earlier.

5.1.1.2 Waveform design via optimization

The concept of waveform design has received considerable attention in the literature of radar signal processing during the last two decades for different applications, i.e., detection, estimation, and tracking. In general, the adaptive transmission technology can be divided into two main different categories according to the degree of the freedom of the transmitter. These categories are *selection* and *design* [4]. Under the selection category, we have the methods that either adaptively select pre-designed signals or select certain parameters of pre-defined signals such as pulse repetition frequency (PRF) and the signal pulse width. On the other hand, the design category includes the methods that either arbitrarily design the transmit waveform or design the aforementioned parameters of an existing waveform. The discussion presented in this chapter deals with the waveform design category. More specifically, the proposed methods or algorithms here design suitable transmit waveforms to compensate for the changes in the radar's surrounding environment. Furthermore, it is assumed that no specific structure for the transmit waveform. In other words, in the design process, each time sample in the desired waveform is considered to be freely available to be designed and constructed.

Mostly, this type of design problem is treated under the *numerical constrained optimization* paradigm where the radar performance metric of interest will be optimized as an *objective function* over the transmitted waveforms while considering some practical/hardware requirements that are imposed by the system as *constraints*. In general, the objective functions and the constraints depend on the task of the radar system, i.e., detection, estimation, and tracking tasks. In the following subsections, different common performance metrics and practical constraints will be discussed in more detail.

5.1.1.3 Radar performance metrics (objective functions)

As mentioned earlier, the transmit waveforms from the radar can be designed by solving optimization problems. The objective functions in these problems usually represent some performance measures as figure of merits. In this part, our goal is to highlight some of these merits.

Signal-to-interference-plus-noise ratio (SINR)

Most of the waveform design approaches aim to enhance the detection ability of radar systems. Enhancing this ability is equivalent to maximizing the radar SINR. SINR has a proportional influence on the probability of detection, and maximizing SINR is equivalent to directly maximizing the probability of detection and, hence, the detection ability [6–8] and the references therein.

Transmit beampattern

Under excessive clutter and/or interference disturbance the improvement in SINR can be indirectly achieved by considering different metrics. For instance, maximizing the energy of the target return (by focusing the transmit power to the expected target location) and reducing it for the unwanted returns has been conducted in the literature by controlling *the transmit beampattern* [9–11] and the references therein. The main idea here is to design a set of waveforms such that the transmitted beampattern matches certain specifications, e.g., a desired beampattern, by minimizing the deviation between the produced and the desired beampatterns. Another approach for transmitting beampattern design considers the properties of autocorrelation and cross-correlation since a waveform with good correlation properties enables good parameter estimation and anti-jamming ability. The goal is to suppress the peak sidelobe level (PSL) [12–14] and the integrated sidelobe level (ISL) [14–16].

Ambiguity function

Another indirect SINR improvement can be achieved by designing a transmit waveform that has a specific *ambiguity* shape [11,17]. This design problem is known in literature as *the ambiguity function* (AF) shaping. The AF represents the range–Doppler response at the output of a filter matched to the transmitted signal when it arrives with a time delay (represents the range) and uncompensated Doppler shift. The transmit waveform has a significant impact on this response, and, hence, designing this waveform can be employed to control the AF for radar systems. Chapter 6 "Cognition-enabled waveform design for ambiguity function shaping" discusses more details about waveform design algorithms using this metric for cognitive radar.

5.1.1.4 Practical constraints in the waveform design process

Furthermore, besides improving the performance metric using the optimal waveform, many hardware limitations imposed by system components must be considered during the design process. These limitations are reflected in the design process as constraints on the designed waveform. Many practical constraints are used in literature. Salient examples include peak-to-average-power-ratio (PAPR) and energy constraints (EC), similarity constraint (SC), spectral constraint (SpecC), and constant modulus constraint (CMC). In the optimization problems, these constraints were used alone or multiple constraints have been simultaneously exploited [18]. In the following, we provide brief descriptions of some of the intensively used/studied constraints.

PAPR and the EC

PAPR and ECs belong to the family of the modulus constraints [4] and usually are imposed on the transmitted waveform to maximize the efficiency of the transmitter hardware [19,20]. Since the CMC results in a hard non-convex optimization problem, PAPR and the EC are employed as a relaxation of the CMC in the literature.

SC

The SC is used to produce a waveform that has some of the desirable properties of a reference signal. The advantage of imposing this constraint has been reported in different works in the literature such as in [8,21]. In these works, without enforcing SC the produced waveforms suffer from undesirable effects in pulse compression, and AF properties as shown in Figure 5.3(a).

Spectral constraint (SpecC)

This constraint has been introduced in the literature of radar signal processing to ensure the co-existence among radar and communication systems, in a spectrally crowded environment [22,23]. The need for this co-existence appears when both systems occupy the same frequency band. For an instant, in wideband transmission, radar systems occupy a large bandwidth and, hence, overlap with the spectrum for other radiators could arise. The term co-existence implies that the frequency signature

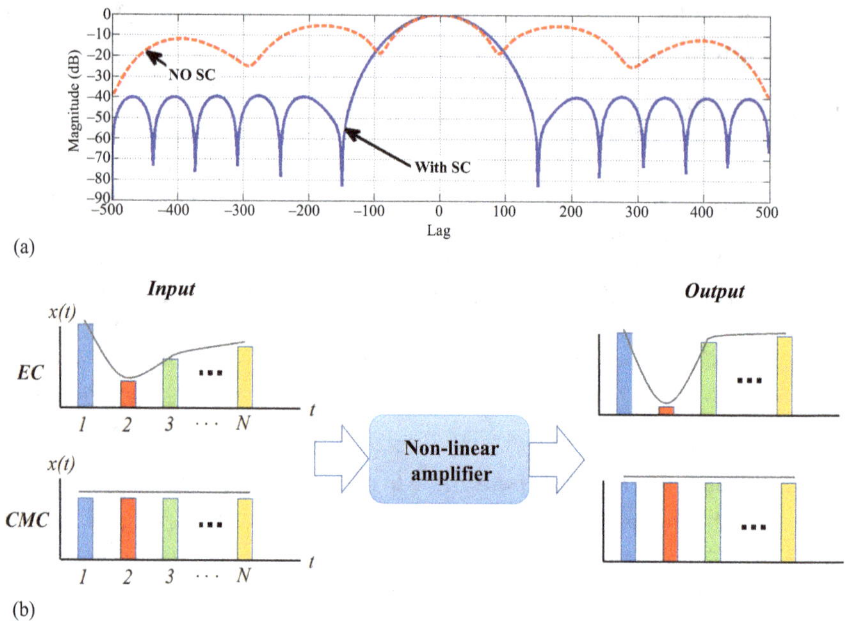

Figure 5.3 Necessity and importance of practical constraints on radar transmit waveform: (a) the similarity constraint and (b) the constant modulus constraint

of the transmit waveforms from the radar system exhibits nulls in the frequency bands of the communication systems.

Orthogonality constraint

Imposing orthogonality across antennas has been shown to be particularly meritorious. Orthogonal MIMO waveforms enable the radar system to achieve an increased virtual array and, hence, leads to many practical benefits [24,25]. A compelling practical challenge is that the "directional knowledge" of target and interference sources utilized in specifying the desired beampattern may not be perfect. In such scenarios, it has been shown in [24,25] that the gain loss in the transmit–receive patterns for orthogonal waveform transmission is very small under target direction mismatch.

Constant modulus constraint (CMC)

The CMC aims to ensure that the waveform's envelope have a constant amplitude. The CMC is crucial in the design process due to the presence of non-linear amplifiers in radar systems [6]. These components are known to operate in saturation mode, and CMC is required to maximize their efficiency. Due to this saturation mode, radar transmitters have a peak power level that cannot be exceeded. Therefore, if the peak amplitude of the waveform exceeds this upper level, it will be clipped, and, hence, the transmitted power will not be fully utilized as shown in Figure 5.3(b). Consequently, less than the expected power is carried to the target, and thus system performance will be degraded.

5.2 Background and motivation

Convex optimization has a long history in signal processing applications including control, circuit design, economics and finance, statistics and machine learning, and radar applications [26] and has emerged as a major signal processing tool that has made a significant impact on numerous problems previously considered intractable [27]. The classical simplex algorithm was developed during the World War II by Dantzig to solve linear programming problems. It begins at a starting extreme point and moves along the edges of the feasible region until it reaches the vertex of the optimal solution. It showed a great improvement over earlier methods, however, it takes a long time to converge and the computational complexity is exponential time at the worst case [28]. In 1980s, Karmarka developed Karmarka's algorithm [29] which has been proven to be four times faster than the simplex method, particularly in polynomial time, by reinventing the interior-point method. It reaches a best solution by traversing the interior of the feasible region contrary to the simplex method. This recognition of the interior-point methods stimulated huge interest in new classes of convex optimization problems such as semidefinite programs and second-order cone programs and enabled to solve the problems as easily as linear programs [30]. The recent advances in processor power dramatically reduce solution time and even accelerated usage of convex optimization in numerous applications.

In this section, we introduce fundamental principles of convex optimization and briefly discuss challenges on applying optimization problems to cognitive radar. Though most of the practical constraints used in the waveform/beampattern design problem are non-convex constraints, it is important to understand principles of convex optimization since many non-convex optimization problem can be solved using the principles such as convex relaxation.

5.2.1 Principles of convex optimization

We first introduce definition of the terminology of convex optimization including convex sets, convex functions, and convex optimization problem with examples of canonical convex optimization problems and then the numerical and analytical approaches to solve convex optimization problems are provided. Lastly, we discuss two popular approaches to non-convex optimization problems using the principles of convex optimization, convex relaxation, and transformation of variables.

5.2.1.1 Convex optimization problem

Constrained optimization problem

Many estimation and design problems in signal processing, particularly, radar applications, can be posed as a constrained optimization problem which has the form

$$\begin{aligned} \underset{\boldsymbol{x}}{\text{minimize}} \quad & f(\boldsymbol{x}) \\ \text{subject to} \quad & g_i(\boldsymbol{x}) \leq 0 \\ & h_i(\boldsymbol{x}) = 0 \end{aligned} \tag{5.1}$$

where the vector $\boldsymbol{x}$ is the optimization variable of the problem, the function f is the objective function or the cost function we desire to minimize, the functions g_i are the inequality constraint functions, and the functions h_i are the equality constraint functions. The optimal solution of the optimization problem, $\boldsymbol{x}^\star$, is a vector that achieves the smallest objective value of the cost function among all vectors that satisfy the constraints. $\boldsymbol{x}^\star$ can be expressed by

$$\boldsymbol{x}^\star = \arg\min_{\boldsymbol{x}\in\{\{\boldsymbol{x}|g_i(\boldsymbol{x})\leq 0\}\cap\{\boldsymbol{x}|h_i(\boldsymbol{x})=0\}\}} f(\boldsymbol{x}) \tag{5.2}$$

where the set $\{\{\boldsymbol{x}|g_i(\boldsymbol{x}) \leq 0\} \cap \{\boldsymbol{x}|h_i(\boldsymbol{x}) = 0\}\}$ is called the constraint set or the domain of the problem. In general, the objective function f is a non-convex function and the problem may have many local optima. Therefore, it is challenging to obtain the global optimal solution of the problem using numerical algorithms or even impossible to find the solution. A specific class of optimization problems is called convex optimization problems if the cost function f and the inequality constraint functions g_i are convex and the equality constraint functions h_i are affine. For convex optimization problems, any local optimum of the problem is the global minimum for convex optimization problems.

Convex sets

A set S is called an affine set if it contains the line through any two points in the set. If S is an affine set, for any two points $\boldsymbol{x}$ and $\boldsymbol{y}$ in the set S, every linear combination of them is in the set S:

$$\boldsymbol{x},\boldsymbol{y} \in S \;\Rightarrow\; \lambda\boldsymbol{x} + (1-\lambda)\boldsymbol{y} \in S \;\text{ for }\; \lambda \in \mathbb{R} \tag{5.3}$$

In addition, a set S is convex if it contains every line segment between any two points in the set. In other words, if S is a convex set, for any two points $\boldsymbol{x}$ and $\boldsymbol{y}$ in the set S,

$$\boldsymbol{x},\boldsymbol{y} \in S \;\Rightarrow\; \lambda\boldsymbol{x} + (1-\lambda)\boldsymbol{y} \in S \;\text{ for }\; 0 \leq \lambda \leq 1 \tag{5.4}$$

The weighted average shown in (5.4) can be generalized to more than two points. We refer to a point of the form $\lambda_1\boldsymbol{x}_1 + \lambda_2\boldsymbol{x}_2 + \cdots + \lambda_k\boldsymbol{x}_k$, where $\lambda_1 + \lambda_2 + \cdots + \lambda_k = 1$ and $\lambda_i \geq 0$ for all i, to a convex combination. Then it can be also shown that a set is convex if and only if it contains every convex combination of points in the set. We can easily know that an affine set is a subset of a convex set and every affine set is a convex set.

The convex hull is defined on a set S as the set of all convex combinations of points in the set S:

$$\textbf{conv}\, S = \{\lambda_1\boldsymbol{x}_1 + \lambda_2\boldsymbol{x}_2 + \cdots + \lambda_k\boldsymbol{x}_k | x_i \in S, \lambda_i \geq 0, \lambda_1 + \lambda_2 + \cdots + \lambda_k = 1\} \tag{5.5}$$

The convex hull is always convex and it is the smallest convex set that contains S. For example, if S is a set of two points, **conv** S is a line segment between the two points. If S is a set of three points, **conv** S is a triangle that is formed by the three points. If a set C that contains S is given, then **conv** $S \subseteq C$.

There are some more examples of convex sets that are widely used in the optimization problems. A hyperplane defined by $\{\boldsymbol{x}|\boldsymbol{a}^T\boldsymbol{x} = b\}$ is affine and convex. A Euclidean ball with the center at $\boldsymbol{x}_c$ and the radius r given by

$$B(\boldsymbol{x}_c, r) = \{\|\boldsymbol{x} - \boldsymbol{x}_c\|_2 \leq r\} \tag{5.6}$$

is a special case of a convex ellipsoid which has the form $\{\boldsymbol{x}|(\boldsymbol{x} - \boldsymbol{x}_c)^T\boldsymbol{A}^{-1}(\boldsymbol{x} - \boldsymbol{x}_c) \leq 1\}$ where $\boldsymbol{A}$ is symmetric and positive definite. A norm ball is one of the mostly used convex sets in the optimization problem, which is defined by

$$\{\boldsymbol{x}|\,\|\boldsymbol{x} - \boldsymbol{x}_c\| \leq r\}. \tag{5.7}$$

Note that a kind of norm is not specified. Any function that is positive definite and absolutely homogeneous and satisfies the triangle inequality is a norm.

Convexity is preserved under intersection and an affine transformation, namely, the intersection of convex sets is convex and the image and inverse image of a convex set on an affine function is convex. For example, a polyhedron which is defined as the solution set of a finite number of linear equalities and inequalities, in other words, the intersection of a finite number of half-spaces and hyperplanes, $\{\boldsymbol{x}|\boldsymbol{A}\boldsymbol{x} \preceq \boldsymbol{b}, \boldsymbol{C}\boldsymbol{x} = \boldsymbol{d}\}$, is a convex set.

Convex functions

A function f is convex if the domain of f is a convex set and

$$f(\lambda\boldsymbol{x} + (1-\lambda)\boldsymbol{y}) \leq \lambda f(\boldsymbol{x}) + (1-\lambda)f(\boldsymbol{y}) \tag{5.8}$$

for all $\boldsymbol{x}, \boldsymbol{y} \in \mathbf{dom}\ f$ and $0 \leq \lambda \leq 1$. This inequality geometrically means that the line segment between $(\boldsymbol{x}, f(\boldsymbol{x}))$ and $(\boldsymbol{y}, f(\boldsymbol{y}))$ lies above the graph of f. For example, an affine function always holds the equality in (5.8) and, therefore, all affine functions are convex. If $-f$ is convex, f is called a concave function. An affine function is also a concave function and any function that is convex and concave is affine.

A convex function can be defined using the α-sublevel set of a function f which is defined as

$$S_\alpha \triangleq \{\boldsymbol{x} \in \mathbf{dom}\ f \mid f(\boldsymbol{x}) \leq \alpha\} \tag{5.9}$$

If f is convex, then its sublevel sets S_α are convex sets for any value of α. Note that the converse is not true. For example, a non-convex monotonically decreasing function has convex sublevel sets. In addition, if a function is differentiable or twice differentiable, in other words, its gradient or Hessian exists at each point in $\mathbf{dom} f$, then f is convex if and only if $\mathbf{dom} f$ is convex and

$$f(\boldsymbol{y}) \geq f(\boldsymbol{x}) + \nabla f(\boldsymbol{x})^T(\boldsymbol{y} - \boldsymbol{x}) \qquad \text{for all } \boldsymbol{x}, \boldsymbol{y} \in \mathbf{dom} f \tag{5.10}$$

$$\nabla^2 f(\boldsymbol{x}) \succeq 0 \qquad \text{for all } \boldsymbol{x} \in \mathbf{dom} f \tag{5.11}$$

Note that the right term of (5.10) represents the first-order Taylor approximation of f at $\boldsymbol{x}$ and the inequality (5.10) implies that the first-order Taylor approximation of a convex function is a global underestimator of the function. The inequality (5.11) shows that the Hessian of a convex function is positive semidefinite.

Convex optimization problem

For the standard form of the optimization problem in (5.1), an optimization problem is a convex optimization problem if the objective function f and the inequality constraint functions g_i are convex and the equality constraint function h_i are affine. Therefore, the standard form of the convex optimization problem can be given as

$$\begin{aligned} \underset{\boldsymbol{x}}{\text{minimize}} \quad & f(\boldsymbol{x}) \\ \text{subject to} \quad & g_i(\boldsymbol{x}) \leq 0 \\ & \boldsymbol{A}\boldsymbol{x} = \boldsymbol{b} \end{aligned} \tag{5.12}$$

Since the equality constraint functions are affine, the equality constraint can be expressed by $h_i(\boldsymbol{x}) = \boldsymbol{a}_i^T\boldsymbol{x} - b_i$ and rewritten as $\boldsymbol{A}\boldsymbol{x} = \boldsymbol{b}$ by stacking all the equality constraints. Convex optimization problems facilitate solving optimization problems since any local optimum of the convex optimization problem is necessarily a global optimum. Therefore, there are many efficient numerical solution methods available that can handle very large problems with high-dimensional variables and a number of constraints.

Some examples of canonical convex optimization problems are useful to introduce. When the objective function and constraint functions are all affine, the problem is called a linear programming (LP) which has the form

$$\begin{aligned} \underset{\boldsymbol{x}}{\text{minimize}} \quad & \boldsymbol{c}^T\boldsymbol{x} \\ \text{subject to} \quad & \boldsymbol{G}\boldsymbol{x} \preceq \boldsymbol{h} \\ & \boldsymbol{A}\boldsymbol{x} = \boldsymbol{b} \end{aligned} \tag{5.13}$$

where $\boldsymbol{G} \in \mathbb{R}^{m\times n}$ and $\boldsymbol{A} \in \mathbb{R}^{p\times n}$, m is the number of the inequality constraints, p is the number of the equality constraints, and n is the number of the optimization variables. The feasible set of the LP forms a polyhedron. LPs are applicable to a number of fields and applications such as finding Chebyshev center of a polyhedron and dynamic activity planning.

A quadratically constrained quadratic program (QCQP) is also a kind of convex optimization problems that is widely used in applications. It has a convex quadratic objective function and convex quadratic inequality constraint functions.

$$\begin{aligned} \underset{\boldsymbol{x}}{\text{minimize}} \quad & \boldsymbol{x}^T\boldsymbol{R}\boldsymbol{x} + \boldsymbol{c}^T\boldsymbol{x} + d \\ \text{subject to} \quad & \boldsymbol{x}^T\boldsymbol{P}_i\boldsymbol{x} + \boldsymbol{q}_i^T\boldsymbol{x} + r_i \leq \boldsymbol{0} \qquad i = 1, \ldots, m \\ & \boldsymbol{A}\boldsymbol{x} = \boldsymbol{b} \end{aligned} \tag{5.14}$$

where $\boldsymbol{R}$ and $\boldsymbol{P}_i$ are all positive semidefinite matrices. In QCQP, each inequality constraint forms an ellipsoid. In the case that $\boldsymbol{P}_i = 0$, the optimization problem becomes a quadratic program (QP) which has a convex quadratic objective function and affine constraint functions. It also includes linear programs when $\boldsymbol{R} =$ and $\boldsymbol{P}_i = 0$. The examples of the QP include least squares approximation and Markowitz portfolio optimization which is a classical portfolio problem.

A semidefinite programming (SDP) has the form

$$\begin{aligned} \underset{X}{\text{minimize}} \quad & \operatorname{tr}(\boldsymbol{R}\boldsymbol{X}) \\ \text{subject to} \quad & \operatorname{tr}(\boldsymbol{P}_i\boldsymbol{X}) \leq 0 \qquad i = 1, \ldots, m \\ & \operatorname{tr}(\boldsymbol{Q}_i\boldsymbol{X}) = 0 \qquad i = 1, \ldots, p \\ & \boldsymbol{X} \succeq 0 \end{aligned} \tag{5.15}$$

where $\boldsymbol{X} \succeq 0$ represents that $\boldsymbol{X}$ is a positive semidefinte matrix. Note that it has a nonnegativity constraint in addition to a linear objective function and linear constraint functions of $\boldsymbol{X}$. Many optimization problems including a matrix norm minimization problem [30], moment problems [31,32], and a fastest mixing Markov chain problem [33] can be cast to a SDP.

5.2.1.2 Solving convex optimization problems

We discuss two main categories for solving convex optimization problems, i.e., obtaining the optimal solution of optimization problems, analytical approaches, and numerical approaches.* Analytical approaches enable a closed-form solution but are not available to all the optimization problems. For those problems where an analytical approach is not available, a solution can be found via numerical approaches.

Analytical approaches

Lagrangian associated with the problem (5.1) is defined as

$$L(\boldsymbol{x}, \boldsymbol{\lambda}, \boldsymbol{\nu}) = f(\boldsymbol{x}) + \sum_{i=1}^{m} \lambda_i g_i(\boldsymbol{x}) + \sum_{i=1}^{p} \nu_i h_i(\boldsymbol{x}) \tag{5.16}$$

where $\boldsymbol{\lambda}$ and $\boldsymbol{\nu}$ are the Lagrange multipliers. The Lagrange dual function is the minimum value of the Lagrangian over $\boldsymbol{x}$

$$g(\boldsymbol{\lambda}, \boldsymbol{\nu}) = \inf_{\boldsymbol{x} \in D} L(\boldsymbol{x}, \boldsymbol{\lambda}, \boldsymbol{\nu}) \tag{5.17}$$

where D is the domain of the optimization problem. The dual function gives the information about the optimal value $p^{\star}$ of the problem. For any $\boldsymbol{\lambda} \succeq 0$ and any $\boldsymbol{\nu}$,

$$g(\boldsymbol{\lambda}, \boldsymbol{\nu}) \leq p^{\star} \tag{5.18}$$

The inequality implies that the Lagrange dual function is a lower bound on the optimal value $p^{\star}$. Then the best lower bound of the optimum value can be obtained by maximizing the Lagrangian dual function:

$$\begin{aligned} \text{maximize} \quad & g(\boldsymbol{\lambda}, \boldsymbol{\nu}) \\ \text{subject to} \quad & \boldsymbol{\lambda} \succeq 0 \end{aligned} \tag{5.19}$$

This is the Lagrange dual problem of (5.1). It is always a convex optimization problem regardless of convexity of (5.1) since the dual function is concave and the constraint

*Apart from analytical and numerical approaches, such convex optimization problems could be solved using learning theory as well.

is convex. Let $d^\star$ denote the optimal value of (5.19). The following weak duality inequality always holds even if the original problem is not convex:

$$d^\star \leq p^\star. \tag{5.20}$$

In the case that the problem is convex and there exists a strictly feasible point, strong duality holds,

$$d^\star = p^\star. \tag{5.21}$$

This means if the dual optimal points $\boldsymbol{\lambda}^\star$ and $\boldsymbol{\nu}^\star$ can be obtained the optimality conditions of the optimization problem can be derived. Let $\boldsymbol{x}^\star$ be the optimal point of the optimization problem (5.1). If strong duality holds,

$$\begin{aligned} f(\boldsymbol{x}^\star) &= g(\boldsymbol{\lambda}^\star, \boldsymbol{\nu}^\star) && (5.22)\\ &\leq f(\boldsymbol{x}^\star) + \sum_{i=1}^{m} \lambda_i^\star g_i(\boldsymbol{x}^\star) + \sum_{i=1}^{p} \nu_i^\star h_i(\boldsymbol{x}^\star) && (5.23)\\ &\leq f(\boldsymbol{x}^\star) && (5.24) \end{aligned}$$

From the equality and inequalities above, the Karush–Kuhn–Tucker (KKT) conditions are derived as follows:

$$\begin{aligned} & g_i(\boldsymbol{x}^\star) \leq 0 && \text{Primal feasibility} \quad (5.25)\\ & h_i(\boldsymbol{x}^\star) = 0 && \text{Primal feasibility} \quad (5.26)\\ & \lambda_i^\star \geq 0 && \text{Dual feasibility} \quad (5.27)\\ & \lambda_i^\star g_i(\boldsymbol{x}^\star) = 0 && \text{Complementary slackness} \quad (5.28)\\ & \nabla f(\boldsymbol{x}^\star) + \sum_{i=1}^{m} \lambda_i^\star \nabla g_i(\boldsymbol{x}^\star) + \sum_{i=1}^{p} \nu_i^\star \nabla h_i(\boldsymbol{x}^\star) = 0 && \text{Gradient condition} \quad (5.29) \end{aligned}$$

If the objective function and constraint functions of the problem are differentiable and strong duality holds, the optimal and the dual optimal points must satisfy the KKT conditions. Moreover, the KKT conditions are sufficient for convex optimization problems, which means any points $\boldsymbol{x}$, $\boldsymbol{\lambda}$, and $\boldsymbol{\nu}$ that satisfy the KKT conditions are primal and dual optimal. One simple example of the optimization problem that can be solved by the KKT conditions is an equality constrained convex quadratic minimization problem which is given by

$$\begin{aligned} \underset{x}{\text{minimize}} \ & (1/2)\boldsymbol{x}^T \boldsymbol{P}\boldsymbol{x} + \boldsymbol{q}^T \boldsymbol{x} + r\\ \text{subject to} \ & \boldsymbol{A}\boldsymbol{x} = \boldsymbol{b} \end{aligned} \tag{5.30}$$

The KKT conditions for this problem are

$$\boldsymbol{A}\boldsymbol{x}^\star = \boldsymbol{b}, \qquad \boldsymbol{P}\boldsymbol{x}^\star + \boldsymbol{q} + \boldsymbol{A}^T \boldsymbol{\nu}^\star = 0 \tag{5.31}$$

$\boldsymbol{x}^\star$ and $\boldsymbol{\nu}^\star$ can be obtained by solving the following linear equation

$$\begin{bmatrix} \boldsymbol{P} & \boldsymbol{A}^T \\ \boldsymbol{A} & \boldsymbol{0} \end{bmatrix} \begin{bmatrix} \boldsymbol{x}^\star \\ \boldsymbol{\nu}^\star \end{bmatrix} = \begin{bmatrix} -\boldsymbol{q} \\ \boldsymbol{b} \end{bmatrix} \tag{5.32}$$

Numerical approaches

The KKT conditions enable an analytical solution of optimization problems, however, they are applicable to a limited number of convex optimization problems. In most cases, it is required to find the solution by numerical algorithms. First consider a simple unconstrained convex optimization problem,

$$\underset{\boldsymbol{x}}{\text{minimize}}\, f(\boldsymbol{x}) \tag{5.33}$$

A necessary and sufficient condition for the optimal point $\boldsymbol{x}^\star$ is $\nabla f(\boldsymbol{x}^\star) = 0$. Starting from an initial point $\boldsymbol{x}_0$, a sequence of $\boldsymbol{x}_n$ is selected by

$$\boldsymbol{x}_{n+1} = \boldsymbol{x}_n + t\boldsymbol{d} \tag{5.34}$$

where t and $\boldsymbol{d}$ are the step size and the search direction, respectively. This is called a descent method if $f(\boldsymbol{x}_{n+1}) < f(\boldsymbol{x}_n)$ holds for every pair of $\boldsymbol{x}_n$ and $\boldsymbol{x}_{n+1}$ in the sequence. For any decent methods for the convex objective function, the search direction $\boldsymbol{d}$ must be a descent direction which satisfies

$$\nabla f(\boldsymbol{x})^T \boldsymbol{d} < 0 \tag{5.35}$$

The step size t can be determined by the exact line search or the inexact line search methods, for example, the backtracking line search. Different choice of the search direction for the descent methods results in various kinds of the descent methods. A descent method which takes the negative gradient as the search direction is called the gradient descent method. For the gradient method, the search direction is chosen by

$$\boldsymbol{d} = -\nabla f(\boldsymbol{x}) \tag{5.36}$$

The gradient descent method shows approximately linear convergence and the convergence rate highly depends on the condition number of the Hessian. Though it is such a simple descent method, the convergence is very slow with the high condition number of the Hessian.

A descent method with a fixed step size $t = 1$ and the search direction

$$\boldsymbol{d} = -\nabla^2 f(\boldsymbol{x})^{-1} \nabla f(\boldsymbol{x}) \tag{5.37}$$

is called Newton's method. It is motivated by the fact that $\boldsymbol{d}$ in (5.37) is the minimizer of the second-order approximation of f at $\boldsymbol{x}$, i.e.,

$$\boldsymbol{d} = \arg\min \hat{f}(\boldsymbol{x} + \boldsymbol{d}) = \arg\min f(\boldsymbol{x}) + \nabla f(\boldsymbol{x})^T \boldsymbol{d} + \frac{1}{2}\boldsymbol{d}^T \nabla^2 f(\boldsymbol{x}) \boldsymbol{d} \tag{5.38}$$

Newton's method shows faster convergence than the gradient descent method and it converges quadratically near $\boldsymbol{x}^\star$. However, it involves computation of the Hessian and the Newton step at every iteration, which requires solving a set of linear equations.

An equality constrained minimization problem can be solved by the decent methods for unconstrained minimization problems described earlier after eliminating the equality constraint and simplified to an unconstrained optimization problem. Consider the following equality constrained minimization problem:

$$\begin{aligned} &\underset{\boldsymbol{x}}{\text{minimize}}\ f(\boldsymbol{x}) \\ &\text{subject to } \boldsymbol{A}\boldsymbol{x} = \boldsymbol{b} \end{aligned} \tag{5.39}$$

After finding a particular solution $\hat{x}$ of the equality constraint $Ax = b$, the equivalent unconstrained optimization problem can be written by

$$\min g(z) = f(Fz + \hat{x}) \tag{5.40}$$

where $F \in \mathbb{R}$ is a matrix whose range is the nullspace of A, i.e., $AF = 0$. This problem can be solved by the descent methods for unconstrained optimization problems. After obtaining $z^\star$, the solution of (5.39) can be easily found by

$$x^\star = Fz^\star + \hat{x} \tag{5.41}$$

Newton's method for the inequality constrained minimization problem is also available. Taking the second-order Taylor approximation of (5.39) yields

$$\begin{aligned} &\underset{x}{\text{minimize}}\ \hat{f}(x + d) = f(x) + \nabla f(x)^T d + \tfrac{1}{2} d^T \nabla^2 f(x) d \\ &\text{subject to}\ A(x + d) = b \end{aligned} \tag{5.42}$$

Recall that the search direction is the minimizer of the above problem and since it is an equality constrained quadratic minimization problem as (5.30), d can be obtained by the KKT condition,

$$\begin{bmatrix} \nabla^2 f(x) & A^T \\ A & 0 \end{bmatrix} \begin{bmatrix} d \\ v \end{bmatrix} = \begin{bmatrix} -\nabla f(x) \\ 0 \end{bmatrix} \tag{5.43}$$

Lastly, a key idea for solving inequality constrained minimization problems is approximately to formulate an inequality constrained problem which can be solved by the methods described earlier. Consider the following equality constrained optimization problem

$$\begin{aligned} &\underset{x}{\text{minimize}}\ f(x) + \textstyle\sum_{i=1}^{m} I_{-}(g_i(x)) \\ &\text{subject to}\ Ax = b \end{aligned} \tag{5.44}$$

where I_{-} is the indicator function of $\mathbb{R}_{-}$ such that $I_{-}(u) = 0$ if $u \leq 0$ and $I_{-}(u) = \infty$ otherwise. Though (5.44) has no equality constraint and it is equivalent to (5.12), Newton's method cannot be applied since $I_{-}(u)$ is not differentiable. The indicator function can also be approximated by the logarithmic barrier function which is differentiable and given by

$$\hat{I}(u) = -(1/t)\log(-u) \tag{5.45}$$

Then the inequality constrained problem can be approximated by the following equality constrained problem:

$$\begin{aligned} &\underset{x}{\text{minimize}}\ f(x) - (1/t)\textstyle\sum_{i=1}^{m} \log(-g_i(x)) \\ &\text{subject to}\ Ax = b \end{aligned} \tag{5.46}$$

This problem is convex and the objective function is differentiable. Therefore, it can be solved by Newton's method.

5.2.1.3 Approaches for non-convex optimization problems

Convex problems can be solved by the analytical and numerical approaches described in the previous sections; however, many of the optimization problems in practice are non-convex. In general, it is difficult to find a solution of the non-convex optimization problem since it has a lot of local minima and maxima and saddle points. We introduce mathematical techniques that help to find optimal or sub-optimal solution of the non-convex problem, convex relaxation/approximation, and transformation of variables with examples.

A compressive sensing signal recovery problem is one of the most actively studied areas in signal-processing applications. The optimization problem is given by an l_0 norm minimization problem,

$$\hat{x} = \min_{x} \|\boldsymbol{x}\|_0 \qquad \text{subject to} \qquad \boldsymbol{y} = \boldsymbol{A}\boldsymbol{x} \tag{5.47}$$

Since the objective function $\|\boldsymbol{x}\|_0$ is non-convex, it can be relaxed as l_1 norm which is convex

$$\hat{x} = \min_{x} \|\boldsymbol{x}\|_1 \qquad \text{subject to} \qquad \boldsymbol{y} = \boldsymbol{A}\boldsymbol{x} \tag{5.48}$$

This problem is a convex optimization problem. Though the objective function is still not differentiable at $\boldsymbol{x} = \boldsymbol{0}$, it can be solved by efficient solvers such as the interior point method and least angle regression (LARS). In radar applications, the CMC is used in many waveform design problems. The SINR maximization subject to the CMC is given by

$$\begin{aligned} &\underset{x}{\text{minimize}} \ \boldsymbol{x}^H \boldsymbol{\Phi} \boldsymbol{x} \\ &\text{subject to} \ |\boldsymbol{x}| = \mathbf{1}/n. \end{aligned} \tag{5.49}$$

where $\boldsymbol{\Phi}$ is an SINR matrix. The CMC is a non-convex constraint and very hard to exploit in the optimization problem. The EC is used as a relaxation of the CMC in many problems

$$\begin{aligned} &\underset{x}{\text{minimize}} \ \boldsymbol{x}^H \boldsymbol{\Phi} \boldsymbol{x} \\ &\text{subject to} \ \|\boldsymbol{x}\|_2 - 1. \end{aligned} \tag{5.50}$$

A non-convex problem can be converted to a convex problem by transformation of variables. Consider a geometric programming which is given by

$$\begin{aligned} &\underset{x}{\text{minimize}} \ \sum_{k=1}^{K_0} e^{\boldsymbol{a}_{0_k}^T \boldsymbol{x} + b_{0_k}} \\ &\text{subject to} \ \sum_{k=1}^{K_i} e^{\boldsymbol{a}_{i_k}^T \boldsymbol{x} + b_{i_k}} \leq 1, \qquad i = 1, \ldots, m \\ &\qquad\qquad e^{\boldsymbol{g}_i^T \boldsymbol{x} + h_i} = 1, \qquad i = 1, \ldots, p \end{aligned} \tag{5.51}$$

This problem is non-convex because of non-convexity of the equality constraint. Taking the logarithm on the objective and constraint functions gives

$$\begin{aligned} &\underset{x}{\text{minimize}} \ \log\left(\sum_{k=1}^{K_0} e^{\boldsymbol{a}_{0_k}^T \boldsymbol{x} + b_{0_k}}\right) \\ &\text{subject to} \ \log\left(\sum_{k=1}^{K_i} e^{\boldsymbol{a}_{i_k}^T \boldsymbol{x} + b_{i_k}}\right) \leq 0, \qquad i = 1, \ldots, m \\ &\qquad\qquad \boldsymbol{g}_i^T \boldsymbol{x} + h_i = 0, \qquad i = 1, \ldots, p \end{aligned} \tag{5.52}$$

The objective and equality constraint functions are convex and the equality constraint function is affine, and, therefore, the problem is convex.

5.2.2 Challenges of optimization problems for cognitive radar

Convex optimization also benefits cognitive radar in many ways and has been applied to essential problems for cognitive radar, including but not limited to target detection and estimation, channel estimation, waveform design, beampattern design, and resource allocation. In practice, however, the optimization problems become more challenging when practical constraints induced by the environment and radar physics are incorporated into the optimization problems. For example, the waveform design problem that maximizes a radar output SINR under the EC, $\boldsymbol{x}^H\boldsymbol{x} = 1$, can be easily solved by the eigenvalue problem and the optimal waveform is given as the eigenvector corresponding to the greatest eigenvalue of the SINR matrix [34]. However, the waveform optimization problem becomes onerous and the solution is difficult to obtain by conventional numerical approaches when the CMC which is practically essential is employed in the optimization problem. Additional constraints such as the SC, the SpecC, and the interference constraint make the optimization problem even more challenging. Table 5.1 provides the objective functions and the constraints commonly used

Table 5.1 Common objective functions and constraints for cognitive radar problems

Objective functions	**Constraints**	**Optimization algorithms**
SNR or SINR (quadratic function)	SC & CMC	Randomization [7,21,35], semidefinite relaxation (SDR) [21,35], Successive QCQP refinement (SQR) [8], block coordinate descent (BCD) [36]
	EC, ISL, PSL	Randomization [13]
	EC, SpecC, SC	SDR & rank-one decomposition [22,36]
Beampattern error	CMC	Sequence of closed form solution (SCF) [10], alternating optimization [37], Complex circle manifold [38]
	EC	Barrier method [9]
	PAR	Cyclic algorithm [19], obtain unconstrained solution and enforce PAR [20]
	CMC & SpecC	SCF [23]
Ambiguity function (quartic)	CMC	Quadratic gradient descent [11], maximum block improvement [17], Phase-only conjugate gradient[39]
	PAR & EC	Majorization–minimization (MM) [40]
ISL	CMC	Iterative min–max problems [12], singular value decomposition [15], MM [16,19]
Mutual information	EC	Eigendecomposition [41]
Radiation power	CMC	Gradient descent [42]
Bayesian CRB	EC	Semidefinite programming [43]

in cognitive radar problems and the approaches to solve the optimization problems for each objective function and constraints.[†]

5.3 Constrained optimization for cognitive radar

In this section, we introduce three successful waveform design algorithms that solve non-convex optimization problems that optimize SINR, the beampattern, and the AF under challenging constraints. They exploit the CMC in common and have different additional constraints depending on the purpose of waveform design. Though the CMC cannot be achievable by conventional numerical approaches, the solutions of these algorithms approach the constant modulus in iterative ways where a convex optimization problem is solved at each iteration step. It has been proven that the solutions converge and achieve the CMC at convergence.

5.3.1 SINR maximization

We consider a collocated narrow-band MIMO radar system with N_T transmit antennas and N_R receive antennas. The received signal in this model is given by

$$\boldsymbol{r} = \alpha_0 \boldsymbol{U}(\theta_0)\boldsymbol{x} + \sum_{k=1}^{K} \alpha_k \boldsymbol{U}(\theta_k)\boldsymbol{x} + \boldsymbol{n} \tag{5.53}$$

where n is a circular complex Gaussian noise vector with zero mean and covariance matrix $\sigma^2\boldsymbol{I}$, α_0, α_k, θ_0 and θ_k denote the complex amplitudes and the angle of the target and the kth clutter source, respectively, and $\boldsymbol{U}(\theta)$ is the steering matrix of a uniform linear array (ULA) antenna with half-wavelength separation between the antennas. The most common criterion in waveform design involves SINR maximization, which involves joint optimization of the transmit waveform and the receive filter. In particular, the receive filter is assumed to be a linear finite impulse response filter $\boldsymbol{w}$. The output is given by

$$r_f = \boldsymbol{w}^H\boldsymbol{r} = \alpha_0 \boldsymbol{w}^H \boldsymbol{U}(\theta_0)\boldsymbol{x} + \sum_{k=1}^{K} \alpha_k \boldsymbol{w}^H \boldsymbol{U}(\theta_k)\boldsymbol{x} + \boldsymbol{w}^H\boldsymbol{n} \tag{5.54}$$

Then, the output SINR can be expressed as

$$\text{SINR} = \frac{\sigma|\boldsymbol{w}^H \boldsymbol{U}(\theta_0)\boldsymbol{x}|^2}{\boldsymbol{w}^H \boldsymbol{\Sigma}(\boldsymbol{x})\boldsymbol{w} + \boldsymbol{w}^H\boldsymbol{w}} \tag{5.55}$$

where $\sigma = E[|\alpha_0|^2]/\sigma_n^2$, $\boldsymbol{\Sigma}(\boldsymbol{x}) = \sum_{k=1}^{K} I_k \boldsymbol{U}(\theta_k)\boldsymbol{x}\boldsymbol{x}^H \boldsymbol{U}^H(\theta_k)$ and $I_k = E[|\alpha_k|^2]/\sigma_n^2$.

[†]There are numerous existing work for cognitive radar problems other than the methods shown in the table. We only include the methods referred in this chapter.

5.3.1.1 Problem formulation

The objective is to design the optimal waveform which maximizes the SINR subject to the CMC and the SC, i.e., to solve the following optimization problem:

$$\begin{aligned}
&\underset{\boldsymbol{w},\boldsymbol{x}}{\text{maximize}} \ \frac{\sigma|\boldsymbol{w}^H \boldsymbol{U}(\theta_0)\boldsymbol{x}|^2}{\boldsymbol{w}^H \boldsymbol{\Sigma}(\boldsymbol{x})\boldsymbol{w}+\boldsymbol{w}^H\boldsymbol{w}} \\
&\text{subject to} \ \|\boldsymbol{x}-\boldsymbol{x}_0\|_\infty \leq \epsilon \\
&\qquad\qquad |\boldsymbol{x}(k)| = 1/\sqrt{N_T N}
\end{aligned} \tag{5.56}$$

It is a joint problem with respect to $\boldsymbol{x}$ and $\boldsymbol{w}$; however, it is separable and an unconstrained optimization problem of $\boldsymbol{w}$. The optimal $\boldsymbol{w}$ can be obtained from the well-known MVDR problem [44]. By substituting the optimal $\boldsymbol{w}$ into (5.50), we obtain an equivalent problem

$$\begin{aligned}
&\underset{\boldsymbol{x}}{\text{maximize}} \ \boldsymbol{x}^H \boldsymbol{\Phi}(\boldsymbol{x})\boldsymbol{x} \\
&\text{subject to} \ \|\boldsymbol{x}-\boldsymbol{x}_0\|_\infty \leq \epsilon \\
&\qquad\qquad |\boldsymbol{x}(k)| = 1/\sqrt{N_T N}
\end{aligned} \tag{5.57}$$

where $\boldsymbol{\Phi}(\boldsymbol{x}) = \boldsymbol{U}^H(\theta_0)[\boldsymbol{\Sigma}(\boldsymbol{x})+\boldsymbol{I}]^{-1}\boldsymbol{U}(\theta_0)$. Since the dependence of $\boldsymbol{\Phi}(\boldsymbol{x})$ on the waveform $\boldsymbol{x}$ makes the optimization problem onerous, it has been solved iteratively assuming $\boldsymbol{\Phi}(\boldsymbol{x}) = \boldsymbol{\Phi}$ for a fixed $\boldsymbol{x}$ and repeatedly optimizing $\boldsymbol{x}$ with a new $\boldsymbol{\Phi}$ till convergence [8,21,45]. However, even for a fixed $\boldsymbol{\Phi}$, the optimization of $\boldsymbol{x}$ is a hard non-convex problem for which approaches to solve the problem involves SDR with randomization [21,46] or MM [36].

5.3.1.2 Successive QCQP refinement

The optimization problem (5.56) with signal-independent clutter (*i.e.*, $\boldsymbol{\Phi}(\boldsymbol{x}) = \boldsymbol{\Phi}$) is equivalent to the following non-convex problem:

$$\begin{aligned}
&\underset{\boldsymbol{x}}{\text{maximize}} \ \boldsymbol{x}^H(\boldsymbol{\Phi}-\lambda\boldsymbol{I})\boldsymbol{x} \\
&\text{subject to} \ \arg\boldsymbol{x}(k) \in [\gamma_k, \gamma_k+\delta] \\
&\qquad\qquad |\boldsymbol{x}(k)| = 1/\sqrt{N_T N}
\end{aligned} \tag{5.58}$$

where λ is a constant greater than the largest eigenvalue of $\boldsymbol{\Phi}$ so that $\boldsymbol{\Phi}-\lambda\boldsymbol{I}$ is negative semidefinite. Due to the CMC, the SC can be expressed by the phase only where $\gamma_k = \arg\boldsymbol{x}_0(k) - \arccos(1-\epsilon^2/2)$ and $\delta = 2\arccos(1-\epsilon^2/2)$. Though it maximizes a concave quadratic function with a negative semidefinite matrix, it is still non-convex due to the feasible set. The key idea of the successive QCQP refinement is to solve the non-convex optimization problem (5.58) by solving a sequence of convex QCQP problems such that in each iteration of the sequence the designed waveform satisfies the SC and the constant modulus is successively achieved at convergence. The CMC enforces the modulus of every element of $\boldsymbol{x}$ to be a constant $(1/\sqrt{N_T N})$, in other words, every element of $\boldsymbol{x}$ should be located on a scaled unit circle in the complex plane. The SC restricts the phase of each element as shown in (5.58). Based on this observation, the successive QCQP refinement method finds the solution by solving the sequence of convex QCQP problems for which the feasible set gets smaller and closer to the constant modulus circle.

Consider the following convex optimization problem which is a relaxation of the non-convex problem (5.58):

$$\begin{aligned} \underset{x}{\text{maximize}} \ & \boldsymbol{x}^H \boldsymbol{Q} \boldsymbol{x} \\ \text{subject to} \ & a_k \Re\{\boldsymbol{x}(k)\} + b_k \Im\{\boldsymbol{x}(k)\} \geq c_k \\ & |\boldsymbol{x}(k)|^2 \leq 1/\sqrt{N_T N} \end{aligned} \tag{5.59}$$

where $\boldsymbol{Q} = \boldsymbol{\Phi} - \lambda \boldsymbol{I}$ is a negative semidefinite matrix, $\Re\{\boldsymbol{x}\}$ and $\Im\{\boldsymbol{x}\}$ represent the real and imaginary parts of a complex vector $\boldsymbol{x}$, respectively. Note that the CMC and the SC are relaxed to the convex quadratic inequality constraint and the affine inequality constraint, respectively. Therefore, (5.59) is a convex optimization problem. The parameters a_k, b_k, and c_k represent the line that intersects with the constant modulus at the interval $[\gamma_k, \gamma_k + \delta_k]$. The tighter the SC of (5.59) (which implies the smaller δ_k), the closer (5.59) to (5.58). For instance, if $\delta = \pi/2$, the feasible value of $|\boldsymbol{x}(k)|$ lies between $1/\sqrt{2}$ and 1 and $|\boldsymbol{x}(k)|$ approaches 1 as δ_k reduces as shown in Figure 5.4.

Figure 5.4 illustrates of the successive QCQP refinement algorithm. At the first iteration, a_k, b_k, and c_k are calculated from the SC parameters, γ_k and δ_k. The feasible set forms a circular segment of the unit circle with a radius $1/N_T N$ as shown in Figure 5.4(a). Then $\boldsymbol{x}^{\star}(k)$ is obtained by solving the problem (5.59). Denote the solution at the first iteration by $\boldsymbol{x}^{(0)}$. The basic idea of updating the feasible set such

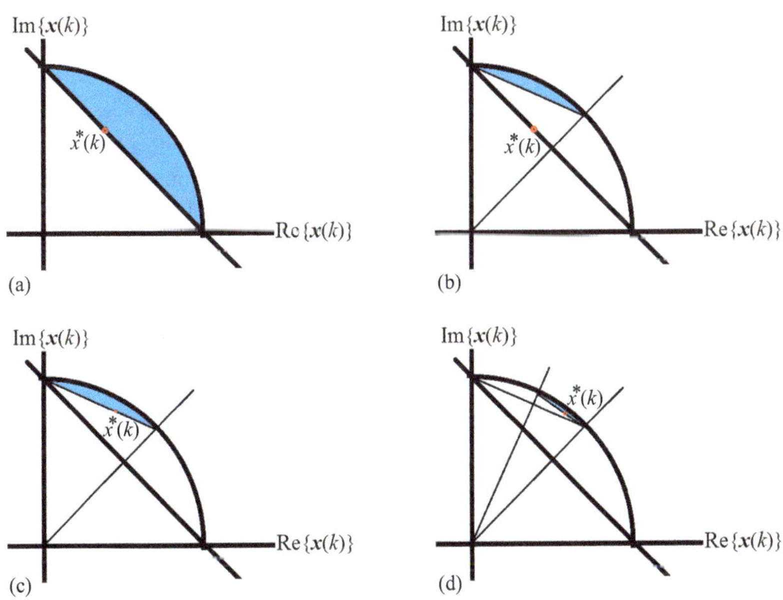

Figure 5.4 Illustration of the successive approximation of problem (5.58). *(a) The convex hull of the feasible set of* (5.58) *is the blue area. (b) The solution point of the convex problem in red. Now we consider only the upper half of the SC and solve again. (c) Second refinement (d) Third refinement, here solution in the third refinement is very close to unity.*

that a new feasible set becomes closer to the constant modulus circle at every iteration is to choose a half of the current feasible set which the solution $\boldsymbol{x}^{(0)}$ belongs to. Specifically, if arg $\boldsymbol{x}^{(0)}(k) \geq \gamma_k + \delta/2$, we set a new SC as $[\gamma_k + \delta/2, \gamma_k + \delta]$, i.e., the new constraint angles become $\gamma_k^{(1)} = \gamma_k + \delta/2$ and $\delta^{(1)} = \delta/2$. If arg $\boldsymbol{x}^{(0)}(k) < \gamma_k + \delta/2$, $\gamma_k^{(1)} = \gamma_k$ and $\delta^{(1)} = \delta/2$. In the same way, the problem (5.59) is solved in the next refinement with the updated γ_k and δ. Repeating the refinements for $n = 2, 3, \ldots$, the interval $[\gamma_k^{(n)}, \gamma_k^{(n)} + \delta/2^n]$ gets smaller and smaller and eventually the modulus of $\boldsymbol{x}^{(n)}(k)$ will converge to the constant modulus circle as shown in Figure 5.4.

5.3.1.3 Analytical and experimental results

Complexity analysis

Based on the computational complexity of a QCQP [47] in each refinement, the overall computational complexity of SQR is $\mathcal{O}(FN_T^{3.5}N^{3.5})$ where F is the total number of refinements. In comparison, SDR with randomization has a computational complexity of $\mathcal{O}(N_T^{3.5}N^{3.5}) + \mathcal{O}(LN_T^2N^2)$ [35]. It is shown that the number of required refinements F is independent of N_TN and in fact $F \ll N_TN$ [8] and the SQR algorithm typically has much lower complexity. The SDR with randomization invariably needs a large number of randomization trials L [21] which makes the term $\mathcal{O}(LN_T^2N^2)$ much larger.

Convergence analysis

It is shown that the SINR of the SQR algorithm is non-decreasing with each refinement, which means the following inequality always holds at each refinement:

$$\text{SINR}^{n-1} = \boldsymbol{x}^{(n-1)^H}\boldsymbol{\Phi}\boldsymbol{x}^{(n-1)} \leq \boldsymbol{x}^{(n)^H}\boldsymbol{\Phi}\boldsymbol{x}^{(n)} = \text{SINR}^n \tag{5.60}$$

The inequality implies that the SQR algorithm improves the SINR after each refinement. Furthermore, it is also shown that the sequence SINR^n is bounded and, therefore, it converges to a finite value $\text{SINR}^\star$ since a function which is monotonically non-decreasing and bounded converges to a finite value from the monotone convergence criterion [48].

The convergence rate of the algorithm highly depends on the number of refinements. The total number of refinements F can be determined from $i(n)$, the maximum improvement of the new iteration $n+1$ over n, which is given by

$$i(n) = \max\{\text{SINR}^{n+1} - \text{SINR}^n\} = \lambda_{\max}(1 - \beta^2(n)) \tag{5.61}$$

where $\beta(n) = \cos\left(\frac{\pi}{2^{n+1}}\right)$ and $\lambda_{\max}$ is the largest eigenvalue of $\boldsymbol{\Phi}$. Figure 5.5 plots $i(n)$ when $\lambda_{\max} = 1$. It is shown that at most only 2% of $\lambda_{\max}$ of improvement on the SINR value is expected after 5 refinements, which means the SINR nearly converges after the small number of refinements.

Experimental results

The numerical simulations are provided to show the performance of the SQR binary search (SQR-BS) method and the sequential optimization algorithm 1 (SOA1) [21]. For experimental setup, transmit and receive antennas have $N_T = 4$ and $N_R = 8$ elements, respectively, and the orthogonal linear frequency modulation (LFM) waveform is considered as the reference waveform $\boldsymbol{x}_0$. The number of randomization trials used in

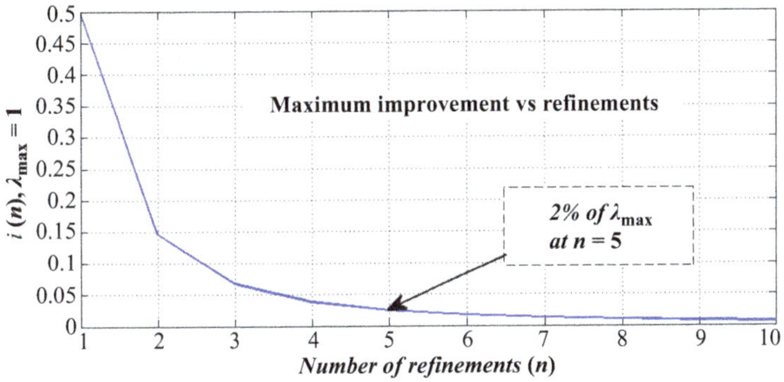

Figure 5.5 $i(n)$ versus the number of refinements for $\lambda_{\max} = 1$

SOA1 is 20,000 and the SQR method involves four refinement steps, i.e., $F = 4$. The target is located at an angle $\theta_0 = 15°$ and three interference sources is at $\theta_1 = -50°$, $theta_2 = -10°$ and $\theta_3 = 40°$.

Figure 5.6 shows the SINR improvement in each iteration in (a) and the beampattern in (b) for the SOA1 and SQR-BS. It is shown that SQR-BS achieves an SINR 1.59 dB higher than SOA1 for the same SC and exhibits much better suppression performance at $\theta = -10°$ and $\theta = 40°$ when compared to SOA1. A plot of the converged SINR value versus the SC parameter ϵ is shown in Figure 5.7. The SOA1 increases approximately linearly with ϵ while the SQR-BS exhibits a superlinear increase.

5.3.2 Spatio-spectral radar beampattern design

In practice, the transmit beampattern design is more challenging for two reasons. The first reason is the requirement of the CMC on the radar transmit waveform, i.e. a constant envelope transmits signal [6]. The second reason is the requirement of spectral compatibility of radar and telecommunication systems, which demands a SpecC on the radar waveform spectral shape. Designing the MIMO radar beampattern in the simultaneous presence of constant modulus and SpecCs remains a stiff open challenge.

5.3.2.1 Problem formulation

Consider a wideband MIMO radar with a ULA of M antennas and equal spacing distance of d as shown in Figure 5.8. The signal transmitted from the mth element is denoted by $z_m(t)$. Let $z_m(t) = x_m(t)e^{j2\pi f_c t}$ where $x_m(t)$ is the baseband signal and f_c is the carrier frequency. We assume that the spectral support of $x_m(t)$ is within the interval $[-B/2, B/2]$ where B is the bandwidth in Hz. The sampled baseband signal transmitted by the mth element is denoted by $x_m(n) \triangleq x_m(t = nT_s)$, $n = 0, \ldots, N-1$

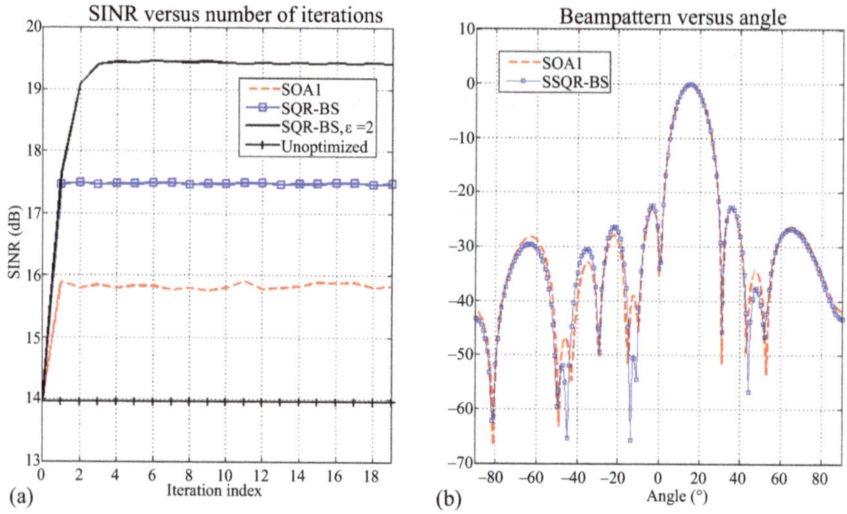

Figure 5.6 (a) The SINR values at each iteration and (b) the beampattern for SOA1 and SQR-BS algorithms with $\epsilon = 0.7$

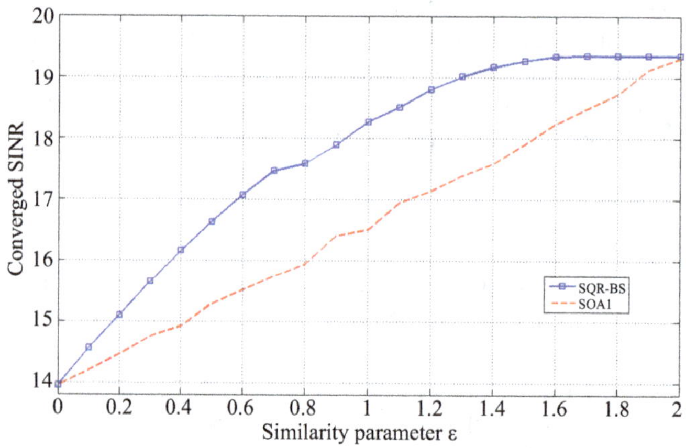

Figure 5.7 The SINR values at convergence of SQR-BS and SOA1 versus ϵ

with N being the number of time samples and $T_s = 1/B$ is the sampling rate. The discrete Fourier transform (DFT) of $x_m(n)$ is denoted by $y_m(p)$ and it is given by

$$y_m(p) = \sum_{n=0}^{N-1} x_m(n) e^{-j2\pi \frac{np}{N}}, \quad p = -\frac{N}{2}, \ldots, 0, \ldots, \frac{N}{2} - 1 \tag{5.62}$$

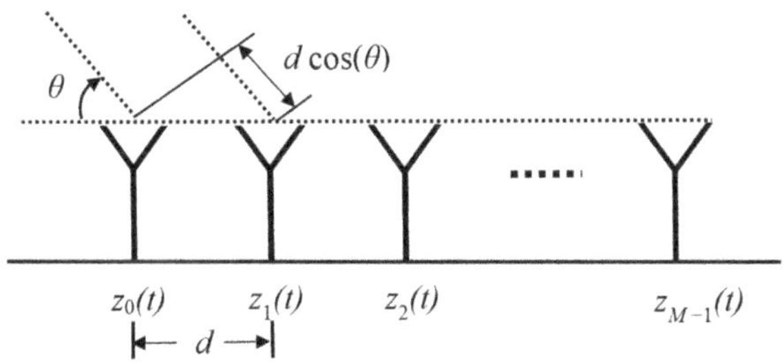

Figure 5.8 Configuration of ULA antenna

where N is assumed to be even in (5.62). If N is odd, then $p = -(N-1)/2, \ldots, 0, \ldots, (N-1)/2$.

The beampattern can be given by the following discrete angle-frequency grid [20]:

$$P_{kp} = |\boldsymbol{a}_{kp}^H \boldsymbol{y}_p|^2 = |\boldsymbol{a}_{kp}^H \boldsymbol{W}_p \boldsymbol{x}|^2 \tag{5.63}$$

where $\boldsymbol{x} \in \mathbb{C}^{MN}$ is the concatenated vector, i.e. $\boldsymbol{x} = [\boldsymbol{x}_0^T \quad \boldsymbol{x}_1^T \quad \cdots \quad \boldsymbol{x}_{M-1}^T]^T$, $\boldsymbol{a}_{kp} = \begin{bmatrix} 1 & e^{j2\pi(\frac{p}{NT_s}+f_c)\frac{d\cos\theta_k}{c}} & \ldots & e^{j2\pi(\frac{p}{NT_s}+f_c)\frac{(M-1)d\cos\theta_k}{c}} \end{bmatrix}^T$, and $\boldsymbol{W}_p \in \mathbb{C}^{M\times MN}$ is given by

$$\boldsymbol{W}_p = \boldsymbol{I}_M \otimes \boldsymbol{e}_p^H \tag{5.64}$$

where $\boldsymbol{e}_p^H = \begin{bmatrix} 1 & e^{-j2\pi\frac{p}{N}} & \ldots & e^{-j2\pi\frac{(N-1)p}{N}} \end{bmatrix} \in \mathbb{C}^N$ and $\boldsymbol{I}_M$ is an $M \times M$ identity matrix.

The problem of spectral co-existence has been of great interest recently [22,41,43] and involves minimization of interference caused by radar transmission at victim communication receivers operating in the same frequency band. In this case, the beampattern of the transmit waveform is required to have nulls in these bands to prevent interference. For J communication receivers, we suppose that the jth communication receiver operating on a frequency band $B_j = [p_l^j, p_u^j]$, where p_l^j and p_u^j are the lower and upper normalized frequencies, respectively. We denote the desired (discrete) spectrum shape by $\hat{\boldsymbol{y}} = [\hat{y}_{-\frac{N}{2}}, \hat{y}_{-\frac{N}{2}+1}, \ldots, \hat{y}_{\frac{N}{2}-1}] \in \mathbb{C}^{N\times 1}$ defined as

$$\hat{y}_p = \begin{cases} 0 & \text{for } p \in B_j = [p_l^j, p_u^j], \qquad j = 1, 2, \ldots, J \\ \gamma & \text{otherwise.} \end{cases} \tag{5.65}$$

where γ is a scalar such that $\hat{\boldsymbol{y}}^H \boldsymbol{F}\boldsymbol{F}^H \hat{\boldsymbol{y}} = N$ and $\boldsymbol{F}$ is the DFT matrix. In SHAPE algorithm proposed by Rowe *et al.* [37], a least-squares fitting approach for the spectral shaping problem for SISO has been formulated by minimizing the following cost function:

$$\|\boldsymbol{F}^H \boldsymbol{x} - \hat{\boldsymbol{y}}\|_2^2 \tag{5.66}$$

We extend (5.66) for MIMO radar and employ it as a constraint in the optimization problem as follows:

$$\|(\boldsymbol{I}_M \otimes \boldsymbol{F}^H)(\boldsymbol{1}_M \otimes \hat{\boldsymbol{y}}) - \boldsymbol{x}\|_2^2 = \|\bar{\boldsymbol{F}}^H \bar{\boldsymbol{y}} - \boldsymbol{x}\|_2^2 \leq E_R \tag{5.67}$$

where $\boldsymbol{1}_M = [1, 1, \ldots, 1] \in \mathbb{R}^{M\times 1}$, $\bar{\boldsymbol{F}} = \boldsymbol{I}_M \otimes \boldsymbol{F}^H$, and $\bar{\boldsymbol{y}} = \boldsymbol{1}_M \otimes \hat{\boldsymbol{y}}$, and E_R is the maximum tolerable spectral error.

5.3.2.2 Beampattern design with interference control (BIC) under constant modulus

The optimization problem for beampattern design under the CMC and the SpecC can be formulated as the following matching problem:

$$\begin{aligned} \underset{\boldsymbol{x}}{\text{minimize}} \quad & \textstyle\sum_{k=1}^{K} \sum_{p=-\frac{N}{2}}^{\frac{N}{2}-1} [d_{kp} - |\boldsymbol{a}_{kp}^H \boldsymbol{W}_p \boldsymbol{x}|]^2 \\ \text{subject to } & |x_m(n)| = 1, \text{ for } m = 1, 2, \ldots, M \text{ and} \\ & \qquad n = 0, 1, \ldots, N-1 \\ & \|\bar{\boldsymbol{F}}^H \bar{\boldsymbol{y}} - \boldsymbol{x}\|_2^2 \leq E_R \end{aligned} \tag{5.68}$$

where $d_{kp} \in \mathbb{R}$ is the desired beampattern. These constraints are neither convex nor linear and it is well known in the literature that (5.68) is a hard non-convex problem even without the SpecC. The objective function can be rewritten as [20]

$$\sum_{k=1}^{K} \sum_{p=-\frac{N}{2}}^{\frac{N}{2}-1} |d_{kp} e^{j\phi_{kp}} - \boldsymbol{a}_{kp}^H \boldsymbol{W}_p \boldsymbol{x}|^2 \tag{5.69}$$

where $\phi_{kp} = \arg\{\boldsymbol{a}_{kp}^H \boldsymbol{W}_p \boldsymbol{x}\}$. This objective function can be optimized by an iterative method [20,49,50] which first optimizes $\boldsymbol{x}$ for a fixed ϕ_{kp} and then finds the optimal ϕ_{kp} for the fixed $\boldsymbol{x}$. For a fixed ϕ_{kp}, the objective function can be further simplified in a quadratic form as follows [10]:

$$\sum_{k=1}^{K} \sum_{p=-\frac{N}{2}}^{\frac{N}{2}-1} |d_{kp} e^{j\phi_{kp}} - \boldsymbol{a}_{kp}^H \boldsymbol{W}_p \boldsymbol{x}|^2 = \sum_{p} \|\boldsymbol{d}_p - \boldsymbol{A}_p \boldsymbol{W}_p \boldsymbol{x}\|_2^2 \tag{5.70}$$

$$= \boldsymbol{x}^H \boldsymbol{P} \boldsymbol{x} - \boldsymbol{q}^H \boldsymbol{x} - \boldsymbol{x}^H \boldsymbol{q} + r \tag{5.71}$$

Moreover, using $\boldsymbol{x}^H \boldsymbol{x} = 2L$, the SpecC can also be simplified as

$$\Re\{\bar{\boldsymbol{y}}^H \bar{\boldsymbol{F}} \boldsymbol{x}\} \geq (1 - E_R/2)L \tag{5.72}$$

The optimization problem (5.68) for a fixed ϕ_{kp} is equivalent to the following problem in real variables:

$$\begin{aligned} \underset{\boldsymbol{s}}{\text{minimize}} \quad & \boldsymbol{s}^T (\boldsymbol{R} + \lambda \boldsymbol{I}) \boldsymbol{s} \\ \text{subject to } & \boldsymbol{s}^T \boldsymbol{E}_l \boldsymbol{s} = 1, \quad l = 1, 2, \ldots, L \\ & \bar{\boldsymbol{s}}^T \boldsymbol{s} \geq (1 - E_R/2)L \end{aligned} \tag{5.73}$$

where λ is an arbitrary positive number, $\bar{\boldsymbol{s}} = [\Re\{\bar{\boldsymbol{F}}^H \bar{\boldsymbol{y}}\}^T \ \Im\{\bar{\boldsymbol{F}}^H \bar{\boldsymbol{y}}\}^T \ 0]^T$, $\boldsymbol{R} = \begin{bmatrix} \boldsymbol{G} & -\boldsymbol{t} \\ -\boldsymbol{t}^T & r \end{bmatrix}$, $\boldsymbol{s} = [\Re\{\boldsymbol{x}^T\} \ \Im\{\boldsymbol{x}^T\} \ 1]^T$, and $\boldsymbol{t} = [\Re\{\boldsymbol{q}^T\} \ \Im\{\boldsymbol{q}^T\}]^T$.

Sequence of closed form solutions

Though (5.73) is a minimization problem with a convex objective function since $\boldsymbol{R}$ is positive semi-definite, it is still non-convex because of the CMC, $\boldsymbol{s}^T\boldsymbol{E}_l\boldsymbol{s} = 1$. It can also be solved by a sequential approach which involves solving a sequence of convex problems. Let us consider the following sequence of constrained quadratic programming (QP) where the nth QP is given by

$$(CP)^{(n)} \begin{cases} \underset{\boldsymbol{s}}{\text{minimize}} & \boldsymbol{s}^T(\boldsymbol{R} + \lambda\boldsymbol{I})\boldsymbol{s} \\ \text{subject to} & \boldsymbol{B}^{(n)}\boldsymbol{s} = \boldsymbol{1} \\ & \bar{\boldsymbol{s}}^{(n)T}\boldsymbol{s} \geq (1 - E_R/2)L \end{cases} \tag{5.74}$$

where $\bar{\boldsymbol{s}}^{(n)}$ is given by:

$$\bar{\boldsymbol{s}}^{(n)} = \begin{bmatrix} \Re\{(\bar{\boldsymbol{F}}^H\bar{\boldsymbol{y}}) \odot e^{\{j\arg(\boldsymbol{x}^{(n-1)}) - \arg(\bar{\boldsymbol{F}}^H\bar{\boldsymbol{y}})\}}\} \\ \Im\{(\bar{\boldsymbol{F}}^H\bar{\boldsymbol{y}}) \odot e^{\{j\arg(\boldsymbol{x}^{(n-1)}) - \arg(\bar{\boldsymbol{F}}^H\bar{\boldsymbol{y}})\}}\} \\ 0 \end{bmatrix} \tag{5.75}$$

and $\boldsymbol{B}^{(n)} = [\boldsymbol{b}_1^{(n)}, \boldsymbol{b}_2^{(n)}, \ldots, \boldsymbol{b}_{L+1}^{(n)}]^T \in \mathbb{R}^{(L+1)\times(2L+1)}$ such that the line defined by $\boldsymbol{b}_l^{(n)T}\boldsymbol{s} = 1$ is a tangent to the circle $\boldsymbol{s}^T\boldsymbol{E}_l\boldsymbol{s} = 1$ for $l = 1, 2, \ldots, L$. Specifically, $\boldsymbol{b}_l$ is given by

$$\boldsymbol{b}_l^{(n)}(i) = \begin{cases} \cos(\gamma_l^{(n)}) & \text{if } i = l \\ \sin(\gamma_l^{(n)}) & \text{if } i = l + L \\ 0 & \text{otherwise.} \end{cases} \tag{5.76}$$

for $l = 1, \ldots, L$ and $\boldsymbol{b}_{L+1}^{(n)} = [0, \ldots, 0, 1]^T$ where $\gamma_l^{(n)} = 2\arg(x_l^{(n-1)}) - \gamma_l^{(n-1)}$ and $x_l^{(n)}$ is the lth elements of $\boldsymbol{x}^{(n)}$ which is the complex version of the optimal solution of (5.74), $\boldsymbol{s}^{(n)}$, that is, $x_l^{(n)} = s_l^{(n)} + js_{l+L}^{(n)}$ and conversely $\boldsymbol{s}^{(n)} = [\Re\{\boldsymbol{x}^{(n)}\}^T \Im\{\boldsymbol{x}^{(n)}\}^T\ 1]^T$. Note that, the term $e^{\{j\arg(\boldsymbol{x}^{(n-1)}) - \arg(\bar{\boldsymbol{F}}^H\bar{\boldsymbol{y}})\}}$ in (5.75) depends on the argument $\boldsymbol{x}^{(n-1)}$, which changes $\boldsymbol{s}^{(n)}$ in each iteration.

Although the problem (5.74) does not result in a constant modulus solution, a sequence of such problems (in the index n) ensures a non-increasing sequence of cost function values, such that the sequence of the corresponding optimal solutions converges to constant modulus for large enough λ [51]. It is shown that the constraints of $CP^{(n)}$ in (5.74) are adjusted so that the feasible set of $CP^{(n)}$ includes $\boldsymbol{x}^{(n-1)}$ [23]. This means that the feasible set of each iteration is updated such that it contains the optimal solution of the optimization problem at the previous iteration step. If $|\boldsymbol{x}^{(n)}| = \boldsymbol{1}$, then the constraints of the next problem $CP^{(n+1)}$ are the same as problem $CP^{(n)}$, which means $\boldsymbol{x}^{(n+1)} = \boldsymbol{x}^{(n)}$ and, hence, the algorithm converges. It has been also shown that the cost function sequence is in fact non-increasing and converges. This procedure is visually illustrated in Figure 5.9.

The optimization problem (5.74) is a convex quadratic minimization with linear equality constraints. Using the optimality conditions for problem (5.74), the sufficient and necessary KKT conditions [30] of (5.74) give the following:

$$2(\boldsymbol{R} + \lambda\boldsymbol{I})\boldsymbol{s}^{(n)} + \boldsymbol{B}^{(n)T}\boldsymbol{\nu}^{(n)} - \mu^{(n)}\bar{\boldsymbol{s}} = 0 \tag{5.77}$$

$$\boldsymbol{B}^{(n)}\boldsymbol{s}^{(n)} = \boldsymbol{1} \tag{5.78}$$

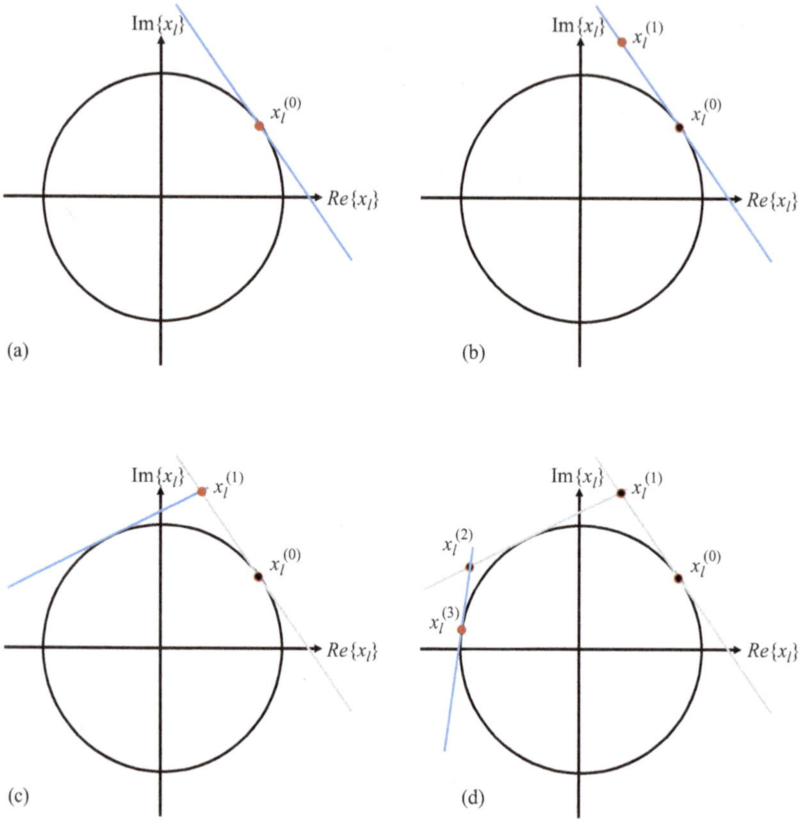

Figure 5.9 Illustration of the successive solutions of (5.74) *for the lth element of the vector* $\boldsymbol{x}^{(n)}$*, i.e.,* $x_l^{(n)}$*. The current feasible set is shown via a blue line. (a) The initial problem* $CP^{(1)}$*, the initial feasible set is the blue line. (b) Solution of problem* $CP^{(1)}$ *lies on the initial feasible set. (c) The new adjusted feasible set (contains* $x_l^{(1)}$*) in blue, the previous feasible set in gray. (d) The converged solution now lies on the constant modulus.*

$$\mu^{(n)}\left(\bar{\boldsymbol{s}}^{(n)T}\boldsymbol{s}^{(n)} - (1 - E_R/2)L\right) = 0 \tag{5.79}$$

$$\bar{\boldsymbol{s}}^{(n)T}\boldsymbol{s}^{(n)} - (1 - E_R/2)L \geq 0 \tag{5.80}$$

$$\mu^{(n)} \geq 0 \tag{5.81}$$

Solving the equations and the inequality above gives a closed form solution as following:

$$\boldsymbol{s}^{(n)} = \begin{cases} \bar{\boldsymbol{R}}^{-1}\boldsymbol{B}^{(n)T}\left(\boldsymbol{B}^{(n)}\bar{\boldsymbol{R}}^{-1}\boldsymbol{B}^{(n)T}\right)^{-1}\mathbf{1} & \text{if } \bar{\boldsymbol{s}}^{(n)T}\boldsymbol{s}^{(n)} - (1 - E_R/2)L \geq 0 \\ \mu^{(n)}\bar{\boldsymbol{R}}^{-1}(\boldsymbol{I} - \boldsymbol{B}^{(n)T}\hat{\boldsymbol{R}}\boldsymbol{B}^{(n)}\bar{\boldsymbol{R}}^{-1})\bar{\boldsymbol{s}}^{(n)} + \hat{\boldsymbol{s}}^{(n)} & \text{otherwise} \end{cases} \tag{5.82}$$

where $\hat{\boldsymbol{R}} = \left(\boldsymbol{B}^{(n)}\bar{\boldsymbol{R}}^{-1}\boldsymbol{B}^{(n)^T}\right)^{-1}$, $\mu^{(n)} = \frac{1}{\alpha^{(n)}}\left(\bar{\boldsymbol{s}}^{(n)T}\hat{\boldsymbol{s}}^{(n)} - (1 - E_R/2)L\right)$, and

$$\alpha^{(n)} = -\begin{bmatrix}\bar{\boldsymbol{s}}^{(n)}\\ \boldsymbol{0}\end{bmatrix}^T \begin{bmatrix}\bar{\boldsymbol{R}} & \boldsymbol{B}^{(n)^T}\\ \boldsymbol{B}^{(n)} & \boldsymbol{0}\end{bmatrix}^{-1} \begin{bmatrix}\bar{\boldsymbol{s}}^{(n)}\\ \boldsymbol{0}\end{bmatrix} \tag{5.83}$$

Nullforming beampattern design

The beampattern design method described earlier can also be applied the nullforming beampattern design problem which is a special case of the full beampattern design. The goal of nullforming beampattern design is to form a beampattern with nulls in desired directions and the optimization problem is given by

$$\begin{aligned} \underset{\boldsymbol{x}}{\text{minimize}} \quad & \boldsymbol{x}^H \boldsymbol{V}\boldsymbol{x} \\ \text{subject to} \quad & |\boldsymbol{x}(k)|^2 \leq 1/(MN) \\ & |x_m(n)| = 1, \text{ for } m = 1, 2, \ldots, M \text{ and } n = 0, 1, \ldots, N-1 \\ & \|\bar{\boldsymbol{F}}^H\bar{\boldsymbol{y}} - \boldsymbol{x}\|_2^2 \leq E_R \end{aligned} \tag{5.84}$$

where $\boldsymbol{V} = \sum_{p=-\frac{N}{2}}^{\frac{N}{2}-1} \boldsymbol{W}_p^H \boldsymbol{A}_p^H \boldsymbol{A}_p \boldsymbol{W}_p$. Since $\boldsymbol{V}$ is positive semidefinite and there are no linear terms in the objective function, the solution can be obtained by the algorithm.

5.3.2.3 Analytical and experimental results

Complexity and convergence analysis

Based on the computational cost of each iteration, the overall computational complexity is $\mathcal{O}(FL^{2.373}) - \mathcal{O}(FL^3)$ [52] where F is the total number of iterations. It also converges to a finite value $\boldsymbol{s}^\star$ since the sequence $\{g(\boldsymbol{s}^{(n)})\}_{n=0}^{\infty}$ is non-increasing and bounded $g(\boldsymbol{s}) \geq 0$ where $g(\boldsymbol{s}) = \boldsymbol{s}^T(\boldsymbol{R} + \lambda\boldsymbol{I})\boldsymbol{s}$. It can be easily shown by the following inequality:

$$\boldsymbol{s}^{(n)T}(\boldsymbol{R} + \lambda\boldsymbol{I})\boldsymbol{s}^{(n)} \leq \boldsymbol{s}^{(n-1)T}(\boldsymbol{R} + \lambda\boldsymbol{I})\boldsymbol{s}^{(n-1)} \tag{5.85}$$

The inequality always holds since $\boldsymbol{s}^{(n-1)}$ belongs to the feasible set of $CP^{(n)}$ and $\boldsymbol{s}^{(n)}$ is the optimal solution of $CP^{(n)}$ which is a convex problem.

Numerical results

Figure 5.10 shows the results for nullforming beampattern of BIC, POVMM [42], SHAPE [37], and JDO SSPARC [53]. POVMM performs nullforming beampattern design by optimizing phases of the waveform under the CMC but no SpecC is involved. The SHAPE algorithm is a computationally efficient method of designing sequences with desired spectrum shapes. In particular, the spectral shape is optimized as a cost function subject to the CMC but the resulting beampattern is an outcome (not explicitly controlled). JDO SSPARC is an approach for beamforming that maximizes the signal power through the forward channels while simultaneously minimizes the response at the co-channels. JDO SSPARC does not control the spectral shape of the waveform in the frequency domain. It is shown that BIC, POVMM, and JDO SSPARC achieve nulls in the desired angles, 10°, 40°, and 120°, i.e., desired spatial control in Figure 5.10(a). SHAPE lacks a spatial control component by virtue of its design. Note that the forward channel for JDO SSPARC is set to be θ = [80° to 100°]; however,

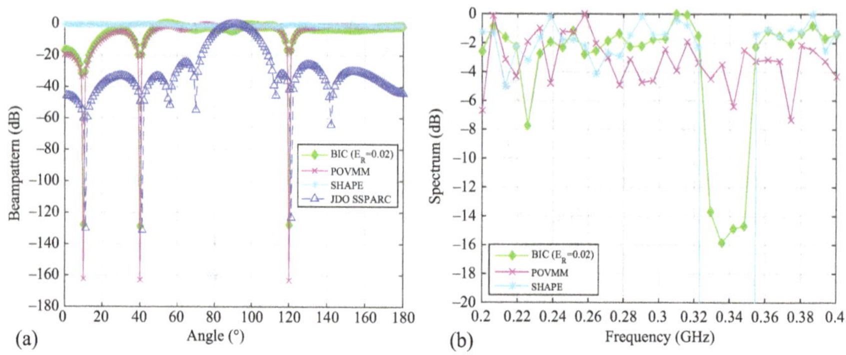

Figure 5.10 Nullforming beampattern design. (a) Beampattern versus angle. (b) Spectrum versus frequency

unlike the other methods, the resulting waveform is non-constant modulus. On the other hand, Figure 5.10(b) plots the spectrum versus the frequency. Here, BIC and SHAPE effectively suppress the energy in the frequency bands where the transmission should be mitigated. Unsurprisingly, POVMM do not provide the desired suppression in the frequency bands of interest because it is not designed for the same. In summary, only the BIC enables the desired spatio-spectral control.

For wideband beampattern design, we place a notch in the band 910–932 MHz and consider the following desired transmit beampattern

$$d(\theta,f) = \begin{cases} 1 & \theta = [95^\circ, 120^\circ] \\ 0 & \text{Otherwise.} \end{cases} \tag{5.86}$$

in Figure 5.11. It shows the angle–frequency plot of the beampattern for WBFIT method [20] (no SpecC) and BIC with the SpecC ($E_R = 0.01$). The BIC method is able to keep the energy of the waveform in particular frequency band low enough as well as achieve higher suppression at the undesired angles compared to WBFIT.

5.3.3 *Quartic gradient descent for tractable radar ambiguity function shaping*

The problem of minimizing the disturbance power at the output of the matched filter in a single antenna cognitive radar set-up improves the SINR for the radar. The aforementioned disturbance power can be shown to be an expectation of the slow-time ambiguity function (STAF) of the transmitted waveform over range–Doppler bins of interest. The design problem is known to be a non-convex quartic function of the transmit radar waveform. This STAF shaping problem becomes even more challenging in the presence of practical constraints on the transmit waveform such as the

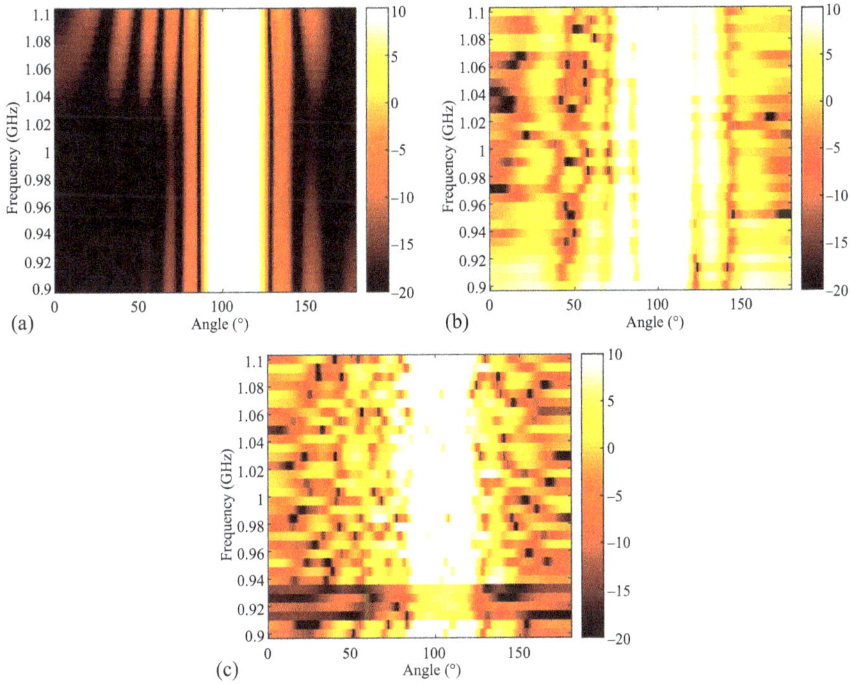

Figure 5.11 *Plot of the beampattern. (a) Unconstrained, (b) WBFIT method, and (c) BIC.*

CMC. Most existing approaches address the aforementioned challenges by suitably modifying or relaxing the design cost function and/or the CMC.

In this part, we discuss a solution that involves direct optimization over the non-convex complex circle manifold, i.e., the CMC set. This solution uses a new update strategy (quartic-gradient-descent (QGD)) that computes an *exact* gradient of the quartic cost and invokes principles of optimization over manifolds towards an iterative procedure with guarantees of monotonic cost function decrease and convergence [38]. Experimentally, QGD outperforms state-of-the-art approaches for shaping the AF under the CMC while being computationally less expensive.

5.3.3.1 System model and problem formulation

We consider a monostatic single-input–single-output (SISO) radar system which transmits a coherent burst of slow-time‡ coded pulses dented by

$$\boldsymbol{x} = [x(0), x(1), \ldots, x(N-1)]^T \in \mathbb{C}^N \tag{5.87}$$

‡For more details about the slow-time and fast-time coding, the reader is advised to see [11,25,54].

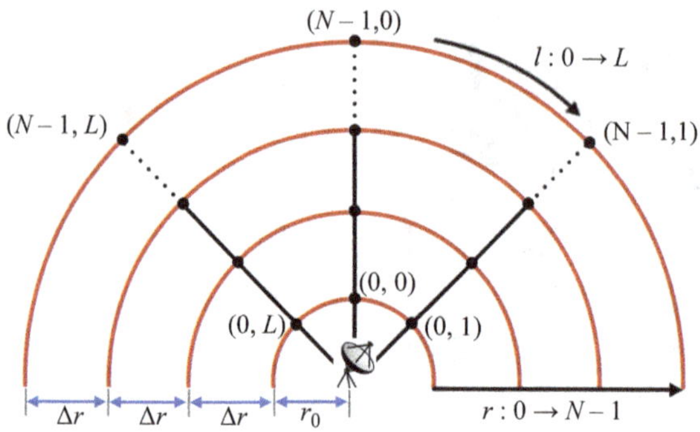

Figure 5.12 Range-azimuth bins, where the target is located at the point $(0,0)$*. The distance* r_0 *is assumed to be* $r_0 \leq r_{max} = \frac{cT_r}{2}$*, where* r_{max} *is the maximum unambiguous range defines the maximum distance to locate a target, and* $\Delta r = r_{max}$*.*

The radar system illuminates the environment by sending N coherent burst of slow-time coded pulses $\boldsymbol{x}$. The signal at the receiver is down-converted to baseband, undergoes a pulse-matched filtering operation, and then is sampled. The received vector $\boldsymbol{v} = [v(0), v(1), \ldots, v(N-1)]^T \in \mathbb{C}^N$ of observations from the range-azimuth cell under consideration is given by

$$\boldsymbol{v} = \alpha_T \boldsymbol{x} \odot \boldsymbol{p}(\nu_{d_T}) + \boldsymbol{d}(\boldsymbol{x}) + \boldsymbol{n} \tag{5.88}$$

where α_T is a complex parameter accounting for channel propagation and back-scattering effects from the target within the range-azimuth bin of interest, $\boldsymbol{p}(\nu_{d_T}) = [1, e^{j2\pi\nu_{d_T}}, \ldots, e^{j2\pi(N-1)\nu_{d_T}}]^T$, ν_{d_T} is the normalized target Doppler frequency, $\boldsymbol{d}(\boldsymbol{x})$ is the vector of interfering echo samples, and $\boldsymbol{n}$ is the filtered noise vector with $\mathbb{E}[\boldsymbol{n}] = 0$ and $\mathbb{E}[\boldsymbol{n}\boldsymbol{n}^H] = \sigma_n^2 \boldsymbol{I}$. According to [17], the vector $\boldsymbol{d}(\boldsymbol{x})$ captures the returns from different N_t interfering scatterers located at different range-azimuth bins[§] (r,l), where $r \in \{0, 1, \ldots, N-1\}$, $l \in \{0, 1, \ldots, L\}$ (as illustrated in Figure 5.12) and $L+1$ is the number of discrete azimuth sectors. This vector $\boldsymbol{d}(\boldsymbol{x})$ can be expressed as

$$\boldsymbol{d}(\boldsymbol{x}) = \sum_{i=1}^{N_t} \rho_i \boldsymbol{J}^{r_i} \left(\underbrace{\boldsymbol{x} \odot \boldsymbol{p}(\nu_{d_i})}_{\boldsymbol{c}_{\nu_{d_i}}} \right) = \sum_{i=1}^{N_t} \rho_i \boldsymbol{J}^{r_i} \boldsymbol{c}_{\nu_{d_i}} \tag{5.89}$$

where $r_i \in \{0, 1, \ldots, N-1\}$ is the range position, ρ_i is the echo complex amplitude, ν_{d_i} and $\boldsymbol{c}_{\nu_{d_i}} = \boldsymbol{x} \odot \boldsymbol{p}(\nu_{d_i})$ are the normalized Doppler frequency and the signature[‖]

[§]Without loss of generality, the target of interest can be assumed to be located at the range-azimuth bin $(0,0)$ and scatterers are located at further range bins [55,56].

[‖]This vector $\boldsymbol{c}_\nu$ will be used in this chapter to represent the signature of any object with a Doppler frequency ν, i.e., $\boldsymbol{c}_\nu = \boldsymbol{x} \odot \boldsymbol{p}(\nu)$.

of the ith scatterers, respectively. $\boldsymbol{J}^r$ is an $N \times N$ shift matrix and $\forall\, r \in \{-N+1, \ldots, 0, \ldots, N-1\}$ it is denoted as:

$$\boldsymbol{J}^r(l_1, l_2) = \begin{cases} 1, & \text{if } l_1 - l_2 = r \\ 0, & \text{otherwise} \end{cases} \quad (l_1, l_2) \in \{1, \ldots, N\}^2 \tag{5.90}$$

with $\boldsymbol{J}^{-r} = (\boldsymbol{J}^r)^T$. Combining Equations (5.88) and (5.89), the output of the matched filter to the target signature $\boldsymbol{c}_{\nu_{d_T}} = \boldsymbol{x} \odot \boldsymbol{p}(\nu_{d_T})$ is given by

$$\boldsymbol{c}_{\nu_{d_T}}^H \boldsymbol{v} = \alpha_T \|\boldsymbol{x}\|_2^2 + \text{Dist}\,(\boldsymbol{x}, \nu, r) \tag{5.91}$$

where Dist $(\boldsymbol{x}, \nu, r)$ represents the disturbance at the output of the match filter, i.e.,

$$\text{Dist}\,(\boldsymbol{x}, \nu, r) = \underbrace{\boldsymbol{c}_{\nu_{d_T}}^H \boldsymbol{n}}_{\text{noise}} + \underbrace{\sum_{i=1}^{N_t} \rho_i \boldsymbol{c}_{\nu_{d_T}}^H \boldsymbol{J}^{r_i} \boldsymbol{c}_{\nu_{d_i}}}_{\text{interference}} \tag{5.92}$$

Assuming that $\boldsymbol{n}$ is uncorrelated with $\boldsymbol{d}(\boldsymbol{x})$, the energy of the disturbances in the match filter output can be expressed as:

$$\mathbb{E}[|\text{Dist}\,(\boldsymbol{x}, \nu, r)\,|^2] = \mathbb{E}\left[\left|\boldsymbol{c}_{\nu_{d_T}}^H \boldsymbol{n}\right|^2\right] + \mathbb{E}\left[\left|\sum_{i=1}^{N_t} \rho_i \boldsymbol{c}_{\nu_{d_T}}^H \boldsymbol{J}^{r_i} \boldsymbol{c}_{\nu_{d_i}}\right|^2\right] \tag{5.93}$$

Problem formulation

In [17], the normalized Doppler frequencies ν_{d_i} are expressed in terms of the difference with respect to ν_{d_T}, and the normalized Doppler frequency range $[-\frac{1}{2}, \frac{1}{2}]$ is divided into N_ν bins. Consequently, the normalized frequencies ν_{d_i} can be represented by the discrete frequencies $\nu_h = -\frac{1}{2} + \frac{h}{N_\nu}$, $h = 0, \ldots, N_\nu - 1$. Using this representation and approximating the statistical expectation in (5.93) by the sample mean, the disturbances in the match filter output will be[¶]

$$\mathbb{E}\left[|\text{Dist}(\boldsymbol{x}, \nu, r)|^2\right] = \sum_{r=0}^{N-1} \sum_{h=0}^{N_\nu - 1} p(r, h) \|\boldsymbol{x}\|_2^2 g_{\boldsymbol{x}}(r, \nu_h) + \sigma_n^2 \|\boldsymbol{x}\|_2^2 \tag{5.94}$$

where $p(r, h)$ is interference map for the range–Doppler bin (r, ν_h) and $g_{\boldsymbol{x}}\,(r, \nu_h)$ is the STAF of the transmitted code $\boldsymbol{x}$ defined as:

$$g_{\boldsymbol{x}}(r, \nu_h) = \frac{1}{\|\boldsymbol{x}\|_2^2} \left|\boldsymbol{x}^H \boldsymbol{J}^r \boldsymbol{c}_\nu\right|^2 \tag{5.95}$$

with $r \in \{0, 1, \ldots, N-1\}$ and $\nu_h = -\frac{1}{2} + \frac{h}{N_\nu}$, $h = 0, \ldots, N_\nu - 1$. Given a (r, ν_h) pair, the STAF $g_{\boldsymbol{x}}(r, \nu_h)$ gives the range–Doppler response from an interfering patch corresponding with a Doppler frequency of ν_h located r time-lag away. As mentioned earlier, the goal is to design a suitable radar waveform $\boldsymbol{x}$ in order to shape the STAF to match a desired range–Doppler response (*shaping the STAF is equivalent to minimizing the disturbances in the match filter output in (5.94)* [17]) under the CMC, i.e.,

[¶]A detailed derivation for (5.94) can be found in [17].

$|x(n)| = 1,\ n = 1, 2, \ldots, N$. With this constraint, the quantity $\|\boldsymbol{x}\|_2^2$ in (5.94) is constant and, hence, the disturbance in the output of the matched filter will be minimized using the following cost function [11]:

$$\phi(\boldsymbol{x}) = \sum_{i=1}^{M} \boldsymbol{x}^H \boldsymbol{C}_i \boldsymbol{x}\boldsymbol{x}^H \boldsymbol{C}_i^H \boldsymbol{x}. \tag{5.96}$$

where $M = N \times N_\nu$, and i is a one-to-one mapping index with (r, h), i.e., for each pair $(r, h) \in \{0, \ldots, N-1\} \times \{0, \ldots, N_\nu - 1\}$, we have $i = rN_\nu + h \in \{1, \ldots, M = N \times N_\nu\}$ and the matrix $\boldsymbol{C}_i$ is defined as $\boldsymbol{C}_i = \boldsymbol{C}_{(r,h)} = \sqrt{p(r,h)}\boldsymbol{J}^r \text{diag}(\boldsymbol{p}(\nu_h))$. Then the optimization problem for shaping the STAF under the CMC will be the following complex quartic minimization problem:

$$\begin{aligned} &\underset{\boldsymbol{x}}{\text{minimize}}\ \ \phi(\boldsymbol{x}) = \textstyle\sum_{i=1}^{M} \boldsymbol{x}^H \boldsymbol{C}_i \boldsymbol{x}\boldsymbol{x}^H \boldsymbol{C}_i^H \boldsymbol{x} \\ &\text{subject to} \qquad\qquad \boldsymbol{x} \in \mathcal{S}^N \end{aligned} \tag{5.97}$$

where $\mathcal{S}^N$ is the complex circle manifold (formal manifold terminology for the CMC set) defined as $\mathcal{S}^N \triangleq \{\boldsymbol{x} \in \mathbb{C}^N : |x(n)| = 1,\ n = 1, 2, \ldots, N\}$.

It has been shown in [17] that the optimization problem (5.97) is NP-hard. The authors in [17] approach this problem via a polynomial time waveform optimization procedure based on maximum block improvement (MBI) method. In their work, the CMC is enforced by employing a randomization strategy [57] which leads to an effective solution but one that has high computational complexity. In [40], a combination of MM (to majorize the quartic cost by a quadratic) and coordinate descent methods is used. Also, a related approach for a unimodular sequence design to minimize the ISL based on the phase-only conjugate gradient and phase-only Newton's method is proposed in [39]. In general, the CMC is extracted in different parts of the optimization but a direct optimization over the non-convex CMC remains elusive.

We invoke principles of optimization over non-convex manifolds to address this open challenge. Our focus is on developing a gradient-based method, which can enable descent on the complex circle manifold while maintaining feasibility. First, the cost function in (5.97) can be altered by adding the term $\gamma \boldsymbol{x}^H \boldsymbol{x}\boldsymbol{x}^H \boldsymbol{x}$, i.e.,

$$\begin{aligned} &\underset{\boldsymbol{x}}{\text{minimize}}\ \ \bar{\phi}(\boldsymbol{x}) = \textstyle\sum_{i=1}^{M} \boldsymbol{x}^H \boldsymbol{C}_i \boldsymbol{x}\boldsymbol{x}^H \boldsymbol{C}_i^H \boldsymbol{x} + \gamma \boldsymbol{x}^H \boldsymbol{x}\boldsymbol{x}^H \boldsymbol{x} \\ &\text{subject to} \qquad\qquad \boldsymbol{x} \in \mathcal{S}^N \end{aligned} \tag{5.98}$$

where $\gamma \geq 0$ (it will be used later in Lemma 1 to control convergence). Since the problem (5.98) enforces the CMC, the term $\gamma \boldsymbol{x}^H \boldsymbol{x}\boldsymbol{x}^H \boldsymbol{x}$ is constant (i.e. $\gamma \boldsymbol{x}^H \boldsymbol{x}\boldsymbol{x}^H \boldsymbol{x} = \gamma N^2$). Hence, the optimal solution of the problem (5.97) and the optimal solution of the problem (5.98) are identical for any $\gamma \geq 0$.

5.3.3.2 QGD algorithm

The goal is to find an efficient method to deal with the non-convex feasible set of the problem (5.97) (or (5.98)), i.e., the complex circle manifold. Many classical line-search algorithms from unconstrained nonlinear optimization in $\mathbb{C}^N$ such as gradient descent can be used in optimization over manifolds but with some modifications. In general, line-search methods in $\mathbb{C}^N$ are based on the following update formula [58]:

$$\boldsymbol{x}_{k+1} = \boldsymbol{x}_k + \beta_k \boldsymbol{\eta}_k \tag{5.99}$$

where $\boldsymbol{\eta}_k \in \mathbb{C}^N$ is the *search direction* and $\beta_k \in \mathbb{R}$ is the *step size*. The most obvious choice for the search direction is the steepest descent direction which is the negative gradient of $\bar{\phi}(\boldsymbol{x})$ at the point $\boldsymbol{x}_k$, i.e., $\boldsymbol{\eta}_k = -\nabla_{\boldsymbol{x}}\bar{\phi}(\boldsymbol{x}_k)$ [58,59]. In the literature [60,61], the following high-level structure is suggested:

- The descent will be performed on the manifold itself rather than in the Euclidean space by means of the *intrinsic* search direction. The *intrinsic* search direction is a vector in the tangent space $\mathcal{T}_{\boldsymbol{x}_k}\mathcal{M}$ to the manifold $\mathcal{M}$ at the point $\boldsymbol{x}_k \in \mathcal{M}$. This *intrinsic* search direction can be obtained by projecting the standard search direction $\boldsymbol{\eta}_k = -\nabla_{\boldsymbol{x}}\bar{\phi}(\boldsymbol{x}_k)$ onto $\mathcal{T}_{\boldsymbol{x}_k}\mathcal{M}$ by means of a projection operator $\mathrm{Proj}_{\mathcal{T}_{\boldsymbol{x}_k}\mathcal{M}}\left(\boldsymbol{\eta}_k\right)$.
- The update is performed on the tangent space along the direction of $\mathrm{Proj}_{\mathcal{T}_{\boldsymbol{x}_k}\mathcal{M}}\left(\boldsymbol{\eta}_k\right)$ with a step β, i.e.,

$$\bar{\boldsymbol{x}}_k = \boldsymbol{x}_k + \beta \mathrm{Proj}_{\mathcal{T}_{\boldsymbol{x}_k}\mathcal{M}}(\boldsymbol{\eta}_k) \in \mathcal{T}_{\boldsymbol{x}_k}\mathcal{M}$$

- Since $\bar{\boldsymbol{x}}_k \notin \mathcal{M}$, it will be mapped back to the manifold by the means of a *retraction* operator, $\boldsymbol{x}_{k+1} = \mathrm{Ret}\left(\bar{\boldsymbol{x}}_k\right)$.

For many manifolds, the projection $\mathrm{Proj}_{\mathcal{T}_{\boldsymbol{x}_k}\mathcal{M}}(.)$ and retraction $\mathrm{Ret}(.)$ operators admit a closed form. Interested readers may refer to [60] for more details.

For the manifold under interest, i.e., the complex circle manifold, [11] developed a framework for the optimization over this manifold. Consequently, the problem of shaping the STAF over CMC defined in (5.98) can be solved by utilizing the aforementioned framework. Precisely, at the kth iteration, (5.98) will be solved iteratively using the following steps (illustrated visually in Figure 5.13):

1. A projection of the search direction $\boldsymbol{\eta}_k = -\nabla_{\boldsymbol{x}}\bar{\phi}(\boldsymbol{x}_k)$ onto the tangent space of the manifold at the point $\boldsymbol{x}_k$, $\mathcal{T}_{\boldsymbol{x}_k}\mathcal{S}^N$, using

$$\boldsymbol{P}_{\mathcal{T}_{\boldsymbol{x}_k}\mathcal{S}^N}(\boldsymbol{\eta}_k) = \boldsymbol{\eta}_k - \mathrm{Re}\{\boldsymbol{\eta}_k^* \odot \boldsymbol{x}_k\} \odot \boldsymbol{x}_k \tag{5.100}$$

2. A descent on this tangent space to update the current value of $\boldsymbol{x}_k$ on the tangent space $\mathcal{T}_{\boldsymbol{x}_k}\mathcal{S}^N$ as

$$\bar{\boldsymbol{x}}_k = \boldsymbol{x}_k + \beta \boldsymbol{P}_{\mathcal{T}_{\boldsymbol{x}_k}\mathcal{S}^N}\left(\boldsymbol{\eta}_k\right) \tag{5.101}$$

3. A retraction of this update to $\mathcal{S}^N$ by using $\boldsymbol{R}(\boldsymbol{w}) = \boldsymbol{w} \odot \frac{1}{|\boldsymbol{w}|}$ as

$$\boldsymbol{x}_{k+1} = \boldsymbol{R}\left(\bar{\boldsymbol{x}}_k\right) \tag{5.102}$$

where $\odot$ is the element-wise product and $|\boldsymbol{x}_k|$ is a vector of element-wise absolute values of $\boldsymbol{x}_k$, i.e., $|\boldsymbol{x}_k| = [|x_k(1)|\ |x_k(2)|\ \ldots\ |x_k(N)|]^T$.

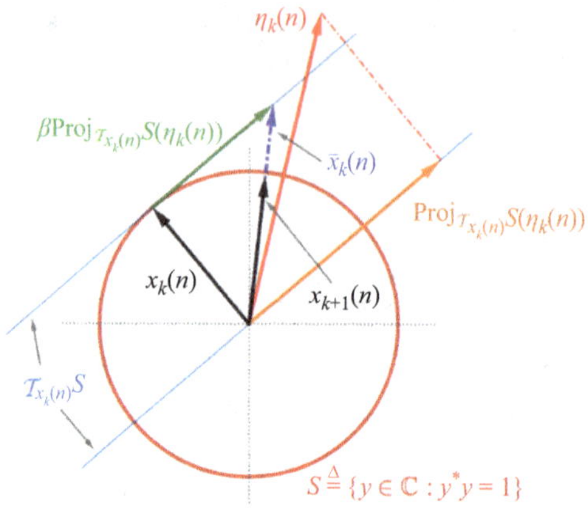

Figure 5.13 Illustration of the update of $x_{k+1}(n)$ starting from $x_k(n)$, where $x_k(n)$ and $\eta_k(n)$ are the nth elements of the vectors $\boldsymbol{x}_k$ and $\boldsymbol{\eta}_k$, respectively

The algorithm utilizing these steps to solve P_2 is named as QGD. It is shown in [11] that the gradient of the quartic cost in (5.98) is

$$\nabla(\bar{\phi}(\boldsymbol{x})) = 2\left(\sum_{i=1}^{M} \boldsymbol{C}_i \boldsymbol{x}\boldsymbol{x}^H \boldsymbol{C}_i^H \boldsymbol{x} + \boldsymbol{C}_i^H \boldsymbol{x}\boldsymbol{x}^H \boldsymbol{C}_i \boldsymbol{x}\right) + 4\gamma N \boldsymbol{x} \tag{5.103}$$

Using optimization over the complex circle manifold along with the gradient in Equation (5.103), the QGD algorithm with Armijo line search method is formally described in Algorithm 1.

5.3.3.3 Analytical and experimental results

Convergence analysis

The cost function in (5.98) is quartic w.r.t. $\boldsymbol{x}$ and, hence, finding a condition on the step size μ to ensure the monotonic decrease in the cost function during the descent step on the tangent space $\mathcal{T}_{\boldsymbol{x}_k}\mathcal{S}^N$ of $\mathcal{S}^N$ at $\boldsymbol{x}_k$ is a challenging task. Empirically, a small step size shows good results (monotone decrease in the cost) during this step. Instead of working with a fixed step size, we can employ well-known backtracking line search methods that produce a variable step size that ensures the reduction in the cost. One of these methods is Armijo line search [60,62]. In [62], Proposition 1.2.1 states that for a gradient method (such as steepest descent method) with a step size chosen by the Armijo method, every limit point of the generated sequence is a stationary point. In the setup, the Armijo line search will be used to ensure the reduction of the tangent space (an affine set) and, hence, the result from the aforementioned proposition can be utilized here. In other words, using Armijo line search method will produce a point on the tangent space of $\mathcal{S}^N$ at $\boldsymbol{x}_k$ with an improvement on the cost, i.e., $\bar{\phi}(\boldsymbol{x}_k) \geq \bar{\phi}(\bar{\boldsymbol{x}}_k)$. Now, the point $\bar{\boldsymbol{x}}_k$ will be on the tangent space and it will be retracted to the complex circle manifold, hence, we need to investigate the effect of this operator on the change

---

Algorithm 1: QGD with Armijo line search

1: **Inputs:** The interference map $p(r, h)$, $\boldsymbol{x}_0 \in \mathcal{S}^N$, scalars $\tau > 0, \beta \in (0, 1), \sigma \in (0, 1)$ and a pre-defined threshold value ϵ.
2: **Output:** A solution $\boldsymbol{x}^\star$ for optimizing $\bar{\phi}(\boldsymbol{x})$ over the complex circle manifold $\mathcal{S}^N$.
3: $i = 1$
4: **for** each $(r, h) \in \{0, \ldots, N-1\} \times \{0, \ldots, N_\nu - 1\}$ **do**
5: Compute $\boldsymbol{J}^r$ as (5.90)
6: Compute ν_h and $\boldsymbol{p}(\nu_h)$
7: $\boldsymbol{C}_{(r,h)} = \sqrt{p(r,h)}\boldsymbol{J}^r \mathrm{diag}(\boldsymbol{p}(\nu_h))$
8: $i \leftarrow i + 1$
9: **end for**
10: Set $k = 0$.
11: Compute the search direction using the gradient in (5.103) as $\boldsymbol{\eta}_k = -\nabla(\bar{\phi}(\boldsymbol{x}_k))$.
12: Compute the projection of the $\boldsymbol{\eta}_k$ onto the tangent space according to (5.100), and let $\boldsymbol{z} = \boldsymbol{P}_{\mathcal{T}_{\boldsymbol{x}_k}\mathcal{S}^N}(\boldsymbol{\eta}_k)$.
13: (**Armijo line search**) Find the smallest integer $m \geq 0$ such that

$$\bar{\phi}(\boldsymbol{x}_k) - \bar{\phi}(\boldsymbol{x}_k + \tau\beta^m \boldsymbol{\eta}_k) \geq \sigma\tau\beta^m \boldsymbol{z}^H \boldsymbol{z}$$

14: Compute the update of $\boldsymbol{s}_k$ onto $\mathcal{T}_{\boldsymbol{s}_k}\mathcal{S}^N$ as

$$\bar{\boldsymbol{x}}_k = \boldsymbol{x}_k + \tau\beta^m \boldsymbol{\eta}_k \tag{5.104}$$

15: Compute the next iterate $\boldsymbol{x}_{k+1}$ by retracting $\bar{\boldsymbol{x}}_k$ to the complex circle manifold by using the retraction formula $\boldsymbol{x}_{k+1} = \boldsymbol{R}(\bar{\boldsymbol{x}}_k)$.
16: **if** $\|\boldsymbol{x}_{k+1} - \boldsymbol{x}_k\|_2^2 < \epsilon$ **then**
17: STOP.
18: **else**
19: k = k + 1.
20: GOTO step (11).
21: **end if**
22: **Output:** $\boldsymbol{x}^\star = \boldsymbol{x}_k$

---

in the cost. The following lemma establishes that the cost function $\bar{\phi}(\boldsymbol{x})$ will be non-increasing through the retraction step given that the positive scalar γ satisfies a certain condition.

Lemma 1. *Let $\lambda_{\boldsymbol{B}}$ denote the largest eigenvalue of the matrix*

$$\boldsymbol{B} = \sum_{i=1}^{M} \left(vec(\boldsymbol{C}_i)vec(\boldsymbol{C}_i)^H\right)$$

If $\gamma \geq \frac{N^2}{8}\lambda_{\boldsymbol{B}}$ then $\bar{\phi}(\bar{\boldsymbol{x}}_k) \geq \bar{\phi}(\boldsymbol{x}_{k+1})$.

Proof. See [11]. □

Enabled by monotonic cost function decrease in both the Armijo line search and retraction steps (Lemma above), we have $\bar{\phi}(\boldsymbol{x}_k) \geq \bar{\phi}(\boldsymbol{x}_{k+1}) \Rightarrow \bar{\phi}(\boldsymbol{x}_k) - \bar{\phi}(\boldsymbol{x}_{k+1}) \geq 0 \Rightarrow \phi(\boldsymbol{x}_k) - \phi(\boldsymbol{x}_{k+1}) \geq 0$. Then the sequence $\{\phi(\boldsymbol{x}_k)\}_{k=0}^{\infty}$ is non-increasing and since $\phi(\boldsymbol{x}_k) \geq 0, \ \forall \boldsymbol{x}$ (bounded below), converges to a finite value ϕ^* is guaranteed.

Experimental results

We show the performance of AF shaping algorithms, the QGD algorithm, the MBI method with a quadratic improvement (MBIQ) [17], and the coordinate iteration for ambiguity function iterative shaping (CIAFIS) [40]. Consistent with existing work [17], the number of bins N_ν in the normalized Doppler frequency axis is set to 50 which produces the discrete frequencies $\nu_h = -\frac{1}{2} + \frac{h}{N_\nu}$, $h = 0, \ldots, 49$. The desired response $p(r, h) = 1$ for $(r, h) \in \{2, 3, 4\} \times \{35, 36, 37, 38\}$, $(r, h) \in \{3, 4\} \times \{18, 19, 20\}$, and 0 otherwise (see Figure 5.14). The signal-to-interference-ratio (SIR) provides a numerical assessment of all algorithms and is defined as

$$\text{SIR} = \frac{N^2}{\sum_{r=1}^{N} \sum_{h=1}^{N_\nu} p(r, h) \|\boldsymbol{x}\|_2^2 g_{\boldsymbol{x}}(r, \nu_h)} \tag{5.105}$$

In Figure 5.15(a)–(c), 2D plots for the STAFs are shown for QGD, MBIQ, and CIAFIS for $N = 25$. From these figures, it is evident that the response for the QGD waveform is the closest one to the desired one (the unwanted range–Doppler responses

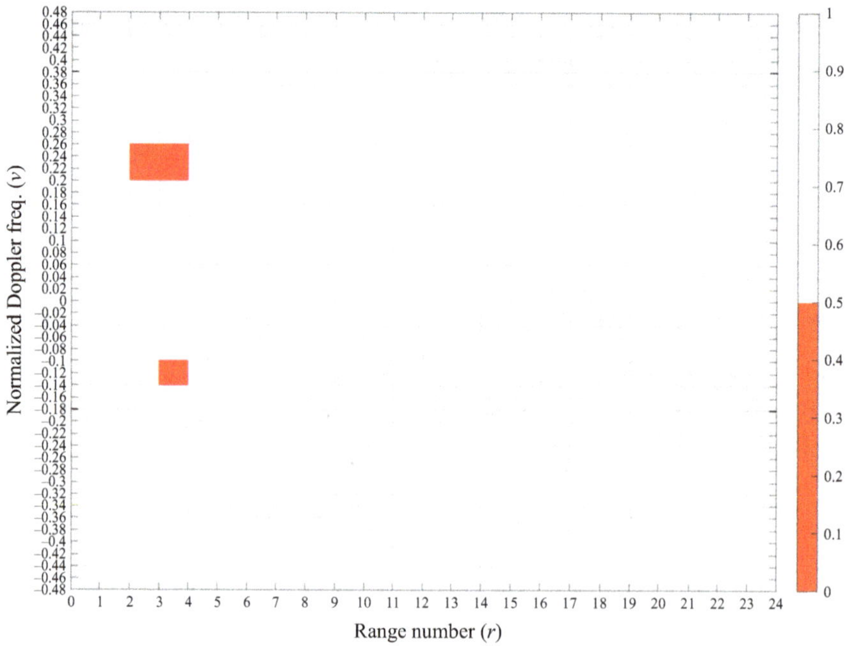

Figure 5.14 The desired STAF (p(r, h))

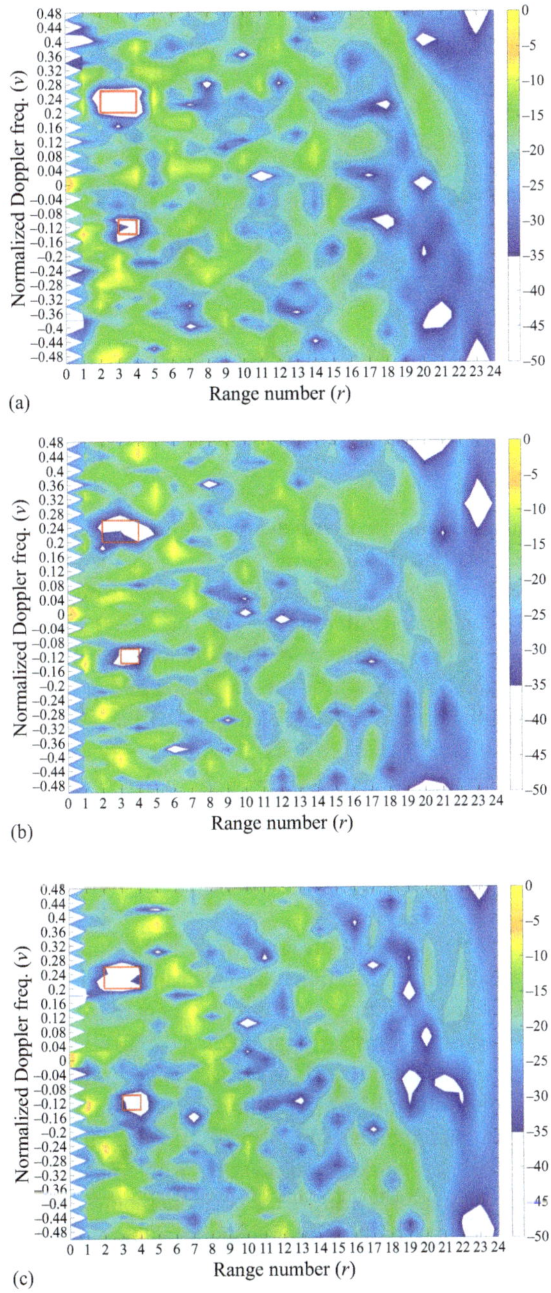

Figure 5.15 STAF for (a) QGD, (b) MBIQ, and (c) CIAFIS for $N = 25$

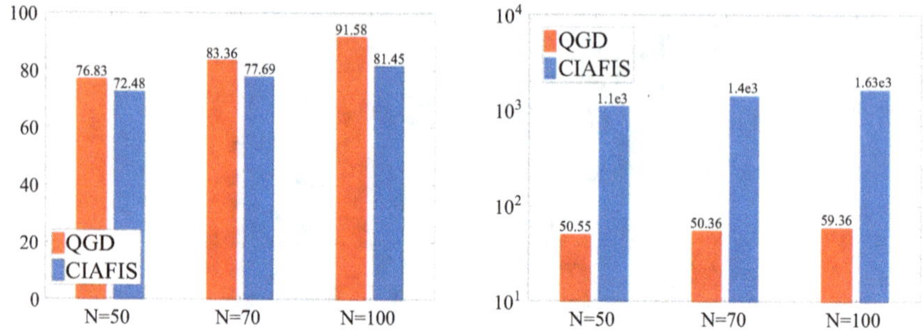

Figure 5.16 *(a) SIR average values, and (b) average simulation times for QGD and CIAFIS for* $N = 50, 70,$ *and* 100. *Each value is averaged over 100 random trials.*

in the red rectangles are suppressed with average values around −45 dB). In Figure 5.16(a) and (b), QGD is compared against the CIAFIS method for different large values for N varying from 50 to 100. We focus on comparisons only against CIAFIS because [40], it has been reported that the results for MBIQ [17] for relatively large values of N, i.e., beyond $N = 25$ take prohibitively long to generate. Figure 5.16(a) shows that both QGD and CIAFIS exhibit the expected average SIR gains with increasing N but QGD can still outperform CIAFIS by 4–10 dB as N varies from 50 to 100. On the other hand, as Figure 5.16(b) reveals, the complexity of QGD increases more gracefully (slowly) with increasing N as compared to that of CIAFIS.**

5.4 Summary

In this chapter, we have reviewed the principles of convex optimization and its application to transmit waveform/beampattern design problems. Though many practical optimization problems for cognitive radar are not convex, convex optimization is still useful since non-convex optimization problems can be exactly or approximately solved using convex optimization skills such as convex relaxation or a sequence of convex optimization problems. We first introduced popular practical constraints for cognitive radar and fundamentals of convex optimization. Next, three successful examples that solve hard non-convex optimization problems using convex optimization are presented. All the methods achieve the constant modulus waveform by solving a sequence of convex optimization problems and simulation results verify that they outperform state-of-the-art algorithms.

**The CIAFIS can be significantly accelerated by the squared iterative method (SQUAREM) [63] which is used in general as an accelerator for the MM algorithms. It was shown that the computational time can be reduced by 10 times via SQUAREM in [40].

References

[1] Guerci JR, Guerci RM, Ranagaswamy M, *et al.* CoFAR: Cognitive Fully Adaptive Radar. In: *IEEE Radar Conference*; 2014. p. 984–989.

[2] Haykin S. Cognitive Dynamic Systems. *Proc IEEE*. 2006;94(11): 1910–1911.

[3] Richards M, Scheer J, Holm W, *et al.* Principles of Modern Radar. Citeseer; 2010.

[4] Patton LK. *On the Satisfaction of Modulus and Ambiguity Function Constraints in Radar Waveform Optimization for Detection.* Wright State University; 2009.

[5] Trees HLV. Optimum Signal Design and Processing for Reverberation-Limited Environments. *IEEE Trans Military Electron.* 1965;9(3):212–229.

[6] Patton L and Rigling BD. Modulus Constraints in Adaptive Radar Waveform Design. In: *IEEE Radar Conference*; 2008. p. 1–6.

[7] Maio AD, Nicola SD, Huang Y, *et al.* Design of Phase Codes for Radar Performance Optimization with a Similarity Constraint. *IEEE Trans Signal Process.* 2008;57(2):610–621.

[8] Aldayel O, Monga V, and Rangaswamy M. Successive QCQP Refinement for MIMO Radar Waveform Design under Practical Constraints. *IEEE Trans Signal Process.* 2016;64(14):3760–3773.

[9] San Antonio G and Fuhrmann DR. Beampattern Synthesis for Wideband MIMO Radar Systems. In: *1st IEEE International Workshop on Computational Advances in Multi-Sensor Adaptive Processing*; 2005. p. 105–108.

[10] Aldayel O, Monga V, and Rangaswmay M. Tractable Transmit MIMO Beampattern Design under a Constant Modulus Constraint. *IEEE Trans Signal Process.* 2017;35(2):237–246.

[11] Alhujaili KA, Monga V, and Rangaswamy M. Quartic Gradient Descent for Tractable Radar Slow-Time Ambiguity Function (STAF) Shaping. *IEEE Trans Aerosp Electron Syst.* 2020;56(2):1474–1489.

[12] Kerahroodi MA, Aubry A, Maio AD, *et al.* A Coordinate-Descent Framework to Design Low PSL/ISL Sequences. *IEEE Trans Signal Process.* 2017;65(22):5942–5956.

[13] Imani S, Nayebi MM, and Ghorashi SA. Colocated MIMO Radar SINR Maximization Under ISL and PSL Constraints. *IEEE Signal Process Lett.* 2018;25(3):422–426.

[14] He H, Stoica P, and Li J. Designing Unimodular Sequence Sets with Good Correlations - Including an Application to MIMO Radar. *IEEE Trans Signal Process.* 2009;57(11):4391–4405.

[15] Stoica P, He H, and Li J. New Algorithms for Designing Unimodular Sequences with Good Correlations Properties. *IEEE Trans Signal Process.* 2009;57(4):1415–1425.

[16] Song J, Badu P, and Palomar DP. Optimization Methods for Designing Sequences with Low Autocorrelation Sidelobes. *IEEE Trans Signal Process.* 2015;63(15):3998–4009.

[17] Aubry A, De Maio A, Jiang B, *et al.* Ambiguity Function Shaping for Cognitive Radar via Complex Quartic Optimization. *IEEE Trans Signal Process.* 2013;61(22):5603–5619.
[18] Wu Z, Xu T, Zhou Z, *et al.* Fast Algorithms for Designing Complementary Sets of Sequences under Multiple Constraints. *IEEE Access*. 2019;7:50041–50051.
[19] Stoica P, Li J, and Zhu X. Waveform Synthesis for Diversity-Based Transmit Beampattern Design. *IEEE Trans Signal Process*. 2008;56(6):2593–2598.
[20] He H, Stoica P, and Li J. Wideband MIMO Systems: Signal Design for Transmit Beampattern Synthesis. *IEEE Trans Signal Process*. 2011;59(2):618–628.
[21] Cui G, Li H, and Rangaswamy M. MIMO Radar Waveform Design with Constant Modulus and Similarity Constraints. *IEEE Trans Signal Process.* 2014;62(2):343–353.
[22] Aubry A, Maio AD, and Farina A. Radar Waveform Design in a Spectrally Crowded Environment Via Nonconvex Quadratic Optimization. *IEEE Trans Aerosp Electron Syst*. 2014;50(2):1138–1152.
[23] Kang B, Aldayel O, Monga V, *et al.* Spatio-Spectral Radar Beampattern Design for Coexistence With Wireless Communication Systems. *IEEE Trans Aerosp Electron Syst*. 2019;55(2):644–657.
[24] Bekkerman I and Tabrikian J. Target Detection and Localization Using MIMO Radars and Sonars. *IEEE Trans Signal Process*. 2006;54(10):3873–3883.
[25] Li J and Stoica P. *MIMO Radar Signal Processing*. Wiley Online Library; 2009.
[26] Mattingley J and Boyd S. Real-Time Convex Optimization in Signal Processing. *IEEE Signal Process Mag*. 2010;27(3):50–61.
[27] Eldar YC, Luo Z, Ma W, *et al.* Convex Optimization in Signal Processing. *IEEE Signal Processing Mag*. 2010;27(3):19,145.
[28] Klee V and Minty G. How good is the simplex algorithm? In O. Shisha (ed.), *Inequalities*, III. Academic Press, New York, NY: 1972.
[29] Karmarka N. A New Polynomial-Time Algorithm for Linear Programming. *Combinatorica*. 1984;4(4):373–395.
[30] Boyd S and Vandenberghe L. *Convex Optimization*, 2nd ed. Cambridge: Cambridge University Press; 2004.
[31] Bertsimas D and Sethuraman J. Moment Problems and Semidefinite Optimization. In: Wolkowicz H, Saigal R, Vandenberghe L, editors. *Handbook of Semidefinite Programming*. Dordrecht: Kluwer; 2000. p. 469–510.
[32] Nesterov Y. Squared Functional Systems and Optimization Problems. In: Frenk J, Roos C, Terlaky T, *et al.*, editors. *High Performance Optimization Techniques*. Dordrecht: Kluwer; 2000. p. 405–440.
[33] Boyd S, Diaconis P, and Xiao L. Fastest Mixing Markov Chain on a Graph. *SIAM Rev*. 2004;46(4):667–689.
[34] Guerci JR, Bergin JS, Guerci RJ, *et al.* A New MIMO Clutter Model for Cognitive Radar. In: *2016 IEEE Radar Conference (RadarConf)*; 2016. p. 1–6.
[35] Aubry A, Maio AD, Piezzo M, *et al.* Cognitive Design of the Receive Filter and Transmitted Phase Code in Reverberating Environment. *IET Radar Sonar Navig*. 2012;6(9):822–833.

[36] Wu L, Badu P, and Palomar DP. Transmit Waveform/Receive Filter Design for MIMO Radar with Multiple Waveform Constraints. *IEEE Trans Signal Process*. 2018;66(6):1526–1540.

[37] Rowe W, Stoica P, and Li J. Spectrally Constrained Waveform Design. *IEEE Signal Process Mag*. 2014;31(3):157–162.

[38] Alhujaili K, Monga V, and Rangaswamy M. Transmit MIMO Radar Beampattern Design via Optimization on the Complex Circle Manifold. *IEEE Trans Signal Process*. 2019;67(13):3561–3575.

[39] Zhang J, Qiu X, Shi C, *et al*. Cognitive Radar Ambiguity Function Optimization for Unimodular Sequence. *EURASIP J Adv Signal Process*. 2016; 2016(1):31.

[40] Wu L, Babu P, and Palomar DP. Cognitive radar-based sequence design via SINR maximization. *IEEE Trans Signal Process*. 2017;65(3):779–793.

[41] Tang B, Tang J, and Peng Y. MIMO Radar Waveform Design in Colored Noise Based on Information Theory. *IEEE Trans Signal Process*. 2010;58(9): 4684–4697.

[42] Guo L, Deng H, Himed B, *et al*. Waveform Optimization for Transmit Beamforming with MIMO Radar Antenna Arrays. *IEEE Trans Antennas Propagat*. 2015;63(2):543–552.

[43] Huleihel W, Tabrikian J, and Shavit R. Optimal Adaptive Waveform Design for Cognitive MIMO Radar. *IEEE Trans Signal Process*. 2013;61(20):5075–5089.

[44] Capon J. High Resolution Frequency-Wavenumber Spectrum Analysis. *Proc IEEE*. 1969;57(8):1408–1418.

[45] Friedlander B. Waveform Design for MIMO Radars. *IEEE Trans Aerosp Electron Syst*. 2007;43(3):1227–1238.

[46] Luo Z, Ma W, So A, *et al*. Semidefinite Relaxation of Quadratic Optimization Problems. *IEEE Signal Process Mag*. 2010;27(3):20–34.

[47] Lobo MS, Vandenberghe L, Boyd S, *et al*. Applications of Second-Order Cone Programming. *Linear Algebra Appl*. 1998;284(1):193–228.

[48] Royden H and Fitzpatrick P. Real Analysis, 4th ed. Englewood Cliffs, NJ: Prentice Hall; 2010.

[49] Sussman SM. Least-Square Synthesis of Radar Ambiguity Functions. *IRE Trans Inform Theory*. 1962;8(3):246–254.

[50] Gerchberg RW and Saxton WO. A Practical Algorithm for the Determination of Phase from Image and Diffraction Plane Pictures. *Optik*. 1972;35(2):237–246.

[51] Aldayel O, Kang B, Monga V, *et al*. *Technical Report: Spatio-Spectral Radar Beampattern Design for Co-existence with Wireless Communication Systems*. The Pennsylvania State University; 2017. http://www.personal.psu.edu/osa105/BICTechReport.pdf.

[52] Williams VV. Multiplying Matrices Faster than Coppersmith-Winograd. In: *Proceedings of the Forty-Fourth Annual ACM Symposium on Theory of Computing*; 2012. p. 887–898.

[53] Guerci JR, Guerci RM, Lackpour A, *et al*. Joint Design and Operation of Shared Spectrum Access for Radar and Communications. In: *IEEE Radar Conference (RadarCon)*; 2015. p. 761–766.

[54] Zheng L, Lops M, Eldar YC, *et al.* Radar and Communication Coexistence: An Overview: A Review of Recent Methods. *IEEE Signal Process Mag.* 2019;36(5):85–99.

[55] DeLong D and Hofstetter E. The Design of Clutter-Resistant Radar Waveforms with Limited Dynamic Range. *IEEE Trans Inform Theory.* 1969;15(3):376–385.

[56] Gregers-Hansen V. Clutter Suppression Using Amplitude Weighted Waveforms. In: *Radar 97* (Conf. Publ. No. 449); 1997. p. 797–801.

[57] Zhang S and Huang Y. Complex Quadratic Optimization and Semidefinite Programming. *SIAM J Optim.* 2006;16(3):871–890.

[58] Sayed AH. *Adaptive Filters.* New York, NY: John Wiley & Sons; 2011.

[59] Nocedal J and Wright SJ. *Numerical Optimization*, 2nd ed. New York, NY: Springer; 2006.

[60] Absil PA, Mahony R, and Sepulchre R. *Optimization Algorithms on Matrix Manifolds.* Princeton, NJ: Princeton University Press; 2009.

[61] Kovnatsky A, Glashoff K, and Bronstein M. MADMM: A Generic Algorithm for Non-smooth Optimization on Manifolds. In: *European Conference on Computer Vision.* Berlin: Springer; 2016. p. 680–696.

[62] Bertsekas DP. *Nonlinear Programming.* Belmont, MA: Athena Scientific; 1999.

[63] Varadhan R and Roland C. Simple and Globally Convergent Methods for Accelerating the Convergence of Any EM Algorithm. *Scand J Statist.* 2008;35(2):335–353.

Part II

Design methodologies

Chapter 6

Cognition-enabled waveform design for ambiguity function shaping

Linlong Wu[1] *and Daniel P. Palomar*[2]

One distinguished feature of cognitive radar is the ability of intelligent sensing, which relies much on the transmit waveform in a self-perpetuating manner. On the one hand, the transmit waveform affects significantly on the quality of the backscatter echoes, from which the environmental parameters are inferred by estimation and learning techniques. On the other hand, the waveform design based on the extracted information will further strengthen the radar performance in the next illumination. This chapter focuses on the latter aspect to illustrate how to design waveforms under specific circumstances by exploiting the prior knowledge obtained by a cognitive radar. Two waveform design problems for different application scenarios are presented in a unified waveform design pattern from the perspective of ambiguity function. The first problem is in fact to shape the ambiguity function by making use of the prior information on the scatters. The second problem is to design a waveform with the desired spectral shape for coexistence by leveraging on the information of stopbands and passbands.

6.1 Introduction

The concept of cognitive radar was first introduced by Simon Haykin in his seminal paper [1] in 2006. Unlike conventional adaptive radars, cognitive radar is distinguished for its dynamic feedback loop encompassing the transmitter, environment, and receiver. The extracted information on the receiver will be exploited by the transmitter to further adjust on-the-fly the waveform for a better illumination of the surrounding environment. This new paradigm of radar systems has been motivating many studies to consolidate the framework. To name a few, the authors of [2] adopted the Bayesian approach for target detecting and tracking, the authors of [3] used machine learning techniques to decide the detection threshold, the authors of [4] designed the transmit waveform sequentially by leveraging the cognition, and the authors of [5]

[1]Signal Processing Applications in Radar and Communications (SPARC) group, Interdisciplinary Center for Security, Reliability and Trust (SnT), University of Luxembourg, Luxembourg
[2]Department of Electronic and Computer Engineering and Department of Industrial Engineering & Decision Analytics at the Hong Kong University of Science and Technology (HKUST), Hong Kong

built the sub-Nyquist prototype to validate its performance. In light of the fact that the transmit waveform is the only means to interact with the ambient environment, waveform design has always been a key problem as that throughout the history of active sensing systems [6,7]. It is worth noting that, different from the conventional radar systems, more advanced waveform designs can be conducted based on the cognitive radar system by leveraging on the provided cognition ability. For example, the radio environment map (REM) is an integrated database storing and updating the available electromagnetic information, which can be used to infer a multitude of environmental characteristics including but not limited to transmitter locations, propagation conditions, spectrum usage, and clutter properties [8]. Hence, by exploiting the equipped REM, the cognitive radar is aware of the surrounding environment, and the transmit waveform can be designed intelligently to adapt the actual operative scenario [9].

In essence, waveform design can be interpreted as shaping the associated ambiguity function (AF) [10–13], which is a major tool to analyze the ability of a radar waveform to distinguish targets on the range-Doppler frequency plane [14]. It is also worth mentioning that the band-limited signal can be recovered from its AF [15,16]. The ideal AF is thumbtack-like, with the peak corresponding to the range–Doppler bin of the target of interest. However, it is physically unrealistic to implement due to the finite peak value and the constant volume of the AF [17]. More practically, it is still achievable to shape the AF so that the response from the unwanted range–Doppler regions is suppressed. But we are often in such a dilemma that although we understand the significance of the AF and are able to shape it, we do not know or have no prior information about what its desired shape is, except for the lower sidelobes of the autocorrelation. Recall that the cognitive radar system capitalizes on the cognitive information provided by the platform like a REM, we are aware of the surrounding environment to some degree, which make it viable to describe the desired AF shape. To be more specific, the response at some range–Doppler bins corresponding to the known or predicted scatters reduces to be as small as possible, while the response at the corresponding range–Doppler bin of the target of interest is maintained at a relatively high level.

In this chapter, we investigate two waveform design problems from the perspective of AF shaping under the valid assumption that the environment information is available based on the cognitive radar system. We hope this self-contained chapter can not only introduce the cognition-enabled waveform design topic but also present some useful optimization methods to solve the related problems. The rest of the chapter is organized as follows. Section 6.2 serves as preliminaries on AF and optimization methods. In Section 6.3, we design a waveform to shape the range–Doppler AF with the PAR constraint. In Section 6.4, the spectral shaping problem is investigated by designing waveform with desired spectrum properties and corresponding algorithms are proposed. Finally, conclusions are given in Section 6.5.

6.2 Preliminaries to AF and optimization methods

In this section, we will first present the concept of AF and its role in waveform design of cognitive radar. Then, the Dinkelbach's algorithm and the majorization–minimization (MM) method will be introduced, which are the major optimization tools used in this chapter.

6.2.1 Ambiguity function and its shaping

Imagine a simple scenario where the Doppler-shifted complex envelope of the received signal is

$$u(t) = s(t)\, e^{j2\pi\nu t}. \tag{6.1}$$

The matched filter is designed to the nominal values (without loss of generality, zero delay, and zero Doppler frequency are assumed) so that its envelope is

$$h(t) = s^*(t). \tag{6.2}$$

Then, the complex envelope of the output of the matched filter is given by [14]

$$u_D(t,\nu) = \int_{-\infty}^{\infty} s(\tau)\, e^{j2\pi\nu\tau} s^*(\tau - t)\, d\tau. \tag{6.3}$$

By exchanging τ and t, we obtain $\mathscr{R}(\tau,\nu)$ defined as

$$\mathscr{R}(\tau,\nu) = \int_{-\infty}^{\infty} s(t)\, s(t-\tau)\, e^{j2\pi\nu t} dt. \tag{6.4}$$

The expression of $\mathscr{R}(\tau,\nu)$ has a well-understood physical meaning. It describes the response of the matched filter to a signal delayed in the time domain by τ and shifted in the Doppler domain by ν. In radar signal processing, $\mathscr{R}(\tau,\nu)$ is referred to as the AF, which is the major tool to study and analyze radar signals. Similarly, if a baseband signal is modulated as a pulse-coded signal

$$u(t) = \sum_{n=1}^{N} s(n)\, p_n(t), \tag{6.5}$$

where $\{s(n)\}_{n=1}^{N}$ is the code sequence to be designed and $p_n(t)$ is the ideal rectangular function, we also have the discrete AF, defined as [6]

$$\mathscr{R}(k,p) = \sum_{n=1}^{N} s(n)\, s^*(n-k)\, e^{j2\pi\frac{(n-k)}{N}p}, \tag{6.6}$$

where $k = -N+1, \ldots, N-1$, and $p = -\frac{N}{2}, \ldots, \frac{N}{2}$ for an even N or $p = -\frac{N-1}{2}, \ldots, \frac{N-1}{2}$ for an odd N. For example, Figure 6.1 shows $|\mathscr{R}(\tau,\nu)|$ corresponding to a linear frequency modulated (LFM) pulse waveform.

The AF has several useful properties as follows [14]:

- Maximum at nominal origin: $|\mathscr{R}(\tau,\nu)| \leq |\mathscr{R}(0,0)|$.
- Constant volume: $\int_{-\infty}^{\infty}\int_{-\infty}^{\infty} |\mathscr{R}(\tau,\nu)|^2 d\tau d\nu = 1$.
- Symmetry to the origin: $|\mathscr{R}(\tau,\nu)| = |\mathscr{R}(-\tau,-\nu)|$.

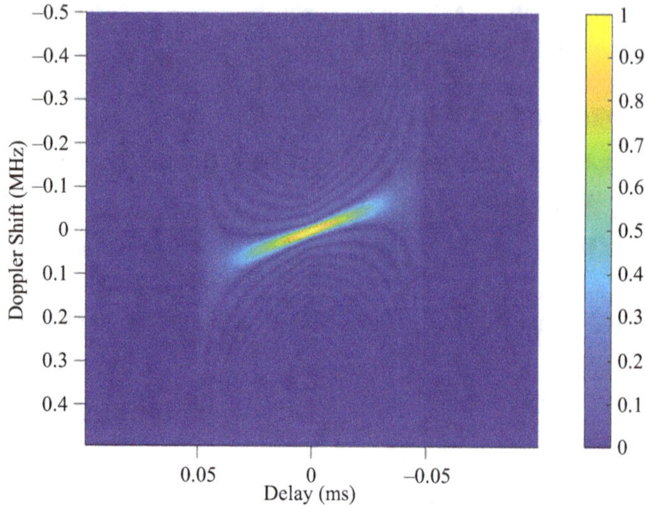

Figure 6.1 $|\mathscr{R}(\tau, \nu)|$ *of an LFM pulse waveform*

For example, based on the property of constant volume, shaping the AF is essentially a reassignment of the values at different range–Doppler bins with a fixed total volume. It is worth noting that the above concepts of AF are for narrowband signals and SISO radar systems. But it has been successfully extended to the MIMO and wideband cases, which is out of the scope of this chapter. Interested readers may refer to [18–20] and references therein for more details.

From the definition of the AF, the transmit waveform $s(t)$ is the variable to shape the AF. In fact, many waveform design problems can be interpreted from the perspective of shaping the AF. For many SNR/SINR maximization problems, if the matched filter is deployed, then these problems are equivalent to the AF shaping by definition. For the problems of spectrum and/or auto-correlation shaping problems [21–24], due to Wiener–Khinchin theorem [25], we can interpret them as shaping the zero-Doppler cut of the AF. The block diagram of waveform design in cognitive radar is illustrated in Figure 6.2.

6.2.2 MM and Dinkelbach's algorithm

In this subsection, we will introduce two useful optimization methods, i.e., MM and Dinkelbach's algorithm. The two methods will be deployed to solve the waveform design problems considered in the book chapter.

6.2.2.1 MM

The MM method is a powerful optimization scheme, especially when the problem is hard to tackle directly. The idea behind the MM algorithm is to convert the original problem into a sequence of simpler problems to be solved until convergence.

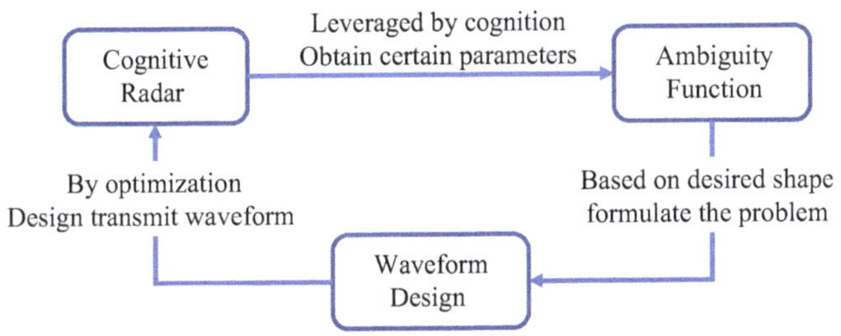

Figure 6.2 Graphical illustration of waveform design for AF shaping in cognitive radar

Consider a general optimization problem

$$\begin{aligned} &\underset{\mathbf{x}}{\text{minimize}} \quad f(\mathbf{x}) \\ &\text{subject to} \quad \mathbf{x} \in \mathscr{X}. \end{aligned} \tag{6.7}$$

Suppose the objective function is hard to directly minimized. Following the general MM idea at the ℓth iteration, we first construct $u(\mathbf{x}, \mathbf{x}_\ell)$, the so-called majorizer of $f(\mathbf{x})$, satisfying the following two requirements at the point $\mathbf{x}_\ell$:

$$u(\mathbf{x}, \mathbf{x}_\ell) \geq f(\mathbf{x}), \text{ for all } \mathbf{x} \in \mathscr{X} \tag{6.8}$$

$$u(\mathbf{x}_\ell, \mathbf{x}_\ell) = f(\mathbf{x}_\ell). \tag{6.9}$$

Then, the MM update is given by

$$\mathbf{x}_{\ell+1} = \underset{\mathbf{x} \in \mathscr{X}}{\text{argmin}}\, u(\mathbf{x}, \mathbf{x}_\ell). \tag{6.10}$$

it means that at each iteration of MM, the function $u(\mathbf{x}, \mathbf{x}_\ell)$, instead of $f(\mathbf{x})$, will be minimized for $\mathbf{x} \in \mathscr{X}$.

One interesting and useful property of MM-based methods is monotonicity:

$$f(\mathbf{x}_{\ell+1}) \leq u(\mathbf{x}_{\ell+1}, \mathbf{x}_\ell) \leq u(\mathbf{x}_\ell, \mathbf{x}_\ell) = f(\mathbf{x}_\ell), \tag{6.11}$$

where the first inequality follows from (6.8), the second one follows from (6.10), and the last equality follows from (6.9). Note that based on (6.11), even if $\mathbf{x}_{\ell+1}$ is not the minimizer of $u(\mathbf{x}, \mathbf{x}_\ell)$, the monotonicity can still be guaranteed as long as it improves the function $u(\mathbf{x}_{\ell+1}, \mathbf{x}_\ell) \leq u(\mathbf{x}_\ell, \mathbf{x}_\ell)$, where the equality means the algorithm has already found a stationary point $\mathbf{x}_{\ell+1}$. Thus, the convergence is guaranteed because $f(\mathbf{x}_\ell)$ is nonincreasing after each iteration. For more details about the convergence of $\{f(\mathbf{x}_\ell)\}$ and $\{\mathbf{x}_\ell\}$, interested readers may refer to [26,27]. The counterpart for maximization problems is referred to as MM, of which the key step is to construct a so-called minorizer. The analysis is similar to that of MM and thus omitted here.

6.2.2.2 Dinkelbach's algorithm

The Dinkelbach's algorithm, first proposed in [28], is a powerful optimization scheme dealing with nonlinear fractional programming problems, which has already been studied in many applications [29]. The idea behind it is to convert, by introducing an auxiliary variable, the original nonlinear fractional problem into a sequence of non-fractional problems to be solved until convergence.

Consider a general fractional programming problem

$$\begin{aligned} \underset{\mathbf{x}}{\text{minimize}} \quad & \frac{f_1(\mathbf{x})}{f_2(\mathbf{x})} \\ \text{subject to} \quad & \mathbf{x} \in \mathscr{X}. \end{aligned} \tag{6.12}$$

where $f_2(\mathbf{x}) > 0$ for $\mathbf{x} \in \mathscr{X}$. Suppose the problem is hard to directly minimize. Following the general idea of the Dinkelbach's algorithm, we need to solve the following problem at the kth iteration,

$$\begin{aligned} \underset{\mathbf{x}}{\text{minimize}} \quad & f_1(\mathbf{x}) - y_k f_2(\mathbf{x}) \\ \text{subject to} \quad & \mathbf{x} \in \mathscr{X}, \end{aligned} \tag{6.13}$$

where y_k is the auxiliary variable updated as

$$y_k = \frac{f_1(\mathbf{x}_k)}{f_2(\mathbf{x}_k)}. \tag{6.14}$$

Assume the optimal solution of problem (6.13) is $\mathbf{x}_{k+1}$. One advantage of the Dinkelbach's algorithm is the guarantee of monotonicity of the sequence $\left\{\frac{f_1(\mathbf{x}_{k+1})}{f_2(\mathbf{x}_{k+1})}\right\}$. Since

$$f_1(\mathbf{x}_{k+1}) - y_k f_2(\mathbf{x}_{k+1}) \leq f_1(\mathbf{x}_k) - y_k f_2(\mathbf{x}_k) = 0, \tag{6.15}$$

we have

$$y_{k+1} = \frac{f_1(\mathbf{x}_{k+1})}{f_2(\mathbf{x}_{k+1})} \leq y_k = \frac{f_1(\mathbf{x}_k)}{f_2(\mathbf{x}_k)}. \tag{6.16}$$

Thus, by alternatively solving problem (6.13) and updating y_k by (6.14), the convergence is guaranteed because y_k is nonincreasing. Also note that the monotonicity can still be guaranteed as long as $f_1(\mathbf{x}_{k+1}) - y_k f_2(\mathbf{x}_{k+1}) \leq 0$ is satisfied even if $\mathbf{x}_{k+1}$ is not the optimal solution of problem (6.13). Specially, if $f_1(\mathbf{x})$ is convex and $f_2(\mathbf{x})$ is concave in the convex set $\mathscr{X}$, the overall iterative algorithm will converge to the global optimum solution of problem (6.12) [30]. For more details about convergence, interested readers may refer to [31,32].

6.3 Waveform design for AF shaping via SINR maximization

In this section, we consider the AF shaping problem with respect to the peak-to-average-power ratio (PAR) constraint, which can be derived from the maximization of the SINR [11]. Recently, the authors of [12] have applied the MM method successfully on this problem, on which the content of this section is based. Interested readers may refer to [12] and references therein for more details. The rest of this section is organized as follows. We first introduce the system model and formulate the problem.

Then we derive the general algorithm within the MM framework and consider several constraints on the problem. Finally, we evaluate the performance of the proposed algorithms via numerical experiments.

6.3.1 System model and problem formulation

Consider a monostatic radar system transmitting a coherent burst of coded pulses, with the N-dimensional vector of observations modeled as [11]:

$$\mathbf{v} = \alpha \mathbf{s} \odot \mathbf{p}(\nu_d) + \mathbf{d}(\mathbf{s}) + \mathbf{n}, \tag{6.17}$$

where α is a complex parameter accounting for channel propagation and backscattering effects, $\mathbf{s}$ is the vector of coded elements, $\mathbf{p}(\nu_d) = \left[1, e^{j2\pi\nu_d}, \ldots, e^{j2\pi(N-1)\nu_d}\right]^T$ is the temporal steering vector, ν_d is the normalized Doppler frequency of the target of interest, $\mathbf{d}(\mathbf{s})$ is the vector of interfering samples, and $\mathbf{n}$ is the vector of the noise samples following the normal distribution $\mathcal{N}\left(\mathbf{0}, \sigma_n^2\mathbf{I}\right)$ uncorrelated with $\mathbf{d}(\mathbf{s})$.

Note that the interfering vector $\mathbf{d}(\mathbf{s})$ accounts for the clutter returns, which can be expressed as [11]:

$$\mathbf{d}(\mathbf{s}) = \sum_{i=1}^{N_t} \rho_i \mathbf{J}^{r_i}(\mathbf{s} \odot \mathbf{p}(\nu_i)), \tag{6.18}$$

where N_t is the total number of interfering scatterers, $r_k \in \{0, 1, \ldots N-1\}$, ρ_i and ν_i are, respectively, the range position, the echo complex amplitude, and the normalized Doppler frequency of the ith scatterer, and $\mathbf{J}^{r_i}$, $r_i \in \{-N+1, \ldots, 0, \ldots, N-1\}$ is the $N \times N$ shift matrix given by

$$\mathbf{J}^{r_i}(m, n) = \begin{cases} 1, & m - n = r_i \\ 0, & m - n \neq r_i. \end{cases} \tag{6.19}$$

Once a target is assumed to be threatening, a track file in the search-and-track modality is opened and continuously updated [11]. This track file usually contains several target parameters, including Doppler velocity measurements [33]. For the details of how the Doppler shift is measured, refer to Chapter 17 in [34]. Thus, we reasonably assume that the Doppler frequency of the target of interest ν_d is known. The output of the matched filter to the echo is given by

$$(\mathbf{s} \odot \mathbf{p}(\nu_d))^H \mathbf{v} = \alpha \|\mathbf{s}\|^2 + (\mathbf{s} \odot \mathbf{p}(\nu_d))^H \mathbf{d}(\mathbf{s}) + (\mathbf{s} \odot \mathbf{p}(\nu_d))^H \mathbf{n}, \tag{6.20}$$

where the last two terms are the disturbance to the target detection. Consequently, the disturbance power after matched filtering is

$$\begin{aligned} &\mathbb{E}\left[\left|(\mathbf{s} \odot \mathbf{p}(\nu_d))^H \mathbf{d}(\mathbf{s}) + (\mathbf{s} \odot \mathbf{p}(\nu_d))^H \mathbf{n}\right|^2\right] \\ &= (\mathbf{s} \odot \mathbf{p}(\nu_d))^H \mathbb{E}\left[\mathbf{d}(\mathbf{s})\mathbf{d}(\mathbf{s})^H\right](\mathbf{s} \odot \mathbf{p}(\nu_d)) + \sigma_n^2 \|\mathbf{s}\|^2, \end{aligned} \tag{6.21}$$

and the signal-to-interference plus noise ratio (SINR) is

$$\text{SINR} = \frac{|\alpha|^2 \|\mathbf{s}\|^4}{(\mathbf{s} \odot \mathbf{p}(\nu_d))^H \mathbb{E}\left[\mathbf{d}(\mathbf{s})\mathbf{d}(\mathbf{s})^H\right](\mathbf{s} \odot \mathbf{p}(\nu_d)) + \sigma_n^2 \|\mathbf{s}\|^2}. \tag{6.22}$$

Considering the constant power constraint $\|\mathbf{s}\|^2 = N$, the SINR maximization problem can be equivalently expressed as

$$\begin{aligned} \underset{\mathbf{s}}{\text{minimize}} \quad & (\mathbf{s} \odot \mathbf{p}(\nu_d))^H \mathbb{E}\left[\mathbf{d}(\mathbf{s})\,\mathbf{d}(\mathbf{s})^H\right](\mathbf{s} \odot \mathbf{p}(\nu_d)) \\ \text{subject to} \quad & \text{PAR}(\mathbf{s}) \leq \gamma \\ & \|\mathbf{s}\|^2 = N. \end{aligned} \tag{6.23}$$

where $\text{PAR}(\mathbf{s}) = N \max_{n=1,\ldots,N} \left\{|s_n|^2\right\} / \sum_{n=1}^{N} |s_n|^2$, and the parameter γ controls the acceptable level of PAR with $1 \leq \gamma \leq N$.

In [11], the normalized Doppler frequency ν_i of the ith clutter is modeled as a uniformly distributed random variable. After discretizing the normalized Doppler interval $\left[-\frac{1}{2}, \frac{1}{2}\right]$ into N_ν bins and approximating the expectation with the sample mean, the objective of problem (6.23) can be approximately expressed as

$$\begin{aligned} & (\mathbf{s} \odot \mathbf{p}(\nu_d))^H \mathbb{E}\left[\mathbf{d}(\mathbf{s})\,\mathbf{d}(\mathbf{s})^H\right](\mathbf{s} \odot \mathbf{p}(\nu_d)) \\ & \approx \sum_{r=0}^{N-1} \sum_{h=0}^{N_\nu-1} p(r,h) \left|\mathbf{s}^H \mathbf{J}^r \text{Diag}(\mathbf{p}(\nu_h))\,\mathbf{s}\right|^2, \end{aligned} \tag{6.24}$$

where $\nu_h = -\frac{1}{2} + \frac{h}{N_\nu}$, $h = 0, 1, \ldots, N_\nu$, is the discrete normalized Doppler frequency and the target Doppler frequency is set as $\nu_h = 0$ without loss of generality; $p(r, h)$ is the interference power for the range–Doppler bin (r, ν_h).

For the objective function of problem (6.23), there always exists a one-to-one mapping $k \in \{1, 2, \ldots, NN_\nu\} \rightarrow (r, h) \in \{0, 1, \ldots, N-1\} \times \{0, 1, \ldots, N_\nu - 1\}$. In the rest of the chapter, k is used to represent the corresponding (r, h) unless otherwise specified. Let $\mathbf{A}_k = \mathbf{J}^r\text{Diag}(\mathbf{p}(\nu_h))$. Then problem (6.23) can be rewritten as

$$\begin{aligned} \underset{\mathbf{s}}{\text{minimize}} \quad & \sum_{k=1}^{NN_\nu} p_k \left|\mathbf{s}^H \mathbf{A}_k \mathbf{s}\right|^2 \\ \text{subject to} \quad & \text{PAR}(\mathbf{s}) \leq \gamma \\ & \|\mathbf{s}\|^2 = N. \end{aligned} \tag{6.25}$$

Before proceeding with the design of the solution to problem (6.25), we address something of our problem formulation:

- $\left|\mathbf{s}^H \mathbf{A}_k \mathbf{s}\right|$ and $\{p_k\}$ are the modulus of the AF of $\mathbf{s}$ and the clutter information at the range–Doppler bin (r, h), respectively. Problem (6.25) can be interpreted as follows: after perceiving the environment by cognitive approaches, the clutter information is incorporated into $\{p_k\}$. By multiplying p_k with the square modulus of the AF at the corresponding range–Doppler bin (r, h) and then minimizing the sum of the products, the designed AF is expected to have low responses in the range–Doppler bins corresponding to high values of p_k.
- Our problem becomes the common ISL problem [21,22] if we let $\nu_h = 0$ for all h and $p_k = 1$ for all k, which means all the scatters have the same Doppler frequency

or we only focus on a specific Doppler frequency of interest. Besides this, if we let $\nu_h = 0$ and choose different values of p_k's, the problem becomes a weighted ISL problem [35,36].

- Note that γ is in the range $[1, N]$. When $\gamma = 1$, the PAR constraint, together with the constant energy constraint, becomes the unit-modulus constraint, which is widely considered in the literature.

6.3.2 Waveform design via MM

The objective function of problem (6.25) can be equivalently reformulated as

$$\sum_{k=1}^{NN_\nu} p_k \left|\mathbf{s}^H \mathbf{A}_k \mathbf{s}\right|^2 = \sum_{k=1}^{NN_\nu} p_k \left|\operatorname{tr}\left(\mathbf{A}_k \mathbf{S}\right)\right|^2 = \sum_{k=1}^{NN_\nu} p_k \operatorname{vec}\left(\mathbf{S}\right)^H \mathbf{B} \operatorname{vec}\left(\mathbf{S}\right), \tag{6.26}$$

where $\mathbf{B} = \sum_{k=1}^{NN_\nu} p_k \operatorname{vec}\left(\mathbf{A}_k\right) \operatorname{vec}\left(\mathbf{A}_k\right)^H$ and $\mathbf{S} = \mathbf{s}\mathbf{s}^H$. Note that

$$\operatorname{tr}\left(\mathbf{B}\right) = \sum_{k=1}^{NN_\nu} p_k \operatorname{tr}\left(\mathbf{A}_k^H \mathbf{A}_k\right) = \sum_{k=1}^{NN_\nu} p_k \left(N - r\right). \tag{6.27}$$

Due to $\operatorname{vec}\left(\mathbf{S}\right)^H \operatorname{vec}\left(\mathbf{S}\right) = \operatorname{tr}\left(\mathbf{S}\mathbf{S}\right) = \operatorname{tr}\left(\mathbf{s}\mathbf{s}^H\mathbf{s}\mathbf{s}^H\right) = N^2$, we further have the following equivalent problem:

$$\begin{aligned} \underset{\mathbf{S},\mathbf{s}}{\text{minimize}} \quad & \operatorname{vec}\left(\mathbf{S}\right)^H \left(\mathbf{B} - \lambda_u\left(\mathbf{B}\right)\mathbf{I}\right) \operatorname{vec}\left(\mathbf{S}\right) \\ \text{subject to} \quad & |s_n| \leq \sqrt{\gamma},\ n = 1, 2, \ldots, N \\ & \|\mathbf{s}\|^2 = N, \mathbf{S} = \mathbf{s}\mathbf{s}^H, \end{aligned} \tag{6.28}$$

where $\lambda_u\left(\mathbf{B}\right) = \sum_{k=1}^{NN_\nu} p_k \left(N - r\right)$ is an upper bound of the eigenvalues of $\mathbf{B}$.

Note that the objective function of problem (6.28) is concave now. We can construct the surrogate function of the objective function of (6.28) by the first-order approximation. Given $\mathbf{S}^{(\ell)} = \mathbf{s}^{(\ell)} \left(\mathbf{s}^{(\ell)}\right)^H$, the first-order approximation is

$$\begin{aligned} u_1\left(\mathbf{S}, \mathbf{S}^{(\ell)}\right) =& \operatorname{vec}\left(\mathbf{S}^{(\ell)}\right)^H \left(\lambda_u\left(\mathbf{B}\right)\mathbf{I} - \mathbf{B}\right) \operatorname{vec}\left(\mathbf{S}^{(\ell)}\right) \\ & + 2\operatorname{Re}\left(\operatorname{vec}\left(\mathbf{S}\right)^H \left(\mathbf{B} - \lambda_u\left(\mathbf{B}\right)\mathbf{I}\right) \operatorname{vec}\left(\mathbf{S}^{(\ell)}\right)\right). \end{aligned} \tag{6.29}$$

Ignoring the constant terms of (6.29), the majorized problem of (6.28) at the point $\mathbf{s}^{(\ell)}$ is given by

$$\begin{aligned} \underset{\mathbf{S},\mathbf{s}}{\text{minimize}} \quad & \operatorname{Re}\left(\operatorname{vec}\left(\mathbf{S}\right)^H \left(\mathbf{B} - \lambda_u\left(\mathbf{B}\right)\mathbf{I}\right) \operatorname{vec}\left(\mathbf{S}^{(\ell)}\right)\right) \\ \text{subject to} \quad & |s_n| \leq \sqrt{\gamma},\ n = 1, 2, \ldots, N \\ & \|\mathbf{s}\|^2 = N, \mathbf{S} = \mathbf{s}\mathbf{s}^H. \end{aligned} \tag{6.30}$$

We can now undo the change of variable $\mathbf{S} = \mathbf{s}\mathbf{s}^H$ in the objective function of (6.30):

$$\begin{aligned}
&\operatorname{Re}\left(\operatorname{vec}(\mathbf{S})^H(\mathbf{B}-\lambda_u(\mathbf{B})\mathbf{I})\operatorname{vec}\left(\mathbf{S}^{(\ell)}\right)\right)\\
=&\operatorname{Re}\left(\operatorname{vec}(\mathbf{S})^H\left(\sum_{k=1}^{NN_v}p_k\operatorname{vec}(\mathbf{A}_k)\operatorname{vec}(\mathbf{A}_k)^H\right)\operatorname{vec}\left(\mathbf{S}^{(\ell)}\right)\right)\\
&-\operatorname{Re}\left(\lambda_u(\mathbf{B})\operatorname{vec}(\mathbf{S})^H\operatorname{vec}\left(\mathbf{S}^{(\ell)}\right)\right)\\
=&\operatorname{Re}\left(\operatorname{tr}\left(\left(\sum_{k=1}^{NN_v}p_k\operatorname{tr}(\mathbf{A}_k^H\mathbf{S}^{(\ell)})\mathbf{A}_k-\lambda_u(\mathbf{B})\mathbf{S}^{(\ell)}\right)\mathbf{S}\right)\right)\\
=&\operatorname{Re}\left(\mathbf{s}^H\left(\sum_{k=1}^{NN_v}p_k\left(\mathbf{s}^{(\ell)}\right)^H\mathbf{A}_k^H\mathbf{s}^{(\ell)}\mathbf{A}_k-\lambda_u(\mathbf{B})\mathbf{s}^{(\ell)}\left(\mathbf{s}^{(\ell)}\right)^H\right)\mathbf{s}\right),
\end{aligned}\tag{6.31}$$

and then problem (6.30) becomes

$$\begin{aligned}
\underset{\mathbf{s}}{\text{minimize}}\quad & \operatorname{Re}\left(\mathbf{s}^H\left(\mathbf{R}-\lambda_u(\mathbf{B})\mathbf{s}^{(\ell)}\left(\mathbf{s}^{(\ell)}\right)^H\right)\mathbf{s}\right)\\
\text{subject to}\quad & |s_n|\le\sqrt{\gamma},\ n=1,2,\ldots,N\\
& \|\mathbf{s}\|^2=N,
\end{aligned}\tag{6.32}$$

where $\mathbf{R}=\sum_{k=1}^{NN_v}p_k\left(\mathbf{s}^{(\ell)}\right)^H\mathbf{A}_k^H\mathbf{s}^{(\ell)}\mathbf{A}_k$.

By defining $\mathbf{P}=\frac{1}{2}\left(\mathbf{R}+\mathbf{R}^H\right)$, we have $\operatorname{Re}\left(\mathbf{s}^H\mathbf{R}\mathbf{s}\right)=\frac{1}{2}\left(\mathbf{s}^H\mathbf{R}\mathbf{s}+\mathbf{s}^H\mathbf{R}^H\mathbf{s}\right)=\mathbf{s}^H\mathbf{P}\mathbf{s}$. Then problem (6.32) can be rewritten as

$$\begin{aligned}
\underset{\mathbf{s}}{\text{minimize}}\quad & \mathbf{s}^H\left(\mathbf{P}-\lambda_u(\mathbf{B})\mathbf{s}^{(\ell)}\left(\mathbf{s}^{(\ell)}\right)^H\right)\mathbf{s}\\
\text{subject to}\quad & |s_n|\le\sqrt{\gamma},\ n=1,2,\ldots,N\\
& \|\mathbf{s}\|^2=N.
\end{aligned}\tag{6.33}$$

The objective function of problem (6.33) is quadratic in $\mathbf{s}$, but it is still hard to solve directly because the matrix $\mathbf{P}$ may be indefinite. Thus, we can majorize the objective function of problem (6.33) at $\mathbf{s}^{(\ell)}$ again to further simplify the problem. Similar to the construction of $u_1\left(\mathbf{S},\mathbf{S}^{(\ell)}\right)$, we need to find an upper bound of the matrix $\left(\mathbf{P}-\lambda_u(\mathbf{B})\mathbf{s}^{(\ell)}\left(\mathbf{s}^{(\ell)}\right)^H\right)$. Before we find the upper bound, let us introduce a useful theorem regarding the bounds of extreme eigenvalues of a Hermitian matrix [37].

Lemma 1. *Let* $\mathbf{M}$ *be an* $n\times n$ *complex matrix with real eigenvalues* $\lambda(\mathbf{M})$, *and* $m=\frac{tr(\mathbf{M})}{n}$ *and* $s^2=\frac{tr(\mathbf{M}^2)}{n-m^2}$. *Then*

$$m-s\sqrt{n-1}\le\lambda_{min}(\mathbf{M})\le m-\frac{s}{\sqrt{n-1}},\tag{6.34}$$

$$m+\frac{s}{\sqrt{n-1}}\le\lambda_{max}(\mathbf{M})\le m+s\sqrt{n-1}.\tag{6.35}$$

We define

$$\mathbf{P}_k = \left(\mathbf{s}^{(\ell)}\right)^H \mathbf{A}_k^H \mathbf{s}^{(\ell)} \mathbf{A}_k + \mathbf{A}_k^H \left(\mathbf{s}^{(\ell)}\right)^H \mathbf{A}_k \mathbf{s}^{(\ell)}, \forall k = 1, 2, \ldots, NN_v \tag{6.36}$$

Each $\mathbf{P}_k$ is Hermitian and, thus, has real eigenvalues. By using Lemma 1, we have the following result about $\mathbf{P}_k$.

Lemma 2. *Let* $\mathbf{P}_k$ *be the matrix defined in (6.36). Then*

$$\lambda_{max}(\mathbf{P}_k) \le \begin{cases} \sqrt{\frac{2(N-r)(N-1)}{N}} \left| \left(\mathbf{s}^{(\ell)}\right)^H \mathbf{A}_k \mathbf{s}^{(\ell)} \right|, & for\ r \neq 0 \\ 2N, & for\ r = 0, \end{cases}$$

where r represents the range and $r = \left\lfloor \frac{k}{N_v} \right\rfloor$.

Proof. See Appendix A.1. □

Let $\mathbf{P} = \sum_{k=1}^{NN_v} \frac{p_k}{2} \mathbf{P}_k$ and we have $\lambda_{max}(\mathbf{P}) \le \sum_{k=1}^{NN_v} \frac{p_k}{2} \lambda_{max}(\mathbf{P}_k)$, the upper bound of the eigenvalues of $\mathbf{P}$ can be expressed as

$$\lambda_u(\mathbf{P}) = \sum_{k=N_v+1}^{NN_v} p_k \sqrt{\frac{(N-r)(N-1)}{2N}} \left| \left(\mathbf{s}^{(\ell)}\right)^H \mathbf{A}_k \mathbf{s}^{(\ell)} \right| + \sum_{k=1}^{N_v} p_k N, \tag{6.37}$$

which is also an upper bound of the eigenvalues of $\left(\mathbf{P} - \lambda_u(\mathbf{B})\, \mathbf{s}^{(\ell)} \left(\mathbf{s}^{(\ell)}\right)^H\right)$. Thus, problem (6.33) is equivalent to

$$\begin{aligned} \underset{\mathbf{s}}{\text{minimize}} \quad & \mathbf{s}^H \left(\mathbf{P} - \lambda_u(\mathbf{B})\, \mathbf{s}^{(\ell)} \left(\mathbf{s}^{(\ell)}\right)^H - \lambda_u(\mathbf{P})\, \mathbf{I}\right) \mathbf{s} \\ \text{subject to} \quad & |s_n| \le \sqrt{\gamma},\ n = 1, 2, \ldots, N \\ & \|\mathbf{s}\|^2 = N. \end{aligned} \tag{6.38}$$

The objective function of (6.38) can also be majorized by the first-order approximation

$$\begin{aligned} u_2\left(\mathbf{s}, \mathbf{s}^{(\ell)}\right) =& 2\mathrm{Re}\left(\mathbf{s}^H \left(\mathbf{P} - \lambda_u(\mathbf{B})\, \mathbf{s}^{(\ell)} (\mathbf{s}^{(\ell)})^H - \lambda_u(\mathbf{P})\, \mathbf{I}\right) \mathbf{s}^{(\ell)}\right) \\ &+ \left(\mathbf{s}^{(\ell)}\right)^H \left(\lambda_u(\mathbf{P})\, \mathbf{I} - \mathbf{P} + \lambda_u(\mathbf{B})\, \mathbf{s}^{(\ell)} \left(\mathbf{s}^{(\ell)}\right)^H\right) \mathbf{s}^{(\ell)} \\ =& 2\mathrm{Re}\left(\mathbf{s}^H \left(\mathbf{P} - \left(\lambda_u(\mathbf{B})\, N + \lambda_u(\mathbf{P})\right) \mathbf{I}\right) \mathbf{s}^{(\ell)}\right) \\ &+ \left(\mathbf{s}^{(\ell)}\right)^H \left(\lambda_u(\mathbf{P})\, \mathbf{I} - \mathbf{P} + \lambda_u(\mathbf{B})\, \mathbf{s}^{(\ell)} \left(\mathbf{s}^{(\ell)}\right)^H\right) \mathbf{s}^{(\ell)}. \end{aligned} \tag{6.39}$$

Ignoring the constant terms and the scalar of (6.39), the majorized problem of (6.33) is

$$\begin{aligned} \underset{\mathbf{s}}{\text{minimize}} \quad & \mathrm{Re}\left(\mathbf{s}^H \mathbf{z}\right) \\ \text{subject to} \quad & |s_n| \le \sqrt{\gamma},\ n = 1, 2, \ldots, N \\ & \|\mathbf{s}\|^2 = N, \end{aligned} \tag{6.40}$$

where

$$\mathbf{z} = \left(\mathbf{P} - \left(\lambda_u\left(\mathbf{B}\right)N + \lambda_u\left(\mathbf{P}\right)\right)\mathbf{I}\right)\mathbf{s}^{(\ell)}. \tag{6.41}$$

The following lemma gives an optimal solution of problem (6.40), for which the detailed proof can be found in [38].

Lemma 3. *An optimal solution to (6.40) is given by*

$$\mathbf{s} = \mathscr{P}_{\mathscr{S}}\left(\mathbf{z}\right), \tag{6.42}$$

where

$$\begin{aligned}\mathscr{P}_{\mathscr{S}}\left(\cdot\right) = &-\left(\mathbf{1}_{\mathbb{R}_0^+}\left(N - m\gamma\right)\right)\sqrt{\gamma}\mathbf{u}_m \odot e^{jarg(\cdot)}\\ &-\left(\mathbf{1}_{\mathbb{R}^-}\left(N - m\gamma\right)\right) min\{\beta|\mathbf{z}|, \sqrt{\gamma}\mathbf{1}\} \odot e^{jarg(\cdot)},\end{aligned} \tag{6.43}$$

$min\{\cdot,\cdot\}$, $|\cdot|$ and $e^{jarg(\cdot)}$ are element-wise operations,

$$\mathbf{1}_A\left(x\right) = \begin{cases} 1, & if\, x \in A,\\ 0, & otherwise,\end{cases} \tag{6.44}$$

$$\mathbf{u}_m = [\underbrace{1,\ldots 1}_{m}, \underbrace{\sqrt{\frac{N - m\gamma}{N\gamma - m\gamma}},\ldots,\sqrt{\frac{N - m\gamma}{N\gamma - m\gamma}}}_{N-m}]^T, \tag{6.45}$$

and

$$\beta \in \left\{\beta \middle| \sum_{n=1}^{N} min\left\{\beta^2\left|z_n\right|^2, \gamma\right\} = N, \beta \in [0, \frac{\sqrt{\gamma}}{min\{|z_n| \mid |z_n| \neq 0\}}]\right\}. \tag{6.46}$$

Note that even though we derive the objective function of (6.40) through two majorization steps, we can merge the two steps into one and obtain the final surrogate function of the objective function of (6.25) given by

$$\begin{aligned}&u\left(\mathbf{s}, \mathbf{s}^{(\ell)}\right)\\ =&2u_2\left(\mathbf{s}, \mathbf{s}^{(\ell)}\right) + 2\lambda_u(\mathbf{P})N + 2\lambda_u(\mathbf{L})N^2 - \text{vec}(\mathbf{S}^{(\ell)})^H\mathbf{L}\text{vec}(\mathbf{S}^{(\ell)})\\ =&4\text{Re}\left(\mathbf{s}^H\mathbf{P}\mathbf{s}^{(\ell)} - \left(\lambda_u(\mathbf{P}) + \lambda_u\left(\mathbf{L}\right)N\right)\mathbf{s}^H\mathbf{s}^{(\ell)}\right)\\ &-\left(2(\mathbf{s}^{(\ell)})^H\mathbf{P}\mathbf{s}^{(\ell)} + \text{vec}(\mathbf{S}^{(\ell)})^H\mathbf{L}\text{vec}(\mathbf{S}^{(\ell)})\right) + 4N\left(\lambda_u(\mathbf{P}) + \lambda_u\left(\mathbf{L}\right)N\right)\\ =&4\text{Re}\left(\mathbf{s}^H\left(\mathbf{P} - \left(\lambda_u\left(\mathbf{B}\right)N + \lambda_u\left(\mathbf{P}\right)\right)\mathbf{I}\right)\mathbf{s}^{(\ell)}\right) + \text{constant}.\end{aligned} \tag{6.47}$$

The complete description of the overall algorithm is given in Algorithm 1.

6.3.3 Convergence analysis and accelerations

The objective function of problem (6.25) is bounded by 0 and the MM method guarantees the monotonicity. Thus, the sequence $\left\{f\left(\mathbf{s}^{(\ell)}\right)\right\}$ generated by MIAFIS is guaranteed to converge to a finite value. In addition, we have the following lemma about the convergence of the sequence $\left\{\mathbf{s}^{(\ell)}\right\}$ generated by MIAFIS.

Algorithm 1: MIAFIS—Majorized iteration for ambiguity function iterative shaping

Require: Initial waveform $\mathbf{s}^{(0)}$
Ensure: Designed waveform $\mathbf{s}$

$\lambda_u(\mathbf{B}) = \sum_{k=1}^{NN_v} p_k(N - r)$
repeat
  $\mathbf{P} = \sum_{k=1}^{NN_v} \frac{p_k}{2}(tr(\mathbf{A}_k^H \mathbf{S}^{(\ell)})\mathbf{A}_k + tr(\mathbf{A}_k \mathbf{S}^{(\ell)})\mathbf{A}_k^H)$
  Calculate $\lambda_u(\mathbf{P})$ according to (32)
  $\mathbf{z} = (\mathbf{P} - (\lambda_u(\mathbf{B}) N + \lambda_u(\mathbf{P}))\mathbf{I})\mathbf{s}^{(\ell)}$
  $\mathbf{s}^{(\ell+1)} = \mathscr{P}_{\mathscr{S}}(\mathbf{z})$
  $\ell \leftarrow \ell + 1$
until convergence

Lemma 4. *Let* $\{\mathbf{s}^{(\ell)}\}$ *be the sequence generated by MIAFIS. Then every limit point of* $\{\mathbf{s}^{(\ell)}\}$ *is a stationary point of problem (6.25).*

Proof. See Appendix A.2. □

For the MM algorithm, the convergence speed is mainly determined by the majorized function. In our case, since the surrogate function might be relatively loose due to the two majorization steps, some acceleration techniques will adopted in the case of slow convergence speed.

6.3.3.1 Acceleration via SQUAREM

SQUAREM [39] refers to the squared iterative method and can be easily implemented as an "off-the-shelf" accelerator for the MM algorithm. Let $\mathbf{F}_{MM}(\cdot)$ denote the nonlinear fixed-point iteration map of the MIAFIS algorithm with $\mathbf{s}^{(\ell+1)} = \mathbf{F}_{MM}\left(\mathbf{s}^{(\ell)}\right)$. The detailed implementation of the proposed algorithm accelerated via SQUAREM is shown in Algorithm 2. Note that applying SQUAREM may cause two potential problems. First, SQUAREM may violate the PAR and constant energy constraints. Second, it may violate the monotonicity of the proposed MM algorithm. For the first problem, we project the infeasible point back to the constraint set by $-\mathscr{P}_{\mathscr{S}}(\cdot)$. For the second problem, a strategy based on backtracking is adopted to preserve the monotonicity, which repeatedly halves the distance between -1 and α: $\alpha = (\alpha - 1)/2$ until the monotonicity is achieved. Note that when $\alpha = -1$, $\mathbf{s}^{(\ell)} - 2\alpha\mathbf{q} + \alpha^2\mathbf{v} = \mathbf{s}_2$. According to the monotonicity of the MM algorithm, $f(\mathbf{s}_1) \leq f(\mathbf{s}_2)$. Thus, it is clear that the monotonicity will finally be achieved when the value of α is approaching -1.

6.3.3.2 Acceleration via local majorization

The potential slowness of the convergence is mainly caused by the double majorization, which might lead to a loose approximation of the original objective function. Besides, we use the upper bound of $\mathbf{B}$ and that of $\left(\mathbf{P} - \lambda_u(\mathbf{B})\mathbf{s}^{(\ell)}\left(\mathbf{s}^{(\ell)}\right)^H\right)$, which could make the approximation even looser. Apart from the SQUAREM scheme, which still

Algorithm 2: MIAFIS acceleration via SQUAREM

Require: Initial waveform $\mathbf{s}^{(0)}$
Ensure: Designed waveform $\mathbf{s}$

```
repeat
    s_1 = F_MM(s^(ℓ))
    s_2 = F_MM(s_1)
    q = s_1 − s^(ℓ)
    v = s_2 − s_1 − q
    α = −||q|| / ||v||
    s^(ℓ+1) = −P_S(s^(ℓ) − 2αq + α²v)
    while f(s^(ℓ+1)) > f(s^(ℓ)) do
        α = (α − 1)/2
        s^(ℓ+1) = −P_S(s^(ℓ) − 2αq + α²v)
    end while
    ℓ ← ℓ + 1
until convergence
```

uses the same surrogate function, a natural idea to accelerate the MM algorithm is to find a better surrogate of the original quartic objective function at every iteration. Note that the monotonicity of the MM algorithm only requires that $u\left(\mathbf{s}, \mathbf{s}^{(\ell)}\right) \geq f\left(\mathbf{s}\right)$ at $\mathbf{s} = \mathbf{s}^{(\ell+1)}$. In other words, $u\left(\mathbf{s}, \mathbf{s}^{(\ell)}\right)$ does not have to be a global upper bound of $f\left(\mathbf{s}\right)$ on the whole domain.

Algorithm 3: MIAFIS acceleration via local majorization

Require: Initial waveform $\mathbf{s}^{(0)}$
Ensure: Designed waveform $\mathbf{s}$

```
λ_u(B) = Σ_{k=1}^{NN_v} p_k(N − r)
repeat
    P = Σ_{k=1}^{NN_v} (p_k/2)(tr(A_k^H S^(ℓ))A_k + tr(A_k S^(ℓ))A_k^H)
    Calculate λ_u(P) according to (32)
    repeat for m in {0, 1, ..., N − 1}
        t = (λ_u(P) + λ_u(B)N) / 2^(N−m)
        s_t* = P_S((P − tI)s^(ℓ))
        m ← m + 1
    until u_t(s_t*, s^(ℓ)) > f(s_t*)
    s_n^(ℓ+1) = s_t*
    ℓ ← ℓ + 1
until convergence
```

Recall the surrogate function of the original objective at the point $\mathbf{s}^{(\ell)}$ in (6.47). The term $(\lambda_u(\mathbf{P}) + \lambda_u(\mathbf{L})N)$ makes the bound globally loose and will influence the convergence speed. By tuning this term, we can achieve a tighter local upper bound of the original objective function at $\mathbf{s}^{(\ell)}$, although it may not be a global upper bound. Thus, we consider the following local upper bound of $f(\mathbf{s})$:

$$\begin{aligned} & u_t\left(\mathbf{s}, \mathbf{s}^{(\ell)}\right) \\ =& -\left(2\left(\mathbf{s}^{(\ell)}\right)^H \mathbf{P}\mathbf{s}^{(\ell)} + \mathrm{vec}\left(\mathbf{S}^{(\ell)}\right)^H \mathbf{B}\mathrm{vec}\left(\mathbf{S}^{(\ell)}\right)\right) \\ & + 4\mathrm{Re}\left(\mathbf{s}^H\mathbf{P}\mathbf{s}^{(\ell)} - t\mathbf{s}^H\mathbf{s}^{(\ell)}\right) + 4Nt \\ =& 4\mathrm{Re}\left(\mathbf{s}^H\mathbf{P}\mathbf{s}^{(\ell)} - t\mathbf{s}^H\mathbf{s}^{(\ell)}\right) + 4Nt - 2\left(\mathbf{s}^{(\ell)}\right)^H \mathbf{P}\mathbf{s}^{(\ell)} - f\left(\mathbf{s}^{(\ell)}\right), \end{aligned} \tag{6.48}$$

where t needs to be chosen such that $u_t\left(\mathbf{s}, \mathbf{s}^{(\ell)}\right) \geq f(\mathbf{s})$ at the minimizer of $u_t\left(\mathbf{s}, \mathbf{s}^{(\ell)}\right)$ over the constraint set, which is

$$\mathbf{s}_t^\star = \mathscr{P}_{\mathscr{S}}\left((\mathbf{P} - t\mathbf{I})\,\mathbf{s}^{(\ell)}\right). \tag{6.49}$$

The complete description of MIAFIS acceleration via local majorization is given in Algorithm 3.

6.3.4 Numerical experiments

The range–Doppler interference scenario is shown in Figure 6.3, where the waveform length is $N = 25$, the red blocks represent the regions of unwanted range–Doppler returns, and the normalized Doppler frequency axis is discretized into $N_\nu = 50$ bins with the discrete Doppler frequency $\nu_h = -\frac{1}{2} + \frac{h}{50}, \forall h = 0, 1, \ldots, N_\nu - 1$. A uniform interference power is assumed among the interference bins, i.e.,

$$p(r,h) = \begin{cases} 1 & (r,h) \in \{2,3,4\} \times \{35,36,37,38\} \\ 1 & (r,h) \in \{3,4\} \times \{18,19,20\} \\ 1 & (r,h) \in \{1,2,\ldots,\ N-1\} \times \{25\} \\ 0 & \text{otherwise.} \end{cases} \tag{6.50}$$

Note that in this interference map, we not only suppress the unwanted range–Doppler returns but also control the ISL over all the lags of the autocorrelation of the transmitted waveform. Also, weighted ISL control can be readily incorporated by letting the $p(r,h)$ corresponding to some particular sidelobes be zero. In the following, all the simulations are based on the above scenario (6.50) unless otherwise specified. All experiments were implemented in MATLAB® R2014b and performed on a PC with a 3.30 GHz i5-4950 CPU and 8 GB RAM.

In Figure 6.4, we plot the convergence curves of the objective value with respect to the number of iterations for the above scenario. A randomly generated waveform is used as the initial one, and the PAR parameter here is $\gamma = 4$. From this figure, we can see clearly that the two accelerated algorithms require far fewer iterations (around 2–3 orders of magnitude less).

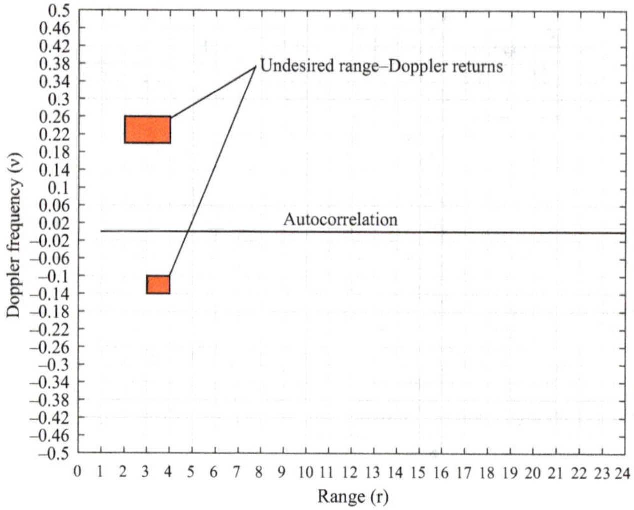

Figure 6.3 Range–Doppler interference scenario for test

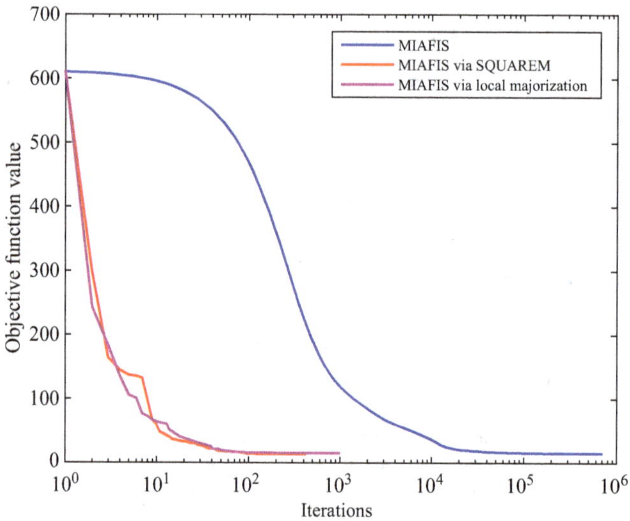

Figure 6.4 Convergence of MIAFIS algorithms for $N = 25$

Recall that the squared magnitude of the AF of the matched filter of the radar waveform $\mathbf{s}$ after normalization is given by

$$g_{\mathbf{s}}(r, v) = \frac{1}{\|\mathbf{s}\|^2} \left| \mathbf{s}^H \mathbf{J}^r \mathrm{diag}\left(\mathbf{p}(v)\right) \mathbf{s} \right|^2 . \tag{6.51}$$

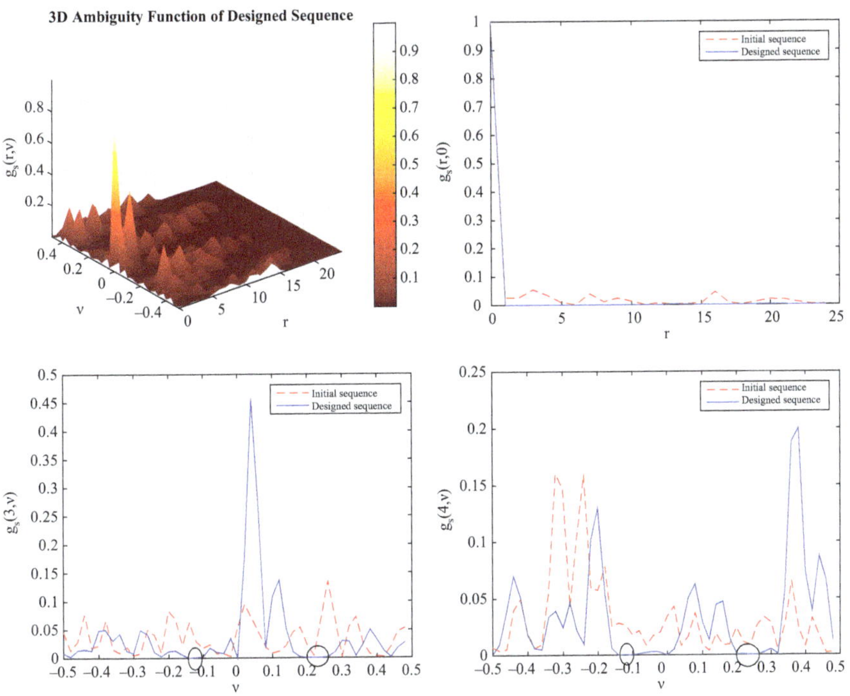

Figure 6.5 Designed ambiguity function and its range/Doppler cuts. Top left: AF. Top right: AF cut at $v = 0$*. Bottom left: AF cut at* $r = 2$*. Bottom right: AF cut at* $r = 3$*.*

In Figure 6.5, we demonstrate the designed AF and its range and Doppler cuts for the range–Doppler bins of interest. The unwanted range–Doppler responses in the two red blocks are suppressed to a very low level and the ISL is significantly improved, which indicates that the proposed algorithms shape the range–Doppler response as expected.

6.4 Waveform design via minimization of regularized spectral level ratio

In the previous section, we tackle the AF shaping problem for considering both range delay and Doppler frequency. In some scenarios, we are more interested in the spectral shape or the zero-Doppler cut of the AF. The motivation behind this can be well understood when considering spectrum sharing among radar and other RF systems. Due to the ever-growing demand for both wireless communication services and accurate sensing capabilities, the amount of desired bandwidth is increasing. Consequently, spectral sharing among radar and telecommunications becomes a solution to this significant issue [40,41]. In order to reduce the mutual interference, it is desired or required that the waveform has deep notches on some particular frequency intervals, in which the prior knowledge of the frequencies is available in the cognitive system. In this section, we focus on solving the spectral shaping problem. The content of this section is based on [24]. Interested readers may refer to [24] and references therein for more details. The rest of this section is organized as follows. We first illustrate the regularized spectral level ratio (RSLR) and the corresponding problem formulation. Then propose two algorithms to solve the problem, followed by the numerical experiments for evaluation.

6.4.1 Regularized SLR and problem formulation

We aim to design a transmit radar waveform $\mathbf{x} = [x_1, \ldots, x_N]^T \in \mathbb{C}^N$ with length being N, which should have a desired spectrum and satisfy a specific PAR level. Let $\mathscr{S}$ and $\mathscr{P}$ denote the stopband and the passband frequency grid set of interest, respectively, which satisfy $\mathscr{S} \cup \mathscr{P} \subseteq \{0, 1, \ldots, N-1\}$ and $\mathscr{S} \cap \mathscr{P} = \emptyset$. Denote the discrete Fourier transform (DFT) matrix by $\mathbf{F}_{DFT} = [\mathbf{f}_0, \ldots \mathbf{f}_{N-1}] \in \mathbb{C}^{N\times N}$, where $\mathbf{f}_\omega = \frac{1}{\sqrt{N}}\left[1, e^{j2\pi\omega/N}, \ldots, e^{j2\pi\omega(N-1)/N}\right]^T \in \mathbb{C}^N$ for $\omega = 0, \ldots, N-1$. The minimal passband level and the maximal stopband level can be expressed by $\min\left\{\left|\mathbf{f}_\omega^H\mathbf{x}\right|^2 |\omega \in \mathscr{P}\right\}$ and $\max\left\{\left|\mathbf{f}_\omega^H\mathbf{x}\right|^2 |\omega \in \mathscr{S}\right\}$, respectively. In [42], the spectral level ratio (SLR) is defined as

$$\mathrm{SLR} = \frac{\max\left\{\left|\mathbf{f}_\omega^H\mathbf{x}\right|^2 |\omega \in \mathscr{S}\right\}}{\min\left\{\left|\mathbf{f}_\omega^H\mathbf{x}\right|^2 |\omega \in \mathscr{P}\right\}}, \tag{6.52}$$

and the problem is formulated as

$$\begin{aligned} \underset{\mathbf{x}}{\text{minimize}} \quad & \mathrm{SLR} \\ \text{subject to} \quad & |x_n| = 1, \forall n = 1, \ldots, N. \end{aligned} \tag{6.53}$$

Intuitively, SLR should be minimized so that $\max\left\{\left|\mathbf{f}_\omega^H\mathbf{x}\right|^2 |\omega \in \mathscr{S}\right\}$ becomes as small as possible and $\min\left\{\left|\mathbf{f}_\omega^H\mathbf{x}\right|^2 |\omega \in \mathscr{P}\right\}$ becomes as large as possible. From the perspective of optimization, it is obvious that problem (6.53) is optimally solved once $\max\left\{\left|\mathbf{f}_\omega^H\mathbf{x}\right|^2 |\omega \in \mathscr{S}\right\} = 0$. Correspondingly, the optimal solution to problem (6.53) is $\mathbf{x}^\star \in \mathscr{N}ull\,(\mathbf{f}_\omega|\omega \in \mathscr{S})$, i.e., the null space of the subspace spanned by $\{\mathbf{f}_\omega\}_{\omega\in\mathscr{S}}$. This cannot guarantee the denominator to be well processed. An extreme example is that $\mathbf{x} = \mathbf{f}_\omega$ for $\forall\omega \in \mathscr{P}$ is also an optimal solution, for which the denominator $\min\left\{\left|\mathbf{f}_\omega^H\mathbf{x}\right|^2 |\omega \in \mathscr{P}\right\}$ might be very small. Note that if $\{\mathbf{f}_\omega\}_{\omega\in\mathscr{S}\cup\mathscr{P}}$ are not the columns of the DFT matrix $\mathbf{F}_{DFT}$, then SLR is still a good optimization metric for spectral shaping. Note that the above case of $\max\left\{\left|\mathbf{f}_\omega^H\mathbf{x}\right|^2 |\omega \in \mathscr{S}\right\} = 0$ only happens for the N point DFT case. If frequency oversampling (more than N frequency samples) is considered for the passbands and stopbands, then the proposed SLR is suitable for optimization.

In order to make the SLR more suitable for optimization, we propose the RSLR as follows:

$$\mathrm{RSLR} = \frac{\max\left\{\left|\mathbf{f}_\omega^H\mathbf{x}\right|^2 \,|\omega \in \mathscr{S}\right\} + c}{\min\left\{\left|\mathbf{f}_\omega^H\mathbf{x}\right|^2 \,|\omega \in \mathscr{P}\right\}}, \tag{6.54}$$

where c is a positive constant.* Therefore, the problem of interest is formulated as

$$\begin{aligned} \underset{\mathbf{x}}{\text{minimize}} \quad & \mathrm{RSLR} \\ \text{subject to} \quad & \|\mathbf{x}\|_2^2 = N, \\ & |x_n| \le \sqrt{\gamma},\ \forall n = 1, \ldots, N, \end{aligned} \tag{6.55}$$

where γ represents the PAR parameter. For simplicity of notation, problem (6.55) will be expressed as

$$\underset{\mathbf{x}\in\mathscr{X}}{\text{minimize}} \quad \frac{\max\left\{\mathbf{x}^H\mathbf{F}_i\mathbf{x}|i \in \mathscr{S}\right\} + c}{\min\left\{\mathbf{x}^H\mathbf{F}_i\mathbf{x}|i \in \mathscr{P}\right\}}, \tag{6.56}$$

where $\mathbf{F}_i = \mathbf{f}_i\mathbf{f}_i^H$ and $\mathscr{X} \triangleq \left\{\mathbf{x}|\ \|\mathbf{x}\|_2^2 = N, |x_n| \le \sqrt{\gamma}, \forall n = 1, \ldots, N\right\}$.

Before proceeding with the design of the algorithm for problem (6.56), we make some comments about this problem formulation:

- Compared with the existing approaches, the highlight of this formulation is that it does not require any spectral settings in advance except $\mathscr{S}$ and $\mathscr{P}$.
- The constraint set is more general than the unit modulus constraint, which is a special case when $\gamma = 1$. In addition, when $\gamma = N$, only the first constraint $\|\mathbf{x}\|_2^2 = N$ takes effect. By increasing the value of γ, we are in fact extending the feasible set, and the optimal objective value should be nonincreasing.
- Generally, the formulated problem is challenging in three aspects: (1) the objective function is fractional; (2) both $\max\left\{\mathbf{x}^H\mathbf{F}_i\mathbf{x}|i \in \mathscr{S}\right\}$ and $\min\left\{\mathbf{x}^H\mathbf{F}_i\mathbf{x}|i \in \mathscr{P}\right\}$ are nondifferentiable; (3) both the objective function and the constraint set are highly nonconvex.

6.4.2 Approximate iterative method for spectrum shaping

In this section, based on the introduced algorithmic frameworks, an iterative method is proposed to solve problem (6.56). At the end of this section, we will summarize the derived method and analyze its complexity and convergence.

*Note that $\mathrm{RSLR} = \frac{\max\left\{|\mathbf{f}_\omega^H\mathbf{x}|^2|\omega\in\mathscr{S}\right\}+c}{\min\left\{|\mathbf{f}_\omega^H\mathbf{x}|^2|\omega\in\mathscr{P}\right\}} = \frac{\max\left\{|\mathbf{f}_\omega^H\mathbf{x}|^2|\omega\in\mathscr{S}\right\}}{\min\left\{|\mathbf{f}_\omega^H\mathbf{x}|^2|\omega\in\mathscr{P}\right\}} + \frac{c}{\min\left\{|\mathbf{f}_\omega^H\mathbf{x}|^2|\omega\in\mathscr{P}\right\}}$. For $\mathbf{x} \in Null\,(\mathbf{f}_\omega|\omega \in \mathscr{S})$, the first term becomes 0. Thus, no matter what value c is, the optimal solution is the one which maximizes $\min\left\{\left|\mathbf{f}_\omega^H\mathbf{x}\right|^2 |\omega \in \mathscr{P}\right\}$ with $\mathbf{x} \in \mathscr{N}ull\,(\mathbf{f}_\omega|\omega \in \mathscr{S})$.

6.4.2.1 Approximation of the point-wise maximum

At the kth iteration of the Dinkelbach's algorithm, we have the following problem:

$$\underset{\mathbf{x}\in\mathcal{X}}{\text{minimize}} \quad \max\left\{\mathbf{x}^H\mathbf{F}_i\mathbf{x}|i\in\mathcal{S}\right\} - y_k\min\left\{\mathbf{x}^H\mathbf{F}_i\mathbf{x}|i\in\mathcal{P}\right\}. \tag{6.57}$$

Due to $y_k = \frac{\max\left\{\mathbf{x}_k^H\mathbf{F}_i\mathbf{x}_k|i\in\mathcal{S}\right\}+c}{\min\left\{\mathbf{x}_k^H\mathbf{F}_i\mathbf{x}_k|i\in\mathcal{P}\right\}} \geq 0$, problem (6.57) is equivalent to

$$\underset{\mathbf{x}\in\mathcal{X}}{\text{minimize}} \quad \max\left\{\mathbf{x}^H\mathbf{F}_i\mathbf{x}|i\in\mathcal{S}\right\} + y_k\max\left\{-\mathbf{x}^H\mathbf{F}_i\mathbf{x}|i\in\mathcal{P}\right\}. \tag{6.58}$$

The objective function is nonconvex and nondifferentiable.

Lemma 5. *Denote the objective function of problem (6.58) by $f(\mathbf{x})$. Then $f(\mathbf{x})$ can be approximated by*

$$g(\mathbf{x}) \approx \alpha log\sum_{i\in\mathcal{S}} exp\left(\frac{\mathbf{x}^H\mathbf{F}_i\mathbf{x}}{\alpha}\right) + \alpha y_k log\sum_{i\in\mathcal{P}} exp\left(-\frac{\mathbf{x}^H\mathbf{F}_i\mathbf{x}}{\alpha}\right) \tag{6.59}$$

with $f(\mathbf{x}) \leq g(\mathbf{x}) \leq f(\mathbf{x}) + \alpha\left(log|\mathcal{S}| + y_k log|\mathcal{P}|\right)$, where $\alpha > 0$ is a constant.

Proof. See Appendix A.3 □

Note that Lemma 5 provides a differentiable approximation of the objective function, and the degree of this approximation can be adjusted by α. Figure 6.6 shows a toy example for intuitive illustration of this approximation. It is clear that the smaller the value of α, the better the approximation.

By using Lemma 5 and ignoring the constant, the approximate problem is given by

$$\underset{\mathbf{x}\in\mathcal{X}}{\text{minimize}} \quad \log\sum_{i\in\mathcal{S}} \exp\left(\frac{\mathbf{x}^H\mathbf{F}_i\mathbf{x}}{\alpha}\right) + y_k\log\sum_{i\in\mathcal{P}} \exp\left(-\frac{\mathbf{x}^H\mathbf{F}_i\mathbf{x}}{\alpha}\right), \tag{6.60}$$

where the objective function is now differentiable but still non-convex. In the next two subsections, we will solve problem (6.60) by applying the MM method.

6.4.2.2 Majorizer construction

For the first term $\log\sum_{i\in\mathcal{S}}\exp\left(\frac{\mathbf{x}^H\mathbf{F}_i\mathbf{x}}{\alpha}\right)$ in problem (6.60), we have

$$\begin{aligned}
&\log\sum_{i\in\mathcal{S}} \exp\left(\frac{\mathbf{x}^H\mathbf{F}_i\mathbf{x}}{\alpha}\right) \\
=&\log\sum_{i\in\mathcal{S}} \exp\left(\frac{\mathbf{x}^H\left(\mathbf{F}_i - (1+\varepsilon)\mathbf{I}\right)\mathbf{x}}{\alpha} + \frac{(1+\varepsilon)\mathbf{x}^H\mathbf{x}}{\alpha}\right) \\
=&\frac{(1+\varepsilon)N}{\alpha} + \log\sum_{i\in\mathcal{S}} \exp\left(-\frac{\mathbf{x}^H\left((1+\varepsilon)\mathbf{I} - \mathbf{F}_i\right)\mathbf{x}}{\alpha}\right),
\end{aligned} \tag{6.61}$$

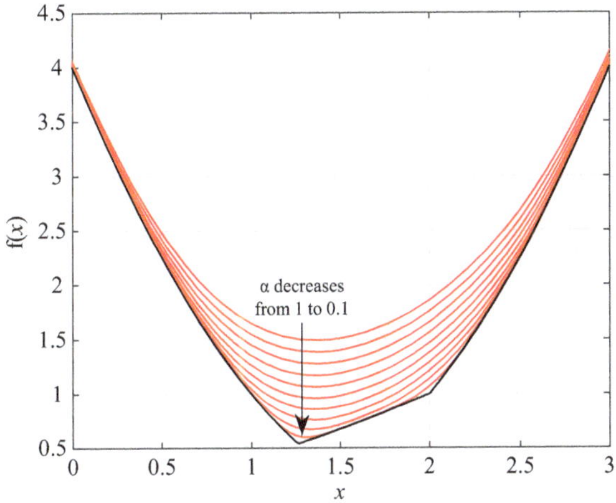

Figure 6.6 Approximation of the point-wise maximum with respect to different α. *Black:* $f(x) = max\{f_i(x) \,|\, i = 1, 2, 3\}$; *Red:* $g(x) = \alpha log \sum_{i=1}^{3} exp\left(\frac{f_i(x)}{\alpha}\right)$, *where* $f_1(x) = x^2 - 2x + 1$, $f_2(x) = x^2 - 4x + 4$ *and* $f_3(x) = 0.1x^2 + 0.3x$.

where ε is a small positive value and we set $\varepsilon = 1 \times 10^{-3}$ hereafter. Similarly, for the second term, we have

$$\log \sum_{j \in \mathscr{P}} \exp\left(-\frac{\mathbf{x}^H \mathbf{F}_j \mathbf{x}}{\alpha}\right) = \frac{\varepsilon N}{\alpha} + \log \sum_{i \in \mathscr{S}} \exp\left(-\frac{\mathbf{x}^H (\mathbf{F}_i + \varepsilon \mathbf{I})\, \mathbf{x}}{\alpha}\right). \tag{6.62}$$

Thus, by defining $\tilde{\mathbf{F}}_i = \frac{1}{\alpha}((1+\varepsilon)\mathbf{I} - \mathbf{F}_i) \succ \mathbf{0}$ and $\hat{\mathbf{F}}_j = \frac{1}{\alpha}(\mathbf{F}_i + \varepsilon \mathbf{I}) \succ \mathbf{0}$, problem (6.60) is equivalent to

$$\underset{\mathbf{x} \in \mathscr{X}}{\text{minimize}} \quad \log \sum_{i \in \mathscr{S}} \exp\left(-\mathbf{x}^H \tilde{\mathbf{F}}_i \mathbf{x}\right) + y_k \log \sum_{j \in \mathscr{P}} \exp\left(-\mathbf{x}^H \hat{\mathbf{F}}_j \mathbf{x}\right), \tag{6.63}$$

Since both terms of the objective function of problem (6.63) have the same structure, we focus on constructing the majorizer of $\log \sum_{i \in \mathscr{S}} \exp\left(-\mathbf{x}^H \tilde{\mathbf{F}}_i \mathbf{x}\right)$ for illustration.

Lemma 6. *At the* ℓ*th iteration,* $log \sum_{i \in \mathscr{S}} exp\left(-\mathbf{x}^H \tilde{\mathbf{F}}_i \mathbf{x}\right)$ *can be majorized by*

$$log \sum_{i \in \mathscr{S}} exp\left(-\mathbf{x}^H \tilde{\mathbf{F}}_i \mathbf{x}\right) \leq 2Re\left[\left(\sum_{i \in \mathscr{S}} \mathbf{A}_i^\ell \mathbf{x}_\ell\right)^H \mathbf{x}\right] + constant \tag{6.64}$$

with

$$\mathbf{A}_i^{\ell} = -\frac{exp\left(-\mathbf{x}_\ell^H\tilde{\mathbf{F}}_i\mathbf{x}_\ell\right)\tilde{\mathbf{F}}_i + \frac{1}{\alpha^2}\left((1+\varepsilon)^2 N - 2\varepsilon - 1\right)\mathbf{x}_\ell\mathbf{x}_\ell^H}{\sum_{i\in\mathscr{S}} exp\left(-\mathbf{x}_\ell^H\tilde{\mathbf{F}}_i\mathbf{x}_\ell\right)}. \tag{6.65}$$

The equality is achieved when $\mathbf{x} = \mathbf{x}_\ell$.

Proof. See Appendix A.4. □

The same techniques can be applied to the second term of the objective function of problem (6.63). We have

$$\log\sum_{i\in\mathscr{P}}\exp\left(-\mathbf{x}^H\hat{\mathbf{F}}_i\mathbf{x}\right) \leq 2\mathrm{Re}\left[\left(\sum_{i\in\mathscr{P}}\mathbf{B}_i^\ell\mathbf{x}_\ell\right)^H\mathbf{x}\right] + \text{constant}, \tag{6.66}$$

where the equality is achieved when $\mathbf{x} = \mathbf{x}_\ell$, and

$$\mathbf{B}_i^{\ell} = -\frac{\exp\left(-\mathbf{x}_\ell^H\hat{\mathbf{F}}_i\mathbf{x}_\ell\right)\hat{\mathbf{F}}_j + \frac{1}{\alpha^2}\left(\varepsilon^2 N + 2\varepsilon + 1\right)\mathbf{x}_\ell\mathbf{x}_\ell^H}{\sum_{i\in\mathscr{P}} \exp\left(-\mathbf{x}_\ell^H\hat{\mathbf{F}}_i\mathbf{x}_\ell\right)}. \tag{6.67}$$

Therefore, the final majorized problem of problem (6.63) is

$$\begin{aligned}
&\underset{\mathbf{x}}{\text{minimize}} && \mathrm{Re}\left(\mathbf{p}_\ell^H\mathbf{x}\right)\\
&\text{subject to} && \|\mathbf{x}\|_2^2 = N.\\
& && |x_n| \leq \sqrt{\gamma}, \forall n = 1,\ldots,N,
\end{aligned} \tag{6.68}$$

where

$$\mathbf{p}_\ell = \left(\sum_{i\in\mathscr{S}}\mathbf{A}_i^\ell + y_k\sum_{i\in\mathscr{P}}\mathbf{B}_i^\ell\right)\mathbf{x}_\ell. \tag{6.69}$$

A closed-form solution to problem (6.68) is given by

$$\mathbf{x}^\star = \mathscr{A}_{\mathscr{X}}\left(\mathbf{p}_\ell\right) \tag{6.70}$$

where $\mathscr{A}_{\mathscr{X}}(\cdot)$ is the same as (6.43) with some notation modifications. For the special case where $\gamma = 1$, the constraint set is reduced to the unit-modulus constraint. Then the optimal solution is $\mathbf{x}^\star = -e^{j\arg(\mathbf{p}_\ell)}$. For the special case where $\gamma = N$, only the constraint $\|\mathbf{x}\|_2^2 = N$ takes effect, and the optimal solution is $\mathbf{x}^\star = -\frac{\sqrt{N}\mathbf{p}}{\|\mathbf{p}\|_2}$.

6.4.2.3 Complexity and convergence analysis

The complete description of the proposed algorithm named as **A**pproximate **I**terative Method for **S**pectrum **S**haping (AISS) is shown in Algorithm 4. It is clear that the main computation of each iteration is the calculation of $\mathbf{p}_\ell$, which consists of $\mathbf{A}_i^\ell\mathbf{x}_\ell$ for all $i\in\mathscr{S}$ and $\mathbf{B}_i^\ell\mathbf{x}_\ell$ for all $i\in\mathscr{P}$. Note that both $\mathbf{A}_i^\ell\mathbf{x}_\ell$ and $\mathbf{B}_i^\ell\mathbf{x}_\ell$ include $\mathbf{f}_\omega^H\mathbf{x}_\ell$ for $\omega\in\mathscr{S}\cup\mathscr{P}$, which can be implemented via the fast Fourier transform (FFT). Thus, the computation cost per iteration is $\mathscr{O}\left(NlogN\right)$.

Algorithm 4: AISS

Require: Initial waveform $\mathbf{s}_{(0)}$, stopband $\mathscr{S}$ and passband $\mathscr{P}$
Ensure: Designed waveform $\mathbf{s}$

repeat
  Set $\ell = 0, \mathbf{s}_\ell = \mathbf{x}_k$
  $y_k = \frac{\max\{\mathbf{x}_k^H \mathbf{F}_i \mathbf{x}_k | i \in \mathscr{S}\} + c}{\min\{\mathbf{x}_k^H \mathbf{F}_i \mathbf{x}_k | i \in \mathscr{P}\}}$
  repeat
    Calculate $\mathbf{f}_\omega^H \mathbf{s}_\ell$ for all $\omega \in \mathscr{S} \cup \mathscr{P}$
    $\sum_{i\in\mathscr{S}} \mathbf{A}_i^\ell \mathbf{s}_\ell = \frac{\sum_{i\in\mathscr{S}} \exp\left(\frac{1}{\alpha}|\mathbf{f}_i^H \mathbf{s}_\ell|^2\right) \mathbf{f}_i^H \mathbf{s}_\ell}{\alpha \sum_{i\in\mathscr{S}} \left(\exp\left(\frac{1}{\alpha}|\mathbf{f}_i^H \mathbf{s}_\ell|^2\right)\right)} \mathbf{f}_i - \left(\frac{1+\varepsilon}{\alpha} + \frac{((1+\varepsilon)^2 N - 2\varepsilon - 1) N |\mathscr{S}| \exp\left(\frac{1+\varepsilon}{\alpha} N\right)}{\alpha^2 \sum_{i\in\mathscr{S}} \exp\left(\frac{1}{\alpha}|\mathbf{f}_i^H \mathbf{s}_\ell|^2\right)} \right) \mathbf{s}_\ell$
    $\sum_{i\in\mathscr{P}} \mathbf{B}_i^\ell \mathbf{s}_\ell = -\frac{\sum_{i\in\mathscr{P}} \exp\left(-\frac{1}{\alpha}|\mathbf{f}_i^H \mathbf{s}_\ell|^2\right) \mathbf{f}_i^H \mathbf{s}_\ell}{\alpha \sum_{i\in\mathscr{P}} \left(\exp\left(-\frac{1}{\alpha}|\mathbf{f}_i^H \mathbf{s}_\ell|^2\right)\right)} \mathbf{f}_i - \left(\frac{\varepsilon}{\alpha} + \frac{(\varepsilon^2 N + 2\varepsilon + 1) N |\mathscr{P}| \exp\left(\frac{\varepsilon}{\alpha} N\right)}{\alpha^2 \sum_{i\in\mathscr{P}} \exp\left(-\frac{1}{\alpha}|\mathbf{f}_i^H \mathbf{s}_\ell|^2\right)} \right) \mathbf{s}_\ell$
    $\mathbf{p}_\ell = \left(\sum_{i\in\mathscr{S}} \mathbf{A}_i^\ell \mathbf{s}_\ell + y_k \sum_{i\in\mathscr{P}} \mathbf{B}_i^\ell \mathbf{s}_\ell\right)$
    Obtain $\mathbf{x}_{\ell+1}$ according to (6.70).
    $\ell \leftarrow \ell + 1$
  until convergence
  $\mathbf{x}_{k+1} = \mathbf{s}_\ell$
  $k \leftarrow k + 1$
until convergence

As illustrated in the preliminary part, both the Dinkelbach's algorithm and the MM method can guarantee the monotonicity of the sequence of the objective values. However, at the kth iteration of the Dinkelbach's algorithm, we in fact solve an approximate problem instead of the standard one. Thus, the existing result about the monotonicity cannot be applied directly. In the following lemma, we analyze the monotonicity of the proposed AISS.

Lemma 7. *For the generated sequence* $\{y_k\}$, *we have*

$$y_{k+1} - y_k \leq \alpha \frac{log\,|\mathscr{S}| + y_k log\,|\mathscr{P}|}{min\left\{\mathbf{x}_{k+1}^H \mathbf{F}_i \mathbf{x}_{k+1} | i \in \mathscr{P}\right\}}. \tag{6.71}$$

Proof. Let $f_1(\mathbf{x}) = \max\left\{\mathbf{x}^H \mathbf{F}_i \mathbf{x} | i \in \mathscr{S}\right\}$, $f_2(\mathbf{x}) = \min\left\{\mathbf{x}^H \mathbf{F}_i \mathbf{x} | i \in \mathscr{P}\right\}$, and $h(\mathbf{x}) = \log \sum_{i\in\mathscr{S}} \exp\left(\mathbf{x}^H \mathbf{F}_i \mathbf{x}/\alpha\right) + y_k \log \sum_{i\in\mathscr{P}} \exp\left(-\mathbf{x}^H \mathbf{F}_i \mathbf{x}/\alpha\right)$. According to Lemma 5, we have the following two inequalities:

$$f_1(\mathbf{x}) \leq \alpha \log \sum_{i\in\mathscr{S}} \exp\left(\mathbf{x}^H \mathbf{F}_i \mathbf{x}/\alpha\right) \leq f_1(\mathbf{x}) + \alpha \log |\mathscr{S}|, \tag{6.72}$$

$$-f_2(\mathbf{x}) \leq \alpha \log \sum_{i\in\mathscr{P}} \exp\left(-\mathbf{x}^H \mathbf{F}_i \mathbf{x}/\alpha\right) \leq -f_2(\mathbf{x}) + \alpha \log |\mathscr{P}|. \tag{6.73}$$

Thus,

$$f_1(\mathbf{x}) - y_k f_2(\mathbf{x}) \leq \alpha h(\mathbf{x}) \leq f_1(\mathbf{x}) - y_k f_2(\mathbf{x}) + \alpha\left(\log|\mathscr{S}| + y_k \log|\mathscr{P}|\right). \tag{6.74}$$

Recall that at the kth iteration, the initial point is $\mathbf{x}_k$ and $y_k = \frac{f_1(\mathbf{x}_k)}{f_2(\mathbf{x}_k)}$. Assume that the output of the kth iteration is $\mathbf{x}_{k+1}$. We have two possible situations for $\mathbf{x}_{k+1}$:

1. $h(\mathbf{x}_{k+1}) \leq 0$. Then $f_1(\mathbf{x}_{k+1}) - y_k f_2(\mathbf{x}_{k+1}) \leq \alpha h(\mathbf{x}_{k+1}) \leq 0$. So $\frac{f_1(\mathbf{x}_{k+1})}{f_2(\mathbf{x}_{k+1})} = y_{k+1} \leq y_k$;
2. $h(\mathbf{x}_{k+1}) > 0$. Then $f_1(\mathbf{x}_{k+1}) - y_k f_2(\mathbf{x}_{k+1}) \leq \alpha h(\mathbf{x}_{k+1})$, which is equivalent to

$$y_{k+1} = \frac{f_1(\mathbf{x}_{k+1})}{f_2(\mathbf{x}_{k+1})} \leq y_k + \frac{\alpha h(\mathbf{x}_{k+1})}{f_2(\mathbf{x}_{k+1})}. \tag{6.75}$$

Since $\mathbf{x}_k$ is the input for the kth iteration and we are using the MM method which guarantees the monotonicity, we have $h(\mathbf{x}_{k+1}) \leq h(\mathbf{x}_k)$. Besides, we have $\alpha h(\mathbf{x}_k) \leq f_1(\mathbf{x}_k) - y_k f_2(\mathbf{x}_k) + \alpha\left(\log|\mathscr{S}| + y_k \log|\mathscr{P}|\right)$, which is based on Lemma 5. Thus, (6.75) can be further relaxed to (also using the equation $f_1(\mathbf{x}_k) - y_k f_2(\mathbf{x}_k) = 0$)

$$y_{k+1} \leq y_k + \frac{\alpha\left(\log|\mathscr{S}| + y_k \log|\mathscr{P}|\right)}{\min\left\{\mathbf{x}_{k+1}^H \mathbf{F}_i \mathbf{x}_{k+1} | i \in \mathscr{P}\right\}}. \tag{6.76}$$

The proof is complete. □

Lemma (7) provides an upper bound of $y_{k+1} - y_k$ as a function of α. Specifically, the smaller the value of α, the smaller the upper bound. In the extreme, $y_{k+1} \leq y_k$ is always guaranteed if $\alpha \to 0$. It is intuitive that when α becomes smaller, the approximate function becomes closer to the original one. If $h(\mathbf{x}_{k+1}) \leq 0$, then $y_{k+1} \leq y_k$. But even when $h(\mathbf{x}_{k+1}) > 0$, $y_{k+1} \leq y_k$ can probably still hold. Empirically, the sequence of $\{y_k\}$ is generally decreasing and finally converges to a small value for a small value of α.

Remark 6.1. *The algorithm can be implemented more efficiently in practice. The inner loop of the proposed algorithm has no need to run until convergence. In fact, we can stop the inner loop as long as* $f_1(\mathbf{x}) - y_k f_2(\mathbf{x}) \leq 0$ *is satisfied.*

6.4.3 Monotonic iterative method for spectrum shaping

In the previous section, we have derived an algorithm named AISS to solve problem (6.17) and analyzed that the monotonicity of AISS can be guaranteed if $\alpha \to 0$. However, since α is always a nonzero value in practice, the monotonicity has no theoretical guarantee although it usually converges empirically. Thus, we derive another algorithm with the guarantee of strict monotonicity in this section, which will be used to provide a good initial point for the iteration of AISS.

6.4.3.1 Minorizer construction of the max–min problem

Recall that the objective function of problem (6.21) can be rewritten as follows:

$$
\begin{aligned}
&\max\left\{\mathbf{x}^H\mathbf{F}_i\mathbf{x}|i\in\mathscr{S}\right\}-y_k\min\left\{\mathbf{x}^H\mathbf{F}_i\mathbf{x}|i\in\mathscr{P}\right\}\\
&=-\left(-\max\left\{\mathbf{x}^H\mathbf{F}_i\mathbf{x}|i\in\mathscr{S}\right\}+y_k\min\left\{\mathbf{x}^H\mathbf{F}_i\mathbf{x}|i\in\mathscr{P}\right\}\right)\\
&=-\left(\min\left\{-\mathbf{x}^H\mathbf{F}_i\mathbf{x}|i\in\mathscr{S}\right\}+y_k\min\left\{\mathbf{x}^H\mathbf{F}_i\mathbf{x}|i\in\mathscr{P}\right\}\right)\\
&=-y_k\left(\min\left\{-\frac{1}{y_k}\mathbf{x}^H\mathbf{F}_i\mathbf{x}|i\in\mathscr{S}\right\}+\min\left\{\mathbf{x}^H\mathbf{F}_i\mathbf{x}|i\in\mathscr{P}\right\}\right).
\end{aligned}
\tag{6.77}
$$

Thus, problem (6.21) is equivalent to

$$
\underset{\mathbf{x}\in\mathscr{X}}{\text{maximize}}\quad \min\left\{\mathbf{x}^H\mathbf{F}_i\mathbf{x}|i\in\mathscr{P}\right\}+\min\left\{-y_k\mathbf{x}^H\mathbf{F}_i\mathbf{x}|i\in\mathscr{S}\right\},\tag{6.78}
$$

where $\widehat{y}_k=\frac{1}{y_k}=\frac{\min\left\{\mathbf{x}_{k-1}^H\mathbf{F}_i\mathbf{x}_{k-1}|i\in\mathscr{P}\right\}}{\max\left\{\mathbf{x}_{k-1}^H\mathbf{F}_i\mathbf{x}_{k-1}|i\in\mathscr{S}\right\}+c}$.

Furthermore, by introducing an auxiliary variable $\mathbf{p}\in\mathbb{R}^{|\mathscr{S}|}$, we have

$$
\min\left\{-\widehat{y}_k\mathbf{x}^H\mathbf{F}_i\mathbf{x}|i\in\mathscr{S}\right\}=\min_{\mathbf{p}\in\mathscr{S}_1}\left\{\sum_{i\in\mathscr{S}}p_i\left(-\widehat{y}_k\mathbf{x}^H\mathbf{F}_i\mathbf{x}\right)\right\}\tag{6.79}
$$

with $\mathscr{S}_1\triangleq\left\{\mathbf{p}|\mathbf{1}^T\mathbf{p}=1,\mathbf{p}\geq 0\right\}$. The optimal $\mathbf{p}$ has only one element being 1 corresponding to the minimal value of $\left\{\mathbf{x}^H\mathbf{F}_i\mathbf{x}\right\}_{i=1}^{|\mathscr{P}|}$ and the rest elements are zeros. For the other term, we also have

$$
\min\left\{\mathbf{x}^H\mathbf{F}_i\mathbf{x}|i\in\mathscr{P}\right\}=\min_{\mathbf{q}\in\mathscr{S}_2}\left\{\sum_{i\in\mathscr{P}}q_i\mathbf{x}^H\mathbf{F}_i\mathbf{x}\right\}.\tag{6.80}
$$

with $\mathscr{S}_2\triangleq\left\{\mathbf{q}|\mathbf{1}^T\mathbf{q}=1,\mathbf{q}\succeq 0\right\}$. Therefore, problem (6.78) can be equivalently rewritten as

$$
\underset{\mathbf{x}\in\mathscr{X}}{\text{maximize}}\quad\min_{\mathbf{p}\in\mathscr{S}_1,\mathbf{q}\in\mathscr{S}_2}\left\{\sum_{i\in\mathscr{P}}q_i\mathbf{x}^H\mathbf{F}_i\mathbf{x}-\widehat{y}_k\sum_{i\in\mathscr{S}}p_i\mathbf{x}^H\mathbf{F}_i\mathbf{x}\right\}.\tag{6.81}
$$

A minorizer of $\min_{\mathbf{p}\in\mathscr{S}_1,\mathbf{q}\in\mathscr{S}_2}\left\{\sum_{i\in\mathscr{P}}q_i\mathbf{x}^H\mathbf{F}_i\mathbf{x}-\widehat{y}_k\sum_{i\in\mathscr{S}}p_i\mathbf{x}^H\mathbf{F}_i\mathbf{x}\right\}$ is provided by the following lemma.

Lemma 8. *At the ℓth iteration of the MM method, a minorizer of the objective function of problem (6.81) is given by*

$$
\ell(\mathbf{x})=\underset{\mathbf{p}\in\mathscr{S}_1,\mathbf{q}\in\mathscr{S}_2}{min}\left\{Re\left\{\mathbf{a}_\ell^H\mathbf{x}\right\}+u_\ell(\mathbf{p},\mathbf{q})\right\},\tag{6.82}
$$

where

$$
\mathbf{a}_\ell=2\left(\sum_{i\in\mathscr{P}}q_j\mathbf{F}_j-\widehat{y}_k\sum_{i\in\mathscr{S}}p_i\left(\mathbf{F}_i-\mathbf{I}\right)\right)\mathbf{x}_\ell\tag{6.83}
$$

and

$$u_\ell(\mathbf{p},\mathbf{q}) = \widehat{y}_k \sum_{i\in\mathscr{S}} p_i\left(\mathbf{x}_\ell^H \mathbf{F}_i \mathbf{x}_\ell - 2N\right) - \sum_{i\in\mathscr{P}} q_i \mathbf{x}_\ell^H \mathbf{F}_i \mathbf{x}_\ell. \tag{6.84}$$

Proof. See Appendix A.5. □

Therefore, the minorized problem of (6.81) is

$$\begin{aligned} &\underset{\mathbf{x}}{\text{maximize}} && \min_{\mathbf{p}\in\mathscr{S}_1,\mathbf{q}\in\mathscr{S}_2} \left\{\text{Re}\left\{\mathbf{a}_\ell^H \mathbf{x}\right\} + u_\ell(\mathbf{p},\mathbf{q})\right\} \\ &\text{subject to} && \|\mathbf{x}\|_2^2 = N. \\ & && |x_n| \le \sqrt{\gamma} \text{ for } n = 1,\ldots,N. \end{aligned} \tag{6.85}$$

The lemma below converts problem (6.85) to an equivalent problem, which is relatively easier to solve.

Lemma 9. *Solving problem (6.85) is equivalent to solving the following problem:*

$$\begin{aligned} &\underset{\mathbf{p},\mathbf{q}}{\textit{minimize}} && \left(\max_{\|\mathbf{x}\|_2^2\le N, |x_n|\le\sqrt{\gamma}} \textit{Re}\left\{\mathbf{a}_\ell^H \mathbf{x}\right\}\right) + u_\ell(\mathbf{p},\mathbf{q}) \\ &\textit{subject to} && \mathbf{p}\in\mathscr{S}_1, \mathbf{q}\in\mathscr{S}_2. \end{aligned} \tag{6.86}$$

Proof. See Appendix A.6. □

Problem (6.86) can be solved via the projected subgradient method, which finds an ε-suboptimal point within a finite number of iterations [43]. Since this method is well established and the application on problem (6.86) is very straightforward, the details are omitted. In fact, when applying this method, we can stop running the projected subgradient method once it makes $\widehat{y}_{k+1} \ge \widehat{y}_k$, which still guarantees the monotonicity of the whole algorithm.

6.4.3.2 Two special cases

The constant energy constraint

If $\gamma = N$, then the inner problem of problem (6.86) becomes

$$\begin{aligned} &\underset{\mathbf{x}}{\text{maximize}} && \text{Re}\left\{\mathbf{a}_\ell^H \mathbf{x}\right\} \\ &\text{subject to} && \|\mathbf{x}\|_2^2 = N, \end{aligned} \tag{6.87}$$

which has a closed-form solution given by

$$\mathbf{x}^\star = \frac{\sqrt{N}\mathbf{a}_\ell}{\|\mathbf{a}_\ell\|_2} \tag{6.88}$$

Substituting (6.88) back into problem (6.86), we have

$$\begin{aligned} &\underset{\mathbf{p},\mathbf{q}}{\text{minimize}} && \sqrt{N}\,\|\mathbf{a}_\ell\|_2 + u_\ell(\mathbf{p},\mathbf{q}) \\ &\text{subject to} && \mathbf{p}\in\mathscr{S}_1, \mathbf{q}\in\mathscr{S}_2, \end{aligned} \tag{6.89}$$

which can be rewritten as

$$\begin{aligned} &\underset{\mathbf{p},\mathbf{q}}{\text{minimize}} \quad 2\sqrt{N}\left\|\mathbf{A}_\ell\mathbf{q}-\mathbf{B}_\ell\mathbf{p}\right\|_2-\mathbf{c}_\ell^H\mathbf{q}-\mathbf{d}_\ell^H\mathbf{p} \\ &\text{subject to} \quad \mathbf{p}\in\mathscr{S}_1, \mathbf{q}\in\mathscr{S}_2. \end{aligned} \tag{6.90}$$

where

$$\mathbf{A}_\ell=\left[\mathbf{F}_1\mathbf{x}_\ell,\mathbf{F}_2\mathbf{x}_\ell,\ldots,\mathbf{F}_{|\mathscr{P}|}\mathbf{x}_\ell\right], \tag{6.91}$$

$$\mathbf{B}_\ell=\left[y_k\mathbf{F}_1\mathbf{x}_\ell-y_k\mathbf{x}_\ell,\ldots,y_k\mathbf{F}_{|\mathscr{S}|}\mathbf{x}_\ell-\mathbf{x}_\ell\right], \tag{6.92}$$

$$\mathbf{c}_\ell=\left[\mathbf{x}_\ell^H\mathbf{F}_1\mathbf{x}_\ell,\ldots,\mathbf{x}_\ell^H\mathbf{F}_{|\mathscr{P}|}\mathbf{x}_\ell\right]^T, \tag{6.93}$$

$$\mathbf{d}_\ell=\begin{bmatrix} \widehat{y}_k\left(2N-\mathbf{x}_\ell^H\mathbf{F}_1\mathbf{x}_\ell\right) \\ \vdots \\ \widehat{y}_k\left(2N-\mathbf{x}_\ell^H\mathbf{F}_{|\mathscr{S}|}\mathbf{x}_\ell\right) \end{bmatrix}. \tag{6.94}$$

Problem (6.90) can be rewritten in a second-order cone programming (SOCP) form and solved efficiently by any off-the-shelf solver like SeDuMi, SDPT3, or Mosek.

The unit-modulus constraint
If $\gamma=1$, then the inner problem of problem (6.86) becomes

$$\begin{aligned} &\underset{\mathbf{x}}{\text{maximize}} \quad \text{Re}\left\{\mathbf{a}_\ell^H\mathbf{x}\right\} \\ &\text{subject to} \quad |x_n|=1 \text{ for } n=1,\ldots,N. \end{aligned} \tag{6.95}$$

which has a closed-form solution given by

$$\mathbf{x}^\star=e^{j\arg(\mathbf{a}_\ell)} \tag{6.96}$$

with $e^{j\arg(\cdot)}$ being an elementwise operation. Substituting (6.96) back into problem (6.86), we have

$$\begin{aligned} &\underset{\mathbf{p},\mathbf{q}}{\text{minimize}} \quad \left\|\mathbf{a}_\ell\right\|_1+u_\ell\left(\mathbf{p},\mathbf{q}\right) \\ &\text{subject to} \quad \mathbf{p}\in\mathscr{S}_1, \mathbf{q}\in\mathscr{S}_2, \end{aligned}$$

which can also be rewritten as

$$\begin{aligned} &\underset{\mathbf{p},\mathbf{q}}{\text{minimize}} \quad 2\left\|\mathbf{A}_\ell\mathbf{q}-\mathbf{B}_\ell\mathbf{p}\right\|_1-\mathbf{c}_\ell^H\mathbf{q}-\mathbf{d}_\ell^H\mathbf{p} \\ &\text{subject to} \quad \mathbf{p}\in\mathscr{S}_1, \mathbf{q}\in\mathscr{S}_2. \end{aligned} \tag{6.97}$$

Problem (6.97) is convex and can be solved efficiently by solvers.

6.4.3.3 Complexity and convergence analysis

The DFT matrix is decomposed into two submatrices: the passband DFT matrix $\mathbf{F}_{\mathscr{P}}=\left[\mathbf{f}_1,\ldots,\mathbf{f}_{|\mathscr{P}|}\right]$ and the stopband DFT matrix $\mathbf{F}_{\mathscr{S}}=\left[\mathbf{f}_1,\ldots,\mathbf{f}_{|S|}\right]$. The complete description of the proposed algorithm, named as **M**onotonic **I**terative method for **S**pectrum **S**haping (MISS), is given in Algorithm 5.

From the pseudo-code of MISS, the main computations include $\mathbf{F}_{\mathscr{P}}^H \mathbf{x}_\ell$, $\mathbf{F}_{\mathscr{S}}^H \mathbf{x}_\ell$ and solving problem (6.90) or (6.97). The first four computations can be implemented via FFT and thus require $\mathscr{O}\,(NlogN)$ flops. Assume that both problems (6.90) and (6.97) are solved by the solver CVX, which will adopt a primal-dual interior point method with the worst-case computational complexity being $\mathscr{O}\left(N^{3.5}\right)$. Therefore, in the worst case, the complexity of each iteration of MISS is $\mathscr{O}\left(N^{3.5}\right)$. Given the monotonicity of both the Dinkelbach's algorithm and MM, the monotonicity of MISS can be guaranteed. Note that the monotonicity of the outer Dinkelbach's algorithm is still guaranteed as long as $f_1(\mathbf{x}) - \widehat{y}_k f_2(\mathbf{x}) \leq 0$, which means that the inner MM method can be run for only several or even one iteration.

Algorithm 5: Monotonic iterative method for spectrum shaping (MISS)

Require: Initial waveform $\mathbf{s}_{(0)}$, stopband $\mathscr{S}$ and passband $\mathscr{P}$
Ensure: Designed waveform $\mathbf{s}$

repeat
- $\widehat{y}_k = \frac{\max\left\{\mathbf{x}_k^H \mathbf{F}_i \mathbf{x}_k | i \in \mathscr{S}\right\}}{\min\left\{\mathbf{x}_k^H \mathbf{F}_i \mathbf{x}_k | i \in \mathscr{P}\right\} + c}$
- Set $\ell = 0$, $\mathbf{s}_\ell = \mathbf{x}_k$
- **repeat**
  - $\mathbf{A}_\ell = \mathbf{F}_{\mathscr{P}} \mathrm{Diag}\left(\mathbf{F}_{\mathscr{P}}^H \mathbf{s}_\ell\right)$
  - $\mathbf{B}_\ell = \widehat{y}_k \left(\mathbf{F}_S \mathrm{Diag}\left(\mathbf{F}_{\mathscr{S}}^H \mathbf{s}_\ell\right) - \mathbf{1}_{|\mathscr{S}|}^T \otimes \mathbf{s}_\ell\right)$
  - $\mathbf{c}_\ell = \left[\mathrm{abs}\left(\mathbf{F}_{\mathscr{P}}^H \mathbf{s}_\ell\right)\right]^2$
  - $\mathbf{d}_\ell = \widehat{y}_k \left(2N\mathbf{1} - \left[\mathrm{abs}\left(\mathbf{F}_{\mathscr{S}}^H \mathbf{s}_\ell\right)\right]^2\right)$
  - Obtain $(\mathbf{p}_{\ell+1}, \mathbf{q}_{\ell+1})$ by solving problem (6.90) or (6.97)
  - $\mathbf{a}_{\ell+1} = 2\left(\mathbf{A}_\ell \mathbf{q}_{\ell+1} - \mathbf{B}_\ell \mathbf{p}_{\ell+1}\right)$
  - $\mathbf{s}_{\ell+1} = \frac{\sqrt{N}\mathbf{a}_{\ell+1}}{\|\mathbf{a}_{\ell+1}\|_2}$ or $\mathbf{s}_{\ell+1} = e^{j\arg(\mathbf{a}_{\ell+1})}$
  - $\ell \leftarrow \ell + 1$
- **until** convergence
- $\mathbf{x}_{k+1} = \mathbf{s}_\ell$
- $k \leftarrow k + 1$

until convergence

6.4.4 Numerical experiments

In this section, we conduct numerical experiments to evaluate the performance of the two proposed methods and compare them with the existing benchmark. Assume that

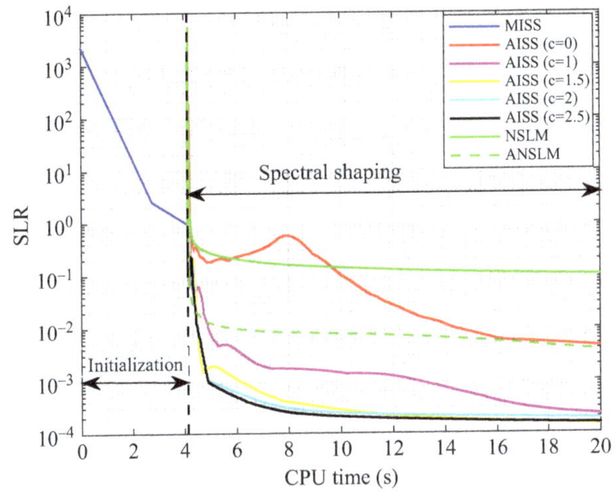

Figure 6.7 Convergence plot of objective value versus CPU time for $\gamma = 1$ and $\alpha = 1 \times 10^{-10}$

the transmitted waveform has the length $N = 162$. This waveform is transmitted in multiple electromagnetic service environment, where the stopbands are given by

$$\begin{aligned}\mathscr{S} = & [0, 0.0617] \cup [0.0988, 0.2469] \cup [0.2593, 0.2840] \cup [0.3086, 0.3827] \\ & \cup [0.4074, 0.4938] \cup [0.5185, 0.5558] \cup [0.9383, 1],\end{aligned} \tag{6.98}$$

where the frequency is normalized so that the total range is [0, 1]. The passbands consist of the complementary sets of $\mathscr{S}$ within [0, 1]. The benchmark method for this design problem is the No-Spectral-Level-Mask (NSLM) method [42], which is for the unit-modulus case. The parameter settings of NSLM follow the suggestions of [42]. Unless otherwise specified, all the parameters are the same in the numerical experiments. All experiments were carried out on a Window desktop PC with a 3.30 GHz i5-4950 CPU and 8 GB RAM.

Since MISS guarantees the monotonicity strictly but at the cost of high computational complexity, it can be used to provide a good initialization for AISS and NSLM. Figure 6.7 shows the curves of the SLR along the CPU time, where the initial point for AISS and NSLM is provided by MISS after a few iterations. From the figure, we can see clearly that AISS decreases much faster and achieves better SLR than NSLM. In addition, although neither AISS nor NSLM has monotonicity guarantee, both can still decrease the objective value along iterations generally. In fact, by choosing a small α, we can reasonably expect that AISS performs well with only small fluctuations or even no fluctuations. The spectra of the designed waveforms are shown in Figure 6.8. The AISS can shape deep notches in these stopbands.

Table 6.1 shows the comparison of average performance between AISS and NSLM. For each value of N, we conduct 50 random trails. In both columns of SLR and CPU time, each presented value is the average of the 50 outcomes. In the last column

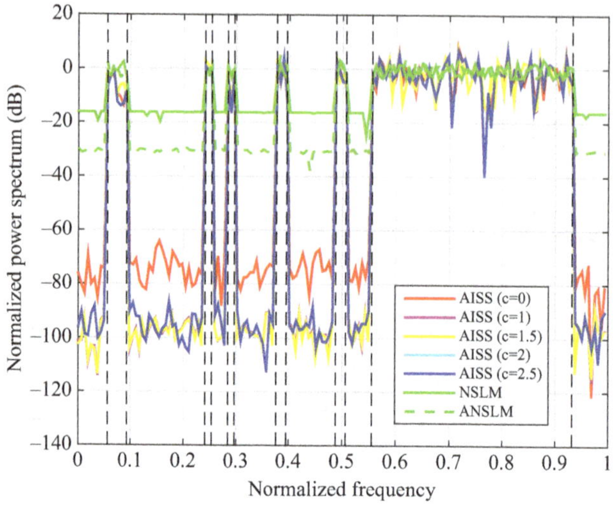

Figure 6.8 Comparison of the spectra of the designed waveforms for $\gamma = 1$ and $\alpha = 1 \times 10^{-10}$

Table 6.1 Performance evaluation of AISS and NSLM

Length	Method	SLR (dB)	CPU time (s)	Exhaustion
$N = 50$	AISS	−2.8243	**14.1050**	76.67%
	NSLM	**−8.8741**	34.4193	100.00%
$N = 100$	AISS	**−11.2808**	**36.7730**	86.67%
	NSLM	−4.4524	66.5400	100.00%
$N = 150$	AISS	**−16.7633**	**67.4600**	100.00%
	NSLM	−0.3881	115.6813	100.00%
$N = 200$	AISS	**−17.9391**	**117.7323**	100.00%
	NSLM	0.1919	179.1473	100.00%
$N = 250$	AISS	**−22.2534**	**160.5987**	100.00%
	NSLM	6.9316	250.7623	100.00%
$N = 300$	AISS	**−16.2970**	**240.3300**	100.00%
	NSLM	8.9960	370.6253	100.00%

named exhaustion, the value represents the percentage of occurrence of the algorithm stopped by meeting the maximum number of iterations. Note that the stopping criterion for both AISS and MISS is $\left(\|\mathbf{x}_{k+1} - \mathbf{x}_k\|_2 / \|\mathbf{x}_k\|_2 \leq \varepsilon \text{ or } k \geq K\right)$, where $\varepsilon = 1 \times 10^{-8}$, $K = 5 \times 10^3$ for AISS and $\varepsilon = 1 \times 10^{-8}$, $K = 5 \times 10^4$ for NSLM. From Table 6.1, we can see that AISS is better than NSLM in terms of both CPU time and SLR (only except the case $N = 50$).

Compared with MISS and NSLM, AISS can deal with the general PAR constraint. Figure 6.9 shows the effect of different values of γ on the SLR performance, where 50

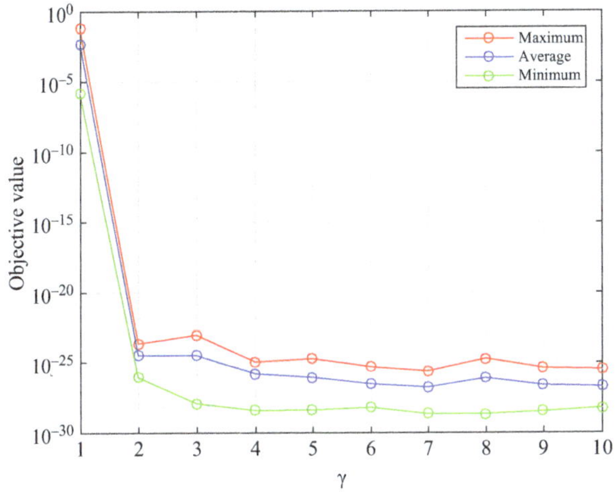

Figure 6.9 Effect of different γ on the objective value for AISS over 50 trials

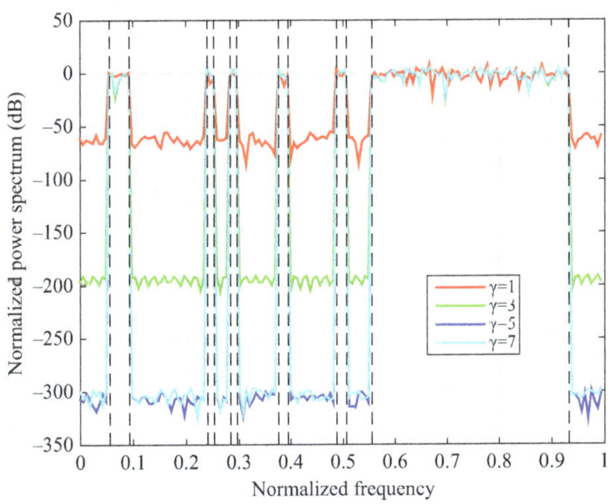

Figure 6.10 Comparison of spectra for different PAR levels of AISS. $\alpha = 1 \times 10^{-7}$

random trials are conducted for each γ. From this figure, we can see that the objective value is generally decreasing as the value of γ increases. From the perspective of optimization, it is reasonable because as γ increases, the feasible set extends so that the achieved objective value probably becomes smaller and smaller. However, the improvement of the averaged objective value is very significant when γ is changed from 1 to 2. After 6, the averaged objective value does not decrease too much. In Figure 6.10, we show the spectra of the designed waveform for different values of γ, where the AISS uses a randomly generated initial waveform. It is clear to see that there are notches in the stopands and these notches become deeper when the value of γ increases. But for the cases $\gamma = 5$ and $\gamma = 7$, the spectra are generally the same, which is consistent with the information provided by Figure 6.9.

6.5 Conclusions

In this chapter, from the perspective of AF shaping, we proposed several optimization algorithms to solve waveform design problems, in which the formulations and corresponding algorithms leverage the capability of cognitive radar. In the first problem, we have interpreted the Doppler-considered SINR maximization as a shaping of the AF of the waveform. The weight for each range–Doppler bin can be obtained within the cognitive radar. An efficient MM-based algorithm named MIAFIS has been derived for this problem. In the case of the ill-construction of the majorization function, two acceleration schemes have been considered. Numerical experiments show the efficiency of the proposed algorithms in shaping a desired AF. In the second problem, the minimization of the regularized SLR is formulated for waveform design. The goal of this problem to obtain a waveform with a desired spectrum, which is in fact a desired zero-Doppler cut of the AF. We have derived two algorithms, AISS and MISS, based on the combination of the Dinkelbach's algorithm and the MM method, where the difference is that AISS approximates the iterative subproblem of the Dinkelbach's framework while MISS solves that directly. Consequently, the AISS has a lower computational complexity but has no strict guarantee of monotonicity, while the MISS is on the contrary. In the numerical experiments, the combination of MISS and AISS is verified and AISS shows better performance than the benchmark in terms of both SLR and running time.

Appendix

A.1 Proof of Lemma 2

Proof. First, $\text{tr}(\mathbf{P}_k) = 2\text{Re}\left(\text{tr}\left(\mathbf{A}_k^H \mathbf{S}^{(\ell)}\right)\text{tr}\left(\mathbf{A}_k\right)\right)$ and $\text{tr}\left(\mathbf{P}_k^2\right) = \text{vec}\left(\mathbf{P}_k\right)^H \text{vec}\left(\mathbf{P}_k\right)$. If $r \neq 0$, $\text{Tr}\left(\mathbf{A}_k\right) = 0$. Thus, $\text{Tr}\left(\mathbf{P}_k\right) = 0$ and

$$
\begin{aligned}
&\text{tr}(\mathbf{P}_k^2) \\
&= \left|\left(\mathbf{s}^{(\ell)}\right)^H \mathbf{A}_k \mathbf{s}^{(\ell)}\right|^2 \text{vec}\left(\mathbf{A}_k\right)^H \text{vec}\left(\mathbf{A}_k\right) \\
&\quad + \left|\left(\mathbf{s}^{(\ell)}\right)^H \mathbf{A}_k \mathbf{s}^{(\ell)}\right|^2 \text{vec}\left(\mathbf{A}_k^H\right)^H \text{vec}\left(\mathbf{A}_k^H\right) \\
&\quad + \left(\mathbf{s}^{(\ell)}\right)^H \mathbf{A}_k \mathbf{s}^{(\ell)} \left(\mathbf{s}^{(\ell)}\right)^H \mathbf{A}_k \mathbf{s}^{(\ell)} \text{vec}\left(\mathbf{A}_k\right)^H \text{vec}\left(\mathbf{A}_k^H\right) \\
&\quad + \left(\mathbf{s}^{(\ell)}\right)^H \mathbf{A}_k^H \mathbf{s}^{(\ell)} \left(\mathbf{s}^{(\ell)}\right)^H \mathbf{A}_k^H \mathbf{s}^{(\ell)} \text{vec}\left(\mathbf{A}_k^H\right)^H \text{vec}\left(\mathbf{A}_k\right) \\
&= 2\left(N - r\right)\left|\left(\mathbf{s}^{(\ell)}\right)^H \mathbf{A}_k \mathbf{s}^{(\ell)}\right|^2,
\end{aligned}
\tag{6.99}
$$

where the last equality holds because $\text{vec}(\mathbf{A}_k)^H \text{vec}(\mathbf{A}_k) = \text{vec}\left(\mathbf{A}_k^H\right)^H \text{vec}\left(\mathbf{A}_k^H\right) = N - r$ and $\text{vec}(\mathbf{A}_k)^H \text{vec}\left(\mathbf{A}_k^H\right) = 0$.

Thus, according to Lemma 1, we have

$$m = 0,\ s^2 = \frac{2(N-r)}{N} \left|\left(\mathbf{s}^{(\ell)}\right)^H \mathbf{A}_k \mathbf{s}^{(\ell)}\right|^2 \tag{6.100}$$

and

$$\lambda_{max}(\mathbf{P}_k) \leq \sqrt{\frac{2(N-r)(N-1)}{N}} \left|\left(\mathbf{s}^{(\ell)}\right)^H \mathbf{A}_k \mathbf{s}^{(\ell)}\right|. \tag{6.101}$$

If $r = 0$, then $\mathbf{P}_k$ is a diagonal matrix. We have $\text{Tr}\left(\mathbf{A}_k \mathbf{S}^{(\ell)}\right) = \sum_{i=0}^{N-1} e^{j2\pi i v_h}$. Let $\mathbf{p} = \left[1, e^{j2\pi v_h}, \ldots, e^{j2\pi(N-1)v_h}\right]^T$, and then

$$\begin{aligned}\mathbf{P}_k &= \left(\sum_{i=0}^{N-1} e^{j2\pi i v_h}\right) \text{Diag}(\mathbf{p})^H + \left(\sum_{i=0}^{N-1} e^{-j2\pi i v_h}\right)^H \text{Diag}(\mathbf{p}) \\ &= \text{Diag}\left(\left\{\sum_{i=0}^{N-1} 2\cos\left(2\pi\,(i-d)\,v_h\right)\right\}_{d=0}^{N-1}\right)\end{aligned} \tag{6.102}$$

with $\sum_{i=0}^{N-1} 2\cos\left(2\pi\,(i-d)\,v_h\right) \leq 2N, \forall d = 0, \ldots, N-1$.

Thus, $\lambda_{max}(\mathbf{P}_k) \leq 2N$. The proof is complete. □

A.2 Proof of Lemma 4

Proof. First, every point of the sequence $\left\{\mathbf{s}^{(\ell)}\right\}$ is bounded with $0 \leq \left|\mathbf{s}^{(\ell)}\right| \leq \sqrt{\gamma}$. According to Theorem 2.17 in [44], at least one limit point must exist.

Denote the objective function of problem (6.25) by $f(\mathbf{s})$ and the feasible set by $\mathcal{S}$. Consider a limit point $\mathbf{z}$ and the corresponding subsequence $\left\{\mathbf{s}^{(\ell_i)}\right\}$. We have

$$\begin{aligned}\left(\mathbf{s}^{(\ell_{i+1})}, \mathbf{s}^{(\ell_{i+1})}\right) &= f\left(\mathbf{s}^{(\ell_{i+1})}\right) \leq f\left(\mathbf{s}^{(\ell_i+1)}\right) \\ &\leq u\left(\mathbf{s}^{(\ell_i+1)}, \mathbf{s}^{(\ell_i)}\right) \leq u\left(\mathbf{s}, \mathbf{s}^{(\ell_i)}\right), \forall \mathbf{s} \in \mathcal{S}.\end{aligned} \tag{6.103}$$

Letting $i \to \infty$, we obtain

$$u(\mathbf{z},\ \mathbf{z}) \leq u(\mathbf{s},\ \mathbf{z}),\ \forall \mathbf{s} \in \mathcal{S}, \tag{6.104}$$

which implies

$$\nabla u(\mathbf{z},\ \mathbf{z})^T \begin{bmatrix} \mathbf{s} - \mathbf{z} \\ (\mathbf{s} - \mathbf{z})^* \end{bmatrix} \geq 0,\ \forall \mathbf{s} \in \mathcal{S}, \tag{6.105}$$

where $\nabla u(\mathbf{z},\ \mathbf{z}) = \left[\begin{array}{cc} \frac{\partial u}{\partial \mathbf{s}} & \frac{\partial u}{\partial \mathbf{s}^*} \end{array}\right]_{(\mathbf{s},\mathbf{s}^*)=(\mathbf{z},\mathbf{z}^*)}$. From the deviation of the majorization function (6.47) of the objective of problem (6.25), we can see clearly that

$$\nabla f(\mathbf{z}) = \nabla u(\mathbf{z}, \mathbf{z}). \tag{6.106}$$

Therefore, $\mathbf{z}$ is a stationary point for problem (6.25). □

A.3 Proof of Lemma 5

Proof. According to the log-sum-exp approximation [45],

$$\begin{aligned}\max\left\{\frac{\mathbf{x}^H\mathbf{F}_i\mathbf{x}}{\alpha}|i\in\mathscr{S}\right\}&\le\log\sum_{i\in\mathscr{S}}\exp\left(\frac{\mathbf{x}^H\mathbf{F}_i\mathbf{x}}{\alpha}\right)\\&\le\log|\mathscr{S}|+\max\left\{\frac{\mathbf{x}^H\mathbf{F}_i\mathbf{x}}{\alpha}|i\in\mathscr{S}\right\},\end{aligned}\tag{6.107}$$

which is further equivalent to

$$\begin{aligned}\max\left\{\mathbf{x}^H\mathbf{F}_i\mathbf{x}|i\in\mathscr{S}\right\}&\le\alpha\log\sum_{i\in\mathscr{S}}\exp\left(\frac{\mathbf{x}^H\mathbf{F}_i\mathbf{x}}{\alpha}\right)\\&\le\alpha\log|\mathscr{S}|+\max\left\{\mathbf{x}^H\mathbf{F}_i\mathbf{x}|i\in\mathscr{S}\right\}.\end{aligned}\tag{6.108}$$

Similarly, for the term $\max\left\{-\mathbf{x}^H\mathbf{F}_j\mathbf{x}|j\in\mathscr{P}\right\}$, we have

$$\begin{aligned}\max\left\{-\mathbf{x}^H\mathbf{F}_j\mathbf{x}|j\in\mathscr{P}\right\}&\le\alpha\log\sum_{j\in\mathscr{P}}\exp\left(-\frac{\mathbf{x}^H\mathbf{F}_j\mathbf{x}}{\alpha}\right)\\&\le\alpha\log|\mathscr{P}|+\max\left\{-\mathbf{x}^H\mathbf{F}_j\mathbf{x}|j\in\mathscr{P}\right\}.\end{aligned}\tag{6.109}$$

The objective function of problem (6.58) is approximated by $\alpha\log\sum_{i\in\mathscr{S}}\exp\left(\frac{\mathbf{x}^H\mathbf{F}_i\mathbf{x}}{\alpha}\right)+\alpha y_k\log\sum_{j\in\mathscr{P}}\exp\left(-\frac{\mathbf{x}^H\mathbf{F}_j\mathbf{x}}{\alpha}\right)$ with the error bounded by $\alpha\left(\log|\mathscr{S}|+y_k\log|\mathscr{P}|\right)$. □

A.4 Proof of Lemma 6

Proof. At the ℓth iteration of the MM method, by using the concavity of logarithm, we have

$$\begin{aligned}&\log\sum_{i\in\mathscr{S}}\exp\left(-\mathbf{x}^H\tilde{\mathbf{F}}_i\mathbf{x}\right)\\&\le\frac{\sum_{i\in\mathscr{S}}\exp\left(-\mathbf{x}^H\tilde{\mathbf{F}}_i\mathbf{x}\right)}{\sum_{i\in\mathscr{S}}\exp\left(-\mathbf{x}_\ell^H\tilde{\mathbf{F}}_i\mathbf{x}_\ell\right)}+\log\sum_{i\in\mathscr{S}}\exp\left(-\mathbf{x}_\ell^H\tilde{\mathbf{F}}_i\mathbf{x}_\ell\right)-1\end{aligned}\tag{6.110}$$

with the equality achieved when $\mathbf{x}=\mathbf{x}_\ell$.

The function $f(x)=e^{-x}, x\in(0,+\infty)$ is β-smooth (i.e., the derivative of $f(x)$ is Lipschitz continuous) with $\beta=1$ because $|f''(x)|=e^{-x}<1$ for $x\in(0,+\infty)$. Thus, for $x,y\in(0,+\infty)$, $f(x)$ is upper bounded by a quadratic function given by

$$f(x)\le f(y)+\nabla f(y)^T(x-y)+\frac{1}{2}||x-y||^2\tag{6.111}$$

with the equality achieved when $x = y$. Substituting $x = \mathbf{x}^H\tilde{\mathbf{F}}_i\mathbf{x}$ and $y = \mathbf{x}_\ell^H\tilde{\mathbf{F}}_i\mathbf{x}_\ell$ into (6.111), we have

$$\begin{aligned}
&\exp\left(-\mathbf{x}^H\tilde{\mathbf{F}}_i\mathbf{x}\right)\\
\leq&\exp\left(-\mathbf{x}_\ell^H\tilde{\mathbf{F}}_i\mathbf{x}_\ell\right) - \exp\left(-\mathbf{x}_\ell^H\tilde{\mathbf{F}}_i\mathbf{x}_\ell\right)\left(\mathbf{x}^H\tilde{\mathbf{F}}_i\mathbf{x} - \mathbf{x}_\ell^H\tilde{\mathbf{F}}_i\mathbf{x}_\ell\right)\\
&+\frac{1}{2}\left\|\mathbf{x}^H\tilde{\mathbf{F}}_i\mathbf{x} - \mathbf{x}_\ell^H\tilde{\mathbf{F}}_i\mathbf{x}_\ell\right\|_2^2\\
=&\frac{1}{2}\mathbf{x}^H\tilde{\mathbf{F}}_i^H\mathbf{x}\mathbf{x}^H\tilde{\mathbf{F}}_i\mathbf{x} - \left(\mathbf{x}_\ell^H\tilde{\mathbf{F}}_i^H\mathbf{x}_\ell + \exp\left(-\mathbf{x}_\ell^H\tilde{\mathbf{F}}_i\mathbf{x}_\ell\right)\right)\mathbf{x}^H\tilde{\mathbf{F}}_i\mathbf{x} + \text{constant}
\end{aligned} \tag{6.112}$$

with the equality achieved when $\mathbf{x} = \mathbf{x}_\ell$.

By combining (6.110) and (6.112), $\log\sum_{i\in\mathscr{S}}\exp\left(-\mathbf{x}^H\tilde{\mathbf{F}}_i\mathbf{x}\right)$ can be majorized as

$$\log\sum_{i\in\mathscr{S}}\exp\left(-\mathbf{x}^H\tilde{\mathbf{F}}_i\mathbf{x}\right) \leq \sum_{i\in\mathscr{S}}\left(\frac{1}{2a^\ell}\mathbf{x}^H\tilde{\mathbf{F}}_i^H\mathbf{x}\mathbf{x}^H\tilde{\mathbf{F}}_i\mathbf{x} - \frac{b_i^\ell}{a^\ell}\mathbf{x}^H\tilde{\mathbf{F}}_i\mathbf{x}\right) + \text{constant}, \tag{6.113}$$

where $a^\ell = \sum_{i\in\mathscr{S}}\exp\left(-\mathbf{x}_\ell^H\tilde{\mathbf{F}}_i\mathbf{x}_\ell\right)$ and $b_i^\ell = \mathbf{x}_\ell^H\tilde{\mathbf{F}}_i^H\mathbf{x}_\ell + \exp\left(-\mathbf{x}_\ell^H\tilde{\mathbf{F}}_i\mathbf{x}_\ell\right)$.

Next, the majorizer of $\frac{1}{2a^\ell}\mathbf{x}^H\tilde{\mathbf{F}}_i^H\mathbf{x}\mathbf{x}^H\tilde{\mathbf{F}}_i\mathbf{x} - \frac{b_i^\ell}{a^\ell}\mathbf{x}^H\tilde{\mathbf{F}}_i\mathbf{x}$ will be constructed. Let $\mathbf{X} = \mathbf{x}\mathbf{x}^H$, then

$$\mathbf{x}^H\tilde{\mathbf{F}}_i^H\mathbf{x}\mathbf{x}^H\tilde{\mathbf{F}}_i\mathbf{x} = \text{vec}\left(\mathbf{X}\right)^H\text{vec}\left(\tilde{\mathbf{F}}_i\right)\text{vec}\left(\tilde{\mathbf{F}}_i\right)^H\text{vec}\left(\mathbf{X}\right). \tag{6.114}$$

The largest eigenvalue of $\bar{\mathbf{F}}_i = \text{vec}\left(\tilde{\mathbf{F}}_i\right)\text{vec}\left(\tilde{\mathbf{F}}_i\right)^H$ is

$$\lambda_{max}\left(\bar{\mathbf{F}}_i\right) = \text{vec}\left(\tilde{\mathbf{F}}_i\right)^H\text{vec}\left(\tilde{\mathbf{F}}_i\right) = \frac{1}{\alpha^2}\left((1+\varepsilon)^2N - 2(1+\varepsilon) + 1\right) \tag{6.115}$$

According to [22, Lemma 1], we have

$$\begin{aligned}
&\mathbf{x}^H\tilde{\mathbf{F}}_i^H\mathbf{x}\mathbf{x}^H\tilde{\mathbf{F}}_i\mathbf{x}\\
=&\text{vec}\left(\mathbf{X}\right)^H\text{vec}\left(\tilde{\mathbf{F}}_i\right)\text{vec}\left(\tilde{\mathbf{F}}_i\right)^H\text{vec}\left(\mathbf{X}\right)\\
\leq&2\text{Re}\left(\text{vec}\left(\mathbf{X}_\ell\right)^H\left(\bar{\mathbf{F}}_i - \lambda_{max}\left(\bar{\mathbf{F}}_i\right)\mathbf{I}\right)\text{vec}\left(\mathbf{X}\right)\right) + \text{constant}\\
=&2\mathbf{x}^H\left(\mathbf{x}_\ell^H\tilde{\mathbf{F}}_i\mathbf{x}_\ell\tilde{\mathbf{F}}_i - \lambda_{max}\left(\bar{\mathbf{F}}_i\right)\mathbf{X}_\ell\right)\mathbf{x} + \text{constant}.
\end{aligned} \tag{6.116}$$

Thus, we have

$$\frac{1}{2a^\ell}\mathbf{x}^H\tilde{\mathbf{F}}_i^H\mathbf{x}\mathbf{x}^H\tilde{\mathbf{F}}_i\mathbf{x} - \frac{b_i^\ell}{a^\ell}\mathbf{x}^H\tilde{\mathbf{F}}_i\mathbf{x} \leq \mathbf{x}^H\mathbf{A}_i^\ell\mathbf{x} + \text{constant}, \tag{6.117}$$

where $\mathbf{A}_i^\ell$ is defined by (6.65). Due to $\mathbf{A}_i^\ell \preceq \mathbf{0}$, the concave term $\mathbf{x}^H\mathbf{A}_i^\ell\mathbf{x}$ can be further majorized by its first-order Taylor expansion given by

$$\mathbf{x}^H\mathbf{A}_i^\ell\mathbf{x} \leq 2\text{Re}\left(\mathbf{x}_\ell^H\mathbf{A}_i^\ell\mathbf{x}\right) + \text{constant}. \tag{6.118}$$

We have

$$\frac{\mathbf{x}^H\tilde{\mathbf{F}}_i^H\mathbf{x}\mathbf{x}^H\tilde{\mathbf{F}}_i\mathbf{x}}{2a^\ell} - \frac{b_i^\ell}{a^\ell}\mathbf{x}^H\tilde{\mathbf{F}}_i\mathbf{x} \le 2\mathrm{Re}\left(\mathbf{x}_\ell^H\mathbf{A}_i^\ell\mathbf{x}\right) + \text{constant}, \tag{6.119}$$

Therefore, by combining (6.113) and (6.119), we have

$$\log\sum_{i\in\mathcal{S}}\exp\left(-\mathbf{x}^H\tilde{\mathbf{F}}_i\mathbf{x}\right) \le 2\mathrm{Re}\left[\left(\sum_{i\in\mathcal{S}}\mathbf{A}_i^\ell\mathbf{x}_\ell\right)^H\mathbf{x}\right] + \text{constant}.$$

□

A.5 Proof of Lemma 8

Proof. Define $f_1(\mathbf{x}) = \sum_{j\in\mathcal{P}} q_j\mathbf{x}^H\mathbf{F}_j\mathbf{x}$ and $f_2(\mathbf{x}) = \sum_{i\in\mathcal{S}} p_i\mathbf{x}^H\mathbf{F}_i\mathbf{x}$. The convex function $f_1(\mathbf{x})$ can be lower bounded by its first-order Taylor expansion as follows:

$$\sum_{j\in\mathcal{P}} q_j\mathbf{x}^H\mathbf{F}_j\mathbf{x} \ge \sum_{j\in\mathcal{P}} q_j\left[\mathbf{x}_\ell^H\mathbf{F}_j\mathbf{x}_\ell + 2\mathrm{Re}\left(\mathbf{x}_\ell^H\mathbf{F}_j(\mathbf{x}-\mathbf{x}_\ell)\right)\right]. \tag{6.120}$$

According to [22, Lemma 1], for each $i\in\mathcal{S}$, we have

$$\mathbf{x}^H\mathbf{F}_i\mathbf{x} \le \mathbf{x}^H\mathbf{x} + 2\mathrm{Re}\left(\mathbf{x}_\ell^H(\mathbf{F}_i-\mathbf{I})\mathbf{x}\right) + \mathbf{x}_\ell^H(\mathbf{I}-\mathbf{F}_i)\mathbf{x}_\ell, \tag{6.121}$$

where $\lambda_u(\mathbf{F}_i)$ is an upper bound of the eigenvalues of $\mathbf{F}_i$.

By combining (6.120) and (6.121) and doing some algebra manipulations, we have

$$\begin{aligned}&\sum_{j\in\mathcal{P}} q_j\mathbf{x}^H\mathbf{F}_j\mathbf{x} - y_k\sum_{i\in\mathcal{S}} p_i\mathbf{x}^H\mathbf{F}_i\mathbf{x}\\ \ge&\mathrm{Re}\left[\mathbf{x}_\ell^H\left(\sum_{j\in\mathcal{P}}2q_j\mathbf{F}_j - y_k\sum_{i\in\mathcal{S}}2p_i(\mathbf{F}_i-\mathbf{I})\right)^H\mathbf{x}\right]\\ &-y_k\sum_{i\in\mathcal{S}}p_i\left(N+\mathbf{x}_\ell^H(\mathbf{I}-\mathbf{F}_i)\mathbf{x}_\ell\right) - \sum_{j\in\mathcal{P}}q_j\mathbf{x}_\ell^H\mathbf{F}_j\mathbf{x}_\ell.\end{aligned} \tag{6.122}$$

by defining $\mathbf{a}_\ell$ and $u_\ell(\mathbf{p},\mathbf{q})$ as (6.48) and (6.84), we have

$$\begin{aligned}&\min_{\mathbf{p}\in\mathcal{S}_1,\mathbf{q}\in\mathcal{S}_2}\left\{\sum_{j\in\mathcal{P}}q_j\mathbf{x}^H\mathbf{F}_j\mathbf{x} - y_k\sum_{i\in\mathcal{S}}p_i\mathbf{x}^H\mathbf{F}_i\mathbf{x}\right\}\\ \ge&\min_{\mathbf{p}\in\mathcal{S}_1,\mathbf{q}\in\mathcal{S}_2}\left\{\mathrm{Re}\left\{\mathbf{a}_\ell^H\mathbf{x}\right\}+u_\ell(\mathbf{p},\mathbf{q})\right\}\end{aligned} \tag{6.123}$$

with equality achieved when $\mathbf{x}=\mathbf{x}_\ell$. □

A.6 Proof of Lemma 9

Proof. First, problem (6.85) is equivalent to

$$\begin{aligned} \underset{\mathbf{x}}{\text{maximize}} \quad & \min_{\mathbf{p}\in\mathscr{S}_1,\mathbf{q}\in\mathscr{S}_2} \left\{ \mathrm{Re}\left\{ \mathbf{a}_\ell^H \mathbf{x} \right\} + u_\ell(\mathbf{p},\mathbf{q}) \right\} \\ \text{subject to} \quad & \|\mathbf{x}\|_2^2 \le N. \\ & |x_n| \le \sqrt{\gamma} \text{ for } n = 1,\ldots,N. \end{aligned} \tag{6.124}$$

The optimal solution to problem (6.124) should satisfy $\|\mathbf{x}^\star\|_2^2 = N$. Otherwise, we can always scale up some elements of $\mathbf{x}^\star$ with a larger objective value.

For problem (6.124), the objective function is bilinear in $\mathbf{x}$ and $(\mathbf{p},\mathbf{q})$, and the constraint sets for $\mathbf{x}$ and $(\mathbf{p},\mathbf{q})$ are both compact convex. According to the minimax theorem [46–48], the equality is achieved so that max and min can be exchanged. Thus, we have the following equivalent problem:

$$\begin{aligned} \underset{\mathbf{p},\mathbf{q}}{\text{minimize}} \quad & \left(\max_{\|\mathbf{x}\|_2^2\le N, |x_n|\le\sqrt{\gamma}} \mathrm{Re}\left\{ \mathbf{a}_\ell^H \mathbf{x} \right\} \right) + u_\ell(\mathbf{p},\mathbf{q}) \\ \text{subject to} \quad & \mathbf{p}\in\mathscr{S}_1, \mathbf{q}\in\mathscr{S}_2. \end{aligned} \tag{6.125}$$

□

References

[1] Haykin S. Cognitive radar: a way of the future. *IEEE Signal Processing Magazine*. 2006;23(1):30–40.

[2] Bell KL, Baker CJ, Smith GE, *et al.* Cognitive radar framework for target detection and tracking. *IEEE Journal of Selected Topics in Signal Processing*. 2015;9(8):1427–1439.

[3] Metcalf J, Blunt SD, and Himed B. A machine learning approach to cognitive radar detection. In: *2015 IEEE Radar Conference (RadarCon)*. Piscataway, NJ: IEEE; 2015. p. 1405–1411.

[4] Turlapaty A and Jin Y. Bayesian sequential parameter estimation by cognitive radar with multiantenna arrays. *IEEE Transactions on Signal Processing*. 2014;63(4):974–987.

[5] Mishra KV, Shoshan E, Namer M, *et al.* Cognitive sub-Nyquist hardware prototype of a collocated MIMO radar. In: *2016 4th International Workshop on Compressed Sensing Theory and its Applications to Radar, Sonar and Remote Sensing (CoSeRa)*. Piscataway, NJ: IEEE; 2016. p. 56–60.

[6] He H, Li J, and Stoica P. *Waveform Design for Active Sensing Systems: A Computational Approach*. Cambridge: Cambridge University Press; 2012.

[7] Blunt SD and Mokole EL. Overview of radar waveform diversity. *IEEE Aerospace and Electronic Systems Magazine*. 2016;31(11):2–42.

[8] Zhao Y, Gaeddert J, Bae KK, *et al.* Radio environment map enabled situation-aware cognitive radio learning algorithms. In: *Software Defined Radio Forum (SDRF) Technical Conference*; 2006.

[9] Gini F, De Maio A, and Patton L. *Waveform Design and Diversity for Advanced Radar Systems*. London: Institution of Engineering and Technology; 2012.

[10] Chen CY and Vaidyanathan P. MIMO radar ambiguity properties and optimization using frequency-hopping waveforms. *IEEE Transactions on Signal Processing*. 2008;56(12):5926–5936.

[11] Aubry A, De Maio A, Jiang B, *et al.* Ambiguity function shaping for cognitive radar via complex quartic optimization. *IEEE Transactions on Signal Processing*. 2013;61(22):5603–5619.

[12] Wu L, Babu P, and Palomar DP. Cognitive radar-based sequence design via SINR maximization. *IEEE Transactions on Signal Processing*. 2016;65(3):779–793.

[13] Jing Y, Liang J, Tang B, *et al.* Designing unimodular sequence with low peak of sidelobe level of local ambiguity function. *IEEE Transactions on Aerospace and Electronic Systems*. 2018;55(3):1393–1406.

[14] Levanon N and Mozeson E. *Radar Signals*. New York, NY: John Wiley & Sons; 2004.

[15] Pinilla S, Mishr KV, Sadler BM, *et al.* Banraw: band-limited radar waveform design via phase retrieval. In: *ICASSP 2021-2021 IEEE International Conference on Acoustics, Speech and Signal Processing (ICASSP)*. Piscataway, NJ: IEEE; 2021. p. 5449–5453.

[16] Pinilla S, Mishra KV, and Sadler B. WaveMax: FrFT-based convex phase retrieval for radar waveform design. In: *2021 IEEE International Symposium on Information Theory (ISIT)*. Piscataway, NJ: IEEE; 2021. p. 2387–2392.

[17] Price R and Hofstetter E. Bounds on the volume and height distributions of the ambiguity function. *IEEE Transactions on Information Theory*. 1965;11(2):207–214.

[18] San Antonio G, Fuhrmann DR, and Robey FC. MIMO radar ambiguity functions. *IEEE Journal of Selected Topics in Signal Processing*. 2007;1(1): 167–177.

[19] Abramovich YI and Frazer GJ. Bounds on the volume and height distributions for the MIMO radar ambiguity function. *IEEE Signal Processing Letters*. 2008;15:505–508.

[20] Li Y, Vorobyov SA, and Koivunen V. Ambiguity function of the transmit beamspace-based MIMO radar. *IEEE Transactions on Signal Processing*. 2015;63(17):4445–4457.

[21] Stoica P, He H, and Li J. New algorithms for designing unimodular sequences with good correlation properties. *IEEE Transactions on Signal Processing*. 2009;57(4):1415–1425.

[22] Song J, Babu P, and Palomar DP. Optimization methods for designing sequences with low autocorrelation sidelobes. *IEEE Transactions on Signal Processing*. 2015;63(15):3998–4009.

[23] Kerahroodi MA, Aubry A, De Maio A, *et al.* A coordinate-descent framework to design low PSL/ISL sequences. *IEEE Transactions on Signal Processing.* 2017;65(22):5942–5956.

[24] Wu L and Palomar DP. Sequence design for spectral shaping via minimization of regularized spectral level ratio. *IEEE Transactions on Signal Processing.* 2019;67(18):4683–4695.

[25] Cohen L. The generalization of the Wiener–Khinchin theorem. In: *Proceedings of the 1998 IEEE International Conference on Acoustics, Speech and Signal Processing, ICASSP'98* (Cat. No. 98CH36181). vol. 3. Piscataway, NJ: IEEE; 1998. p. 1577–1580.

[26] Razaviyayn M, Hong M, and Luo ZQ. A unified convergence analysis of block successive minimization methods for nonsmooth optimization. *SIAM Journal on Optimization.* 2013;23(2):1126–1153.

[27] Sun Y, Babu P, and Palomar DP. Majorization–minimization algorithms in signal processing, communications, and machine learning. *IEEE Transactions on Signal Processing.* 2017;65(3):794–816.

[28] Dinkelbach W. On nonlinear fractional programming. *Management Science.* 1967;13(7):492–498.

[29] Fan W, Liang J, and Li J. Constant modulus MIMO radar waveform design with minimum peak sidelobe transmit beampattern. *IEEE Transactions on Signal Processing.* 2018;66(16):4207–4222.

[30] Shen K and Yu W. Fractional programming for communication systems – Part I: power control and beamforming. *IEEE Transactions on Signal Processing.* 2018;66(10):2616–2630.

[31] Borde J and Crouzeix JP. Convergence of a Dinkelbach-type algorithm in generalized fractional programming. *Zeitschrift für Operations Research.* 1987;31(1):A31–A54.

[32] Schaible S. Fractional programming. II: on Dinkelbach's algorithm. *Management Science.* 1976;22(8):868–873.

[33] Mahafza BR. *Introduction to Radar Analysis.* Boca Raton, FL: CRC Press; 1998.

[34] Richards MA, Scheer JA, Holm WA, *et al.* *Principles of Modern Radar.* Stevenage: SciTech Publishing; 2010.

[35] He H, Stoica P, and Li J. Waveform design with stopband and correlation constraints for cognitive radar. In: *2010 2nd International Workshop on Cognitive Information Processing (CIP).* Piscataway, NJ: IEEE; 2010. p. 344–349.

[36] Song J, Babu P, and Palomar DP. Sequence Design to Minimize the Weighted Integrated and Peak Sidelobe Levels. arXiv preprint arXiv:150604234. 2015. http://arxiv.org/abs/1506.04234.

[37] Wolkowicz H and Styan GP. Bounds for eigenvalues using traces. *Linear Algebra and Its Applications.* 1980;29:471–506.

[38] Tropp JA, Dhillon IS, Heath RW, *et al.* Designing structured tight frames via an alternating projection method. *IEEE Transactions on Information Theory.* 2005;51(1):188–209.

[39] Varadhan R and Roland C. Simple and globally convergent methods for accelerating the convergence of any EM algorithm. *Scandinavian Journal of Statistics*. 2008;35(2):335–353.

[40] Wicks M. Spectrum crowding and cognitive radar. In: *2010 2nd International Workshop on Cognitive Information Processing (CIP)*. Piscataway, NJ: IEEE; 2010. p. 452–457.

[41] Davis ME. Frequency allocation challenges for ultra-wideband radars. *IEEE Aerospace and Electronic Systems Magazine*. 2013;28(7):12–18.

[42] Jing Y, Liang J, Zhou D, *et al.* Spectrally constrained unimodular sequence design without spectral level mask. *IEEE Signal Processing Letters*. 2018;25(7):1004–1008.

[43] Bertsekas DP. Incremental gradient, subgradient, and proximal methods for convex optimization: a survey. *Optimization for Machine Learning*. 2011;2010(1-38):3.

[44] Ponnusamy S and Silverman H. *Complex Variables with Applications*. Berlin: Springer Science & Business Media; 2007.

[45] Boyd S and Vandenberghe L. *Convex Optimization*. Cambridge: Cambridge University Press; 2004.

[46] Palomar DP, Cioffi JM, and Lagunas MA. Uniform power allocation in MIMO channels: a game-theoretic approach. *IEEE Transactions on Information Theory*. 2003;49(7):1707–1727.

[47] Scutari G, Palomar DP, and Barbarossa S. Competitive design of multiuser MIMO systems based on game theory: a unified view. *IEEE Journal on Selected Areas in Communications*. 2008;26(7):1089–1103.

[48] Scutari G, Palomar DP, and Barbarossa S. Cognitive MIMO radio. *IEEE Signal Processing Magazine*. 2008;25(6):46–59.

Chapter 7

Training-based adaptive transmit–receive beamforming for MIMO radars

Mahdi Shaghaghi[1], *Raviraj S. Adve*[1] *and George Shehata*[1]

7.1 Introduction

To detect the presence of a target, a pulsed surveillance radar repeatedly transmits radar waveforms and processes the received returns. Given an antenna array, the transmitter uses transmit beamforming to focus its available power toward a chosen angle, most commonly known as the look direction. If the return signal is only corrupted by white noise, statistically uncorrelated across space and time, the optimal receiver implements matched filtering, i.e., the receive filter is matched to the transmitted waveform. Matched filtering maximizes the output signal-to-noise ratio (SNR); the SNR is further enhanced by receive beamforming also matched to the look direction. The multiple transmissions (pulses) can be processed to obtain the Doppler (target speed) information.

This basic approach requires significant revision in the case where the target returns are buried in interference, e.g., the clutter seen in an airborne radar. In this case, the optimal approach under Gaussian interference is the *adaptive*-matched filter which linearly combines the signals across antennas and pulses in a manner such that the signal-to-interference-plus-noise ratio (SINR) is maximized [1]. It is worth noting that other than a few works, e.g., [2], for the most part, researchers implementing such space–time adaptive processing (STAP) have assumed Gaussian interference. While the STAP approach is fairly simple to derive mathematically, its implementation is significantly complicated by the fact that the result depends on knowledge of the interference covariance matrix, which, in practice, is invariably *a priori* unknown. Effective estimation of the required covariance matrix has kept many researchers busy for many years [3]. A new related estimation approach entails channel estimation for a fully adaptive radar [4]. The need for this sort of estimation will underline this chapter as well as we extend receive processing to adapting transmissions as well.

As mentioned, a traditional radar system would repeatedly transmit the same waveform, possibly with a phase shift to achieve beamsteering [5,6]. More recently, however, it has become possible to create waveforms "in software," i.e., to design

[1]Department of Electrical and Computer Engineering, University of Toronto, Toronto, ON, Canada

waveforms in real-time to achieve some operational purpose, usually, again to maximize discrimination between signal and interference. This flexibility now allows each element in a transmit array to transmit a different waveform, i.e., the array would have multiple inputs. Crucially, this would allow the multiple receivers in an array to distinguish the impact of the individual transmissions; indeed, recently, such multiple-input multiple-output (MIMO) radar systems have received significant attention [7].

7.1.1 Background

The possibility of real-time waveform design forms the background for this chapter wherein we extend the receive adaptive processing scheme to adaptive transmissions in the context of a cognitive radar. At a receiver, the linear combination across antennas and pulses requires the scaling of each contribution by a complex scalar. This STAP-based approach can now also be used at the transmitter, i.e., each transmit element can scale and phase shift a template waveform before transmission. These scalars play a role in transmission similar to STAP weights for reception. Importantly, adaptive transmit processing and adaptive receive processing can complement each other to truly maximize the SINR. As with receive processing, mathematically deriving the optimal transmit weights is relatively straightforward. The fundamental question of this chapter is how to obtain, in practice, the needed information to execute the theory.

It is worth discussing the relationship between adaptive transmissions and cognitive radar. While many interpret a cognitive radar to require machine learning techniques, as stated in [8, Chapter 3], "The ability to feed measurements back into a system computer that will use these measurements and changing external conditions to optimize future measurements is a unique feature that helps to distinguish cognitive radar." As stated in [9], "The key strength of [a cognitive] system is its ability to learn the channel or target environment and then adapt both the transmitter and receiver to provide an enhanced performance." Several others have investigated this aspect of cognitive radar [10]. It is in this context that we develop real-time transmit adaptivity wherein the transmitter changes the weights of the template waveform to adapt to an *a priori* unknown environment; an environment that is probed and the resulting information is used for the adaptation process.

The weights used at the transmitter have also been referred to as the transmit code. In [11], Friedlander introduces the notion of transmit waveform design for MIMO radars. In recent times, a key step in this direction is the work of De Maio *et al.* who introduce the use of optimization theory in this application; specifically, the authors optimize the weights to maximize detection performance while maintaining similarity to a pre-chosen "good" code. Since these works, it has been shown in many different works that detection performance can be significantly improved by the *joint design* of the transmit code and the receive filter [12–21]. We also drop the reader's attention to the fundamental work in [22]. Here, we will assume a fixed waveform to be scaled by a factor to be determined. The scaling factor across space and time will be referred to as a code. The resulting optimization problems are generally not jointly convex in both the transmit and receive weights and, so, are usually solved by iterating between optimizing the transmit code and the receive weights or using biquadratic optimization [23]. The non-convexity of the overall problem was proven empirically in [24] and more formally using linear algebra considerations in [25], where it was shown that the weight vector on receive had to simultaneously satisfy two competing conditions and, hence, the non-convexity results.

As we will see, as with receive-only STAP techniques, the optimization problems rely on an assumed knowledge of the second-order statistics of the interference. In these works, this crucial information is either assumed known directly *a priori* or enough is known about the environment such that this information can be derived. While this sort of approach has shown some promise in STAP applications [26], assuming *a priori* knowledge of either the statistics or the environment seems restricted to special scenarios. This chapter will develop alternative approaches wherein the *transmit covariance matrix* is estimated from the *received data*.

7.1.2 Contributions

While the approach we develop can be used in most of the approaches mentioned, we will apply the estimation approach in the context of a novel optimization framework. As mentioned, a phased array radar searches for a target in a chosen look direction; however, the target's relative speed is unknown. In many cases, the optimization framework has focused on a single target speed (Doppler bin). Given the time available to complete the detection tasks, since the transmit codeword must be optimized for each case, it is unlikely that we can interrogate each possible Doppler bin individually. In this chapter, we consider all Doppler bins simultaneously; in this regard, we may wish to maximize the average SINR across all bins or the minimum SINR across all Doppler bins. Here, we choose to maximize the minimum SINR as in [20,21,27,28]. The key difference is that we now consider the practical case of unknown interference statistics.

It is worth commenting on why transmit adaptivity requires a new approach: first, as mentioned, the required statistics at the transmitter must be estimated using receive data, a seemingly non-causal process since the data is received after transmission. Second, interference sources such as clutter are, by definition, dependent on the transmit waveform. Any optimization formulation must, therefore, account for this dependence. To deal with the non-causal nature of the problem and the dependence of the statistics on the transmit waveform, we develop a *training-based* approach to probe the environment and obtain the required information. Importantly, in practice, this training has to be done only once since each subsequent pulse can be used for both detection and to obtain the next (set of) pulse(s). This capability would be useful in scenarios where the interference is dynamic; our motivating example is ionospheric clutter; the characteristics of ionospheric clutter change through a day [29].

Our discussion so far has assumed that we can jointly process all elements in the array and pulses in a coherent pulse interval (CPI). Unfortunately, the limited training available makes this almost impossible in practice—even for receive-only processing. This is because, in practice, the available training data is limited. In developing a practical approach to transmit adaptivity, we must also develop techniques for *reduced-dimension* adaptive techniques. On the receiving side, the usual approach is to reduce the number of *adaptive* degrees of freedom, requiring a fewer training samples. As with the fully-adaptive case, reduced-dimension transmit adaptivity is very different from the receive-only cases (though, our approach borrows from joint domain localized processing developed for receive processing [30,31]). It is worth emphasizing that while reduced-dimension techniques have the added benefit of reduced computational complexity, it is the limited available training samples that fundamentally drive the investigation of reduced-dimension techniques.

In summary, this chapter will introduce:

- an optimization problem to estimate the required "transmit" second-order statistics using received data, and
- reduced-dimension transmit adaptive processing.
  Both these items are, we believe, essential to be able to implement transmit adaptive processing.
- extensions of the transmit adaptivity problem to the max–min SINR case where the analysis covers multiple Doppler bins

Our numerical examples will cover two very different radar systems. The first is a collocated MIMO radar system in which the propagation suffers from random phase changes [29,32]. In such a radar, the interference that limits the detection is the clutter induced by the transmitted signal [20]. Consequently, the clutter statistics, specifically its covariance matrix, depend on the transmitted signal. The second case is an airborne radar system, which has a similar formulation to the random phase radar, with the addition of jamming signals in the interference component of the received signal [1]. Importantly, the jamming is independent of the transmitted signals.

7.2 System model

Consider a collocated MIMO radar system with N_T transmit and N_R receive antennas. The transmitted waveform from the nth antenna ($1 \leq n \leq N_T$) at time t is given by

$$u_n(t) = \sum_{m=1}^{M} C_{nm} s(t - mT), \tag{7.1}$$

where $s(t)$ is a template pulse shape common to all transmitters, T is the pulse repetition interval (PRI, the slow-time interval), M denotes the number of pulses that form a CPI, and C_{nm} denotes the amplitude of the mth transmitted pulse from the nth antenna. The template pulse $s(t)$ has unit energy, i.e., the energy in the transmitted signal $u_n(t)$ is determined by the amplitude term, C_{nm}. We define the code matrix $\mathbf{C} \in \mathbb{C}^{N_T \times M}$ by setting its (n, m)th element to C_{nm}. Transmit adaptivity implies optimizing this code matrix

In fast time, we have L range samples per PRI. Overall, the received signals form, therefore, an $N_R \times M \times L$ radar data cube; however, for our purposes, for each range bin, ℓ, it is more convenient to stack the received data from the M pulses and N_R receivers into a length-N_RM vector, $\mathbf{x}_\ell$. Our convention is such that the $((n -$

$1)M + m$)th element of $\mathbf{x}_\ell$ corresponds to the sample from the nth receive antenna ($1 \le n \le N_R$) of the mth slow-time pulse.

This data vector can be written as a combination of the contributions from a target $\mathbf{t}_\ell$ (possibly), clutter $\mathbf{q}_\ell$, and noise $\mathbf{w}_\ell$:

$$\mathbf{x}_\ell = \mathbf{t}_\ell + \mathbf{q}_\ell + \mathbf{w}_\ell. \tag{7.2}$$

In the following, for convenience, we drop the subscript ℓ. We begin by briefly reviewing the formulation of the target, clutter, and noise components in the context of a radar undergoing phase perturbations during transmission [33], with comments on how to change the model for an airborne radar.

7.2.1 Target contribution

The target is assumed to be a far-field point source with radar cross-section (RCS) α_t moving with normalized Doppler frequency f_t at azimuth angle ϕ_t (look angle-Doppler of (ϕ_t, f_t)). The normalization here is with respect to the pulse repetition frequency (PRF). We define the following vectors and matrices: let $\mathbf{a}_R(\phi) \in \mathbb{C}^{N_R \times 1}$ and $\mathbf{a}_T(\phi) \in \mathbb{C}^{N_T \times 1}$ be the receive and transmit steering vectors, respectively. For an inter-element spacing of d, the rth and nth elements of $\mathbf{a}_R(\phi)$ and $\mathbf{a}_T(\phi)$ are given by $\exp(j2\pi(d/\lambda)(r-1)\cos(\phi))$ and $\exp(j\pi(d/\lambda)(n-1)\cos(\phi))$, respectively, where λ denotes the operating wavelength. We define the Doppler vector $\mathbf{a}_D(f) \in \mathbb{C}^{M \times 1}$ such that its mth element is given by $\exp(j2\pi(m-1)f)$.

The steering vectors can be combined into the matrix $\check{\mathbf{\Theta}}(\phi, f) \in \mathbb{C}^{N_R M \times N_R M N_T}$ as

$$\check{\mathbf{\Theta}}(\phi, f) = \operatorname{diag}(\mathbf{a}_R(\phi)) \otimes \operatorname{diag}(\mathbf{a}_D(f)) \otimes (\mathbf{a}_T(\phi))^T, \tag{7.3}$$

where $\operatorname{diag}(\mathbf{a})$ represents a square matrix with its diagonal elements equal to the vector $\mathbf{a}$, and $\otimes$ and $(\cdot)^T$ represent the Kronecker product and the transpose operator, respectively.

Let the block diagonal matrix $\bar{\mathbf{C}} \in \mathbb{C}^{MN_T \times M}$ be formed by placing the columns of $\mathbf{C}$ as its diagonal blocks. Now, define matrix $\check{\mathbf{C}} \in \mathbb{C}^{N_R M N_T \times N_R M}$ as

$$\check{\mathbf{C}} = \mathbf{I}_{N_R} \otimes \bar{\mathbf{C}}, \tag{7.4}$$

where $\mathbf{I}_{N_R}$ is the $N_R \times N_R$ identity matrix. We assume that the signal received from the target by the rth antenna at the mth PRI is perturbed by the random phase φ_{rm}. Define the phase perturbation vector $\mathbf{p}_t \in \mathbb{C}^{N_R M \times 1}$ such that its $((r-1)M + m)$th element is equal to $\exp(j\varphi_{rm})$.

Given these settings, the target vector can be expressed as

$$\mathbf{t} = \alpha_t \check{\mathbf{\Theta}}_t \check{\mathbf{C}} \mathbf{p}_t, \tag{7.5}$$

where $\check{\mathbf{\Theta}}_t$ is shorthand for $\check{\mathbf{\Theta}}(\phi_0, f_0)$. Importantly, $\check{\mathbf{\Theta}}_t \check{\mathbf{C}}$ is a diagonal matrix. To show this, we use the fact that for arbitrary matrices $\mathbf{A}$, $\mathbf{B}$, $\mathbf{C}$, and $\mathbf{D}$ with consistent sizes, we have: $(\mathbf{A} \otimes \mathbf{B})(\mathbf{C} \otimes \mathbf{D}) = (\mathbf{AC}) \otimes (\mathbf{BD})$. Using this fact, and the expressions in (7.3) and (7.4), we have

$$\check{\mathbf{\Theta}}_t \check{\mathbf{C}} = \left(\operatorname{diag}(\mathbf{a}_R(\phi))\, \mathbf{I}_{N_R}\right) \otimes \left(\left(\operatorname{diag}(\mathbf{a}_D(f)) \otimes (\mathbf{a}_T(\phi))^T\right) \bar{\mathbf{C}}\right).$$

Clearly, diag $(\mathbf{a}_R(\phi))\,\mathbf{I}_{N_R} = \text{diag}\,(\mathbf{a}_R(\phi))$ is a diagonal matrix. As the Kronecker product of a diagonal matrix and a row vector, diag $(\mathbf{a}_D(f)) \otimes (\mathbf{a}_T(\phi))^T$ is a block diagonal matrix, where each block is of size $1 \times N_T$. The matrix $\bar{\mathbf{C}}$ is also block diagonal wherein each block is to length-N_T column vector. Thus, $\left(\text{diag}\,(\mathbf{a}_D(f)) \otimes (\mathbf{a}_T(\phi))^T\right)\bar{\mathbf{C}}$ is a diagonal matrix. Finally, $\breve{\boldsymbol{\Theta}}_t\breve{\mathbf{C}}$ is diagonal since the Kronecker product of two diagonal matrices is also diagonal.

This analysis is important because, now, the target vector can also be written as

$$\mathbf{t} = \alpha_t \breve{\mathbf{C}}^T \breve{\boldsymbol{\Theta}}_t^T \mathbf{p}_t. \tag{7.6}$$

Later in this chapter, we will discuss the model appropriate for the target RCS. Finally, to extend this model to the case of airborne radar, we just have to remember that propagation is the line of sight, i.e., there is no phase perturbation. The random phase φ_{rm} is, therefore, zero, and, consequently, all the elements of $\mathbf{p}_t$ are equal to one.

7.2.2 Clutter contribution

As is common, the clutter model is based on the target model [1]. The clutter at any range bin is represented as a superposition of V rays incident from azimuth angles ϕ_v $(1 \le v \le V)$ with RCS α_v and normalized Doppler frequency $f_v(\phi_v)$. Similar to (7.6), the clutter vector is given by

$$\mathbf{q} = \sum_{v=1}^{V} \alpha_v \breve{\mathbf{C}}^T \breve{\boldsymbol{\Theta}}_v^T \mathbf{p}_v, \tag{7.7}$$

where $\breve{\boldsymbol{\Theta}}_v$ is shorthand for $\breve{\boldsymbol{\Theta}}(\phi_v, f_v)$ and $\mathbf{p}_v$ is the phase perturbation vector of ray v.

As we will see, the clutter covariance matrix is particularly important. It is defined as $\mathbf{R}_q = \mathbb{E}\left\{\mathbf{q}\mathbf{q}^H\right\}$, where $\mathbb{E}\{\cdot\}$ and $(\,\cdot\,)^H$ stand for the expectation and Hermitian operators, respectively. Using (7.7), $\mathbf{R}_q$ is given by

$$\mathbf{R}_q = \breve{\mathbf{C}}^T \mathbf{R}_\phi \breve{\mathbf{C}}^*, \tag{7.8}$$

where $(\,\cdot\,)^*$ represents the complex conjugate operator and $\mathbf{R}_\phi \in \mathbb{C}^{N_R M N_T \times N_R M N_T}$.

Importantly, (7.8) expresses the clutter covariance matrix into the impact of the code matrix (in $\breve{\mathbf{C}}$) and another covariance matrix, $\mathbf{R}_\phi$, that is *independent of the transmit code*. For the model in (7.7), $\mathbf{R}_\phi$ is given by

$$\mathbf{R}_\phi = \sum_{v=1}^{V}\sum_{u=1}^{V} \mathbb{E}\left\{\alpha_v \alpha_u^* \breve{\boldsymbol{\Theta}}_v^T \mathbf{p}_v \mathbf{p}_u^H \breve{\boldsymbol{\Theta}}_u^*\right\}. \tag{7.9}$$

In the case of an airborne radar, homogeneous clutter is modeled using clutter rays incident from all azimuth angles. The RCS of the clutter patches are independent and, as with the target model, there are no phase perturbations. The Doppler associated with a clutter ray at azimuth ϕ_v is given by [1]

$$f_v(\phi_v) = \frac{2v_p T}{\lambda} \cos(\phi_v)$$

where v_p is the velocity of the aircraft. For the airborne case, therefore, (7.9) must be replaced with

$$\mathbf{R}_\phi = \sum_{v=1}^{V} \mathbb{E}[|\alpha_v|^2] \breve{\mathbf{\Theta}}_v^T \mathbf{1}_{N_R M} \mathbf{1}_{N_R M}^T \breve{\mathbf{\Theta}}_v^*. \tag{7.10}$$

where the double summation is replaced by a single summation since the clutter rays are assumed to be zero-mean and independent and the phase perturbation vector is replaced with the all-ones vector $\mathbf{1}_{N_R M}$, of length $N_R M$. The average power in the vth clutter patch, $\mathbb{E}[|\alpha_v|^2]$, can be obtained by the element beampattern, transmit power, and other system parameters [1].

7.2.3 Noise model

We distinguish noise from clutter in terms of key characteristics—noise is independent of the transmitted signal. The two components of noise we consider here are thermal noise (also known as white noise) and jamming, modeled here as a white noise jammer.

Thermal noise is modeled as zero-mean circularly-symmetric complex Gaussian $\mathcal{CN}(\mathbf{0}, \mathbf{R}_n)$ where $\mathbf{R}_n = \sigma_n^2 \mathbf{I}_{N_R M}$. The model uses a scaled identity matrix, i.e., the thermal noise components in space and time are both uncorrelated and have equal variance (white noise model).

Jamming is particularly important in the case of an airborne radar. In this case, noise vector has an additional component modeled as a barrage noise jammer arising from angle ϕ_j. The covariance matrix of this specific component is given by [1]

$$\mathbf{R}_j = \sigma_j^2 \mathbf{a}_R(\phi_j) \left(\mathbf{a}_R(\phi_j)\right)^H \otimes \mathbf{I}_M, \tag{7.11}$$

where σ_j^2 is the jammer power. In this case, the overall noise covariance $\mathbf{R}_w$ comprises the covariance matrices of the white noise and the jamming signal given by

$$\mathbf{R}_w = \mathbf{R}_n + \mathbf{R}_j. \tag{7.12}$$

7.3 Adaptive beamforming

Having developed our data model, we are now ready to develop the optimization problem at hand—our objective to maximize is the SINR; our optimization variables are the receive and transmit beamformers. When the covariance matrices are known, this is now a solved problem [13,16,33]. The main contribution here is to consider what can be done when the interference statistics are unknown. As we will see, the receive beamforming is essentially the same as in the vast body of literature on STAP [3] (and the references within).

Transmit beamforming refers to choosing the transmit code matrix $\mathbf{C}$, while receive beamforming refers to choosing a combining vector $\mathbf{h}$ applied to the receive data vector $\mathbf{x}$ (here, the vector $\mathbf{x}$ refers to the received signal within the range cell of interest). The output of the receive beamformer given by $z = \mathbf{h}^H \mathbf{x}$ is used to determine the presence of a target at a specified look azimuth angle, ϕ_0, and normalized Doppler frequency, f_0, for a given range bin. The figure of merit and our objective function are the SINR given by

$$\mathrm{SINR} = \frac{\mathbb{E}\left\{|\mathbf{h}^H\mathbf{t}(\phi_0,f_0)|^2\right\}}{\mathbb{E}\left\{|\mathbf{h}^H\mathbf{q}|^2\right\}+\mathbb{E}\left\{|\mathbf{h}^H\mathbf{w}|^2\right\}}, \tag{7.13}$$

where $\mathbf{t}(\phi_0,f_0)$ denotes the steering vector corresponding to the look angle–Doppler and is given by (7.5). Importantly, in this equation, the look angle–Doppler may not match the angle–Doppler of a true target, i.e., it is not necessary that $(\phi_0,f_0) = (\phi_t,f_t)$. This is an important consideration—ideally, a target would be detected only when $(\phi_0,f_0) = (\phi_t,f_t)$. Indeed, when $(\phi_0,f_0) \neq (\phi_t,f_t)$, the target acts as discrete interference! In what follows, for the ease of exposition, where the correspondence is clear, we will drop the specification of the look angle–Doppler (ϕ_0,f_0). The transmit code and receive weights are, unless otherwise specified, designed to maximize the SINR corresponding to the look angle–Doppler.

Using (7.5), the numerator in (7.13) is given by

$$\mathbb{E}\left\{|\mathbf{h}^H\mathbf{t}|^2\right\} = |\alpha_t|^2\mathbf{h}^H\check{\mathbf{\Theta}}_0\check{\mathbf{C}}\mathbf{R}_0^{\phi}\check{\mathbf{C}}^H\check{\mathbf{\Theta}}_0^H\mathbf{h} \tag{7.14}$$

where $\mathbf{R}_0^{\phi} = \mathbb{E}\left\{\mathbf{p}_0\mathbf{p}_0^H\right\}$ and where $\check{\mathbf{\Theta}}_0$ is shorthand for $\check{\mathbf{\Theta}}(\phi_0,f_0)$. Since it only represents a scale factor, we can set $\alpha_t = 1$.* The denominator of (7.13) can be written as

$$\mathbb{E}\left\{|\mathbf{h}^H\mathbf{q}|^2\right\}+\mathbb{E}\left\{|\mathbf{h}^H\mathbf{w}|^2\right\} = \mathbf{h}^H\mathbf{R}_{qw}\mathbf{h} \tag{7.15}$$

where $\mathbf{R}_{qw} = \mathbf{R}_q + \mathbf{R}_w$.

Optimizing the SINR over $\mathbf{h}$ requires the knowledge of $\mathbf{R}_0^{\phi}$ and $\mathbf{R}_{qw}$. In practice, these matrices are unknown and need to be estimated from the received data. The matrix $\mathbf{R}_0^{\phi}$, the covariance matrix of the target phase perturbations, is particularly difficult to estimate since the target is assumed to be only present in a specific range bin, azimuth angle, and Doppler frequency. Unless a phase perturbation model is available from the underlying propagation physics, it is essentially impossible to estimate. We note that even a physics-based model includes parameters that must be estimated. We propose that, *for the purposes of the optimization*, we eliminate phase perturbations, i.e., replace $\mathbf{R}_0^{\phi}$ with the all-ones matrix to simplify the problem. This is equivalent to assuming the random phases in the target vector to be zero. In the airborne radar system, the random phases are already zero.

Based on this approximation and (7.13), (7.14), and (7.15), the SINR can be written as

$$\mathrm{SINR} = \frac{\left|\mathbf{h}^H\check{\mathbf{\Theta}}_0\check{\mathbf{C}}\mathbf{1}_{N_RM}\right|^2}{\mathbf{h}^H\mathbf{R}_{qw}\mathbf{h}}. \tag{7.16}$$

Given that they contribute to both the numerator and the denominator in a specific manner, the objective function at hand is a non-convex function of the optimization variables, the transmit code matrix $\mathbf{C}$ and the weight vector $\mathbf{h}$, making it difficult

*It is worth commenting that in the vast majority of look angle and Doppler bins, there is no target present, i.e., $\alpha_0 = 0$. The SINR being optimized, therefore, is the *potential* SINR if a target were to be present.

to optimize. Generally, the approach taken is to iterate between transmit and receive beamforming [33]. We start by transmitting an initial code matrix $\mathbf{C}$ and obtain the best combining vector $\mathbf{h}$ corresponding to this transmission. Now, given this combining vector $\mathbf{h}$, we can obtain the best transmit code matrix to maximize the SINR. To meet a constraint on the available power, we set $\|\mathbf{C}\|_F^2 \leq 1$ where $\|\cdot\|_F^2$ indicates the Frobenius norm of a matrix. This iterative procedure is repeated.

7.3.1 Receive beamforming

Optimizing the receive beamformer, $\mathbf{h}$, for a fixed transmit code matrix, $\mathbf{C}$, is essentially the well-established STAP approach. We begin by noting that

$$\check{\boldsymbol{\Theta}}_0\check{\mathbf{C}}\mathbf{1}_{N_RM} = \boldsymbol{\Theta}_0\mathbf{c}, \tag{7.17}$$

where $\mathbf{c} \in \mathbb{C}^{MN_T\times 1}$ is the transmit code vector obtained by stacking the columns of matrix $\mathbf{C}$, the matrix $\boldsymbol{\Theta}(\phi,f) \in \mathbb{C}^{N_RM\times MN_T}$ is defined as

$$\boldsymbol{\Theta}(\phi,f) = \mathbf{a}_R(\phi) \otimes \mathrm{diag}\,(\mathbf{a}_D(f)) \otimes (\mathbf{a}_T(\phi))^T\,. \tag{7.18}$$

Using (7.17), the SINR can be rewritten as

$$\mathrm{SINR} = \frac{\left|\mathbf{h}^H\boldsymbol{\Theta}_0\mathbf{c}\right|^2}{\mathbf{h}^H\mathbf{R}_{qw}\mathbf{h}}. \tag{7.19}$$

As mentioned, we follow an iterative optimization procedure; assuming the transmit code vector $\mathbf{c}$ is known and the objective is to design the receive filter only $\mathbf{h}$ such that the SINR, as given in (7.19), is maximized. The solution to maximizing the SINR is the well-known minimum variance distortionless response (MVDR) beamformer [34] given by

$$\mathbf{h}_o = \frac{\mathbf{R}_{qw}^{-1}\boldsymbol{\Theta}_0\mathbf{c}}{\mathbf{c}^H\boldsymbol{\Theta}_0^H\mathbf{R}_{qw}^{-1}\boldsymbol{\Theta}_0\mathbf{c}}. \tag{7.20}$$

Finally, the matrix $\mathbf{R}_{qw}$ can be estimated using secondary data samples from adjacent range bins which are assumed to be target free [1]:

$$\widehat{\mathbf{R}}_{qw} = \frac{1}{K}\sum_{\ell=1}^{K}\mathbf{x}_\ell\mathbf{x}_\ell^H, \tag{7.21}$$

where $\mathbf{x}_\ell, \ell = 1,\ldots,K$ denote the K secondary data samples.

This estimation process assumes the K samples are statistically homogeneous and the Reed–Mallet–Brennan (RMB) rule suggests that K be greater than twice the adaptive degrees of freedom [3]. Here, since $\mathbf{h}$ is a length-N_RM vector, we would need K on the order of $2N_RM$. It may not be possible to obtain such a large number of homogeneous samples in practice. There is a wealth of literature on how to deal with non-homogeneous clutter and processing with a reduced number of degrees of freedom. We refer the reader to [3] and the references therein. Later in this chapter, we present one such approach.

7.3.2 Transmit beamforming: known covariance

Having developed an approach to optimize the receive beamformer, given the code vector $\mathbf{c}$, we now consider the converse problem: how to optimize the transmit beamformer assuming the receive beamformer $\mathbf{h}$ is assumed known (from the previous transmission and using (7.20).) Using (7.6) and also noting that $|\mathbf{h}^H\mathbf{t}|^2$ is real, (7.14) can be rewritten as

$$\mathbb{E}\left\{|\mathbf{h}^H\mathbf{t}|^2\right\} = |\alpha_0|^2\mathbf{h}^T\check{\mathbf{C}}^H\check{\mathbf{\Theta}}_0^H\left(\mathbf{R}_t^\phi\right)^*\check{\mathbf{\Theta}}_0\check{\mathbf{C}}\mathbf{h}^*. \tag{7.22}$$

Using (7.18), we have $\mathbf{h}^T\check{\mathbf{C}}^H\check{\mathbf{\Theta}}_0^H = \mathbf{c}^H\mathbf{\Theta}_0^H\operatorname{diag}(\mathbf{h})$. We can, therefore, rewrite the numerator in the SINR, given in (7.22), as

$$\mathbb{E}\left\{|\mathbf{h}^H\mathbf{t}|^2\right\} = |\alpha_0|^2\mathbf{c}^H\mathbf{\Theta}_0^H\mathbf{H}\left(\mathbf{R}_t^\phi\right)^*\mathbf{H}^H\mathbf{\Theta}_0\mathbf{c} \tag{7.23}$$

where $\mathbf{H} = \operatorname{diag}(\mathbf{h})$. Next, using (7.8) and also noting that $|\mathbf{h}^H\mathbf{q}|^2$ is real, we have

$$\mathbb{E}\left\{|\mathbf{h}^H\mathbf{q}|^2\right\} = \mathbf{h}^T\check{\mathbf{C}}^H\mathbf{R}_\phi^*\check{\mathbf{C}}\mathbf{h}^*. \tag{7.24}$$

Rewriting $\mathbf{h}^T\check{\mathbf{C}}^H$ as $\mathbf{h}^T\check{\mathbf{C}}^H = \mathbf{c}^H\check{\mathbf{H}}$ by defining the $MN_T \times N_RMN_T$ matrix $\check{\mathbf{H}}$ as

$$\check{\mathbf{H}} = \left(\mathbf{1}_{N_R}^T \otimes \mathbf{I}_{MN_T}\right)\operatorname{diag}\left(\mathbf{h}\otimes\mathbf{1}_{N_T}\right), \tag{7.25}$$

we have a useful expression for the denominator of the SINR given as

$$\mathbb{E}\left\{|\mathbf{h}^H\mathbf{q}|^2\right\} = \mathbf{c}^H\check{\mathbf{H}}\mathbf{R}_\phi^*\check{\mathbf{H}}^H\mathbf{c}. \tag{7.26}$$

Similar to the receive beamforming case, we replace $\mathbf{R}_t^\phi$ with the all-ones matrix and the matrices, $\mathbf{H}$ and $\check{\mathbf{H}}$ is based on the optimized receive beamforming vector $\mathbf{h}_o$ obtained in the previous section, denoted as $\mathbf{H}_o$ and $\check{\mathbf{H}}_o$, respectively. The power constraint on the transmit code matrix $\mathbf{C}$ given by $\|\mathbf{C}\|_F^2 \le 1$ can be rewritten as $\|\mathbf{c}\|^2 \le 1$.

With these settings, the transmit beamforming problem can be cast as the following optimization problem:

$$\begin{aligned}\mathbf{c}_o &= \arg\max_{\mathbf{c}} \frac{\left|\mathbf{c}^H\mathbf{\Theta}_0^H\mathbf{H}_o\mathbf{1}_{N_RM}\right|^2}{\mathbf{c}^H\check{\mathbf{H}}_o\mathbf{R}_\phi^*\check{\mathbf{H}}_o^H\mathbf{c} + \mathbf{h}_o^H\mathbf{R}_w\mathbf{h}_o}\\ &= \arg\max_{\mathbf{c}} \frac{\left|\mathbf{c}^H\mathbf{\Theta}_0^H\mathbf{h}_o\right|^2}{\mathbf{c}^H\check{\mathbf{H}}_o\mathbf{R}_\phi^*\check{\mathbf{H}}_o^H\mathbf{c} + \mathbf{h}_o^H\mathbf{R}_w\mathbf{h}_o}\\ &\qquad \text{subject to } \|\mathbf{c}\|^2 \le 1.\end{aligned} \tag{7.27}$$

Comparing (7.19) and (7.27), we see that the two expressions are very similar; what makes the transmit problem different is the power constraint. We use the work in [22] where the authors showed that the solution to (7.27) is the normalized version of the solution of the following unconstrained problem [22]:

$$\mathbf{c}_* = \arg\max_{\mathbf{c}} \frac{\left|\mathbf{c}^H\mathbf{\Theta}_0^H\mathbf{h}_o\right|^2}{\mathbf{c}^H\check{\mathbf{H}}_o\mathbf{R}_\phi^*\check{\mathbf{H}}_o^H\mathbf{c} + \mathbf{h}_o^H\mathbf{R}_w\mathbf{h}_o.\mathbf{c}^H\mathbf{c}} \tag{7.28}$$

i.e., $\mathbf{c}_o = \mathbf{c}_*/\|\mathbf{c}_*\|$. Note that $\mathbf{c}_*$ is not orthogonal to $\mathbf{\Theta}_t^H\mathbf{h}_o$, since it would result in the minimum SINR value of zero. Furthermore, the SINR value does not depend on the norm of $\mathbf{c}$. Therefore, to get a unique solution, we can add the constraint $\mathbf{c}^H\mathbf{\Theta}_t^H\mathbf{h}_o = 1$ to (7.28) without affecting the maximization. Defining the $MN_T \times MN_T$ matrices $\mathbf{\Gamma}_q$, $\mathbf{\Gamma}_w$, $\mathbf{\Gamma}_{qw}$ as

$$\mathbf{\Gamma}_q = \check{\mathbf{H}}_o\mathbf{R}_\phi^*\check{\mathbf{H}}_o^H, \tag{7.29}$$

$$\mathbf{\Gamma}_w = \mathbf{h}_o^H\mathbf{R}_w\mathbf{h}_o\mathbf{I}_{MN_T}, \tag{7.30}$$

$$\mathbf{\Gamma}_{qw} = \mathbf{\Gamma}_q + \mathbf{\Gamma}_w., \tag{7.31}$$

the solution to (7.28), $\mathbf{c}_*$, is given by

$$\mathbf{c}_* \propto \mathbf{\Gamma}_{qw}^{-1}\mathbf{\Theta}_0^H\mathbf{h}_o. \tag{7.32}$$

Finally, $\mathbf{c}_o$ is obtained by normalizing $\mathbf{c}_*$ as

$$\mathbf{c}_o = \frac{\mathbf{\Gamma}_{qw}^{-1}\mathbf{\Theta}_0^H\mathbf{h}_o}{\|\mathbf{\Gamma}_{qw}^{-1}\mathbf{\Theta}_0^H\mathbf{h}_o\|}. \tag{7.33}$$

7.3.3 Transmit BF: estimating the required covariance matrix

As is clear from (7.33), as with the case of receive beamforming, transmit beamforming requires knowledge of the second-order statistics in $\mathbf{R}_\phi$ and, more importantly, $\mathbf{R}_w$. Assuming these are known, the development has followed [16,33]. Let us now consider the missing piece of the puzzle, viz., how to enable transmit beamforming using *estimated* covariance matrices. This problem is different from receive beamforming because the "transmit" covariance matrix is to be estimated using *only received data.* Furthermore, for the case of receive beamforming, the processor requires only an estimate of the sum of the clutter and noise covariance matrices; in contrast, as is clear from (7.29) and (7.30), for transmit beamforming, the two covariance matrices, $\mathbf{R}_\phi$ and $\mathbf{R}_w$ are involved in two different functions before addition.

To obtain estimates of $\mathbf{R}_\phi$ and $\mathbf{R}_w$, we propose that the radar system begin with a few training CPIs; in these training CPIs, the code matrices are pre-selected, i.e., they are known. Specifically, we assume N_{tr} transmissions with code matrices $\mathbf{C}_1, \mathbf{C}_2, \cdots, \mathbf{C}_{N_{tr}}$ (equivalently, the code vectors $\mathbf{c}_1, \mathbf{c}_2, \cdots, \mathbf{c}_{N_{tr}}$). As in (7.21), for the ith transmission ($1 \le i \le N_{tr}$), the estimated clutter-plus-noise covariance matrix $\widehat{\mathbf{R}}_{qw}^{(i)}$ is obtained using the target-free received data vectors $\mathbf{x}_\ell^{(i)}$ from K secondary range bins as

$$\widehat{\mathbf{R}}_{qw}^{(i)} = \frac{1}{K}\sum_{\ell=1}^{K}\mathbf{x}_\ell^{(i)}\left(\mathbf{x}_\ell^{(i)}\right)^H. \tag{7.34}$$

Recalling (7.8), in the ith CPI, the true clutter-plus-noise covariance matrix $\mathbf{R}_{qw}^{(i)}$ is given by

$$\mathbf{R}_{qw}^{(i)} = \mathbf{R}_q^{(i)} + \mathbf{R}_w = \check{\mathbf{C}}_i^T\mathbf{R}_\phi\check{\mathbf{C}}_i^* + \mathbf{R}_w \tag{7.35}$$

where $\check{\mathbf{C}}_i$ is obtained from $\mathbf{C}_i$ as defined in (7.4). To estimate $\mathbf{R}_\phi$ and $\mathbf{R}_w$, we chose these estimates to minimize the resulting difference between $\widehat{\mathbf{R}}_{qw}^{(i)}$ and $\mathbf{R}_{qw}^{(i)}$. Importantly, $\mathbf{R}_\phi$ is positive semidefinite (denoted as $\mathbf{R}_\phi \succeq \mathbf{0}$) and, in the absence of jamming $\mathbf{R}_w = \sigma_n^2 \mathbf{I}_{N_R M}$ with unknown σ_n^2.

The estimation problem can, therefore, be cast as

$$
\begin{aligned}
(\widehat{\mathbf{R}}_\phi, \widehat{\mathbf{R}}_w) = \arg\min_{\mathbf{R}_\phi, \mathbf{R}_w} \sum_{i=1}^{N_{tr}} \left\| \check{\mathbf{C}}_i^T \mathbf{R}_\phi \check{\mathbf{C}}_i^* + \mathbf{R}_w - \widehat{\mathbf{R}}_{qw}^{(i)} \right\|_F \\
\text{subject to } \mathbf{R}_\phi \succeq \mathbf{0}, \\
\mathbf{R}_w = \gamma \mathbf{I}_{N_R M}, \\
\gamma \geq 0. \qquad (7.36)
\end{aligned}
$$

The optimization problem in (7.36) is convex and can, therefore, be efficiently solved using standard optimization tools. For an airborne radar system, where a jamming signal may be present, $\mathbf{R}_w$ is given by (7.12). In this case, instead of a diagonal matrix, the constraint on $\mathbf{R}_w$ is only that it should be positive semi-definite, i.e., the second constraint in (7.36) is replaced with $\mathbf{R}_w \succeq \mathbf{0}$. Importantly, estimating the transmit covariance using received data is very different from traditional STAP.

In summary, the steps of our proposed algorithm are:

1. Transmit a series of N_{tr} training codes.
2. For each transmission, estimate the receive beamformer using (7.20) in conjunction with (7.21).
3. Estimate the transmit covariance matrix by solving (7.36).
4. Use this estimated covariance matrix in conjunction with (7.33) to optimize the transmit code matrix. Use this code matrix for the next CPI.
5. Iterate steps 2–4 using the most recent N_{tr} transmissions.

To the best of our knowledge, this is the first formulation to adaptively design the transmit code, in real time, using training data. We emphasize that the formulation makes no assumptions on the structure of the covariance matrices involved (other than them being positive semi-definite). Clearly, if a structure is known (such as $\mathbf{R}_w$ being diagonal as in (7.36)), this can be incorporated into the formulation to improve the estimate.

It is worth asking the question why this process could not be done using the covariance matrix estimated for receive beamforming. Essentially, why is this training phase required? This is because to obtain good estimates of the covariance matrices, $\mathbf{R}_\phi$ and $\mathbf{R}_w$, we need multiple *different* samples of the covariance matrix (with a known transmit codeword). If we did not use this training phase, and transmitted the same codeword, we would obtain essentially the same estimated covariance providing poor estimates of $\mathbf{R}_\phi$ and $\mathbf{R}_w$. This issue arises because the *transmit* covariance matrices are estimated using *received* data. If we were to transmit the same signal (i.e., the same training), we would get essentially the same returns (subject to the randomness int environment).

7.4 Reduced-dimension transmit beamforming

In the previous section, we extended adaptive transmit beamforming to the case of unknown covariance matrices. We now consider an extension in a different direction: reducing the number of adaptive degrees of freedom. As mentioned, since the number of training samples required is at least twice the number of adaptive degrees of freedom, the fully adaptive process is usually impossible to implement in practice [3]. Here, we develop reduced-dimension processing algorithm based on the JDL approach of [30,31] algorithm. Reducing the number of adaptive degrees of freedom reduces the required training. It is worth emphasizing that the loss in performance from using fully adaptive processing is, often, not as large as would be expected—this is because, most often, the clutter covariance matrix, $\mathbf{R}_q$ is low-rank and, hence, reducing the number of adaptive degrees of freedom is possible.

Dimension reduction can be implemented as a linear transformation with any data vector $\mathbf{x}$ replaced with the reduced-dimension $\tilde{\mathbf{x}} = \mathbf{T}_R^H\mathbf{x}$, where $\mathbf{T}_R \in \mathbb{C}^{N_RM\times D}$ (the subscript $_R$ indicates receive processing). Usually, $D \ll N_RM$. The covariance matrix $\mathbf{R}_{qw}$ is replaced by the $D \times D$ matrix $\tilde{\mathbf{R}}_{qw} = \mathbf{T}_R^H\mathbf{R}_{qw}\mathbf{T}_R$.[†]

Given the transformation matrix $\mathbf{T}_R$, (7.15) is replaced with $\tilde{\mathbf{h}}^H\tilde{\mathbf{R}}_{qw}\tilde{\mathbf{h}}$ and the resulting SINR can be written as

$$\mathrm{SINR} = \frac{\left|\tilde{\mathbf{h}}^H\mathbf{T}_R^H\boldsymbol{\Theta}_0\mathbf{c}\right|^2}{\tilde{\mathbf{h}}^H\tilde{\mathbf{R}}_{qw}\tilde{\mathbf{h}}}. \tag{7.37}$$

The reduced-dimension MVDR beamformer is then given by

$$\tilde{\mathbf{h}}_o = \frac{\tilde{\mathbf{R}}_{qw}^{-1}\mathbf{T}_R^H\boldsymbol{\Theta}_0\mathbf{c}}{\mathbf{c}^H\boldsymbol{\Theta}_0^H\mathbf{T}_R\tilde{\mathbf{R}}_{qw}^{-1}\mathbf{T}_R^H\boldsymbol{\Theta}_0\mathbf{c}}. \tag{7.38}$$

The detection statistic z is formed as $\tilde{\mathbf{h}}_o^H\tilde{\mathbf{x}}$, i.e., this reduced-dimension beamforming vector, $\tilde{\mathbf{h}}_o$ corresponds to a full-size vector $\mathbf{h}_o = \mathbf{T}_R\tilde{\mathbf{h}}_o$.

Finally, the matrix $\tilde{\mathbf{R}}_{qw}$ can be estimated using a limited number of secondary target-free data from range bins close to the range bin of interest as

$$\hat{\tilde{\mathbf{R}}}_{qw} = \frac{1}{K}\sum_{\ell=1}^{K}\tilde{\mathbf{x}}_\ell\tilde{\mathbf{x}}_\ell^H \tag{7.39}$$

where K, as before, is the number of range bins used for covariance matrix estimation and $\tilde{\mathbf{x}}_\ell = \mathbf{T}_R^H\mathbf{x}_\ell$. The key difference is that now K is on the order of $2D$.

While several reduced-dimension STAP algorithms can be fit into the framework above, in the JDL algorithm, the columns of $\mathbf{T}_R$ are chosen to be a few steering vectors around the look angle–Doppler (ϕ_0,f_0). A popular choice is to choose vectors associated with 3 angle and 3 Doppler bins centered at the look angle–Doppler bin, leading to $D = 9$. Specifically, let ϕ_0^-, ϕ_0^+, f_0^-, and f_0^+ represent the neighboring grid points

[†]We expand on the choice of D later in this document.

to ϕ_0 and f_0, respectively. Define the matrices $\mathbf{A}_R(\phi_0) = \left[\mathbf{a}_R(\phi_0^-), \mathbf{a}_R(\phi_0), \mathbf{a}_R(\phi_0^+)\right]$ and $\mathbf{A}_D(f_0) = \left[\mathbf{a}_D(f_0^-), \mathbf{a}_D(f_0), \mathbf{a}_D(f_0^+)\right]$. Then, $\mathbf{T}_R$ is given by

$$\mathbf{T}_R = \mathbf{A}_R(\phi_0) \otimes \mathbf{A}_D(f_0). \tag{7.40}$$

Note that the transformation matrix, $\mathbf{T}_R$, is a function of the look angle–Doppler (ϕ_0, f_0).‡

We now consider how to extend this reduced-dimension receive beamforming to the transmit case. The development in Section 7.3.3 suffers two drawbacks. First, as in the receive case, the estimation of full-size matrix $\mathbf{R}_{qw}$ (to estimate $\mathbf{R}_\phi$ and $\mathbf{R}_w$) requires large sample support which is usually not available in practice. Second, the computational burden of the optimization problem in (7.36) can quickly become prohibitive for a system with a large number of antennas or a large number of pulses per CPI. Finally, in the expected dynamic scenarios, the required transmit covariance matrices need to be updated. These issues motivate reduced-dimension methods *for the transmitter* as well. Unfortunately, extending JDL to the transmit case is not straightforward.

With the assumption that, at each PRI, there is at least one active transmit antenna, the matrix $\check{\mathbf{C}}$ is full-rank. Then, it can be seen from (7.8) that if the clutter covariance matrix $\mathbf{R}_q$ is low-rank, the matrix $\mathbf{R}_\phi$ is also low-rank. As a result, the matrix $\boldsymbol{\Gamma}_{qw}$ as given in (7.31) is a summation of the low-rank matrix $\boldsymbol{\Gamma}_q$ and the diagonal matrix $\boldsymbol{\Gamma}_w$. Based on the work in Section 4.3 in [1], it can be shown that with such a structure, the vector $\mathbf{c}_*$ as given in (7.32) lies entirely in a low-dimensional subspace.

Specifically, assume the rank of $\boldsymbol{\Gamma}_q$ is r, and let the eigenvectors corresponding to its nonzero eigenvalues be arranged as the columns of matrix $\mathbf{E} \in \mathbb{C}^{MN_T \times r}$. Furthermore, define the adaptive steering vector for the transmitter as

$$\mathbf{s}(\phi, f) = (\boldsymbol{\Theta}(\phi, f))^H \mathbf{h}_o. \tag{7.41}$$

Then, it can be shown that

$$\mathbf{c}_* \subset \text{span}\{[\mathbf{s}_0, \mathbf{E}]\} \tag{7.42}$$

where $\mathbf{s}_0$ denotes $\mathbf{s}(\phi_0, f_0)$.

Theorem 1. *Let $\mathbf{T}_T \in \mathbb{C}^{MN_T \times J}$ be the transmitter dimensionality-reduction matrix, where $(r+1) \leq J \ll MN_T$. If $\mathbf{T}_T$ is designed to satisfy*

$$\text{span}\{[\mathbf{s}_0, \mathbf{E}]\} \subset \text{span}\{\mathbf{T}_T\} \tag{7.43}$$

then, the transmit code vector $\mathbf{c}_o$ is equal to

$$\mathbf{c}_o = \frac{\mathbf{T}_T \tilde{\mathbf{c}}_*}{\|\mathbf{T}_T \tilde{\mathbf{c}}_*\|} \tag{7.44}$$

where $\tilde{\mathbf{c}}_* \in \mathbb{C}^{J \times 1}$ is the reduced-dimension transmit code vector given by

$$\tilde{\mathbf{c}}_* = \left(\mathbf{T}_T^H \boldsymbol{\Gamma}_{qw} \mathbf{T}_T\right)^{-1} \mathbf{T}_T^H \boldsymbol{\Theta}_t^H \mathbf{h}_o. \tag{7.45}$$

‡It is worth mentioning that dimensionality reduction via random projections has been suggested. This is a potential research topic and could build upon the work in [35].

Proof. The proof follows the same lines as the proof of Theorem 2 in [1]. Here, J plays the role of the number of dimensions after dimensionality reduction. □

We should note that, so far, our derivation is self-referential in that designing $\mathbf{T}_T$ requires an estimate of $\mathbf{E}$, which in turn requires an estimate of $\boldsymbol{\Gamma}_q$. But, $\boldsymbol{\Gamma}_q$ is a large matrix (effectively, it has many degrees of freedom); attempting to estimate $\boldsymbol{\Gamma}_q$ would require as much training as before. We circumvent this problem using an approach based on the JDL algorithm as described earlier.

To form the transmit dimensionality reduction matrix, $\mathbf{T}_T$, we choose as its columns the adaptive transmit steering vectors as given in (7.41) around the look angle–Doppler. Specifically, let $\phi_1 = \phi_0^-$, $\phi_2 = \phi_0$, $\phi_3 = \phi_0^+$, $f_1 = f_0^-$, $f_2 = f_0$, and $f_3 = f_0^+$. Then, the jth column of $\mathbf{T}_T$ is chosen as $\mathbf{s}(\phi_l, f_m)$, where $j = 3(m-1) + l$ and $1 \le l, m \le 3$. In this case, we have $J = 9$. Note that the order of the columns of $\mathbf{T}_T$ does not make a difference.§ Furthermore, note that $\mathbf{s}(\phi_l, f_m)$ depends on the receive weight vector $\mathbf{h}_o$. Therefore, matrix $\mathbf{T}_T$ needs to be updated at every transmission. Computing $\mathbf{T}_T$ can be simplified using the fact that $\mathbf{h}_o = \mathbf{T}_R\tilde{\mathbf{h}}_o$, and that $\boldsymbol{\Theta}(\phi, f)$ and $\mathbf{T}_R$ are fixed for all transmissions.

Define $\tilde{\boldsymbol{\Gamma}}_{qw} \in \mathbb{C}^{J\times J}$ as

$$\tilde{\boldsymbol{\Gamma}}_{qw} = \mathbf{T}_T^H \boldsymbol{\Gamma}_{qw} \mathbf{T}_T. \tag{7.46}$$

Note that $\tilde{\boldsymbol{\Gamma}}_{qw}$ is a $J \times J$ matrix irrespective of the number of antennas or pulses. From (7.45), we require an estimate of this lower dimension matrix thereby reducing both the computation complexity and, most importantly, the required sample support. The remaining issue with our reduced-dimension approach is to estimate this matrix using the received data only.

Similar to the case of fully adaptive transmit beamforming, we transmit preselected using N_{tr} initial training codes $\mathbf{c}_1, \mathbf{c}_2, \cdots, \mathbf{c}_{N_{tr}}$. Define the matrices $\tilde{\boldsymbol{\Gamma}}_q$, $\tilde{\boldsymbol{\Gamma}}_w \in \mathbb{C}^{J\times J}$ as

$$\tilde{\boldsymbol{\Gamma}}_q = \mathbf{T}_T^H \boldsymbol{\Gamma}_q \mathbf{T}_T = \mathbf{T}_T^H \check{\mathbf{H}}_o \mathbf{R}_\phi^* \check{\mathbf{H}}_o^H \mathbf{T}_T \tag{7.47}$$

$$\tilde{\boldsymbol{\Gamma}}_w = \mathbf{T}_T^H \boldsymbol{\Gamma}_w \mathbf{T}_T = \mathbf{h}_o^H \mathbf{R}_w \mathbf{h}_o \mathbf{T}_T^H \mathbf{T}_T. \tag{7.48}$$

We wish to formulate a problem which is only dependent on matrices of size J. We must, therefore, relate $\mathbf{R}_{qw}$ to $\tilde{\boldsymbol{\Gamma}}_{qw}$. This requires us to estimate the larger matrix $\mathbf{R}_\phi$ from the smaller matrix $\tilde{\boldsymbol{\Gamma}}_q$. This results in an underdetermined system of equations, given in (7.47); we are forced to use least squared (LS). Using (7.47), the LS solution for $\mathbf{R}_\phi$ is given by

$$\mathbf{R}_\phi^* = \left(\mathbf{B}^\dagger\right)^H \tilde{\boldsymbol{\Gamma}}_q \mathbf{B}^\dagger, \tag{7.49}$$

§ Since the Doppler information is obtained from the slow-time pulses, and indeed angle information from the antennas, using a fast Fourier transform (FFT), the choice of frequencies can wrap around. This is because a length-N FFT repeats every N frequency samples.

where $\mathbf{B} = \check{\mathbf{H}}_o^H \mathbf{T}_T$ and $\mathbf{B}^\dagger$ denotes the pseudo-inverse of $\mathbf{B}$. We must acknowledge that the LS squares solution is one of the infinitely many possibilities; however, as we will see, this solution provides excellent performance.

We are now ready to introduce the procedure to obtain the reduced-dimension estimate, $\tilde{\boldsymbol{\Gamma}}_q$. Using (7.49) and the definition of $\mathbf{R}_{qw}^{(i)}$ in (7.35), we obtain the relation

$$\left(\mathbf{R}_{qw}^{(i)}\right)^* = \check{\mathbf{C}}_i^H \left(\mathbf{B}^\dagger\right)^H \tilde{\boldsymbol{\Gamma}}_q \mathbf{B}^\dagger \check{\mathbf{C}}_i + \mathbf{R}_w^*. \tag{7.50}$$

We can process this further to directly relate reduced-dimension matrices. To this end, define the reduced-dimension clutter and noise covariance matrix during the ith training CPI as, $\tilde{\mathbf{R}}_{qw}^{(i)} = \mathbf{T}_R^H \mathbf{R}_{qw}^{(i)} \mathbf{T}_R$. Then, using (7.50), we have

$$\left(\tilde{\mathbf{R}}_{qw}^{(i)}\right)^* = \mathbf{D}_i^H \tilde{\boldsymbol{\Gamma}}_q \mathbf{D}_i + \sigma_n^2 \mathbf{T}_R^T \mathbf{T}_R^*, \tag{7.51}$$

where $\mathbf{D}_i$ is a reduced-dimension $J \times D$ matrix given by $\mathbf{D}_i = \mathbf{B}^\dagger \check{\mathbf{C}}_i \mathbf{T}_R^*$. Note that we have set the noise covariance matrix to $\mathbf{R}_w = \sigma_n^2 \mathbf{I}_{N_R M}$. Crucially, (7.51) specifies an approximate (due to the LS solution as described earlier) relationship between matrices $\tilde{\boldsymbol{\Gamma}}_q$ and the covariance matrix $\tilde{\mathbf{R}}_{qw}^{(i)}$, both of reduced dimension. The matrix on the LHS of this equation can be estimated using received data.

During the ith training CPI, as done for receive processing, $\tilde{\mathbf{R}}_{qw}^{(i)}$ can be estimated using K secondary data vectors $\mathbf{x}_\ell^{(i)}$ received at range bins close to the range bin of interest as

$$\widehat{\tilde{\mathbf{R}}}_{qw}^{(i)} = \frac{1}{K} \sum_{\ell=1}^{K} \tilde{\mathbf{x}}_\ell^{(i)} \left(\tilde{\mathbf{x}}_\ell^{(i)}\right)^H, \tag{7.52}$$

where $\tilde{\mathbf{x}}_\ell^{(i)} = \mathbf{T}_R^H \mathbf{x}_\ell^{(i)}$. Note that the value of K used here can be much smaller (on the order of $2D$) than in the case of fully adaptive STAP.

As in the case of fully adaptive transmit processing, we must obtain an estimate of $\tilde{\boldsymbol{\Gamma}}_q$ and σ_n^2 such that the model for the covariance matrix in (7.51) is consistent with the estimated covariance matrix in (7.52). This is the reduced-dimension equivalent of the optimization problem in (7.36) and can be formulated as

$$\begin{aligned} \left(\widehat{\tilde{\boldsymbol{\Gamma}}}_q, \widehat{\sigma}_w^2\right) &= \arg\min_{\tilde{\boldsymbol{\Gamma}}_q, \gamma} \sum_{i=1}^{N_{tr}} \left\| \mathbf{D}_i^H \tilde{\boldsymbol{\Gamma}}_q \mathbf{D}_i + \gamma \mathbf{T}_R^T \mathbf{T}_R^* - \left(\widehat{\tilde{\mathbf{R}}}_{qw}^{(i)}\right)^* \right\|_F \\ \text{subject to} \quad & \tilde{\boldsymbol{\Gamma}}_q \succeq \mathbf{0} \\ & \gamma \geq 0. \end{aligned} \tag{7.53}$$

We emphasize that the purpose of this seemingly complicated derivation is to reduce significantly the size of the matrices involved, in turn reducing the required number of samples in estimating the covariance matrix $\mathbf{R}_{qw}^{(i)}$. The matrices involved here are of size $J \times J$ (as opposed to size $N_R M \times N_R M$) in (7.36)).

The procedure to be followed, therefore, is essentially the same as in the case of fully adaptive transmit processing. The estimates $\tilde{\boldsymbol{\Gamma}}_q$ and σ_n^2 are obtained after a training phase of N_{tr} CPIs. Within each CPI, the received signals at the N_R antennas

M pulses are transformed to a smaller dimension D using the transformation matrix $\mathbf{T}_r$. The required estimates are obtained by solving (7.53). As before, after an initial phase of N_{tr} CPIs, for each additional CPI, the signals from the previous N_{tr} CPIs can be used.

7.5 Transmit BF for multiple Doppler bins

To build on the available literature, so far, we have focused on optimizing the transmissions for a single look angle–Doppler bin. Such an approach is, unfortunately, not feasible in practice. While a radar interrogates specific regions of space (angle bin or look direction) in each CPI covering M pulses, this would have to be repeated for every possible target speed (Doppler) massively increasing the time required to interrogate each Doppler bin. In the case of receive-only adaptive processing, this is not a significant issue since the received data can be processed independently for each Doppler bin. For transmit beamforming, however, we must choose the transmission *a priori*. We now, therefore, extend the techniques developed in the previous sections to cover multiple Doppler bins simultaneously.

Denote as $f_1, \cdots, f_{N_D}$ as the N_D Doppler bins of interest. To this list, we add Doppler bins f_0 and f_{N_D+1}. These act as the neighboring Doppler bins to f_1 and f_{N_D}, respectively, and will be used to form a proposed extended transformation matrix. Furthermore, consider the angle bins ϕ_0^-, ϕ_0, and ϕ_0^+, where, as before, ϕ_0 is the look angle. We propose that the code vector $\mathbf{c} \in \mathbb{C}^{MN_T \times 1}$ be written using the *extended* transformation matrix $\mathbf{T}_e \in \mathbb{C}^{MN_T \times P}$ as

$$\mathbf{c} = \mathbf{T}_e \tilde{\mathbf{c}} \tag{7.54}$$

where $\tilde{\mathbf{c}} \in \mathbb{C}^{P \times 1}$ is the reduced-dimension transmit code vector, and $P = 3(N_D + 2)$.

This approach is similar to the development in Section 7.4 where we used a transmit transformation matrix $\mathbf{T}_T$ based on angle and Doppler bins near the angle–Doppler bin of interest. Here, we extend this choice of Doppler bins such that the columns of $\mathbf{T}_e$ are the adaptive steering vectors covering the three angle bins and, to cover the entire Doppler space, the $(N_D + 2)$ Doppler bins $(f_0, f_1, \ldots, f_{N_D}, f_{N_D+1})$. In particular, the jth column of $\mathbf{T}_e$ is chosen as $\mathbf{s}(\phi_l, f_m)$, where $j = 3m + l$, $1 \le l \le 3$, and $0 \le m \le N_D + 1$. Adaptive steering vectors were specified in (7.41). Here, we must make one change from that specification: the receive adaptive beamformer $\mathbf{h}_o$ depends on the Doppler bin, i.e., we change and we use the N_D receive beamforming vectors corresponding to the N_D Doppler bins of interest. For the bins f_0 and f_{N_D+1}, we use the weight vector corresponding to f_1 and f_{N_D}, respectively. Note that for $N_D = 1$, as expected, the matrices $\mathbf{T}_T$ and $\mathbf{T}_e$ will be identical.

As before, we wish to maximize the SINR; here, since each Doppler bin receives a different SINR, we optimize the transmission to maximize the minimum SINR across the N_D Doppler bins of interest. Let SINR_m denote the SINR value at the look angle

ϕ_0 and the mth Doppler bin ($1 \leq m \leq N_D$). Then, the optimization problem can be cast as

$$\begin{aligned} \mathbf{c}_o &= \arg\max_{\mathbf{c}} \min_{m} \text{SINR}_m \\ &\quad \text{subject to } \|\mathbf{c}\|^2 \leq 1. \end{aligned} \tag{7.55}$$

Keeping with the theme of this chapter, unlike the work in [21] (and references therein) we do not assume knowledge of the interference statistics. We base our solution on the estimation approach described in Sections 7.3 and 7.4. Specifically, we rewrite (7.55) as

$$\begin{aligned} \tilde{\mathbf{c}}_o &= \arg\max_{\tilde{\mathbf{c}}} \min_{m} \text{SINR}_m \\ &\quad \text{subject to } \|\mathbf{T}_e\tilde{\mathbf{c}}\|^2 \leq 1, \end{aligned} \tag{7.56}$$

and use $\mathbf{c}_o = \mathbf{T}_e\tilde{\mathbf{c}}_o$.

The optimization problem in (7.56) is non-convex and difficult to solve; we use semidefinite relaxation to obtain a sub-optimal solution. We start by revisiting the SINR function. Using (7.23) and (7.26), and replacing $\mathbf{R}_t^{\phi}$ with the all-ones matrix, the SINR for the mth Doppler bin can be written as

$$\text{SINR}_m = \frac{\mathbf{c}^H \boldsymbol{\Theta}_{0,m}^H \mathbf{h}_m \mathbf{h}_m^H \boldsymbol{\Theta}_{0,m} \mathbf{c}}{\mathbf{c}^H \check{\mathbf{H}}_m \mathbf{R}_{\phi}^* \check{\mathbf{H}}_m^H \mathbf{c} + \mathbf{h}_m^H \mathbf{R}_w \mathbf{h}_m} \tag{7.57}$$

where $\check{\mathbf{H}}_m$ is obtained from $\mathbf{h}_m$ using (7.25) and $\boldsymbol{\Theta}_{0,m}$ denotes $\boldsymbol{\Theta}(\phi_0, f_m)$. Now, using (7.54), and defining the rank-one matrix $\tilde{\mathbf{C}} = \tilde{\mathbf{c}}\tilde{\mathbf{c}}^H$, we have

$$\text{SINR}_m = \frac{\mathbf{h}_m^H \boldsymbol{\Theta}_{t,m} \mathbf{T}_e \tilde{\mathbf{C}} \mathbf{T}_e^H \boldsymbol{\Theta}_{t,m}^H \mathbf{h}_m}{\text{Tr}\left\{\check{\mathbf{H}}_m \mathbf{R}_{\phi}^* \check{\mathbf{H}}_m^H \mathbf{T}_e \tilde{\mathbf{C}} \mathbf{T}_e^H\right\} + \mathbf{h}_m^H \mathbf{R}_w \mathbf{h}_m} \tag{7.58}$$

where $\text{Tr}\{\cdot\}$ stands for the trace operator. Since $\tilde{\mathbf{C}}$ is ranked one and is positive semidefinite, the optimization problem in (7.56) can be rewritten as

$$\begin{aligned} \tilde{\mathbf{C}}_o &= \arg\max_{\tilde{\mathbf{C}}} \min_{m} \text{SINR}_m \\ &\quad \text{subject to } \text{Tr}\left\{\mathbf{T}_e \tilde{\mathbf{C}} \mathbf{T}_e^H\right\} \leq 1 \\ &\quad \tilde{\mathbf{C}} \succeq \mathbf{0} \\ &\quad \text{rank}\{\tilde{\mathbf{C}}\} = 1. \end{aligned} \tag{7.59}$$

The constraint on the rank of $\tilde{\mathbf{C}}$ is non-convex making (7.59) difficult to solve. One approach to deal with this problem is to relax the constraint on the rank of $\tilde{\mathbf{C}}$, which results in the following problem:

$$\begin{aligned}\tilde{\mathbf{C}}_* &= \arg\max_{\tilde{\mathbf{C}}} \min_m \mathrm{SINR}_m \\ &\quad \text{subject to } \mathrm{Tr}\left\{\mathbf{T}_e\tilde{\mathbf{C}}\mathbf{T}_e^H\right\} \le 1 \\ &\quad \tilde{\mathbf{C}} \succeq \mathbf{0}.\end{aligned} \tag{7.60}$$

If the solution of (7.60), $\tilde{\mathbf{C}}_*$, is a rank-one matrix, it is also the solution of (7.59). Otherwise, a suboptimal solution to (7.59) can be obtained from the solution to (7.60). Specifically, let $\mathbf{u}$ be the eigenvector corresponding to the largest eigenvalue of $\tilde{\mathbf{C}}_*$. Then, the optimal reduced-dimension transmit code vector can be approximated as

$$\tilde{\mathbf{c}}_o \approx \frac{\mathbf{u}}{\|\mathbf{T}_e\mathbf{u}\|}. \tag{7.61}$$

Consider rewriting (7.60) as

$$\begin{aligned}\tilde{\mathbf{C}}_* &= \arg\max_{\tilde{\mathbf{C}}}\{\gamma\} \\ &\quad \text{subject to } \mathrm{Tr}\left\{\mathbf{T}_e\tilde{\mathbf{C}}\mathbf{T}_e^H\right\} \le 1 \\ &\quad \tilde{\mathbf{C}} \succeq \mathbf{0} \\ &\quad \mathrm{SINR}_m \ge \gamma \text{ for } 1 \le m \le N_D.\end{aligned} \tag{7.62}$$

Then, (7.62) can be solved using a bisection search algorithm to find a value of γ for which the solution is feasible.

The final step to complete our proposed algorithm of adaptive transmit beamforming for multiple Doppler bins is to consider a practical method to estimate the required covariance matrices at the transmitter. Specifically, evaluating the SINR objective function requires an estimate for $\mathbf{R}_\phi$ and $\mathbf{R}_w$. W build on our work in the previous section.

For the mth ($1 \le m \le N_D$) Doppler bin, an estimate for $\tilde{\mathbf{\Gamma}}_q$ and σ_n^2 can be found using (7.53) and the last N_{tr} received signals as

$$\begin{aligned}\left(\widehat{\tilde{\mathbf{\Gamma}}}_q^{(m)}, \widehat{\sigma}_w^{2(m)}\right) &= \arg\min_{\tilde{\mathbf{\Gamma}}_q,\gamma} \sum_{i=1}^{N_{tr}} \Bigg\| \left(\mathbf{D}_i^{(m)}\right)^H \tilde{\mathbf{\Gamma}}_q \mathbf{D}_i^{(m)} \\ &\qquad + \gamma\left(\mathbf{T}_R^{(m)}\right)^T\left(\mathbf{T}_R^{(m)}\right)^* - \left(\widehat{\tilde{\mathbf{R}}}_{qw}^{(i)}\right)^* \Bigg\|_F \\ \text{subject to} &\quad \tilde{\mathbf{\Gamma}}_q \succeq \mathbf{0} \\ &\quad \gamma \ge 0\end{aligned} \tag{7.63}$$

where $\mathbf{D}_i^{(m)} = \left(\mathbf{B}^{(m)}\right)^\dagger \check{\mathbf{C}}_i \left(\mathbf{T}_R^{(m)}\right)^*$, and $\left(\mathbf{B}^{(m)}\right)^\dagger$ is the pseudoinverse of $\mathbf{B}^{(m)} = \check{\mathbf{H}}_m^H\mathbf{T}_T^{(m)}$. Here, $\check{\mathbf{H}}_m$ is obtained using (7.25) and the receive weight vector for the mth Doppler bin, i.e., $\mathbf{h}_m$. The receive transformation matrix $\mathbf{T}_R^{(m)}$ is obtained using (7.40), and f_{m-1}, f_m, and f_{m+1}. Furthermore, $\mathbf{T}_T^{(m)}$ is formed using a subset of the columns of the extended matrix $\mathbf{T}_e$. Specifically, columns of $\mathbf{T}_e$ which correspond to the (ϕ_0, f_m) angle–Doppler bin and also the adjacent bins are selected. Based on our definition of $\mathbf{T}_e$, matrix $\mathbf{T}_T^{(m)}$ can be formed by taking the $(1 + 3(m - 1))$th to the $(9 + 3(m - 1))$th column of $\mathbf{T}_e$.

Given $\widehat{\tilde{\boldsymbol{\Gamma}}}_q^{(m)}$, the estimate $\widehat{\mathbf{R}}_\phi^{(m)}$ can be obtained using $\mathbf{B}^{(m)}$ and (7.49). Finally, estimations for $\mathbf{R}_\phi$ and σ_w^2 are obtained by averaging over the estimates $\widehat{\mathbf{R}}_\phi^{(m)}$ and $\widehat{\sigma}_w^{2(m)}$ made at different Doppler bins.

7.6 Numerical results

Having developed the theory of estimating the required second-order statistics for transmit adaptivity and then extend this notion to the case of reduced-dimension processing and a max–min problem across Doppler bins, we now illustrate the efficacy of the proposed methods. The results match the developed theory in that we use scenarios with random phase perturbations [17,33] and airborne radar [1].

In all of the examples, unless otherwise specified, the number of secondary range bins used to estimate the clutter-plus-noise covariance matrix is set to $K = 20$. Furthermore, the number of training transmissions is set to $N_{tr} = 8$. The first N_{tr} transmissions are chosen to be random Gaussian codes, independently and identically drawn from a circularly symmetric Gaussian distribution with an identity covariance matrix. To meet the power constraint, the Gaussian code vector is normalized to unit length. As mentioned earlier, we only need to "start" the system with these N_{tr} training transmissions; after these initial training CPIs, the most recent N_{tr} transmissions are used to estimate the covariance matrices needed to optimize the transmission. If required, the probability of false alarm is set to 0.01. In all examples, the noise is assumed to be white with unit power. The numerical results presented are the result of 10,000 Monte Carlo trials.

In all figures, the dashed curves represent the scenarios where the required covariance matrices are available at either the transmitter or at the receiver. These results are not realizable, since the covariance matrices are unknown in practice, i.e., these curves represent our performance benchmark (the performance of a clairvoyant receiver/transmitter.)

The curves marked with triangles represent either the case where only one transmit antenna is used (with the amplitude of the pulses kept the same) or the case where a conventional beamformer is used at the transmitter. The former is denoted as "isotropic Tx," whereas the latter is denoted by "directed Tx" in the figures. In the case of isotropic Tx, there is no transmit beamforming. For the curves marked with triangles, known covariance matrices are used at the receiver to obtain the optimal receive weight vectors.

The curves marked with squares correspond to the case when the true covariance matrices are available at both the transmitter and the receiver and are used to obtain the optimal transmit code vector and receive filters. The curves marked with diamonds represent the case when the transmission is isotropic (or directed) and the covariance matrices are estimated at the receiver. The curves marked with circles depict the results for transmit beamforming with known and receive weight vector design with estimated covariance matrices. Finally, in the case that at both the transmitter and the receiver, the covariance matrices are estimated to obtain the optimal transmit and receive filters, the curves are marked with pentagrams.

7.6.1 Random phase radar signals

For the case of random phase radar signals, we consider two scenarios: optimizing the transmit code vector for a single look angle–Doppler bin or, alternatively, optimizing the code word for a range of Doppler bins using the max–min method in Section 7.5. In both cases, the transmit and receive antennas are uniform linear arrays (ULA) with half-wavelength inter-element spacing of 30 m. The PRI is set to $T = 1/40$ s. The overall number of range bins is $L = 21$. A target is present in the 11th range bin at azimuth angle $\phi_0 = 50°$ at a speed of 310 m/s ($f_0 = 0.2583$). There are $V = 31$ clutter rays incident uniformly distributed from azimuth angles 30° to 60°.

Single-look angle–Doppler

In our first experiment, we consider transmit code design for a specific look angle–Doppler bin for a radar system with $N_T = 4$ transmit and $N_R = 4$ receive antennas. The number of pulses per CPI is set to $M = 6$. For such a small problem we are able to use full-dimension processing as a baseline.

Figure 7.1 plots the output SINR for different target RCS values to illustrate the benefits of transmit beamforming. The process of estimating the receive covariance matrix (the curve with diamonds) represents receive-only STAP. With respect to receive-only STAP, performing transmit beamforming provides a gain of about 6 dB in the output SINR. Furthermore, as the plot shows, when compared to the case where all the covariance matrices are known, estimating the transmit and receive matrices suffers a 3 dB loss. This is consistent with the RMB which states that keeping this loss to less than 3 dB requires that the number of samples used to estimate the covariance matrices must be twice the number of adaptive degrees of freedom.

Figure 7.2 presents a similar comparison in terms of the probability of detection. The performance comparisons using this figure are consistent with those in Figure 7.1. The curves marked with squares in Figures 7.1 and 7.2 correspond to the results presented for the clairvoyant receiver/transmitter in [33] (and also the methods introduced in [16] when written for random phase radar signals). Note that these results are obtained using the true covariance matrices at both the transmit and receive sides. In practice, we need to use estimates of these matrices.

In the second experiment, we consider the number of transmit antennas to be the same as the first example, i.e., $N_T = 4$. However, the number of receive antennas and the pulses are increased to $N_R = 16$ and $M = 32$, respectively. These numbers necessitate reduced-dimension processing; here we set $D = J = 9$. The neighboring

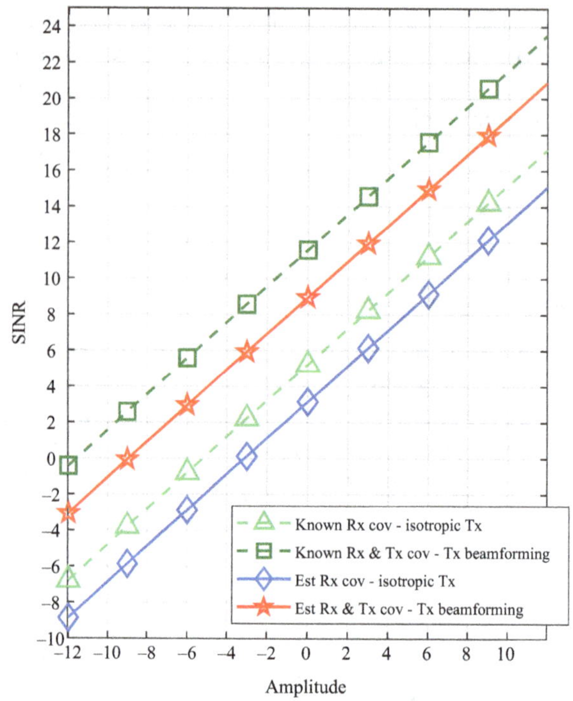

Figure 7.1 Output SINR versus the target RCS for the case of transmit beamforming for a single angle–Doppler bin with full-dimension processing

angle–Doppler bins to form the reduced-dimension matrices are chosen from the 2-dimensional discrete Fourier transform (2-D DFT) grid as $\phi_0^- = 41.4°$, $\phi_0^+ = 60°$, $f_0^- = 0.22$, and $f_0^+ = 0.28$.

Figure 7.3 plots the output SINR for different target RCS values. As is clear from the figure, using known covariance matrices at the transmitter, optimizing the transmit code allows for a 3 dB gain in the output SINR. In practice, when the estimated covariance matrices are used at the transmitter, a gain of about 2 dB can be achieved. As with the case of full-dimensional processing above, the corresponding probability of detection curves in Figure 7.4 shows comparable performance.

Multiple Doppler bins

In the third experiment, we design the transmit code vectors in max–min sense for $N_D = 3$ Doppler bins which include the target Doppler bin. The radar system has $N_T = 4$ transmit and $N_R = 8$ receive antennas. The number of pulses per CPI is set to $M = 16$. Reduced-dimension processing with $D = 9$ and $P = 15$ elements is performed at the receiver and the transmitter, respectively. The neighboring angles are selected from the DFT grid as $\phi_0^- = 0°$, $\phi_0^+ = 60°$ (a consequence of the small

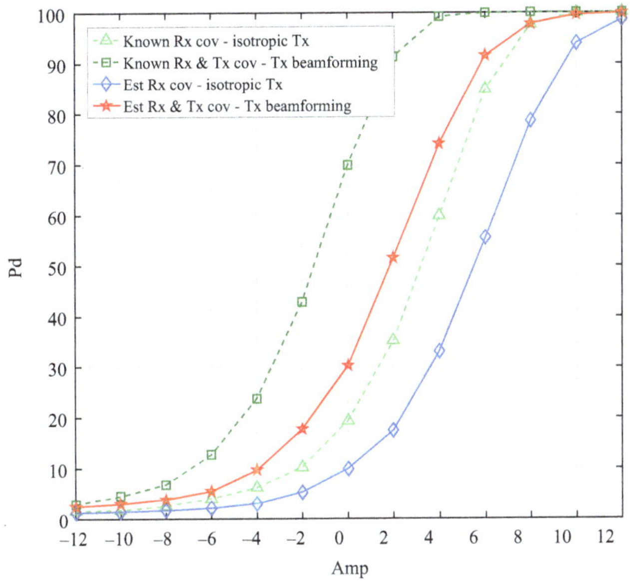

Figure 7.2 Probability of detection versus the target RCS for the case of transmit beamforming for a single angle–Doppler bin with full-dimension processing

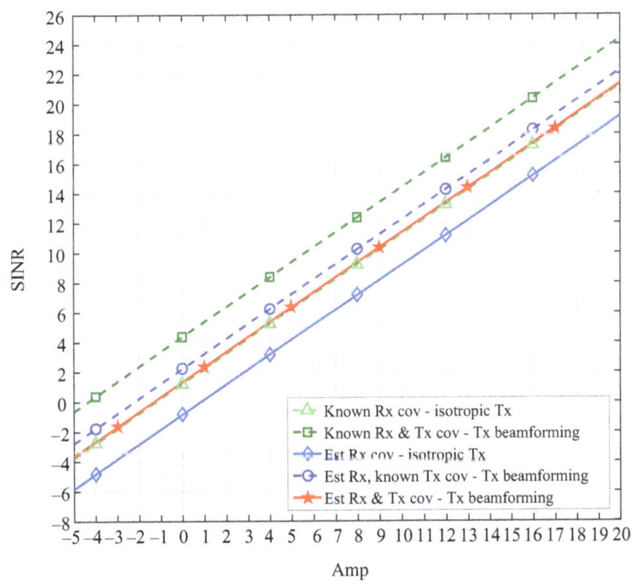

Figure 7.3 Output SINR versus the target RCS for the case of transmit beamforming for a single angle–Doppler bin with reduced-dimension processing

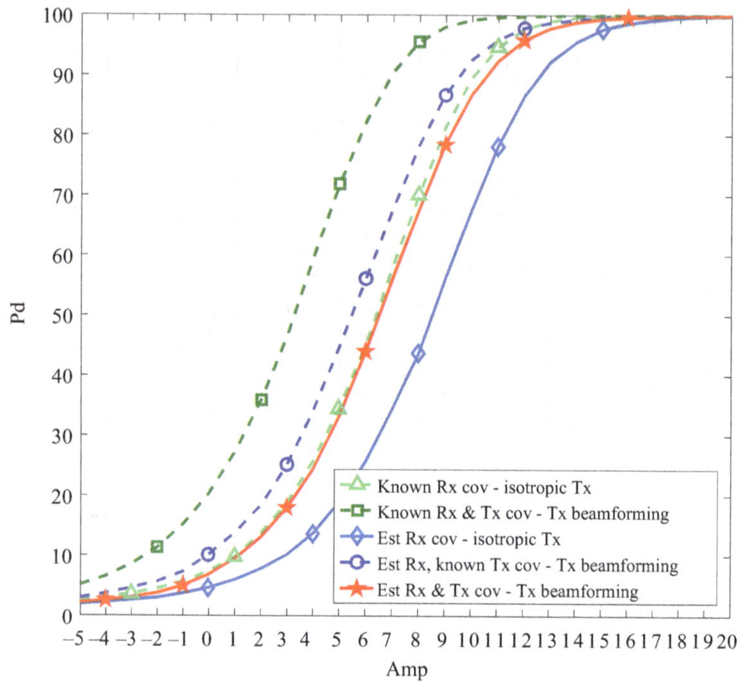

Figure 7.4 Probability of detection versus the target RCS for the case of transmit beamforming for a single angle–Doppler bin with reduced-dimension processing

number of antennas). The neighboring Doppler bins are also taken from the DFT grid as $f_0 = 0.125$, $f_1 = 0.188$, $f_2 = 0.25$, $f_3 = 0.313$, and $f_4 = 0.375$. The output SINR (at the target Doppler bin) and probability of detection versus the target RCS values are depicted in Figures 7.5 and 7.6, respectively. The figures show that a gain of about 2 dB can be achieved by performing transmit beamforming with estimated covariance matrices.

7.6.2 *Airborne radar*

The second set of results are for an airborne radar. We consider two scenarios: in the first case, there is no jamming signal, and in the second one, the interference includes a jamming component (which is independent of the transmit signal). In both cases, the transmit and receive antennas are ULAs with half-wavelength inter-element spacing of 1/ m. The PRI is set to $T = 1/300$ s. The platform velocity is set to $v_p = 30$ m/s. A target is present at azimuth angle $\phi_0 = 90°$ at a speed of 10 m/s with respect to the platform ($f_0 = 0.1$). There are $N_T = 4$ transmit and $N_R = 16$ receive antennas. The number of pulses per CPI is set to $M = 64$. We set $J = D = 9$.

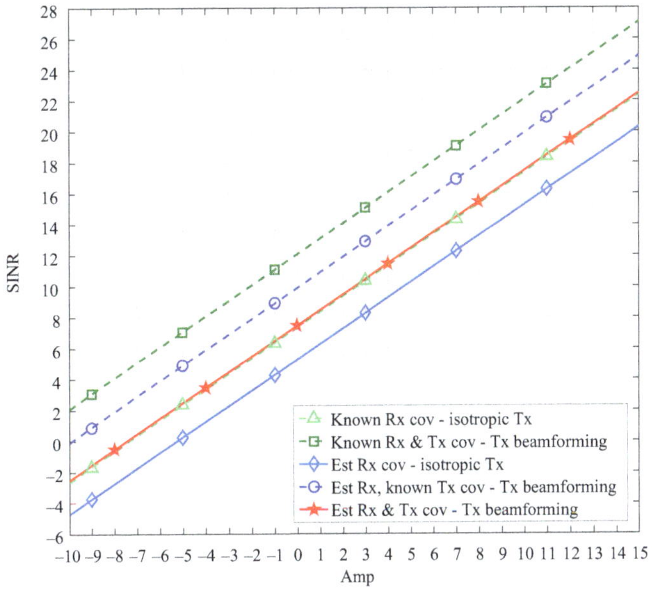

Figure 7.5 *Output SINR versus the target RCS for the case of transmit beamforming with the max–min method*

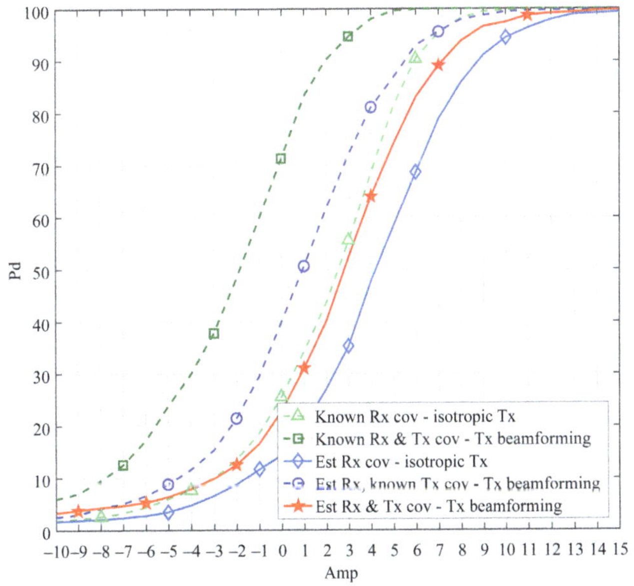

Figure 7.6 *Probability of detection versus the target RCS for the case of transmit beamforming with the max–min method*

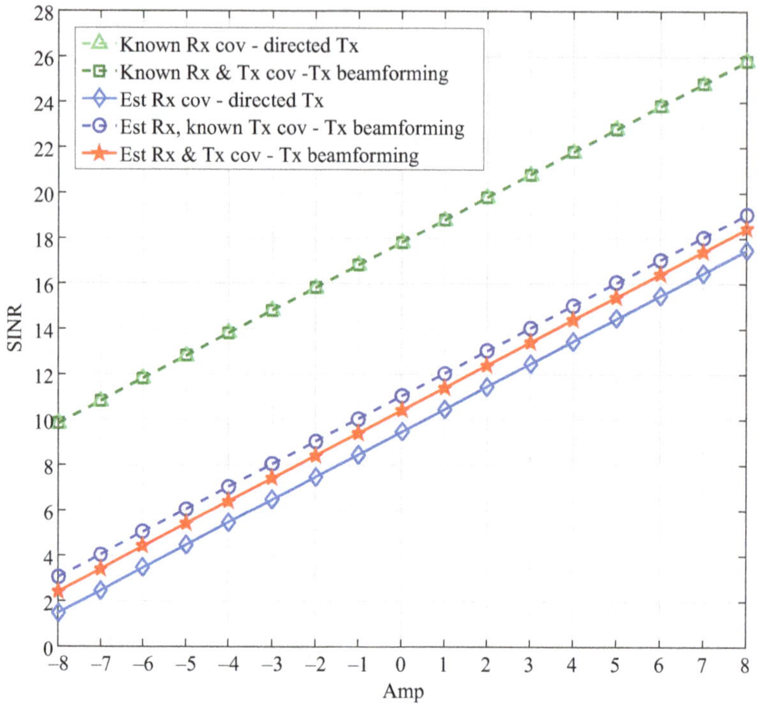

Figure 7.7 Output SINR versus the target RCS for the case of airborne system without jamming

Figures 7.7 and 7.8 present the results for the first scenario (without jamming). The curves marked "directed Tx" indicate the case with non-adaptive transmissions. In this example, the neighboring angle–Doppler bins, used to form the transformation matrices for dimension reduction, are chosen from the 2-D DFT grid points around the look angle–Doppler bin. Again, the proposed adaptive transmit beamforming outperforms the conventional beamformer by about 2 dB in terms of the output SINR and about 1 dB in terms of the probability of detection. Interestingly, the curves show that, in this case, transmit adaptivity only provides a small gain with respect to receive-only STAP.

Finally, we consider an example of the airborne radar system with a jammer (at $\phi_j = 92°$). In this experiment, we have chosen the neighboring angles (ϕ_0^- and ϕ_0^+) 7.24° away from the target and the neighboring Doppler bins (f_0^- and f_0^+) 0.04 from the target. Although the noise covariance matrix is no longer diagonal (due to the presence of the jamming signal), we have used the same processing as done for the case that there is no jamming signal. The results are presented in Figures 7.9 and 7.10. In this case, transmit beamforming provides a significant performance gain over receive-only STAP. The output SINR is improved by about 4 dB as depicted in

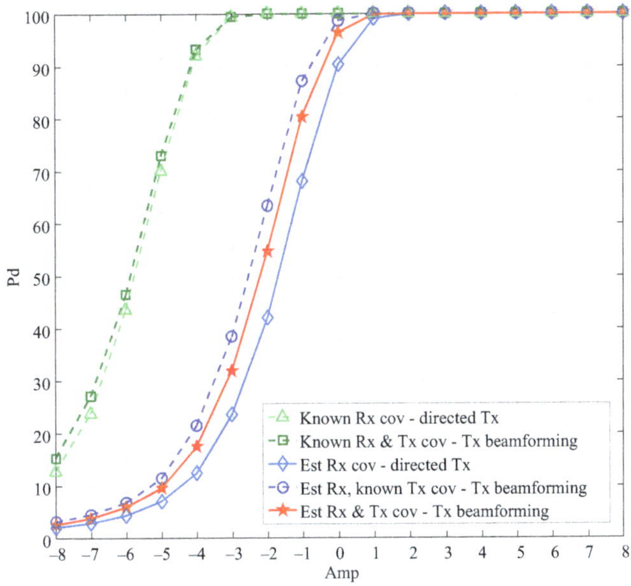

Figure 7.8 Probability of detection versus the target RCS for the case of airborne system without jamming

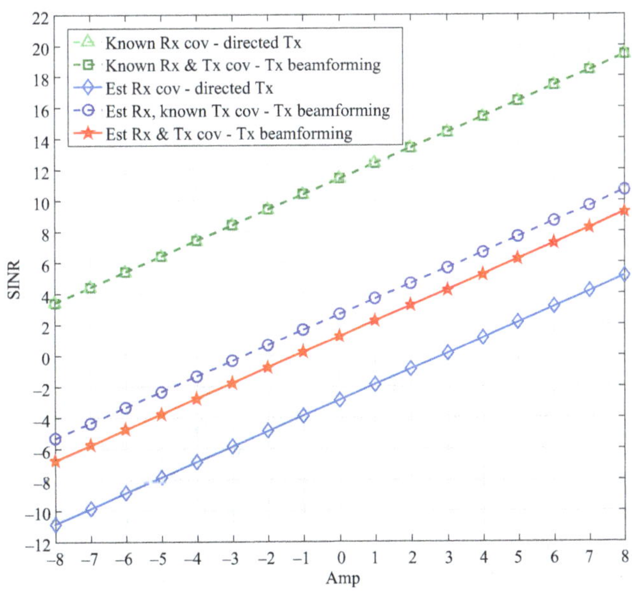

Figure 7.9 Output SINR versus the target RCS for the case of airborne system with jamming

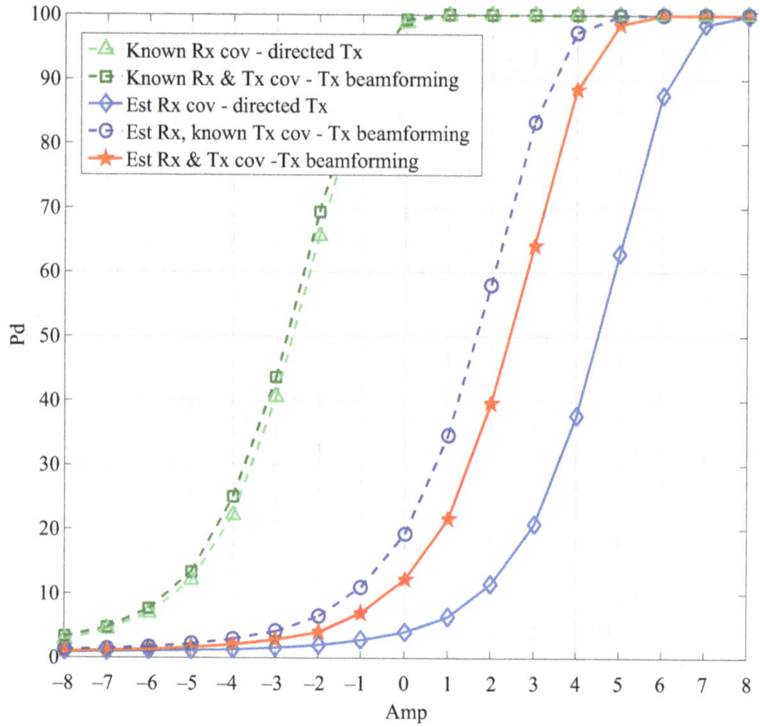

Figure 7.10 Probability of detection versus the target RCS for the case of airborne system with jamming

Figure 7.9, and the probability of detection is improved by about 2 dB as seen in Figure 7.10.

7.7 Conclusion

The advent of digital control of transmitted signals has opened up the possibility that each element in a transmit array can be independently controlled. The natural use of this capability is to optimize transmissions to maximize detection probability when dealing with interference. In this chapter, we have investigated how a cognitive radar might acquire the information needed to implement transmit adaptivity. Specifically, we developed a training model to obtain the needed second-order statistics.

An important implication of our work is that while transmit adaptivity is conceptually similar to receive adaptivity, at an implementation level, they are quite different. This is because transmit characteristics must be derived from the receiver. In this chapter, we developed one computationally challenging approach to achieve this; specifically we used a sequence of training transmissions to interrogate the environment—enabling the perception cycle in a cognitive radar. We hope this work sparks interest in exploring other possible approaches.

In this chapter, we also discussed two other topics: reduced-dimension processing with reduced required training and a max–min approach to cover multiple range bins. Both these are essential for any practical implementation of transmit adaptivity. Our results show that using the techniques developed does not only suffer an (expected) performance loss but also improve on non-adaptive transmission approaches.

There are many open questions to be answered some of which we have touched on in this chapter. How to ensure the appropriate choices of J and D, the reduced dimensionality. Other works have considered using the rank of the clutter matrix, but this has not been extended to transmit covariance matrix estimation. Similarly, the role of a changing environment and whether the real-time computation load.

Acknowledgment

This work is supported by Raytheon Canada, the Natural Sciences and Engineering Research Council (NSERC) of Canada, and the Defence Research and Development Canada (DRDC).

References

[1] J. Ward, "Space–time adaptive processing for airborne radar," MIT Lincoln Laboratory, Lexington, MA, Tech. Rep. 1015, 1994.

[2] M. Rangaswamy and F. C. Lin, "Performance analysis of the namf test in heterogeneous non-gaussian radar clutter scenarios," in *2007 Conference Record of the 41st Asilomar Conference on Signals, Systems and Computers*, 2007, pp. 706–710.

[3] M. C. Wicks, M. Rangaswamy, R. S. Adve, and T. B. Hale, "Space-time adaptive processing: a knowledge-based perspective," *IEEE Signal Process. Mag.*, vol. 23, no. 1, pp. 51–65, 2006.

[4] B. Kang, S. Gogineni, M. Rangaswamy, J. Guerci, and E. Blasch, "Adaptive channel estimation for cognitive fully adaptive radar," *IET Trans. Radar, Sonar Navig.*, vol. 16, pp. 720–734, 2021.

[5] R. Adve, "A brief review of array theory." https://www.comm.utoronto.ca/~rsadve/Notes/ArrayTheory.pdf

[6] C. Balanis, *Antenna Theory: Analysis and Design*. John Wiley, New York, NY, 1997.

[7] J. Li and P. Stoica, "MIMO radar with colocated antennas," *IEEE Signal Process. Mag.*, vol. 24, no. 5, pp. 106–114, 2007.

[8] N. Goodman, in R. Chellappa and S. Theodoridis, Eds., *Academic Press Library in Signal Processing*, vol. 7, Academic Press, London, 2018.

[9] K. V. Mishra, M. R. B. Shankar, and B. Ottersten, "Toward metacognitive radars: concept and applications," in *2020 IEEE International Radar Conference (RADAR)*, 2020, pp. 77–82.

[10] X. Zhang and X. Liu, “Adaptive waveform design for cognitive radar in multiple targets situation,” *Entropy*, vol. 20, p. 114, 2018.

[11] B. Friedlander, “Waveform design for MIMO radars,” *IEEE Trans. Aerosp. Electron. Syst.*, vol. 43, no. 3, pp. 1227–1238, 2007.

[12] P. Stoica, J. Li, and M. Xue, “Transmit codes and receive filters for radar,” *IEEE Signal Process. Mag.*, vol. 25, no. 6, pp. 94–109, 2008.

[13] A. Aubry, A. De Maio, A. Farina, and M. Wicks, “Knowledge-aided (potentially cognitive) transmit signal and receive filter design in signal-dependent clutter,” *IEEE Trans. Aerosp. Electron. Syst.*, vol. 49, no. 1, pp. 93–117, 2013.

[14] F. Gini, A. De Maio, and L. Patton, *Waveform Design and Diversity for Advanced Radar Systems*. Inst. Eng. Technol. (IET), Series 22, 2012.

[15] P. Stoica, H. He, and J. Li, “Optimization of the receive filter and transmit sequence for active sensing,” *IEEE Trans. Signal Process.*, vol. 60, no. 4, pp. 1730–1740, 2012.

[16] J. Liu, H. Li, and B. Himed, “Joint optimization of transmit and receive beamforming in active arrays,” *IEEE Signal Process. Lett.*, vol. 21, no. 1, pp. 39–42, 2014.

[17] A. A. Gorji and R. S. Adve, “Waveform optimization for random-phase radar signals with PAPR constraints,” in *Proceedings of the IEEE International Radar Conference*, Lille, Oct. 2014, pp. 1–5.

[18] B. Tang and J. Tang, “Joint design of transmit waveforms and receive filters for MIMO radar space-time adaptive processing,” *IEEE Trans. Signal Process.*, vol. 64, no. 18, pp. 4707–4722, 2016.

[19] A. Aubry, A. De Maio, M. Piezzo, A. Farina, and M. Wicks, “Cognitive design of the receive filter and transmitted phase code in reverberating environment,” *IET Radar, Sonar, Navig.*, vol. 6, no. 9, pp. 822–833, 2012.

[20] M. M. Naghsh, M. Soltanalian, P. Stoica, M. Modarres-Hashemi, A. De Maio, and A. Aubry, “A Doppler robust design of transmit sequence and receive filter in the presence of signal-dependent interference,” *IEEE Trans. Signal Process.*, vol. 62, no. 4, pp. 772–785, 2014.

[21] A. Aubry, A. De Maio, and M. M. Naghsh, “Optimizing radar waveform and Doppler filter bank via generalized fractional programming,” *IEEE J. Sel. Topics Signal Process.*, vol. 9, no. 8, pp. 1387–1399, 2015.

[22] C. Y. Chen and P. P. Vaidyanathan, “MIMO radar waveform optimization with prior information of the extended target and clutter,” *IEEE Trans. Signal Process.*, vol. 57, no. 9, pp. 3533–3544, 2009.

[23] S. M. O’Rourke, P. Setlur, M. Rangaswamy, and A. L. Swindlehurst, “Relaxed biquadratic optimization for joint filter-signal design in signal-dependent stap,” *IEEE Trans. Signal Process.*, vol. 66, no. 5, pp. 1300–1315, 2018.

[24] P. Setlur and M. Rangaswamy, “Waveform design for radar stap in signal dependent interference,” *IEEE Trans. Signal Process.*, vol. 64, no. 1, pp. 19–34, 2016.

[25] S. M. O’Rourke, P. Setlur, M. Rangaswamy, and A. L. Swindlehurst, “Relaxed biquadratic optimization for joint filter-signal design in signal-dependent stap,” *IEEE Trans. Signal Process.*, vol. 66, no. 5, pp. 1300–1315, 2018.

[26] J. S. Bergin, C. M. Teixeira, P. M. Techau, and J. R. Guerci, "Improved clutter mitigation performance using knowledge-aided space-time adaptive processing," *IEEE Trans. Aerospace Electron. Syst.*, vol. 42, no. 3, pp. 997–1009, 2006.

[27] A. De Maio, Y. Huang, and M. Piezzo, "A Doppler robust max-min approach to radar code design," *IEEE Trans. Signal Process.*, vol. 58, no. 9, pp. 4943–4947, 2010.

[28] M. M. Naghsh, M. Soltanalian, P. Stoica, and M. Modarres-Hashemi, "Radar code design for detection of moving targets," *IEEE Trans. Aerosp. Electron. Syst.*, vol. 50, no. 4, pp. 2762–2778, 2014.

[29] M. Ravan, R. J. Riddolls, and R. S. Adve, "Ionospheric and auroral clutter models for HF surface wave and over the horizon radar systems," *Radio Sci.*, vol. 47, pp. 1–12, 2012.

[30] H. Wang and L. Cai, "On adaptive spatial–temporal processing for airborne surveillance radar systems," *IEEE Trans. Aerosp. Electron. Syst.*, vol. 30, no. 3, pp. 660–670, 1994.

[31] R. S. Adve, T. B. Hale, and M. C. Wicks, "Joint domain localized adaptive processing in homogeneous and non-homogeneous environments. Part I: homogeneous environments," *IEE Proc. Radar Sonar Navig.*, vol. 147, no. 2, pp. 57–65, 2000.

[32] M. Shaghaghi and R. S. Adve, "Training-based adaptive transmit–receive beamforming for random phase radar signals," in *Proceedings of the IEEE International Radar Conference*, Philadelphia, PA, May 2016, pp. 1–5.

[33] A. A. Gorji, R. J. Riddolls, M. Ravan, and R. S. Adve, "Joint waveform optimization and adaptive processing for random phase radar signals," *IEEE Trans. Aerosp. Electron. Syst.*, vol. 51, no. 4, pp. 2627–2640, 2015.

[34] J. Capon, "High-resolution frequency–wavenumber spectrum analysis," *Proc. IEEE*, vol. 57, no. 8, pp. 1408–1418, 1969.

[35] O. Saleh, M. Ravan, R. Riddolls, and R. Adve, "Fast fully adaptive processing: a multistage stap approach," *IEEE Trans. Aerosp. Electron. Syst.*, vol. 52, no. 5, pp. 2168–2183, 2016.

Chapter 8

Random projections and sparse techniques in radar

Pawan Setlur[1,*]

It is anticipated that cognitive radars will be able to leverage autonomy via feedback-enabled resource scheduling, waveform design, beam-shaping and beam-steering, autonomous tracking, etc. Radar sensors sample data at high rates, and, with new technologies like cognition in radar, we envision a sensor that is inundated with data. We, therefore, investigate random projections and sparse techniques in radar to reduce the dimensionality of the data.

In this chapter, first, we revisit the underpinnings of Compressive Sensing (CS) theory, namely random projections [1] which is the pre-cursor to CS theory [2]. Using random projections, we take a critical look at sub-sampling claims in CS theory in the analog domain.

Additionally, we also consider random projections on radar space–time adaptive processing (STAP). In radar STAP, training data from neighboring range cells is limited. This precludes the implementation of the full-dimensional adaptive detectors, i.e., the minimum variance distortion-less response (MVDR) filter. In that regard, we reduce the dimension of the problem by random sampling, i.e., by projecting the data into a random d-dimensional subspace. This offers two advantages, first, it permits the implementation of classical detectors in the limited sample size regime. Second, it offers significant computational savings permitting possible real-time solutions. Both these advantages are, however, at the cost of reducing the *output SINR* for radar STAP. In STAP, the cell under test is assumed to have known desired spatial and temporal responses. To ameliorate over this signal-to-noise and interference ratio (SINR) loss from random projections, we propose other techniques where the lower dimension subspace is not entirely random, but is decomposed into both random and deterministic parts. The family of random and random type projections we develop here, are either l_2 norm preserving or contraction mappings in statistical expectation.

[1]Northrop Grumman Corporation, Xetron, Cincinnati, OH
*This work was performed when the author was with United States Air Force (USAF), WPAFB, USA and in the author's personal capacity and no DoD or USAF endorsement is and was assumed or acknowledged. The views expressed in this chapter are the author's own and are not endorsed by the DoD or USAF, or Northrop Grumman Corporation in anyway whatsoever.

8.1 Introduction

Recently, there has been a spate of technical literature in a field called compressive sensing (CS). Numerous PhD theses and technical literature devoted to this subject have been published and are too numerous to cite here. The field of CS grew popular because practitioners in this field claimed that this theory can reconstruct (sparse) signals by under-sampling or sampling below the Nyquist rate.

Other sub-sampling techniques: Prony's method is an example which has existed for 200 years and reconstructs a k sparse signal with the first $2k$ Fourier coefficients, this is less than using the entire discrete Fourier transform [3,4].

In radar, frequency-modulated pulses are stretch processed and sampled at rates much less than the Nyquist rate of the original modulated pulse. This is possible because stretch processing converts a linear frequency modulated pulse to a sinusoid at the expense of range–Doppler coupling. Such techniques have existed since 1960s.

The original papers in CS theory [5,6] all used discrete signals and have never investigated how CS would actually sub-sample an analog signal, except in [7–10]. Our analysis leads to conclusions which are in direct contrast to [7–10]. Other approaches to sampling sparse signals have been proposed, see [11] *et al.* and are not discussed here.

In [7], the author proposes a data acquisition architecture and implies that random projections may be implemented via random convolution. Plainly speaking, random projections involve taking an inner product of one vector with many other random vectors. Clearly on L_2, this inner product is in fact an integral. Further elementary signal-processing concepts tell us that the inner product on vectors (e.g. in $\mathbb{R}^N$), $< \mathbf{x}, \mathbf{y} >= \sum_{n=0}^{N-1} x(n)y(n)$ is in l_2 and analogously on signal spaces, for signals time limited in $[0, T]$ this inner product is also defined as $< \mathbf{x}, \mathbf{y} >= \int_0^T x(t)y(t)\mathrm{d}t$ is in L_2. Random projections were implemented as an inner product in L_2 in [8,9] but in [7], a random convolution was used to replicate random projections. It is also unclear why linear convolution is implemented as a circular convolution using a random waveform in [7]. The main analytical theorems in [7] also resort to circular convolution for their proofs. It is unclear in [7], how a waveform can be random and then a circular convolution operation can then be used on it. If a waveform is random, then it has no pattern; circular convolution is used for periodic extensions, so the assumptions made in [7] are untenable.

Circular convolution theory is appropriate for finite-length discrete signals, see [12, Chapter 8]. Further, in [12, Section 8.7.2 p. 577] "circular convolution as linear convolution with aliasing," hence, arguments put forth on sampling below the Nyquist are contradictory in [7]. It becomes evident that the analysis in [7] are for signals that have already been sampled and are discrete.

In the real world, signals are analog and they are sampled and digitized by an analog-to-digital (A/D) converter. Therefore, any claims of sub-sampling must be investigated from first principles in an analog to digital framework; our goal is to address this issue.

We consider an architecture that implements random projections in an analog fashion. An analog signal is projected, i.e., mixed with many random basis signals, then integrated, and then sampled at some CS rate lower than the Nyquist rate of the original analog signal. By analyzing the signals passing through the analog chain, we show that the final signal in this chain is altered spectrally. Our architecture is similar to that proposed in [8–10] but our analysis and results are different.

We use elementary signal processing concepts, this is intended to enable the reader to verify our analysis and results readily. Recall that multiplying signals is convolution of their respective spectra. Therefore, the spectrum of the signals after mixing with the random basis functions in the technique of random projections is actually spread.

A spread signal has a Nyquist sampling frequency greater than its un-spread counterpart. The integrators further act as smoothers or low-pass filters (LPF) and, hence, attenuate high-frequency components. Our analysis demonstrates that if these spread signals are sampled below their Nyquist rate (which is greater than or equal to the Nyquist sampling frequency of the original un-spread signal), then they have an irrecoverable loss of spectral information, are aliased, and result in an SNR loss. The claims of CS then sub-sampling below the Nyquist rate of the original signal itself is called into question.

Of course there is a silver lining, like in the computer science literature [1,13–16], random projections in fact may be used on the digitized signals to reduce the dimensionality of the signal processing problem and computational complexity of algorithms by operating on the lower dimensional data sets. In the rest of the chapter, we focus on random projections for signals that have already been sampled, i.e. digitized signals.

A powerful result by Johnson and Lindenstrauss (JL) [17] states that a higher dimensional data set can be operated upon by a Lipschitz function and the resulting *lower dimensional* output data set has similar pairwise distances as the original data set. Random projections employ this concept, but project the data into a random subspace spanned by d-random vectors. It is noticeable that in a vector space, the number of "*nearly orthogonal vectors*" exponentially increases as the dimension increases. This is exploited by the random projections principle, and it was shown that this technique preserved the pairwise distances of the data after projection via statistical expectations, with a prescribed but tolerable variance [1,13,16].

Multisensor radar data such as those from multi-spectral imaging, electro-optical/infrared (EO/IR), and RF sensor data suffer from the curse of dimensionality. In particular, for radar, STAP data becomes heterogeneous within a few cells from the cell under test [18]. Therefore, sufficient training data is unavailable to implement the full dimension adaptive filter. This problem is further exacerbated for multistatic radar STAP since the statistics depend on the placement geometry of the transmit and receive arrays [19]. Further, the STAP filter in the original higher dimension setting is computationally expensive due to the inversion of large, almost always *non-sparse* matrices.

Random projections for STAP are appealing from two perspectives. First, it reduces the dimension of the problem leading to faster computations in inverting the covariance matrices. Second, in the reduced dimension setting, a fewer samples are needed to generate representative sample covariance matrices. However, these gains come with a price of SINR loss [20]. Random projections afford a significant computational cost reduction for radar STAP [21].

To ameliorate over the loss of SINR, we consider another technique, which uses both a random and a deterministic lower dimensional subspace. We call this localized random projections [22]. The idea behind it is simple, in radar and especially STAP, detection is performed for a cell having a specific angle and Doppler. Therefore, for the deterministic subspace, we choose the steering vector for the corresponding angle, and other column vectors from its orthogonal subspace. Now since the target may have any Doppler, we consider its corresponding subspace to be random. The resulting lower dimensional subspace consists of a deterministic part as well as a random part. Using this technique, we demonstrate in general that it performs better than traditional random projections with respect to SINR.

Prior work: Researchers from theoretical computer science were the first to use random projections but their motivation was different from CS [1,13]. Specifically, they wanted to speed up the algorithms by retrieving smaller representative dataset(s) rather than retrieving the entire dataset, and also preserving memory for other data-intensive problems. Our motivation is dimensionality reduction from random projections as a solution for the scarcity of training data problem in radar STAP [23], and of course as a straightforward computational cost reducer.

Random projections are closely related and could be considered as a predecessor to CS and the older "sketching" or "streaming" techniques (see [15,24]) in the computer science literature. However, random projections do not deal with signal recovery or estimation as CS. Random projections are the first stage in CS where the signal is sampled or projected onto random basis functions. Other reduced dimension STAP algorithms exist (see [25–28] and references therein). Unlike the random projections technique, some of these approaches make parametric assumptions and operate with different dimension-reducing transformations for different cells under test (c.u.t.) but for the same data set and may have implications on constant false alarm rates (CFAR).

8.2 A critical perspective on sub-sampling claims in compressive sensing theory

A system is shown in Figure 8.1 which emulates and implements random projections in the analog domain, followed by an A/D which converts the analog signal to a discrete signal. The resulting discrete samples are furnished to the l_1 optimization routines which implement signal reconstruction. The l_1 optimization algorithms are implemented in software and are, therefore, not shown in Figure 8.1. We consider examples of the reconstruction from l_1 optimization algorithms subsequently. The crux of the matter here is whether the analog signal, $s(t)$, may be reconstructed by sampling at a rate below the Nyquist rate by the system as shown in Figure 8.1.

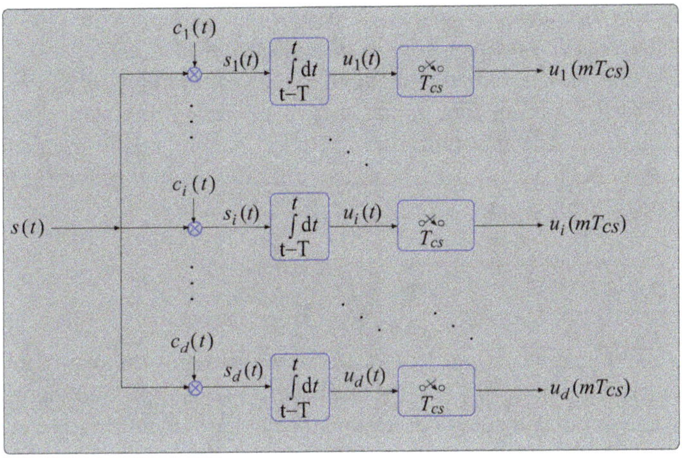

Figure 8.1 Architecture implementing compressive sensing in the analog domain. Like any practical A/D, an anti-aliasing filter (not shown here but assumed) is used in the A/D and filters before sampling.

Table 8.1 Table of some important parameters and their definitions used in Section 8.2 and rest of this chapter

Parameter	Definition
T	Integration time in random projections (directly effects LPF cut-off)
F_{CS}	CS sampling rate, assumed to be lower than the traditional Nyquist rate
T_{CS}	CS sampling period inverse of CS sampling rate F_{CS}
T_s	(over) sampling period inverse of (over) sampling rate F_s, used in simulations
D	A larger dimension
d	A smaller dimension than D
T_c	Chip length in random basis
t_o	A delay random variable
T_o	Maximum support of distribution of t_o
T_r	Rectangular pulse width
*	Complex conjugate
$\circledast$	Linear convolution
$\mathscr{F}^{-1}$	Fourier inverse operation

SWAP-C: Practical implementation of sensing systems try to minimize size, weight, aperture, power and, therefore, cost, this is termed as SWAP-C. Clearly, from Figure 8.1, the SWAP-C is increased. The CS analog implementation in this figure will require d analog mixers, d integrators, and d A/Ds as compared to just one A/D in a traditional Nyquist sampling scheme.

Before we delve into details, we provide a small table (see Table 8.1) that defines some important parameters used in Section 8.2.

We describe this analog system next. Assume that the input signal $s(t)$ is a sparse signal which we need to recover from its discrete samples, $u_i(mT_{CS}), i = 1, 2, \ldots, d$. The analog signal, $s(t)$, is passed as an input to the system in Figure 8.1.

Multiplication/mixing: The signal, $s(t)$, is multiplied by many random basis, $c_i(t), i = 1, 2, \ldots, d$. The multiplication operation may be readily implemented by an analog device such as a mixer. The output after this multiplication (or mixing) is denoted as, $s_i(t)$, expressed as,

$$s_i(t) = s(t) \times c_i(t), i = 1, 2, \ldots, d. \tag{8.1}$$

Recall that multiplication in time domain is equivalent to convolution in the frequency domain. Hence, it is noteworthy now to examine that $s_i(t)$ is a spectrally spread version of $s(t)$. The amount of spreading depends on the random waveform $c_i(t)$, or rather its power spectral density (PSD). We will revisit this spreading behavior as we consider some examples subsequently.

We claim that the argument of CS being able to sub-sample below the Nyquist rate is weakened already for the following reasons:

1. Each spectrally spread signal, $s_i(t), i = 1, 2, \ldots, d$ in general has a different Nyquist rate than the Nyquist rate of another spectrally spread signal, $s_j(t), j \neq i$, as well as different from the Nyquist rate of the original, un-spread signal, $s(t)$.
2. if $s_i(t), i = 1, 2, \ldots, d$ are non-stationary (which we will see is true by default in CS implementation, see Section 8.2.1, then the Nyquist rate comparison is itself untenable.

Sub-sampling the random waveforms? Additionally, note that the random waveforms, $c_i(t)$, are analog. The l_1 reconstruction algorithms are sensitive to the sensing matrices. If the sensing matrices are not representative of the random waveforms, $c_i(t)$, this will affect reconstruction. Herein lies a central problem. Take note that for l_1 CS reconstruction, the sensing matrices consist of rows of the sampled random waveforms. Thus far, no CS literature discusses how these random waveforms, $c_i(t), i = 1, 2, \ldots, d$ are themselves sampled in the first place. Would they require above Nyquist sampling rates? Would compressing sensing be used on the random waveforms themselves? The former would imply that the Nyquist sampling theorem would then be a basis for CS. The latter would imply a circular argument. In both cases, the questions are valid and have been unanswered by the CS community. Presently, we and the CS community are at a loss to answer these questions.

Integration as low pass filtering: The signals, $s_i(t), i = 1, 2, \ldots, d$ are passed to the integrator, denote the output signals after integration as, $u_i(t)$, expressed as:

$$u_i(t) = \int_{t-T}^{t} s_i(t) \mathrm{d}t \tag{8.2}$$

The integrator $\int_{t-T}^{t}$ is special, in that it is a causal system and after an initial delay of T secs, it operates in real time. It may be shown readily that the integrator in (8.2) functions as a LPF with transfer function,

$$H(f) = T\text{sinc}(fT)\exp(-j\pi fT) \tag{8.3}$$

where, $\text{sinc}(x) := \sin(\pi x)/\pi x$, and $\text{sinc}(0) := 1$.

Analog to digital conversion: The analog signals, $u_i(t), i = 1, 2 \ldots, d$, are now passed through d A/D converter which samples the signal at sampling rate, T_{CS}. The resulting discrete signals are $u_i(mT_{CS}),\ m = 0, 1, 2, \ldots,, i = 1, 2, \ldots, d$.

Anti-aliasing filter: Practical A/Ds have an anti-aliasing filter. The anti-aliasing filter is used prior to the A/D actually sampling the signal, i.e. filter before actual sampling. This filter is used to suppress strong out-of-band signals. Without the anti-aliasing filter, the out-of-band interference will alias and distort the signal spectrum that is being sampled and observed. In essence, the anti-alias filter band-limits the spectra and attenuates the signals being aliased for some set sample rate.

For example, let us assume a sampling rate of 40 kHz and no anti-aliasing filter. The signal of interest is at 10 kHz. if we have a strong interference signal at 55 kHz, the interference is aliased and will show up at 15 kHz. Now suppose an anti-aliasing filter is used prior to sampling by the A/D and its cut-off band is limited to 50 kHz. The interference is suppressed before sampling and, therefore, has no opportunity to distort the spectrum after sampling.

Note that the integrator may itself function as an anti-aliasing filter if the signal of interest was near DC and in the main lobe response of the integrator transfer function, but not otherwise. For example, if the processing is done at intermediate frequency (IF), then an anti-aliasing filter has to be used. If the signal is spread across many bands, the anti-aliasing filter will surely remove some spectral content and, therefore, would affect the signal reconstruction afterwards. These are practical design challenges inherent to CS sampling architectures.

The system described so far is emulative of how CS would be implemented in the analog domain. It is also worth noting that the architecture described thus far is causal and operates in real-time.

8.2.1 General issues of non-stationarity

Statistics of wide sense stationary processes have a time invariant mean and the auto-correlation is a function of the time lag (differences) and not the particular time instants. Stationary processes passed through linear systems remain stationary. Wide sense stationary stochastic processes have a deterministic and time invariant PSD.

The auto-correlation and cross-correlation function of the signals, $s_i(t) = s(t)c_i(t), i = 1, 2, \ldots, d$ are:

$$\begin{aligned} R_i(t,\tau) &:= \mathrm{E}\{s(t)c_i(t)s^*(t+\tau)c_i^*(t+\tau)\} \\ R_{ik}(t,\tau) &:= \mathrm{E}\{s(t)c_i(t)s^*(t+\tau)c_k^*(t+\tau)\}, i \neq k \end{aligned} \tag{8.4}$$

Assuming that the random basis waveforms, $c_i(t), i = 1, 2, \ldots, d$, are stationary, then there is absolutely no guarantee that signals, $s_i(t), = s(t)c_i(t)i = 1, 2, \ldots, d$ result in wide sense or strict sense stationary processes. This is because any product of a deterministic signal or stationary random process with another random process is not guaranteed to be stationary.

Passing non-stationary input random processes through the integrators results in non-stationary processes at the outputs. Herein lies another practical problem with CS implemented in an analog system.

We consider an example: assume a linear frequency modulated signal (LFM), $s(t) = \exp(j\alpha 2\pi t^2 + j\theta)$, where θ is a random variable with some arbitrary distribution and independent from the random basis, $c_i(t)$ for all $i = 1, 2, \ldots, d$. Clearly then, $s_i(t) = s(t)c_i(t)$ is non-stationary since $\mathrm{E}\{s_i(t)s_i^*(t+\tau)\}$ is a function of time. LFM signals are always used in radar, but with the CS analog architecture as shown in Figure 8.1, the signals at the outputs after multiplication (or mixing) and integration are neither cyclo-stationary nor stationary.

Indeed, if $c_i(t)$ are zero mean and white, then the underlying waveforms, $s_i(t)$ are all zero mean and stationary with delta-type autocorrelation function. Therefore, note that $s_i(t)$ is spread over an infinite bandwidth and spectral information will be lost when these signals are passed through the integrator. Therefore, this strongly suggests that choosing $c_i(t), i = 1, 2, , \ldots, d$ being zero mean and white is impractical; an assumption, however, always made in the CS literature, see e.g. [5,10]. To reduce spectral spreading, we chip the random waveforms as seen subsequently.

We now take specific examples of the CS analog sampling system and see how it performs for different signal classes. For the ease of exposition, we focus on a single family of random basis, e.g., a BPSK waveform, analogous to a Bernoulli sensing matrix in CS literature. Explicitly,

$$c_i(t) = \sum_{n=0}^{\infty} a_{ni}\mathrm{rect}\left(\frac{t - nT_c}{T_c}\right), i = 1, 2, \ldots, d \tag{8.5}$$

where T_c is the chip width, and $\mathrm{rect}(t/T_c)$ is a rectangular waveform defined as,

$$\mathrm{rect}\left(\frac{t}{T_c}\right) := \begin{cases} 1 & 0 \le t \le T_c \\ 0 & \text{otherwise.} \end{cases}$$

Additionally, in (8.5), a_{ni} is a random variable which is ± 1 with equal probability. The random variables, a_{ni}, are independent, identically distributed across both dimensions indexed by (n, i). That is, random variables, $a_{n_1 i}$ and $a_{n_2 i}$ are mutually independent with $(n_1, n_2) \in 0, 1, 2, \ldots$ and $n_1 \neq n_2$. Likewise, random variables, a_{ni} and a_{nj} are also mutually independent with $(i,j) \in 1, 2, \ldots, d$ and $i \neq j$.

We note again that the random basis waveforms, $c_i(t)$ are analogous to Bernoulli sensing matrices used in CS theory. Bernoulli sensing matrices strongly satisfy RIP with high probability, see [29].

The random basis waveforms in (8.5) have an auto-correlation function which is cyclo-stationary with period, T_c, and, therefore, has an (average) power spectral density, $C_i(f)$ and average auto-correlation function over one period, $R_{ci}(\tau)$ expressed as [30],

$$\begin{aligned} C_i(f) &= T_c \text{sinc}^2(fT_c) \\ R_{ci}(\tau) &= \mathscr{F}^{-1}(C_i(f)), i = 1, 2, \ldots, d \end{aligned} \tag{8.6}$$

where $\mathscr{F}^{-1}$ is the inverse Fourier transform. We note that the PSDs, $C_i(f)$, are all identical, for all $i = 1, 2, \ldots, d$.

With the analog architecture described thus far, we consider specific sparse signals of interest and describe the signals $u_i(mT_{CS}), m = 0, 1, 2, \ldots, i = 1, 2, \ldots, d$ in the spectral domain.

8.2.2 Sparse signal in intermediate frequency (IF)

Assume $s(t) = \exp(j\omega_c t + j\theta)$, where ω_c is the carrier frequency, θ is a random variable uniformly distributed in $[-\pi, \pi]$ and independent of the random basis functions, $c_i(t)$ for all $i = 1, 2, \ldots, d$. In typical RF systems including radar and communications, signals are sampled at an IF, ω_{IF}. By convention, $\omega_{IF} < \omega_c$. Thus, assuming down conversion, the new signal is $s(t) := \exp(j\omega_{IF} t + j\theta)$. This is now passed into the sampling architecture as described earlier in Figure 8.1.

In this case, the sinusoid, $\exp(j2\pi f_{IF} t + j\theta)$ is stationary, but $s_i(t) = s(t)c_i(t), i = 1, 2, \ldots, d$ are cyclo-stationary since $c_i(t)$ is cyclo-stationary. The corresponding average auto-correlation and average PSD of $s_i(t)$ are expressed as:

$$\begin{aligned} R_i(\tau) &= \exp(j2\pi f_{IF}\tau) R_{ci}(\tau) \\ S_i(f) &= \delta(f - f_{IF}) \circledast C_i(f) = C_i(f - f_{IF}), i = 1, 2, \ldots, d \end{aligned} \tag{8.7}$$

where $\circledast$ is the convolution operation.

The signals, $s_i(t), i = 1, 2 \ldots, d$ are passed through the integrators, the outputs are $u_i(t)$. The integrators are equivalent to LPFs. Obviously, from (8.7) and (8.6), the LPF will reject the IF if its cut-off frequency is much smaller than the IF. In fact the integration time, T, must be chosen such that it would allow at least the main lobe of $\text{sinc}^2((f - f_{IF})T_c)$ to pass through with little distortion. For that to happen,

$$\frac{1}{T} \geq f_{IF} + \frac{1}{T_c}. \tag{8.8}$$

An SNR loss is still incurred since the original signal was spread. Also note that the cutoff of the LPF (equivalently integrator) must be higher than the IF, this poses unnecessary demands on filter design. Furthermore such high cutoffs would allow frequencies from DC to IF pass un-distorted and, therefore, also reduce system SNR.

We take an example to make our point. Let IF be 1 MHz, assume that $\omega_c = 2\pi \times 1$ GHz. If we assume a nominal $1/T_c = 1$ MHz, then the integration filter has a zero crossing at $1/T = 2$ MHz. Therefore, from (8.8) all frequencies from DC to 2 MHz will be passed by the integration filter. Note, however, that this is unacceptable since the noise bandwidth is increased and any or all signals in the pass band will be transmitted un-distorted. A few numerical simulations of IF processing are shown subsequently.

We note two other important points, $s(t)$, were sparse in the frequency domain. The signals in the CS analog scheme, $s_i(t), i = 1, 2, \ldots, d$ are no longer sparse in the spectral domain. In fact, its PSD is spread around the IF. The Nyquist rate of the original signal is $1/(2f_{IF})$. Even if we assume the integrator allows the main-lobe of $\mathrm{sinc}^2((f - f_{IF})T_c)$ to pass un-distorted, then it is obvious that sampling at a rate below Nyquist $T_{CS} > 1/(2f_{IF})$, or $F_{CS} < 2f_{IF}$ will result in the discrete signals, $u_i(mT_{CS})$ aliased and no longer a true representation of $s_i(t)$ and, therefore, also of $s(t)$.

8.2.3 Temporally sparse signal in baseband

We consider a signal that is typical in radar, a rectangular pulse with an unknown and random delay, t_o. We assume that the random delay, t_o is uniformly distributed in $[0, T_o]$. Let the pulse width of the rectangular waveform be $T_r < T_o$. The signal, $s(t)$ is then,

$$s(t) = \mathrm{rect}\left(\frac{t - t_o}{T_r}\right). \tag{8.9}$$

Clearly, $s(t)$ is a random process, we may show readily that it is wide sense stationary, with mean and auto-correlation, expressed as:

$$\begin{aligned}\mu_s &= \mathrm{E}\{s(t)\} = \frac{T_r}{T_o}\\ R_s(\tau) &= \mathrm{E}\{s(t)s^*(t+\tau)\}\\ &= \frac{1}{T_o}\int_0^{T_o} \mathrm{rect}\left(\frac{t - t_o}{T_o}\right)\mathrm{rect}\left(\frac{t - t_o + \tau}{T_o}\right)\mathrm{d}t_o\\ &= \frac{T_r}{T_o}\mathrm{tri}\left(\frac{\tau}{T_r}\right)\end{aligned} \tag{8.10}$$

where,

$$\mathrm{tri}(x) = \begin{cases}(1 - |x|) & |x| < 1\\ 0 & \text{otherwise.}\end{cases}$$

The random process $s(t)$ is also not surprisingly ergodic in both mean and auto-correlation.

Further, in this case, $s_i(t) = s(t)c_i(t)$ is cyclo-stationary since $c_i(t)$ is cyclo-stationary. Like before, the mean and the auto-correlation are periodic with period T_c. The average PSD and average auto-correlation functions for $s_i(t)$ are expressed as:

$$\begin{aligned}S_i(f) &= \frac{T_r^2}{T_o}\mathrm{sinc}^2(fT_r) \circledast T_c\mathrm{sinc}^2(fT_c)\\ &= \frac{T_r}{T_o}g_1(f) \circledast g_2(f)\\ R_i(\tau) &= \mathcal{F}^{-1}\{S_i(f)\}\end{aligned} \tag{8.11}$$

where $g_1(f) := T_r \text{sinc}^2(fT_r)$ and $g_2(f) := T_c \text{sinc}^2(fT_c)$. Instead of analytically simplifying (8.11), we take a few numerical examples to demonstrate the loss of SNR and the loss of spectral content by sub-sampling below the Nyquist rate in the analog scheme implementing CS.

Consider, Figure 8.2(a)–(c). Here we assumed a rectangular pulse of width, $T_r = 0.01$s, the chip width, T_c, of the random basis functions as in (8.5), is varied. In Figure 8.2(a)–(c), $Tc = 0.005$s, $T_c = 0.002$s, and $T_c = 0.05$s, respectively. The PSD of the original signal ($s(t)$) is shown in blue, the average PSD of the random basis function ($c_i(t)$) is shown in black. The PSD after multiplying the two, i.e. $s_i(t) = s(t)c_i(t)$ is shown in red in these figures. In Figure 8.2(c), $Tc > T_r$ and, hence, we see minimal spreading this is because one chip captures the entire rectangular pulse having width T_r. In other cases, the spectral spread is significant. In all these figures, the Nyquist rate is 200 Hz. Sampling below Nyquist rate, i.e. $T_{CS} > 1/200$ s or $F_{CS} < 200$ Hz results in spectral content being lost as well as an SNR loss.

8.3 Random projections STAP model

Until now, we have discussed the CS analog sensing framework and demonstrated the deleterious effects of spreading and low-pass filtering (integration) on the signal. In essence, we argued that CS techniques prior to sampling are not well suited for radar problems.

Herein, we characterize random projections for radar STAP. Note that this is applied for radar data after sampling. Random projections are used for minimizing the digital computations.

The STAP model is well known and is described succinctly, as follows. We consider an airborne radar looking at the ground. The radar consists of M calibrated sensor elements and transmits a burst of N pulses. The objective is to test for the presence or absence of a target. Assuming narrowband operation and a receive filter vector $\mathbf{w} \in \mathbb{C}^{MN}$, the hypotheses may be formulated as

$$\begin{aligned} \mathscr{H}_0 &: \ \mathbf{y} = \mathbf{c} + \mathbf{n} + \mathbf{i} \\ \mathscr{H}_1 &: \ \mathbf{y} = \gamma \mathbf{x} + \mathbf{c} + \mathbf{n} + \mathbf{i} \end{aligned} \tag{8.12}$$

where $\mathbf{y} \in \mathbb{C}^{MN}$ is the space–time snapshot measurement, $\mathbf{x} \in \mathbb{C}^{MN}$ is the deterministic and known spatio-temporal response of the target, and γ is a complex amplitude. The clutter, noise, and interference responses are the vectors, $\mathbf{c}, \mathbf{n}, \mathbf{i}$, respectively, assumed each to be zero mean, and mutually independent in the statistical sense [23]. Assume that the random vectors in (8.12) could be modeled with a covariance matrix, denoted as $\mathbf{R} \in \mathbb{C}^{MN \times MN} = \mathbf{R_c} + \mathbf{R_n} + \mathbf{R_i}$, where each matrix is the respective covariance of the clutter, noise, and interference. Then, the *clairvoyant* SINR (assumes knowledge of covariance matrix), denoted as ($\mathtt{SINR_c}$) is,

$$\mathtt{SINR_c} = \frac{|\gamma \mathbf{w}^H \mathbf{x}|^2}{\mathbf{w}^H \mathbf{R} \mathbf{w}}, \quad \mathbf{w_o} = \mathbf{R}^{-1}\mathbf{x} \tag{8.13a}$$

$$\mathtt{SINR_c^o} = \mathtt{SINR_c}(\mathbf{w_o}) = |\gamma|^2 \mathbf{x}^H \mathbf{R}^{-1} \mathbf{x}. \tag{8.13b}$$

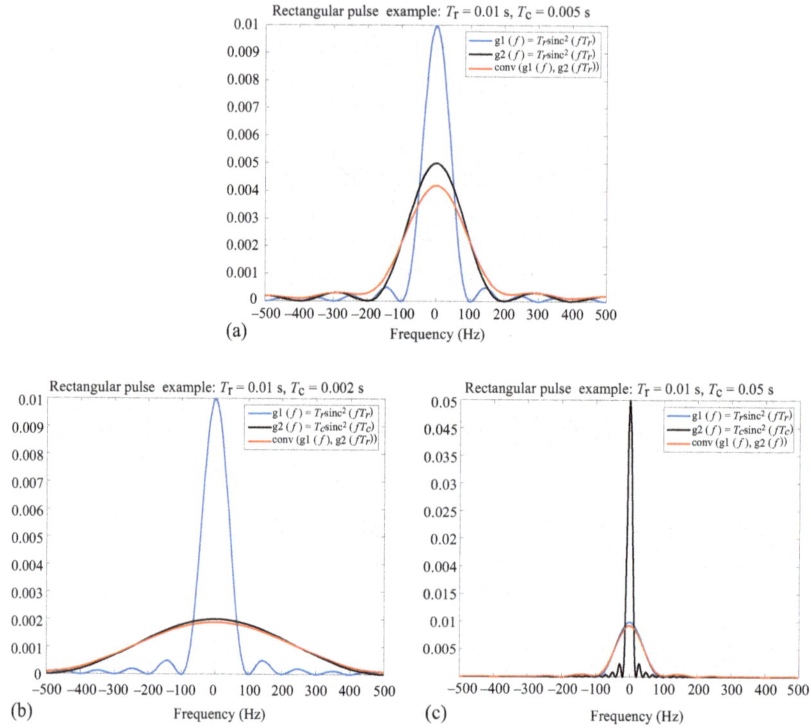

Figure 8.2 *Rectangular pulse ($T_r = 0.01$ s) example as in Section 8.2.3 and* (8.11)*: (a) $T_c = 0.005$ s, (b) $T_c = 0.002$ s, and (c) $T_c = 0.05$ s. Original PSD (blue), random basis PSD (black), PSD arising from multiplying random basis with signal in red, see also* (8.11)*. The Nyquist rate is 200 Hz, so under sampling ($\frac{1}{T_{CS}} = F_{CS} < 200$ Hz) the signals (in red) in (a), (b), and (c) will result in an irrecoverable loss of spectral content, aliasing and an SNR loss. This is true even if the integration limit, T (related to the LPF cut off) is chosen to preserve a significant portion of the signals of interest. Note that in (a), (b), and (c), the original signal spectrum (blue) after multiplying with random basis spectrum (black) is spread (in red).*

where in (8.13a), the (optimum) weight vector, $\mathbf{w_o}$, maximizes the SINR. As evident from (8.12) and (8.13a), the weight vector is essentially a whitening match filter. In (8.13b), the clairvoyant SINR evaluated at the optimum weight vector in (8.13a) is denoted as $\mathtt{SINR}_\mathtt{c}^o$.

8.3.1 Computational complexity and a "small" data problem

From (8.12) and (8.13a), we see that for every spatio-temporal cell, the weight vector involves inverting matrices. The complexity is then $O(M^3N^3)$ making real-time implementation of STAP computationally prohibitive. Furthermore, we note that in practical cases, only partial knowledge of $\mathbf{R}$ is available, and, hence, the sample covariance matrix, $\hat{\mathbf{R}}$, is used as surrogate instead of $\mathbf{R}$ in the STAP implementation. The sample covariance matrix is constructed in the usual maximum-likelihood way for Gaussian statistics, by averaging (outer products of) responses from nearby cells in the vicinity of the c.u.t. Unfortunately, the radar scene is homogeneous for a few cells near the c.u.t. [23]. Therefore, we do not have enough samples to form the sample covariance matrix, and guarantee its invertibility. This is a well-known problem in radar STAP.

Another SINR metric useful when dealing with the sample covariance matrix is the normalized SINR, denoted as

$$\mathtt{SINR}_{\mathtt{n}} = \frac{(\mathbf{x}^H\hat{\mathbf{R}}^{-1}\mathbf{x})^2}{(\mathbf{x}^H\hat{\mathbf{R}}^{-1}\mathbf{R}\hat{\mathbf{R}}^{-1}\mathbf{x})(\mathbf{x}^H\mathbf{R}^{-1}\mathbf{x})}. \tag{8.14}$$

The normalized SINR is always between (0, 1] and measures a loss of performance of the system when using the sample matrix instead of the true covariance matrix.

Motivation: Assume that we have $L \leq MN$ samples to construct the sample covariance matrix. To motivate a solution to our problem, recall the Johnson–Lindenstrauss theorem.

Theorem 1. (*Johnson–Lindenstrauss (JL) [17,31]*) For some $0 < \epsilon_o < 1$, and for any set $\mathscr{X} \subset \mathrm{R}^{MN}$ of $L \leq MN$ data points in R^{MN}, there exists a Lipschitz mapping, $f(\cdot) : \mathrm{R}^{MN} \rightarrow \mathrm{R}^d$ such that for all $\mathbf{u}, \mathbf{v} \in \mathscr{X}$

$$(1-\epsilon_o)||\mathbf{u}-\mathbf{v}||^2 \leq ||f(\mathbf{u})-f(\mathbf{v})||^2 \leq (1+\epsilon_o)||\mathbf{u}-\mathbf{v}||^2$$

where $d > \lceil \frac{\kappa \ln(L)}{g(\epsilon_o)} \rceil$, κ, is a universal constant, and $g(\epsilon_o)$ is an arbitrary function of ϵ_o, and $f(\cdot)$ may be found in randomized polynomial time.

We note that the reduced dimension, d, is independent of the original dimension of the problem, i.e. MN, but is dependent only on L. For specific types of $g(\epsilon_o)$, see [17,31], and references therein. The JL theorem is unique in that it states that by projecting the data onto a lower dimension, the pairwise distances are only *slightly perturbed*. The original theorem was proved from the real domain but can be readily extended to the complex domain as well.

The idea then is to reduce the dimension of the spatio-temporal measurement vector $\mathbf{y}$ in (8.12) to a lower dimension $d << MN$, with the JL theorem providing guarantees that the newly projected data is representative of its original higher dimension counterpart. In other words, if the c.u.t. has no target, then its lower dimension projection also is similar (in a l_2 norm sense) and is comprised of clutter, noise, and interference like components. Likewise, if the c.u.t. has a target embedded in clutter, noise, and interference, its lower dimension counterpart is also representative of the same (in a l_2 norm sense). The reduced dimension problem's covariance is now $d \times d$ and if $d < L$, the lower dimension covariance is invertible. The inverse matrix computation required for the new weight vector is also reduced from $O(M^3N^3)$ to $O(d^3)$ flops with $d << MN$.

8.3.2 Random projections

The JL theorem is general and does not specify the type of the lower dimension Lipschitz mapping, $f(\cdot)$. Furthermore, the fidelity constant ϵ_o is difficult to ascertain readily and depends on d, the chosen dimension via a function $g(\epsilon_o)$. A solution is then to project the data into a lower d-dimensional subspace spanned by d random vectors. It was shown that, with some prescribed variance, on an average, the pairwise distances between the data points after projection are preserved [13].

Consider such a lower dimension transformation denoted by matrix $\mathbf{T} \in \mathbb{C}^{MN\times d}$, where $d \leq MN$. The new null hypothesis is then, $\mathscr{H}_0 : \mathbf{y_t} = \mathbf{T}^H(\mathbf{c}+\mathbf{n}+\mathbf{i})$ and the new alternate hypothesis is then, $\mathscr{H}_1 : \mathbf{y_t} = \mathbf{T}^H(\gamma\mathbf{x}+\mathbf{c}+\mathbf{n}+\mathbf{i})$. Likewise, the reduced dimension optimum weight vector ($\mathbf{w_t} \in \mathbb{C}^d$), transformed covariance matrix ($\mathbf{R_t} \in \mathbb{C}^{d\times d}$), reduced dimension clairvoyant SINR at this new optimum weight vector ($\mathtt{SINR}^o_{\mathtt{ct}}$) and new normalized SINR ($\mathtt{SINR}_{\mathtt{nt}}$) are expressed as:

$$\mathbf{w_t} = \mathbf{R_t}^{-1}\mathbf{x_t},\ \mathbf{R_t} = \mathbf{T}^H\mathbf{R}\mathbf{T},\ \mathtt{SINR}^o_{\mathtt{ct}} = |\gamma|^2\mathbf{x_t}^H\mathbf{R_t}^{-1}\mathbf{x_t} \tag{8.15a}$$

$$\mathtt{SINR}_{\mathtt{nt}} = \frac{(\mathbf{x_t}^H\hat{\mathbf{R}}_\mathbf{t}^{-1}\mathbf{x_t})^2}{(\mathbf{x_t}^H\hat{\mathbf{R}}_\mathbf{t}^{-1}\mathbf{R_t}\hat{\mathbf{R}}_\mathbf{t}^{-1}\mathbf{x_t})(\mathbf{x_t}^H\mathbf{R_t}^{-1}\mathbf{x_t})}. \tag{8.15b}$$

where as usual, in the lower dimension setting, $\mathbf{x_t} = \mathbf{T}^H\mathbf{x}$. We note that as long as $\mathrm{rank}(\mathbf{T}) = d$, it may be shown readily that the transformed covariance matrix is also full rank. The number of samples L is unchanged with this transformation and when $d \leq L$ the resulting sample matrix $\hat{\mathbf{R}}_\mathbf{t} = \mathbf{T}^H\hat{\mathbf{R}}\mathbf{T}$ is invertible.

Conditional statistics: The appeal of using random projections is that the matrix $\mathbf{T}$ is held constant for the entire data set including all the cells under test. Hence, as is assumed usually, invoking that the distributions in (8.12) are complex normal, let $f(\mathbf{y}|\mathscr{H}_0) = \mathscr{CN}(\mathbf{0},\mathbf{R})$ and $f(\mathbf{y}|\mathscr{H}_1) = \mathscr{CN}(\alpha\mathbf{x},\mathbf{R})$. Then, under the random projections paradigm, we work with the conditional distributions in the new reduced dimension hypotheses (since $\mathbf{T}$ is constant for all cells under test), as $f(\mathbf{y}|\mathscr{H}_0,\mathbf{T}) = \mathscr{CN}(\mathbf{0},\mathbf{R_t})$ and $f(\mathbf{y_t}|\mathscr{H}_1,\mathbf{T}) = \mathscr{CN}(\alpha\mathbf{x_t},\mathbf{R_t})$. Indeed we are also interested in the behavior of random projections for all other instances of $\mathbf{T}$, and this analysis is not tractable in closed form but can be shown via Monte-Carlo simulation subsequently.

8.3.2.1 Distributions

In general, we want the columns of:

$$\mathbf{T} = \frac{1}{\sqrt{d}}[\mathbf{t}_1,\mathbf{t}_2,\cdots,\mathbf{t}_d]$$

to be i.i.d., zero mean and have constant variance to guarantee the pairwise norm preservation in the data on the average.

(*Normal*) Typically, $\mathbf{t}_i, i = 1,2,\ldots,d$ are chosen from $\mathscr{CN}(\mathbf{0},\mathbf{I})$ or when $\mathbf{T}$ is real, $\mathbf{t}_i, i = 1,2,\ldots,d$ are sampled from $\mathscr{N}(\mathbf{0},\mathbf{I})$, where $\mathbf{I}$ is the identity matrix. When the columns of $\mathbf{T}$ are normal (real or complex) the random projections is *non-sparse*. Other sparse versions of random projections exist, and for $\mathbf{T}$ complex, the real and complex elements of $\mathbf{t}_i, i = 1, 2, \ldots, d$ are sampled independently from the discrete distribution given below

$$\sqrt{\frac{p}{2}}\begin{cases} +\ 1 & \text{with prob.} \quad 1/2p \\ \ \ \ 0 & \text{with prob.} \quad 1 - 1/p\ . \\ -\ 1 & \text{with prob.} \quad 1/2p \end{cases} \tag{8.16}$$

When $\mathbf{T}$ is purely real, then elements of the columns are sampled from a discrete distribution similar to (8.16) but with $\sqrt{p}$ instead of $\sqrt{p/2}$.

(*Achiloptas [1]*) For $p = 1$ or $p = 3$, results in a sparse random projection (more zeros, leading to sampling fewer data-points).

(*Li-Has)* [13]) For $p = \sqrt{d}$, or $p = d/\ln(d)$, with $p >> 3$ in both cases, these distributions result in *very* sparse random projections [13]. We do not recommend using $p = d/\ln(d)$ because several of our Monte-Carlo instantiations, resulted in non-full-rank covariance matrices for particular values of d. The choice of $p = d/\ln(d)$ is aggressive and is motivated from the exponential tail error bounds (Chernoff, Hoefdding, etc.) of several distributions.

(*Bernoulli*) Another simple but sometimes overlooked distribution in the random projections literature is the Bernoulli with equi-probable symbols ± 1 and scaled appropriately so that each element in $\mathbf{T}$ has zero mean and unit variance. These distributions are used in our numerical experiments to generate elements of $\mathbf{T}$ for both complex and real cases.

8.3.3 Localized random projections

In localized random projections, the angle information is incorporated to generate the resulting transformation matrix. At the target range cell, consider the transformation matrix

$$\mathbf{T} = \mathbf{T}_{\vartheta} \otimes \mathbf{T}_S(\theta) \tag{8.17}$$

where $\otimes$ is the Kronecker product, $\mathbf{T}_{\vartheta} \in C^{N \times d_1}$ is the Doppler component, and $\mathbf{T}_S \in C^{M \times d_2}$ is the angle component, with $d_2 \leq M$ being the number of vectors in the angle space and is completely deterministic. We note that $d_1 \times d_2 = d \leq MN$.

$$\mathbf{T}_S(\theta) = \left[\frac{\mathbf{a}(\theta)}{||\mathbf{a}(\theta)||}, \mathbf{t}_1^{\theta}, \mathbf{t}_2^{\theta}, \cdots, \mathbf{t}_{d_2-1}^{\theta}\right] \tag{8.18}$$

where $\mathbf{a}(\theta)$ is the steering vector of the angle under test. The rest of the vectors, $\mathbf{t}_i^{\theta}, i = 1, 2, \ldots d_2 - 1$, are chosen randomly without replacement from the columns of vectors spanning the orthogonal space spanned by $\mathbf{a}(\theta)$, i.e. from the columns of $\mathbf{I} - \frac{\mathbf{a}(\theta)\mathbf{a}^H(\theta)}{||\mathbf{a}(\theta)||^2}$. Since there is no prior information about the target Doppler, the matrix $\mathbf{T}_{\vartheta}$ is a random matrix, given by

$$\mathbf{T}_{\vartheta} = \frac{1}{K_1}[\mathbf{t}_1, \mathbf{t}_2, \cdots, \mathbf{t}_{d_1}] \tag{8.19}$$

where the constant K_1 will be determined subsequently.

Rank considerations: As before, we require $\mathbf{T}^H\mathbf{RT}$ to be full rank for matrix inversion. In addition, Rank($\mathbf{A} \otimes \mathbf{B}$) = Rank($\mathbf{A}$) $\otimes$ Rank($\mathbf{B}$). This imposes several constraints on d, we know from (8.18) that $d_2 \leq M$ and, hence, Rank($\mathbf{T}_S(\theta)$) = d_2, likewise, the maximum rank of $\mathbf{T}_\vartheta$ is N for $d_1 \geq N$ and d_1 for $d_1 \leq N$. Hence, $d \leq d_2N$. For example, $M = 10, N = 32, d_2 = 3$ then $d \leq 96$. Random projections have no such restrictions on d, except $d \leq MN$.

8.3.4 Semi-random localized projection

In this variant, the matrix $\mathbf{T}$ is decomposed as a Kronecker product of the spatial and Doppler projection matrices. The Doppler projection matrix is random and similar to the Localized random projections as in (8.19). The spatial projection, however, is different and is expressed as,

$$\mathbf{T}_S(\theta) = \frac{1}{\sqrt{2}}\left[\frac{\mathbf{a}(\theta)}{||\mathbf{a}(\theta)||}, \mathbf{T}_{S_1}\right] \tag{8.20}$$

where $\mathbf{T}_{S_1} = \frac{1}{K_2}[\mathbf{t}_1, \mathbf{t}_2, \ldots, \mathbf{t}_{d_2-1}]$, and K_2 will be determined subsequently.

8.4 Statistical analysis

We explore the statistics of the data after random and localized random projection. In particular, the mean of the l_2 norms before and after projection.

Random projections: Let the high-dimensional vectors be denoted as $\mathbf{x}_i \in \mathrm{C}^D, i = 1, 2$ and the random transformation is $\mathbf{T} = \frac{1}{K}[\mathbf{t}_1, \mathbf{t}_2, \cdots, \mathbf{t}_d] \in \mathrm{C}^{D\times d}$. Assume that the random variables comprising the rows and columns of $\mathbf{T}$ is zero mean, and has a variance σ^2. The distribution is immaterial here, but the random variables are assumed independent and identically distributed. We have,

$$E\{||\mathbf{y}_i||^2\} = \frac{1}{K^2}E\{\mathbf{x}_i^H\mathbf{T}\mathbf{T}^H\mathbf{x}_i\} = \frac{d\sigma^2||\mathbf{x}_i||^2}{K^2}, i = 1, 2. \tag{8.21}$$

where $E\{\cdot\}$ is the statistical expectation operator. From above, it is then clear that to preserve distance on the average $K = \sqrt{d\sigma^2}$. The closed form expression for the variance may also be derived in a straightforward manner but is omitted here. However, it may be shown that the variance, $Var\{||\mathbf{y}_i||^2\} = O(1/d)$.

Localized random projection: In a similar fashion for the localized random projections, if we consider $\mathbf{x}_i = \mathbf{x}_{i1} \otimes \mathbf{x}_{i2}$ of appropriate dimensions, then,

$$E\{||\mathbf{y}_i||^2\} = E\{\mathbf{x}_{i1}^H\mathbf{T}_\vartheta\mathbf{T}_\vartheta^H\mathbf{x}_{i1}\} \otimes \mathbf{x}_{i2}^H\mathbf{T}_S(\theta)\mathbf{T}_S^H(\theta)\mathbf{x}_{i2} \tag{8.22}$$

We note that in (8.22), the first part is identical to the analysis as the random projections method, hence, $K_1 = \sqrt{d_1\sigma^2}$. However, we also note that the transformation $\mathbf{T}_S(\theta)$ operating on $\mathbf{x}_{i2}$ is not distance preserving, i.e. $||\mathbf{T}_S^H(\theta)\mathbf{x}_{i2}||^2 \neq ||\mathbf{x}_{i2}||^2$ but since,

$$||\mathbf{T}_S^H(\theta)\mathbf{x}_{i2}||^2 \leq ||\mathbf{T}_S^H(\theta)||^2||\mathbf{x}_{i2}||^2 = ||\mathbf{x}_{i2}||^2$$

where $||\mathbf{T}_S^H(\theta)||$ is the induced two norms on this matrix. Now with $K_1 = \sqrt{d_1\sigma^2}$, we can readily see that

$$E\{||\mathbf{y}_i||^2\} \leq ||\mathbf{x}_{i1}||^2||\mathbf{x}_{i2}||^2 = ||\mathbf{x}_i||^2, i = 1, 2 \tag{8.23}$$

Therefore, localized random projections do not preserve distance but rather reduces it, and in the best case preserves it.

Semi-random localized projection: Using the same notation as before, and with the obvious choice of $K_2 = \sqrt{(d_2 - 1)\sigma^2}$, we can readily show that,

$$\begin{aligned} E\{||\mathbf{y}_i||^2\} &= ||\mathbf{x}_{i1}||^2 \times \frac{1}{2}\mathbf{x}_{i2}^H(\frac{\mathbf{a}(\theta)}{||\mathbf{a}(\theta)||}\frac{\mathbf{a}^H(\theta)}{||\mathbf{a}(\theta)||} + \mathbf{I})\mathbf{x}_{i2} \\ &\leq ||\mathbf{x}_{i1}||^2 \times \frac{1}{2}||\mathbf{x}_{i2}||^2(\mathrm{Tr}(\frac{\mathbf{a}(\theta)}{||\mathbf{a}(\theta)||}\frac{\mathbf{a}^H(\theta)}{||\mathbf{a}(\theta)||}) + 1) \\ &= ||\mathbf{x}_{i1}||^2||\mathbf{x}_{i2}||^2 = ||\mathbf{x}_i||^2, i = 1, 2. \end{aligned} \tag{8.24}$$

Therefore, like the localized random projection technique, the semi-random localized projection also reduces the distance and in the best case preserves it.

8.4.1 Probabilistic bounds

Before we delve into the main results, we present some preliminaries. Assume a matrix $\mathbf{A} \in \mathbb{C}^{N\times N}$ is positive definite and has eigenvalues, $\lambda_1 \geq \lambda_2 \geq \cdots \geq \lambda_N$.

We are now interested in deriving probabilistic upper and lower bounds for $\mathtt{SINR}^o_{\mathtt{ct}}$, when the transformation matrix $\mathbf{T}$ is from the complex normal distribution. Without loss of generality assume $\gamma = 1$. Before that let us define these events and their associated probabilities. Let the event $E_i = \{||\mathbf{t}_i||^2 \leq x_1\}$, with probability p_1, then $\Pr\{\bigcap_{i=1}^{d} E_i\} = p_1^d$. Now consider $\mathbf{x_t} = \mathbf{T}^H\mathbf{x}$, it is easy to see that $\mathbf{x_t}/||\mathbf{x}||$ is the standard multivariate complex normal, hence, $||\mathbf{x_t}||^2/||\mathbf{x}||^2$ is chi-squared distributed with $2d$ degrees of freedom. From this fact, we can readily establish $\Pr\{||\mathbf{x_t}||^2 \leq x_2\}$, denote this probability as p_2.

Also, define the following events, with their respective probabilities,

$$A_i = x_3 \leq ||\mathbf{t_i}||^2 \leq x_4, \Pr\{A_i\} = p_3 \tag{8.25}$$

$$A_{ij} = |\mathbf{t_i}^H\mathbf{t_j}| \leq \frac{x_5}{(d-1)\mathrm{Tr}\{\mathbf{R}\}}, \Pr\{A_{ij}\} = p_4 \tag{8.26}$$

$$B = ||\mathbf{z}||^2 \leq x_6, \Pr\{B\} = p_5. \tag{8.27}$$

Theorem 2 (Proof in [32]). When $T = \frac{1}{\sqrt{d\sigma^2}}[\mathbf{t}_1, \mathbf{t}_2, \ldots, \mathbf{t}_d]$ and the columns are chosen independently from a zero mean complex Gaussian with variance $\sigma^2\mathbf{I}$, we have

1. $\mathbf{x_t}^H\mathbf{R_t}^{-1}\mathbf{x_t} \geq \frac{x_2}{d\lambda_{\max}(\mathbf{R})x_1}$ with probability less than or equal to $\min(p_1^d, p_2)$ at most and with probability greater than or equal to $\max(0, p_1^d + p_2 - 1)$ at least.

2. $\mathbf{x}_t^H \mathbf{R}_t^{-1} \mathbf{x}_t \leq (\frac{dx_4}{d-1})^{d-1} \frac{x_6}{(\lambda_{\min}(\mathbf{R})x_3 - x_5)^d}$ with probability less than or equal to $\min(p_3^d, p_4, p_5)$ at most and with probability greater than or equal to $\max(0, p_3^d + p_4 d(d-1)/2 + p_5 - d(d-1)/2 - 1)$ at least.

Note: For the above upper bound on $\mathtt{SINR}_{\mathtt{ct}}^{\circ}$ to be relevant, we need the lower bound on the determinant to be positive, and ideally small, i.e., $(\lambda_{\min}(\mathbf{R})x_3 - x_5) = \epsilon > 0$.

For localized random projections, an analogue to Theorem 2 is readily derived, replacing d with d_1 along with some other minor modifications. This is because, in localized random projections, $\mathbf{T}_\nu$ is random but $\mathbf{T}_S(\theta)$ is completely deterministic. We may further decompose the original steering vectors and the original covariance matrices as,

$$\mathbf{x} = \mathbf{x}_1 \otimes \mathbf{x}_2$$
$$\mathbf{R} = \mathbf{R}_1 \otimes \mathbf{R}_2$$

where these decompositions are temporal and spatial, i.e., $\mathbf{x}_1 \in \mathbb{C}^N$, $\mathbf{x}_2 \in \mathbb{C}^M$. Likewise, $\mathbf{R}_1 \in \mathbb{C}^{N \times N}$ and $\mathbf{R}_2 \in \mathbb{C}^{M \times M}$. With these decompositions and the Kronecker mixed properties, the analogue to Theorem 2 may be readily derived.

For the semi-random localized projection, a similar approach may be followed to establish the analogue of Theorem 2. However, we take note that now both the spatial and temporal transformations are random. Nonetheless, with a bit of algebra, such an analogue may also be readily derived, but not shown here.

(**Normalized SINR distribution**) The distribution of the normalized SINR i.e., for *all families of random projections* $\mathtt{SINR}_{\mathtt{nt}}$ is as described in the next theorem.

Theorem 3 (Normalized SINR, proof in [32], derived from [33]). If $\mathbf{y}_l \sim CN(\mathbf{0}, \mathbf{R}), l = 1, \ldots L$ and $\hat{\mathbf{R}} = \sum_{l=1}^{L} \mathbf{y}_l \mathbf{y}_l^H / L$, then for $L \geq d, 0 \leq \mathtt{SINR}_{\mathtt{nt}} \leq 1$, we have $p(\mathtt{SINR}_{\mathtt{nt}})$ is distributed as,

$$p(\mathtt{SINR}_{\mathtt{nt}}) = \frac{L!}{(d-2)!(L+1-d)!}(1 - \mathtt{SINR}_{\mathtt{nt}})^{d-2}(\mathtt{SINR}_{\mathtt{nt}})^{L+1-d}$$

Remark: Theorem 3 is quite powerful and they apply to all families of random projections. From the properties of the Wishart distribution and using a sequence of transformations as in the seminal [33], the distribution conditioned on $\mathbf{T}$ is independent of $\mathbf{R}_t$ and, therefore, also independent of $\mathbf{T}$.

8.5 Simulations

In this section, we consider numerical examples and simulations validating the theory presented previously.

8.5.1 Integration as low-pass filtering

Simulations demonstrating that the integration in (8.2) is equivalent to LPF are shown. Indeed, for many readers, this is a trivial exercise but an important one since we are emphasizing here that the integrators in Figure 8.1 act as LPFs and, hence, these will remove any high-frequency components in the CS scheme. So any spectrally spread signal will be altered by the LPF and frequency information is lost and cannot be recovered.

In our simulations shown in Figure 8.3, we consider a signal comprised of two sinusoids at frequencies 250 Hz and 350 Hz. The signal is now integrated with different integration times and the results are shown in Figure 8.3(b) and (d). The impulse responses in the frequency domain of the filters are shown in Figure 8.3(a) and (c). In Figure 8.3(a), the integration time is $T = 0.011$ s, in Figure 8.3(c), the integration

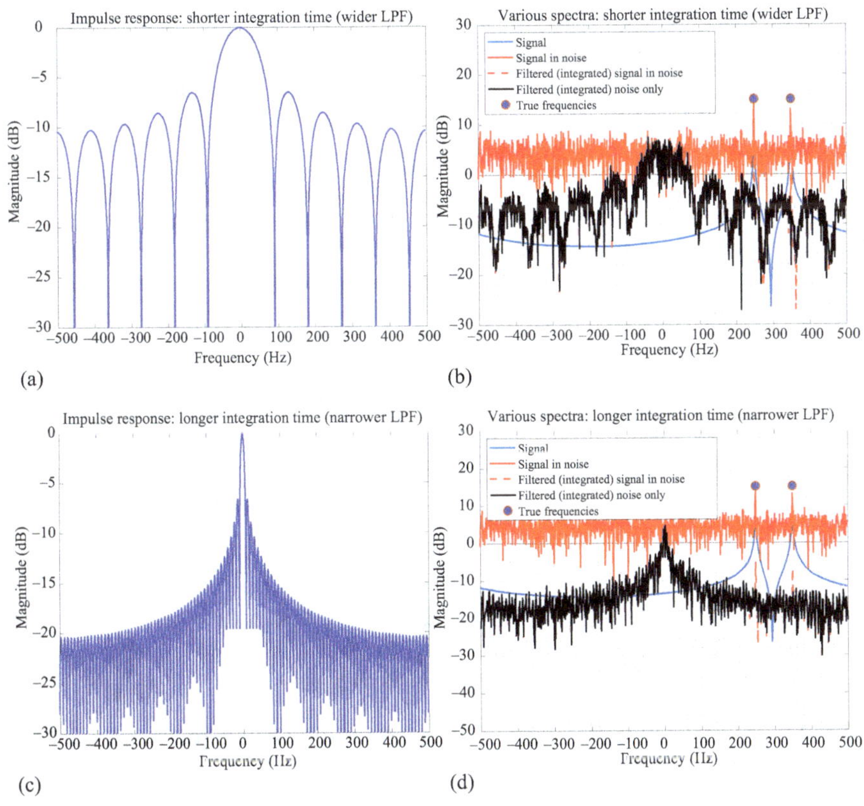

Figure 8.3 Integration as low-pass filtering example: (shorter) integration time $T = 0.011$ s: (a) impulse response and (b) sinusoidal signals filtered by integrators both noise free and noiseless. (Longer) Integration time, $T = 0.11$ s, (c) impulse response and (d) sinusoidal signals filtered by integrators both noise free and noiseless.

time is $T = 0.11$ s. Recognize that the integration operation in (8.2) is similar to smoothing. A smoothing operation removes high-frequency components.

In Figure 8.3(b) and(d), the noise is added at an SNR of 10 dB. The results after integration of the noise free and the noisy signal are shown. The integrators have filtered the sinusoids and they are below the noise floor. These results are trivial but they give a perspective that integration operation in Figure 8.1 acts as low-pass filtering.

8.5.2 CS: sinusoid in IF example

We consider a sinusoidal signal at 1 kHz. The signal is noise free and is sparse in the spectral domain. The random basis functions as seen in (8.5) are used, we consider two different chip lengths, $T_c = 0.0027$ s (shorter chip length) and $T_c = 0.0138$ s (longer chip length).

The spectrum of the sinusoidal signal is shown in Figure 8.4(a), this signal is multiplied by the random basis signal with the shorter chip length ($T_c = 0.0027$ s). The resulting signal's spectrum is shown in Figure 8.4(b), clearly the sinusoid is now spread. The shorter chip length random basis signal causes the sinusoid at 1 kHz to spread significantly. In Figure 8.4(c), the spectra, PSD estimate computed via the periodogram and the PSD from theory are all shown for the shorter chip length random basis signal.

In Figure 8.4(d), the longer chip length random basis waveform is multiplied with the sinusoid at 1 kHz, the resulting signal's spectrum is shown. The sinusoid at 1 kHz is spread but not as significantly as in Figure 8.4(b). Nonetheless, energy originally confined to one band has now been spread to several bands. In Figure 8.4(e), the spectra, PSD estimate computed via the periodogram, and the PSD from theory are all shown for the longer chip length random basis signal.

The resulting signals in Figure 8.4(b) and (d) would be low-pass filtered and even if the LPF cut off was above 1 kHz, any sub-sampling would alias the signal and it is no longer a true representation of the sinusoidal signals.

8.5.3 CS: rectangular pulse example

In this numerical experiment, we consider a rectangular pulse, this signal is sparse in the time domain. Before we analyze what happens to this signal by the analog scheme implementing CS in Figure 8.1, we evaluate the mean and the auto-correlation of this signal for random time shifts via Monte-Carlo simulations.

8.5.3.1 Monte-Carlo simulation for rectangular pulse mean and auto-correlation

In this example, we consider a rectangular pulse of width $T_r = 0.2$ s. For the Monte-Carlo simulations, we consider random delays uniformly distributed in $[0, 5]$ s. The rectangular pulse is shown in Figure 8.5(a). For 10,000 Monte-Carlo trials, the mean and the auto-correlation function are shown in Figure 8.5(b) and (c), respectively. It is clearly seen that both the mean and the auto-correlation are time invariant; the auto-correlation is a function of the lag (τ) alone. These simulations validate (8.10).

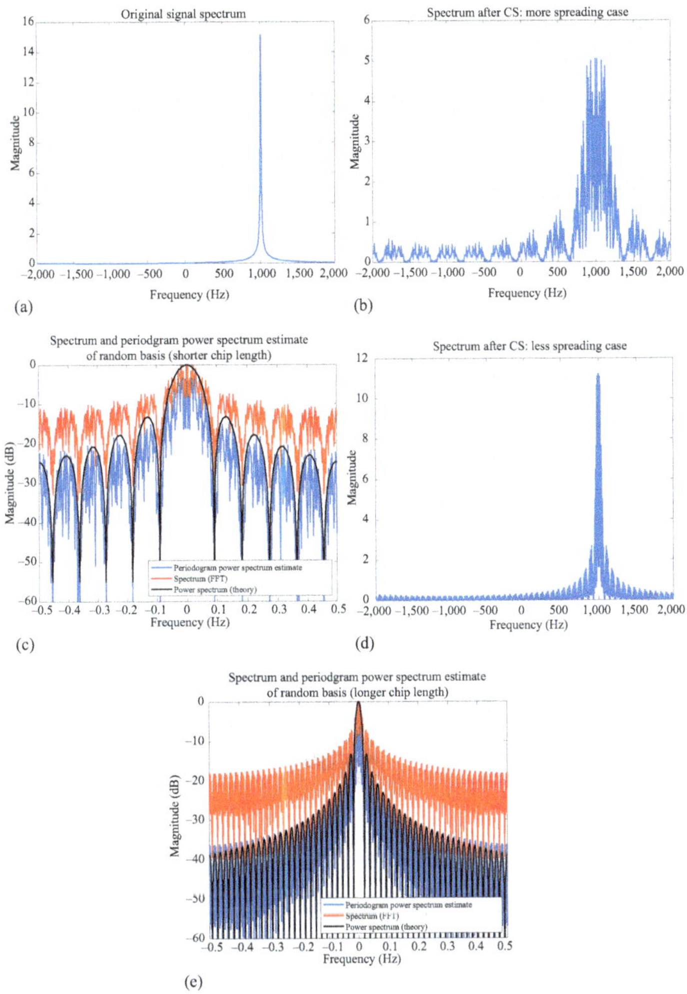

Figure 8.4 CS sinusoid case: (a) sinusoid, a sparse signal in frequency domain and (b) spectrum of signal resulting from multiplying sinusoid with random basis function. Random basis function specified in (8.5) *with* $T_c = 0.0027$ *s (shorter chip length). In (c), the PSD (from theory), PSD estimate from periodogram and the spectrum of the (shorter chip length) random basis signal are shown. In (d), the spectrum of signal resulting from multiplying sinusoid with random basis function. Random basis function specified in* (8.5) *with* $T_c = 0.0138$ *s (longer chip length). In (e), the PSD (from theory), PSD estimate from periodogram, and the spectrum of the (longer chip length) random basis signal are shown.*

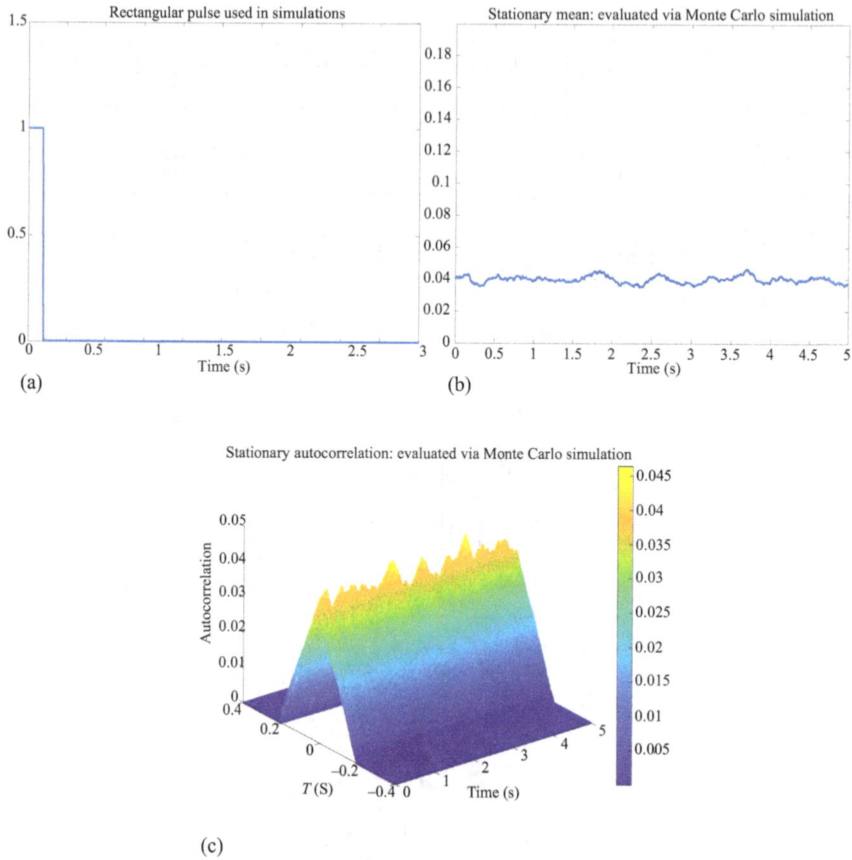

Figure 8.5 Stationarity of mean and auto-correlation for rectangular pulse with random delay using Monte-Carlo simulation, (a) pulse, (b) mean, and (c) auto-correlation

8.5.3.2 CS baseband rectangular pulse example

In Figure 8.6, we consider a rectangular pulse of width $T_r = 1$ ms. The signal is noise free and is sparse in the time domain. The random basis functions as seen in 8.5 are used, we consider two different chip lengths, $T_c = 0.00011$ s (shorter chip length) and $T_c = 0.00044$ s (longer chip length).

The spectrum of the rectangular pulse is shown in Figure 8.6(a), this signal is multiplied by the random basis signal with the shorter chip length ($T_c = 0.00011$ s). The resulting signal's spectrum is shown in Figure 8.6(b), clearly the pulse is now spread. The shorter chip length random basis signal causes the rectangular pulse to spread significantly in frequency. In Figure 8.6(c), the spectra, PSD estimate computed via the periodogram, and the PSD from theory is shown for the shorter chip length random basis signal.

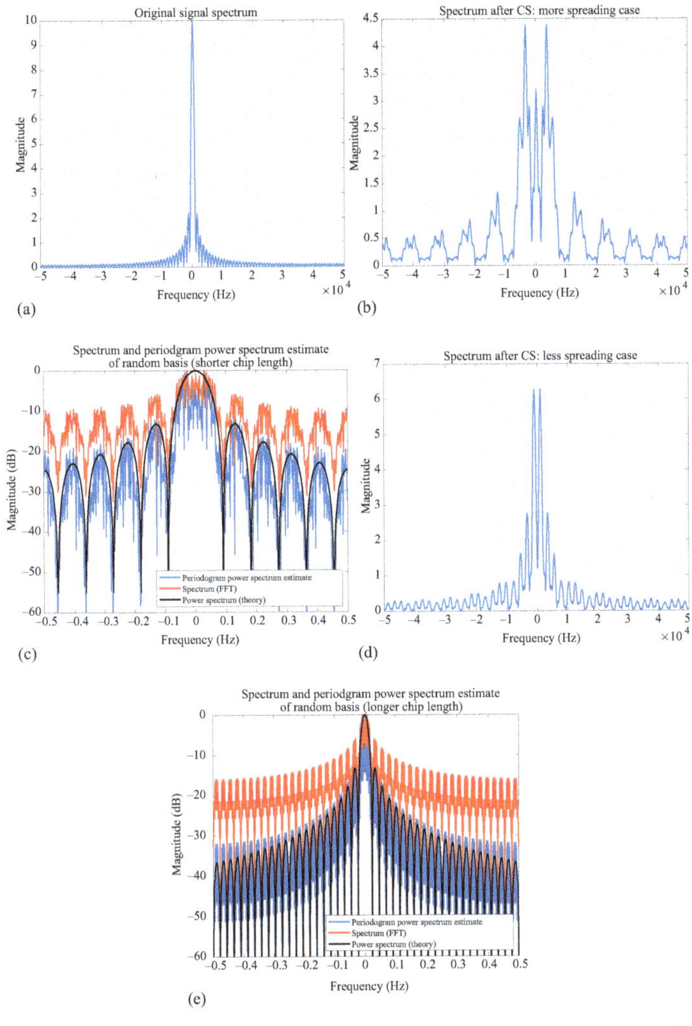

Figure 8.6 CS rectangular pulse case: (a) spectrum of rectangular pulse, a sparse signal in time domain, (b) spectrum of signal resulting from multiplying pulse with random basis function. Random basis function specified in (8.5) *with* $T_c = 0.00011$ *s (shorter chip length). In (c), the PSD (from theory), PSD estimate from periodogram, and the spectrum of the (shorter chip length) random basis signal are shown. In (d), the spectrum of signal resulting from multiplying pulse with random basis function. Random basis function specified in* (8.5) *with* $T_c = 0.00044$ *s (longer chip length). In (d), the PSD (from theory), PSD estimate from periodogram, and the spectrum of the (longer chip length) random basis signal are shown.*

In Figure 8.6(d), the longer chip length random basis waveform is multiplied with the rectangular pulse, the resulting signal's spectrum is shown. The rectangular pulse is spread but not as significantly as in Figure 8.6(b). In Figure 8.6(e), the spectra, PSD estimate computed via the periodogram, and the PSD from theory are all shown for the longer chip length random basis signal.

The resulting signals in Figure 8.6(b) and (d) would be integrated or equivalently low-pass filtered, and if even if the LPF cut off was above 1 kHz, any sub-sampling would alias the signal and it is no longer a true representation of the rectangular pulse.

8.5.4 Realistic examples of CS reconstructions

So far, we have not considered l_1 reconstructions of sparse signals. Our intent now is to demonstrate l_1 reconstructions incorporating the effects of integration. We also consider noisy signals.

Unlike significant literature on CS, we consider and add noise before compression. That is, the random projections are made on noisy signals instead of random projections on noise-free signals and then adding noise, as seen in [10, see e.g. (7)].

Although noise is added, in some simulations, we choose very high SNRs to isolate the problems in l_1 reconstruction mitigating the effect of noise completely. This is intentional to show that the effect of integration filters has on the signal gains after reconstruction.

Experimental procedure: Our procedure to reconstruct sparse signals is as follows. Since we use a computer, we cannot generate an analog signal. But, however, to a certain degree, we can simulate the effects of the analog signal as it is passing through the RF chain digitally. In that regard, we design an arbitrary sparse signal in frequency domain. Noise is then added to this signal. The signal is originally highly over-sampled at 2,000 Hz, we then generate $d = 100$ random vectors from a standard normal. The original signal is of length $N = 8{,}001$ samples or is of 4 s in duration. The signal is then multiplied with $d = 100$ random vectors and then integrated. The length of the integration filter is N_f samples. The resulting signal is then under-sampled, there are now $d = 100$ under-sampled signals. We now use l_1 magic [34] and then reconstruct the signal. We vary the filter lengths, N_f, and vary the undersampling rates of the signal and show the l_1 reconstruction in the frequency domain.

This systematic procedure is illustrated in Figure 8.7. In the figure, recall that the original signal, $s(t) = s(mT_s)$, used is discrete and is oversampled. Therefore, the integrator can be replaced by a Riemann summation as shown below:

$$\int_{t-T}^{t} f(t)\mathrm{d}t = \lim_{\delta t \to 0} \sum_{n=1}^{N_f} f(t_n)\delta t \tag{8.28}$$

where $t_n :\in [n\delta t, (n+1)\delta t]$ for a highly oversampled signal as we use here, $\delta t = T_s$, i.e. the oversampling period. It is no surprise that the summation in (8.28) can be efficiently implemented as filter whose b (numerator coefficients) is $[1, 1, \ldots N_f \text{ times }]T_s$ and a (denominator coefficient) is 1. The filter gain at DC

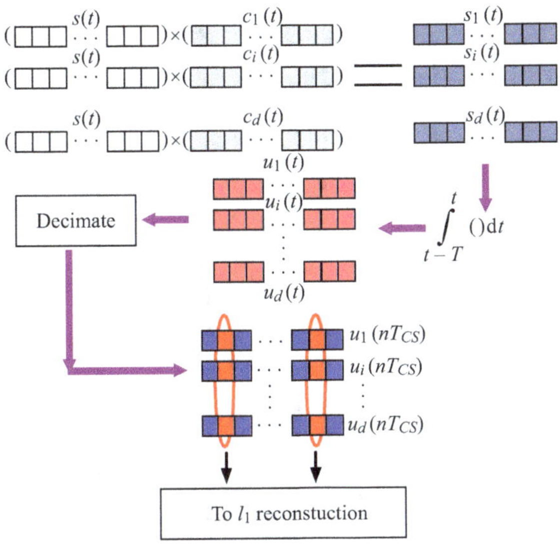

Figure 8.7 *Procedure to. employ l_1 reconstruction using the sampling architecture in Figure 8.1 for highly oversampled signals. The variable t is replaced by mT_s where, T_s is the original (over) sampling period. The integral can be efficiently implemented by a discrete filter see* (8.28). *After decimation the red samples are passed to l_1 reconstruction and the signal is reconstructed batch-wise. The red samples are inner products of random vectors with the original signal for the chosen filter length, T along with the gain due to the integrator filter. The blue samples are not utilized for reconstruction.*

is $N_f T_s = T$ as expected. The filter command in MATLAB® can now be used to implement (8.28).

The sparse signal along-with the impulse response of the integrators are seen in Figure 8.8. We know integration is similar to a smoother. Certain samples of the sub-sampled signal may now be used for l_1 reconstruction. We elaborate, for example, if $N_f = 501$ samples, then with subsampling rate of 20, the 26th sample of the $d = 100$ subsampled signals may be used to re-construct the first 501 samples of the original signal in l_1 magic. Likewise, the 51st sample of the $d = 100$ sub sampled signals may be used to reconstruct the samples, 501 to 1,001 in the original signal. This process is continued till we reconstruct the full signal batch-wise. Clearly we see that this procedure is tedious and is unrealistic in a practical sampling system and poses unattainable requirements on system design and timing when compared to traditional ADC. We however do it still so as to not disadvantage the l_1 reconstruction algorithm. We also note that many samples in the under-sampled signals are never utilized in l_1 reconstruction resulting in inefficiency.

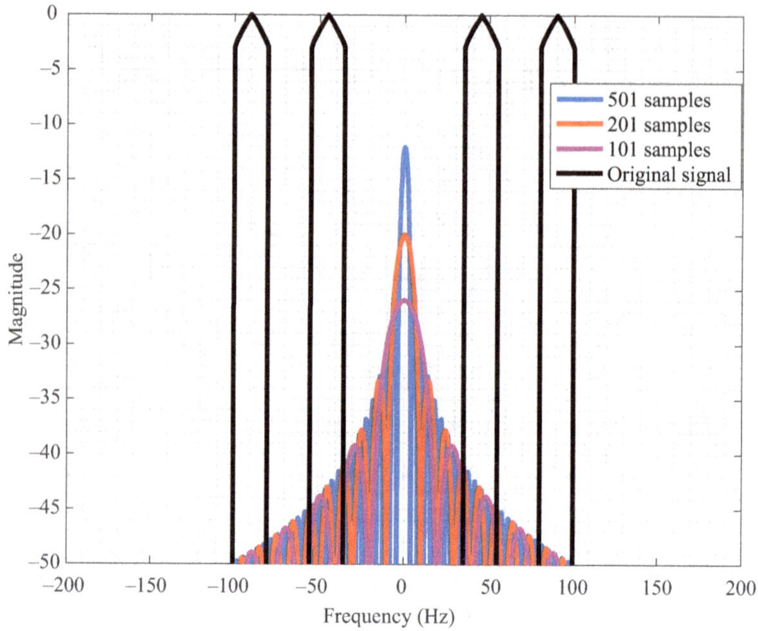

Figure 8.8 Original signal spectrum and impulse response of integration filters used. The filter length is shown in the figure legend in samples. The filter lengths are 501, 201, *and* 101 *samples in length. Original sampling frequency is* 2, 000 *Hz and is 10 times the Nyquist rate of* 200 *Hz.*

For $N_f = 501$, we have 16 batches of signal reconstructed in 501 samples resulting in complete $N = 8001$ sample reconstruction. Similarly for $N_f = 101$, we have 80 batches of originals signal reconstruction with each batch being 101 samples in length. This is the lower limit of the integration filter length. **Note**: Any integration length less than 101 samples will result in creating more batches of sub-sampled signal resulting in oversampling the original signal. For example, if the integration filter length was $N_f = 51$ samples, then we have 160 batches of $d = 100$ sub-sampled signals. This is effectively 160 × 100 CS samples which is twice that of the original signal length of $N = 8,001$ samples.

Integration filter lengths: In Figure 8.9, the results of l_1 reconstructions are shown for varying integration filter lengths. We sample the signal at the Nyquist rate that is 200 Hz. We used the primal-dual linear programming problem to solve for reconstruction. The tolerance for convergence was set at $1e - 06$ duality gap. The reconstructions are inferior across all integration filter lengths as seen in Figure 8.9. As noted earlier, it is meaningless to have integration filter lengths less than the number of random vectors used, i.e. $N_f \nleq d$. From many of the practical challenges delineated before in this chapter, it is not surprising that the results of reconstruction are poor. The results speak for themselves.

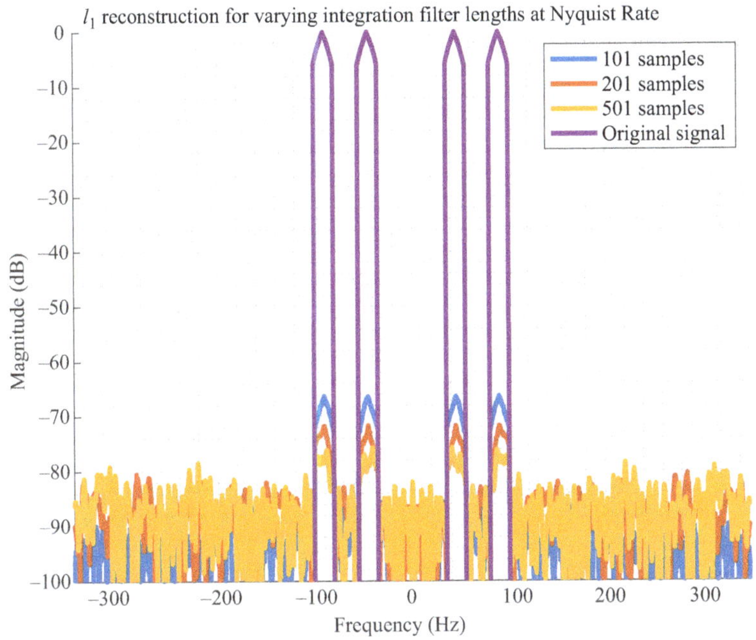

Figure 8.9 *Original signal spectrum and spectrum of reconstructions for varying integration filter lengths corresponding to* 501, 201, *and* 101 *samples. The l_1 reconstruction was used on samples at the Nyquist rate at* 200 *Hz. The primal-dual linear programming optimization was used in l_1 magic [34] with set convergence tolerance of* 1e − 06 *duality gap. SNR used was extremely high, set at an impractical* 100 *dB. The reason to do this was to analyze reconstruction mitigating the effect of noise but to also demonstrate that the integration filters will reduce the gain of the signal being reconstructed.*

In practice, if a signal spreads in the RF chain, the analog components downstream must have reasonable gain at the new spread bandwidth at either RF, IF, or base-band. Without this, further SNR losses are incurred and spectral information is lost. It is impractical to increase bandwidth for some components in the analog RF chain while not doing so for the rest. Additionally, increasing the system bandwidth in the middle of the RF chain is not good design practice since it will also increase the noise bandwidth. These effects were not simulated and, therefore, practically we expect the results of reconstruction in Figure 8.9 to be worse than they already are.

Undersampling rates: In Figure 8.10, we vary the undersampling rates. Recall, from Figure 8.1, the undersampling happens after integration filtering and we then reconstruct the signal. The decimate function with default parameters in MATLAB was used to undersample the signal before using the l_1 reconstruction. In Figure 8.10, the SNR used was 20 dB. We see that the results are unacceptable.

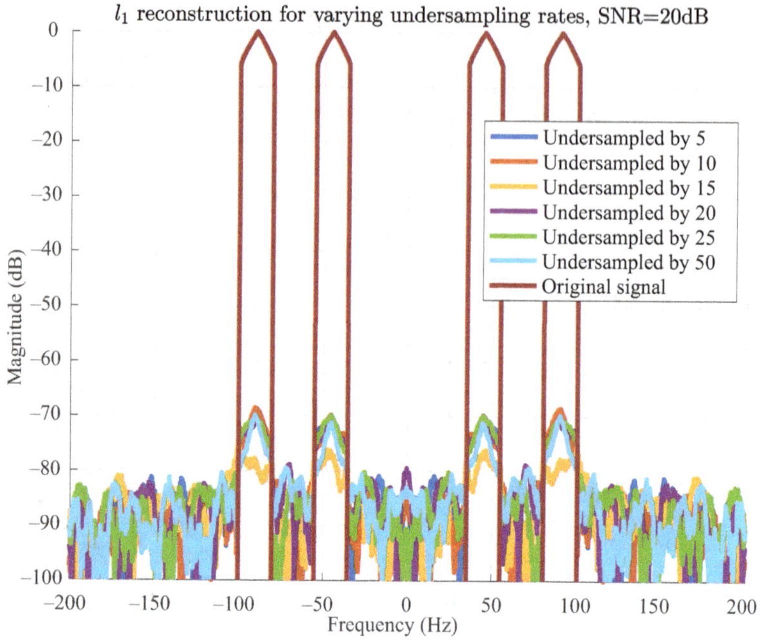

Figure 8.10 Original signal spectrum and spectrum of reconstructions for varying sub-sampling rates. The primal-dual linear programming optimization was used in l_1 magic [34] with set convergence tolerance of 1e − 06 *duality gap. SNR used was* 20 *dB.*

Not true undersampling? Although we highlight the results in Figure 8.10 as under-sampled data which it is due to decimation. However, note that the samples used for reconstruction are judiciously selected and contain the inner products of the random vectors with the signal to advantage l_1 reconstruction . Furthermore, many samples after decimation are not utilized. Therefore, it is a misnomer to qualify them as true undersampling effects on l_1 reconstruction.

8.5.5 Random projections with different distributions

A linear array geometry is assumed, with 10 sensors transmitting 16 pulses. The data is matched filtered and the random transformation is applied for a particular cell under test. The covariance model is similar to that used in [19,35], and is, therefore, known. Representative data from neighboring cells is then generated from this covariance matrix class by using the Cholesky decomposition and the multivariate standard normal distribution.

In the next simulation in Figure 8.11, we use the sample covariance matrix instead of the actual covariance matrix. We compare the normalized SINR for the original full-dimensional problem as in (8.14) and normalized SINR for the dimension reduced problem as in (8.15b). The number of samples used to generate the data is fixed as $L = 2d$. A similar number of samples are also used for the original STAP problem (full dimensional) and compared with the dimension-reduced random projection technique. Complex normal distributions were used in generating the transformation matrices, and were identical for the 500 Monte-Carlo trials but were different for different d's. The mean normalized SINR is shown in Figure 8.11. For the original problem, when $L \leq 160, L = 2d$ the resulting sample covariance matrix is rank deficient, and diagonal loading was used. When $L \geq 160$ i.e. $d \geq 80$, the sample matrices are full rank, and, hence, diagonal loading was not used. Results with a load factor equal to 50 times the minimum eigenvalue of the true covariance matrix, and 100 times the minimum eigenvalue are shown in Figure 8.11. We note that, in practice, the true covariance is never known, so the load factor has to be obtained from other techniques, for example, optimizing some function of the SINR with the sample covariance matrix instead, indeed at the cost of increased computations.

We see that random projections performs well and is close to the RMB predicted rule (black dashed line in Figure 8.11) [23,36], when sample covariance matrices are used. A significant computational reduction is afforded by the random projections. We note that the normalized SINR from random projection is also slightly higher than the RMB prediction for small subspace dimension, d. This is a Monte-Carlo effect, and we observed very high variance of the SINR for small dimensions of random projections. From this phenomena, it is erroneous to conclude that working in the lower dimension subspaces leads to gains in normalized SINR. On the contrary, however, comparing the clairvoyant SINR, (8.13a), of the full dimensional problem to the clairvoyant SINR of the reduced dimension problem, (8.15a), and as seen in Figure 8.12, we see that the overall clairvoyant SINR of the dimension reduced problem also decreases when d decreases. The choice of the distribution used for generating $\mathbf{T}$ is irrelevant here, as seen in Figure 8.12. For generating the results in Figure 8.12, we used 500 Monte-Carlo trials, and, for each trial, we used different random projection matrices from various real and complex families of distributions. The clairvoyant SINR for the reduced dimension problem as in (8.15a) is, therefore, also random since $\mathbf{T}$ varies for each trial. The mean and variance of the clairvoyant SINR of the dimension reduced problem is depicted in Figure 8.12. We observe from Figure 8.12, the mean normalized SINR is nearly identical for both real distributions and complex distributions. The variance, however, is slightly different, but nonetheless follows similar trends among the distributions.

8.5.6 Random and random-type projections

Next, we present the radar simulations. A linear array geometry is assumed, with $M = 10$ sensors transmitting $N = 16$ pulses. We compare both the random and localized random projections technique. The covariance model is similar to that used in [19,35] and is, therefore, known. The clairvoyant SINR as in (8.13b) is shown, this is the full dimensional SINR evaluated at the full-dimensional optimum weight vector. This is the upper bound on the SINR given a particular covariance matrix. The columns of the random matrix in the random projections and the random-type projections

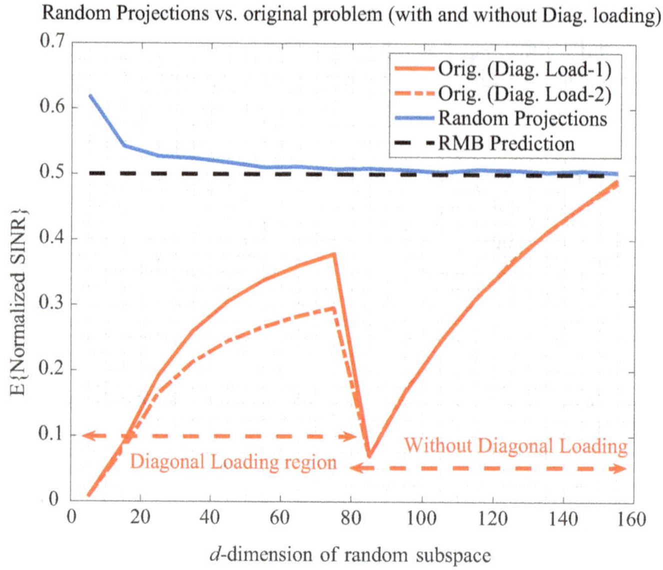

Figure 8.11 Sample covariance: varying d for random projections, the number of samples, L = 2d. The expected value of SINR of the original full dimensional problem (blue) is compared with and without diagonal loading (red) along-with the Reed–Mallet–Brennan [23,36]. prediction (black). Diagonal loading with load factors equal to 50 times (red solid) and 100 times (red dashed dotted) the minimum eigenvalue of the true covariance matrix is shown. Figure reproduced from [20] with permission from IEEE.

are from the complex normal family. The column dimensions of the deterministic spatial transformation matrix, $\mathbf{T}_S(\theta)$ is d_2. Three different scenarios, $d_1 = 3, 5, 7$, are simulated with 500 Monte-Carlo trials for each simulation, and the results are shown in Figure 8.13. Due to the constraints on d_2 and, therefore, on the rank of $\mathbf{T} = \mathbf{T}_\vartheta \otimes \mathbf{T}_S(\theta)$, $d \leq 40$, $d \leq 80$, $d \leq 112$ for the three different values of d_2. For each of these simulations, d was varied from one to its corresponding maximum.

In Figure 8.13, the error-bars (mean $\pm$ standard deviation) are shown for all the techniques. Two observations are immediate, localized random projections perform better than random projections but has a higher variance. The semi-random projections perform the best but at the cost of higher variance. The clairvoyant SINR for the localized random projections starts coinciding with clairvoyant SINR for random projections as we increase d_1. This may be explained because as we are increasing d_1 other angles spanning other subspaces are included in $\mathbf{T}_S(\theta)$ which have no bearing with the angle under test, therefore, decreasing the SINR. Furthermore, we note that these subspaces are also spanned by the columns of the random projections

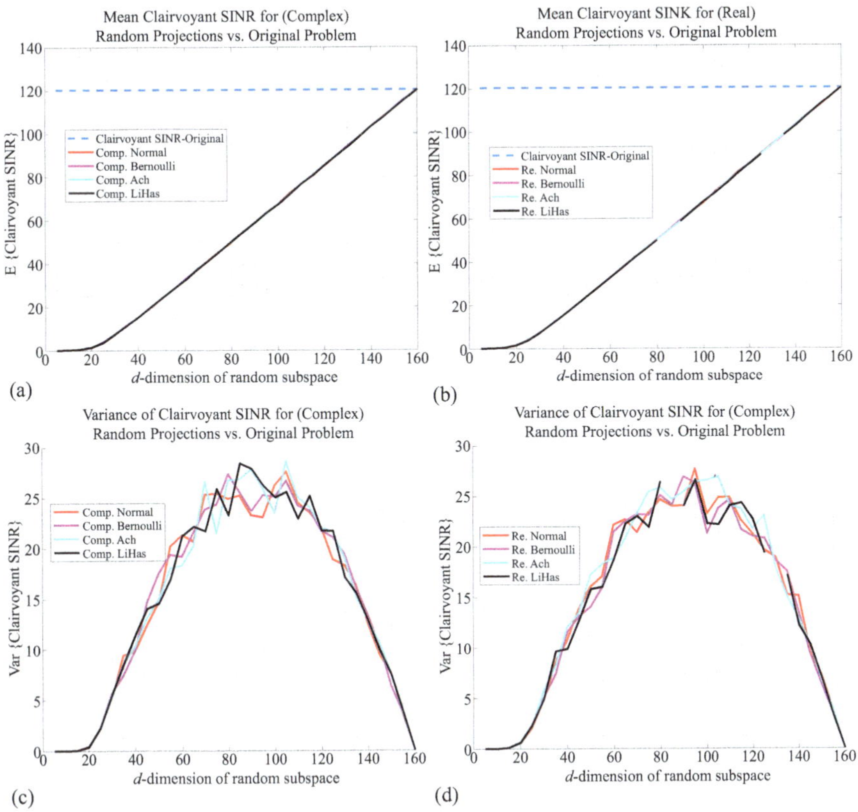

Figure 8.12 True covariance: random projections for various distributions used in generating elements of **T** *(see Section 8.3.2.1) and clairvoyant SINR of original problem vs d-dimension of the random subspace: (a) complex matrices, (b) real matrices, (c) variance in complex case, and (d) variance in real case. Figure reproduced from [20] with permission from IEEE.*

transformation matrix. Therefore, it is natural to expect that the clairvoyant SINR for localized random projections will start coinciding with the clairvoyant SINR of the random projections technique.

8.6 Discussion and conclusions

The author acknowledges that the reference list of CS literature and applications is thoroughly incomplete. There have been manifold spin-off literature by many renowned academics on CS and its applications to various fields in engineering based

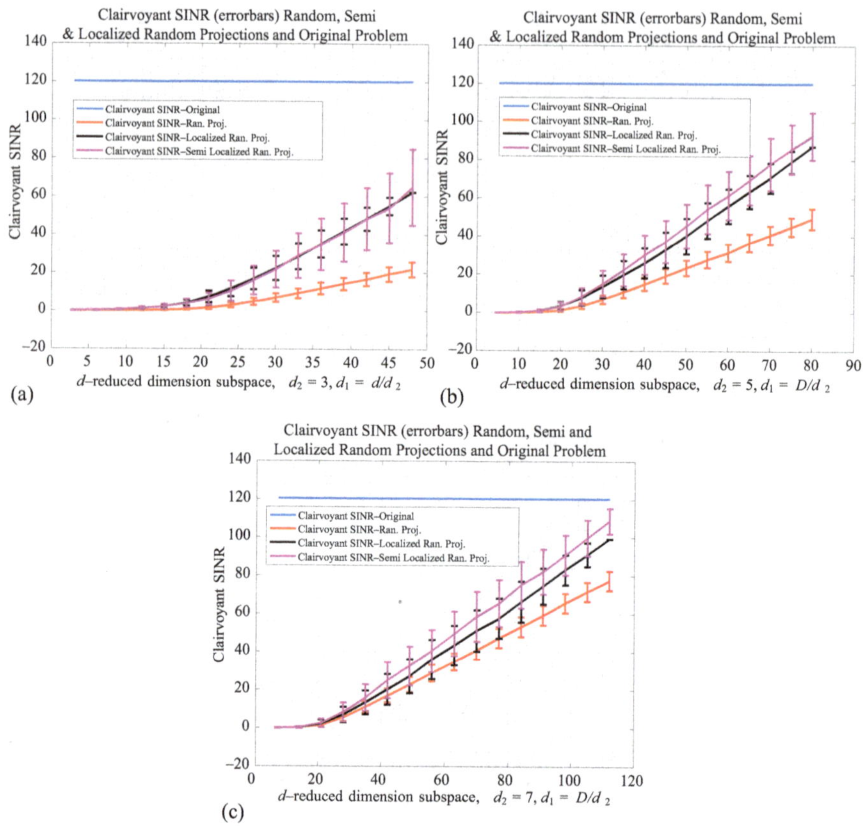

Figure 8.13 True covariance: Clairvoyant SINR error bar for random projections, localized random projections, semi-random localized projection and original problem, (a) $d_2 = 3$, (b) $d_2 = 5$, and (c) $d_2=7$. Figure reproduced from [21] with permission from IEEE.

on the sub-Nyquist sampling claims. The reader is instructed to visit their favorite scientific literature repository for latest trends on CS.

For more than 15 years, CS theorists have claimed that their techniques can sample signals below the Nyquist rate. We addressed those claims meticulously from an analog sensing framework. This chapter is certainly not giving any weight to the sub-sampling claims made in the literature but rather it demonstrated unsatisfactory CS results. To the many practical engineers who read this chapter and design actual systems, the results and analysis here should not be surprising and may already be well known.

MIMO communication theory and CS theory are mostly contemporaneous [5,37,38] *et al.* On the one hand, MIMO communication hardware have proliferated our everyday lives. One may go to a neighborhood electronic store and buy a high performing MIMO router for a few dollars. On the other hand, the same cannot be said for CS-related hardware. The author is currently ignorant of any A/D converter manufacturer who has adopted and designed a CS enabled A/D converter for commercial use. This strongly indicates how the for-profit industry judges one theory over the other.

From a cognitive sensing perspective, we foresee a deluge of radar data due to high sampling rates and large dimensional radar problems. It is the opinion that random projections may be used to reduce dimensionality of the cognitive radar problem for almost real-time processing. However, for radar STAP, the problem is the opposite. We seldom have full rank training data for processing. Therefore, nonetheless we have to reduce dimensionality of the radar problem as well as the radar data to perform the usually mundane radar tasks such as detection and tracking. In both cases, random projections serve to alleviate the computational burden dealing with large dimensional data sets and large dimensional radar problems.

We addressed the scarcity of training data in radar STAP by reducing the problem dimension. Specifically, we analyzed random projections for STAP and demonstrated that this technique suffers from a loss of SINR after dimensionality reduction. To improve this SINR loss, we investigated two techniques, where the lower dimension transformation was decomposed into random and deterministic counterparts. Statistical analysis via probabilistic bounds were derived to quantify performance of these techniques.

Numerical experiments on the analog architecture implementing CS were also shown along with l_1 reconstruction. The results were unsatisfactory, and contrary to that claimed in the literature. We believe the Nyquist sampling theory will always prevail for many years to come in communications and RF sensing.

Acknowledgment

The author thanks United States Air Force Personnel, Mr Frank Scenna, Mr Nathan Wilson, and Dr Christopher Sigler. The author thanks Dr Tariq Qureshi, Intel Labs, OR, USA for many enlightening discussions on random projection variants and application to radar. The author also likes to thank the editors for providing this opportunity to present an alternate narrative to CS.

References

[1] Achlioptas D. Database-friendly random projections: Johnson-Lindenstrauss with binary coins. *Journal of Computer and System Sciences*. 2003;66(4):671–687. Special Issue on {PODS} 2001.

[2] Baraniuk R, Davenport M, DeVore R, *et al.* A simple proof of the restricted isometry property for random matrices. *Constructive Approximation*. 2008;28(3):253–263.

[3] Sauer T. Prony's method: an old trick for new problems. *Snapshots of Modern Mathematics from Oberwolfach*. 2018;(4).
[4] Recht B. CS838 Topics in Optimization: Convex Geometry in High-Dimensional Data Analysis; 2010. [Online; accessed 4-Oct-2022]. https://pages.cs.wisc.edu/ brecht/cs838docs/Lecture06.pdf.
[5] Candès E, Romberg J, and Tao T. Stable signal recovery from incomplete and inaccurate measurements. *Communications on Pure and Applied Mathematics*. 2006;59:1207–1223.
[6] Donoho DL. Compressed sensing. *IEEE Transactions on Information Theory*. 2006;52(4):1289–1306.
[7] Romberg J. Compressive sensing by random convolution. *SIAM Journal on Imaging Sciences*. 2009;2(4):1098–1128.
[8] Tropp JA, Laska JN, Duarte MF, *et al.* Beyond Nyquist: efficient sampling of sparse bandlimited signals. *IEEE Transactions on Information Theory*. 2010;56(1):520–544.
[9] Yazicigil RT, Haque T, Kinget PR, *et al.* Taking compressive sensing to the hardware level: breaking fundamental radio-frequency hardware performance tradeoffs. *IEEE Signal Processing Magazine*. 2019;36(2):81–100.
[10] Candes EJ and Wakin MB. An introduction to compressive sampling. *IEEE Signal Processing Magazine*. 2008;25(2):21–30.
[11] Vetterli M, Marziliano P, and Blu T. Sampling signals with finite rate of innovation. *IEEE Transactions on Signal Processing*. 2002;50(6): 1417–1428.
[12] Oppenheim AV, Schafer RW, and Buck JR. *Discrete-time Signal Processing*, 2nd ed. Englewood Cliffs, NJ: Prentice Hall; 1999.
[13] Li P, Hastie TJ, and Church KW. Very sparse random projections. In: *Proceedings of the ACM SIGKDD International Conference on Knowledge Discovery and Data Mining. KDD'06*. New York, NY: ACM; 2006. pp. 287–296.
[14] Muthukrishnan S. Data streams: algorithms and applications. In: *Proceedings of the Annual ACM-SIAM Symposium on Discrete Algorithms. SODA'03*. Philadelphia, PA: Society for Industrial and Applied Mathematics; 2003. pp. 413–413.
[15] Indyk P, Koudas N, and Muthukrishnan S. Identifying representative trends in massive time series data sets using sketches. In: *Proceedings of the International Conference on Very Large Data Bases. VLDB'00*. San Mateo, CA: Morgan Kaufmann Publishers Inc.; 2000. pp. 363–372.
[16] Vempala SS. *The Random Projection Method. Vol. 65 of DIMACS Series in Discrete Mathematics and Theoretical Computer Science*. Providence, RI: American Mathematical Society; 2004.
[17] Johnson W and Lindenstrauss J. Extensions of Lipschitz mappings into a Hilbert space. In: *Conference in Modern Analysis and Probability* (New Haven, CT, 1982). Vol. 26 of *Contemporary Mathematics*. Providence, RI: American Mathematical Society; 1984. pp. 189–206.
[18] Setlur P and Rangaswamy M. *Joint Filter and Waveform Design for Radar STAP in Signal Dependent Interference*. Dayton, OH: US Air

Force Res. Lab., Sensors Directorate, WPAFB; 2014. *DTIC*, available at: https://arxiv.org/abs/1510.00055.

[19] Qureshi TR, Rangaswamy M, and Bell KL. Reducing the effects of training data heterogeneity in multistatic MIMO radar. In: *2015 49th Asilomar Conference on Signals, Systems and Computers*; 2015. pp. 589–593.

[20] Setlur P, Qureshi T, and Rangaswamy M. Random projections for reduced dimension radar space-time adaptive processing. In: *2017 51st Annual Conference on Information Sciences and Systems (CISS)*; 2017. pp. 1–6.

[21] Setlur P and Rangaswamy M. A family of random and random type projections for radar STAP. In: *2018 IEEE Radar Conference (RadarConf18)*; 2018. pp. 0856–0861.

[22] Qureshi TR, Setlur P, and Rangaswamy M. Localized random projections for space-time adaptive processing. In: *2017 IEEE Radar Conference (RadarConf)*; 2017. pp. 1413–1418.

[23] Setlur P and Rangaswamy M. Waveform design for radar STAP in signal dependent interference. *IEEE Transactions on Signal Processing*. 2016;64(1): 19–34.

[24] Woodruff DP. Sketching as a tool for numerical linear algebra. *Found Trends in Theoretical Computer Science*. 2014;10:1–57.

[25] Ward J. *Space-time Adaptive Processing for Airborne Radar*. Tec. Rep. Massachusetts Institute of Technology, Lincoln Laboratory; 1994.

[26] Klemm R. *Principles of Space-Time Adaptive Processing*. London: Institution of Electrical Engineers; 2002.

[27] Wang H and Cai L. On adaptive spatial-temporal processing for airborne surveillance radar systems. *IEEE Transactions on Aerospace and Electronic Systems*. 1994 ;30(3):660–670.

[28] Guerci JR. *Space-Time Adaptive Processing for Radar*. Boston, MA: Artech House; 2003.

[29] Baraniuk RG, Davenport MA, DeVore RA, *et al.* A simple proof of the restricted isometry property for random matrices. *Constructive Approximation*. 2008;28(3):253–263.

[30] Proakis JG and Salehi M. *Digital Communications*, 5th ed. Chicago, IL: McGraw-Hill Higher Education; 2008.

[31] Dasgupta S and Gupta A. An elementary proof of a theorem of Johnson and Lindenstrauss. *Random Structures & Algorithms*. 2003;22(1): 60–65.

[32] Setlur P, Qureshi T, and Rangaswamy M. Random and localized random projections for radar: Statistical and performance analysis. In: *2017 51st Asilomar Conference on Signals, Systems, and Computers*; 2017. pp. 392–397.

[33] Reed IS, Mallett JD, and Brennan LE. Rapid convergence rates in adaptive arrays. *IEEE Transactions on Aerospace and Electronic Systems*. 1973;AES-10(6):853–863.

[34] Candes EJ and Justin R. l_1 Magic: Recovery of Sparse Signals via Convex Programming; 2005. [Online; accessed 2-Oct-2022]. https://github.com/scgt/l1magic.

[35] Qureshi TR, Rangaswamy M, and Bell KL. Improving multistatic MIMO radar performance in data-limited scenarios. In: *2014 48th Asilomar Conference on Signals, Systems and Computers*; 2014. pp. 1423–1427.

[36] Brennan LE and Reed IS. Theory of adaptive radar. *IEEE Transactions on Aerospace and Electronic Systems*. 1973;AES-9(2):237–252.

[37] Foschini GJ. Layered space-time architecture for wireless communication in a fading environment when using multi-element antennas. *Bell Labs Technical Journal*. 1996;1(2):41–59.

[38] Alamouti SM. A simple transmit diversity technique for wireless communications. *IEEE Journal on Selected Areas in Communications*. 1998;16(8): 1451–1458.

Chapter 9

Fully adaptive radar resource allocation for tracking and classification

Kristine Bell[1], Christopher Kreucher[2], Aaron Brandewie[3] and Joel Johnson[3]

9.1 Introduction

Modern digital radars offer unprecedented flexibility in their transmitted waveforms, radar parameter settings, and transmission schemes in order to support multiple system objectives including target detection, tracking, classification, and other functions. This flexibility provides the potential for improved system performance, but requires a closed-loop sense and respond approach to realize that potential. The concept of fully adaptive radar (FAR), also called cognitive radar [1–5], is to mimic the perception–action cycle (PAC) of cognition [6] to adapt the radar sensor in this closed-loop manner. In this work, we apply the FAR concept to the radar resource allocation (RRA) problem to decide how to allocate finite radar resources such as time, bandwidth, and antenna beamwidth to multiple competing radar system tasks and to decide the transmission parameters for each task so that radar resources are used efficiently and system performance is optimized.

A number of perception–action approaches to RRA have been proposed, including [7–19]. Recent work in this area has been referred to as cognitive radar resource management [16–19], while older related work has been referred to as simply sensor management and/or resource allocation [7–15]. These algorithms rely on two fundamental steps. First, they capture (perceive) the state of the surveillance area probabilistically. Next, they use this probabilistic description to select future sensing actions by determining which actions are expected to maximize utility. Another approach to RRA is the game-theoretic approach [20–23], which is the subject of Chapter 11 "Applications of game theory in cognitive radar."

A key challenge of any RRA algorithm is to balance the multiple competing objectives of target detection, tracking, classification, and other radar tasks. This is addressed through the objective function used in the optimization step to select the

[1]Metron, Inc., Reston, VA, USA
[2]KBR Government Solutions, Ann Arbor, MI, USA
[3]Department of Electrical and Computer Engineering and ElectroScience Laboratory, The Ohio State University, Columbus, OH, USA

next radar actions. Objective functions are also referred to as payoff, criteria, value, or cost functions. Articulating the system goals in a mathematical form suitable for optimization is thus critical to the operation of a fully adaptive radar resource allocation (FARRA) system. As the number of parameters available for adaptation and the number of radar system tasks grow, this becomes increasingly difficult. There are two basic approaches to this optimization: task-driven [19] and information-driven [10].

In the task-driven approach, performance quality of service (QoS) requirements are specified for each task, such as the expected time to detect a target or the tracking root mean square error (RMSE), and a composite objective function is constructed by weighting the utility of various tasks. This has the benefit of being able to separately control task performance and lay out the relative importance of the tasks. However, it requires significant domain knowledge and judgment on the part of the user to specify task requirements and sensor costs and to construct cost/utility functions and weightings for combining disparate task performance metrics [19,24,25].

In the information-driven approach, a global information measure is optimized. Common measures of information include entropy, mutual information (MI), Kullback–Leibler divergence (KLD), and Rényi (alpha) divergence [8,26–29]. Information metrics implicitly balance different types of information that a radar may acquire. This has the desirable property of a common measuring stick (information flow) for all tasks [12], but does not explicitly optimize a task criterion such as RMSE. As such, the information theoretic measures can be difficult for the end-user to understand and attribute to specific operational goals [30]. Furthermore, without additional ad-hoc weighting, they do not allow for separate control of tasks and may produce solutions that over-emphasize some tasks at the expense of others or select sensor actions that provide only marginal gain when judged by user preference.

In this work, we consider a radar system performing concurrent tracking and classification of multiple targets. The FAR framework developed in [18,31], which is based on stochastic optimization [32] (see also Chapter 10 "Stochastic control for cognitive radar"), provides the structure for our PAC. A similar framework is used in Chapter 2 "Adversarial radar inference: inverse tracking, identifying cognition, and designing smart interference." We develop and compare task and information-driven FARRA algorithms for allocating system resources and setting radar transmission parameters, and illustrate the performance on a simulated airborne radar scenario and on the Cognitive Radar Engineering Workspace (CREW) laboratory testbed at The Ohio State University (OSU). More details on the CREW and other experimental testbeds can be found in Chapter 17 "Advances in cognitive radar experiments." This work combines and extends our previous work in sensor management [8–14] and FAR [18,25,31,33–35]. A preliminary version was published in [36]. The results show that the task and information-driven algorithms have similar performance but select different actions to achieve their solutions. We show that the task and information-driven algorithms are actually based on common information-theoretic quantities, so the distinction between them is in the granularity of the metrics used and the degree to which the metrics are weighted.

This chapter is organized as follows. In Section 9.2, we provide an overview of the FAR framework and, in Section 9.3, we develop the multitarget multitask FARRA system model by specifying the components of the FAR framework for this problem. In Section 9.4, we describe the perceptual and executive processors that make up the FARRA PAC, including the task and information-based objective functions we employ. In Section 9.5, we provide airborne radar simulation results comparing the optimization approaches, and, in Section 9.6, we show CREW testbed results. Finally, Section 9.7 presents the conclusions from this effort.

9.2 Fully adaptive radar framework

The FAR framework for a single PAC was developed in [18,31] and is summarized here. A system block diagram is shown in Figure 9.1. The PAC consists of the perceptual processor and the executive processor. The PAC interacts with the external environment through the hardware sensor and with the radar system through the perceptual and executive processors. The perceptual processor receives data from the hardware sensor and processes it into a perception of the environment. The perception is passed to the radar system in order to accomplish system objectives and to the executive processor to decide the next action. The executive processor receives the perception from the perceptual processor along with requirements from the radar system and solves an optimization problem to determine the next sensor action. The executive processor informs the hardware sensor of the settings for the next observation, the sensor collects the next set of data, and the cycle repeats.

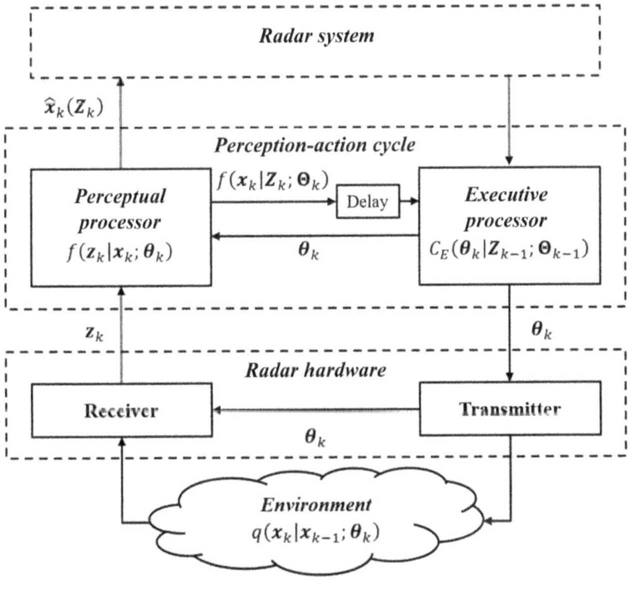

Figure 9.1 Single PAC FAR framework

To develop the mathematical model of the PAC, we assume that the objective of the FAR system is to estimate the state of a target (or targets) at time t_k, denoted as $\boldsymbol{x}_k$. The time-varying nature of the target state is characterized by the state transition (motion) model, which is assumed to be a first-order Markov model with initial target state probability density function (PDF) $q(\boldsymbol{x}_0)$ and transition PDF $q(\boldsymbol{x}_k|\boldsymbol{x}_{k-1};\boldsymbol{\theta}_k)$, which represents the probability that a target in state $\boldsymbol{x}_{k-1}$ will evolve to state $\boldsymbol{x}_k$. The transition density may depend on the sensor parameters $\boldsymbol{\theta}_k$; this will occur, for example, when the choice of sensor parameters affects the time difference $t_k - t_{k-1}$. The hardware sensor observes the environment and produces a measurement vector z_k that depends on the target state $\boldsymbol{x}_k$ and the sensor parameters $\boldsymbol{\theta}_k$. The measurement model is described by the conditional PDF, or likelihood function, $f(z_k|\boldsymbol{x}_k;\boldsymbol{\theta}_k)$.

The perceptual processor processes the data and produces a perception of the target state in the form of a posterior PDF $f(\boldsymbol{x}_k|\boldsymbol{Z}_k;\boldsymbol{\Theta}_k)$ and a target state estimate $\hat{\boldsymbol{x}}_k(\boldsymbol{Z}_k)$, where $\boldsymbol{Z}_k \doteq \{z_1, z_2, \cdots, z_k\}$ denotes the measurements up to time t_k and $\boldsymbol{\Theta}_k \doteq \{\boldsymbol{\theta}_1, \boldsymbol{\theta}_2, \cdots, \boldsymbol{\theta}_k\}$ denotes the sensor parameters up to time t_k. For the Markov motion model, the posterior PDF of $\boldsymbol{x}_k$ given $\boldsymbol{Z}_k$ can be obtained from the Bayes–Markov recursion:

$$f^+(\boldsymbol{x}_0) = q(\boldsymbol{x}_0) \tag{9.1}$$

$$f^-(\boldsymbol{x}_k) \doteq f(\boldsymbol{x}_k|\boldsymbol{Z}_{k-1};\boldsymbol{\Theta}_k) = \int q(\boldsymbol{x}_k|\boldsymbol{x}_{k-1};\boldsymbol{\theta}_k) f^+(\boldsymbol{x}_{k-1}) d\boldsymbol{x}_{k-1} \tag{9.2}$$

$$f^-(z_k) \doteq f(z_k|\boldsymbol{Z}_{k-1};\boldsymbol{\Theta}_k) = \int f(z_k|\boldsymbol{x}_k;\boldsymbol{\theta}_k) f^-(\boldsymbol{x}_k) d\boldsymbol{x}_k \tag{9.3}$$

$$f^+(\boldsymbol{x}_k) \doteq f(\boldsymbol{x}_k|\boldsymbol{Z}_k;\boldsymbol{\Theta}_k) = \frac{f(z_k|\boldsymbol{x}_k;\boldsymbol{\theta}_k) f^-(\boldsymbol{x}_k)}{f^-(z_k)}, \tag{9.4}$$

where $f^-(\boldsymbol{x}_k)$ is the motion-updated predicted density and $f^+(\boldsymbol{x}_k)$ is the information-updated posterior density. The state estimation performance is characterized by the posterior Bayes risk, which is the expected value of the perceptual processor error function $\epsilon\left(\hat{\boldsymbol{x}}(\boldsymbol{Z}_k), \boldsymbol{x}_k\right)$ with respect to the posterior PDF,

$$R^+(\boldsymbol{Z}_k;\boldsymbol{\Theta}_k) = E^+\left\{\epsilon\left(\hat{\boldsymbol{x}}(\boldsymbol{Z}_k), \boldsymbol{x}_k\right)\right\}, \tag{9.5}$$

where $E_k^+\{\cdot\}$ denotes expectation with respect to $f^+(\boldsymbol{x}_k)$. The state estimate is found by minimizing the posterior Bayes risk:

$$\hat{\boldsymbol{x}}_k(\boldsymbol{Z}_k) = \underset{\hat{\boldsymbol{x}}(\boldsymbol{Z}_k)}{\operatorname{argmin}}\; R^+(\boldsymbol{Z}_k;\boldsymbol{\Theta}_k). \tag{9.6}$$

The goal of the executive processor is to find the next set of sensor parameters to optimize the performance of the state estimator that will include the next observation z_k as well as the previously received observations $\boldsymbol{Z}_{k-1}$. We define the joint conditional PDF of $\boldsymbol{x}_k$ and z_k conditioned on $\boldsymbol{Z}_{k-1}$ as

$$f^\uparrow(\boldsymbol{x}_k, z_k) \doteq f(\boldsymbol{x}_k, z_k|\boldsymbol{Z}_{k-1};\boldsymbol{\Theta}_k) = f(z_k|\boldsymbol{x}_k;\boldsymbol{\theta}_k) f(\boldsymbol{x}_k|\boldsymbol{Z}_{k-1};\boldsymbol{\Theta}_k). \tag{9.7}$$

Using the definitions in (9.2) and (9.4), this can also be written as:

$$f^\uparrow(\boldsymbol{x}_k, z_k) = f^+(\boldsymbol{x}_k) f^-(z_k). \tag{9.8}$$

We define the predicted conditional (PC)-Bayes risk by taking the expectation of the error function with respect to the joint conditional PDF,

$$R^{\uparrow}(\boldsymbol{\theta}_k|\boldsymbol{Z}_{k-1};\boldsymbol{\Theta}_{k-1}) = E_k^{\uparrow}\{\epsilon(\hat{\boldsymbol{x}}(\boldsymbol{Z}_k),\boldsymbol{x}_k)\}, \tag{9.9}$$

where $E_k^{\uparrow}\{\cdot\}$ denotes expectation with respect to $f^{\uparrow}(\boldsymbol{x}_k, \boldsymbol{z}_k)$. Using (9.5) and (9.8), we can also write the PC-Bayes risk as the expectation of the posterior Bayes risk with respect to $f^{-}(\boldsymbol{z}_k)$, i.e.,

$$R^{\uparrow}(\boldsymbol{\theta}_k|\boldsymbol{Z}_{k-1};\boldsymbol{\Theta}_{k-1}) = E_{\boldsymbol{z}_k}^{-}\{R^{+}(\boldsymbol{Z}_k;\boldsymbol{\Theta}_k)\}, \tag{9.10}$$

where $E_{\boldsymbol{z}_k}^{-}\{\cdot\}$ denotes expectation with respect to $f^{-}(\boldsymbol{z}_k)$. In many applications, the PC-Bayes risk may be difficult to compute and in general will not have a closed form analytical expression. To overcome this difficulty, information-theoretic surrogate functions that are analytically tractable and provide a good indication of the quality of the target state estimate are often substituted. The next set of sensor parameters are chosen to minimize an executive cost (or objective) function $C_E(\boldsymbol{\theta}_k|\boldsymbol{Z}_{k-1};\boldsymbol{\Theta}_{k-1})$. In the task-driven approach, the executive cost function is a scalar function that incorporates the processor performance, derived from the PC-Bayes risk or a surrogate, with system requirements and the cost of obtaining measurements. In the information-driven approach, the executive cost function is an information theoretic measure. The executive processor optimization problem is then

$$\boldsymbol{\theta}_k = \underset{\boldsymbol{\theta}}{\operatorname{argmin}} \;\; C_E(\boldsymbol{\theta}|\boldsymbol{Z}_{k-1};\boldsymbol{\Theta}_{k-1}). \tag{9.11}$$

In the next two sections, we specialize the general FAR framework for the multitarget multitask RRA problem.

9.3 Multitarget multitask FARRA system model

The multitarget multitask FARRA system model is shown in Figure 9.2. There is a single PAC with a perceptual processor that consists of M tasks and an executive processor that allocates system resources to the M tasks and specifies the next sequence of transmissions of the radar.

9.3.1 Radar resource allocation model

We define a resource allocation frame as an interval of fixed length T_F and let k denote the frame (time) index. We assume that there are M variable length dwells in the kth frame, corresponding to M different tasks, where M is fixed and known. In the examples in this chapter, the tasks are tracking and classification of individual targets, and each dwell corresponds to one coherent processing interval (CPI) for the given task. During each task dwell, we assume that the radar can transmit nothing (taking up no time) or one of the L waveforms from a waveform library. Let $a_l; l = 0, \ldots, L$ denote each of the possible actions (waveforms), where a_0 is the action of no transmission, and let $A = \{a_0, \ldots, a_L\}$ denote the set of actions. We assume the chosen waveform is fixed during the entire task dwell/CPI.

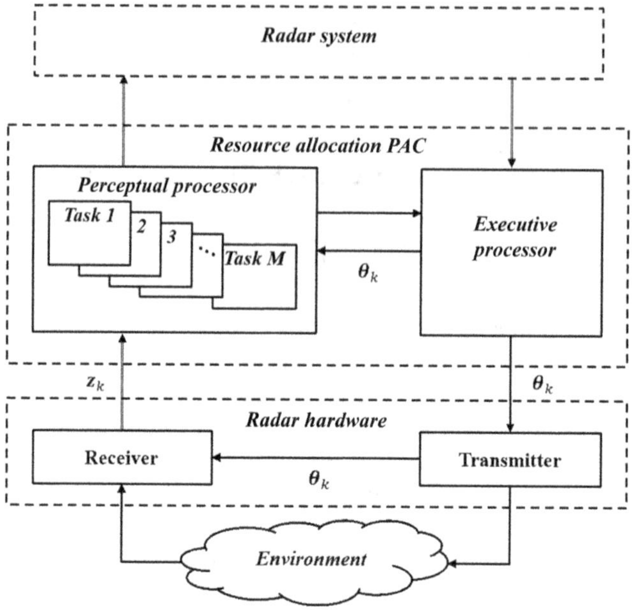

Figure 9.2 System model for multiple task FARRA

9.3.2 Controllable parameters

The radar resource parameter vector, or action vector, for the kth frame is defined as the $M \times 1$ vector

$$\boldsymbol{\theta}_k = \left[\theta_{1,k}\ \theta_{2,k}\ \cdots\ \theta_{M,k}\right]^T, \tag{9.12}$$

where $\theta_{m,k} \in A$ is the action in the mth dwell of the kth frame. The objective of the FARRA executive processor is to determine the best action vector for the next one or more frames.

9.3.3 State vector

We assume that there are N targets, where N is fixed and known. Following the model in [10–12], the multitarget state vector has the form:

$$\boldsymbol{x}_k = \left[\boldsymbol{x}_{1,k}^T\ \boldsymbol{x}_{2,k}^T\ \cdots\ \boldsymbol{x}_{N,k}^T\right]^T, \tag{9.13}$$

where $\boldsymbol{x}_{n,k}$ is the state vector for the nth target and contains components relevant to the tasks at hand. Here we consider multitarget tracking and classification, therefore, each target's state is composed of a tracking state vector $\boldsymbol{y}_{n,k}$ and a classification state variable $c_{n,k}$:

$$\boldsymbol{x}_{n,k} = \left[\boldsymbol{y}_{n,k}^T\ c_{n,k}\right]^T. \tag{9.14}$$

The tracking state vector $\boldsymbol{y}_{n,k}$ consists of kinematic variables (position, velocity, and possibly acceleration) and the received signal-to-noise ratio (SNR). The tracking state variables are continuous random variables, while the target class is a discrete random variable that takes on one of a discrete set of values. As such, the various PDFs defined in Section 9.2 become a combination of a PDF for the continuous components and a probability mass function (PMF) for the discrete components.

9.3.4 Transition model

The transition model consists of a prior PDF/PMF $q(\boldsymbol{x}_0)$ and a transition PDF/PMF $q(\boldsymbol{x}_k|\boldsymbol{x}_{k-1};\boldsymbol{\theta}_k)$. We assume that the target transition models are independent across targets and that for each target, the tracking and classification transition models are independent, therefore, the joint tracking and classification transition model has the form:

$$q(\boldsymbol{x}_0) = \prod_{n=1}^{N} q(\boldsymbol{x}_{n,0}) = \prod_{n=1}^{N} q(\boldsymbol{y}_{n,0})q(c_{n,0}) \tag{9.15}$$

and

$$q(\boldsymbol{x}_k|\boldsymbol{x}_{k-1};\boldsymbol{\theta}_k) = \prod_{n=1}^{N} q(\boldsymbol{x}_{n,k}|\boldsymbol{x}_{n,k-1};\boldsymbol{\theta}_k) = \prod_{n=1}^{N} q(\boldsymbol{y}_{n,k}|\boldsymbol{y}_{n,k-1};\boldsymbol{\theta}_k)q(c_{n,k}|c_{n,k-1}). \tag{9.16}$$

In this model, we assume that the tracking transition model depends on the sensor parameters but the classification transition model does not, as explained below.

Let the random vector $\boldsymbol{y}$ have a multivariate Gaussian distribution with mean $\boldsymbol{\mu}$ and covariance matrix $\boldsymbol{\Sigma}$. We use the notation $N(\boldsymbol{y};\boldsymbol{\mu},\boldsymbol{\Sigma})$ to denote the multivariate Gaussian PDF for the random variable $\boldsymbol{y}$, i.e.,

$$N(\boldsymbol{y};\boldsymbol{\mu},\boldsymbol{\Sigma}) \doteq \frac{1}{\sqrt{|2\pi\boldsymbol{\Sigma}|}} \exp\left\{-\frac{1}{2}[\boldsymbol{y}-\boldsymbol{\mu}]^T\boldsymbol{\Sigma}^{-1}[\boldsymbol{y}-\boldsymbol{\mu}]\right\}. \tag{9.17}$$

For each target, we assume an initial tracking state distribution that is multivariate Gaussian with mean $\boldsymbol{\mu}_{n,0}$ and covariance matrix $\boldsymbol{\Sigma}_{n,0}$, therefore,

$$q(\boldsymbol{y}_{n,0}) = N\left(\boldsymbol{y}_{n,0};\boldsymbol{\mu}_{n,0},\boldsymbol{\Sigma}_{n,0}\right). \tag{9.18}$$

Let $\Delta t(\boldsymbol{\theta}_k) = t_k - t_{k-1}$. We assume a linear motion model of the form:

$$\boldsymbol{y}_{n,k} = \boldsymbol{F}_n(\Delta t(\boldsymbol{\theta}_k))\boldsymbol{y}_{n,k-1} + \boldsymbol{e}_{n,k}, \tag{9.19}$$

where $\boldsymbol{F}_n(\Delta t(\boldsymbol{\theta}_k))$ is the state transition matrix and $\boldsymbol{e}_{n,k}$ is zero-mean additive white Gaussian noise (AWGN) with covariance matrix $\boldsymbol{Q}_n(\Delta t(\boldsymbol{\theta}_k))$. The transition PDF is then:

$$q(\boldsymbol{y}_{n,k}|\boldsymbol{y}_{n,k-1};\boldsymbol{\theta}_k) = N\left(\boldsymbol{y}_{n,k};\boldsymbol{F}_n(\Delta t(\boldsymbol{\theta}_k))\boldsymbol{y}_{n,k-1},\boldsymbol{Q}_n(\Delta t(\boldsymbol{\theta}_k))\right). \tag{9.20}$$

We assume that the target class $c_{n,k}$ takes on one of a discrete set of N_c values in the set C,

$$c_{n,k} \in C = \{1, 2, \ldots, N_c\}. \tag{9.21}$$

For each target, the prior distribution is characterized by the PMF $q(c_{n,0})$, which is represented by the $N_c \times 1$ vector $\boldsymbol{q}_n$, which consists of the N_c probabilities

$$[\boldsymbol{q}_n]_i = P(c_{n,0} = i); \qquad i = 1, \ldots, N_c. \tag{9.22}$$

The transition model $q(c_{n,k}|c_{n,k-1})$ is represented by the $N_c \times N_c$ transition matrix $\boldsymbol{\Upsilon}_n$, where

$$[\boldsymbol{\Upsilon}_n]_{ij} = P(c_{n,k} = i|c_{n,k-1} = j); \qquad i,j = 1, \ldots, N_c. \tag{9.23}$$

Depending on the model, the target may or may not be able to switch classes. For example, if the class represents a target behavior class, then switching can occur, however, if the class represents a type of vehicle or aircraft, then switching cannot occur and $\boldsymbol{\Upsilon}_n = \boldsymbol{I}$. We assume that the classification transition model does not depend on the time between updates or the sensor parameters $\boldsymbol{\theta}_k$.

9.3.5 Measurement model

We assume that each target is allocated one dwell per frame, thus the number of tasks (dwells), M, is equal to the number of targets, N. In each dwell, we might obtain both a tracking measurement and a classification measurement, a tracking measurement only, a classification measurement only, or no measurement. Let $\boldsymbol{z}_{n,k}$ and $\boldsymbol{\xi}_{n,k}$ denote the tracking and classification measurement vectors, respectively, for the nth target during the kth frame. Either or both of these may be empty if there is no measurement of that type. Thus we have $2N$ possible measurements from $M = N$ dwells and the measurement vector has the form:

$$\boldsymbol{z}_k = \left[\boldsymbol{z}_{1,k}^T \ \boldsymbol{\xi}_{1,k}^T \ \boldsymbol{z}_{2,k}^T \ \boldsymbol{\xi}_{2,k}^T \ \cdots \ \boldsymbol{z}_{N,k}^T \ \boldsymbol{\xi}_{N,k}^T\right]^T. \tag{9.24}$$

We assume that the measurements are independent, thus the likelihood function has the form:

$$f(\boldsymbol{z}_k|\boldsymbol{x}_k;\boldsymbol{\theta}_k) = \prod_{n=1}^{N} f(\boldsymbol{z}_{n,k}|\boldsymbol{y}_{n,k};\boldsymbol{\theta}_k) f(\boldsymbol{\xi}_{n,k}|c_{n,k};\boldsymbol{\theta}_k). \tag{9.25}$$

For tracking, we assume that measurements are received with false alarm probability P_F and detection probability $P_D\left(\boldsymbol{y}_{n,k};\boldsymbol{\theta}_k\right)$. The detection probability is determined by the received SNR, which is a function of the target state and sensor parameters, and the detection threshold, which is a function of the required P_F. When measurements are received, we assume they follow a nonlinear, AWGN measurement model of the form:

$$\boldsymbol{z}_{n,k} = \boldsymbol{h}_{n,k}\left(\boldsymbol{y}_{n,k}\right) + \boldsymbol{n}_{n,k}, \tag{9.26}$$

where $\boldsymbol{h}_{n,k}\left(\boldsymbol{y}_{n,k}\right)$ is a nonlinear transformation from the target state space to the radar measurement space and $\boldsymbol{n}_{n,k}$ is the measurement error, which is modeled as a zero-mean Gaussian random vector with covariance matrix $\boldsymbol{R}_{n,k}\left(\boldsymbol{\theta}_k\right)$. The single target likelihood function is then:

$$f(\boldsymbol{z}_{n,k}|\boldsymbol{y}_{n,k};\boldsymbol{\theta}_k) = N\left(\boldsymbol{z}_{n,k};\boldsymbol{h}_{n,k}\left(\boldsymbol{y}_{n,k}\right), \boldsymbol{R}_{n,k}\left(\boldsymbol{\theta}_k\right)\right). \tag{9.27}$$

For classification, we consider two measurement models: a discrete class measurement model and a continuous Gaussian feature vector model. In the discrete class model, we assume that the sensor makes a discrete valued measurement of target class, i.e., $\xi_{n,k} \in C = \{1, 2, \ldots, N_c\}$ and the likelihood function $f(\xi_{n,k}|c_{n,k}; \boldsymbol{\theta}_k)$ is represented by the $N_c \times N_c$ likelihood matrix $\boldsymbol{L}_n(\boldsymbol{\theta}_k)$, where

$$[\boldsymbol{L}_n]_{ij}(\boldsymbol{\theta}_k) = P(\xi_{n,k} = i|c_{n,k} = j; \boldsymbol{\theta}_k); \qquad i,j = 1, \ldots, N_c. \tag{9.28}$$

In the Gaussian feature vector model, we assume that the sensor makes a continuous valued measurement of a feature vector $\boldsymbol{\xi}_{n,k}$ and the likelihood function is a Gaussian density with mean and covariance matrix determined by the target class. For the ith class, the likelihood function is:

$$f(\boldsymbol{\xi}_{n,k}|c_{n,k} = i; \boldsymbol{\theta}_k) = N\left(\boldsymbol{\xi}_{n,k}; \boldsymbol{\mu}_{n,i}(\boldsymbol{\theta}_k), \boldsymbol{\Sigma}_{n,i}(\boldsymbol{\theta}_k)\right); \quad i = 1, \ldots, N_c. \tag{9.29}$$

9.4 FARRA PAC

9.4.1 Perceptual processor

Since the motion and measurement models developed in Section 9.3 are independent across targets and tasks, the Bayes–Markov recursions decouple and can be computed separately for the tracking and classification tasks for each target. While the Bayes–Markov recursion expressions in (9.1)–(9.4) appear straightforward, it is usually analytically and/or computationally infeasible to evaluate them exactly. One exception is in the tracking problem when the motion and measurement models are linear with AWGN, and the transition density, likelihood function, predicted density, and posterior density are all Gaussian. In this case, the exact solution is given by the linear Kalman filter (KF), and the motion and measurement updates consist of explicit calculations of the mean vectors and covariance matrices that characterize the Gaussian densities. For the general tracking problem, approximate and suboptimal implementations include the extended Kalman filter (EKF), unscented Kalman filter (UKF), particle filters, and many others. Our tracking model includes a nonlinear AWGN measurement model, and we will use the EKF as an approximate solution to the Bayes–Markov recursion. The EKF reduces to the exact KF if the measurement model is in fact linear. Another exception is in the classification problem when the target state is one of a discrete set of classes, the motion model is specified by a transition matrix, and the likelihood function has a closed form analytical expression. Our classification models meet these requirements.

For the tracking tasks, the predicted and posterior PDFs are computed using the EKF. The PDFs are presumed to be Gaussian of the form:

$$f^-(\boldsymbol{y}_{n,k}) = N\left(\boldsymbol{y}_{n,k}; \boldsymbol{\mu}_{n,k}^-, \boldsymbol{P}_{n,k}^-\right) \tag{9.30}$$

$$f^+(\boldsymbol{y}_{n,k}) = N\left(\boldsymbol{y}_{n,k}; \boldsymbol{\mu}_{n,k}^+, \boldsymbol{P}_{n,k}^+\right). \tag{9.31}$$

The EKF is initialized with:

$$\boldsymbol{\mu}^{+}_{n,0} = \boldsymbol{\mu}_{n,0} \tag{9.32}$$

$$\boldsymbol{P}^{+}_{n,0} = \boldsymbol{\Sigma}_{n,0} \tag{9.33}$$

and the recursions have the form:

$$\boldsymbol{\mu}^{-}_{n,k} = \boldsymbol{F}_n\left(\Delta t(\boldsymbol{\theta}_k)\right)\boldsymbol{\mu}^{+}_{n,k-1} \tag{9.34}$$

$$\boldsymbol{P}^{-}_{n,k} = \boldsymbol{F}_n\left(\Delta t(\boldsymbol{\theta}_k)\right)\boldsymbol{P}^{+}_{n,k-1}\boldsymbol{F}_n\left(\Delta t(\boldsymbol{\theta}_k)\right)^T + \boldsymbol{Q}_n\left(\Delta t(\boldsymbol{\theta}_k)\right) \tag{9.35}$$

$$\boldsymbol{H}_{n,k} = \tilde{\boldsymbol{H}}_{n,k}(\boldsymbol{\mu}^{-}_{n,k}) \tag{9.36}$$

$$\boldsymbol{K}_{n,k} = \boldsymbol{P}^{-}_{n,k}\boldsymbol{H}^{T}_{n,k}\left[\boldsymbol{H}_{n,k}\boldsymbol{P}^{-}_{n,k}\boldsymbol{H}^{T}_{n,k} + \boldsymbol{R}_{n,k}(\boldsymbol{\theta}_k)\right]^{-1} \tag{9.37}$$

$$\boldsymbol{\mu}^{+}_{n,k} = \boldsymbol{\mu}^{-}_{n,k} + \boldsymbol{K}_{n,k}\left[\boldsymbol{z}_{n,k} - \boldsymbol{h}_{n,k}(\boldsymbol{\mu}^{-}_{n,k})\right] \tag{9.38}$$

$$\boldsymbol{P}^{+}_{n,k} = \boldsymbol{P}^{-}_{n,k} - \boldsymbol{K}_{n,k}\boldsymbol{H}_{n,k}\boldsymbol{P}^{-}_{n,k}, \tag{9.39}$$

where $\tilde{\boldsymbol{H}}_{n,k}(\boldsymbol{y})$ is the Jacobian matrix, defined as:

$$\tilde{\boldsymbol{H}}_{n,k}(\boldsymbol{y}) = \left[\nabla_{y}\boldsymbol{h}_{n,k}(\boldsymbol{y})^T\right]^T. \tag{9.40}$$

For the classification tasks, the predicted and posterior PMFs are computed using the exact Bayes–Markov recursions. Let $f_i^{-}(c_{n,k}) \doteq P(c_{n,k} = i|\boldsymbol{Z}_{k-1}; \boldsymbol{\Theta}_{k-1})$ denote the predicted PMF and $f_i^{+}(c_{n,k}) \doteq P(c_{n,k} = i|\boldsymbol{Z}_k; \boldsymbol{\Theta}_k)$ denote the posterior PMF. The recursion is initialized with:

$$f_i^{+}(c_{n,0}) = [\boldsymbol{q}_n]_i\,; \qquad i = 1, \ldots, N_c, \tag{9.41}$$

and the predicted PMF is computed from:

$$f_i^{-}(c_{n,k}) = \sum_{j=1}^{N_c}[\boldsymbol{\Upsilon}_n]_{ij}f_j^{+}(c_{n,k-1}); \qquad i = 1, \ldots, N_c. \tag{9.42}$$

For the discrete class measurement model, the information update has the form:

$$f^{-}(\xi_{n,k}) = \sum_{j=1}^{N_c}[\boldsymbol{L}_n]_{\xi_{n,k},j}(\boldsymbol{\theta}_k)f_j^{-}(c_{n,k}) \tag{9.43}$$

$$f_i^{+}(c_{n,k}) = \frac{[\boldsymbol{L}_n]_{\xi_{n,k},i}(\boldsymbol{\theta}_k)f_i^{-}(c_{n,k})}{f^{-}(\xi_{n,k})}; \quad i = 1, \ldots, N_c, \tag{9.44}$$

while for the Gaussian feature vector measurement model, the information update has the form:

$$f^{-}(\boldsymbol{\xi}_{n,k}) = \sum_{j=1}^{N_c}f(\boldsymbol{\xi}_{n,k}|c_{n,k} = j; \boldsymbol{\theta}_k)f_j^{-}(c_{n,k}) \tag{9.45}$$

$$f_i^{+}(c_{n,k}) = \frac{f(\boldsymbol{\xi}_{n,k}|c_{n,k} = i; \boldsymbol{\theta}_k)f_i^{-}(c_{n,k})}{f^{-}(\boldsymbol{\xi}_{n,k})}; \quad i = 1, \ldots, N_c. \tag{9.46}$$

The posterior Bayes risk for the multitarget tracking and classification state vector is the sum of the traces of the posterior mean square error (MSE) matrices and the posterior probability of incorrect classification across all targets. The solution to (9.6) is the mean of the posterior PDF for the tracking variables and the maximum of the posterior PMF for the classification variables:

$$\hat{\mathbf{y}}_{n,k}(\mathbf{Z}_k) = E_k^+\left\{\mathbf{y}_{n,k}\right\} = \boldsymbol{\mu}_{n,k}^+; \qquad n = 1, \ldots, N \tag{9.47}$$

$$\hat{c}_{n,k}(\mathbf{Z}_k) = \underset{i \in C}{\operatorname{argmax}}\ f_i^+\left(c_{n,k}\right); \qquad n = 1, \ldots, N. \tag{9.48}$$

9.4.2 Executive processor

We consider both task-driven and information-driven methods for specifying the objective function used by the executive processor. It should be noted here that in both approaches, the optimization is in a global sense and may not be the optimal solution for a particular radar task.

9.4.2.1 Task-driven (QoS) approach

Following the development in [19], we assume that there are M tasks. The perceptual processor for the mth task computes a perception of its environment, which may include quantities such as target location, target class, and target SNR. The executive processor analytically evaluates the performance of the perceptual processor for the mth task in terms of a task QoS metric, which is denoted by $G_{m,k}(\boldsymbol{\theta}_k|\mathbf{Z}_{k-1}; \boldsymbol{\Theta}_{k-1})$. The QoS metric for the current frame will in general depend on the perception from the previous frame, the previous sensing actions, and the current sensing action. Each task QoS metric has a task QoS requirement, denoted $\bar{G}_m$. The task QoS metrics and requirements are physically meaningful quantities with appropriate physical units. The task QoS metric and requirement are converted to a task utility $U_{m,k}(\boldsymbol{\theta}_k|\mathbf{Z}_{k-1}; \boldsymbol{\Theta}_{k-1})$, which is a unitless quantity on the interval [0, 1]. It represents the level of satisfaction with the QoS and is determined from the task utility function,

$$U_{m,k}(\boldsymbol{\theta}_k|\mathbf{Z}_{k-1}; \boldsymbol{\Theta}_{k-1}) = u_m(G_{m,k}(\boldsymbol{\theta}_k|\mathbf{Z}_{k-1}; \boldsymbol{\Theta}_{k-1}), \bar{G}_m). \tag{9.49}$$

The executive processor combines and balances the task utilities along with resource constraints to determine the resource allocation for the next frame. The mission utility, or mission effectiveness, is a measure of the radar system's ability to meet all of its requirements. It is a weighted sum of the task utilities, where the task weighting, w_m, represents the relative importance of the mth task to the overall mission, and the weights sum to one. The mission utility is given by

$$U(\boldsymbol{\theta}_k|\mathbf{Z}_{k-1}; \boldsymbol{\Theta}_{k-1})) = \sum_{m=1}^{M} w_m U_{m,k}(\boldsymbol{\theta}_k|\mathbf{Z}_{k-1}; \boldsymbol{\Theta}_{k-1}). \tag{9.50}$$

Constraints on system resources are described by the function $g_c(\boldsymbol{\theta}_k)$, constructed so the constraint may be expressed as the inequality $g_c(\boldsymbol{\theta}_k) \leq 0$. The next action vector is then determined by maximizing the mission utility subject to the constraint

$$\boldsymbol{\theta}_k = \underset{\boldsymbol{\theta}}{\operatorname{argmax}}\ U(\boldsymbol{\theta}|\mathbf{Z}_{k-1}; \boldsymbol{\Theta}_{k-1}), \qquad s.t. g_c(\boldsymbol{\theta}) \leq 0. \tag{9.51}$$

For a tracking task, we use the position and velocity RMSE and the requirement is an upper limit on the RMSE. In most cases, it is not possible to evaluate the RMSE analytically. However, the Bayesian Cramér–Rao lower bound (BCRLB), the inverse of the Bayesian information matrix (BIM), provides a (matrix) lower bound on the MSE matrix of any estimator [37] and is usually analytically tractable. For tracking applications, this yields the posterior Cramér–Rao lower bound (PCRLB) [38,39]. The PCRLB provides a lower bound on the global MSE that has been averaged over $\boldsymbol{x}_k$ and $\boldsymbol{Z}_k$, thus it characterizes tracker performance for all possible data that might have been received. Here we use a predicted conditional BIM (PC-BIM) and a predicted conditional Cramér–Rao lower bound (PC-CRLB) to bound the PC-MSE matrix, which is averaged over the joint density of $\boldsymbol{x}_k$ and $\boldsymbol{z}_k$ conditioned on $\boldsymbol{Z}_{k-1}$. The PC-CRLB differs from the PCRLB in that it characterizes performance conditioned on the actual data that has been received. For our model, the PC-BIM $\boldsymbol{B}^{\uparrow}_{n,k}(\boldsymbol{\theta}_k|\boldsymbol{Z}_{k-1};\boldsymbol{\Theta}_{k-1})$ has the same form as the inverse of the EKF posterior covariance matrix in (9.39), which simplifies to

$$\begin{aligned}\boldsymbol{B}^{\uparrow}_{n,k}(\boldsymbol{\theta}_k|\boldsymbol{Z}_{k-1};\boldsymbol{\Theta}_{k-1}) &= \left[\boldsymbol{P}^{-}_{n,k} - \boldsymbol{K}_{n,k}\boldsymbol{H}_{n,k}\boldsymbol{P}^{-}_{n,k}\right]^{-1} \\ &= \left[\boldsymbol{P}^{-}_{n,k}\right]^{-1} + \boldsymbol{H}^{T}_{n,k}\boldsymbol{R}_{n,k}(\boldsymbol{\theta}_k)^{-1}\boldsymbol{H}_{n,k}.\end{aligned} \tag{9.52}$$

In our model, a detection is obtained with probability $P_D\left(\boldsymbol{y}_{n,k};\boldsymbol{\theta}_k\right)$. When a detection is obtained, the PC-BIM has the form given in (9.52). When a detection is missed, the second term is equal to zero and the PC-BIM is equal to the inverse of the predicted covariance matrix. Using the approach of the information reduction factor bound in [40], and substituting the mean of the predicted density, $\boldsymbol{\mu}^{-}_{n,k}$, for the unknown target state, $\boldsymbol{y}_{n,k}$, the tracking PC-BIM with missed detections is:

$$\tilde{\boldsymbol{B}}^{\uparrow}_{n,k}(\boldsymbol{\theta}_k|\boldsymbol{Z}_{k-1};\boldsymbol{\Theta}_{k-1}) = \left[\boldsymbol{P}^{-}_{n,k}\right]^{-1} + P_D\left(\boldsymbol{\mu}^{-}_{n,k};\boldsymbol{\theta}_k\right)\boldsymbol{H}^{T}_{n,k}\boldsymbol{R}_{n,k}(\boldsymbol{\theta}_k)^{-1}\boldsymbol{H}_{n,k}. \tag{9.53}$$

The tracking PC-CRLB is the inverse of the PC-BIM,

$$\tilde{\boldsymbol{C}}^{\uparrow}_{n,k}(\boldsymbol{\theta}_k|\boldsymbol{Z}_{k-1};\boldsymbol{\Theta}_{k-1}) = \tilde{\boldsymbol{B}}^{\uparrow}_{n,k}(\boldsymbol{\theta}_k|\boldsymbol{Z}_{k-1};\boldsymbol{\Theta}_{k-1})^{-1}. \tag{9.54}$$

Temporarily dropping the conditioning on $\boldsymbol{Z}_{k-1}$ and $\boldsymbol{\Theta}_{k-1}$ to simplify the notation, the QoS metrics are the position and velocity RMSEs obtained from the PC-CRLB as follows:

$$G^{R}_{n,k}(\boldsymbol{\theta}_k) = \sqrt{\left[\tilde{\boldsymbol{C}}^{\uparrow}_{n,k}(\boldsymbol{\theta}_k)\right]_x + \left[\tilde{\boldsymbol{C}}^{\uparrow}_{n,k}(\boldsymbol{\theta}_k)\right]_y + \left[\tilde{\boldsymbol{C}}^{\uparrow}_{n,k}(\boldsymbol{\theta}_k)\right]_z} \tag{9.55}$$

$$G^{V}_{n,k}(\boldsymbol{\theta}_k) = \sqrt{\left[\tilde{\boldsymbol{C}}^{\uparrow}_{n,k}(\boldsymbol{\theta}_k)\right]_{\dot{x}} + \left[\tilde{\boldsymbol{C}}^{\uparrow}_{n,k}(\boldsymbol{\theta}_k)\right]_{\dot{y}} + \left[\tilde{\boldsymbol{C}}^{\uparrow}_{n,k}(\boldsymbol{\theta}_k)\right]_{\dot{z}}}. \tag{9.56}$$

The QoS requirements are the values that we want the RMSEs to be below, denoted as $\bar{G}_n^R$ and $\bar{G}_n^V$. We then define the position and velocity task utility functions to be:

$$U_{n,k}^R(\boldsymbol{\theta}_k) = \begin{cases} \dfrac{\bar{G}_n^R}{G_{n,k}^R(\boldsymbol{\theta}_k)} & G_{n,k}^R(\boldsymbol{\theta}_k) > \bar{G}_n^R \\ 1 & G_{n,k}^R(\boldsymbol{\theta}_k) \leq \bar{G}_n^R \end{cases} \tag{9.57}$$

$$U_{n,k}^V(\boldsymbol{\theta}_k) = \begin{cases} \dfrac{\bar{G}_n^V}{G_{n,k}^V(\boldsymbol{\theta}_k)} & G_{n,k}^V(\boldsymbol{\theta}_k) > \bar{G}_n^V \\ 1 & G_{n,k}^V(\boldsymbol{\theta}_k) \leq \bar{G}_n^V. \end{cases} \tag{9.58}$$

With these utility functions, if the QoS metric is below the required value, the resulting utility is one and there is neither a penalty nor any additional utility for being below the requirement.

For a classification task, the desired QoS metric is the probability of incorrect classification. The posterior probability of incorrect classification is difficult to compute and in general does not have a closed form analytical expression. The PC-probability of incorrect classification is even more difficult to compute since it involves an additional expectation over the next measurement $\boldsymbol{\xi}_{n,k}$. To overcome this difficulty, we substitute an information-theoretic surrogate that is analytically tractable and provides a good indication of the quality of the target class estimate. As in [34–36], we use the entropy, which can be calculated directly from the discrete classification PMF. The entropy of the predicted and posterior PMFs, respectively, are defined as [26]:

$$H^-(c_{n,k}) = -\sum_{i=1}^{N_c} f_i^-(c_{n,k}) \ln f_i^-(c_{n,k}) \tag{9.59}$$

$$H^+(c_{n,k}) = -\sum_{i=1}^{N_c} f_i^+(c_{n,k}) \ln f_i^+(c_{n,k}). \tag{9.60}$$

The entropy has the property $0 \leq H \leq \ln(N_c)$. It is low when the PMF is concentrated on one of the classes and high when the PMF is distributed across the classes. The posterior entropy is a surrogate for the posterior probability of incorrect classification and is used to characterize classification performance after the measurement is received. In order for the executive processor to determine the next sensing action, we also need a surrogate for the PC-probability of incorrect classification, which characterizes the expected performance of the current (next) measurement, given the past measurements that have been observed. If we take the expected value of the posterior entropy with respect to $f^-(\boldsymbol{\xi}_{n,k})$, we obtain the desired surrogate, which is the conditional entropy [26] of $c_{n,k}$ given $\boldsymbol{\xi}_{n,k}$ conditioned on the past measurements $\boldsymbol{Z}_{k-1}$.

For the discrete class measurement model, we must compute $f^-(\xi_{n,k})$ and $f_i^+(c_{n,k})$ for every $\xi_{n,k}$ using (9.43) and (9.44). Using slightly more explicit notation, define $f_{i|j}^+(c_{n,k}|\xi_{n,k}) \doteq P(c_{n,k} = i|\xi_{n,k} = j, \boldsymbol{Z}_{k-1}; \boldsymbol{\Theta}_k)$ and $f_j^-(\xi_{n,k}) \doteq P(\xi_{n,k} = j|\boldsymbol{Z}_{k-1}; \boldsymbol{\Theta}_k)$. The conditional entropy is then computed from:

$$H_n^\dagger(\boldsymbol{\theta}_k|\boldsymbol{Z}_{k-1}; \boldsymbol{\Theta}_{k-1}) = \sum_{j=1}^{N_c} f_j^-(\xi_{n,k}) \left\{ -\sum_{i=1}^{N_c} f_{i|j}^+(c_{n,k}|\xi_{n,k}) \ln f_{i|j}^+(c_{n,k}|\xi_{n,k}) \right\}. \tag{9.61}$$

For the Gaussian feature vector measurement model, we must compute $f^{-}(\boldsymbol{\xi}_{n,k})$ and $f_i^{+}(c_{n,k})$ as a function of $\boldsymbol{\xi}_{n,k}$ using (9.45) and (9.46). Using the notation $f_i^{+}(c_{n,k}|\boldsymbol{\xi}_{n,k}) \doteq P(c_{n,k} = i|\boldsymbol{\xi}_{n,k}, \boldsymbol{Z}_{k-1}; \boldsymbol{\Theta}_k)$, the conditional entropy is then computed from:

$$H_n^{\uparrow}(\boldsymbol{\theta}_k|\boldsymbol{Z}_{k-1};\boldsymbol{\Theta}_{k-1}) = \int f^{-}(\boldsymbol{\xi}_{n,k})\left\{-\sum_{i=1}^{N_c} f_i^{+}(c_{n,k}|\boldsymbol{\xi}_{n,k}) \ln f_i^{+}(c_{n,k}|\boldsymbol{\xi}_{n,k}) d\boldsymbol{\xi}_{n,k}\right\}. \quad (9.62)$$

The integral in (9.62) does not have a closed form expression and must be evaluated numerically or approximated.

The QoS classification accuracy metric is the conditional entropy, $G_{n,k}^{C}(\boldsymbol{\theta}_k) = H_n^{\uparrow}(\boldsymbol{\theta}_k|\boldsymbol{Z}_{k-1};\boldsymbol{\Theta}_{k-1})$, and the QoS requirement is the value that FARRA wants the conditional entropy to be below, denoted as $\bar{G}_n^{C}$. We then define the classification task utility function to be:

$$U_{n,k}^{C}(\boldsymbol{\theta}_k) = \begin{cases} \dfrac{\bar{G}_n^{C}}{G_{n,k}^{C}(\boldsymbol{\theta}_k)} & G_{n,k}^{C}(\boldsymbol{\theta}_k) > \bar{G}_n^{C} \\ 1 & G_{n,k}^{C}(\boldsymbol{\theta}_k) \le \bar{G}_n^{C} \end{cases} \quad (9.63)$$

The mission utility function is obtained by assigning weights to the task utility functions, which we denote as w_n^R, w_n^V, and w_n^C, and computing the weighted sum of the task utilities,

$$U(\boldsymbol{\theta}_k|\boldsymbol{Z}_{k-1};\boldsymbol{\Theta}_{k-1}) = \sum_{n=1}^{N} \left(w_n^R U_{n,k}^R(\boldsymbol{\theta}_k) + w_n^V U_{n,k}^V(\boldsymbol{\theta}_k) + w_n^C U_{n,k}^C(\boldsymbol{\theta}_k)\right). \quad (9.64)$$

9.4.2.2 Information-driven approach

In the information-driven approach, the relative merit of different sensing actions is measured by the corresponding expected gain in information [8,10,13,17,29]. Assume, temporarily, that at time t_k a FARRA strategy has selected action $\boldsymbol{\theta}_k$, it has been executed, and measurement z_k has been received. To judge the value of this action, we compute the information gained by that measurement; specifically the information gain between the predicted PDF on target state before the measurement was taken, $f^{-}(\boldsymbol{x}_k)$, and the posterior PDF after the measurement has been received, $f^{+}(\boldsymbol{x}_k)$. The most popular approach uses the KLD. The KLD between $f^{+}(\boldsymbol{x}_k)$ and $f^{-}(\boldsymbol{x}_k)$ is defined as [26]:

$$D(f^{+}(\boldsymbol{x}_k)||f^{-}(\boldsymbol{x}_k)) \doteq \int f^{+}(\boldsymbol{x}_k) \ln \frac{f^{+}(\boldsymbol{x}_k)}{f^{-}(\boldsymbol{x}_k)} d\boldsymbol{x}_k. \quad (9.65)$$

There are a number of generalizations of the KLD in the literature, including the Rényi divergence, the Arimoto-divergences, and the f-divergence [27–29]. The KLD has a number of desirable theoretical and practical properties, including (a) the ability to compare actions which generate different types of knowledge (e.g., knowledge about target class versus knowledge about target position) using a common measuring stick—information gain; (b) the asymptotic connection between information gain and risk-based optimization; and (c) the avoidance of weighting schemes to value different types of information. Taking the expectation with respect to $f^{-}(z_k)$, we obtain the expected KLD, which is also known as the MI [26]:

$$I_{xz}(\boldsymbol{\theta}_k|\boldsymbol{Z}_{k-1};\boldsymbol{\Theta}_{k-1}) \doteq E^{-}_{z_k}\left\{\int f^{+}(\boldsymbol{x}_k)\ln\frac{f^{+}(\boldsymbol{x}_k)}{f^{-}(\boldsymbol{x}_k)}d\boldsymbol{x}_k\right\}. \tag{9.66}$$

The next action vector is then determined by maximizing the mutual information,

$$\boldsymbol{\theta}_k = \underset{\boldsymbol{\theta}}{\text{argmax}}\ I_{xz}(\boldsymbol{\theta}|\boldsymbol{Z}_{k-1};\boldsymbol{\Theta}_{k-1}). \tag{9.67}$$

For our model, the global MI decomposes into the sum of the tracking and classification MIs, which we denote as $I_{yz;n}(\boldsymbol{\theta}_k|\boldsymbol{Z}_{k-1};\boldsymbol{\Theta}_{k-1})$ and $I_{c\xi;n}(\boldsymbol{\theta}_k|\boldsymbol{Z}_{k-1};\boldsymbol{\Theta}_{k-1})$, respectively,

$$I_{xz}(\boldsymbol{\theta}_k|\boldsymbol{Z}_{k-1};\boldsymbol{\Theta}_{k-1}) = \sum_{n=1}^{N}\left(I_{yz;n}(\boldsymbol{\theta}_k|\boldsymbol{Z}_{k-1};\boldsymbol{\Theta}_{k-1}) + I_{c\xi;n}(\boldsymbol{\theta}_k|\boldsymbol{Z}_{k-1};\boldsymbol{\Theta}_{k-1})\right). \tag{9.68}$$

The tracking MI has the form

$$I_{yz;n}(\boldsymbol{\theta}_k|\boldsymbol{Z}_{k-1};\boldsymbol{\Theta}_{k-1}) = \frac{1}{2}\ln|\boldsymbol{P}^{-}_{n,k}| + \frac{1}{2}\ln\left|\left[\boldsymbol{P}^{-}_{n,k}\right]^{-1} + \boldsymbol{H}^{T}_{n,k}\boldsymbol{R}_{n,k}(\boldsymbol{\theta}_k)^{-1}\boldsymbol{H}_{n,k}\right|. \tag{9.69}$$

The classification MI is the difference between the entropy of the predicted PMF in (9.59) and the conditional entropy in (9.61) or (9.62),

$$I_{c\xi;n}(\boldsymbol{\theta}_k|\boldsymbol{Z}_{k-1};\boldsymbol{\Theta}_{k-1}) = H^{-}(c_{n,k}) - H^{\uparrow}_{n}(\boldsymbol{\theta}_k|\boldsymbol{Z}_{k-1};\boldsymbol{\Theta}_{k-1}). \tag{9.70}$$

Comparing the second term in (9.69) to the expression for the PC-BIM in (9.52), we see that the tracking MI is a function of the determinant of the PC-BIM. Thus, the task-based and information-based methods developed here have at their core the same information theoretic quantities, and the distinction is in the separation and weighting of individual tasks in the task-based method versus a global approach in the information-based method.

9.5 Simulation results

We now demonstrate FARRA algorithm performance for concurrent tracking and classification of multiple targets using a single multimode radar sensor. We consider a scenario consisting of an airborne radar platform and three airborne targets, as illustrated in Figure 9.3. The radar platform is flying with a velocity of 200 m/s at an altitude of 12 km, Target 1 is 375 m/s at 12 km, Target 2 is 300 m/s at 13 km, and Target 3 is 200 m/s at 13 km. The scenario runs for 60 s.

The tracking state vector $\boldsymbol{y}_{n,k}$ is a ten-dimensional vector consisting of the position, velocity, and acceleration $(x_{n,k}, \dot{x}_{n,k}, \ddot{x}_{n,k}, y_{n,k}, \dot{y}_{n,k}, \ddot{y}_{n,k}, z, \dot{z}_{n,k}, \ddot{z}_{n,k})$, and the SNR in decibels, which we denote as $s_{n,k} = 10\log_{10}\zeta_{n,k}$, where $\zeta_{n,k}$ is the SNR in linear scale. The classification state variable $c_{n,k}$ is assumed to be one of $N_c = 5$ classes.

For tracking, we use a Singer model [41] for target motion. For classification, we assume the transition matrix has diagonal entries $[\boldsymbol{\Upsilon}_n]_{ii} = 0.95$ and off-diagonal entries $[\boldsymbol{\Upsilon}_n]_{ij} = 0.0125$.

For the tracking measurement model, we assume the radar transmits a waveform and receives returns through an antenna with a fixed azimuth beamwidth $(\Delta\phi)$

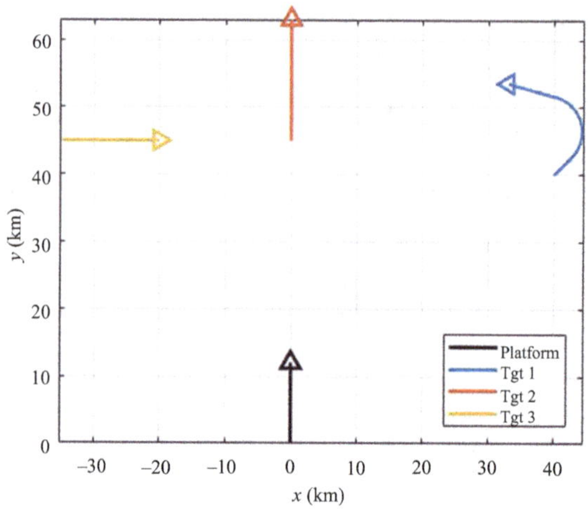

Figure 9.3 Airborne radar simulation scenario with an airborne radar platform and three airborne targets

and elevation beamwidth ($\Delta\theta$). The transmitted waveform is characterized by its center frequency (f_c), pulse bandwidth (B_p), pulse repetition frequency (PRF) (f_p), and number of pulses (N_p). The tracking measurement process results in detections which provide estimates of target range (R), range-rate ($\dot{R}$), azimuth angle (ϕ), elevation angle (θ), and SNR in decibels ($s = 10\log_{10}\zeta$). In this example, we assume f_c, $\Delta\phi$, $\Delta\theta$ are fixed and B_p, f_p, and N_p are adjustable. We also assume that the detection threshold and the P_F are fixed. Let $B_{p;n,k}, f_{p;n,k}$, and $N_{p;n,k}$ denote the parameters of the selected waveform for the nth target. The probability of detection is given by [42]:

$$P_D(\zeta_{n,k};\boldsymbol{\theta}_k) = Q_{MAR}\left(\sqrt{2N_{p;n,k}\zeta_{n,k}}, \sqrt{-2\ln P_F}\right), \tag{9.71}$$

where $Q_{MAR}(a,b)$ is the Marcum Q-function. The estimation covariance matrix is a diagonal matrix whose components are [43,44]:

$$\left[\boldsymbol{R}_{n,k}(\boldsymbol{\theta}_k)\right]_R = \left[2N_{p;n,k}\zeta_{n,k}\left(\frac{2}{c}\right)^2(3B_{p;n,k})^2\right]^{-1} \tag{9.72}$$

$$\left[\boldsymbol{R}_{n,k}(\boldsymbol{\theta}_k)\right]_{\dot{R}} = \left[2N_{p;n,k}\zeta_{n,k}\left(\frac{4\pi f_c}{c}\right)^2\left(\frac{1}{12B_{p;n,k}^2} + \frac{(N_{p;n,k}^2-1)}{12f_{p;n,k}^2}\right)\right]^{-1} \tag{9.73}$$

$$\left[\boldsymbol{R}_{n,k}(\boldsymbol{\theta}_k)\right]_\phi = \left[2N_{p;n,k}\zeta_{n,k}\left(\frac{1.782\pi}{\Delta\phi}\right)^2\right]^{-1} \tag{9.74}$$

$$\left[\boldsymbol{R}_{n,k}(\boldsymbol{\theta}_k)\right]_\theta = \left[2N_{p;n,k}\zeta_{n,k}\left(\frac{1.782\pi}{\Delta\theta}\right)^2\right]^{-1} \tag{9.75}$$

$$\left[\boldsymbol{R}_{n,k}(\boldsymbol{\theta}_k)\right]_s = \left(\frac{10}{\ln(10)}\right)^2, \tag{9.76}$$

where c is the speed of light.

Table 9.1 Waveform parameters and dwell times for airborne radar simulation

Waveform	Parameter(s)			Dwell time
0		N/A		0.0
#	B_p (MHz)	f_p (kHz)	N_p	T (ms)
1,2,3	1,5,10	20	1	0.05
4,5,6	1,5,10	10	1	0.1
7,8,9	1,5,10	20	10	0.5
10,11,12	1,5,10	10	10	1.0
13,14,15	1,5,10	20	20	1.0
16,17,18	1,5,10	10	20	2.0
19,20,21	1,5,10	20	50	2.5
22,23,24	1,5,10	10	50	5.0
#		p_{dc}		T (ms)
25		0.3		1.0
26		0.6		2.5
27		0.75		5.0

For classification, the radar transmits a waveform, processes the received data, and returns a discrete classification call. The transmitted waveform is characterized by the probability that the discrete classification call is correct, denoted by p_{dc}. Let $p_{dc;n,k}$ denote the value corresponding to the selected waveform for the nth target. The classification likelihood matrix has the form:

$$[\boldsymbol{L}_n]_{ij}(\boldsymbol{\theta}_k) = \begin{cases} p_{dc;n,k} & i = j \\ \dfrac{1 - p_{dc;n,k}}{N_c - 1} & i \neq j. \end{cases} \tag{9.77}$$

We assume the RRA frame length is the same as the track update interval, which is $T_F = \Delta t = 100\,\text{ms}$. During the frame, the radar sensor must allocate resources to a surveillance task for detecting new targets and to tracking and classification tasks for each of the known targets. We assume 90 ms are used for surveillance (search) dwells and the remaining 10 ms are for tracking and classification dwells. In this study, we focus on the tracking and classification tasks and consider the surveillance task only by allocating it a fixed amount of the RRA frame time, thus restricting the time available for the tracking and classification tasks.

The available waveforms, their parameters, and dwell times are listed in Table 9.1. Also included is the "do nothing" waveform #0. The fixed tracking waveform parameters are $f_c = 3\text{GHz}$, $\Delta\phi = 2°$, $\Delta\theta = 6°$, and $P_F = 10^{-6}$.

The FARRA algorithm may elect to measure each target during the dwell or any subset of the targets as long as the total measurement time fits into the allocated time budget. For each target, the sensor may select from the following options:

- **Do nothing**: Choose waveform #0. This takes zero time and generates zero utility. It frees up the timeline to dedicate extra dwell time to other targets.

- **Perform a track dwell**: Choose from waveforms #1 – 24. This takes variable time given by N_p/f_p and provides variable utility depending on the waveform parameters.
- **Perform a classification dwell**: Choose from waveforms #25 – 27. This takes variable time and provides variable utility.

For the QoS metric, the position RMSE requirement for Target 1 is $\bar{G}_1^R = 100\,\text{m}$ and the velocity RMSE requirement is $\bar{G}_1^V = 20\,\text{m/s}$. For Targets 2 and 3, the position and velocity requirements are $\bar{G}_2^R = \bar{G}_3^R = 200\,\text{m}$ and $\bar{G}_2^V = \bar{G}_3^V = 60\,\text{m/s}$. The classification goal for Target 1 is that the posterior probability of the correct class is at least 0.6, while the classification goal for Targets 2 and 3 is that this probability is at least 0.8. When there are five target classes, these posterior probability goals roughly correspond to entropy values of 1.2 and 0.8, respectively. Therefore, we set $\bar{G}_1^C = 1.2$ and $\bar{G}_2^C = \bar{G}_3^C = 0.8$. The requirements are equally weighted, $w_n^R = w_n^V = w_n^C = 1/9; n = 1, 2, 3$.

The predicted utility of a sensing action is scored using either the task-driven metric in (9.64) or the information-driven metric in (9.68). The objective is then maximized subject to the timeline constraint.

The simulation is repeated for 1,000 Monte-Carlo trials for each method. The trials have the sensor and target trajectories fixed as shown in Figure 9.3, but a random realization of the measurements is drawn anew each time. This, in turn, affects the adaptive resource allocation calculations leading to different allocations and performance each time.

Figure 9.4 shows the position and velocity RMSEs and Figure 9.5 shows the classification entropy and the posterior probability of the correct class for each method. Also shown are the performance goals used in the task-driven method. These are not used in the information-driven method, but are shown for reference. Figure 9.6 shows the MI for each method. This is not used in the task-driven method, but is shown for reference. Figure 9.7 shows how the resource allocation algorithm selected to use the sensors over time by looking at what fraction of the 10 ms frame is used for tracking and classification dwells for each target at each time.

Figures 9.4 and 9.5 show that the task-driven and information-driven FARRA algorithms produce similar RMSEs and posterior probability of correct classification for the three targets. In the task-driven method, Targets 2 and 3 always meet their performance goals, while Target 1 is only able to meet its classification goal. In the information-driven method, where performance goals are not considered, Target 1 RMSE values are slightly higher than the task-driven method, and Target 2 and 3 RMSE values are considerably lower than the task-driven method. The information-based classification entropies are essentially the same for all three targets. The task-based entropy is higher than the information-based entropy for Target 1 and lower than the information-based entropy for Targets 2 and 3. The information-driven method maximizes the total mutual information and does so by making the mutual information approximately the same for each target, as shown in Figure 9.6. The task-driven method does not consider mutual information and achieves a slightly higher value

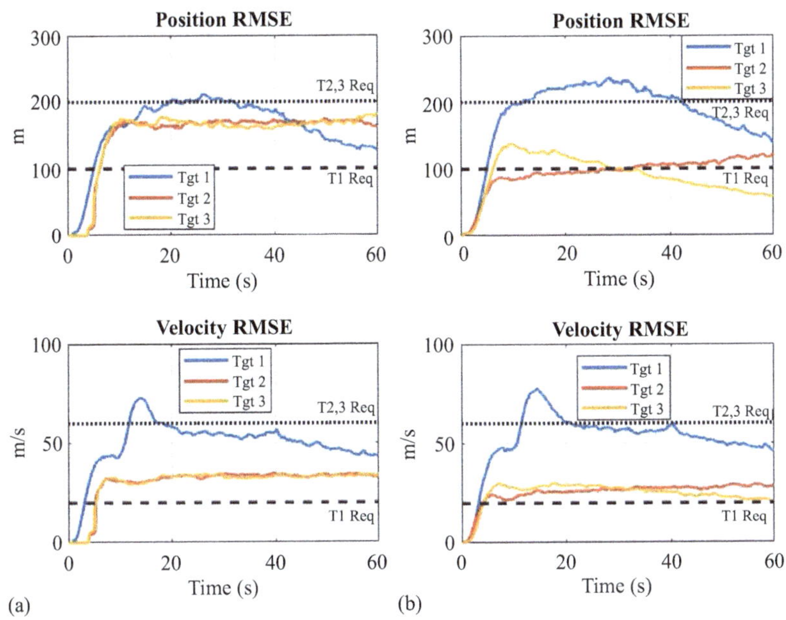

Figure 9.4 Tracking position and velocity RMSE of three targets in airborne radar simulation scenario: (a) task-driven method (left panels) and (b) information-driven method (right panels)

than the information-driven method for Target 1 but lower values (on average) than the information-driven method for Targets 2 and 3.

The methods the scheduling algorithms deploy to reach the roughly equal tracking and classification performance are different, as shown in Figure 9.7. Broadly speaking, we find that both approaches interleave tracking and classification, with more classification dwells during the first portion of the simulation to maintain track accuracy but also to learn about the target class. After that, classification dwells are taken periodically to main classification performance. For tracking, the task-driven method typically measures one target at 5 ms and the other two at 2.5 ms. In contrast, the information-driven approach prefers to make 5 ms dwells. It does this by typically measuring two targets at 5 ms and skipping one target. This generates larger P_D dwells for two targets at the expense of not measuring a third. In the task-driven method, the algorithm is not able to meet the RMSE performance goals for Target 1, but tries very hard by taking tracking measurements about 85% of the time and classification measurements about 15% of the time. For Targets 2 and 3, the RMSE goals are met easily, and less time is spent on tracking measurements. In the information-driven method, the allocation of tracking, classification, and no measurement dwells is roughly the same for all three targets at approximately 60%, 25%, and 20%, respectively.

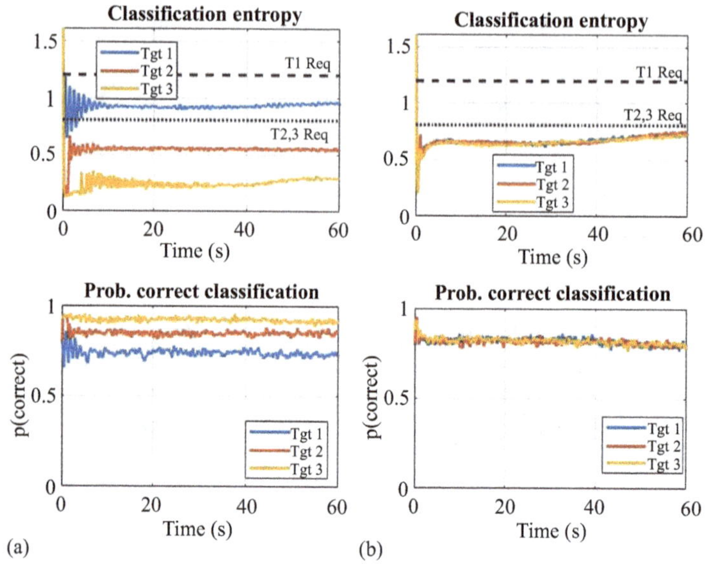

Figure 9.5 Classification entropy and posterior probability of correct classification of three targets in airborne radar simulation scenario: (a) task-driven method (left panels) and (b) information-driven method (right panels)

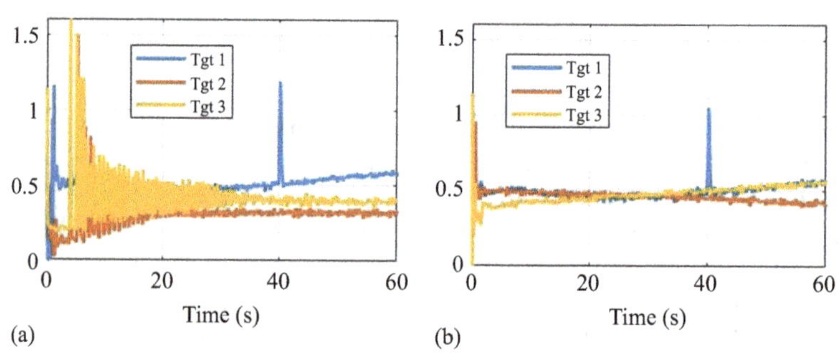

Figure 9.6 Mutual information of three targets in airborne radar simulation scenario: (a) task-driven method (left) and (b) information-driven method (right)

In this example, we have evaluated the performance of task-based and information-based FARRA methods for a scenario involving concurrent tracking and classification of multiple airborne targets using a single airborne radar platform and showed that although the two methods opt for different sensor usage strategies, they in fact have similar performance.

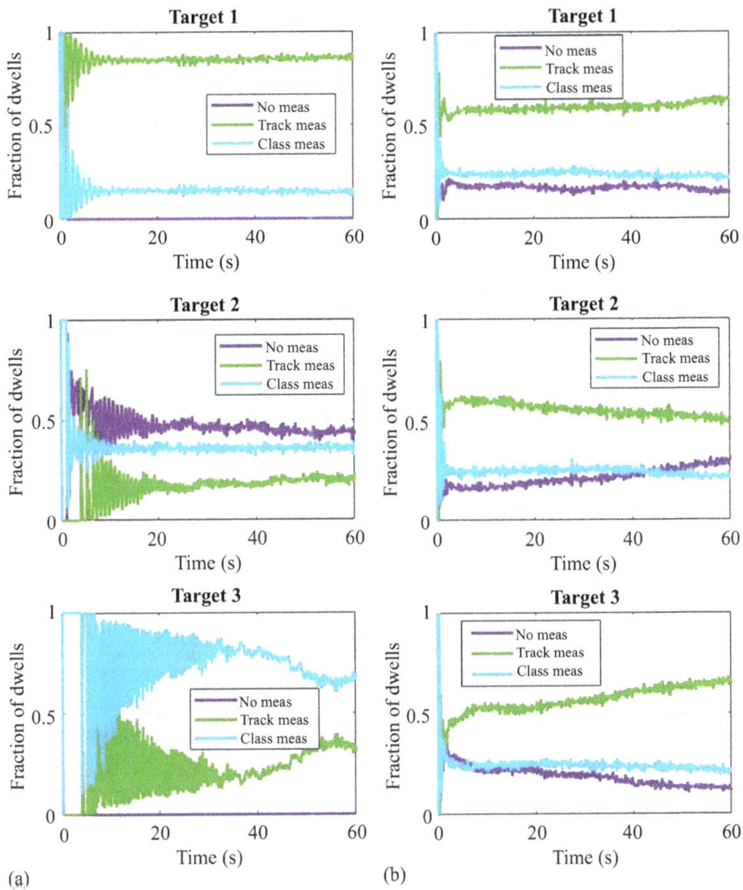

Figure 9.7 *FARRA selection of waveforms by three targets in airborne radar simulation scenario: (a) task-driven method (left panels) and (b) information-driven method (right panels)*

9.6 Experimental results

In this section, we demonstrate FARRA algorithm performance for concurrent tracking and classification of a single target using a single radar sensor in the CREW testbed. The CREW is a waveform agile, millimeter-wavelength, multistatic radar system designed by OSU specifically to test cognitive and FAR principles [33]. The scenario is illustrated in Figure 9.8. As the human target moves back and forth in the laboratory, the tracking task estimates the target's range and velocity and the classification task separately assigns one of the three motion classes: "walking," "jogging/running," and "punching" to the observed target.

The tracking model is the same as in [31]. The state vector $\boldsymbol{y}_k$ is a three-dimensional vector consisting of the target's range (R), velocity (V), and the

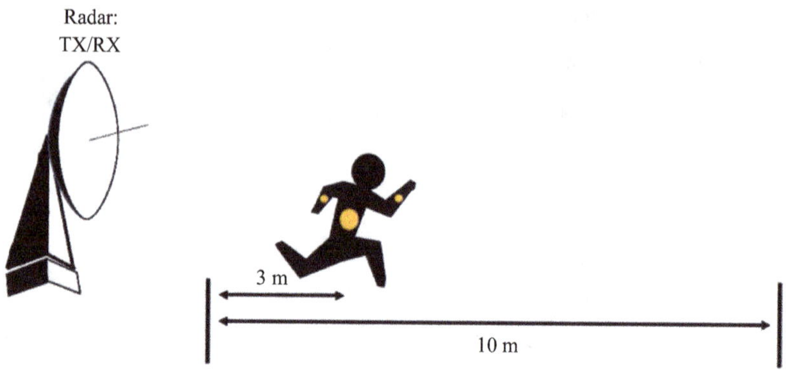

Figure 9.8 CREW experimental scenario with a single human target

Table 9.2 Waveform parameters for CREW experiment

Parameter	Number of values	Adaptive value range
B_p	10	100–1,000 MHz
τ_p	10	0.1–1.0 μs
f_p	10	2–15 kHz
N_p	10	64–640

pulse-integrated SNR in linear scale ($S = N_p\zeta$). We use a nearly constant velocity target motion model in the tracker with an empirically determined process noise covariance matrix [31]. The classification state variable c_k is assumed to be one of the $N_c = 3$ classes described earlier and the transition matrix has diagonal entries $[\boldsymbol{\Upsilon}_n]_{ii} = 0.99$ and off-diagonal entries $[\boldsymbol{\Upsilon}_n]_{ij} = 0.005$.

During each dwell, the CREW radar transmits linear frequency modulation (LFM) waveform pulses, where the available waveforms consist of different combinations of allowable LFM bandwidths (B_p), pulse widths (τ_p), PRFs (f_p), and number of pulses (N_p). The CREW can set these parameters with very fine precision, thus the number of available waveforms is too large to enumerate as we did in Table 9.1. Here we identify the chosen waveform using a sensor parameter vector defined in terms of the four adjustable parameters, i.e.,

$$\boldsymbol{\theta}_k = \left[B_{p;k}\ \tau_{p;k}\ f_{p;k}\ N_{p;k}\right]^T. \tag{9.78}$$

The range of values of the four adjustable parameters is given in Table 9.2. In this experiment, we limited each parameter to ten values. The CREW can also adjust its transmit power (P_t) on each dwell; however, in this example, we hold it fixed. The fixed transmit power is $P_t = 11.5$ dBm and the center frequency is $f_c = 95.5$ GHz.

For the tracking task, the data is range–Doppler processed and detection-level data of target range (R), Doppler frequency (F), and pulse-integrated SNR (S) are obtained. The transmit power is high enough that the probability of detection is equal to one. The estimation covariance matrix is a diagonal matrix whose components are [31]:

$$[\boldsymbol{R}_k(\boldsymbol{\theta}_k)]_R = \left[C_R S_k B_{p;k}^2\right]^{-1} \tag{9.79}$$

$$[\boldsymbol{R}_k(\boldsymbol{\theta}_k)]_F = \left[C_F S_k \left(\frac{N_{p;k}}{f_{p;k}}\right)^2\right]^{-1} \tag{9.80}$$

$$[\boldsymbol{R}_k(\boldsymbol{\theta}_k)]_S = \left[C_S N_{p;k}\right]^{-1}, \tag{9.81}$$

where the constants C_R, C_F, and C_S were determined through empirical data analysis [31].

For the classification task, the data is processed to produce a two-dimensional micro-Doppler feature vector. As described in [35], the classification feature vector is obtained by processing the I/Q data from the CREW sensor to produce a range profile and locate the target range bin. A spectrogram is then computed using the short-time Fourier transform, and a high-dimensional feature vector is formed from 500 normalized spectral samples. This was repeated many times to obtain a set of training data, and multiple discriminant analysis (MDA) was performed to obtain a matrix for projection of the full 500-dimensional feature vector down to a two-dimensional (2D) feature vector. The projection matrix was then stored for later use. This was done for two training datasets collected with two different radar parameter settings, which we denote as waveforms 1C and 2C, shown in Table 9.3.

Figure 9.9 shows the 2D feature vectors obtained after MDA and projection to 2D space for the two training datasets. Note that the feature space vectors are unique up to an arbitrary angular rotation in 2D space, thus this has to be recognized and ignored when comparing the datasets. The plots show that waveform 2C produces much tighter class clusters than waveform 1C, thus it provides a higher level of classification performance, but uses a larger bandwidth and dwell time. In particular, the walking (legs moving) and punching (hands moving) classes of waveform 1C have a large overlap resulting in increased classification uncertainties for these classes.

The means and covariance matrices of the 2D feature vectors shown in Figure 9.9 are determined from the sample mean and covariance of the clusters. The mean is the center of the cluster and the covariance is represented by the 2σ error ellipse

Table 9.3 Waveform parameters and dwell times for CREW classification experiment

Waveform	**Parameter(s)**				**Dwell time**
#	B_p (MHz)	τ_p (μs)	f_p (kHz)	N_p	T (ms)
1C	300	0.5	4.889	64	13.1
2C	1000	0.5	15.0	512	34.1

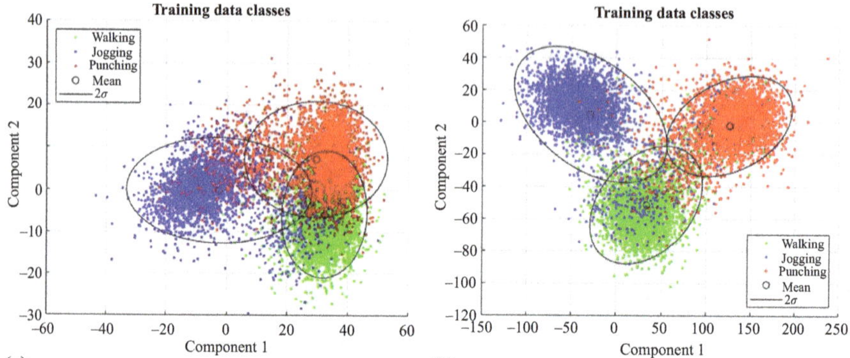

(a) (b)

Figure 9.9 Feature space comparison of CREW measured target returns for two training datasets: (a) waveform 1C and (b) waveform 2C

overlaid on the data. These are the values of $\boldsymbol{\mu}_i(\boldsymbol{\theta}_k)$ and $\boldsymbol{\Sigma}_i(\boldsymbol{\theta}_k)$ used in the likelihood function in (9.29).

The FARRA algorithm must decide the sensor parameters to use during each dwell. The tracking task processes data from every dwell and tracking information updates are always performed regardless of sensor parameter settings used to collect data, however, classification can be performed only when the data is collected using the sensor parameter settings for waveforms 1C and 2C. If it chooses parameters to optimize tracking performance, then no data may be provided to the classification task and there will be no classification information update. If it chooses parameters to optimize classification performance, then data is still provided to the tracker, but it may be of less utility. There is no "do nothing" option in this experiment.

Let $\Delta t(\boldsymbol{\theta}_k)$ denote the measurement update interval. It depends on the dwell time given by N_p/f_p and the processing time, t_{proc},

$$\Delta t(\boldsymbol{\theta}_k) = \frac{N_{p;k}}{f_{p;k}} + t_{proc}. \tag{9.82}$$

For this example, we use the multi-objective optimization cost function approach to develop a task-based FARRA objective function [25]. This approach specifies cost functions rather than utility functions, and the objective function is minimized. The tracking QoS metrics are the range and velocity RMSEs obtained from the PC-CRLB and the classification QoS metric is the conditional entropy:

$$G_k^R(\boldsymbol{\theta}_k) = \sqrt{\left[\tilde{\boldsymbol{C}}_k^{\uparrow}(\boldsymbol{\theta}_k)\right]_R} \tag{9.83}$$

$$G_k^V(\boldsymbol{\theta}_k) = \sqrt{\left[\tilde{\boldsymbol{C}}_k^{\uparrow}(\boldsymbol{\theta}_k)\right]_V} \tag{9.84}$$

$$G_k^C(\boldsymbol{\theta}_k) = H^{\uparrow}(\boldsymbol{\theta}_k|\boldsymbol{Z}_{k-1};\boldsymbol{\Theta}_{k-1}). \tag{9.85}$$

The QoS requirements are the values that we want the RMSEs and conditional entropy to be below, denoted as $\bar{G}^R$, $\bar{G}^V$, and $\bar{G}^C$. We then define the position, velocity, and entropy cost functions to be:

$$C_k^R(\boldsymbol{\theta}_k) = \begin{cases} \dfrac{G_k^R(\boldsymbol{\theta}_k) - \bar{G}^R}{\bar{G}^R} & G_k^R(\boldsymbol{\theta}_k) > \bar{G}^R \\ 0 & G_k^R(\boldsymbol{\theta}_k) \leq \bar{G}^R \end{cases} \tag{9.86}$$

$$C_k^V(\boldsymbol{\theta}_k) = \begin{cases} \dfrac{G_k^V(\boldsymbol{\theta}_k) - \bar{G}^V}{\bar{G}^V} & G_k^V(\boldsymbol{\theta}_k) > \bar{G}^V \\ 0 & G_k^V(\boldsymbol{\theta}_k) \leq \bar{G}^V \end{cases} \tag{9.87}$$

$$C_k^C(\boldsymbol{\theta}_k) = \begin{cases} \dfrac{G_k^C(\boldsymbol{\theta}_k) - \bar{G}^C}{\bar{G}^C} & G_k^C(\boldsymbol{\theta}_k) > \bar{G}^C \\ 0 & G_k^C(\boldsymbol{\theta}_k) \leq \bar{G}^C. \end{cases} \tag{9.88}$$

With these cost functions, if the QoS metric is below the required value, the resulting cost is zero and there is neither a penalty nor any additional utility for being below the requirement. The mission processing cost function is obtained by assigning weights to the task cost functions, which we denote as w_R, w_V, and w_C, and computing the weighted sum of the task costs,

$$C_P(\boldsymbol{\theta}_k) = w_R C_k^R(\boldsymbol{\theta}_k) + w_V C_k^V(\boldsymbol{\theta}_k) + w_C C_k^C(\boldsymbol{\theta}_k). \tag{9.89}$$

In this example, there is no hard constraint on the observation time; however, we define a measurement cost function to characterize user preferences for parameter selections. The preferred sensor parameter values are denoted as $\bar{B}_p$, $\bar{\tau}_p$, $\bar{f}_p$, and $\bar{N}_p$ and the measurement cost weights are denoted as w_B, w_τ, w_f, and w_N. The measurement cost function is defined as:

$$C_M(\boldsymbol{\theta}_k) = w_B \left| \frac{B_{p;k} - \bar{B}_p}{\bar{B}_p} \right| + w_\tau \left| \frac{\tau_{p;k} - \bar{\tau}_p}{\bar{\tau}_p} \right| + w_f \left| \frac{f_{p;k} - \bar{f}_p}{\bar{f}_p} \right| + w_N \left| \frac{N_{p;k} - \bar{N}_p}{\bar{N}_p} \right|. \tag{9.90}$$

Finally, the executive cost function is the sum of the measurement and processor cost functions,

$$C_E(\boldsymbol{\theta}_k | \boldsymbol{Z}_{k-1}; \boldsymbol{\Theta}_{k-1}) = C_M(\boldsymbol{\theta}_k) + C_P(\boldsymbol{\theta}_k). \tag{9.91}$$

The next action vector is then determined by minimizing the executive cost function

$$\boldsymbol{\theta}_k = \underset{\boldsymbol{\theta}}{\operatorname{argmin}} \; C_E(\boldsymbol{\theta} | \boldsymbol{Z}_{k-1}; \boldsymbol{\Theta}_{k-1}). \tag{9.92}$$

For the QoS metric, the range and velocity RMSE requirements are $\bar{G}^R = 0.1$ m and $\bar{G}^V = 0.1$ m/s. The classification requirement is $\bar{G}^C = 0.3$, which corresponds to the posterior probability of the correct class being about 0.93. The weights are $w_R = 1$, $w_V = 1$, and $w_C = 0.9$. Following [25], we chose the measurement goal values and weights to favor timeline minimization. The time required to update a target track is the sum of the dwell time and the processing time, with the dwell time being the dominant factor. We set the goal PRF to the highest value, $\bar{f}_p = 15$ kHz, and the goal number of pulses to the lowest value, $\bar{N}_p = 64$. The processing time is a function of the amount of data the radar has to process. The number of fast-time samples collected with each pulse increases with pulse width so small pulse width is preferable and we set the goal to the lowest value, $\bar{\tau}_p = 0.1\,\mu\text{s}$. Similarly, decreasing the waveform bandwidth decreases the Nyquist sampling rate, so lower sampling frequencies can be used. This decreases the number of samples per measurement, but only if the sampling frequency is adjustable. This is not a feature of the CREW system, but the bandwidth was still included in the optimization to demonstrate how a flexible system might benefit, and the bandwidth goal was set to the lowest value, $\bar{B}_p = 100\,\text{MHz}$. Both pulse width and bandwidth impacted the overall track update time less than the dwell time, so they were given weights $w_B = w_\tau = 0.02$, while PRF and N_p were given weights of $w_f = 0.4$ and $w_N = 0.2$, respectively. The PRF weight was double the N_p weight to reflect an additional preference for higher PRFs to avoid Doppler aliasing.

In this example, we also explored incorporating measurement cost and task weighting in the information-based approach. We created cost functions for the tracking and classification tasks, assigned weights to the tasks which, we denote as w_T and w_C, and summed to form the processor cost function,

$$C_k^T(\boldsymbol{\theta}_k) = -I_{yz}(\boldsymbol{\theta}_k|\boldsymbol{Z}_{k-1};\boldsymbol{\Theta}_{k-1}) \tag{9.93}$$

$$C_k^C(\boldsymbol{\theta}_k) = -I_{c\xi;n}(\boldsymbol{\theta}_k|\boldsymbol{Z}_{k-1};\boldsymbol{\Theta}_{k-1}) \tag{9.94}$$

$$C_P(\boldsymbol{\theta}_k) = w_T C_k^T(\boldsymbol{\theta}_k) + w_C C_k^C(\boldsymbol{\theta}_k). \tag{9.95}$$

Note that the costs defined above can be less than zero. We also incorporate the measurement cost function defined in (9.90), this time weighted by w_M,

$$C_E(\boldsymbol{\theta}_k|\boldsymbol{Z}_{k-1};\boldsymbol{\Theta}_{k-1}) = w_M C_M(\boldsymbol{\theta}_k) + C_P(\boldsymbol{\theta}_k). \tag{9.96}$$

For this example, we chose $w_T = 1$, $w_C = 6$, and $w_M = 0.1$. With these weights, the weighted tracking, classification, and measurement costs are roughly the same order of magnitude.

The predicted cost of a sensing action is scored using either the task-based metric in (9.91) or the information-based metric in (9.96). The cost is then minimized. Our computational approach is to use MATLAB®'s "fmincon" sequential quadratic programming algorithm in the Optimization Toolbox. Despite the discrete set of available parameters in the waveform library, the optimization was solved over a continuous parameter space, and each parameter final solution was rounded up to the nearest available value. The rounding approach results in an overspending of resources, but we preferred this solution because the continuous space optimizations were faster than explicitly solving the discrete parameter problem.

The target in this example initially jogs away from the sensor for 4 s, then stops and punches for 4 s, then walks away from the sensor for 4 s, then reaches the maximum range and stops and punches for 4 s, then jogs toward the sensor for 4 s, then reaches the minimum range and stops and punches for 4 s.

Figure 9.10 shows the measurement cost, processor cost, and executive cost vs. time and Figure 9.11 shows the sensor parameters vs. time. The tradeoff between processor and measurement cost is evident in the plots. Most of the time, the executive

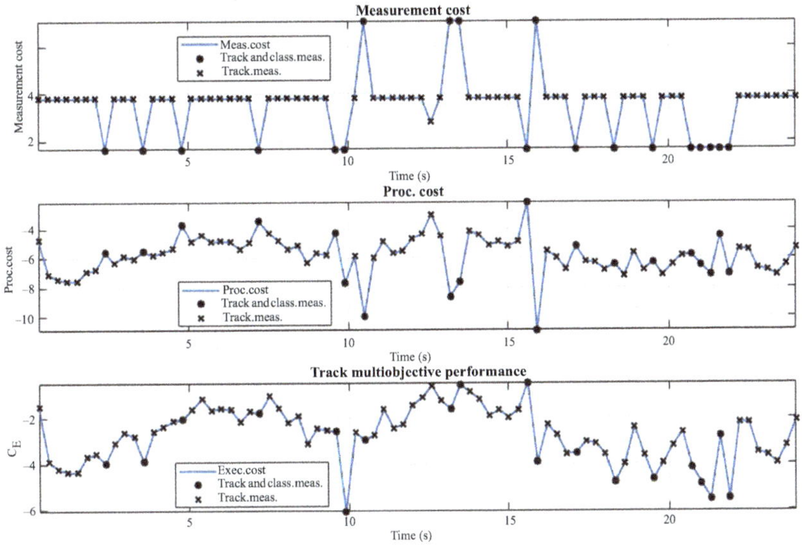

Figure 9.10 Information-based FARRA cost functions in CREW experiment; measurement cost (top panel), processor cost (middle panel), and executive cost (bottom panel)

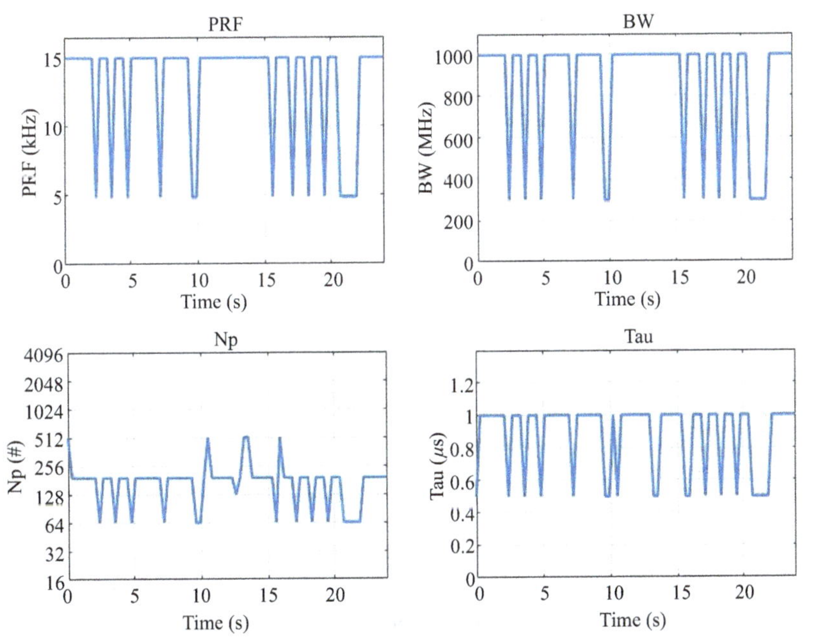

Figure 9.11 Information-based FARRA waveform parameter selections in CREW experiment

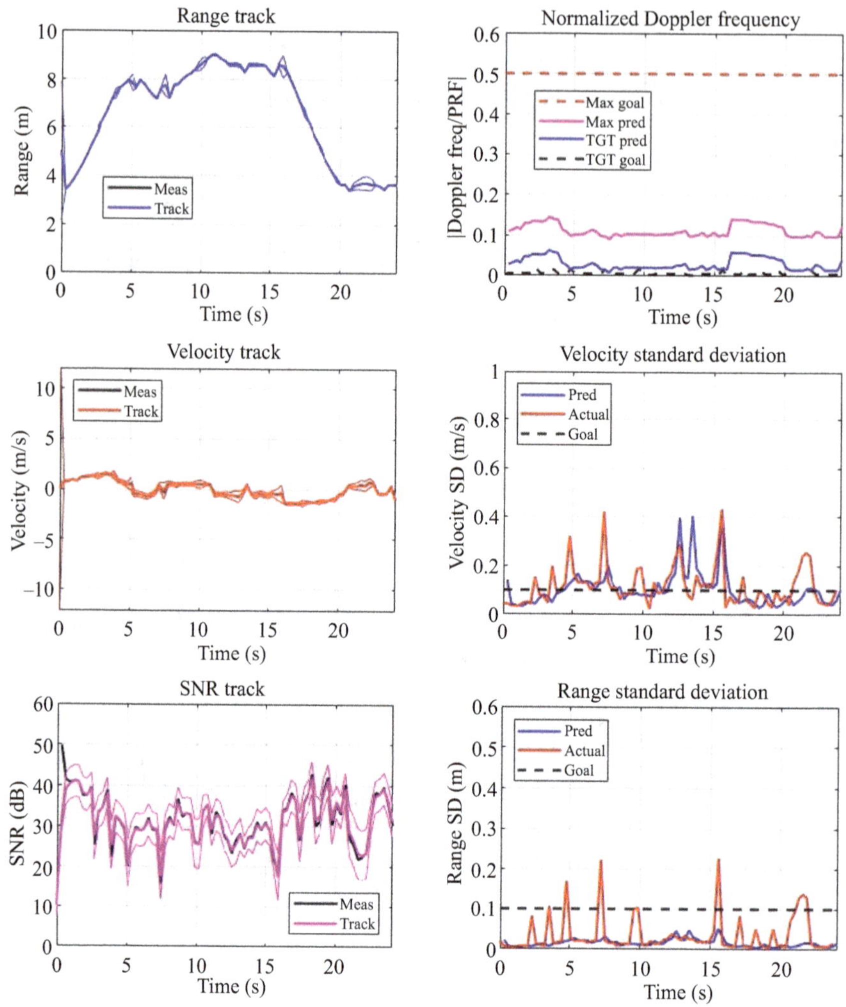

Figure 9.12 Information-based FARRA tracking performance in CREW experiment; range, velocity, and SNR tracks (left panels), Doppler clutter/ambiguity avoidance (top right panel), velocity RMSE (middle right panel), and range RMSE (lower right panel)

cost is minimized by choosing a tracking waveform with moderate measurement cost and moderate processor cost. However, when the posterior entropy in the classification task gets too high, the classification cost becomes the dominant factor and the executive cost is minimized by choosing a classification waveform. When waveform 1C is chosen, the measurement cost is low and the processor cost is high, and when waveform 2C is chosen, the measurement cost is high and the processor cost is low.

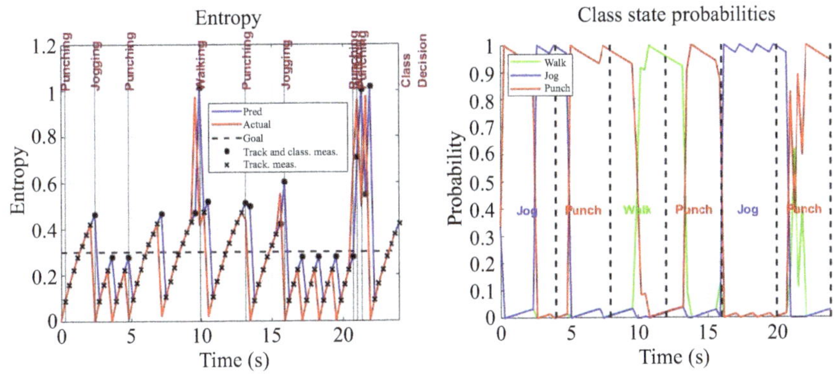

Figure 9.13 Information-based FARRA classification performance in CREW experiment; entropy (left panel) and posterior class probabilities (right panel)

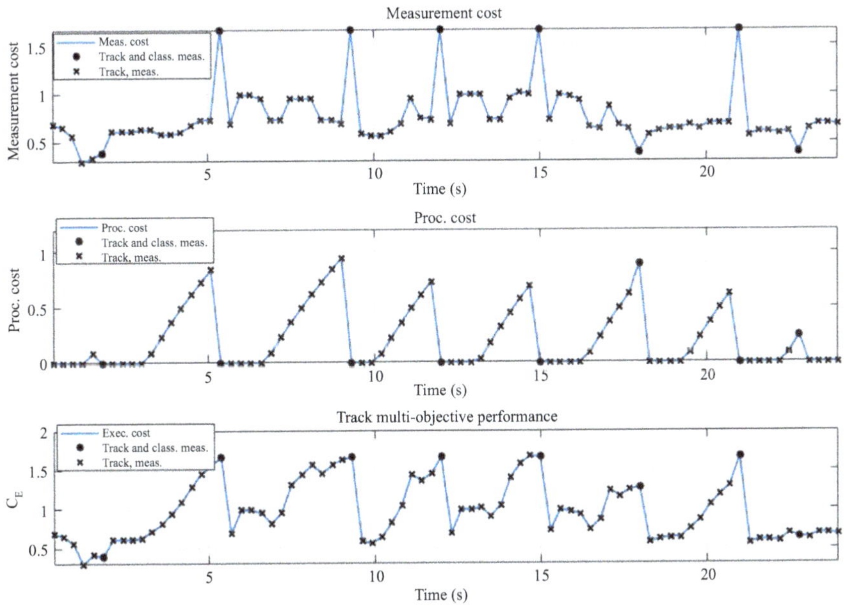

Figure 9.14 Task-based FARRA cost functions in CREW experiment; measurement cost (top panel), processor cost (middle panel), and executive cost (bottom panel)

Figure 9.12 shows the tracking task performance. The three plots in the first column on the left show the range, velocity, and SNR tracks. The three plots in the second column show Doppler clutter/ambiguity avoidance, velocity RMSE compared to the task-based requirement, and range RMSE compared to the task-based requirement.

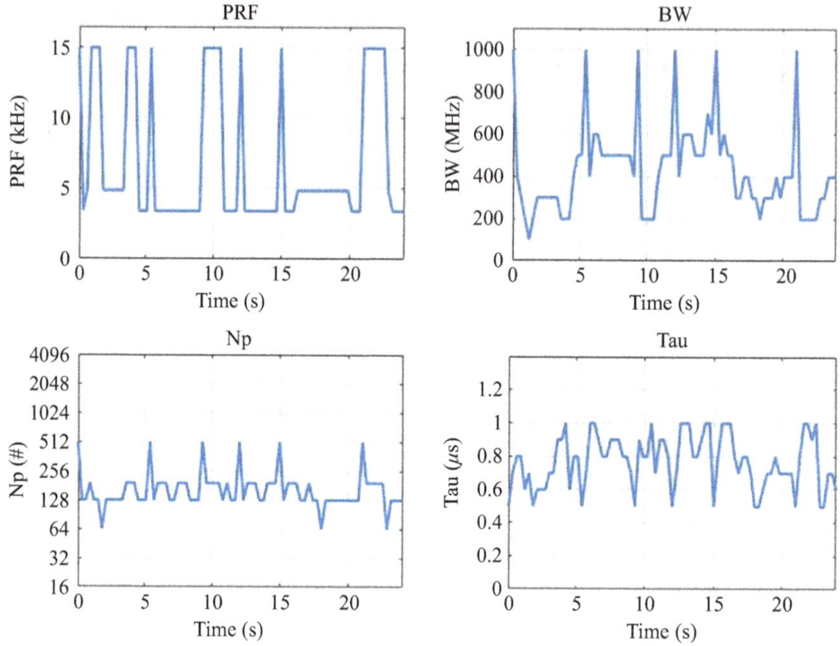

Figure 9.15 Task-based FARRA waveform parameter selections in CREW experiment

Because the information-based FARRA algorithm made no attempt to meet the tracking requirements, the velocity RMSE exceeded the requirement most of the time and the range RMSE was well below the requirement most of the time.

Figure 9.13 shows the classification task performance. The left plot shows the posterior entropy compared to the task-based requirement and the right plot shows the posterior class probabilities. The information-based algorithm makes no attempt to meet the requirement, and we see that the entropy gradually increases until a classification measurement is received, then drops significantly because the classifier is usually able to determine the correct class with high probability with just a single measurement. The correct class probabilities lag the class changes due to the timing of the received measurements. Waveform 1C has very low measurement cost, so it is chosen more often than waveform 2C. When jogging is the previous state, waveform 1C is more likely to be chosen. When walking/punching is previous state, waveform 2C is more likely to be chosen. This is related to the class separations in Figure 9.9. Sometimes when waveform 1C is chosen, the result is uncertain and then waveform 2C is chosen to get a better classification measurement. The frequency of the classification measurements and the waveform selected will be impacted by the weights chosen for the objective function, and different performance can be obtained by varying these weights.

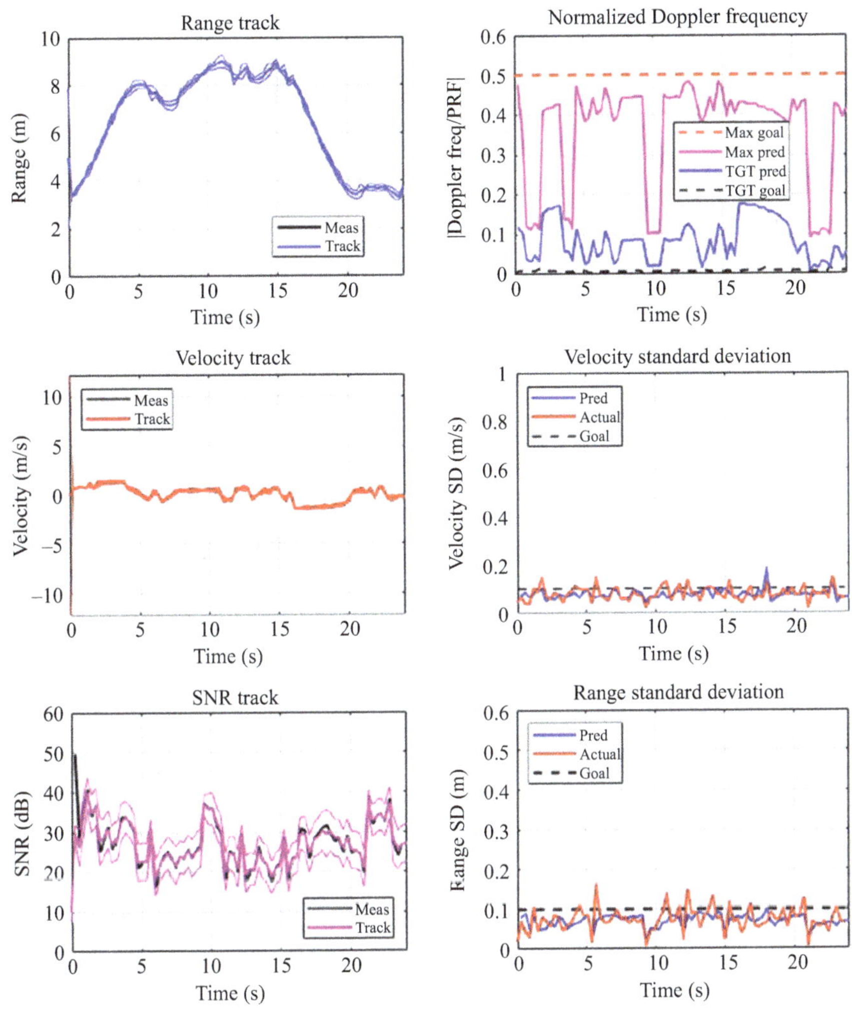

Figure 9.16 Task-based FARRA tracking performance in CREW experiment; range, velocity, and SNR tracks (left panels), Doppler clutter/ambiguity avoidance (top right panel), velocity RMSE (middle right panel), and range RMSE (lower right panel)

In comparison, Figure 9.14 shows the cost functions vs. time for the task-based FARRA algorithm and Figure 9.15 shows the waveform selections. Figure 9.16 shows the tracking task performance and Figure 9.17 shows the classification task performance. Here we see the same tradeoff between processor and measurement cost. Most of the time, the executive cost is minimized by choosing a tracking waveform with low measurement cost and low processor cost. When the posterior entropy in

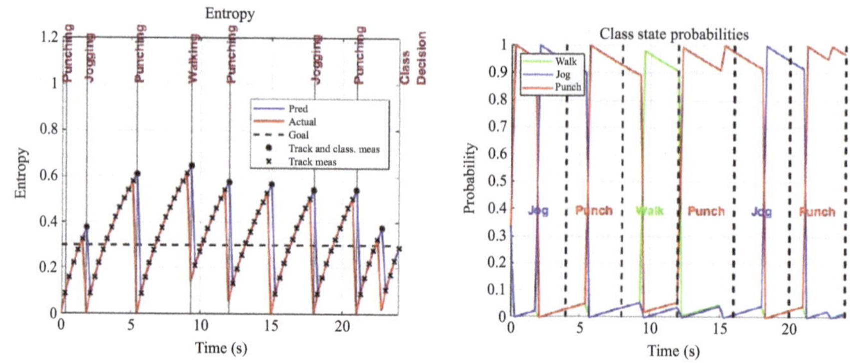

Figure 9.17 Task-based FARRA classification performance in CREW experiment; entropy (left panel) and posterior class probabilities (right panel)

the classification task gets too high, the classification cost becomes the dominant factor and the executive cost is minimized by choosing waveform 2C. In this case, the range and velocity RMSEs are kept very close to the requirements by choosing tracking waveforms most of the time with parameters that vary. Again we see that the entropy gradually increases until a classification measurement is received. Again, the correct class probabilities lag the class changes due to the timing of the received measurements.

In this example, we have evaluated the performance of task-based and information-based FARRA methods for tracking and classification of a single human target using a single radar sensor in the CREW testbed. The two methods use different waveform selection strategies and achieve different performance, especially in the tracking task, where the requirements are closely met in the task-based algorithm.

9.7 Conclusion

In this work, we demonstrated FARRA algorithm performance for concurrent tracking and classification of multiple targets using a single radar sensor. The FARRA approach is based on the PAC of cognition and includes a perceptual processor that performs multiple radar system tasks and an executive processor that allocates system resources to the tasks to decide the next transmission of the radar on a dwell-by-dwell basis. This formulation allowed us to allocate not only radar timeline and power but to adjust waveform transmission parameters on a dwell-by-dwell basis to achieve performance objectives for each task. We used a simulation to model a scenario consisting of an airborne radar platform and multiple airborne targets and the CREW experimental testbed to model a scenario consisting of a single moving target and a single stationary sensor. In both cases, we presented examples to demonstrate the application of task-based and information-based FARRA algorithms to simultaneous tracking and classification.

We showed that the task and information-based algorithms were actually based on the same information-theoretic quantities, and the examples showed that the two methods had similar tracking and classification performance but selected different waveform parameter sets to achieve their solutions. Furthermore, the task-based and information-based algorithms had essentially the same computational complexity since their objective functions were based on the same fundamental quantities and they used the same methods for solving the executive processor optimization problem.

In our scenarios, we assumed that the targets were spatially separated and there was no spectral interference. Future investigations will need to consider spectrum sharing, as discussed in Chapter 14 “Cognitive radar and spectrum sharing.” We also assumed a fixed waveform library with basic LFM waveforms and a steered beam antenna. Future work might also consider more sophisticated adaptive waveform design and beamforming techniques, such as those discussed in Chapters 4–8.

With our current optimization methodology, the complexity of the solution space grows exponentially with the number of tasks. Developing efficient techniques for solving the optimization problem will be of critical importance and particular attention will need to be given to converting the exponential dependence on the number of tasks to a linear dependence if a real-time implementation is to be achieved for the more complex scenarios expected in practice. Reformulating the optimization problem using the techniques discussed in Chapter 5 “Convex optimization for cognitive radar” and using neural networks to perform the computations [45] (see also Chapter 12 “The role of neural networks in cognitive radar”) are under investigation for this purpose.

The work presented in this chapter presents a rigorous methodology for allocating resources and selecting transmission parameters for multiple competing tasks. The examples demonstrate that an effective solution can be achieved in real time for simple scenarios. Many challenges remain before cognitive radar principles will be realized in real-world systems, but the techniques described in this book offer promising solutions to those challenges.

Acknowledgment

This material is based upon work supported by the Air Force Research Laboratory (AFRL) under Contract no. FA8649-20-P-0940. Any opinions, findings, and conclusions or recommendations expressed in this material are those of the authors and do not necessarily reflect the views of AFRL.

References

[1] Haykin S, Xue Y, and Setoodeh M. Cognitive radar: step toward bridging the gap between neuroscience and engineering. *Proceedings of the IEEE*. 2012;100(11):3102–3130.

[2] Haykin S. *Cognitive Dynamic Systems: Perception-Action Cycle, Radar and Radio*. Cambridge: Cambridge University Press; 2012.

[3] Guerci JR. *Cognitive Radar: The Knowledge-Aided Fully Adaptive Approach*. Boston, MA: Artech House; 2010.

[4] Greco M, Gini F, Stinco P, *et al*. Cognitive radars: on the road to reality: progress thus far and possibilities for the future. *IEEE Signal Processing Magazine*. 2018;35(4):112–125.

[5] Gurbuz S, Griffiths H, Charlish A, *et al*. An overview of cognitive radar: past, present, and future. *IEEE Aerospace and Electronic Systems Magazine*. 2019;34(12):6–18.

[6] Fuster JM. *Cortex and Mind: Unifying Cognition*. Oxford: Oxford University Press; 2010.

[7] Hernandez M, Kirubarajan T, and Bar-Shalom Y. Multisensor resource deployment using posterior Cramer–Rao bounds. *IEEE Transactions on Aerospace and Electronic Systems*. 2004;40(2):399–416.

[8] Kreucher C, Kastella K, and Hero AO. Multi-target sensor management using alpha-divergence measures. In: *Proceedings of the Information Processing in Sensor Networks*; 2003. p. 209–222.

[9] Kreucher C, Hero AO, Kastella K, *et al*. Efficient methods of non-myopic sensor management for multitarget tracking. In: *Proceedings of the 43rd IEEE Conference on Decision and Control*; 2004. p. 722–727.

[10] Kreucher C, Hero A, and Kastella K. A comparison of task driven and information driven sensor management for target tracking. In: *Proceedings of the 44th IEEE Conference on Decision and Control*; 2005. p. 4004–4009.

[11] Kreucher CM, Hero AO, Kastella KD, *et al*. Information-based sensor management for simultaneous multitarget tracking and identification. In: *Proceedings of the 13th Conference on Adaptive Sensor Array Processing*; 2005.

[12] Kreucher C and Hero A. Non-myopic approaches to scheduling agile sensors for multistage detection, tracking and identification. In: *Proceedings of the IEEE International Conference on Acoustics, Speech, and Signal Processing*; 2005. p. 885–888.

[13] Kreucher C, Kastella K, and Hero AO. Multi-platform information-based sensor management. *Proceedings of the SPIE*. 2005; 141–151.

[14] Kreucher C, Hero A, Kastella K, *et al*. An information-based approach to sensor management in large dynamic networks. *Proceedings of the IEEE*. 2007;95(5):978–999.

[15] Tharmarasa R, Kirubarajan T, and Hernandez ML. Large-scale optimal sensor array management for multitarget tracking. *IEEE Transactions on Systems, Man, and Cybernetics – Part C: Applications and Reviews*. 2012;60(5): 803–814.

[16] Chavali P and Nehorai A. Scheduling and power allocation in a cognitive radar network for multiple-target tracking. *IEEE Transactions on Signal Processing*. 2012;60(2):715–729.

[17] Romero R and Goodman N. Cognitive radar network: cooperative adaptive beamsteering for integrated search-and-track application. *IEEE Transactions on Aerospace and Electronic Systems*. 2013;49(2):915–931.

[18] Bell K, Baker C, Smith G, *et al.* Cognitive radar framework for target detection and tracking. *IEEE Journal on Selected Topics in Signal Processing.* 2015;9(8):1427–1439.

[19] Charlish A and Hoffmann F. Cognitive radar management. In: *Novel Radar Techniques and Applications Volume 2: Waveform Diversity and Cognitive Radar, and Target Tracking and Data Fusion.* London: Institution of Engineering and Technology; 2017. p. 157–193.

[20] Song X, Willett P, Zhou S, *et al.* The MIMO radar and jammer games. *IEEE Transactions on Signal Processing.* 2012;60(2):687–699.

[21] Deligiannis A, Panoui A, Lambotharan S, *et al.* Game theoretic power allocation and the Nash equilibrium analysis for a multistatic MIMO radar network. *IEEE Transactions on Signal Processing.* 2017;65(24):6397–6408.

[22] Shi C, Wang F, Sellathurai M, *et al.* Game theoretic power allocation for coexisting multistatic radar and communication systems. In: *Proceedings of the 2018 IEEE International Conference on Signal Processing*; 2018. p. 872–877.

[23] Mishra KV, Martone A, and Zaghloul AI. Power allocation games for overlaid radar and communications. In: *Proceedings of the 2019 URSI Asia-Pacific Radio Science Conference*; 2019. p. 1–4.

[24] Nadjiasngar R and Charlish A. Quality of service resource management for a radar network. In: *Proceedings of the 2015 IEEE Radar Conference*; 2015. p. 344–349.

[25] Mitchell AE, Smith GE, Bell KL, *et al.* Cost function design for the fully adaptive radar framework. *IET Radar, Sonar, and Navigation.* 2018;12(12): 1380–1389.

[26] Cover T and Thomas J. *Elements of Information Theory.* New York: Wiley; 1991.

[27] Liese F and Vajda I. On divergences and informations in statistics and information theory. *IEEE Transactions on Information Theory.* 2006;52(10): 4394–4412.

[28] Aughenbaugh JM and LaCour BR. Metric selection for information theoretic sensor management. In: *Proceedings of the 11th International Conference on Information Fusion*; 2008.

[29] Yang C, Kadar I, Blasch E, *et al.* Comparison of information theoretic divergences for sensor management. *Proceedings of the SPIE.* 2011.

[30] Castañón DA, Mahler R, Hintz KJ, *et al.* Issues in resource management with applications to real-world problems. *Proceedings of the SPIE.* 2006.

[31] Mitchell AE, Smith GE, Bell KL, *et al.* Hierarchical fully adaptive radar. *IET Radar, Sonar, and Navigation.* 2018;12(12):1371–1379.

[32] Charlish A, Bell K, and Kreucher K. Implementing perception-action cycles using stochastic optimization. In: *Proceedings of the 2020 IEEE Radar Conference*; 2020.

[33] Smith GE, Cammenga Z, Mitchell AE, *et al.* Experiments with cognitive radar. *IEEE Aerospace and Electronic Systems Magazine.* 2016;31(12): 34–36.

[34] Bell K, Smith GE, Mitchell AE, *et al.* Multiple task fully adaptive radar. In: *Proceedings of the 52nd Asilomar Conference on Signals, Systems, and Computers*; 2018.

[35] Bell K, Smith G, Mitchell A, *et al.* Fully adaptive radar for target classification. In: *Proceedings of the 2019 IEEE Radar Conference*; 2019.

[36] Bell K, Kreucher C, and Rangaswamy M. An evaluation of task and information driven approaches for radar resource allocation. In: *Proceedings of the 2021 IEEE Radar Conference*; 2021.

[37] Van Trees HL, Bell KL, and Tian Z. *Detection, Estimation, and Modulation Theory, Part I.* New York: Wiley; 2013.

[38] Tichavsky P, Muravchik CH, and Nehorai A. Posterior Cramer–Rao bounds for discrete-time nonlinear filtering. *IEEE Transactions on Signal Processing.* 1998;46(5):1386–1396.

[39] Van Trees HL and Bell KL. *Bayesian Bounds for Parameter Estimation and Nonlinear Filtering/Tracking.* New York: Wiley; 2007.

[40] Niu R, Willett PK, and Bar-Shalom Y. Matrix CRLB scaling due to measurements of uncertain origin. *IEEE Transactions on Signal Processing.* 2001;49(7):1325–1335.

[41] Singer RA. Estimating optimal tracking filter performance for manned maneuvering targets. *IEEE Transactions on Aerospace and Electronic Systems.* 1970;6(4):473–483.

[42] Richards M. *Fundamentals of Radar Signal Processing.* New York: McGraw-Hill; 2005.

[43] Van Trees HL. *Optimum Array Processing.* New York: Wiley; 2002.

[44] Dogandzic A and Nehorai A. Cramer–Rao bounds for estimating range, velocity, and direction with an active array. *IEEE Transactions on Signal Processing.* 2001;49(6):1122–1137.

[45] John-Baptiste P, Johnson JT, and Smith GE. Neural network-based control of an adaptive radar. *IEEE Transactions on Aerospace and Electronic Systems.* 2022;58(1):168–179.

Chapter 10
Stochastic control for cognitive radar

Alexander Charlish[1], Folker Hoffmann[1], Kristine Bell[2] and Chris Kreucher[3]

Cognitive radar problems involve the selection of actions based on the uncertain knowledge of a system state that is partially observed through noisy measurements. This process of sequential decision making under uncertainty can be considered as a stochastic optimization problem. This chapter explicitly makes the connection between cognitive radar and stochastic optimization by presenting a framework for describing cognitive radar problems in terms of stochastic optimization, thereby pointing to ways to employ stochastic optimization for designing perception–action cycles in a cognitive radar.

10.1 Introduction

Cognitive radar problems require the selection of actions based on an uncertain perception that is obtained through inexact measurements. There is a broad variety of cognitive radar problems that differ in terms of the relevant perception and the types of actions selected, for example, waveform selection and optimization, measurement scheduling, resource management, detection, tracking, and imaging [1]. A single radar may in fact comprise several individual perception–actions cycles, spread over multiple information abstraction levels [2]. Despite their differences, the variety of cognitive radar problems can be described in terms of a set of similar problem components. Consequently, after identifying the problem components, similar methodologies can be applied for designing perception–action cycles for a cognitive radar.

Cognitive radar problems can be classed as types of stochastic optimization problems. Stochastic optimization is a broad term for techniques that perform decision making under uncertainty, which are currently widely deployed in a range of applications including finance, business, logistics and transportation, and science and engineering. Stochastic optimization methods seek a policy that exploits models to map from a perception, which represents all the available information at the current

[1] Fraunhofer FKIE, Wachtberg, Germany
[2] Metron, Inc., Reston, VA, USA
[3] KBR Government Solutions, Ann Arbor, MI, USA

time, into an optimized action. As this policy is essentially a perception–action cycle, the design of perception–action cycles for cognitive radar can benefit from applying algorithmic strategies for finding policies from the stochastic optimization field.

There are many communities focusing on stochastic optimization problems, who have established a wide variety of algorithmic solutions. These stochastic optimization communities have conducted research covering techniques and applications such as decision trees, stochastic search, optimal stopping, optimal control, (partially observable) Markov decision processes (MDPs/POMDPs), approximate dynamic programming, reinforcement learning, model predictive control, stochastic programming, ranking and selection, and multiarmed bandit problems. It has been shown [3] that these problems can be described in a single stochastic optimization framework, and the respective solution methodologies can be grouped into just four classes.

Some of the work in cognitive radar explicitly refers to these stochastic optimization techniques. For example, multiarmed bandits [4], model predictive control [5], and reinforcement learning [6]. However, for many techniques developed in cognitive radar, the connection is less clear. The primary contribution of this chapter is to directly connect the cognitive radar problem with the large body of work done in stochastic optimization. This connection makes the methodologies developed in the stochastic optimization communities directly applicable to the cognitive radar problem, promising to lead to improved methods for designing perception–action cycles in cognitive radar. This chapter extends our previous work in [7].

10.2 Connection to earlier work

Current approaches to cognitive radar build on and can be traced back to earlier work which was referred to as sensor management [8]. These earlier efforts, while often applied to radar sensing, were ostensibly agnostic to the sensing modality and as such addressed the broad problem of determining the best way to task a sensor or group of sensors when each sensor may have multiple agilities. This section briefly reviews early work in sensor management to give context and connection to the current state of cognitive radar research.

Sensor management research frequently focused on the use case of tasking sensors to deduce the kinematic state (e.g., position and velocity) and identification of a group of targets as well as the number of targets. Applications of sensor management were often military in nature [9], but also included things such as wireless networking [10] and robot path planning [11].

Like cognitive radar, one of the main issues sensor management research addressed is the many competing objectives an automated decision maker may be tuned to meet, e.g., minimization of track loss, probability of new target detection, minimization of track error/covariance, and identification accuracy. Each of these different objectives taken alone may lead to a different sensor allocation strategy [9,12]. Sensor management work was interested in mechanisms for capturing the trade-off between these competing objectives to deliver a measurement strategy that effectively addresses all of the objectives.

Information measures, including entropy reduction, Kullback–Leibler divergence (KLD) and mutual information, were a popular way of capturing the utility of sensing actions in foundational sensor management work and as such were explored by a number of researchers. Hintz [13,14] did early work using the expected change in Shannon entropy when tracking a single target moving in one dimension with Kalman filters. A related approach used discrimination gain based on a measure of relative entropy, the KLD. Schmaedeke and Kastella [15] used the KLD to determine sensor-to-target taskings. Kastella [16,17] used KLD to manage a sensor between tracking and identification mode in the multitarget scenario. Mahler [18] used the KLD as a metric for "optimal" multisensor multitarget sensor allocation. Zhao [19] compared several approaches, including simple heuristics and information-based techniques based on entropy and relative entropy.

For multi-stage planning, sensor management was often formulated as a Partially Observable Markov Decision Process (POMDP) [20,21] and researchers worked to develop approximate solution techniques. For example, Krishnamurthy [22,23] used a multi-arm bandit formulation involving hidden Markov models. In [22], an optimal algorithm was formulated to track multiple targets with an electronically scanned array that has a single steerable beam. Since the optimal approach has prohibitive computational complexity, several suboptimal approximate methods are given and some simple numerical examples involving a small number of targets moving among a small number of discrete states are presented. In [23], the problem was reversed, and a single target was observed by a single sensor from a collection of sensors. Again, approximate methods were formulated due to the intractability of the globally optimal solution.

Bertsekas and Castañon [24] did early work where they formulated heuristics for the solution of a stochastic scheduling problem corresponding to sensor scheduling. They implemented a rollout algorithm based on heuristics to approximate the solution of the stochastic dynamic programming algorithm. Additionally, Castañon [25,26] investigated the problem of classifying a large number of stationary objects with a multi-mode sensor based on a combination of stochastic dynamic programming and optimization techniques. In [27], Malhotra proposed using reinforcement learning as an approximate approach to dynamic programming.

Chhetri [28] approached the long-term scheduling problem for a single target using particle filters and the unscented transform. The method involves drawing samples from the predicted future distribution and minimizing expected future costs. This requires enumeration of the exponentially growing number of possible sensing actions, a very computationally demanding procedure. This is combined with branch and bound techniques which require some restrictive assumptions on additivity of costs. In a series of works, Zhao [10,19,29] investigated sensor management in the setting of a wireless ad hoc network, which involved long-term considerations such as power management.

With those connections as background, we now turn our attention to laying out a general framework for describing cognitive radar problems which makes them amenable to modern solution approaches.

10.3 Stochastic optimization framework

This section presents a framework, inspired by [3],* which enables cognitive radar problems to be described in terms of stochastic optimization problems.

10.3.1 General problem components

As described in [3], all the problems addressed by the stochastic optimization communities comprise the problem components described in this subsection. The next subsection shows how these components can be extended for the case when a system state is partially observed through noisy measurements. As the partially observable case is more relevant to cognitive radar problems, it is used as the focus for the remainder of this chapter.

System state: We are interested in the state of a dynamic system, which can be modeled as a random vector X_k for decision step k. A realization of the random vector at decision step k is denoted $\mathbf{x}_k \in \mathscr{X}$ where $\mathscr{X}$ is the system state space.

Actions and action space: We can select an action or action vector at each decision step k, which influences the transition of the system state between time step k and $k+1$. An instantiation of an action for decision step k is denoted $\mathbf{a}_k \in \mathscr{A}$ where $\mathscr{A}$ is the action space.

Exogenous information: Additional information is revealed at each sequential decision step and can, along with previously revealed information, be used as the basis for the action selection at the current decision step. The information revealed at each time step is modeled as a random vector Z_k and a realization of this random vector is denoted $\mathbf{z}_k$. For completely observable problems, the exogenous information is the system state.

State transition function: Between decision steps, the system state evolves according to a transition function $\mathbf{x}_{k+1} = f_X(\mathbf{x}_k, \mathbf{a}_k, \mathbf{w}_k)$, where $\mathbf{w}_k$ is a realization of the state transition noise (alternatively termed process noise). Due to the state transition noise, the transition can be described probabilistically by the transition probability density $p(\mathbf{x}_{k+1}|\mathbf{x}_k, \mathbf{a}_k)$.

Reward function: At each decision step, a reward is encountered, which is described by the function $r_x(\mathbf{x}_k, \mathbf{a}_k, \mathbf{z}_{k+1})$. Cost can be handled as a negative reward, and, therefore, cost and reward are used interchangeably in this chapter. This objective function is described in more detail in the following sections.

These common components allow the breadth of stochastic optimization problems considered by the stochastic optimization communities to be described. Note that although the system state is completely observable, decision making under uncertainty is present due to the stochastic state transitions. A perception–action cycle using these components is illustrated in Figure 10.1.

*Although we adopt the framework in [3], we use the terminology and notation that is established in the signal processing community.

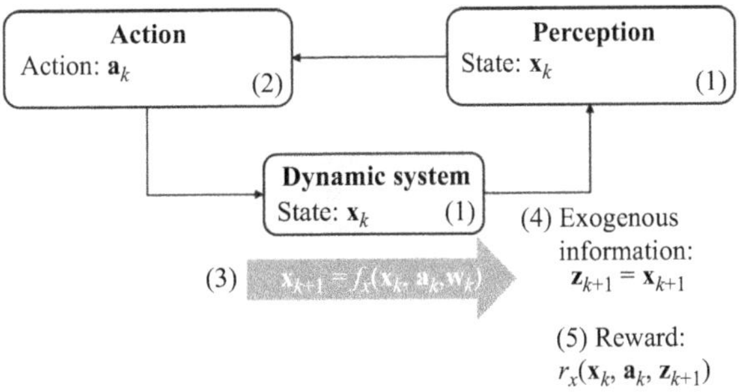

Figure 10.1 General perception–action cycle for a completely observable system using stochastic optimization components. The following repetitive steps occur: (1) the system has a state $\mathbf{x}_k$*, which is completely observed as the perception, (2) an action* $\mathbf{a}_k$ *is selected, (3) the system state transitions to* $\mathbf{x}_{k+1}$*, (4) the system state* $\mathbf{x}_{k+1}$ *is revealed as exogenous information, and (5) a reward is generated.*

10.3.2 Partial observability

A common aspect of cognitive radar problems is that the system state is only *partially observable* through noisy measurements. Therefore, uncertainty is not only present due to stochastic state transitions but also through stochastic measurements. Consequently, we extend and adapt the components described in Section 10.3.1 to the more specific partially observable case, which results in a framework closely resembling a POMDP.

Measurements and measurement space: The exogenous information described in Section 10.3.1 can now be thought of as a noisy measurement of the system state. Now, the random vector Z_k can be defined more exactly as a measurement with realization $\mathbf{z}_k \in \mathscr{Z}$ where $\mathscr{Z}$ is the measurement space.

Measurement-likelihood function: Measurements are related to the system state through the measurement function $\mathbf{z}_k = h(\mathbf{x}_k, \mathbf{a}_{k-1}, \mathbf{v}_k)$ where $\mathbf{v}_k$ is a realization of the measurement noise. Due to the measurement noise, the measurement process can be described by the measurement-likelihood function $\mathscr{L}(\mathbf{x}_k|\mathbf{z}_k, \mathbf{a}_{k-1}) \equiv p(\mathbf{z}_k|\mathbf{x}_k, \mathbf{a}_{k-1})$.

Information state: As the state of the system is not observable, it is necessary to decide on an action based on the information state. The information state is the set of actions and measurements that have occurred prior to the current decision step. The information state for decision step k is denoted $\mathscr{I}_k = (\mathbf{a}_0, \mathbf{z}_1, \ldots, \mathbf{a}_{k-1}, \mathbf{z}_k)$. This information state grows with each time step, i.e., $\mathscr{I}_k = \mathscr{I}_{k-1} \cup (\mathbf{a}_{k-1}, \mathbf{z}_k)$.

Belief state: As the cardinality of the information state grows with each time step, it is generally undesirable to be used as the perception upon which actions are decided. Instead, decisions can be based on a belief state. The belief state is a set of parameters with fixed cardinality that are an (ideally sufficient) statistic of the information state. The belief state at decision step k is modeled as a random vector B_k

and a realization of a belief state at decision step k is denoted $\mathbf{b}_k$. For example, under linear Gaussian assumptions, a sufficient statistic of the information state is the mean and covariance of the posterior PDF, i.e., $p(\mathbf{x}_k|\mathscr{I}_k) \equiv p(\mathbf{x}_k|\mathbf{b}_k)$. Typical belief states are parameters of a Gaussian, a Gaussian sum, or a set of particles. Although this belief state represents imprecise knowledge of the underlying system state, it is itself completely observable. Consequently, by treating this belief state as the system state in Section 10.3.1, a partially observable problem can be handled like a completely observable problem.

Belief state transition function: It is necessary to define a transition function for belief states, analog to the system state transition function. This transition function is denoted $\mathbf{b}_{k+1} = f_B(\mathbf{b}_k, \mathbf{a}_k, \mathbf{z}_{k+1})$. As the belief state can be thought of as parameters of the posterior PDF $p(\mathbf{x}_k|\mathbf{b}_k)$, the transition function represents the standard Bayesian prediction and update steps. As a cognitive radar is an observer, it is often the case that the system state transition is not influenced by the selected sensing action. However, the belief state transition certainly will be influenced by the selected action.

Reward function: A reward function is now defined as a function of the belief state, i.e., $r(\mathbf{b}_k, \mathbf{a}_k, \mathbf{z}_{k+1})$. This differs from the reward function described in Section 10.3.1, which was a function of the system state. The reward function maps to the reward that is associated with the measurement realization $\mathbf{z}_{k+1}$ when the belief state was $\mathbf{b}_k$ and action $\mathbf{a}_k$ was taken. The next subsection describes specific forms of this reward function.

A perception–action cycle for the case of a partially observable system state is illustrated in Figure 10.2. In this figure, $\mathbf{a}_k = A^\pi(\mathbf{b}_k)$ is the policy function that maps from belief states to actions. This policy function is described in detail in Section 10.5.

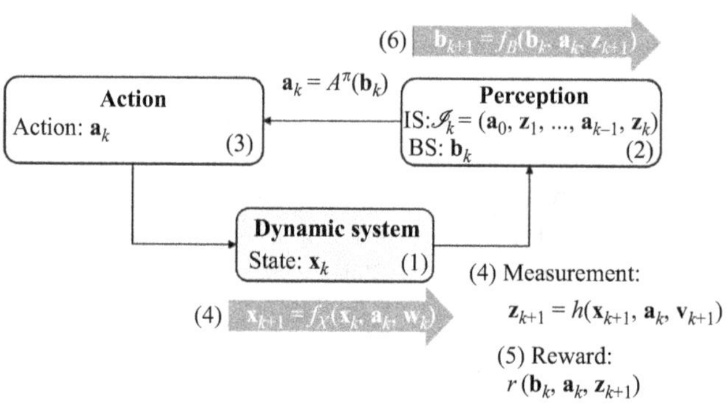

Figure 10.2 Partially observable perception–action cycle using stochastic optimization components. The following iterative steps occur: (1) the system has a state $\mathbf{x}_k$, (2) the perception of the system state is summarized in a belief state $\mathbf{b}_k$, (3) an action $\mathbf{x}_k$ is selected according to the policy function, (4) the system state transitions to $\mathbf{x}_{k+1}$ and a measurement $\mathbf{z}_{k+1}$ is generated, (5) a reward is produced, and (6) the belief state transitions to $\mathbf{b}_{k+1}$.

For the remainder of this chapter, we will assume a partially observable problem. However, a completely observable problem can be recovered by substituting the belief state with the observable system state, considering the likelihood function as a Dirac delta function, and using the state transition function instead of the belief state transition function.

10.4 Objective functions for cognitive radar

The exact form of the reward function $r(\mathbf{b}_k, \mathbf{a}_k, \mathbf{z}_{k+1})$ is crucial, as it must accurately represent the physical problem to be solved. Specifying reward functions for cognitive radar can be loosely categorized into task, information, or utility (quality-of-service) based approaches. However, the separation between the categories is not always distinct and existing approaches form more of a continuum.

10.4.1 Task-based reward functions

Task-based reward functions calculate the cost or reward of an action in terms of a measure that is specific to the task being performed. Relevant task-based metrics include radar timeline or spectrum usage, probability of target detection, detection range for an undetected target density, tracking root mean square error (RMSE), track sharpness, track purity, track continuity, and probability of correct target classification, to name a few.

Each task-based reward function can be regarded as some function $q(\mathbf{b}_k, \mathbf{a}_k, \mathbf{z}_{k+1})$ that is combined in some way to produce a scalar function that maps into the quality space $\mathcal{Q}$. It is often the case that a desired task-based metric is difficult to calculate and is replaced by a surrogate metric such as signal-to-interference plus noise ratio (SINR) or an information theoretic metric.

10.4.2 Information theoretic reward functions

A second class of reward functions used in cognitive radar and related fields is based on information theory. Broadly speaking, an information theoretic function gauges the relative merit of a sensing action in terms of the information flow it provides. While this does not correspond directly to an operational criterion like track hold probability, information flow does capture actions that ultimately lead to good operational performance. A primary motivation for information-based reward functions is the ability to compare actions which generate different types of knowledge (e.g., knowledge about a target class versus knowledge about target position) using a common measuring stick.

A review of the history of information metrics in this context is provided in [8]. Here, we highlight some of the most commonly used reward functions. The most basic information theoretic cost function is the Posterior Shannon Entropy, given as:

$$\mathcal{H}(X_{k+1}|\mathbf{b}_k, \mathbf{a}_k, \mathbf{z}_{k+1}) = \int p(\mathbf{x}_{k+1}|\mathbf{b}_k, \mathbf{a}_k, \mathbf{z}_{k+1}) \ln p(\mathbf{x}_{k+1}|\mathbf{b}_k, \mathbf{a}_k, \mathbf{z}_{k+1}) d\mathbf{x}_{k+1} \tag{10.1}$$

Note that $p(\mathbf{x}_{k+1}|\mathbf{b}_k, \mathbf{a}_k, \mathbf{z}_{k+1}) \equiv p(\mathbf{x}_{k+1}|\mathbf{b}_{k+1})$ in the case that the belief state is a sufficient statistic of the information state.

A related approach computes the information *gain* between densities rather than just the information contained in the posterior. The most popular approach uses the KLD, which is defined using the prior and posterior densities as:

$$\mathscr{D}\left(p(\mathbf{x}_{k+1}|\mathbf{b}_k, \mathbf{a}_k, \mathbf{z}_{k+1})||p(\mathbf{x}_{k+1}|\mathbf{b}_k)\right) = \int p(\mathbf{x}_{k+1}|\mathbf{b}_k, \mathbf{a}_k, \mathbf{z}_{k+1}) \ln \frac{p(\mathbf{x}_{k+1}|\mathbf{b}_k, \mathbf{a}_k, \mathbf{z}_{k+1})}{p(\mathbf{x}_{k+1}|\mathbf{b}_k)} d\mathbf{x}_{k+1} \tag{10.2}$$

The KLD has several desirable properties [30], including its connection to Mutual Information. There are a number of generalizations of the KLD in the literature, including the Rényi Divergence [31], the Arimoto α-divergences, and the f-divergence [32].

A third approach specific to parameter estimation is the Fisher Information Matrix (FIM) and related Bayesian Information Matrix (BIM) [33], which characterize the amount of information that a distribution contains about individual parameters (such as target position or velocity). The inverse of the FIM is the Cramér–Rao Lower Bound (CRLB) and the inverse of the BIM is the Bayesian CRLB, which quantifies the uncertainty in the parameter estimates. The (square root of) the Bayesian CRLB has the property that it is in the units of the parameter being estimated and is a lower bound on the RMSE. Thus, it is often used as a surrogate for the RMSE and categorized as a task-based metric.

The Bayesian CRLB approach is actually closely related to the KLD approach, since the BIM is related to a more general version of the KLD [34], and there is an equivalent Bayesian α-CRLB that is derived from the Bayesian version of the Rényi divergence [35]. Thus, these approaches have at their core the same information theoretic quantities, and the distinction is in the separation and weighting of individual tasks in the task-based Bayesian CRLB method versus a global approach in the information-based KLD method. A comparison between the approaches for fully adaptive radar resource allocation is explored in Chapter 9.

10.4.3 Utility and QoS-based objective functions

Quality-of-service approaches [2,36] differ from task or information-based reward functions in that they optimize the user or operator satisfaction that is derived from a task. A utility function is defined on the task quality space $\hat{u} : \mathscr{Q} \mapsto [0, 1]$ that should accurately describe the satisfaction that is derived from the different possible task quality levels. Combining the quality and utility functions results in a function of the required form $u(\mathbf{b}_k, \mathbf{a}_k, \mathbf{z}_{k+1}) \equiv (\hat{u} \circ q)(\mathbf{b}_k, \mathbf{a}_k, \mathbf{z}_{k+1})$, where $(\hat{u} \circ q)$ is the composite function of $\hat{u}$ following q.

Using utility functions allows a user to specify requirements on task qualities, which are generally tangible to the user. This is very valuable in the context of radar resource management [2] as it enables a radar with limited resources to optimize multiple tasks based on the task quality levels that are required by the mission. Mapping the quality levels of differing radar tasks into the common utility space enables trade-offs between tasks evaluated using differing quality metrics. The global utility across the multiple tasks is typically formed by taking a weighted sum of task utilities. When considering the resource usage, a resource function $g(\mathbf{b}_k, \mathbf{a}_k)$ can be used as a constraint on the permissible actions. This quality-of-service conceptual approach can also be identified in other work [37,38].

10.5 Multi-step objective function

A general objective is to find a policy that determines a feasible action based on the belief state. The policy is a mapping from belief state to action denoted $\mathbf{a}_k = A^{\pi}(\mathbf{b}_k)$, where π carries information about the type of function and its parameters. As the belief state is a set of parameters describing a perception of the system state, the policy can be thought of as the perception–action cycle for a cognitive radar. The policy is not necessarily an analytical function and may actually represent an optimization problem. This section describes how a multi-step objective function is used to define optimal values and policies that are the basis for the design of perception–action cycles in the following section.

10.5.1 Optimal values and policies

The objective of a stochastic optimization problem is to maximize rewards or minimize costs over a time horizon comprising H future decision steps. The expected reward achievable over the current and future decision steps that originate from the current belief state is termed the value of the belief state. Let $V_H^{\pi}(\mathbf{b}_k)$ denote the value of a belief state when following policy A^{π}. It is defined as the expected value of the summed rewards with respect to the set of future measurements $(Z_{k+1}, \ldots, Z_{k+H})$, conditioned on the belief state $\mathbf{b}_k$:

$$V_H^{\pi}(\mathbf{b}_k) = \mathbb{E}\left[\sum_{t=k}^{k+H} r\left(B_t^{\pi}, A^{\pi}(B_t^{\pi}), Z_{t+1}\right) | B_k^{\pi} = \mathbf{b}_k\right] \tag{10.3}$$

where the belief state random variables in the summation evolve according to the belief state transition function when following policy π, i.e., $B_{k+1}^{\pi} = f_B(B_k^{\pi}, A^{\pi}(B_k^{\pi}), Z_{k+1})$. It is common to rewrite (10.3) by splitting it into the expected reward for the current time step and the expected reward for subsequent time steps to give:

$$V_H^{\pi}(\mathbf{b}_k) = R(\mathbf{b}_k, A^{\pi}(\mathbf{b}_k)) + \mathbb{E}\left[V_{H-1}^{\pi}(B_{k+1}^{\pi}) | B_k^{\pi} = \mathbf{b}_k\right] \tag{10.4}$$

where the expectation is taken with respect to the future measurement Z_{k+1}. The single step reward, $R(\mathbf{b}_k, A^{\pi}(\mathbf{b}_k))$, is the expected reward with respect to the future measurement Z_{k+1}:

$$R(\mathbf{b}_k, A^{\pi}(\mathbf{b}_k)) = \mathbb{E}\left[r\left(B_k, A^{\pi}(B_k), Z_{k+1}\right) | B_k = \mathbf{b}_k\right] \tag{10.5}$$

Note that the expectation with respect to the remaining future measurements $(Z_{k+2}, \ldots, Z_{k+H})$ in (10.3) is now contained in the future value term $V_{H-1}^{\pi}(B_{k+1}^{\pi})$

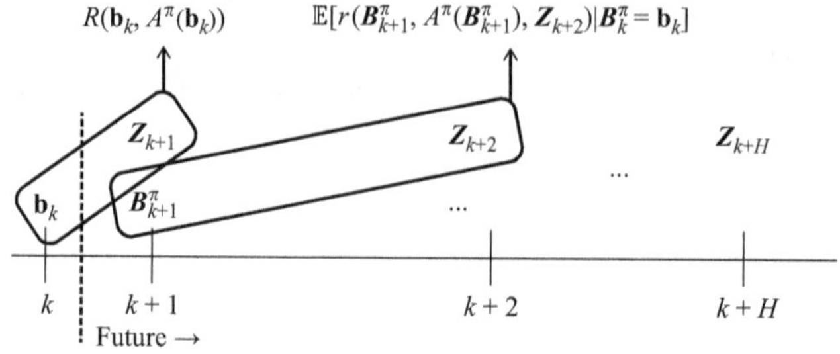

Figure 10.3 Calculation of the value of a belief state when following policy A^π. The single step expected reward is calculated using the current belief state realization $\mathbf{b}_k$ and with respect to the future measurement random variable Z_{k+1}. The expected reward from future times steps is calculated with respect to future belief state and measurement random variables.

in (10.4). Equation (10.4) can be identified as a form of Bellman's equation. The calculation of the value of a belief state when following policy π is illustrated in Figure 10.3.

Similar to the value of a belief state when following policy π, it is possible to define the optimal value of a belief state as:

$$V^*_H(\mathbf{b}_k) = \max_{\mathbf{a}\in\mathscr{A}} \left(R(\mathbf{b}_k, \mathbf{a}) + \mathbb{E}\left[V^*_{H-1}(B^{\mathbf{a}}_{k+1})|B_k = \mathbf{b}_k\right]\right) \tag{10.6}$$

where $B^{\mathbf{a}}_{k+1}$ is a random variable representing the belief state in the next decision step that evolves when taking action $\mathbf{a}$, i.e. $B^{\mathbf{a}}_{k+1} = f_B(B_k, \mathbf{a}, Z_{k+1})$. Using the optimal value function, the optimal policy function can be defined, which is a description of an optimal perception–action cycle:

$$A^*(\mathbf{b}_k) = \arg\max_{\mathbf{a}\in\mathscr{A}} \left(R(\mathbf{b}_k, \mathbf{a}) + \mathbb{E}\left[V^*_{H-1}(B^{\mathbf{a}}_{k+1})|B_k = \mathbf{b}_k\right]\right) \tag{10.7}$$

The first term in (10.7) represents the expected reward associated with the current belief state and the chosen action, and is relatively easy to calculate. However, the second term that represents the expected reward associated with future belief states in the time horizon is very difficult to calculate. Consequently, solving the optimal policy function is generally intractable. The majority of stochastic optimization approaches focus on approximate solutions to this optimal policy function.

Equation (10.3) is a multi-step objective function for the case when it is desired to optimize the expected rewards accumulated over the time horizon. Alternatively, the terminal reward may be of interest at the end of the time horizon. This can be accommodated by using an altered reward function that returns zero except for the last decision step in the time horizon. This section has described a problem with finite horizon H. An infinite horizon problem can be described in the same way, but requires the inclusion of a discounting factor.

10.5.2 Simplified multi-step objective functions

Finding policies that solve (10.7) is very challenging due to the need to evaluate the impact of the current action on expected future rewards, knowing only the current belief state. There are simplifications that are often performed that drastically reduce the complexity of the problem but result in an objective function that does not fully consider the uncertainty present in the problem. These simplifications are often applied in current cognitive radar techniques, as will be shown in Section 10.7.

10.5.2.1 Myopic optimization

If the time horizon is taken as a single step, i.e., $H = 1$, then the problem of evaluating the impact of the action on expected future rewards is removed. Hence, the optimal policy function in (10.7) is significantly simplified to:

$$A^*(\mathbf{b}_k) = \arg\max_{\mathbf{a}\in\mathscr{A}} (R(\mathbf{b}_k, \mathbf{a})) \tag{10.8}$$

This approach is known as myopic or greedy optimization as it focuses on the immediate expected reward and ignores the impact of potential future rewards. This approach can represent a significant simplification of the problem that may result in poor action selection and, hence, a reduced accumulated reward. However, there may be problems in which the optimal myopic policy coincides with the optimal non-myopic policy. In which case, this simplification is completely justified.

10.5.2.2 Deterministic optimization

A second common simplification is to perform a deterministic optimization based on expected values of the belief state and/or future measurements, instead of treating them as random variables and calculating the expected reward. An example of this approach would be to simplify the myopic reward function in (10.5) as:

$$R(\mathbf{b}_k, A^\pi(\mathbf{b}_k)) \approx r\left(\mathbf{b}_k, A^\pi(\mathbf{b}_k), \mathbb{E}\left[Z_{t+1}|B_k = \mathbf{b}_k\right]\right) \tag{10.9}$$

Whereas myopic optimization ignores the propagation of uncertainty into the future, deterministic optimization ignores the uncertainty in the belief state transition and measurement processes. However, by treating the optimization problem as being deterministic, it can be easier to solve. As the reward is now a deterministic mapping from the actions to a real number, standard techniques to optimize functions can be used, for example, numerical optimization methods, metaheuristics such as simulated annealing, or convex optimization.

10.5.2.3 Discussion

Stochastic optimization techniques aim to find a policy that closely matches the optimal policy function and, therefore, perform an action that is optimized considering the uncertainty in the future evolution of the system and the noisy measurement process. However, it should be clear that solving the optimal value and policy functions for realistic problems is intractable. Consequently, existing cognitive radar techniques often simplify the problem by performing myopic or deterministic optimization. However, advances in computational capability combined with the development of new algorithms mean that it is possible to move away from these simplifications and look towards designing perception–action cycles that fully consider the uncertainty in the problem. A subsequent and critical question for any problem is then: *which sources of problem uncertainty have a significant impact on performance and should therefore be incorporated into the optimization process?*

10.6 Policies and perception–action cycles

Solving a stochastic optimization problem involves finding a policy that maps from belief states into actions and hence constitutes a perception–action cycle. This section gives an overview of methods for finding policies that are widely used in stochastic optimization. As mentioned earlier, [3] organizes the methods of finding policies into four classes that cover all approaches in the literature. The first two methods are policy search approaches and are referred to as policy function approximations (PFAs) and cost function approximations (CFAs). The second two approaches are lookahead approaches and are referred to as value function approximations (VFAs) and direct lookahead. We discuss each of these in turn here, with the purpose of showing that established algorithmic strategies from the field of stochastic optimization can be valuable tools for designing perception–action cycles in a cognitive radar. More details on each of these methodologies can be found in [3] and the references therein.

10.6.1 Policy search

The general approach to policy search is to find and tune a policy that matches or approximates the optimal policy function in (10.7). Generally, the optimal policy is unlikely to be found. Instead an approximation to the optimal policy function is sought in the form of PFA or a CFA.

10.6.1.1 Policy function approximations

PFAs attempt to find and tune a function that approximates the optimal policy function in (10.7). For example, we can consider a family of functions $\mathscr{F}$, where a function $f \in \mathscr{F}$ is parameterized by $\theta \in \Theta^f$. Our goal is then to find a function f and parameterization, θ so that the optimal policy function in (10.7) can be approximated as:

$$A_{PFA}^{f,\theta}(\mathbf{b}_k) = f(\mathbf{b}_k;\theta) \tag{10.10}$$

The optimal policy can be found if the optimal policy belongs to the family of functions and the corresponding parameter space. The goal of PFAs is not to find the optimal policy, but to find the best approximation within a class of function approximations. The function class may be any approach for approximating a function, such as an analytic function or a neural network.

An example from the radar literature is the work presented in [39]. Here the problem of optimizing the radar revisit times for a target track is considered. The authors define the concept of a *track sharpness*, which is the major axis of the uncertainty ellipsoid in antenna coordinates (u–v space) relative to the beam width. The general strategy is to schedule a radar dwell to update the track once the track sharpness crosses a given threshold.

It is possible to cast the problem in [39] into the framework components in Section 10.3. The tracker provides some of the parameters for the belief state $\mathbf{b}_k = [r \ \Theta \ \Sigma \ \sigma \ B]^T$, where r is the estimated target range, Θ, Σ are parameters of the Singer target dynamic model, σ is the measurement error standard deviation, and B is the radar half beamwidth. Note that σ and B are sensor parameters that may be dependent on the target kinematic parameters. The action space is a scalar representing the revisit interval time, i.e., $a_k \in \mathbb{R}_+$, where $\mathbb{R}_+$ denotes the positive real numbers. The authors of [39] propose a function for finding the steady-state revisit interval:

$$A_{PFA}^{V_0}(\mathbf{b}_k) = 0.4\left(\frac{r\sigma\sqrt{\Theta}}{\Sigma}\right)^{0.4}\frac{U^{2.4}}{1 + 0.5U^2} \tag{10.11}$$

where $U = BV_0/\sigma$ is the variance reduction ratio. Although it is not stated in [39], this can be considered as a PFA, whereby the function in (10.11) is parameterized by the track sharpness $\theta = V_0$. After proposing the policy function in (10.11), the authors of [39] proceed by finding the function parameterization, i.e. the value of V_0, that minimizes the radar loading while also maintaining track on the target.

10.6.1.2 Cost function approximations

Instead of approximating the entire policy function as with a PFA, a CFA finds a functional approximation for the cost function, which is interchangeable with the reward function described in this paper. Consequently, the optimal policy function in (10.7) is replaced with:

$$A_{CFA}^{\pi,\theta}(\mathbf{b}_k) = \arg\max_{\mathbf{a}\in\mathscr{A}^{\pi}(\theta)} [\tilde{r}^{\pi}(\mathbf{b}_k, \mathbf{a}; \theta)] \tag{10.12}$$

which comprises of the approximation to the cost function $\tilde{r}^{\pi}(\mathbf{b}_k, \mathbf{a}_k; \theta)$ as well as a potentially constrained action space $\mathscr{A}^{\pi}(\theta)$.

An example of such a CFA can be found in the work of [40]. Here the track revisit interval is chosen to minimize the trace of the predicted conditional Bayesian information matrix (PC-BIM). Without any additional constraints such an optimization would always choose the minimum revisit interval. Therefore, a parameter K is introduced, which can be tuned to perform a trade-off between the information gain and the resource usage. Consequently, the following minimization is performed on a CFA:

$$A_{CFA}^{K}(\mathbf{b}_k) = \arg\min_{a\in\mathscr{A}}\left(\mathrm{tr}(\mathbf{P}(a, \mathbf{b}_k)) + \frac{K}{a}\right), \tag{10.13}$$

where $\mathscr{A} = \mathbb{R}_+$ is the set of all possible revisit intervals, and $\mathrm{tr}(\mathbf{P}(a, \mathbf{b}_k))$ is the trace of the PC-BIM after a measurement is produced with a revisit interval of a and conditioned on the current belief state realization $\mathbf{b}_k$.

10.6.2 Lookahead approximations

Lookahead approximations differ from policy search as they attempt to evaluate the influence of an action on future rewards, instead of approximating the policy function. A lookahead approximation can be performed via a VFA or by simulating a direct lookahead.

10.6.2.1 Value function approximations

A VFA uses the optimal policy function in (10.7), but replaces the true optimal value of future belief states $V^*_{H-1}(B^{\mathbf{a}}_{k+1})$ with an approximation $\tilde{V}_{H-1}(B^{\mathbf{a}}_{k+1})$. In some cases, the expectation in (10.7) may be difficult to calculate, in which case a VFA can be used to replace $\mathbb{E}\left[V^*_{H-1}(B^{\mathbf{a}}_{k+1})|B_k = \mathbf{b}_k\right]$.

The resulting policy for a VFA is:

$$X^{\tilde{V}}_{VFA}(\mathbf{b}_k) = \arg\max_{\mathbf{a}\in\mathscr{A}} \left(r(\mathbf{b}_k, \mathbf{a}) + \mathbb{E}\left[\tilde{V}(B^{\mathbf{a}}_{k+1})\right]\right) \tag{10.14}$$

Another variant is the approximation of the action value $\tilde{Q}$, which results in the policy

$$X^{\tilde{Q}}_{VFA}(\mathbf{b}_k) = \arg\max_{\mathbf{a}\in\mathscr{A}} \tilde{Q}(\mathbf{b}_k, \mathbf{a}) \,. \tag{10.15}$$

A famous algorithm from the literature that uses such policies is the Q-learning algorithm [41], whose variants have also been used for radar management [6]. A non-learning example in radar management can be found in [42]. Here the reward is based on the detection range of the radar, and the action $\mathbf{a}_k = [t\ ri]^T$ consists of the revisit interval ri and the dwell duration t. The problem is radar search, where the detection range of the radar should be optimized. To frame this problem in terms of Equation (10.15), $\tilde{Q}(\mathbf{b}_k, \mathbf{a})$ is the expected target detection range. Note that $\mathbf{b}_k$ here does not contain the estimate of a target track, as no track is detected yet. Instead, it contains observable quantities such as the platform's altitude and prior information about the expected target RCS and popup range. In [42], this value is approximated with a lookup table and used in a QoS resource allocation algorithm.

10.6.2.2 Direct lookahead

For the cases when it is not possible to find an accurate VFA, the expected future value can be evaluated by simulating future system evolutions using the available models. As this process is computationally very costly, direct lookahead methods focus on making effective simplifications that still lead to accurate values. Common methods belonging to this class are deterministic lookaheads, Monte-Carlo sampling, rollout policies and Monte-Carlo tree search.

Myopic, single period lookahead policies are variants of this method, which are often used in the radar literature, e.g. [43,44]. A non-myopic lookahead method can be found in [45], which uses a policy rollout method to determine the optimal track revisit intervals. Note that this work also contains components of a CFA, by encoding the performance to resource trade-off with a tune-able parameter in the cost

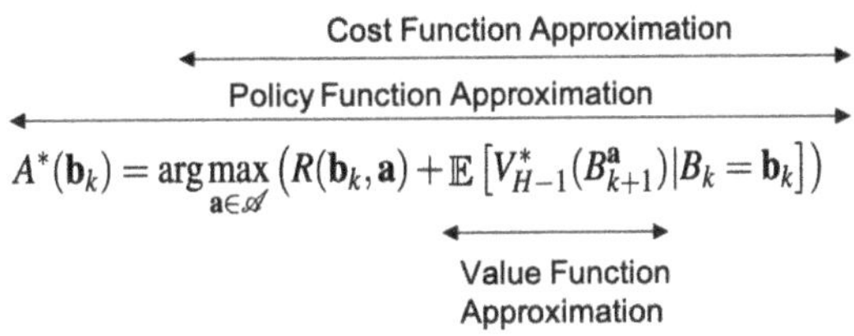

Figure 10.4 Different function approximation types for the optimal policy function

function. Policy rollout is also used in [46] for solving the radar resource management problem.

10.6.3 Discussion

General methodologies for finding policies involve finding a function approximation to either the policy function, the cost function or the value function. The difference between these approaches is simply where the functional approximation is made, as illustrated in Figure 10.4. The effectiveness of these approaches depends on how well a function approximation can capture these respective relationships. All of these methodologies can be implemented with handcrafted models or using machine learning techniques. Although it is typical to perform offline training, these function approximations could be updated online as more data becomes available. Direct lookahead approaches are used when it is not possible to capture the structure of the problem with a function approximation.

10.7 Relationship between cognitive radar and stochastic optimization

In the previous sections, the general framework of stochastic optimization, as well as possible solution techniques are described. This framework models the problem of selecting the best action under uncertainty, to maximize a reward. Cognitive radar is an application domain, which falls under the assumptions of this framework. Knowledge about the true state is only received by noisy measurements, and state and belief state transitions are non-deterministic. A radar controller must select the optimal sensing actions to maximize performance of the system, which can be formalized by a suitable reward function. In the following, this view of the cognitive radar task as a stochastic optimization problem is mapped out.

10.7.1 Problem components

A representative set of cognitive radar problems for different applications can be found in the references. Although it may not always be explicitly stated, these problems can be characterized as stochastic optimization problems that possess the framework components described in Section 10.3. The components are sometimes explicitly stated or can be inferred.

In the case of target tracking [2,37,39,44,47–51], the belief state characterizes a posterior probability density function defined on the system state space. Typical belief states are the mean and covariance matrix of the distribution or a set of particles. The belief state transition function incorporates the Bayesian prediction and update processes. The exogenous information is some noisy function of the system state that maps to radar measurements, thus the system state is partially observable. Often, the likelihood function is a Gaussian approximation of the true measurement errors. Adaptive tracking [2,39,48] methods select actions in the form of revisit interval times as well as the waveform energy for the next measurement, in order to minimize resource usage while maintaining track. An early approach [39] was to use a function that mapped measurement and track accuracies, and Singer maneuver parameters to a revisit interval time. In the context of the methods described in Section 10.6, this can be thought of as an empirically derived PFA. Another strand of work has focused on waveform selection and adaptation [44,47,50], whereby the action space comprised different waveform modulations that were selected in order to minimize track RMSE.

The framework components are easy to identify for tracking problems, because the framework is essentially an extension to the standard Bayesian tracking process. However, other radar functions and applications can also be cast into the framework. For a search problem, the belief state can parameterize an undetected target posterior density. In target detection [52–55], the system state is the state of the clutter, interference, and noise environment. Typical belief states include the clutter, interference, and noise covariance matrix or a posterior distribution on a spectrum occupancy state. For imaging and classification [56] the belief state characterizes a posterior probability mass function. Typical belief states are the pairwise likelihood ratios or the posterior probabilities themselves. Some works also consider a combination of radar functions [43,57,58].

Generally, the action space is some set of parameters that characterize the radar transmission and reception, including transmit and receive sensor selection and scheduling, transmit frequency, bandwidth, time, duration, power, and waveform design. The exogenous information is some noisy function of the system state, thus the system state is partially observable. Generally, reward functions differ widely, but can be categorized according to the classes in Section 10.4.

10.7.2 Typical cognitive radar solution methodologies

A variety of solution methodologies have been applied to cognitive radar problems, which can be compared with the strategies described in Section 10.6. The majority of the reference works formulate myopic optimization problems, which represent a simplification with respect to the general non-myopic multi-step objective function. Depending on the problem, this can be a very valid approach to reduce the complexity of the optimization, especially if it is clear that the current action does not influence future rewards. However, it is worthwhile to explicitly consider how the myopic and non-myopic solutions differ, as there are certainly problems where considering the future rewards associated with the current action can significantly improve performance.

There are also cases in the reference works where an optimization is performed on an expected value of the belief state and/or an expected future measurement, instead of treating the system state and future measurements as random variables and calculating the expected reward. This approach has the benefit of enabling deterministic optimization methods to be applied and is a valid approximation if the reward function is not sensitive in the region of significant probability as described by the posterior and expected measurement PDFs. However, this approach ignores or under-utilizes the uncertainty in the future state evolution and corresponding measurements, which could significantly impact performance.

The cognitive radar methodologies in the reference works generally attempt to solve an optimization problem online by performing numerical optimizations or searches over the action space. However, the strategies described in Section 10.6 first attempt to identify structure in the policy, cost or value function and attempt to use specific models or machine learning to produce a functional approximation. This is a particularly attractive approach because it can reduce the complexity of the online optimization problem, or remove the need to perform an online optimization, depending on the functional approximation type. This approach is underrepresented in the reference works, but can be identified in [51], where a neural network is used to learn the policy function that an optimizer with more complexity would generate.

10.7.3 Cognitive radar objective functions

Although cognitive radar problems and approaches can be cast into the framework described in Section 10.3, there are some key differences. A main difference comes in form of the objective function.

In Section 10.5, the objective of the framework is described as finding a policy with maximal accumulated reward. This is a common formulation in many fields of stochastic optimization, for example, in control theory or reinforcement learning. Almost all papers in the reinforcement learning literature demonstrate the development of these accumulated rewards (or costs) over the time of the training, consequently allowing a summary of the performance of a policy in a single metric. Therefore, it is possible to directly compare two policies and decide which is better.

On the other hand, such a statement is rare in the cognitive radar literature. Typically several metrics are considered in the evaluation. Although these metrics are used in the performance evaluation, they are not always stated explicitly as part of the objective function. We acknowledge that the evaluations yield important insights, however, it is useful to also clearly state the true objective in a single quantifiable way. For example, "in the given evaluation scenario, we want to minimize the sum of the tracking errors over the whole scenario time." A common occurrence in the radar literature is the usage of surrogate functions (e.g., SNR, mutual information, etc.). Generally, it should be clear whether this is the true objective or actually a surrogate for the harder to evaluate true objective function.

We note that finding representative and quantifiable objective functions is a non-trivial challenge, which is a justification for using simplified or surrogate objective functions that are then evaluated against the actual true objectives. For example, it is very challenging to find appropriate objective functions for multi-function radars, which are required to balance the conflicting demands of different functions. Generally, the radar designer may be less interested in the result of optimization as the all-round performance of the radar in realistic situations, which may not be possible to evaluate until after the optimization has been performed. Regardless, differences between the objective function and the actual objective and hence evaluation metrics are an indicator that the objective function may not be truly representative of the problem to be solved. We identify the construction of representative objective functions as an important challenge in cognitive radar research.

10.8 Simulation examples

In this section, we present two simulation examples that demonstrate the influence of different sources of uncertainty in the control process.

10.8.1 Adaptive tracking example

This section presents an adaptive tracking example, whereby it is necessary to decide on the next revisit interval for tracking a target with an agile beam radar. As the radar steers the beam to the estimated target position, a beam positioning loss occurs that is dependent on the difference between the beam pointing direction and the true target direction, which in turn depends on the accuracy of the track. Overall, it is desired to use as few resources as possible to track the target while also aiming to prevent the target from escaping the radar beam.

10.8.1.1 Problem components

This problem can be described in terms of the problem components introduced in Section 10.3.2.

System state and state transition function: The underlying system state is the position and velocity of the target in antenna (i.e., u–v) coordinates. The system state transitions according to a constant velocity, continuous white noise acceleration model, with a specific process noise intensity.

Actions and action space: The action is the revisit interval, which is the interval between the current time and the time of the next track update. Revisit intervals between 0.1 s and 5 s are allowed and this continuous range is discretized into 50 possible revisit interval values.

Measurements and measurement space: The radar produces measurements of the target angle in antenna (i.e., u–v) coordinates. Therefore, $\mathbf{z}_{k+1} \in [-1, 1]^2$. A measurement occurs if the signal amplitude exceeds the detection threshold assuming Swerling 1 radar cross-section fluctuations. The detection and measurement processes depend on an SNR value which is influenced by the beam positioning loss. This beam positioning loss occurs as the beam is directed to the angle given by the estimate in the track, which may differ from the true target angle. The beam positioning loss is modeled as a Gaussian loss function matched to the radar beamwidth.

Measurement-likelihood function: Each measurement dimension is corrupted by independent Gaussian noise with standard deviation depending on the SNR:

$$\sigma_{u,v} = \frac{B}{2\sqrt{SNR}} \tag{10.16}$$

where B is the radar 3-dB beamwidth. Consequently, $\mathscr{L}(\mathbf{x}_k|\mathbf{z}_k, \mathbf{a}_{k-1}) \equiv \mathscr{N}(\mathbf{z}_k; \mathbf{H}\mathbf{x}_k, \mathbf{R}_k)$ where $\mathbf{H}$ is the observation matrix

$$\mathbf{H} = \begin{bmatrix} 1 & 0 & 0 & 0 \\ 0 & 1 & 0 & 0 \end{bmatrix} \tag{10.17}$$

and $\mathbf{R}_k = \text{diag}([\sigma_{u,v}^2 \sigma_{u,v}^2])$ is the measurement error covariance.

Information state: As described in Section 10.3.2, the information state is the collection of previous actions and measurements. As the belief state described next is a sufficient statistic of the information state, it is not necessary to maintain the information state.

Belief state: The belief state comprises an estimate of the target angles in antenna coordinates and the associated covariance matrix. Additionally, the belief state contains the known mean target radar cross-section and the process noise intensity for the system state transition model.

Belief state transition function: The belief state transition function incorporates the standard Kalman filter prediction and update steps.

Reward function: If a detection occurs, then the reward is the revisit interval value that was selected. If no detection occurs, then zero reward is achieved, leading to the reward function:

$$r(\mathbf{b}_k, \mathbf{a}_k, \mathbf{z}_{k+1}) - \begin{cases} 0 & \text{if } \mathbf{z}_{k+1} = \{\} \\ \mathbf{a}_k & \text{otherwise} \end{cases} \tag{10.18}$$

Consequently, the controller is motivated to maximize the revisit interval while also ensuring a detection occurs and hence that the target does not escape the beam. The reward is normalized by the number of actions in the horizon in order to allow an easy comparison between different horizon lengths.

10.8.1.2 Control methods

The problem described above is solved using an exhaustive direct lookahead that evaluates the expected reward for every possible action. For a time horizon of multiple time steps, all possible action sequences are evaluated. The reward function is a function of several random variables that can be considered as sources of uncertainty. The state uncertainty is represented by the belief state, and results in an SNR loss due to the uncertain target angle differing from the estimated target angle. The measurement uncertainty results from Swerling 1 radar cross-section fluctuations that impact on the SNR, the stochastic detection process, and stochastic angular measurement errors.

For this analysis, we use different methods that account for different sources of uncertainty. When a source of uncertainty is considered by the controller, then the expected reward is evaluated using Monte-Carlo sampling for the respective random variable. We compare the following lookahead strategies for evaluating the expected reward:

- *Expected value state/expected value measurement* (EVS/EVM): The expected values of the state and the target radar cross-section are used. The SNR is scaled by the non-zero expected beam positioning loss. An angular measurement is generated with no measurement noise. The reward is scaled by the probability of detection.
- *Randomly sampled state/expected value measurement* (RSS/EVM): Samples of the state are drawn from the belief state, leading to samples of the beam positioning loss and consequently samples of the SNR based on the mean radar cross-section. For each sample, an expected angular measurement is generated with no measurement noise. The reward is scaled by the probability of detection.
- *Expected value state/randomly sampled measurement* (EVS/RSM): The expected value of the state and the beam positioning loss is used. The radar cross-section and hence SNR is sampled and the detection process simulated for each sample. If a detection occurs, noisy angular measurements are generated according to the standard deviation of the respective SNR sample.
- *Randomly sampled state/randomly sampled measurement* (RSS/RSM): Samples of the state and the radar cross-section are drawn, leading to samples of the beam positioning loss and consequently samples of the SNR. The detection process is simulated for each sample. If a detection occurs, noisy angular measurements are generated according to the standard deviation of the respective SNR sample.

As RSS/RSM evaluates the expected reward considering all the modeled sources of uncertainty, it can be considered as the true reward value, under the assumption that the underlying models match to the reality.

10.8.1.3 Results

These results are produced using the parameter values in Table 10.1.

The expected rewards associated with the possible actions using a single step horizon are illustrated in Figure 10.5. It can be seen that the different consideration of the sources of uncertainty in the lookahead strategies lead to different expected rewards. As the action with the greatest expected reward is selected, not considering certain sources of uncertainty can lead to sub-optimal action selections. In this result, measurement sampling (RSM) does not influence the expected reward and instead an expected measurement can be used. This is logical, as the reward function for a one-step horizon is not impacted by the different measurement values or the associated measurement noise covariances. The reward function is impacted by the detection probability, however, it is not necessary to simulate the actual detection process. In Figure 10.5, it can be seen that sampling the state results in significantly different expected rewards. The state uncertainty is the source that has the greatest influence on the expected reward. When considering RSS/RSM to be the true expected reward, not considering the state uncertainty leads to the selection of a 3.1 s revisit interval

Table 10.1 *Simulation parameters unless otherwise stated in the results. SNR is the SNR for the mean radar cross-section and without beam positioning losses.*

Parameter	**Value**
SNR	111 (20 dB)
Probability of false alarm	10^{-6}
Mean RCS	1 m^2
Beamwidth	1 (°)
Track sharpness	0.15 (beamwidths)
Process noise	$(0.004)^2$

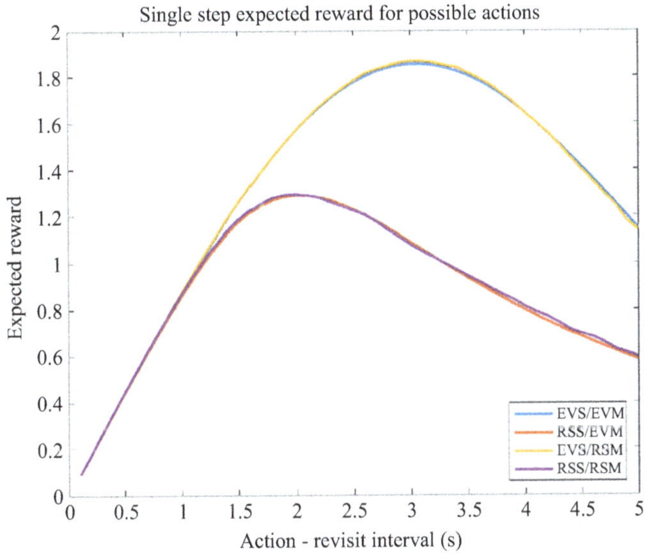

Figure 10.5 *The expected rewards using the different lookahead strategies for a one-step horizon*

instead of 2 s, which results in a 19% reduction of the expected reward from 1.294 to 1.048.

Figure 10.6 shows the best action selected for different values of the process noise intensity and track sharpness. Generally, lower revisit intervals are selected for larger process noise intensities and initial track sharpness. Not considering the state uncertainty leads to sub-optimal action selection, regardless of whether measurement sampling is performed or not. As to be expected based on Figure 10.5, EVS/EVM and EVS/RSM selected optimistically long revisit intervals, because they do not

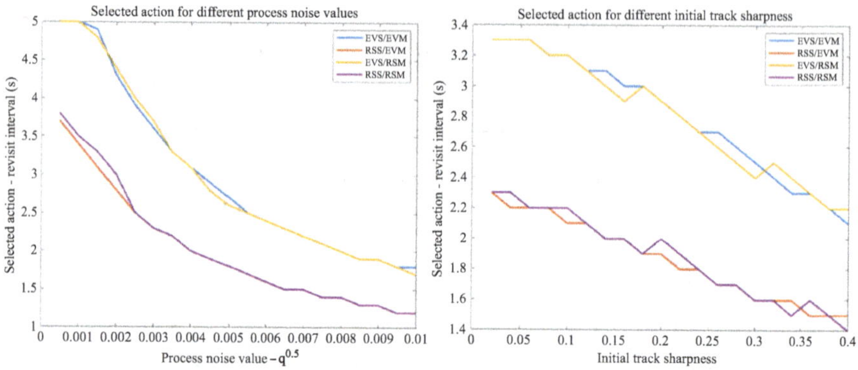

Figure 10.6 The selected actions using the different lookahead strategies for a one-step horizon

adequately evaluate the impact of beam positioning loss on the expected reward. In Figure 10.6, it can be identified that there are simple functional relationships between the parameters of the belief state and the selected action. Consequently, it is possible to perform regression on the results from the exhaustive direct lookahead to produce a PFA that has negligible online computation.

Figure 10.7 illustrates the expected reward of the possible actions using the different lookahead strategies for a two-step horizon length. In contrast to Figure 10.5, all sources of uncertainty impact on the evaluation of the expected reward. Now, measurement sampling in the first step influences the probability of detection and hence the reward in the second step. However, just performing state sampling and not measurement sampling still results in expected rewards that are closer to the optimum of RSS/RSM. By analyzing this figure, it can be seen that EVS/EVM, RSS/EVM, and EVS/RSM result in a loss of expected reward of 22.68%, 1.06%, and 8.35%, respectively. For this example, a single-step horizon instead of a two-step horizon results in a loss of expected reward of 2.33%. Although a longer time horizon improves performance, considering the sources of uncertainty has a greater impact on the expected reward and hence action selection.

Figure 10.8 shows the best action sequence selected for different values of the process noise intensity and track sharpness. Again it can be seen that larger process noise intensities and track sharpness lead to lower revisit intervals. A general recognizable strategy is to schedule a short revisit interval followed by a long revisit interval, especially in the cases of low process noise intensities as well as large initial track sharpness. As seen with the single-step horizon, a basic functional relationship between the belief state parameters and the selected action can be seen. Consequently, a PFA can be produced using regression to approximate the result of this exhaustive direct lookahead, which required significant computation even for this simple example.

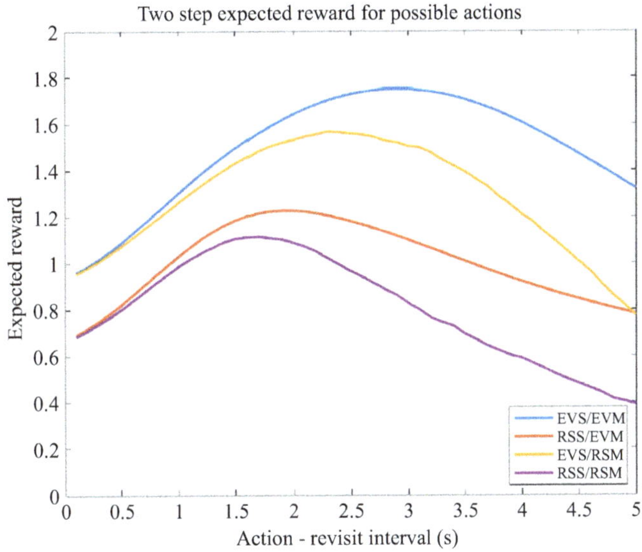

Figure 10.7 *The expected rewards using the different lookahead strategies for a two-step horizon*

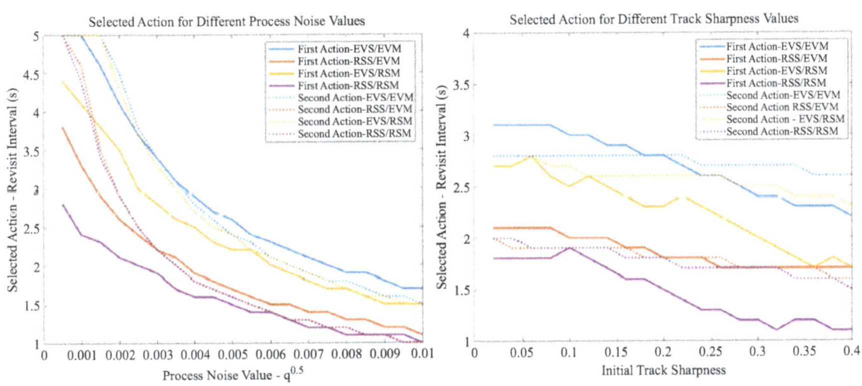

Figure 10.8 *The selected action sequence using the different lookahead strategies for a two-step horizon*

This analysis of the different lookahead strategies assumed that the underlying models match the reality and that the reward function matches the objective of the radar. Although the choice of the reward function is intuitively appealing, a radar engineer is likely to start wondering how this control strategy performs in terms of other performance measures, such as track losses and track accuracies. This highlights the difficulty in creating truly representative reward functions, as discussed in Section 10.7.3. Additionally, RSS/RSM can be considered as the true expected reward when the models are true. However, it is not the case that an object exhibits continuous white noise acceleration motion in antenna coordinates. An evaluation of these strategies on real trajectories will result in different accumulated rewards to those predicted by the lookahead strategies. This motivates the use of learning techniques for learning policy, cost or VFAs based on realistic conditions.

10.8.2 Target resource allocation example

This example uses a myopic lookahead method for allocating radar resources between multiple targets.

10.8.2.1 Scenario

The scenario contains a stationary radar, which tracks three airborne targets. The targets exhibit Swerling I radar cross-section (RCS) fluctuations and follow trajectories 1–3 from [59]. Figure 10.9(a) shows the geometry of the scenario, which has a duration of 140 s. Additional parameters are given in Table 10.2. Nominal radar parameters in Table 10.2 specify the radar performance, the actual SNR is calculated based on the actual parameter values by scaling the nominal SNR.

The radar uses a 2,000-Hz low pulse repetition frequency waveform and allocates a fixed time budget of 10% to the task of tracking the three targets. For every time step of 1,200 ms, it uses 120 ms which is equivalent to 240 pulses for tracking. At each time step, the controller must make the decision of how to allocate those pulses to each of the targets. The targets are tracked using an IMM-EKF tracker, with a nearly constant velocity and a maneuver model. As the simulation focuses on tracking and not search, tracks are initialized with the ground truth state at the beginning of the

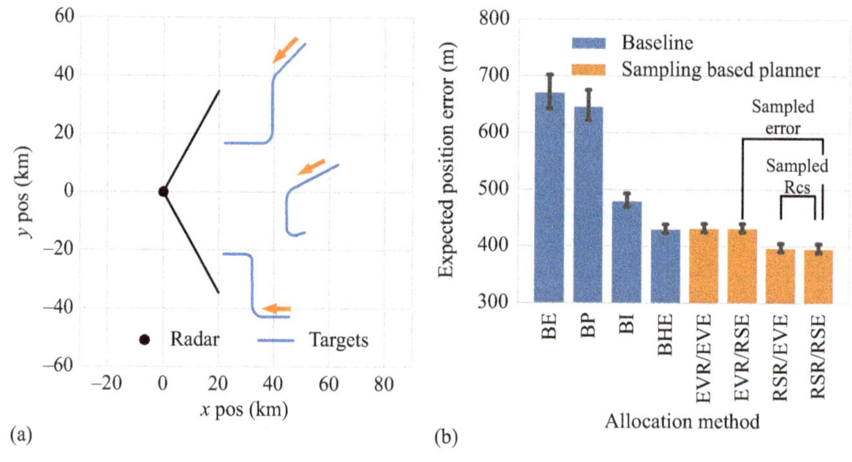

Figure 10.9 Myopic planning example. (a) Scenario. The black lines show the field of view of the radar, and the orange arrows the movement direction of the targets. (b) Expected position error for different allocation methods.

Table 10.2 Simulation parameters

Parameter	Value
Nominal radar range	50 km
Nominal radar SNR	20 (13 dB)
Nominal RCS	1 m^2
Nominal number of pulses	100
Nominal pulse length	2 μs
Pulse length	2 μs
Pulse repetition frequency	2,000 Hz
Signal bandwidth	1 MHz
Probability of a false alarm	10^{-6}
Wavelength	0.03 m
Target RCS	1 m^2
No-maneuver model noise factor q_0	10.0
Maneuver model noise factor q_1	1,000.0
Simulation length	140 s

simulation. The tracker does not drop tracks during the simulation and it uses an ideal measurement to track association.

Since tracks are not dropped, the beam is always steered towards the true position of the target in order to avoid track divergence. While a track might still diverge, it would receive measurements when it gets resources allocated again. This is of course a simplifying assumption, and a real system would need some kind of reacquisition method for lost tracks. However, the implementation of such a method would make the simulation more complex, and distract from its purpose of comparing the sampling uncertainties.

10.8.2.2 Objective

The objective is to minimize the uncertainty of the tracks. We quantify this using just the position part of the covariance matrices for the tracks, leading to a negative reward (cost) at each stage k of

$$r(\mathbf{b}_k, \mathbf{a}_k, \mathbf{z}_{k+1}) = -\sum_{t \in T} \sqrt{\operatorname{tr}\left(\mathbf{P}^{\text{pos}}_{(k+1)t}\right)} \tag{10.19}$$

where T is the set of targets, $\mathbf{P}^{\text{pos}}_{(k+1)t}$ the 3×3 Cartesian position part of the covariance matrix of target t at time step $k+1$, updated with $\mathbf{z}_{k+1}$, and `tr` is the trace function.

10.8.2.3 Control methods

We use four baseline control methods, which use simple heuristics:

- (BE) equally allocates the same amount of pulses to each target.
- (BP) allocates the pulses to targets such that all targets achieve the same SNR.
- (BI) uses all the pulses for a single target and iterates through the targets between decision steps.

- (BHE) uses all pulses on the target with the currently highest expected error, i.e., the target whose contribution to the reward is the highest.

The planner is a myopic one-step lookahead planner. It considers a discrete action set A of possible allocations and selects the action

$$\mathbf{a}_k = \arg\max_{\mathbf{a} \in A} \mathbb{E}\left[r_k(\mathbf{b}_k, \mathbf{a}, Z_{k+1}) \mid \mathbf{a}\right] \tag{10.20}$$

The actions consist of possible allocations of multiples of 60 pulses, leading to 15 different possible actions. For example, 240/0/0 or 60/120/60 are two possible actions. The control algorithm evaluates the expected value using Monte-Carlo sampling. We sample two different random variables in the measurement generation process: the RCS of the targets and the measurement errors. The RCS fluctuation influences the measurement SNR and therefore also the detection probability and measurement covariance. The measurement error is distributed according to the covariance of the measurement and influences not only the point estimate of the target but also the likelihood of the different maneuver models. As we want to compare the influence of these two factors, we either perform Monte-Carlo sampling or take the expected value. Each action is sampled 66 times, leading to a full computing budget of slightly below 1,000 samples. We use common random numbers when comparing the actions in order to reduce the sampling variance.

10.8.2.4 Results

Figure 10.9(b) shows the performance of the different methods after 100 Monte-Carlo runs. The names of the different planner instantiations consist of combinations of randomly sampled (RS) or expected value (EV) of the RCS (R) or the error (E). For example, the rightmost entry results from a randomly sampled RCS and a randomly sampled error.

The results are given as the average from the covariance per target and per decision step. Note that because the tracks are not dropped during the scenario and consequently the number of targets does not change, this scaled representation is proportional to the negative sum of the rewards $r_k(\mathbf{b}_k, \mathbf{a}_k, \mathbf{z}_{k+1})$ over the whole scenario. However, a scaled representation results in a more interpretable error metric.

We can see multiple effects in Figure 10.9(b). First, it seems to pay off to focus the pulses on a single target. Both baseline heuristic methods that spread the pulses (BE and BP) are significantly worse than those that focus them on a single target (BI and BHE). Second, the random variable that is sampled in the measurement process for the planner has a clear influence. When using the expected RCS, the planner is approximately as good as the strongest baseline. However, when sampling the RCS, the planner surpasses the baseline. On the other hand, whether it samples the measurement error or simply takes expected measurement has no significant influence on performance. In this scenario, the RCS sampling has mostly an effect on the topmost target, which is furthest away. When taking the expected value of the RCS, the planner assumes that there is no detection chance at all and never allocates resources to this target for the first 25 s of the scenario. However, when sampling the RCS, it recognizes that the RCS fluctuations provide a chance of a detection.

The keen-eyed reader has likely realized that we did not consider a third source of uncertainty, the actual position of the target. Instead, we only used the expected value of the track. In theory, we have a probability distribution over the target given by the mean and covariance in the track. However, it is very rare that the target is really distributed according to this probability distribution function as the control output of few pilots is truly Gaussian noise, as assumed by many models. A Gaussian process model works well to make the tracker stable, but is not necessarily suited for lookahead planning. In our experiments, sampling the state had sometimes even detrimental effects, as the planner thought the target could be very far away and it would not reach a detection, for example, in situations involving maneuvers that led to a large covariance. This highlights the value of higher fidelity target dynamic models [60] in the context of planning. In the given setup, we could also not find benefits from a non-myopic planning. Note that this is not necessarily surprising, given the performance guarantees of greedy strategies in comparison to non-myopic controllers [61].

This example shows that it is important to consider the source of uncertainty in the planning step. Some sources may have a large impact on performance and some may have little impact. This consideration can be incorporated into the design of a planning algorithm using the stochastic optimization framework described in this chapter. For example, the results above would indicate that replacing the sampling process by just two outcomes—detection and non-detection, weighted by their analytical probabilities, is likely sufficient. The influence of other sources of uncertainties, e.g., the target state, might have an influence; however, care must be taken that the sampling distribution actually corresponds to the true possible states of the targets.

10.9 Conclusion

Many cognitive radar techniques are emerging that tackle different applications or sub-problems in a radar system. This chapter has presented a common framework for describing these cognitive radar problems in terms of a stochastic optimization problem. By doing so, the cognitive radar problem can be addressed using existing algorithmic strategies from the field of stochastic optimization. Specifically, the strategy of finding functional approximations for the optimal policy, cost or value function using machine learning techniques is an attractive approach. In general, learning techniques can be adopted to tackle a stochastic optimization problem, when models in the problem are difficult to describe analytically. Consequently, both control-theoretic methods and learning methods fall under the same framework.

Traditionally, cognitive radar and radar management have performed myopic and deterministic optimizations. However, advances in computing and algorithmic capabilities can enable the more general stochastic optimization problem to be tackled, which fully considers uncertain measurements and state transitions as well as the impact of action selection on future rewards. However, it is important to consider which sources of uncertainty actually impact on the performance. Since increasing the consideration of uncertainty in a planning algorithm ultimately results in increased computation, focusing on just the critical sources of uncertainty enables an efficient and performant algorithm.

References

[1] Gurbuz SZ, Griffiths HD, Charlish A, *et al.* An overview of cognitive radar: past, present, and future. *IEEE Aerospace and Electronic Systems Magazine.* 2019;34(12):6–18.

[2] Charlish A and Hoffmann F. Cognitive radar management. In: *Novel Radar Techniques and Applications. Volume 2: Waveform Diversity and Cognitive Radar, and Target Tracking and Data Fusion.* London: Institution of Engineering and Technology; 2017. p. 157–193.

[3] Powell WB. A unified framework for stochastic optimization. *European Journal of Operational Research.* 2019;275(3):795–821.

[4] Howard WW, Thornton CE, Martone AF, *et al.* Multi-player Bandits for distributed cognitive radar. In: *2021 IEEE Radar Conference (RadarConf21); 2021.*

[5] Boer TD, Schöpe MI, and Driessen H. Radar resource management for multi-target tracking using model predictive control. In: *IEEE 24th International Conference on Information Fusion (FUSION).* Los Altos, CA: International Society of Information Fusion (ISIF); 2021.

[6] Thornton CE, Kozy MA, Buehrer RM, *et al.* Deep reinforcement learning control for radar detection and tracking in congested spectral environments. *IEEE Transactions on Cognitive Communications and Networking.* 2020;6(4):1335–1349.

[7] Charlish A, Bell K, and Kreucher C. Implementing perception-action cycles using stochastic optimization. In: *2020 IEEE Radar Conference (RadarConf20)*; 2020. p. 1–6.

[8] Hero AO, Castañon DA, Cochran D, *et al. Foundations and Applications of Sensor Management.* Berlin: Springer; 2007.

[9] Musick S and Malhotra R. Chasing the elusive sensor manager. In: *Proceedings of NAECON*; 1994 May. p. 606–613.

[10] Liu J, Cheung P, Guibas L, *et al.* A dual-space approach to tracking and sensor management in wireless sensor networks. In: *ACM International Workshop on Wireless Sensor Networks and Applications*; 2002 September.

[11] Lumelsky VJ, Mukhopadhyay S, and Sun K. Dynamic path planning in sensor-based terrain acquisition. *IEEE Transactions on Robotics and Automation.* 1990;6(4):462–472.

[12] Popoli R. The sensor management imperative. In: Bar-Shalom Y, editor. *Multitarget-Multisensor Tracking: Advanced Applications.* vol. II. Boston, MA: Artech House; 1992. p. 325–392.

[13] Hintz KJ and McVey ES. Multi-process constrained estimation. *IEEE Transactions on Man, Systems, and Cybernetics.* 1991;21(1):434–442.

[14] Hintz KJ. A measure of the information gain attributable to cueing. *IEEE Transactions on Systems, Man and Cybernetics*. 1991;21(2):237–244.

[15] Schmaedeke W and Kastella K. Event-averaged maximum likelihood estimation and information-based sensor management. *Proceedings of SPIE*. 1994;2232:91–96.

[16] Kastella K. Discrimination gain for sensor management in multitarget detection and tracking. *IEEE-SMC and IMACS Multiconference CESA*. 1996;1:167–172.

[17] Kastella K. Discrimination gain to optimize classification. *IEEE Transactions on Systems, Man and Cybernetics—Part A: Systems and Humans*. 1997;27(1):112–116.

[18] Mahler R. Global optimal sensor allocation. In: *Proceedings of the Ninth National Symposium on Sensor Fusion*; 1996, vol. 1. p. 167–172.

[19] Zhao F, Shin J, and Reich J. Information-driven dynamic sensor collaboration. *IEEE Signal Processing Magazine*. 2002;p. 61–72.

[20] Chong E, Kreucher C, and Hero A. Partially observable Markov decision process approximations for adaptive sensing. *Discrete Event Dynamic Systems*. 2009;19(3):377–422.

[21] Chong E, Kreucher C, and Hero A. POMDP approximation using simulation and heuristics. In: Hero A, Castañon D, Cochran D, *et al.*, editors. *Foundations and Applications of Sensor Management*. Berlin: Springer; 2008. p. 95–120.

[22] Krishnamurthy V and Evans D. Hidden Markov model multiarm bandits: a methodology for beam scheduling in multitarget tracking. *IEEE Transactions on Signal Processing*. 2001;49(12):2893–2908.

[23] Krishnamurthy V. Algorithms for optimal scheduling and management of hidden Markov model sensors. *IEEE Transactions on Signal Processing*. 2002;50(6):1382–1397.

[24] Bertsekas D and Castañon D. Rollout algorithms for stochastic scheduling problems. *Journal of Heuristics*. 1999;5(1):89–108.

[25] Castañon D. Approximate dynamic programming for sensor management. In: *Proceedings of the 1997 IEEE Conference on Decision and Control*; 1997.

[26] Castañon D. Optimal search strategies for dynamic hypothesis testing. *IEEE Transactions on Systems, Man, and Cybernetics*. 1995;25(7):1130–1138.

[27] Malhotra R. Temporal considerations in sensor management. In: *Proceedings of the IEEE 1995 National Aerospace and Electronics Conference, NAECON 1995*; 1995 May, vol. 1. p. 86–93.

[28] Chhetri A, Morrell D, and Papandreou-Suppappola A. The use of particle filtering with the unscented transform to schedule sensors multiple steps ahead. In: *Proceedings of International Conference on Acoustics, Speech, and Signal Processing 2004*; 2004.

[29] Shin J, Guibas L, and Zhao F. A distributed algorithm for managing multi-target identities in wireless ad-hoc sensor networks. In: *Proceedings of 2nd International Workshop on Information Processing in Sensor Networks*; 2003 April.

[30] Aoki EH, Bagchi A, Mandal P, *et al.* A theoretical look at information-driven sensor management criteria. In: *14th International Conference on Information Fusion*; 2011. p. 1–8.

[31] Sundaresan R. A measure of discrimination and its geometric properties. In: *Proceedings of the 2002 IEEE International Symposium on Information Theory*. IEEE; 2002. p. 264.

[32] Liese F and Vajda I. On divergences and informations in statistics and information theory. *IEEE Transactions on Information Theory*. 2006;52(10):4394–4412.

[33] Van Trees HL and Bell KL, editors. *Bayesian Bounds for Nonlinear Filtering/Tracking*. New York: Wiley; 2007.

[34] Ashok Kumar M and Mishra KV. Information geometric approach to Bayesian lower error bounds. In: *2018 IEEE International Symposium on Information Theory (ISIT)*; 2018. p. 746–750.

[35] Ashok Kumar M and Mishra KV. Cramér–Rao lower bounds arising from generalized Csiszár divergences. *Information Geometry*. 2020;3:33–59.

[36] Hansen JP, Ghosh S, Rajkumar R, *et al.* Resource management of highly configurable tasks. In: *18th International Parallel and Distributed Processing Symposium*. Santa Fe, NM; 2004. p. 116.

[37] Mitchell AE, Smith GE, Bell KL, *et al.* Cost function design for the fully adaptive radar framework. *IET Radar, Sonar, and Navigation*. 2018;12(12):1380–1389.

[38] Yuan Y, Yi W, Kirubarajan T, *et al.* Scaled accuracy based power allocation for multi-target tracking with colocated MIMO radars. *IEEE Journal on Selected Topics in Signal Processing*. 2019;158:227–240.

[39] van Keuk G and Blackman SS. On phased-array radar tracking and parameter control. *IEEE Transactions on Aerospace and Electronic Systems*. 1993;29(1):186–194.

[40] Christiansen JM, Olsen KE, and Smith GE. Fully adaptive radar for track update-interval control. In: *2018 IEEE Radar Conference, RadarConf 2018*. IEEE; 2018. p. 400–404.

[41] Sutton RS and Barto AG. *Reinforcement Learning: An Introduction*, 2nd ed. Cambridge, MA: MIT Press; 2018.

[42] Hoffmann F and Charlish A. A resource allocation model for the radar search function. In: *International Radar Conference*. Lille: IEEE; 2014. p. 1–6.

[43] Bell KL, Baker CJ, Smith GE, *et al.* Cognitive radar framework for target detection and tracking. *IEEE Journal on Selected Topics in Signal Processing*. 2015;9(8):1427–1439.

[44] Sira SP, Li Y, Papandreou-Suppappola A, *et al.* Waveform-agile sensing for tracking. *IEEE Signal Processing Magazine*. 2009;26(1):53–64.

[45] Charlish A and Hoffmann F. Anticipation in cognitive radar using stochastic control. In: *2015 IEEE Radar Conference (RadarCon)*. Arlington, VA: IEEE; 2015. p. 1692–1697.

[46] Schöpe MI, Driessen H, and Yarovoy A. Multi-task sensor resource balancing using Lagrangian relaxation and policy rollout. In: *2020 IEEE 23rd International Conference on Information Fusion (FUSION)*; 2020. p. 1–8.

[47] Kershaw DJ and Evans RJ. Waveform selective probabilistic data association. *IEEE Transactions on Aerospace and Electronic Systems*. 1997;33(4): 1180–1188.

[48] Kirubarajan T, Bar-Shalom Y, Blair WD, *et al.* IMMPDAF for radar management and tracking benchmark with ECM. *IEEE Transactions on Aerospace and Electronic Systems*. 1998;34(4):1115–1134.

[49] Chong EKP, Kreucher CM, and Hero AO. Monte-Carlo-based partially observable Markov decision process approximations for adaptive sensing. In: *9th International Workshop on Discrete Event Systems*; 2008. p. 173–180.

[50] Haykin S. Cognitive Dynamic Systems: Perception-Action Cycle, Radar and Radio. Cambridge: Cambridge University Press; 2012.

[51] John-Baptiste P and Smith GE. Utilizing neural networks for fully adaptive radar. In: *IEEE Radar Conference*; 2019. p. 1–6.

[52] Guerci JR. *Cognitive Radar: The Knowledge-Aided Fully Adaptive Approach*. Boston, MA: Artech House; 2010.

[53] Aubry A, De Maio A, Piezzo M, *et al.* Radar waveform design in a spectrally crowded environment via nonconvex quadratic optimization. *IEEE Transactions on Aerospace and Electronic Systems*. 2014;50(2):1138–1152.

[54] Stinco P, Greco M, and Gini F. Cognitive radars in spectrally dense environments. *IEEE Aerospace and Electronic Systems Magazine*. 2016;31(10): 20–27.

[55] Martone AF, Ranney KI, Sherbondy K, *et al.* Spectrum allocation for noncooperative radar coexistence. *IEEE Transactions on Aerospace and Electronic Systems*. 2018;54(1):90–105.

[56] Goodman NA, Venkata PR, and Neifeld MA. Adaptive waveform design and sequential hypothesis testing for target recognition with active sensors. *IEEE Journal of Selected Topics in Signal Processing*. 2007;1(1):105–113.

[57] Kreucher C, Hero AO, and Kastella K. A comparison of task driven and information driven sensor management for target tracking. In: *44th IEEE Conference on Decision and Control*; 2005. p. 4004–4009.

[58] Charlish A and Katsilieris F. Array radar resource management. In: *Novel Radar Techniques and Applications: Real Aperture Array Radar, Imaging Radar, and Passive and Multistatic Radar*. London: Institution of Engineering and Technology; 2017. p. 135–171.

[59] Blair WD, Watson GA, Kirubarajan T, *et al.* Benchmark for radar allocation and tracking in ECM. *IEEE Transactions on Aerospace and Electronic Systems*. 1998;34(4):1097–1114.

[60] Jung S, Schlangen I, and Charlish A. A mnemonic Kalman filter for non-linear systems with extensive temporal dependencies. *IEEE Signal Processing Letters*. 2020;27:1005–1009.

[61] Williams JL. *Information Theoretic Sensor Management*. Cambridge, MA: Massachusetts Institute of Technology; 2007.

Chapter 11
Applications of game theory in cognitive radar

Chenguang Shi[1], Mathini Sellathurai[2], Fei Wang[1] and Jianjiang Zhou[1]

In this chapter, the game theory is applied to deal with the problem of spectrum sharing between cognitive multistatic radar and a communication system. To be specific, the non-cooperative game theory-based power allocation (NCGT-PA) problem for cognitive multistatic radar systems is studied, which coexists with a communications system in the same frequency band. The key mechanism of the cognitive radar system is to reduce the transmit power of each radar node while satisfying a certain target detection criterion and a maximum tolerable interference constraint for the communication system. Considering the rationality and selfishness of each radar, we adopt the non-cooperative game model to capture the interactions among multiple radars. The utility function of each radar is defined and serves as the optimization criterion for designing the sub-optimal power allocation strategy, taking into account the target detection requirement and the total interference to the communications system. Furthermore, the analytical expression for the Nash equilibrium of the established game model is derived, and the existence and uniqueness of the Nash equilibrium are strictly proved. An efficient iterative power allocation approach that determines the transmit power of each radar is put forward. Finally, several simulation results are provided to demonstrate the theoretical analysis of the Nash equilibrium and the effectiveness of the proposed strategy.

11.1 Introduction

11.1.1 Research background

In the wake of the rapid developments of large bandwidth wireless networks, multi-channel electronic scanning antenna, high-speed low-cost processors, and precise synchronization techniques, it becomes feasible to implement the decentralized cognitive multistatic radar system in practice [1]. Multiple transmitters can simultaneously transmit multiple different and independent waveforms owing to the

[1]Key Laboratory of Radar Imaging and Microwave Photonics, Ministry of Education, Nanjing University of Aeronautics and Astronautics, China
[2]School of Engineering and Physical Sciences, Heriot-Watt University, UK

unique configuration of multistatic system [2]. It has been shown that the cognitive multistatic radar system equipped with multiple transmitters and multiple receivers possesses many advantages when compared with the traditional monostatic radar, due to its spatial and signal diversities. Extensive studies have been conducted to explore the potential utilization of such system in miscellaneous situations such as target detection [3,4], target localization [5], target tracking [1,6], parameter estimation [7,8], adaptive waveform optimization [9,10], sensor assignment [11,12], information extraction [13], and so forth.

Due to the recent advancements in high-bandwidth services and mobile communications, the scarcity of radio frequency (RF) spectrum has become a worldwide problem. One of the solutions to handle such a problem is to enhance the utilization of the existing RF spectrum. Recently, spectrum sharing has been considered as an effective measure to tackle the problem of spectrum congestion [14], which is composed of two or more users, i.e., radar or communications systems, sharing the same frequency band. The authors in [15] propose a dynamic spectrum allocation strategy for a radar system coexisting with a communications base station, in which the problem of joint optimization of transmitted waveform and power spectrum is addressed with a predefined signal-to-interference-plus-noise ratio (SINR) constraint. In [16], the time-delay estimation for cooperative multicarrier radar and communications system is addressed, and it has been illustrated that the radar can improve its parameter estimation accuracy with the target returns due to communication signals. Taking into account the fact that it is impossible to obtain the precise characteristics of target spectra in reality, the authors in [17] investigate the problem of robust orthogonal frequency division multiplexing (OFDM) radar waveform design for the coexisting radar and multiple communications systems with the minimum possible power. In [18], a cooperative spectrum-sharing scheme is proposed, which enhances the detection performance of the radar system with a specified rate constraint for the communications system. The authors in [19] present a mathematical framework for coexisting pulsed radars and communication systems. Considering user quality of service (QoS) requirements, minimum power, and interference limits, the joint user association and power allocation problem is also studied in [20] for millimeter-wave-based ultra-dense networks.

11.1.2 Literature review

Non-cooperative game theory is an effective tool for decentralized optimization problems, because it provides a mathematical framework to analyze the interactions between rational but selfish players [21]. Each player in the model strives to maximize its own utility given the strategies of the other players [22]. Game theoretic models have been extensively studied in various fields, and recently have become a promising tool for the cognitive radar system and signal processing.

11.1.2.1 Anti-jamming design

Many works focus on the applications of game theory in anti-jamming design for a cognitive radar system, in this scenario, the interaction between the radar and the adversary is generally modeled as a non-cooperative game based on the available information and then the established optimization problems are tackled in order to determine the favorable strategies. For example, the authors in [23] exploit a two-person zero-sum (TPZS) game with mutual information (MI) between the received signal and the path gain as utility functions to formulate the interaction between a smart target and a cognitive multiple-input multiple-output (MIMO) radar, where the Stackelberg equilibrium for the hierarchical game is analytically derived and the existence condition of the Nash equilibrium for the symmetric game is also analyzed. Considering the impact of the clutter on the results of the game, the authors in [24] apply the Stackelberg game that captures the interaction between a MIMO radar and a jammer with mutual information as utility functions. The solutions corresponding to weak clutter and strong clutter conditions are analytically derived with a two-step water-filling method. Besides, the impact of the MIMO radar with destroyed antennas on the resulting solutions is also taken into account. Reference [25] looks into the power allocation game between a cognitive radar network and multiple jammers where the primary goal of the radars is to maintain the detection performance for multiple targets with minimum possible power consumption. The existence and uniqueness of Nash equilibrium are analytically proved, and the power allocation strategy of the radar system is derived based on the best response function. In [26], a TPZS game is established with the performance of constant false alarm rate being utility functions and then the strategy selection for both the cognitive radar and the jammer is investigated for three jamming scenarios, i.e., ungated range noise, range-gated noise, and false-target jamming. The authors in [27] formulate the polarization design problem for a cognitive MIMO radar in the presence of deceptive jamming as a TPZS game and propose two design methods based on the solution of the proposed unilateral game and Nash game. In addition, considering the target radar cross-section (RCS) as the incomplete information for a cognitive MIMO radar [28], the interactions between the radar and a jammer are modeled as a TPZS game and a Bayesian game, respectively. The resulting optimization problems are formulated as a power allocation game with MI as utility functions and an iterative algorithm is proposed to achieve the Nash equilibrium. In [29], the power allocation problem for a joint radar and communications system is formulated as a Bayesian game with uncertainty about the capability of a jammer, where the SINR is employed as the performance metric. The analytical expression of the Bayesian Nash equilibrium is derived as a function of probability. The authors in [30] investigate the problem of joint beamforming and power allocation for a radar network in the presence of multiple jammers. The main objective of each radar is to gauge the detection performance, i.e., SINR requirement for the target, and to mitigate the jamming effect from the jammer with minimum power consumption, while the multiple jammers decide their jamming strategies based on the transmit power of the radars. In [31], the electronic counter-countermeasures (ECCM) between a cognitive radar and a jammer are formulated as a Stackelberg game where the radar acts as the leader and the jammer as the follower. Then, the authors design the ECCM strategy for the radar by optimizing the convex utility functions that are a trade-off between the SINR of the target measurement and power consumption. Finally, the conditions that the optimal ECCM strategy is an increasing function of the jamming power injected into the radar receiver are derived.

11.1.2.2 Power control design

Considerable efforts have been devoted to the problem of power allocation for a cognitive radar system under the constraints of predefined performance requirements and limited resource budgets. For instance, a game theoretical power allocation scheme is developed in [32] for distributed radar network, which aims to improve the target detection performance of the underlying system by optimizing the transmit power of each radar. The authors in [33] apply a non-cooperative game to address the power allocation problem in a multistatic MIMO radar network, whose primary goal is to achieve the predetermined SINR threshold for each radar with the minimum power consumption. Based on the works in [32], the existence and uniqueness of the Nash equilibrium are strictly proved with the Karush–Kuhn–Tucker (KKT) conditions and an interesting conclusion that the number of radars illuminating signals is equal to the number of the radars satisfying the SINR requirement with equality is drawn [2]. Besides, the robust power allocation for a multistatic cognitive radar system with the existence of estimation error is investigated in [34]. Taking into account the fact that the knowledge of channel gains could be incomplete, the authors in [35] incorporate a Bayesian game to tackle the power allocation problem for the radar system with the objective to maximize the expected SINR for each radar given the power budgets. In [36], the authors utilize the Stackelberg game to capture the interaction between a cognitive radar network and a hostile intercept receiver, where the intercept receiver acts as the leader who determines the price of transmission for the radar system while the radar system as the follower aiming at minimizing its transmit power under the constraints on the specified SINR threshold. In [37], the problem of joint beamforming and power allocation for a multistatic radar system in the presence of multiple targets is investigated based on the proposed three game models, i.e., a strategic non-cooperative game, a partially cooperative game, and a Stackelberg game with the transmit power of each radar being utility functions and the SINR thresholds for each of the targets being performance constraints. In [38], the robust power allocation problem for a radar network coexisting with a communications system is investigated based on the Stackelberg game, where the communications system is the leader inferring the interference from the radars and sending the price of transmission to the radars while the radars are the follower determining their power allocation according to the price issued by the leader. The resulting optimization problem for the radars is formulated as optimizing the utility function for the worst case under the constraints on the SINR thresholds and the power limits. The authors in [39] investigate the problem of joint beamforming and power allocation for a multistatic MIMO radar network in the presence of multiple targets, where the primary goal of each radar is to guarantee a required SINR threshold for each target while consuming as little power as possible.

11.1.2.3 Waveform design

It has been shown that game theory provides a desired tool for cognitive radar waveform design. For instance, in [40], the problem of polarimetric waveform design is studied for a distributed cognitive MIMO radar, where the transmit polarizations are determined based on the solution of the proposed TPZS game between the radar and the adversary. In [41], the problem of code waveforms design is investigated with non-cooperative games. Specifically, the code design problems are formulated as maximizing the output SINR of a matched filter, a minimum integrated sidelobe level filter, and a minimum peak to sidelobe level filter for each radar. For each case, the existence of the Nash equilibrium is strictly proved with the theory of potential games and the non-cooperative code update algorithm is proposed. Inspired by the works in [41], the authors in [42] exploit the potential game theory to tackle the problem of optimal waveform design for multistatic radar networks where the optimization problem is formulated as maximizing the sum of the denominators of the SINR of all radars under the constraints on available waveforms. The authors in [43] investigate the problem of joint design of amplitudes and frequency-hopping codes from the perspective of game theory for a colocated MIMO radar system, in which two players, with the objectives of minimizing the cost functions corresponding to amplitude design and code design, respectively, compete with each other subject to the energy constraint. Two joint design algorithms, i.e., a non-cooperative scheme and a cooperative scheme, are proposed to achieve the approximated equilibrium. In [44], the problem of waveform design is formulated as a TPZS game considering the conflict between the cognitive monostatic radar and the jammer. Similar to [23], the Stackelberg equilibrium is analytically derived and the condition that there exists the Nash equilibrium is studied. Taking into account the uncertainty of the target spectrum, a Stackelberg game is applied in [45] to design the transmit waveform in the presence of a jammer. In the proposed game, the radar is a leader while the jammer is a follower and the resulting optimization problems are formulated as a TPZS game with one player intending to maximize the MI and the other to minimize the MI.

11.1.3 Motivation

Although the problem of spectrum sharing between cognitive radar and communications systems has been widely studied, there are still a number of aspects that need to be improved: (i) all the existing researches only focus on the traditional monostatic radar system and are not applicable to the cognitive multistatic radar architecture, which is characterized by the huge complicated limitations and calculations. (ii) Even though the non-cooperative game theory has been adopted to guide power resource allocation in multistatic radar, the analytical expression for game theory-based power allocation has not been provided. (iii) The application of a non-cooperative game model for spectral coexistence between cognitive multistatic radar and communications systems has not been considered yet. Furthermore, although the coexistence problem of radar and long-term evolution (LTE) systems is tackled with non-cooperative game theory in [14], the details of the proposed algorithm and the corresponding simulation results are not presented. Particularly, the authors in [46] propose a non-cooperative game-based power allocation and sub-channel assignment algorithm for heterogeneous networks with incomplete channel state information (CSI). Motivated by the analysis in [46], we extend the non-cooperative game to spectral coexistence scenario, where multiple radar nodes and a communications system share the same frequency band. To the best of our knowledge, we are the first to study the problem of game theoretic power allocation of the cognitive multistatic radar system for spectral coexistence.

11.1.4 Major contributions

In this chapter, the problem of power allocation for a cognitive multistatic radar system is investigated, where multiple radar nodes coexist with a communications system in the same RF spectrum. Considering that different radars may not cooperate with each other, we establish a non-cooperative game theoretic model to deal with the above optimization problem. The main objective of the cognitive radar system is to maintain a specified SINR threshold for target detection and guarantee a maximum total interference for the communications system while allocating the minimum possible transmit power for each radar.

The main contributions of this chapter are as follows:

1. We study the problem of non-cooperative game theory-based power allocation (NCGT-PA) for a cognitive multistatic radar system coexisting with a communications system. Specifically, the NCGT-PA strategy aims to minimize the power consumption of each radar node, while satisfying a predefined SINR threshold for target detection and an acceptable interference level for the communications system. Considering the rationality and selfishness of each radar, the non-cooperative game theoretic technique is employed to solve the decentralized power allocation problem. Herein, both the desired SINR requirement and the total received interference power level are incorporated into the utility function of each radar. Thereby, the NCGT-PA strategy is based on the maximization of the utility function, which leads to the sub-optimal transmit power allocation for each radar.
2. The closed-form expression for the Nash equilibrium of the non-cooperative game model is analytically derived by the Lagrangian dual function. In addition, the existence and uniqueness of the Nash equilibrium are strictly testified.
3. A distributed iterative power allocation approach is provided, which is capable of obtaining the Nash equilibrium solutions to the proposed strategy from any feasible initial points. The proposed algorithm significantly reduces the computational complexity and signaling overhead, and it ensures fast convergence in practice.
4. We provide several simulation results to evaluate the performance of the proposed NCGT-PA strategy. It is illustrated that the NCGT-PA strategy can satisfy the specified SINR requirement for target detection and guarantee the maximum tolerable interference for the communications system while assigning the minimum transmit power to each radar. Furthermore, it is also shown that the power allocation results are associated with the target's RCS and the relative geometry between the cognitive multistatic radar system and the target.

11.1.5 Outline of the chapter

The remaining parts of this chapter are organized in the following. Section 11.2 presents the essential assumptions and the spectral coexistence model between a cognitive multistatic radar and a wireless communications system. The game theoretic formulation of the underlying optimization problem is presented in Section 11.3. Section 11.4 focuses on the proof of the existence and uniqueness of Nash equilibrium. Section 11.5 presents a non-cooperative distributed iterative algorithm, which determines the solution to the NCGT-PA model. Simulation results are provided in Section 11.6 to evaluate the performance of the proposed strategy. Finally, Section 11.7 concludes this chapter.

11.2 System and signal models

11.2.1 System model

Consider a cognitive multistatic radar system that consists of N_T radar nodes sharing the same frequency band with a communications system, as illustrated in Figure 11.1. The main idea of the cognitive radar system is to reduce its total power consumption by optimizing the power resource allocation among different radars, subject to the desired SINR threshold for target detection and a maximum tolerable interference for the communications system, respectively. Due to the lack of radar transmission synchronization [47], the signals transmitted from multiple radars might not be orthogonal and consequently leads to extensive inter-radar interference. It is supposed that each

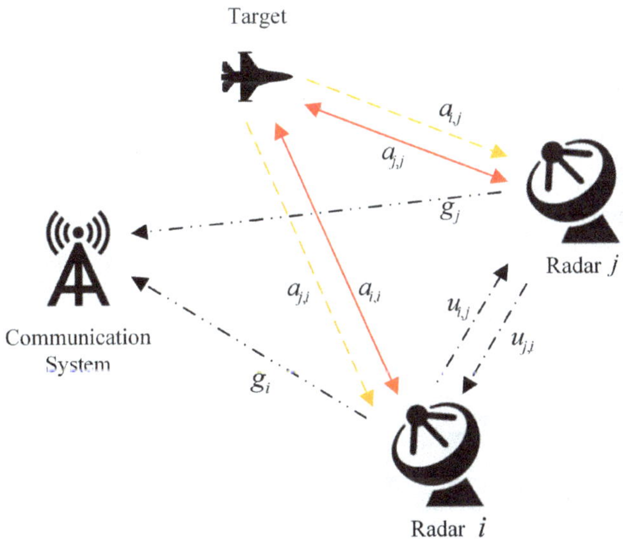

Figure 11.1 Illustration of the system model

radar adopts the successive interference cancellation (SIC) technique [16] to decode the communication signals and to remove the interference induced by both direct and target scattered communication signals. We also assume that the radar signal scattered off the target is extremely weak at the communications system, compared to the signal that comes through the direct path from the radar and hence is neglected for the sake of simplicity.

11.2.2 Signal model

As mentioned earlier, each radar independently optimizes its transmit power to achieve a predefined SINR threshold for target detection in the considered non-cooperative game model. In order to determine the presence of a target, we employ the generalized likelihood ratio test (GLRT) [2] to design an efficient detector for the hypothesis testing. Therefore, the time-domain samples of the received signals for radar i are given by:

$$\text{Target being absent: } \mathbf{x}_i = \sum_{j=1, j\neq i}^{N_\mathrm{T}} \kappa_{i,j}\sqrt{P_j}\mathbf{s}_j + n_i, \tag{11.1}$$

$$\text{Target being present: } \mathbf{x}_i = \xi_i\sqrt{P_i}\mathbf{s}_i + \sum_{j=1, j\neq i}^{N_\mathrm{T}} \kappa_{i,j}\sqrt{P_j}\mathbf{s}_j + n_i, \tag{11.2}$$

where $\mathbf{s}_i = \phi_i\mathbf{a}_i$ represents the emitted signals from radar i, ϕ_i is the predesigned waveform emitted from radar i, $\mathbf{a}_i = [1, e^{j2\pi f_{\mathrm{D},i}}, \cdots, e^{j2\pi(N-1)f_{\mathrm{D},i}}]$ stands for the Doppler steering vector of radar i with respect to the target, $f_{\mathrm{D},i}$ denotes the corresponding Doppler frequency shift with respect to radar i, N is the number of signals received samples during the dwell time, ξ_i is the channel gain incorporating the target's RCS, P_i denotes the transmit power of radar i, $\kappa_{i,j}$ represents the direct cross-channel gain from radar i to radar j, and n_i is white Gaussian noise (WGN) with variance σ_n^2. We suppose that $\xi_i \sim \mathcal{CN}(0, a_{i,i})$, $\kappa_{i,j} \sim \mathcal{CN}(0, c_{i,j}(a_{i,j} + u_{i,j}))$, and $n_i \sim \mathcal{CN}(0, \sigma_n^2)$, where $a_{i,i}$ stands for the variance of the desired channel gain, including the target's RCS, $c_{i,j}a_{i,j}$ is the variance of the target reflection gain at radar j due to the transmission of radar i, $c_{i,j}u_{i,j}$ represents the variance of the direct cross-channel gain from radar i to radar j, and $c_{i,j}$ denotes the cross-correlation factor between radar i and radar j.

The propagation gains of the corresponding paths are defined as:

$$\begin{cases} a_{i,i} = \dfrac{G_\mathrm{t}G_\mathrm{r}\sigma_{i,i}\lambda^2}{(4\pi)^3R_i^4}, \\ a_{i,j} = \dfrac{G_\mathrm{t}G_\mathrm{r}\sigma_{i,j}\lambda^2}{(4\pi)^3R_i^2R_j^2}, \\ u_{i,j} = \dfrac{G_\mathrm{t}'G_\mathrm{r}'\lambda^2}{(4\pi)^2d_{i,j}^2}, \\ g_i = \dfrac{G_\mathrm{t}'G_\mathrm{c}\lambda^2}{(4\pi)^2d_i^2}, \end{cases} \tag{11.3}$$

where $a_{i,i}$ denotes the propagation gain for the path from radar i to target to radar i, $a_{i,j}$ is the propagation gain for the path from radar i to target to radar j, $u_{i,j}$ stands for the propagation gain for the path from radar i to radar j, and g_i represents the propagation gain for the path from radar i to communications system. G_t denotes the main-lobe transmitting antenna gain of each radar, G_r describes the main-lobe receiving antenna gain of each radar, $G_t^{'}$ represents the average side-lobe transmitting antenna gain of each radar, and $G_r^{'}$ is the average side-lobe receiving antenna gain of each radar. G_c denotes the receiving antenna gain of communications system. $\sigma_{i,i}$ stands for the target's RCS with respect to radar i, $\sigma_{i,j}$ represents the target's RCS from radar i to radar j, λ is the wavelength of radar waveform, R_i describes the range between radar i and the target, $d_{i,j}$ is the range from radar i to radar j, and d_i represents the range between radar i and communications system. For simplicity, it is assumed that all the channel gains remain invariant during target detection period.

Here, the GLRT is exploited to design an appropriate detector [2]. The probability of target detection $p_{\mathrm{d},i}(\delta_i, \gamma_i)$ and the probability of false alarm $p_{\mathrm{fa},i}(\delta_i)$ are expressed by:

$$\begin{cases} p_{\mathrm{d},i}(\delta_i, \gamma_i) = \left(1 + \dfrac{\delta_i}{1-\delta_i} \cdot \dfrac{1}{1+N\gamma_i}\right)^{1-N}, \\ p_{\mathrm{fa},i}(\delta_i) = (1-\delta_i)^{N-1}, \end{cases} \tag{11.4}$$

where δ_i denotes the detection threshold and N represents the number of received samples during the dwell time. γ_i is the achievable SINR of radar i, which is written as:

$$\gamma_i = \frac{a_{i,i}P_i}{\sum_{j=1, j\neq i}^{N_\mathrm{T}} c_{i,j}\left(u_{i,j}P_j + a_{i,j}P_j\right) + \sigma_n^2} = \frac{a_{i,i}P_i}{I_{-i}}, \tag{11.5}$$

where I_{-i} represents the interference and noise received at radar i.

In general, the detection probability $p_{\mathrm{d},i}$ and the false alarm probability $p_{\mathrm{fa},i}$ are used to assess the performance of each radar [2]. It has been shown that as the number of samples approaches infinity, the performance of the GLRT detector goes to the Neyman–Pearson detector. Thus, we can determine the detection threshold δ_i given false alarm probability $p_{\mathrm{fa},i}$ and number of samples N, and hereby the desired SINR $\gamma_{\min}$ can be obtained with a given detection probability $p_{\mathrm{d},i}$. In this work, the non-cooperative game model is exploited to capture the interactions among multiple radars and design the power allocation strategy for each radar, as presented in the next section.

11.3 Game theoretic formulation

Technically speaking, the problem of distributed power resource allocation can be formulated as a mathematical optimization model for minimizing the transmit power for each cognitive radar under the constraints of a specified SINR requirement for target detection and a maximum tolerable interference for the communications system.

Due to the rationality and selfishness of each radar in the multistatic system, we adopt the non-cooperative game theory to establish such an optimization model. More specifically, the set of radars $\mathscr{N} = \{1, \cdots, N_T\}$ is considered as the set of players in the game. The strategy set of all players is $\mathscr{P} = \mathscr{P}_1 \times \mathscr{P}_2 \times \cdots \times \mathscr{P}_{N_T}$ with $\mathscr{P}_i = \{P_i \mid 0 \leq P_i \leq P_i^{\max}\}, i \in \mathscr{N}$. The game model is completed by the definition of the utility function $U_i(P_i, \mathbf{P}_{-i}) = \ln(\gamma_i - \gamma_{\min}) - \psi_i g_i P_i$, where $\mathbf{P}_{-i}$ represents the power allocation strategies of all other players except player i, ψ_i is the time-varying price per unit of interference, and $\gamma_{\min}$ denotes the predetermined SINR threshold. Hence, the non-cooperative game model can be formulated as follows:

$$\mathscr{G} = \langle \mathscr{N}, \{\mathscr{P}_i\}_{i \in \mathscr{N}}, \{U_i\}_{i \in \mathscr{N}} \rangle . \tag{11.6}$$

In order to maximize the utility function of the ith player, the NCGT-PA strategy can mathematically be formulated as:

$$\max_{P_i \in \mathscr{P}_i} \; U_i(P_i, \mathbf{P}_{-i}), \tag{11.7a}$$

$$\text{s.t.}: \quad \begin{cases} 0 \leq P_i \leq P_i^{\max}, \\ \gamma_i \geq \gamma_{\min}, \\ \sum_{i=1}^{N_T} g_i P_i \leq T_{\max}, \end{cases} \tag{11.7b}$$

where $T_{\max}$ denotes the maximum interference that can be tolerated by the communications system.

It is crucial to study whether the game $\mathscr{G}$ converges to a feasible solution when it comes to a game theoretic analysis. Nash equilibrium is such a solution where no player in the game achieves more benefits by changing its strategy unilaterally, given the strategies of the other players. As such, the transmit power vector $\mathbf{P}_i^* = (P_i^*, \mathbf{P}_{-i}^*)$ is a Nash equilibrium solution of the proposed game model $\mathscr{G}$ for each player only if:

$$U_i(P_i, \mathbf{P}_{-i}^*) \leq U_i(P_i^*, \mathbf{P}_{-i}^*), \forall P_i \in \mathscr{P}_i. \tag{11.8}$$

11.3.1 Feasible extension

In the previous subsection, the scenario where a cognitive multistatic radar system illuminates one target has been considered. Nevertheless, the NCGT-PA model can also be extended to the scenario that there exist multiple targets. For example, if there exist Q targets, we assume that the transmit waveforms of each radar for different targets are orthogonal with negligible delays and Doppler shifts. Note that such an assumption implies that multiple targets are well separated in range and that the waveforms have good side lobes, which can also be considered from the perspective of joint optimization of beamforming and power allocation for the radar system [37]. Here, we only focus on the problem of power allocation in the presence of multiple targets. In this case, the SINR of target q for radar i can be reformulated as:

$$\gamma_{i,q} = \frac{a_{i,i,q} P_{i,q}}{\sum_{j=1, j\neq i}^{N_T} c_{i,j,q} \left(u_{i,j,q} P_{j,q} + a_{i,j,q} P_{j,q} \right) + \sigma_n^2}, \tag{11.9}$$

where $a_{i,i,q}$, $u_{i,i,q}$, $a_{i,j,q}$ are the corresponding propagation gain for target q, $P_{i,q}$ denotes the transmit power of radar i for target q, and $c_{i,j,q}$ denotes the cross-correlation factor between radar i and radar j. For simplicity, the definition of all above propagation gains are omitted as we focus on extending the NCGT-PA model to multiple targets case. Moreover, the utility function of the ith player can be defined as:

$$U_i^* = \sum_{q=1}^{Q} (\ln(\gamma_{i,q} - \gamma_{\min}) - \psi_{i,q} g_{i,q} P_{i,q}), \tag{11.10}$$

where $\psi_{i,q}$ and $g_{i,q}$ denote the time-varying price per unit of interference and the propagation gain for the path from radar i to communications system. Therefore, the optimization problem for multiple targets case can be written as:

$$\max_{P_{i,q} \in \mathscr{P}_i^*} \ U_i^*, \tag{11.11a}$$

$$\text{s.t.}: \begin{cases} P_i \geq 0, \\ \sum_{q=1}^{Q} P_{i,q} \leq P_i^{\max}, \\ \gamma_{i,q} \geq \gamma_{\min}, \\ \sum_{i=1}^{N_T} \sum_{q=1}^{Q} g_{i,q} P_{i,q} \leq T_{\max}, \forall q \in \mathscr{Q}. \end{cases} \tag{11.11b}$$

where $\mathscr{P}_i^*$ denotes the feasible strategy set for the ith player, and the subscript q represents the corresponding target with $q \in \mathscr{Q} = \{1, 2, \cdots, Q\}$. Hence, through tackling the optimization problem (11.11), we can obtain the power allocation strategy for each radar in the presence of multiple targets.

In the following sections, we will concentrate on the analysis of the existence and uniqueness of the Nash equilibrium that corresponds to the single target scenario presented in (11.7). Nevertheless, the multiple targets model (11.11) can also be investigated in a similar way.

11.4 Existence and uniqueness of the Nash equilibrium

11.4.1 Existence

Lemma 1 (Existence). *There exists at least one Nash equilibrium in the proposed game* $\mathscr{G} = \langle \mathscr{N}, \{\mathscr{P}_i\}_{i \in \mathscr{N}}, \{U_i\}_{i \in \mathscr{N}} \rangle$.

Proof: At least one Nash equilibrium of the game $\mathscr{G}$ exists if the following conditions hold:

(i) For all the players $i \in \mathscr{N}$, the strategy set $\mathscr{P}_i$ is non-empty, compact, and convex.
(ii) The utility function $U_i(P_i, \mathbf{P}_{-i})$ is continuous on $\mathscr{P}$ and quasi-concave in P_i.

In the proposed game model, the set $\mathscr{P}_i = \{P_i \mid 0 \leq P_i \leq P_i^{\max}\}$ is obviously compact and convex, which satisfies condition (i).

Taking the first-order derivative of $U_i(P_i, \mathbf{P}_{-i})$ with respect to P_i, we have:

$$\frac{\partial U_i(P_i, \mathbf{P}_{-i})}{\partial P_i} = \frac{1}{\gamma_i - \gamma_{\min}} \frac{a_{i,i}}{I_{-i}} - \psi_i g_i, \tag{11.12}$$

and then the second-order derivative of $U_i(P_i, \mathbf{P}_{-i})$ with respect to P_i can be written as:

$$\frac{\partial^2 U_i(P_i, \mathbf{P}_{-i})}{\partial P_i^2} = -\frac{(a_{i,i})^2}{I_{-i}^2(\gamma_i - \gamma_{\min})^2} < 0. \tag{11.13}$$

Therefore, $U_i(P_i, \mathbf{P}_{-i})$ is strictly concave in P_i and continuous on $\mathscr{P}$. This concludes the proof that there exist at least one Nash equilibrium in the proposed NCGT-PA model.

11.4.2 Uniqueness

Although the existence of the Nash equilibrium is guaranteed, the equilibrium point is not necessarily unique. Next, we need to demonstrate that the best response strategy of each player is a standard function, which leads to the uniqueness of the Nash equilibrium. Herein, we provide the definition of a standard function in the following:

A function $f(x)$ must possess the following properties so as to be a standard function for all $x \geq 0$:

(i) Positivity: $f(x) > 0$.
(ii) Monotonicity: If $x_1 > x_2$, then $f(x_1) > f(x_2)$.
(iii) Scalability: For all $a \geq 1$, $af(x) > f(ax)$.

In order to obtain the best response function of the ith player, we need to solve the optimization problem (11.7). Note that the problem is typically a convex optimization problem as the objective is rigorously concave and the constraints are linear with respect to P_i after some transformations. In order to gain insights into the power allocation strategy based on Nash equilibrium, we tackle it via Lagrange multipliers. Let the corresponding Lagrangian be:

$$L = -U_i(P_i, \mathbf{P}_{-i}) + \lambda(\sum_{i=1}^{N_T} g_i P_i - T_{\max}), \tag{11.14}$$

where $\lambda > 0$ is the Lagrange multiplier associated with the interference constraint on the communications system. First, by setting the first order derivative of L with respect to P_i equal to zero, we have:

$$\gamma_i = \gamma_{min} + \frac{a_{i,i}}{I_{-i}(\psi_i + \lambda)g_i}. \tag{11.15}$$

Substituting (11.5) and rearranging terms, we can get the transmit power of radar i as:

$$P_i = \frac{I_{-i}}{a_{i,i}}\gamma_{min} + \frac{1}{(\psi_i + \lambda)g_i}. \tag{11.16}$$

According to $\gamma_i = \frac{a_{i,i}}{I_{-i}} P_i$ and (11.16), one has:

$$P_i^{(n+1)} = \frac{P_i^{(n)}}{\gamma_i^{(n)}}\gamma_{\min} + \frac{1}{(\psi_i^{(n)} + \lambda)g_i}, \tag{11.17}$$

where n is the iteration index, and λ is determined by $\sum_{i=1}^{N_T} g_i P_i = T_{\max}$.

Thus, based on the above algebraic manipulations, the iterative function for the transmit power strategy of radar i can be given by:

$$P_i^{(n+1)} = \left[\frac{P_i^{(n)}}{\gamma_i^{(n)}} \gamma_{\min} + \frac{1}{(\psi_i^{(n)} + \lambda) g_i} \right]_0^{P_i^{max}}, \tag{11.18}$$

where $[x]_0^b = \max\{\min(x, b), 0\}$, and $\psi_i^{(n)}$ is the time-varying price per unit of interference designed for increasing the convergence speed, which should satisfy the following conditions:

$$\psi_i^{(n+1)} = \begin{cases} \psi_i^{(n)}, \text{if } \gamma_i^{(n)} \le \gamma_{\min} \\ \psi_i^{(n)} \left(\dfrac{\gamma_i^{(n)}}{\gamma_{\min}} \right)^2, \text{if } \gamma_i^{(n)} > \gamma_{\min} \end{cases} \tag{11.19}$$

It is noteworthy that if $\gamma_i^{(n)} > \gamma_{\min}$, $\psi_i^{(n+1)}$ is increased which imposes the punishment for player i to decrease the transmit power as the SINR requirement is satisfied, whereas if $\gamma_i^{(n)} \le \gamma_{\min}$, $\psi_i^{(n+1)}$ remains unchanged increasing the transmit power and improving detection performance.

Lemma 2 (Uniqueness). *The best response function of each radar in the proposed game $\mathcal{G}$ is standard.*

Proof: From (11.18), we can obtain the best response function of radar i as:

$$f(P_i) = \frac{P_i}{\gamma_i} \gamma_{\min} + \frac{1}{(\psi_i + \lambda) g_i}, \forall P_i \ge 0. \tag{11.20}$$

(i) Positivity: Since

$$f(P_i) = \frac{P_i}{\gamma_i} \gamma_{\min} + \frac{1}{(\psi_i + \lambda) g_i} > 0, \tag{11.21}$$

then we obtain $f(P_i) > 0$.

(ii) Monotonicity: If $P_i^a > P_i^b$, then:

$$f(P_i^a) - f(P_i^b) = \frac{P_i^a - P_i^b}{\gamma_i} \gamma_{min} > 0. \tag{11.22}$$

Thus, we have $f(P_i^a) > f(P_i^b)$.

(iii) Scalability: For arbitrary $a > 1$,

$$af(P_i) - f(aP_i) = \frac{a-1}{(\psi_i + \lambda) g_i} > 0. \tag{11.23}$$

Thus, we get $af(P_i) > f(aP_i)$.

This concludes the proof that the best response function $f(P_i)$ is a standard function. Hence, there exists only one Nash equilibrium for the proposed game model.

11.5 Iterative power allocation method

Based on the analysis of the existence and uniqueness of the Nash equilibrium, we develop a decentralized iterative power allocation method with low complexity and rapid convergence rate, which can achieve the Nash equilibrium point of the proposed game from any feasible starting values. It is noteworthy that each radar independently performs the NCGT-PA strategy in a distributed manner, which satisfies the desired target detection performance and controls the total interference power level to the communications system. Therefore, the proposed NCGT-PA algorithm is attractive to dynamic network architectures that require asynchronous implementation, in which each radar autonomously determines its illumination strategy, given the strategies that all the other radars adopt. In other words, each of the radars senses the knowledge related to the current situation and environment, i.e., the estimated SINR for the target and the interference level to the communications system, through the SINR estimation [2] and communication techniques. Then, the power allocation strategy is determined by leveraging learned knowledge such that the interference to the communications system and mutual interference between the radars are properly alleviated. Moreover, each radar needs to adjust its strategy according to the dynamic environment. Such process constitutes a perception–action cycle of the cognitive radars [48]. The pseudo-code of the iterative power allocation algorithm is presented in Algorithm 1.

Algorithm 1: Detailed steps of iterative power allocation algorithm

---

1: **Input** $P_i^{(0)}$ for $\forall i$, $\gamma_{\min}$, $T_{\max}$, $\psi_i^{(0)}$, L_{max}, and $\varepsilon > 0$, and the corresponding channel gains.
2: **Set** iteration index $n = 0$.
3: **Repeat**
  For $i = 1$ to N_{T} **do**
    (i) Update $P_i^{(n)}$ according to (11.18), and send those values to all the other radars;
    (ii) **If** $\gamma_i^{(n)} > \gamma_{\min}$

$$\psi_i^{(n+1)} \leftarrow \psi_i^{(n)} \left(\frac{\gamma_i^{(n)}}{\gamma_{\min}} \right)^2;$$

    Else

$$\psi_i^{(n+1)} \leftarrow \psi_i^{(n)};$$

    End if
  End for
  Set $n \leftarrow n + 1$.
4: **Until** $\left| P_i^{(n+1)} - P_i^{(n)} \right| < \varepsilon$ or $n = L_{max}$.
5: **Output** $P_i^{(n)}$ for $\forall i$.

11.6 Simulation results and performance evaluation

In this section, extensive numerical examples are conducted to demonstrate the convergence of the NCGT-PA strategy to the Nash equilibrium and evaluate the performance of the proposed strategy compared to the existing power allocation algorithms.

11.6.1 Parameter designation

Here, we consider a cognitive multistatic radar system that consists of four radar nodes, i.e., $N_{\mathrm{T}} = 4$, coexisting with a communications system. It is assumed that the location of communications system is $[-10, 0]$km, while the four radar nodes are located at $[25\sqrt{2}, 25\sqrt{2}]$km, $[-25\sqrt{2}, 25\sqrt{2}]$km, $[-25\sqrt{2}, -25\sqrt{2}]$km and $[25\sqrt{2}, -25\sqrt{2}]$km, respectively. In each time index, we assume that the number of received samples of each radar is $N = 512$. We also set the maximum number of iterations as $L_{max} = 25$ to show the convergence of the proposed game. In order to verify that the power allocation results of the multistatic radar system is associated with the relative geometry between the multistatic system and the target, we consider two different target locations with Location 1 being $[0, 0]$km and Location 2 being $[-\frac{25}{\sqrt{2}}, \frac{25}{\sqrt{2}}]$km. Furthermore, to better reveal the relationship between the target's RCS and power allocation strategy, we consider two different RCS models, where the first RCS model is denoted as $\boldsymbol{\sigma}_1 = [1, 1, 1, 1]\mathrm{m}^2$, while the second one is $\boldsymbol{\sigma}_2 = [1, 0.2, 3, 5]\mathrm{m}^2$.

Before initializing the NCGT-PA model, we should determine the desired SINR threshold γ_{min} for each radar. The desired SINR can be calculated with (11.4) as described in Section 11.2, given the probability of target detection p_{d} and the probability of false alarm p_{fa}. In our simulations, the probabilities of detection and false alarm are set as $p_{\mathrm{d}} = 0.9914$ and $p_{\mathrm{fa}} = 10^{-6}$, respectively, and thus we get the detection threshold and the SINR requirement for each radar, that is, $\delta_i = 0.0267$ and $\gamma_{\min} = 5$ dB, respectively. Finally, the simulation parameters are provided in Table 11.1.

Table 11.1 Simulation parameters

Parameter	Value	Parameter	Value
$T_{\max}$	-110 dBmW	$c_{i,j}$	0.01
λ	0.10 m	$P_i^{\max}$	1,000 W
G_{t}	27 dB	G_{r}	27 dB
$G_{\mathrm{t}}^{'}$	-30 dB	$G_{\mathrm{r}}^{'}$	-30 dB
G_{c}	0 dB	σ_n^2	10^{-18} W
ε	10^{-16}	$\psi_i^{(0)}$	10^{17}

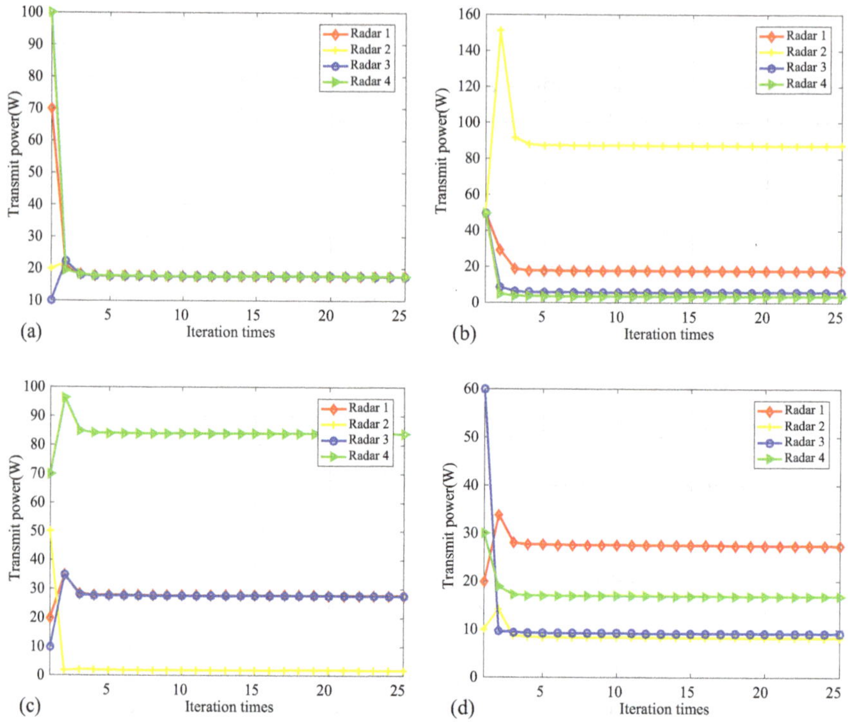

Figure 11.2 Convergence of power allocation of the cognitive multistatic radar system for two target locations with different RCS models: (a) Location 1 with $\boldsymbol{\sigma}_1$, (b) Location 1 with $\boldsymbol{\sigma}_2$, (c) Location 2 with $\boldsymbol{\sigma}_1$, and (d) Location 2 with $\boldsymbol{\sigma}_2$

11.6.2 Numerical results

Figure 11.2 demonstrates the power allocation behaviors of all the radars in the multistatic system with different initial values of transmit power, i.e., $\mathbf{P}^{(0)} = [70, 20, 10, 100]$W, $\mathbf{P}^{(0)} = [50, 50, 50, 50]$W, $\mathbf{P}^{(0)} = [20, 50, 10, 70]$W, and $\mathbf{P}^{(0)} = [20, 10, 60, 30]$W. It is evident that the proposed NCGT-PA scheme converges to a unique solution, regardless of the initial power allocation values. Moreover, the proposed algorithm is highly efficient, as it determines the power allocation strategy within five iterations. These results verify the analysis in Section 11.4, confirming the uniqueness of the Nash equilibrium.

In order to investigate the impacts of the relative geometry between the cognitive radar system and the target and the target reflectivity on the power allocation strategy, we define the transmit power ratio as $\theta_i = \frac{P_i}{\sum_{i=1}^{N_T} P_i}$. Figure 11.3 depicts the transmit power ratio results of multiple radar nodes for two target locations with different RCS models. First, it can be seen from Figure 11.3(a) and (b) that the NCGT-PA

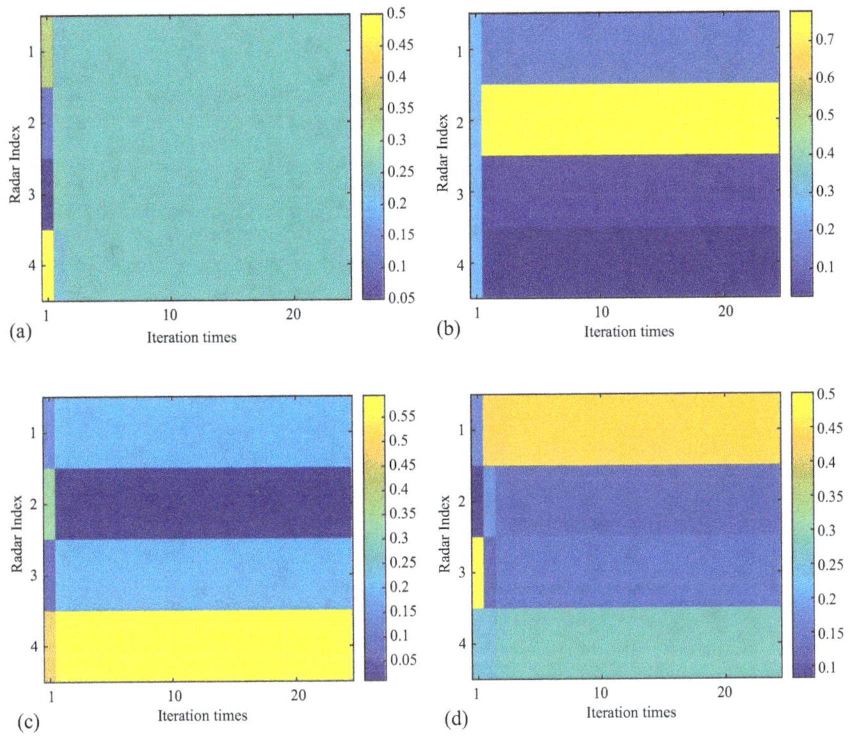

Figure 11.3 The transmit power ratio of the cognitive multistatic radar system for two target locations with different RCS models: (a) Location 1 with $\boldsymbol{\sigma}_1$*, (b) Location 1 with* $\boldsymbol{\sigma}_2$*, (c) Location 2 with* $\boldsymbol{\sigma}_1$*, and (d) Location 2 with* $\boldsymbol{\sigma}_2$

strategy allocates more power to Radar 1 and Radar 2, which is due to the fact that the target reflectivities with respect to these two radars are much weaker than others. Additionally, the relationship between the power allocation results and the geometry between the multistatic radar system and the target is shown in Figure 11.3(c), which presents the power allocation results for Location 2. It is apparent that more power resource is distributed to Radar 4, as the range between Radar 4 and the target is much larger than those of the other three radars. In other words, the radar with a larger range of the target entails more power in order to maintain the specified target detection performance. Therefore, a significantly important conclusion that more power tends to be assigned to the channels with worse conditions can be drawn from Figure 11.3.

Figure 11.4 shows the SINR convergence behaviors of the proposed NCGT-PA algorithm for each radar. One can observe that the SINR values of all radars approach the desired SINR threshold but still exceed the threshold after almost five iterations. Thus, we can conclude that the presented game model is capable of reducing the transmit power as much as possible while satisfying the SINR requirement for target

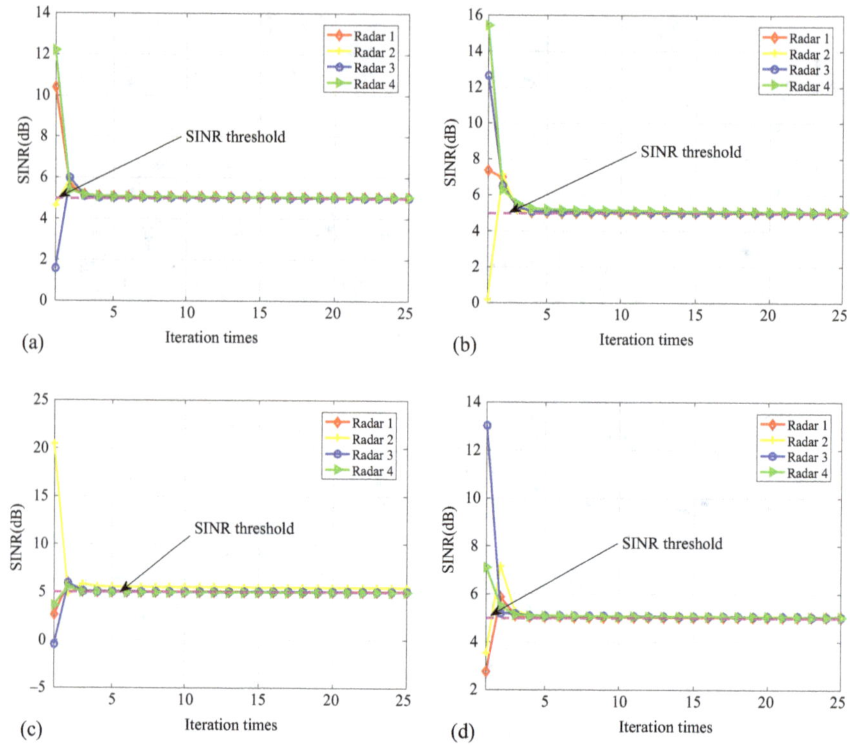

Figure 11.4 Convergence of SINR of the cognitive multistatic radar system for two target locations with different RCS models: (a) Location 1 with $\boldsymbol{\sigma}_1$*, (b) Location 1 with* $\boldsymbol{\sigma}_2$*, (c) Location 2 with* $\boldsymbol{\sigma}_1$*, and (d) Location 2 with* $\boldsymbol{\sigma}_2$

detection. It is worth pointing out that the proposed power allocation method is hugely attractive for the application of target tracking, which requires fine detection performance to achieve the exact location of the target and an efficient power allocation strategy to balance the tracking performance and power consumption for the cognitive radar system.

Figures 11.5 and 11.6 show the comparisons between the proposed NCGT-PA algorithm and the other three algorithms, i.e., the uniform power allocation algorithm, the Koskie and Gajic's (K-G) algorithm presented in [21], and the adaptive non-cooperative power control (ANCPC) algorithm proposed in [49], with respect to the transmit power and the SINR value of each radar for two target locations with different RCS models. By imposing an additional constraint that assigns the transmit power to all the radars in the multistatic system uniformly in the problem (11.7), we can attain the non-cooperative game theory-based uniform power allocation (NCGT-UPA) algorithm. We will show the superiority of the proposed NCGT-PA strategy from the following three perspectives. First, the NCGT-PA strategy is better than the NCGT-UPA strategy in terms of target detection performance, which

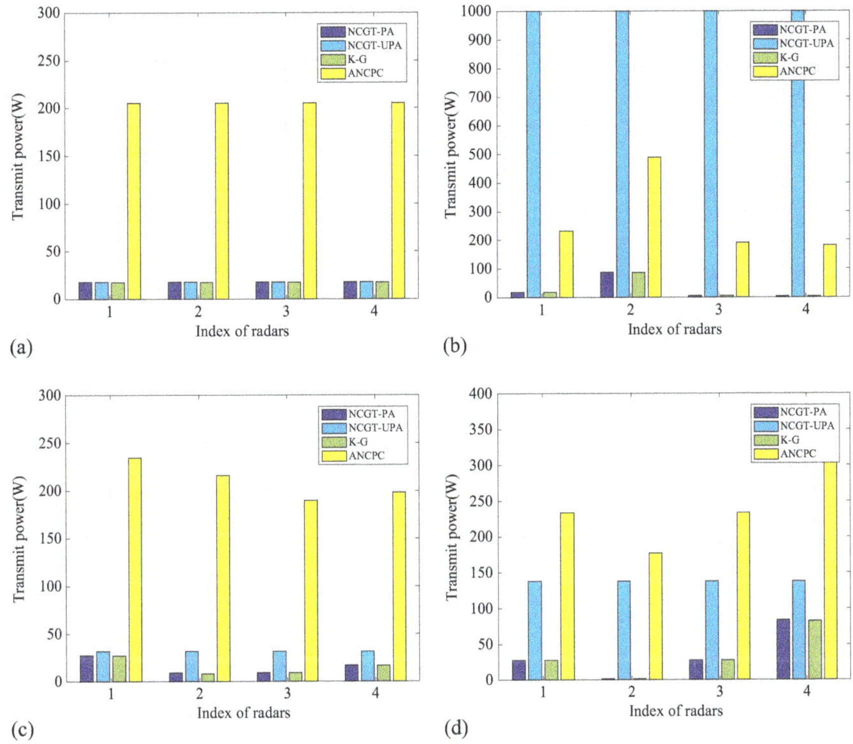

Figure 11.5 Comparisons of the total transmit power of the cognitive multistatic radar system by exploiting various methods for two target locations with different RCS models: (a) Location 1 with $\boldsymbol{\sigma}_1$, (b) Location 1 with $\boldsymbol{\sigma}_2$, (c) Location 2 with $\boldsymbol{\sigma}_1$, and (d) Location 2 with $\boldsymbol{\sigma}_2$

is because the NCGT-UPA strategy cannot satisfy the SINR requirement in all the scenarios. Second, although consuming the least transmit power, the K-G algorithm fails to guarantee the predefined target detection requirement of all the radars, i.e., the specified SINR threshold cannot be achieved. Finally, the transmit power level of the ANCPC approach is much higher than that of the NCGT-PA strategy, which demonstrates its poor power-saving performance.

In order to investigate the effectiveness of different power allocation strategies in the RF spectral coexistence environment, we compare the total interference power levels to the communications system for two target locations with different RCS models, as the histogram shown in Figure 11.7. It can be observed that the proposed NCGT-PA strategy and the K-G strategy can secure the QoS of the communications system, which is due to the fact that the total interference levels are below the predetermined interference threshold, i.e., the achievable interference levels of the above two methods are lower than the maximum tolerable interference $T_{\max}$. Nevertheless,

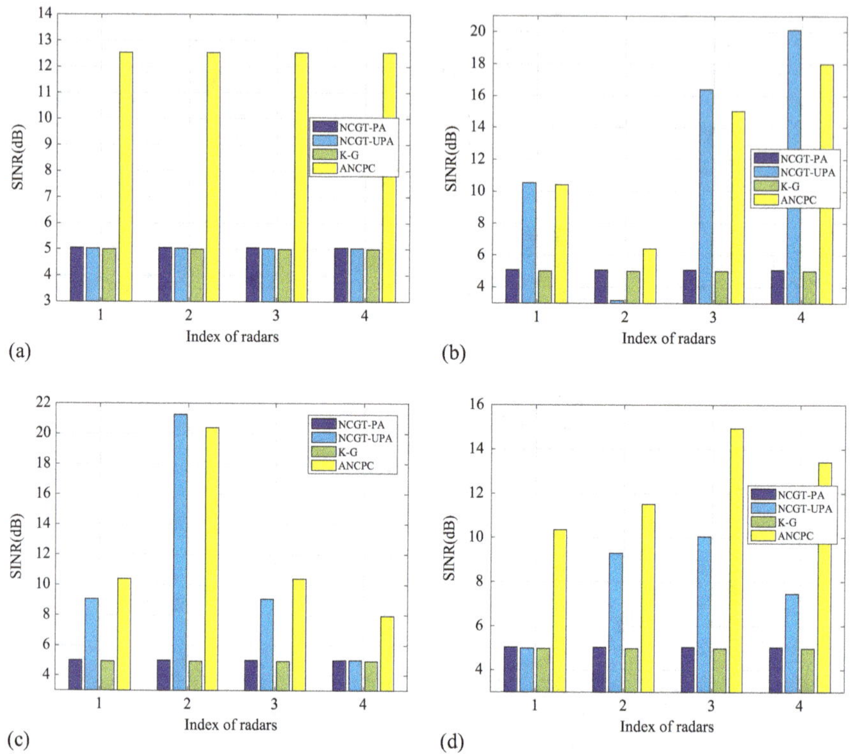

Figure 11.6 Comparisons of the achievable SINR values of the cognitive multistatic radar system by exploiting various methods for two target locations with different RCS models: (a) Location 1 with $\boldsymbol{\sigma}_1$, (b) Location 1 with $\boldsymbol{\sigma}_2$, (c) Location 2 with $\boldsymbol{\sigma}_1$, and (d) Location 2 with $\boldsymbol{\sigma}_2$

the total interference levels of the NCGT-UPA and the ANCPC schemes fail to stay below the interference threshold, which means that the communications system cannot maintain fine QoS when coexisting with the multistatic radar system in such cases. Owing to the failure of achieving the desired SINR target, the K-G approach cannot guarantee the required target detection performance with the minimum transmit power. Hence, the proposed NCGT-PA strategy is the recommended method for spectral coexistence in terms of power-saving, target detection, and spectrum sharing performance between the cognitive multistatic radar and the communications system.

11.7 Conclusion

In this chapter, we have addressed the problem of game theory-based power allocation for the cognitive multistatic radar system in a spectral coexistence scenario,

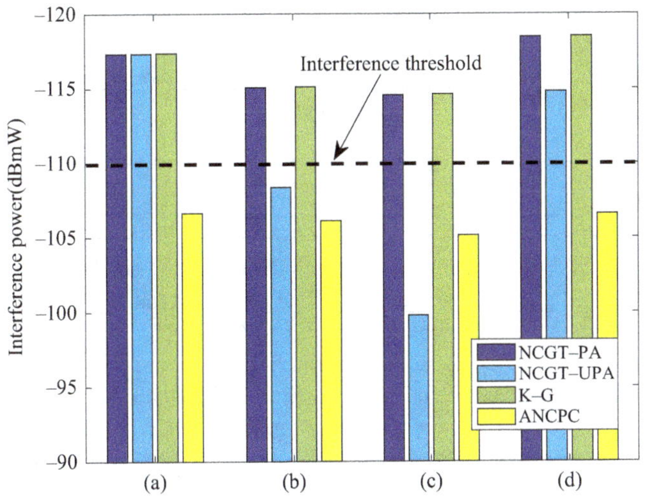

Figure 11.7 Comparisons of the total interference power levels to the communications system by exploiting various methods for two target locations with different RCS models: (a) Location 1 with $\boldsymbol{\sigma}_1$, (b) Location 1 with $\boldsymbol{\sigma}_2$, (c) Location 2 with $\boldsymbol{\sigma}_1$, and (d) Location 2 with $\boldsymbol{\sigma}_2$

where each radar optimizes its transmit power to maintain the target detection performance and control the total interference to a communications system. We built the optimization problem as a non-cooperative game model. Moreover, we calculated the closed-form expression of the Nash equilibrium, and strictly proved the existence and uniqueness of the Nash equilibrium. In order to strengthen the distributed nature of the multistatic radar system, an iterative power allocation algorithm based on the best response function was developed. Finally, extensive simulation results confirmed the superiorities of the presented strategy in terms of power saving, target detection performance, and spectrum sharing performance between the cognitive multistatic radar and the communications system. Particularly, it is also demonstrated that the power allocation results are associated with the target's RCS and the relative geometry between the cognitive radar system and the target.

Although the non-cooperative game is utilized to control mutual interference between the radars and the interference to the communications system in this chapter, there exist wide applications for cognitive radar in other scenarios. A typical application is to model the interaction between a cognitive radar and an adversarial jammer, where both sides need to acquire the knowledge about the environment and the strategy adopted by the other, and then determine and adjust their strategies in order to adapt to the dynamic antagonism environment. Furthermore, most of the existing works focus on the joint design of strategies for multiple systems, which is largely based on the high-quality communication between the systems. Nevertheless, it is hard to maintain real-time and high-quality communication in a complicated electromagnetic environment. In such circumstances, the non-cooperative game would play a key role in independently designing the strategy for each system [2]. Last but not least, game theory can also be a useful tool in improving multiple functions of an individual system, which can be seen as the interaction between multiple players with each player concentrating on one of the functions [43].

References

[1] Yan J, Liu H, Pu W, *et al.* Joint beam selection and power allocation for multiple target tracking in netted colocated MIMO radar system. *IEEE Transactions on Signal Processing*. 2016;64(24):6417–6427.
[2] Deligiannis A, Panoui A, Lambotharan S, *et al.* Game-theoretic power allocation and the Nash equilibrium analysis for a multistatic MIMO radar network. *IEEE Transactions on Signal Processing*. 2017;65(24):6397–6408.
[3] Fishler E, Haimovich A, Blum RS, *et al.* Spatial diversity in radars—models and detection performance. *IEEE Transactions on Signal Processing*. 2006;54(3):823–838.
[4] Naghsh MM, Modarres-Hashemi M, ShahbazPanahi S, *et al.* Unified optimization framework for multi-static radar code design using information-theoretic criteria. *IEEE Transactions on Signal Processing*. 2013;61(21):5401–5416.
[5] Dogancay K. Online optimization of receiver trajectories for scan-based emitter localization. *IEEE Transactions on Aerospace and Electronic Systems*. 2007;43(3):1117–1125.
[6] Godrich H, Tajer A, and Poor HV. Distributed target tracking in multiple widely separated radar architectures. In: *2012 IEEE 7th Sensor Array and Multichannel Signal Processing Workshop (SAM)*; 2012. p. 153–156.
[7] He Q, Blum RS, and Haimovich AM. Noncoherent MIMO radar for location and velocity estimation: more antennas means better performance. *IEEE Transactions on Signal Processing*. 2010;58(7):3661–3680.
[8] Shi CG, Salous S, Wang F, *et al.* Modified Cramér–Rao lower bounds for joint position and velocity estimation of a Rician target in OFDM-based passive radar networks. *Radio Science*. 2017;52(1):15–33.
[9] Nguyen NH, Dogancay K, and Davis LM. Adaptive waveform selection for multistatic target tracking. *IEEE Transactions on Aerospace and Electronic Systems*. 2015;51(1):688–701.
[10] Nguyen NH, Doğançay K, and Davis LM. Joint transmitter waveform and receiver path optimization for target tracking by multistatic radar system. In: *2014 IEEE Workshop on Statistical Signal Processing (SSP)*; 2014. p. 444–447.
[11] Godrich H, Petropulu AP, and Poor HV. Sensor selection in distributed multiple-radar architectures for localization: a Knapsack problem formulation. *IEEE Transactions on Signal Processing*. 2012;60(1):247–260.
[12] Shi C, Wang F, Sellathurai M, *et al.* Transmitter subset selection in FM-based passive radar networks for joint target parameter estimation. *IEEE Sensors Journal*. 2016;16(15):6043–6052.

[13] Song X, Willett P, and Zhou S. Optimal power allocation for MIMO radars with heterogeneous propagation losses. In: *2012 IEEE International Conference on Acoustics, Speech and Signal Processing (ICASSP)*; 2012. p. 2465–2468.

[14] Labib M, Reed JH, Martone AF, *et al.* A game-theoretic approach for radar and LTE systems coexistence in the unlicensed band. In: *2016 USNC-URSI Radio Science Meeting*; 2016. p. 17–18.

[15] Turlapaty A and Jin Y. A joint design of transmit waveforms for radar and communications systems in coexistence. In: *2014 IEEE Radar Conference*; 2014. p. 0315–0319.

[16] Bica M and Koivunen V. Delay estimation method for coexisting radar and wireless communication systems. In: *2017 IEEE Radar Conference (RadarConf)*; 2017. p. 1557–1561.

[17] Shi C, Wang F, Sellathurai M, *et al.* Power minimization-based robust OFDM radar waveform design for radar and communication systems in coexistence. *IEEE Transactions on Signal Processing*. 2018;66(5):1316–1330.

[18] Li B and Petropulu AP. Joint transmit designs for coexistence of MIMO wireless communications and sparse sensing radars in clutter. *IEEE Transactions on Aerospace and Electronic Systems*. 2017;53(6):2846–2864.

[19] Zheng L, Lops M, Wang X, *et al.* Joint design of overlaid communication systems and pulsed radars. *IEEE Transactions on Signal Processing*. 2018;66(1):139–154.

[20] Zhang H, Huang S, Jiang C, *et al.* Energy efficient user association and power allocation in millimeter-wave-based ultra dense networks with energy harvesting base stations. *IEEE Journal on Selected Areas in Communications*. 2017;35(9):1936–1947.

[21] Koskie S and Gajic Z. A Nash game algorithm for SIR-based power control in 3G wireless CDMA networks. *IEEE/ACM Transactions on Networking*. 2005;13(5):1017–1026.

[22] Tsiropoulou EE, Vamvakas P, and Papavassiliou S. Supermodular game-based distributed joint uplink power and rate allocation in two-tier femtocell networks. *IEEE Transactions on Mobile Computing*. 2017;16(9): 2656–2667.

[23] Song X, Willett P, Zhou S, *et al.* The MIMO radar and jammer games. *IEEE Transactions on Signal Processing*. 2012;60(2):687–699.

[24] Lan X, Li W, Wang X, *et al.* MIMO radar and target Stackelberg game in the presence of clutter. *IEEE Sensors Journal*. 2015;15(12):6912–6920.

[25] Deligiannis A, Rossetti G, Panoui A, *et al.* Power allocation game between a radar network and multiple jammers. In: *2016 IEEE Radar Conference (RadarConf)*; 2016. p. 1–5.

[26] Bachmann DJ, Evans RJ, and Moran B. Game theoretic analysis of adaptive radar jamming. *IEEE Transactions on Aerospace and Electronic Systems*. 2011;47(2):1081–1100.

[27] Zhang X, Ma H, Wang J, *et al.* Game theory design for deceptive jamming suppression in polarization MIMO radar. *IEEE Access*. 2019;7:114191–114202.

[28] Gao H, Wang J, Jiang C, *et al.* Equilibrium between a statistical MIMO radar and a jammer. In: *2015 IEEE Radar Conference (RadarCon)*; 2015. p. 0461–0466.

[29] Garnaev A, Trappe W, and Petropulu A. A dual radar and communication system facing uncertainty about a jammer's capability. In: *2018 52nd Asilomar Conference on Signals, Systems, and Computers*; 2018. p. 417–422.

[30] He B and Su H. Game theoretic countermeasure analysis for multistatic radars and multiple jammers. *Radio Science*. 2021;56:1–14.

[31] Gupta A and Krishnamurthy V. Principal agent problem as a principled approach to electronic counter-countermeasures in radar. *IEEE Transactions on Aerospace and Electronic Systems*. 2022;58:1–1.

[32] Bacci G, Sanguinetti L, Greco MS, *et al.* A game-theoretic approach for energy-efficient detection in radar sensor networks. In: *2012 IEEE 7th Sensor Array and Multichannel Signal Processing Workshop (SAM)*; 2012. p. 157–160.

[33] Panoui A, Lambotharan S, and Chambers JA. Game theoretic power allocation technique for a MIMO radar network. In: *2014 6th International Symposium on Communications, Control and Signal Processing (ISCCSP)*; 2014. p. 509–512.

[34] Panoui A, Lambotharan S, and Chambers JA. Game theoretic power allocation for a multistatic radar network in the presence of estimation error. In: *2014 Sensor Signal Processing for Defence (SSPD)*; 2014. p. 1–5.

[35] Deligiannis A and Lambotharan S. A Bayesian game theoretic framework for resource allocation in multistatic radar networks. In: *2017 IEEE Radar Conference (RadarConf)*; 2017. p. 0546–0551.

[36] Shi C, Qiu W, Wang F, *et al.* Stackelberg game-theoretic low probability of intercept performance optimization for multistatic radar system. *Electronics*. 2019;8(4):397.

[37] Deligiannis A, Lambotharan S, and Chambers JA. Game theoretic analysis for MIMO radars with multiple targets. *IEEE Transactions on Aerospace and Electronic Systems*. 2016;52(6):2760–2774.

[38] Shi C, Wang F, Salous S, *et al.* A robust Stackelberg game-based power allocation scheme for spectral coexisting multistatic radar and communication systems. In: *2019 IEEE Radar Conference (RadarConf)*; 2019. p. 1–5.

[39] He B, Su H, and Huang J. Joint beamforming and power allocation between a multistatic MIMO radar network and multiple targets using game theoretic analysis. *Digital Signal Processing*. 2021;115(5):103085.

[40] Gogineni S and Nehorai A. Game theoretic design for polarimetric MIMO radar target detection. *Signal Processing*. 2012;92(5):1281–1289.

[41] Piezzo M, Aubry A, Buzzi S, *et al.* Non-cooperative code design in radar networks: a game-theoretic approach. *Eurasip Journal on Advances in Signal Processing*. 2013;2013(1):63.

[42] Panoui A, Lambotharan S, and Chambers JA. Game theoretic distributed waveform design for multistatic radar networks. *IEEE Transactions on Aerospace and Electronic Systems*. 2016;52(4):1855–1865.

[43] Han K and Nehorai A. Jointly optimal design for MIMO radar frequency-hopping waveforms using game theory. *IEEE Transactions on Aerospace and Electronic Systems*. 2016;52(2):809–820.
[44] Li K, Jiu B, and Liu H. Game theoretic strategies design for monostatic radar and jammer based on mutual information. *IEEE Access*. 2019;7:72257–72266.
[45] Chen X, Song X, Xin F, *et al.* MI-based robust waveform design in radar and jammer games. *Complexity*. 2019;2019:4057849.
[46] Zhang H, Du J, Cheng J, *et al.* Incomplete CSI based resource optimization in SWIPT enabled heterogeneous networks: a non-cooperative game theoretic approach. *IEEE Transactions on Wireless Communications*. 2018;17(3): 1882–1892.
[47] Shi C, Wang F, Sellathurai M, *et al.* Non-cooperative game theoretic power allocation strategy for distributed multiple-radar architecture in a spectrum sharing environment. *IEEE Access*. 2018;6:17787–17800.
[48] Martone AF and Charlish A. Cognitive radar for waveform diversity utilization. In: *2021 IEEE Radar Conference (RadarConf21)*; 2021. p. 1–6.
[49] Wang X, Yang G, Tan X, *et al.* Adaptive power control algorithm in cognitive radio based on game theory. *IET Communications*. 2015 09;9.

Chapter 12
The role of neural networks in cognitive radar

Sevgi Z. Gurbuz[1], Stefan Bruggenwirth[2], Taylor Reininger[3], Ali C. Gurbuz[4] and Graeme E. Smith[3]

The augmentation of engineering systems with some form of "intelligence" has long been a goal of designers to improve robustness and performance. Radar systems typically operate by transmitting a fixed, pre-defined waveform regardless of changes in the environment. Thus, information flow is one way. In contrast, cognitive radar envisions an architecture that has two-way interactions with its surroundings, using this feedback to optimize its performance. Formally, the IEEE defines cognitive radar as *"a radar system that in some sense displays intelligence, adapting its operation and processing in response to a changing environment and target scene. In comparison to active radar, cognitive radar learns to adapt operating parameters as well as processing parameters and may do so over extended time periods"* [1]. Underlying this definition is a vision of artificial emulation of human cognition and cognitive processes in such a way that the radar may "reason" and make decisions on its actions based on information learned from new measurements.

The cognitive neuroscientist Dr Joaquin Fuster [2] has posited that there are five essential cognitive processes: (1) the perception–action cycle (PAC), (2) attention, (3) memory, (4) language, and (5) intelligence. Broadly speaking, common functions of intelligence include perceptual processes (e.g. attention and recognition), long-term and short-term memory, linguistic constructs (e.g. concepts and categories), though processing, problem solving, reasoning, decision making, judgment and anticipation. In many of these processes, learning plays an integral role and has thus driven interest in the potential application of deep neural networks (DNNs) in cognitive radar design. This chapter provides an in-depth look at the use of DNNs in cognitive processing modeling, physics-aware DNN design for incorporating domain knowledge, reinforcement learning as a mechanism for the PAC, and novel approaches for end-to-end DNN design enabling real-time processing. The chapter concludes with a discussion of challenges and future directions.

[1]Department of Electrical and Computer Engineering, The University of Alabama, Tuscaloosa, AL, USA
[2]Department Cognitive Radar, Fraunhofer Institute for High Frequency Physics and Radar Techniques FHR, Wachtberg, Germany
[3]Applied Physics Laboratory, Johns Hopkins University, Laurel, MD, USA
[4]Department of Electrical and Computer Engineering, Mississippi State University, Mississippi State, MS, USA

12.1 Cognitive process modeling with neural networks

12.1.1 Background and motivation

Cognitive (radar) architectures are often inspired by human cognitive performance models and findings from cognitive psychology. Cognitive science investigates the human cognition process. This is necessarily an experimental approach, involving, e.g., MRT-imaging techniques of neuro-science or psychologist investigating human problem-solving strategies in computer-programs: "a cognitive theory should be like a computer program" [3]. For the psychological concept of *intelligence*, several definitions exist:

- "Judgment, otherwise called 'good sense', 'practical sense', 'initiative' - the faculty of adapting one's self to circumstances" [4].
- "a general capacity of an individual consciously to adjust his thinking to new requirements … a general mental adaptability to new problems and conditions of life" [5].
- "The aggregate or global capacity of the individual to act purposefully, to think rationally, and to deal effectively with his environment" [6].
- "Goal-directed adaptive behavior" [7].
- "Intelligence measures an agent's ability to achieve goals in a wide range of environments" [8].

The term cognition comes from the Latin word "cognoscere," which means to conceptualize or to recognize. It is often stated that cognition encompasses an information processing act. While in the early 20th century, behavioristic psychology was dominant, with the "cognitive revolution" around the year 1956, the emphasis shifted towards internal, mental processes. Higher human cognitive capabilities encompass, e.g., situation awareness (SA), attention, problem solving, planning, remembering, learning, and language understanding. In the following 20 years, several cognitive capabilities were analyzed and understood by psychologists using symbol processing computer programs. This "Computer-Metaphor" is based on the physical symbol system hypothesis, which states that "A physical symbol system has the necessary and sufficient means for general intelligent action" [9]. AI software based on symbol-manipulation, such as the "General-Problem-Solver" is nowadays often referred to as "Good Old Fashioned Artificial Intelligence" [10]. Modern software-tools in cognitive psychology hence also include sub-symbolic approaches, e.g. based on activation patterns or neural nets.

12.1.2 Situation awareness and connection to perception–action cycle

SA is a psychological concept, that is closely linked to others like perception, attention, workload. Several definitions exist:

- "Continuous extraction of environmental information, integration of this knowledge to form a coherent mental picture, and the use of that picture in directing further perception and anticipating future events" [11].
- "Just a label for a variety of cognitive processing activities that are critical to dynamic, event-driven and multitask fields of practice" [12].
- "Situation awareness is the perception of the elements in the environment within a volume of time and space, the comprehension of their meaning, and the projection of their status in the near future" [13].

Mica Endsley's definition and the model shown in Figure 12.1 are particularly widespread. She describes three levels of SA, whereas level 1 ("perception of elements in current situation") encompasses all directly perceived objects (e.g. cars, aircrafts, pedestrians) in a scene and their state (e.g. position, dynamics, mode of operation). Level 2 SA ("comprehension of current situation") describes the association between perceived objects towards an abstract description of the situation. For this, an interpretation and assessment of the facts due to a priori knowledge and experience is required. Level 3 ("projection of future status") extrapolates L1 and L2 elements perceived into the future. This represents an even further degree of abstraction and allows statements about future events.

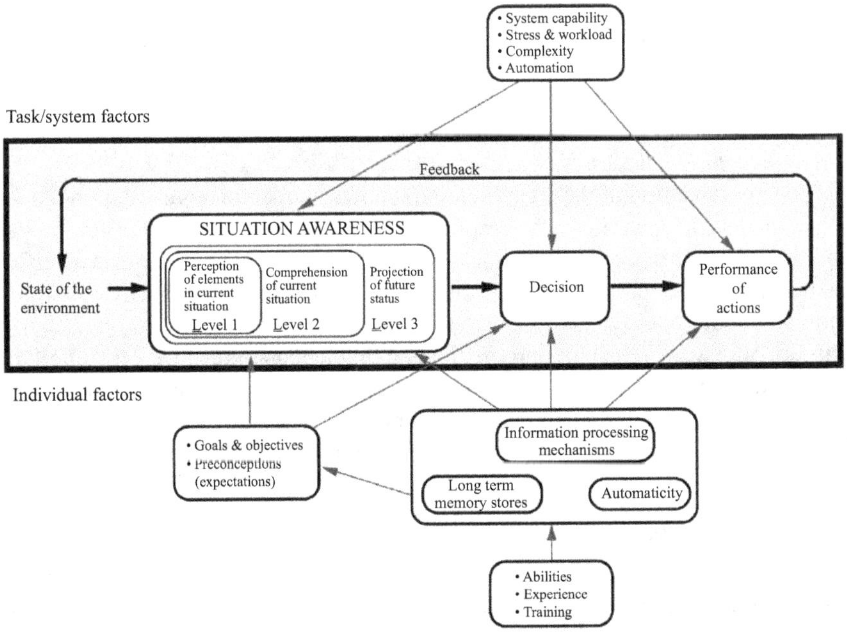

Figure 12.1 Endsley's model of SA

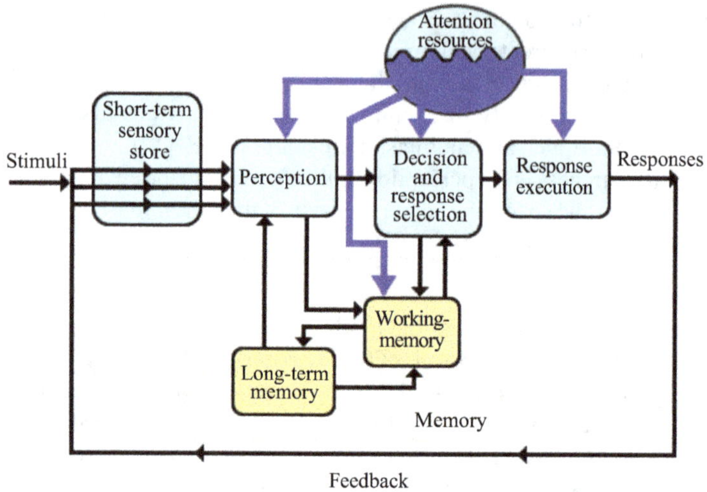

Figure 12.2 Wickens model of information processing

12.1.3 Memory and attention

The information processing scheme according to [14] shown in Figure 12.2 is a standard-model in many cognitive architectures. The approach distinguishes between the pure reception of a stimuli by the receiving organs and the information processing by higher cortical structures in the brain. The reception of the stimuli is represented by the short-term sensory store, which can hold a large amount of data for short time (e.g. 0.1–0.5 s) to provide the incoming signals to the subsequent perception and pattern recognition processes.

The interpretation of the signal into an internal, semantic representation (called mental models) happens in the "perception" block. The "decision and response selection" and "response execution" blocks represent subsequent the human decision making and plan execution stages. The working memory can temporarily hold information for about 20–45 s. According to [15], its capacity is restricted to 7 ± 2 chunks of information (e.g. remembering the digits in a telephone-number). The long-term memory in contrast retains large chunks of information for long-time periods (e.g. a lifetime). In the Wickens model, a limited amount of attention resources is available to be distributed among the different information processing blocks.

12.1.4 Knowledge representation

Most cognitive architectures differentiate between short- and long-term memory, whereas short-term refers to the "working memory" of the system, e.g., its current perceived state or SA. Long-term memory on the other hand commonly refers to the static background or "a priori" knowledge that the system uses in order to make inferences or act upon the current working memory content. Learning usually involves a modification of this a priori knowledge (either offline or online).

A computer implementation of such cognitive algorithms hence is strongly influenced by the type of knowledge representation in the short- and the long-term memory of the architecture. Several formats exist and have been successfully used in various computer science applications.

Logic and fuzzy logic

The philosophical discipline of logic is devoted to reasoning. Classical logic is a calculus whose statements are mapped to one of two truth values (usually true (w) or false (f)) according to the principle of bivalence. are mapped. In propositional calculus, propositions (also called formulas) can be composed of atomic statements (e.g., A: "Socrates is a man," B: "Men are mortal") and junctors (e.g.,$\forall$,$\vee$,$\wedge$, $\Rightarrow$, $\Longleftrightarrow$). Assuming that both statements A and B are true, the truth value of the compound statement $A \wedge B$ can be determined to be true. In the first-order logic, it is also possible to represent the inner structure of propositions that are not propositionally true which cannot be further decomposed in propositional logic. The inner structure of the propositions is represented by predicates and their arguments. The predicate expresses, for example, a property that applies to its argument, or a relation that is relation that exists between its arguments. The membership of a predicate to a set can be answered in classical logic unambiguously and can be answered with true or false.

A generalization of the function to a real-valued number from the interval $[0,1]$ is allowed by fuzzy logic [16]. It thus colloquially allows the modeling of "vagueness" or also "uncertainty" (however, not as a probability value as in the following section following section, but as "the degree to which a property is true"). Fuzzification describes the mapping of descriptive terms like "weak," "medium," "strong" to the real-valued degree of membership of the element. The fuzzy logic describes a logic calculus on this by defining the usual logical operators of conjunction (AND-connection), disjunction (OR-connection), negation (NOT-connection), and implication. Defuzzification transfers the result of a fuzzy inference, i.e., the degree of membership of an element searched for, back into a qualitative description. Although the operators of fuzzy logic are described mathematically unambiguously, the fuzzy modeling process often turns out to be not very systematic. Nevertheless, systems based on fuzzy logic are successfully used in the field of automation, consumer electronics, speech recognition, etc.

Graphs and semantic networks

A graph $G = (V, E)$ denotes a set of vertices and connecting edges. If the edges are characterized by a direction, the graph is called directed. If there are no closed edges (the so-called cycles), the graph is called a tree (Figure 12.3(a)). If there are costs associated with the connections between the nodes, these are noted on the edges and we speak of a weighted graph (cf. Figure 12.3(b)). A semantic network is a graph structure that is used to represent classification hierarchies (so-called taxonomies) and relations (Figure 12.3(c)). Nodes represent either categories or objects, i.e., concrete individuals of a class. Directed edges represent relations between nodes. The type "Instance" denotes a relation between an individual (K, e.g., "UAV4") and the class (H,

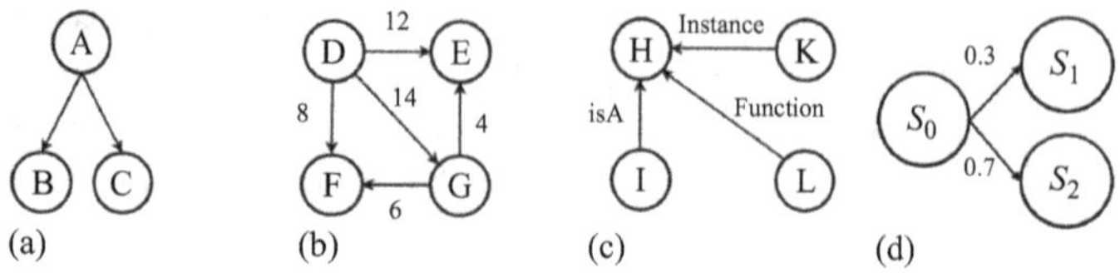

Figure 12.3 (a) Tree, (b) weighted graph, (c) semantic net, and (d) state transition diagram

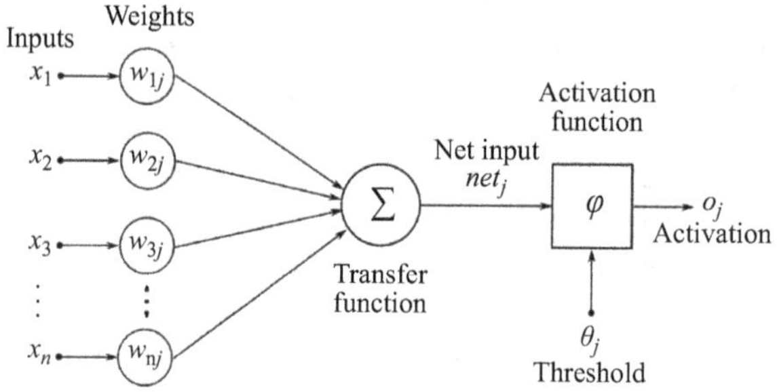

Figure 12.4 Artificial neural network

e.g., “aircraft”). An “isA” relationship describes a subset, or inheritance relationship (e.g., I, “UAV”). Functions describe properties and relationships between nodes (e.g. L, “civil,” edge type “admission”).

Figure 12.3(d) shows a probabilistic state transition diagram. It is well suited to represent dynamic processes in partially observable or stochastic environments. The vertices S_0, S_1, and S_2 represent three different states that, e.g., the environment can be in. The edges represent a transition between the states S_0, S_1, and S_2 with a probability of 0.3 and 0.7, respectively.

Artificial neural networks

Biological studies showed early on that information processing in the human brain emerges as an emergent effect through the interaction of the electrical signals from a large number of simple neurons—an approach that is also referred to as Connectionism. As early as the 1940s, the authors of [17] described networks of artificial neurons that are capable of approximating arbitrary approximate arithmetic functions. In contrast to the symbol-based representation forms mentioned so far in this chapter, artificial neural networks (ANNs) work with subsymbolic real-valued activation potentials. With suitable coding, however, logical functions can also be realized.

Figure 12.4 shows a simple mathematical model for an artificial neuron. The neuron is connected at its input to the activation potential of the outputs of other neurons in the network. These are given weights and form the transfer function by linear combination. From this, the activation function generates the activation potential at the output of the neuron. The activation function is usually a non-linear function (e.g., the sigmoid or Tanh function), which changes its output value abruptly from 0 to 1 when a threshold θ changes abruptly from 0 to 1 (the neuron "fires"). The network structure of artificial neurons is called topology. A perceptron is a simple feedforward network with an input and an output layer. Multilayer feedforward networks contain one or more hidden layers in between. Recurrent networks, in contrast to the feed-forward principle, contain backward edges, which thus enable a memory or feedback mechanism. The creation of the network topology is crucial for the performance and is usually done manually or experimentally. In general, given a topology, the approximation accuracy of the network is improved by adjusting the weights in supervised learning. This requires an annotated set of training data that associates known input vectors with desired outputs. Minimizing an error metric (usually the mean square deviation between the output of the network and the training data) is thus an optimization problem over the weights. Frequently used methods in this context are the delta rule or the backpropagation algorithm. A general problem of stochastic learning methods is overfitting, i.e., the precise replication of the used training data under loss of the generalization ability of the generalization ability of the network to unknown data patterns.

Comparison

Table 12.1 shows a qualitative comparison of the memory representations presented in this chapter, evaluating representation forms of the memories with respect to expressive power (which is usually inversely proportional to the efficiency of the associated inference process), intuitive comprehensibility by a human user and suitable environment classes.

Predicate logic is a widely used and established form of representation in mathematics, philosophy and established form of representation in mathematics, philosophy and computer science. Accordingly, the distribution as well as the comprehensibility by human users and knowledge engineers is high. The expressive power is also high,

Table 12.1 Qualitative comparison of different knowledge representation formats. "++/+" indicate a (strong) advantage of the respective method in this category, while "0" indicates a neutral and "–" a negative ranking.

Name	**Type**	**Expressivity**	**Human readable**	**Environment**	**Implementation**
Predicate logic	Model	0	+	Deterministic	CPU
Graphs	Model	–	+	Deterministic	CPU/GPU
Fuzzy logic	Model	+	+	Partially observable	CPU
ANN/RNN	Data driven	+	–	Partially observable	CPU/GPU
Physics-aware DNN	Model + Data driven	++	–	Partially observable	CPU/GPU
Reinforcement learning	Model	++	–	Partially observable	CPU/GPU

since besides statements and junctors also quantifiers are supported. Taxonomies and set relations cannot be represented natively. Predicate logic is the classic form of representation for symbol-based approaches, and is thus primarily suitable for deterministic, fully observable environments. Fuzzy logic is an extension of classical logic, and is thus at least as expressive as the expressive as the underlying propositional or predicate logic. The comprehensibility is limited because the fuzzy modeling process is not always clear. This, however, makes fuzzy logic applicable to partially observable or stochastic environments.

Graphs are also a common form of knowledge representation in mathematics and computer science. They are also intuitive for human users. However, the expressive power is lower, since natively only relations between nodes in the form of between nodes in the form of directed and weighted edges. Thus a suitability for deterministic, completely observable environments is present. Via modeling techniques, however, these weaknesses can be partially, albeit not natively, compensated.

ANNs are able to represent arbitrary logical and arithmetic functions. However, the efficiency of the approach strongly depends on the chosen network topology. The subsymbolic representation form is, therefore, not intuitively understandable and the topology is often difficult to grasp or to model. Due to the statistical, data-driven learning methods and the generalization capability, the approach the approach is well suited for stochastic and partially observable environments.

12.1.5 A three-layer cognitive architecture

The cognitive radar architecture developed at Fraunhofer FHR is based on the three-layer model of human cognitive performance by Rasmussen. As shown in Figure 12.5, the complex human cognition process is broken down into eight simplified cognitive subfunctions arranged in three behavioral layers of increasing abstraction. The *skill-based* behavior represents subconscious subsymbolic processing of input sensory data inside the gray perception block, and execution of sensory-motor-patterns on

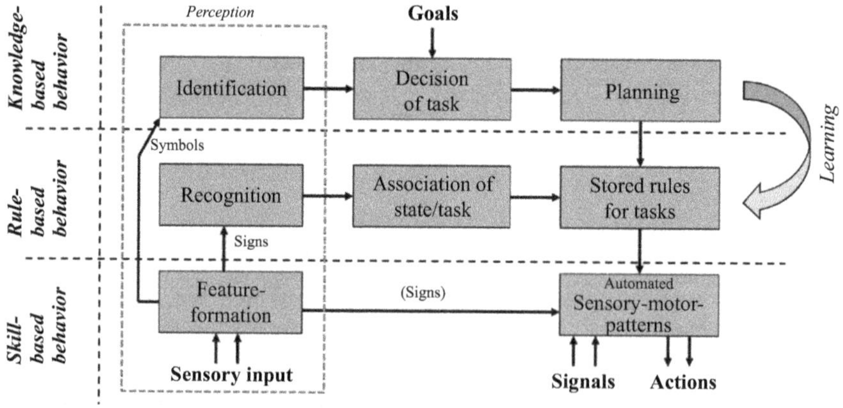

Figure 12.5 Three-layer cognitive architecture

the action side. *Rule-based* behavior allows a human to quickly and consciously react to a known situation by learned procedures. *Knowledge-based* behavior provides the highest flexibility by a semantically rich, symbolic identification of the environment and deliberative task planning. In this concept, *learning* refers to "automating" knowledge-based behavior, by abstracting and compressing the desired behavior into a set of pre-computed rules. In a particular situation, a reactive procedure will immediately be executed instead of the full knowledge-based reasoning process.

The three layers of behavior are associated with different forms of perception and knowledge representation:

Signals are sensory data representing time–space variables from a dynamical spatial configuration in the environment, and they can be processed by the organism as continuous variables.

Signs indicate a state in the environment with reference to certain conventions for acts. Signs are related to certain features in the environment and the connected conditions for action. Signs cannot be processed directly, they serve to activate stored patterns of behavior.

Symbols represent other information, variables, relations, and properties and can be formally processed. Symbols are abstract constructs related to and defined by a formal structure of relations and processes—which by conventions can be related to features of the external world.

12.1.6 Applications of machine learning in a cognitive radar architecture

Figure 12.6 shows the adaption of the generic three-layer architecture from Figure 12.5 to a cognitive radar application domain. The figure also illustrates algorithmic and signal-processing approaches towards the software implementation of the required cognitive subfunctions.

In the field of cognitive radar, several different architectures that integrate self-learning have been proposed (refer for instance to the Chapters 3 and 9 in this book).

In our approach, the transitioning from sub-symbolic, e.g., raw input data on the skill-based layer towards a symbolic representation on the rule-based layer is well suited for machine learning (ML) methods. This bridging of the *semantic gap* allows higher level functions to process the data in a more abstract representation. In Radar, deep learning (DL) methods have been used successfully for the identification of non-cooperative targets (NCTI). In this case, raw input data (e.g., high-resolution range profiles or SAR/ISAR images) are assigned a to a target class, that can be further exploited using knowledge-based reasoning or automated planning functions. Once a discrete state matches a predefined procedure, optimal-control policies are executed on the rule-based layer. The reinforcement learning (RL) technique illustrated in Section 12.3 is well-suited for this. Adaptive signal-processing and

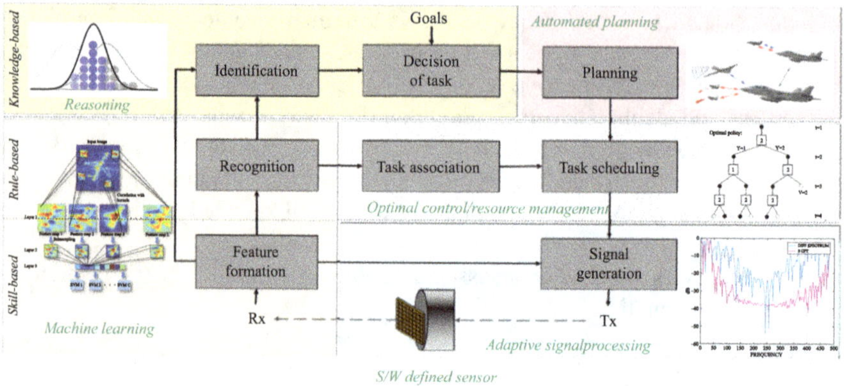

Figure 12.6 Three-layer cognitive radar architecture with algorithms

waveform-generation methods are finally utilized to emit optimized TX-pulses on the raw-data output level.

12.2 Integration of domain knowledge via physics-aware DL

Perception is a key function of cognitive radar that inherently incorporates prior knowledge in the processes of training DNNs for recognition and identification tasks based on the sensor input (see Figure 1.5). Prior knowledge can come in the form of learned information stored in long-term memory or physics-based models. Over the years, a large body of knowledge, e.g., [18–20], has been established on the electromagnetic backscatter from a variety of significant targets, include vehicles, aircraft, drones, people, animals, and clutter—the backscatter from surfaces other than the person of interest, such as the ground, buildings, or trees. These models have formed the basis for advanced radar simulations, which predict the received RF radar return given any antenna-target geometry and clutter environment, as well as the development of radar signal processing algorithms for target detection and tracking. The principal disadvantage of models, however, is that they do not take into account dynamic changes in the environment or target behavior, which could impact the accuracy of models being used. Furthermore, computationally convenient models may not capture the phenomenology of the signal in its entirety. While more complex models surely could be developed to improve accuracy, the dynamic nature of the sensing environment ensures that there will always be some part of the signal that is unknown. Practical examples of this include device-specific sensor artifacts, glitching, RF interference, or terrain with great topographical variation.

DL offers a data-driven approach to learning, which can bridge the gap between models and the real world. However, DNNs require massive amounts of data to accurately learn underlying representations from scratch. However, the application of black-box DNNs has had limited success when applied to sensing problems due to limitations in the amount of type of measured data for training, inability to produce physically consistent results, and difficulty to generalize to out-of-sample scenarios. The acquisition of sensing data can be costly and time-consuming, especially if human subjects are involved. Moreover, it may not be possible to acquire training data corresponding to all possible target profiles or antenna-target geometries, especially when dealing with airborne sensing applications. Because the training of DNNs aims to optimize weights based on a cost function, metrics, and distance measures are statistical in nature, rending the network incapable of recognizing physically impossible samples. Finally, because DNNs are data-driven, they are severely limited when challenged to recognize new samples with significant inter and intra-class variations in comparison to the training samples.

This has motivated research in an emerging domain of ML often known as *physics-aware*, *physics-based*, or *physics-inspired* DL [21] which aims to integrate physics-based models with data-driven DNN architectures in a synergistic manner. The resulting hybrid approach thus optimizes trade-offs between prior versus new knowledge, models vs. data, uncertainty, complexity, and computation time, for greater accuracy and robustness, as summarized in Figure 12.7.

Domain knowledge and physics-based models can be incorporated into any step within an ML approach, starting from the way RF data is presented to a DNN, to how the DNN is trained and structured, and the design of the cost function to be minimized. As a case study through which physics-aware thinking will be illustrated, this section will be using the problem of radar-based human motion recognition using micro-Doppler signatures [22,23]. The need to understand human movements lies at the core of all ambient intelligence applications, including defense and security, remote health monitoring of gait, falls, and fall risk, and human–computer interaction based on gesture or sign language recognition [24]. Human motion recognition is also a great problem where domain knowledge has significant tangible benefits to ML: although we might not know what a specific person is doing at a particular time, we do know in general a lot about how humans move. This is reflected in bio-mechanical models for human gait [25] and the physical constraints of how the parts of the body

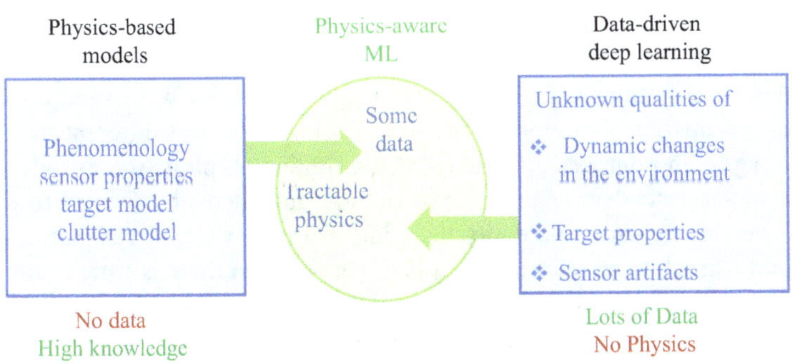

Figure 12.7 Physics-aware DL trade-off

move relative to each other during an activity. But, there is a great degree of variability in the walking styles of different people, and nearly an infinite number of different movements a person can make. This presents great challenges in the training of DNNs, generalization, and open-set classification (testing a model on a class not included in the training data).

The ability to accurately generate synthetic data is critical to the development of cognitive radar for automatic target recognition (ATR) not only because of its essential role in training deep models for ATR (discussed in Section 1.2.1) but also because of its potential role in the Testing and Evaluation (T&E) of cognitive radar systems. It is often not feasible to test cognitive radar across all possible operational scenarios it may encounter. This is because in real-world applications, target behavior and environmental factors are dynamic. In contrast, simulations would provide the ability to fully validate ATR algorithms prior to deployment in operational settings for which real data is not obtainable. This requires not only accurately representing expected target signatures but also site-specific clutter. Currently, there remains a significant gap between measured radar signatures and synthetic datasets, which precludes to the use of simulations for T&E. Advancement of physics-aware synthetic data generation techniques could close this gap and thus fill an important need critical to the advancement of cognitive radar design and development.

12.2.1 Physics-aware DNN training using synthetic data

An important issue in the training of DNNs is the initialization of the network. CNNs, one of the more commonly used DNN architectures, are usually randomly initialized before being trained in a supervised fashion using a training dataset. However, when the amount of available training data is small, this approach may not yield the best possible performance. This is because the objective function of a CNN is highly non-convex, i.e., the parameter space of the model contains many local minima. Thus, training a DNN by randomly initializing model parameters (weights and biases) may not be as effective as gradient-based optimization algorithms, which may converge to a local minima that is not optimal in a global sense [26]. Randomly initialized DNNs require large training sample support to converge to a good solution, which may not be available for RF applications.

Two common alternatives to random initialization are transfer learning [27] and unsupervised pre-training [28]. Transfer learning provides one way to address the limited data problem by pre-training the network first on data from a different source, e.g., optical images. In contrast, unsupervised pre-training exploits the encode–decode structure of autoencoders (AEs) or convolution autoencoders (CAEs) to greedily train the weights to learn an identity mapping. For micro-Doppler signatures, it has been found that while transfer learning is effective when there is truly minimal real data available, CAEs are more effective for moderate amounts of data [29,30]. This is because of the difference in phenomenology between RF micro-Doppler signatures and optical data, which results in different spatial correlations between adjacent pixels. For example, micro-Doppler signatures are bounded by the maximum possible

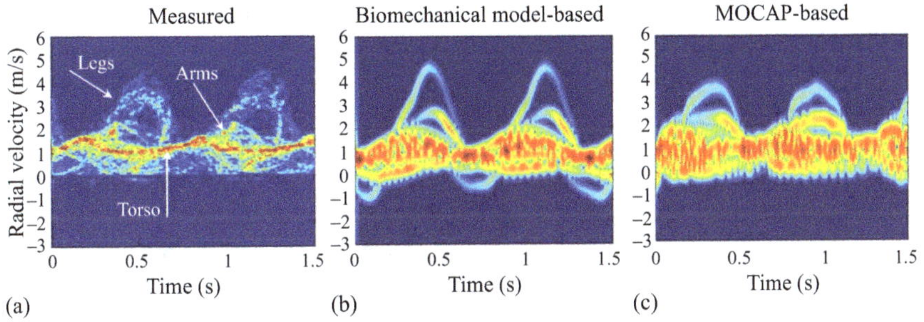

Figure 12.8 Comparison of micro-Doppler signatures for walking: (a) 77 GHz RF data, and synthetic data generated using (b) gait model and (c) MOCAP

velocities of the body, whereas optical imagery has spatial correlations based on the physical location of objects.

As a result, physics-based models have been proposed [31] to generate synthetic data for initialization of DNNs. Both bio-mechanical models and motion capture (MOCAP)-based skeleton tracking have been used to synthesize micro-Doppler signatures, but MOCAP has become more commonly used as it can capture individual variations for almost any activity. As shown in Figure 12.8, while the synthetic micro-Doppler signatures bear good resemblance to that extracted from real RF data, MOCAP-derived signatures are more realistic than that of the biomechanical models because they measure the actual positional variations incurred and do not rely on functional approximations of joint trajectories.

Because MOCAP-based skeleton tracking relies on actual measurements from a human subject, as with radar data, the size of the dataset is still limited by the human effort, time, and cost of data collections. To overcome this limitation, diversification can be applied to generate physically meaningful transformations of the underlying skeletal model. In this way, a small amount of MOCAP data can be leveraged to generate a large number of synthetic micro-Doppler signatures. This is accomplished by scaling the skeletal dimensions to model different body sizes, scaling the time dimension to model different speeds, and perturbing the parameters of a Fourier-based model of joint trajectories to emulate individualized gait style.

When this technique [31] was applied to 55 MOCAP samples (5 samples for 11 activity classes), a total of 32,000 synthetic samples were generated. The synthetic samples were then used to initialize a 15-layer residual neural network, which was then fine-tuned with approximately 40 samples/class. Note that the depth of 15 layers is a significant increase over the 7-layers of a CNN trained with measured data only, and results in an improvement in classification accuracy of over 15%. The overall classification accuracy of 95% surpasses that attained with alternative forms of network initialization [32], such as transfer learning from optical imagery and unsupervised pre-training.

Thus, physics-aware initialization with knowledge transfer from model-based simulations is a powerful technique for overcoming the problem of limited training data and can also improve generalization performance by exploiting the simulation of scenarios for which real data acquisition may impractical.

12.2.2 Adversarial learning for initialization of DNNs

While model-based training data synthesis has been quite effective in the replication of target signatures, it does not account for other sources of noise and interference, such as sensor artifacts and ground clutter. Because interference sources may be device-specific or environment-specific, data-driven methods for data synthesis such as adversarial learning are well-suited to account for such factors. Adversarial learning can be exploited in several different ways to learn and transfer knowledge in offline model training, as illustrated in Figure 12.9; for example,

- To improve realism of synthetic data generated from physics-based models.
- To adapt data from a different source domain to better resemble data from the target domain.
- To directly synthesize both target and clutter components of measured RF data.

The main benefit of using adversarial learning to improve the realism of synthetic images generated from physics-based models is that it preserves the micro-Doppler signature properties that are bound by the physical constraints of the human body and kinematics, while using the adversarial neural network to learn features in the data unrelated to the target model, e.g. sensor artifacts and clutter. The goal for improving realism [33] is to generate training images that better capture the characteristics of each class, and thus improve the resulting test accuracy. However, as the goal of the refiner is merely to improve its similarity to real data, a one-to-one correspondence is maintained between synthetic and refined samples. In other words, however, much data we have at the outset, as generated by physics-based models, is the same amount of data that we have after the refinement process—no additional data is synthesized.

Alternatively, the data from a source domain may be adapted or transformed to resemble real data acquired in the target domain [34]; then, the adapted data is used for network initialization. In this approach, the source domain can be real data acquired using a different RF sensor with different transmit waveform parameters (center frequency, bandwidth, and pulse repetition frequency), while the target domain is that which is to be classified. For example, consider the case where the target domain is RF data acquired with a 77-GHz FMCW automotive radar, but there is insufficient data to adequately train a DNN for classification. Perhaps data from some other sensor, however, is available: this could be data from a publicly released dataset, or data from a different RF sensor. Suppose we have ample real measurements from two other RF sensors—a 10-GHz ultra-wide band impulse radar and a 24-GHz FMCW radar. Although the data from these three RF sensors will be similar for the same activity, there are sufficient differences in the micro-Doppler signatures that direct transfer learning suffers from catastrophic performance degradation. While the classification accuracy of 77 GHz data with training data from the same sensor can be as high as 91%, the accuracy attained when trained on 24 GHz and 10 GHz data is just 27% and 20% [35], respectively. This represents over 65% poorer accuracy. On the other hand,

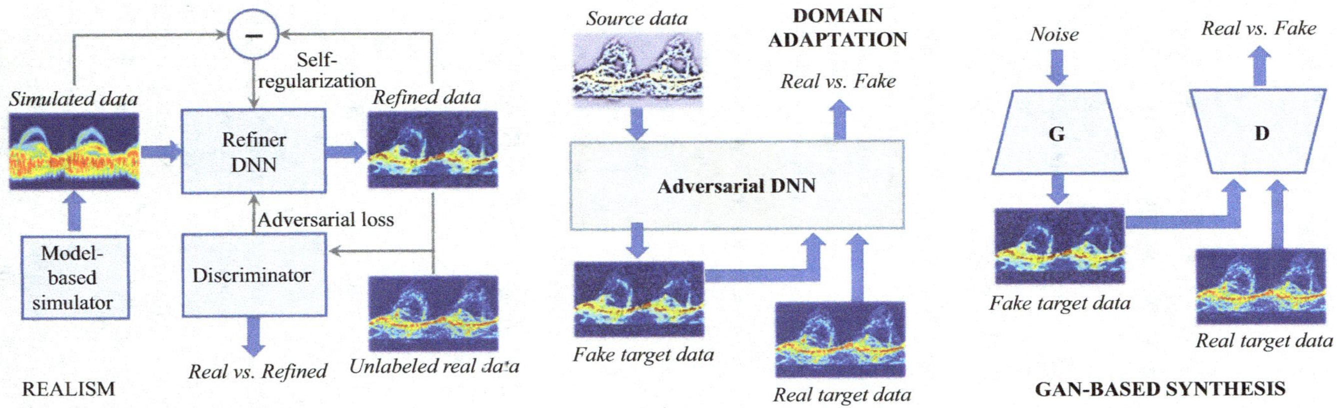

Figure 12.9 Techniques for utilizing adversarial learning for DNN initialization

when adversarial domain adaptation is applied to first transform the 10 GHz and 24 GHz data to resemble that of the target 77 GHz data, classification accuracies that surpass that of training with just real target data can be achieved [36].

A number of image-to-image translation techniques such as Pix2Pix [37] and CycleGAN [38] have been proposed in the literature:

- **Pix2Pix:** Pix2Pix is a type of conditional GAN (cGAN), where the generation of the output image is conditioned on the input; in this case, a source image. The generator of Pix2Pix uses the U-Net [39] architecture. In general, image synthesis architectures take in a random vector as input, project it onto a much higher dimensional vector via a fully connected layer, reshape it, and then apply a series of de-convolutional operations until the desired spatial resolution is achieved. In contrast, the generator of Pix2Pix resembles an auto-encoder. The generator takes in the image to be translated, compresses it into a low-dimensional vector representation, and then learns how to upsample it into the output image. The generator is trained via adversarial loss, which encourages it to generate plausible images in the target domain. The generator is also updated via an l1-loss measured between the generated image and the expected output image. This additional loss encourages the generator model to create plausible translations of the source image.
  The architecture of the discriminator is a PatchGAN/Markovian discriminator [40] that works by classifying individual $(N \cdot N)$ patches in the image as "real vs. fake," as opposed to classifying the entire image. This enforces more constraints that encourage sharp high-frequency detail in the output images. The discriminator is provided both with a source image and the target image and must determine whether the target is a plausible transformation of the source image. One limitation of Pix2Pix is that since it is a paired image-to- image translation method, the total number of synthetic samples generated is identical to the number of real target signatures acquired.
- **CycleGAN:** In contrast to Pix2Pix, CycleGAN is a GAN for unpaired image-to-image translation. Thus, a greater amount of synthetic data can be generated than the real imitation samples used at the input of the network. For two domains A and B, CycleGAN learns two mappings: $G : A \cdot B$ and $F : B \cdot A$. CycleGAN translates an image from a source domain A to a target domain B by forming a series connection between two GANs to form a "cycle": the first GAN tries to synthesize "fake fluent" ASL data from the imitation signing data, while the second GAN works to reconstruct the original sample, synthesizing "fake imitation" ASL samples. Thus, the network tries to minimize the cycle consistency loss, i.e., the difference between the input of the first GAN and the output of second GAN. Each CycleGAN generator is comprised of three sections: an encoder, a transformer, and a decoder. The input image is fed directly into the encoder, which shrinks the representation size while increasing the number of channels. The encoder is composed of three convolution layers. The resulting activation is passed to the transformer, a series of six residual blocks. It is then expanded again by the decoder, which uses two transpose convolutions to enlarge the representation size,

and one output layer to produce the final transformed image. The discriminators are comprised of PatchGANs—fully convolutional neural networks that look at a "patch" of the input image, and output the probability of the patch being "real." This is both more computationally efficient than trying to look at the entire input image and is also more effective since it allows the discriminator to focus on more localized features, like texture.

Domain adaptation techniques have been utilized in several radar applications, including SAR image retrieval [41], cross-target mapping between synthetic and measured data for improved generalization [42], classification in multi-frequency radar networks [43], and American sign language recognition [44] to bridge the difference in fluency across users. In domain adaptation, however, there are two significant sources of errors: first, there is the discrepancy between the target and source domains, which may not be fully compensated for by domain adaptation networks; second, there are kinematic errors generated by the generative process within DNN itself.

DNNs can generate kinematic errors because RF data is not naturally an image, but is actually converted into a 2D format via radar signal processing; in the case of micro-Doppler signatures, time-frequency analysis is employed to extract the micro-Doppler shifts as a function of time. Consequently, spatial correlations are not based on physical proximity (as in optical images), but depend on the distribution of velocity across the target (in this case, human body) and constraints imposed by the physical structure of the target (e.g., human skeleton). generative adversarial network (GAN) architectures [45] are not supplied with any information or metric pertaining to these constraints, resulting in synthetic samples that bear spatial resemblance, but in fact may correspond to physically impossible target movements. Different adversarial networks may have different degrees to which such errors are generated based on the DNN architecture itself. For example, CycleGAN generates significantly more kinematically flawed synthetic data than Pix2Pix, as illustrated in Figure 12.10. Note that neither Pix2Pix nor CycleGAN is able to adequately re-create the impulsive peak in the original data. While Pix2Pix generates a weaker signal with less textural richness relative to the original data, CycleGAN signatures are blurry and even the signal strengths observed in the repetitious portion of the signature are not replicated with a consistent amplitude. This is at least in part because the CycleGAN architecture includes two generators, in contrast to the single generator of Pix2Pix; hence, the greater amount of kinematic errors.

As discrepancies between source and target may remain despite domain adaptation, if a moderate amount of real data is available, it may be preferable to simply directly synthesize target data directly. As GANs are used for direct synthesis as well, the next section discusses in more detail the types of kinematic errors commonly observed and ways a physics-based approach can be used to reduce such errors.

12.2.3 Generative models and their kinematic fidelity

This section considers the kinematic fidelity of synthetic data generated through three well-known networks: the Wasserstein GAN (WGAN), Conditional Variational Autoencoder (CVAE), and Auxiliary Conditional GAN (ACGAN).

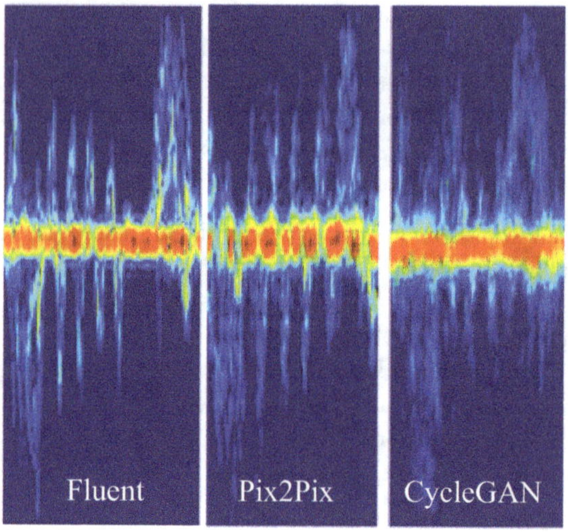

Figure 12.10 Comparison of real micro-Doppler signature for the ASL sign WATER *and the synthetic signatures generated by Pix2Pix and CycleGAN*

WGANs are a popular variant of the GAN architecture, which employs the 1-Wasserstein distance, also known as the Earth-Mover (EM) distance rather than alternative metrics, such as the Kullback–Leibler (KL) divergence or the Jenson–Shannon Divergence (JSD), to quantify the distance between the model and target distributions [46]. The WGAN is advantageous because it provides for a more stable training process, with proven convergence of the loss function, and is less sensitive to model architecture or hyperparameter selection.

Alternatively, conditional generative models, principally the CVAE and ACGAN, allow the generative model to condition on external class labels. This has the benefit of improving the visual accuracy of the synthetic images generated. CVAEs are an extension of the vanilla VAE where the input observations modulate the prior on Gaussian latent variables that generate the outputs. A vanilla VAE consists of an encoder, a decoder, and a loss function. The encoder and the decoder are usually designed as neural networks, and they are given the weights of θ and ϕ, respectively. The encoder takes an input image and outputs a latent representation in lower dimensions. It is important to note that the latent space is stochastic: the encoder outputs parameters to a Gaussian probability density, which can then be sampled to obtain noisy values of the latent representation z. Then, the decoder takes the encoded latent representation as an input and outputs parameters to the probability distribution of the data. Let us denote the encoder and decoder as $q_\theta(\mathrm{z}|\mathrm{x})$ and $p_\phi(\mathrm{x}|\mathrm{z})$, respectively.

The loss function of a vanilla VAE is the negative log-likelihood with a regularizer. It can be decomposed into a single spectrogram image since there are no global connections between images. The loss function l_i for a single image x_i is defined as

$$l_i(\theta, \phi) = -\mathrm{E}_{\mathrm{z}\sim q_\theta(\mathrm{z}|\mathrm{x}_i)}[\log p_\phi(\mathrm{x}_i|\mathrm{z})] + \mathrm{KL}(q_\theta(\mathrm{z}|\mathrm{x}_i)||p(\mathrm{z})) \tag{12.1}$$

where the first and the second term represent the reconstruction error and the regularizer, respectively. The former encourages the decoder network to learn how to reconstruct the input data, while providing the smallest error, as in basic autoencoders. If the decoder is unable to reconstruct the data well enough, then it will incur a high loss function value. The regularizer is the Kullback–Leibler (KL) divergence, which measures how much information is lost when using $q_\theta(\mathrm{z}|\mathrm{x})$ to represent $p(\mathrm{z})$. The regularization term forces the encoder to map images from the same classes onto the same region in the latent space. Moreover, in the VAE, p is specified as the normal distribution with mean zero and variance one ($N(0, 1)$).

Similar to vanilla VAEs, a CVAE consists of an encoder, a decoder, and a loss function. However, in contrast to VAEs, CVAEs have additional input branches called conditions (external class labels) to both the encoder and decoder. Due to embedding of class labels, the encoder is conditioned on the spectrograms and corresponding class labels, whereas the decoder is conditioned on latent variables and class labels. Other than conditional embeddings, CVAEs have the same principle as VAEs, where the encoder takes the spectrograms and class labels (x, y) and outputs a hidden representation z, with the attached weights (θ) and biases (ϕ). Then, the decoder takes z and y as inputs and outputs the parameters to the probability distribution of the data. The CVAE is trained to maximize the conditional log-likelihood. In CVAEs, the empirical lower bound is defined as

$$L_{cvae}(\mathrm{x}, \mathrm{y}; \theta, \phi) = -\mathrm{KL}(\, q_\phi(\mathrm{z}|\mathrm{x}, \mathrm{y}) \,||\, p_\theta(z|\mathrm{x})) + \frac{1}{L}\sum_{l=1}^{L} \log p_\theta(\mathrm{y}|\mathrm{x}, \mathrm{z}^{(l)}), \quad (12.2)$$

where $\mathbf{z}^{(l)} \approx N(0, 1)$, L is the number of samples, $q_\phi(z|x,y)$ is the conditional recognition distribution, and $p_\theta(z|x)$ is the generative distribution.

ACGANs are an extension of the vanilla GAN model that enables the model to be conditioned on external labels to improve the quality of the generated images. One method to produce class conditional samples can be done by supplying both generator and discriminator with class labels as in CVAE. However, the strategy behind the ACGAN is to instead of feeding the class information to the discriminator, one can task the discriminator with reconstructing the label information. This can be done by modifying the discriminator to contain an auxiliary decoder network that outputs the class labels for the training data. In this respect, the objective function of the ACGAN has two parts: the log-likelihood of the correct source, L_S, and the log-likelihood of the correct class, L_y.

$$L_s = \mathrm{E}[\log p(\mathrm{s} = real|\mathrm{x}_{real})] + \mathrm{E}[\log p(\mathrm{s} = fake|\mathrm{x}_{fake})]. \quad (12.3)$$

$$L_\mathrm{y} = \mathrm{E}[\log p(\mathrm{Y} = y|\mathrm{x}_{real})] + \mathrm{E}[\log p(\mathrm{Y} = y|\mathrm{x}_{fake})], \quad (12.4)$$

where s are the generated images. The discriminator is trained in order to maximize the $L_s + L_Y$ while the generator is trained to maximize $L_Y - L_s$.

A sample of some of the spectrograms generated by WGAN, CVAE, and ACGAN are shown in Figure 12.11 in comparison to real RF signatures for the class of walking. At the outset, it may be noticed that the CVAE-generated signatures are almost unrealistically blurry, a feature exhibited across all classes. The main reason

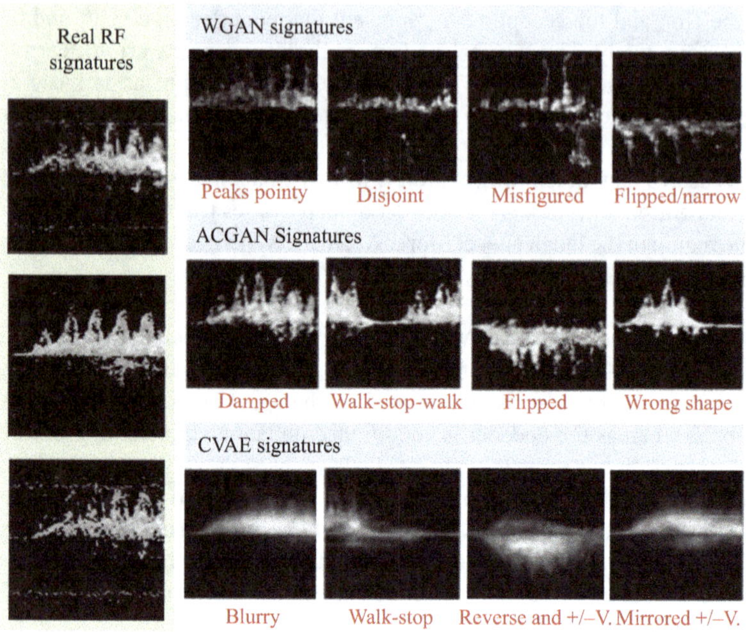

Figure 12.11 Comparison of real micro-Doppler signature for walking and the synthetic signatures generated by WGAN, ACGAN, and CVAE. "+/–V" denotes simultaneous positive and negative velocity components.

for this blurriness stems from the challenge of fitting of the data distribution into a tractable density distribution. Nevertheless, all were found to generate data that exhibits significant discrepancies from that of real RF signatures. Examples include

- *Disjoint components micro-Doppler*: Real micro-Doppler signatures are connected and continuous, because all points on the human body are connected with each other, forming a continuous spread of velocities. This prevents human RF signatures from having disjoint components or regions in the signature.
- *Leakage between target and non-target components*: A benefit of GANs is that sensor-artifacts can also be synthesized, but sometimes this results in leakage (connected segments) between target movements and sensor artifacts or noise, which are not physically possible.
- *Incorrect shape of signature*: When the shape of the micro-Doppler is distorted, with additional peaks, or symmetric reflections about the x-axis, these components correspond to physically impossible movements; e.g., a person whose hand simultaneous moves towards and away from the radar, additional repetitions, or sudden motion back and forth that are not normally part of the sign.

While these erroneous components may not seem significant visually, they ultimately correspond to kinematically impossible articulations, which, when used as training data, incorrectly trains the DNN and significantly degrades classification accuracy. For example, consider the initialization of a DNN for the classification of eight different human activities using 40,000 synthetic samples generated with ACGAN, and fine-tuning the training with just 474 real samples. The use of the ACGAN drastically reduces real training data requirements; however, recognition accuracy is boosted by 10% simply by discarding the 9,000 kinematically impossible samples, which are identified as outliers in the distribution using Principal Component Analysis [47].

Although DNNs are able to learn—given enough data—complex spatiotemporal relationships, the application of DL in radar applications has been limited by three important challenges: (1) far too much data is required for training, which is rarely available for all real scenarios of interest; (2) the learned models can produce physically inconsistent results; and (3) therefore cannot generalize to scenarios not explicitly represented in the training data. While MOCAP-based models allow a physically meaningful way of diversifying target simulations to a wide range of probably kinematic profiles, the GAN-synthesized samples are simply statistically diverse—we have not placed any constraints on the data to ensure their kinematic fidelity nor is it evident what the physical articulation that corresponds to each synthetic sample is.

12.2.4 Physics-aware DNN design

Physics-aware DNN design aims at addressing such challenges via a combination of physics-based modeling with data-driven DL. There are two key considerations: (1) the architecture of the DNN itself, and (2) the loss function we seek to minimize during training. One way to modify the architecture is to minimize the error between the physics-based model and a learned model—a method known as residual modeling [48]. However, because this approach does not explicitly model physical quantities, a limitation of this approach is that it cannot enforce physics-based constraints. Alternatively, signal processing can be used to compute physically meaningful signals and integrate them into the learning process, so that the DNN becomes "aware" of the significance of certain variables and is therefore able to take this into account in feature learning and loss minimization.

For example, let us modify the GAN architecture [49] to now include an auxiliary branch, which extracts features from the 1D envelope of the micro-Doppler signatures, in addition to the main branch, which extract features from the 2D signature itself, as illustrated in Figure 12.12. The envelop of the signature embodies critical kinematic properties because it reflects the highest frequency components that are generated by the motion of the legs. Consequently, it is critical for synthetic signatures to preserve as much as possible the shape and features of the envelope. The MBGAN architecture aims to emphasize envelope consistency by explicitly extracting and deriving features from the envelope. As can be seen from the comparison of the signatures, the MBGAN architecture does indeed provide greater consistency in the envelope profile.

Further improvement to the kinematic fidelity of the synthesized signatures can be made through modification of the loss function [50] used in the GAN. WGANs utilize the Earth Mover's (EM) distance, which is a statistical measure of the distance between two probability distributions. Alternatively, kinematic loss metrics can be added to

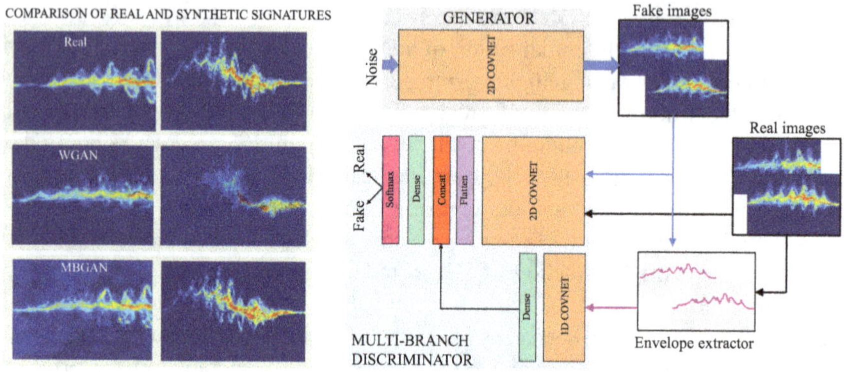

Figure 12.12 Multi-branch GAN (MBGAN) with integrated envelope extraction: architecture and comparison of signatures

the loss function to account for discrepancies in physics-based factors. Hence, we add another term to the discriminator loss function, L_D, which is dependent upon a kinematic metric:

$$L_D = L_C + GP + L_K \tag{12.5}$$

where $L_C = D(x) - D(G(z))$ is the critic loss; GP is the gradient penalty,

$$GP = \lambda\left(||\nabla_{\hat{x}} D(\hat{x})||\right)^2, \tag{12.6}$$

and L_K is the kinematic loss, based on a kinematic metric. In the above expressions, $D(x)$ is the discriminator's estimate of the probability that real data instance x is real; $G(z)$ is the generator's output when given noise z; $D(G(z))$ is the discriminator's estimate of the probability that a fake instance is real.

Some possible kinematic metrics that are computable from the signature envelopes can be based on

1. *Curve matching*: Considering the envelope as a time-series or a curve, the similarity between curves can be measured in a way that takes into account both the location and ordering of the points along the curve [51]. It is noted that different measures for curve matching appear in several application domains, including time series analysis, shape matching, speech recognition, and signature verification. Curve matching has been studied extensively in computational geometry, and many measures of similarity have been examined [52]. Both the Frechet distance and dynamic time wrapping (DTW) distance are considered in this work, which represent two of the most commonly curve matching metrics. The Frechet distance is a maximum measure over a parametrization, whereas DTW is a sum-measure method. In time series analysis, DTW is an algorithm for measuring the similarity between two temporal sequences that may vary in time or speed. For instance, similarities in walking patterns could be detected using DTW, even if

one person was walking faster than the other, or if there were acceleration and deceleration during the course of an observation.

2. *Correlation coefficient*: GANs synthesize the spectrograms by trying to replicate the distribution of the real spectrograms. Thus, measuring the correlation between the distribution of the real samples and the synthetic samples can be useful for quantifying the degree of similarity between the synthetic and real RF data. Pearson correlation coefficients measure the linear correlation between two random variables by computing the covariance of the two variables divided by the product of their standard deviations. It has a value between +1 and −1. A value of +1 is the total positive linear correlation, 0 is no linear correlation, and −1 is the total negative linear correlation.

As a result of this new term in the loss function, when increased mismatch between the envelope of the synthetic and real samples occurs, the generator will be compelled to try to minimize this loss by synthesizing signatures with greater matching with the envelopes, thereby further improving the kinematic fidelity of the loss regularized MB-GAN (LR-MBGAN) synthesized samples. Physics-aware GAN designs have led to increased classification accuracy in both classification of gross motor activities as well as in fine-grain motions, such as sign language recognition [44].

12.2.5 Addressing temporal dependencies in time-series data

In the approaches discussed in the previous sub-sections, the radar data is batch processed for a finite time interval. Micro-Doppler signatures, for example, are an image acquired over a fixed time period which illustrates velocity versus time. Temporal dependencies are implicitly captured via consideration of spatial correlations—not explicitly modeled. As data is acquired, such "snapshots" must be recorded and are independently processed, precluding the consideration of dependencies between snapshots. Because radar data is inherently time dependent, the issues of how to properly account for temporal dependencies and deal with new data as it is acquired gains critical importance.

Recurrent neural networks (RNNs) have been the principle mechanism through which sequential time-series data classification has been approached. RNNs model temporal behavior using connections between nodes that form a directed graph along a sequence. At each time step, an output is produced. Long short-term memory (LSTM) RNNs are able to model longer-term behavior through the inclusion of a memory block that consists of a cell, input, output, and forget gates. On radar datasets, it has become common to combine RNNs with CNNs that take in 2D images, e.g., spectrograms, or 3D range-Doppler-time tensors [53,54]. Gated recurrent units (GRUs) are a variant of RNNs, which have simpler structure and better performance on smaller datasets. In a recent study [55], stacked GRUs were used to classify the concatenation of the micro-Doppler signatures for different activities to form a sequence. Spectrogram concatenation, however, does not accurately represent the overall measurement acquired from continuous observation of a target, because of the transitional motions that occur when changing kinematics. Only a few works have actually considered the sequential classification of continuous data using bi-directional LSTMs [56] or joint domain multi-input multi-task learning (JD-MIMTL) [57].

However, RNNs do not adequately address the problem of time-series radar data classification because they only capture local correlations between samples within a motion class—they do not address the problem of the kinematic dependencies between different classes of target behavior, as constrained by physics and probable target trajectory. Moreover, RNNs also require significant amounts of data for training. Although there has been some work in the broader context of GANs for time-series synthesis [58], this remains an open area of research. Sequential modeling and training both can benefit from physics-aware approaches to supplement DNNs in accurately learning long-term temporal dependences.

Another significant, but related issue, beyond just recognizing sequential data, is also how to update radar operation on both transmit and receive. The development of online training strategies for updating initial batch training can improve recognition of received data, and involves balancing the trade-offs between the benefits of learning from new data, which can provide adaptability to new scenarios, and preventing catastrophic forgetting, which would eliminate any knowledge embodied in initial batch training. An approach that has been in the forefront of research involving adaptation on transmit as well as receive is RL, as discussed in the next section.

12.3 Reinforcement learning

12.3.1 Overview

In many domains, the rise in popularity of artificial intelligence (AI) algorithms has increased the dynamic nature of engineering solutions, while reducing execution time and processing requirements, and often outperforming solutions created by human experts. In March 2016, the AlphaGo algorithm beat 18-time world champion Lee Sedol (4 to 1) at the game of Go, a game many times more complex than chess [59,60]. The AlphaGo RL algorithm created by David Silver and his team at DeepMind showed the advancements made by the field of AI in achieving goals with abstract metrics of success and delayed reward structures. In recent years, Nvidia Corporation's deep learning super sampling (DLSS) algorithm has utilized a convolutional autoencoder AI to drastically reduce the computational demand required to produce high-resolution images on graphical processing units (GPUs) [61]. In almost every field, AI has pushed the boundaries of what is possible and is only growing in relevance. The way radar is designed will also be subject to this trend, adopting AI solutions to advance the state of the art.

The use of AI and ML solutions for radar and communications systems has also grown in relevance in recent years. Selvi *et al.* investigate the use of a Markov decision process to enable a cognitive radar to coexist with communication systems while maintaining a target track [62]. Additionally, Ak *et al.* demonstrated the use of RL for anti-jamming cognitive radar applications and Thornton *et al.* investigate the use of RL for not-cooperative coexistence between cognitive radar and nearby communication systems [63]. Finally, Smith *et al.* explored the use of RL for waveform notching in a phase-coded signal, which will be used as an example here [64].

12.3.2 Basics of reinforcement learning

RL is a branch of ML in which an agent seeks to optimize a reward received from its environment. It involves an exploration phase where the agent gathers data by interacting with its environment to gain new experiences. During the training stage, the agent uses the data gathered through its experiences to improve its performance for future interactions. An agent is an autonomous entity which seeks to achieve a specific goal by taking actions that it believes will produce the highest reward. The agent operates within an environment which controls the rules of operations, including the reward structure. Through an iterative process, the agent attempts to improve its performance through experimentation. A typical RL pipeline at time t would begin with the environment presenting a state (s_t), which the agent observes (o_t). If the agent has perfect observation, o_t will be equal to s_t. For the examples in this chapter, we will assume perfect observation, and will solely use s_t in equations moving forward.

The agent will take an action (a_t) which will be processed by the environment. The environment will then produce a reward (r_t) for this action, and return a new state (s_{t+1}). The reward function is a critical component of any problem space, and special caution must be taken when designing a reward function to avoid unintended consequences, as it is the metric being used during training. This loop is repeated many times until the agent is sufficiently trained on the environment.

Figure 12.13 shows a typical RL pipeline for one state-action-reward cycle.

There are a large number of specific RL algorithms that can be used in a variety of applications. It is important to identify the specific algorithm that best solves the task at hand. Here we cover a small sample of RL algorithms that may be of use in cognitive radar applications.

12.3.3 Q-Learning algorithm

One of the simplest RL algorithms is Q-learning. Q-learning is a method for training an agent to select the best possible action for the current state based on the total reward

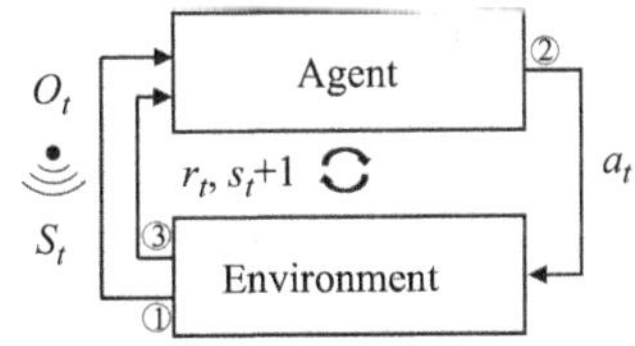

Figure 12.13 Reinforcement learning diagram

Q-Table				
	a_1	...	a_n
s_1	$Q_{1,1}$	...	$Q_{1,n}$
...	...	...	...
s_n	$Q_{n,1}$	...	$Q_{n,n}$

Figure 12.14 Q-Table diagram

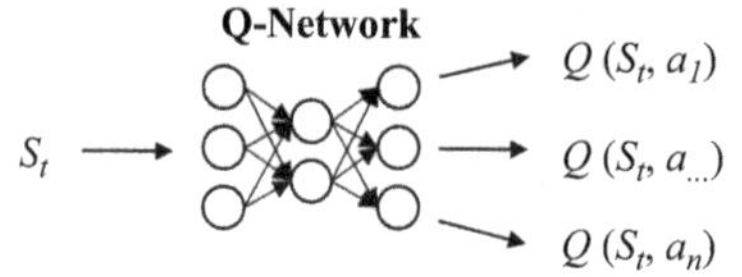

Figure 12.15 Deep Q-network diagram

that it expects to receive in the future. The Bellman equation is used to calculate the quality of each possible discrete action at the current state, called a Q-function:

$$Q(s,a) = r + \gamma \cdot max_{a'}\left(Q(s',a')\right) \tag{12.7}$$

Where r is the reward at the current time step, γ is the discount factor for events in the future, and $Q(s', a')$ is the quality estimate of future state-action pairs. Exploration of the environment through trial and error serves to train the Q-function until it is able to accurately determine the quality value of each action at the current state, at which point the agent only needs to select the action with the highest quality. For finite state and action spaces with discrete values, this process results in the agent producing a lookup table to map states to action qualities.

Q-learning is useful when the environment is simple and there are a limited number of possible states and actions, making it rather limited in its utility for radar problems.

12.3.4 Deep Q-network algorithm

Q-learning begins to suffer when operating in sufficiently large environments, or one that has continuous-valued state spaces. In this case, it is useful to add a DNN to the algorithm to produce a deep Q-network (DQN) [65]. In a DQN, the Q-function is replaced with a Q-network. As in other fields of ML, the DNN has significantly increased algorithms' capabilities by allowing them to generalize across similar states. The same is achieved here, by providing a mechanism to generalize across a large number of possible states.

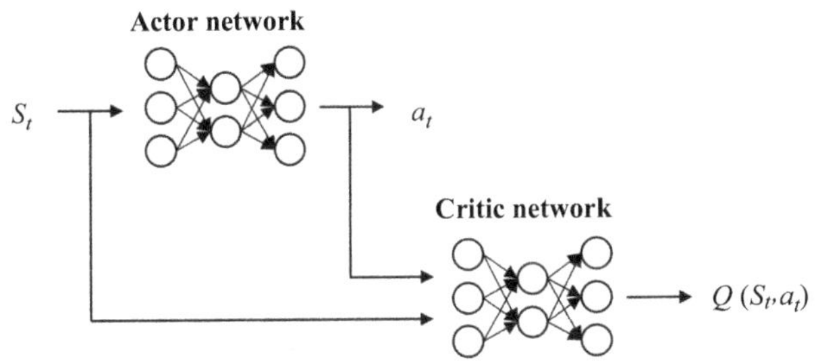

Figure 12.16 Deep deterministic policy gradient diagram

DQN is useful for problems with large state spaces, but a finite number of possible actions. This algorithm may have many applications in developing cognitive radar.

12.3.5 Deep deterministic policy gradient algorithm

For many applications, the algorithm must also accommodate large or continuous action spaces in addition to large or continuous state spaces.

DQN does not allow for large action spaces or continuous values because there is no way to select an action that optimizes the Q-function when there are infinite actions. One method for achieving this is to use two neural networks in combination, namely deep deterministic policy gradient (DDPG). One neural network is tasked with mapping states to actions, and the other neural network is tasked with mapping state-action pairs into quality values. An image depicting the process of the DDPG algorithm is shown here.

In addition to having the ability to handle large and continuous state spaces, the DDPG algorithm is capable of processing large state spaces to produce simultaneous, continuous-valued actions. DDPG may have many applications for cognitive radar, as the algorithm supports large problem spaces, as seen in the radar domain. The main drawback to using DDPG over other RL algorithms is the added complexity of training multiple neural networks simultaneously.

There are, of course, a great many other RL algorithms that could be of use to cognitive radar applications that are not addressed here.

12.3.6 Algorithm selection

There are dozens of RL algorithms currently available in open literature, and many more are being developed. Having knowledge of these algorithms allows the radar designers to effectively identify their needs and potential RL solutions.

In order to select the appropriate RL agent for their application, the designer must identify the state and action space dimensions and if they require continuous values. Designers should attempt to choose the simplest algorithm possible while still solving their intended problem. Example use cases for RL in cognitive radar applications would be planning, task scheduling, and signal generation.

12.3.7 Example reinforcement learning implementation

In order to demonstrate the process of applying an RL algorithm to a radar problem, a highly simplified phase-coded waveform spectral notching problem from [64] is examined to demonstrate a signal generation use case. Spectral notching is the process of designing a waveform that has notches in its power spectrum (PS) to facilitate spectral sharing. The PS can be managed by modulating the waveform using polyphase codes.

Defining an ordered set of N phases, $\Phi = \{\phi_1, \phi_2, \dots \phi_N\} = \{\phi_n\}$ for $n = 1 \dots N$ we can define the baseband radar waveform, $s[n]$, as

$$s[n] = \exp(j\phi_n), \tag{12.8}$$

where j is the imaginary number. It is clear from the form of (12.8) that $s[n]$ will have a constant amplitude envelope, a desirable property for high-power amplifiers. For this simplified example, the only constraint we have is that each ϕ_n must be in the interval $[-180°, 180°]$. One approach to control the shape of the PS of $s[n]$ is using optimization to select the phases, Φ, against some criteria [66].

In order to determine the PS of the resulting signal, we use equation (12.9):

$$P[n] = |\mathscr{F}(s[n])|^2 \tag{12.9}$$

Where $P[n]$ is the power spectrum as a function of frequency bin, $\mathscr{F}$ is the Fourier transform, and $s[n]$ is the complex signal. Here the PS will be represented in the logarithmic domain for convenience.

The spectral notching problem is typically solved through a gradient-based optimizer which determines a set of phases to be used in a phase-coded waveform [67–69]. In order for an optimizer to produce a phase-coded waveform, it will use trial and error to explore a cost function, specified by the designer. The optimizer will attempt to minimize the output value by following the gradient of the cost function through a series of small changes to the input variables. While the quality of notches synthesized this way is high, their iterative nature can lead to lengthy execution times. This is the most common limitation of applying optimization algorithms to design problems.

Alternatively, optimized solutions can be synthesized offline to be stored in a lookup table for future use. By doing so, high-quality solutions can be produced without the time required to run an optimizer. This method is, however, limited in its flexibility since a limited number of solutions can be stored in a lookup table. Additionally, determining which solution should be selected from the lookup table becomes its own design problem.

For these reasons, it is desirable to look at alternative solutions, which are becoming viable due to recent advancements in ML and AI. Specifically, we show an approach to solving this problem with the DDPG algorithm outlined above.

The major components of a DDPG algorithm are listed here:

- The *environment*: the simulation responsible for executing the logic that the AI operates within.
- The *state*: the observable space presented by the environment at a given time step.
- The *action*: the set of values created by the AI in response to a given *state*.

- The *reward function*: the equation that quantifies success for any *action* taken within the *environment* at the applicable *state*.
- The *actor* network: the neural network responsible for taking actions.
- The *critic* network: the neural network responsible for directing the behavior of the *actor* during training, based on its learned approximation of the *reward function*.
- The *quality* metric: the value that the *critic* attributes to any given *action*.

DDPG was selected to solve the radar notching problem due to the fact that it supports multiple, continuous actions, and because it accommodates abstract metrics of success.

For this example, we use an environment which consists of an extremely simplified waveform notching problem. The environment produces a randomly generated notch mask, a series of length N consisting of ones and negative ones. The indices with a value of one represent the desired transmit regions and the indices with a value of negative one represent the desired attenuation regions. The notching region has width of M and is constrained so that the notched region is not positioned near the beginning or end of the spectrum. Initially, we start with $N = 16$ and $M = 3$, a simple case which has a limited number of unique states. Figure 12.17 shows every possible state for this environment.

The actor network interacts with the environment by responding to the observed state with an action, a set of continuous phase values, Φ_n. The phases are to be used

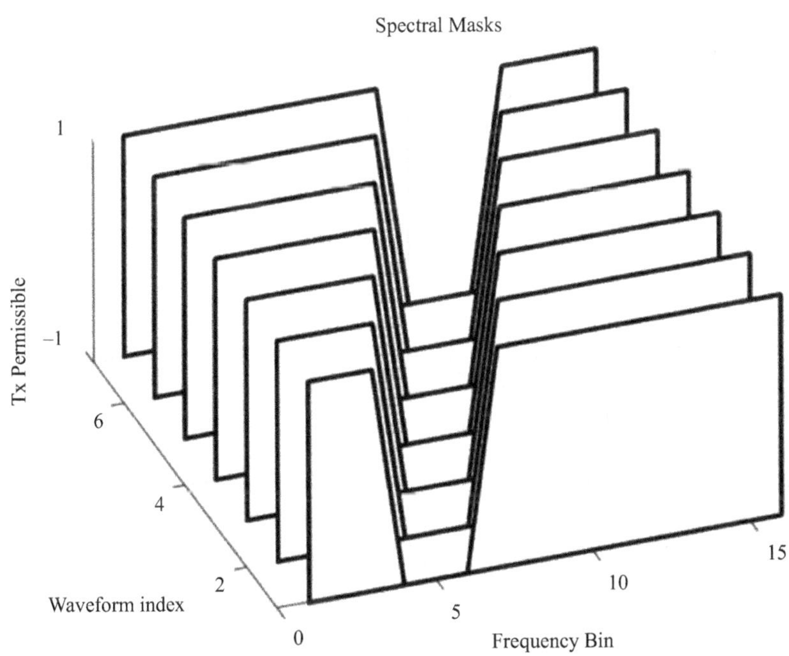

Figure 12.17 Example masks presented by notching environment

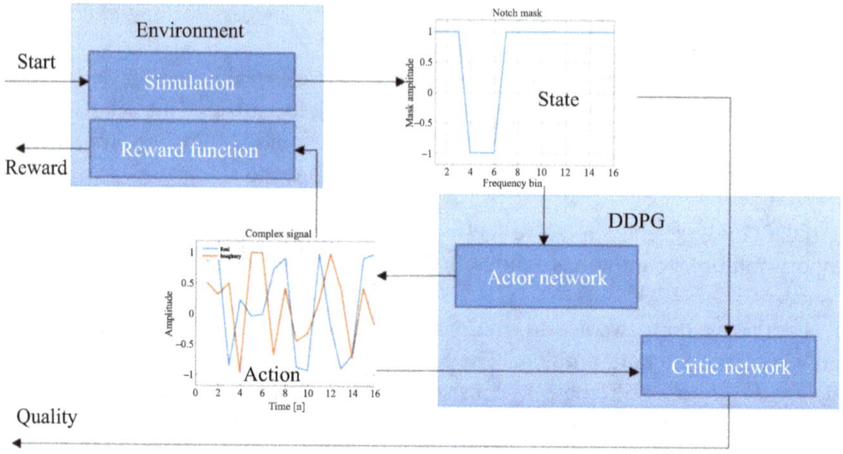

Figure 12.18 Block diagram DDPG implementation

in the phase-coded waveform in equation (12.8). The environment then processes the state-action pair to determine the reward, based on the reward function designed by the designer. The critic network takes the state-action pair and approximates the quality of the action given the state. The delta between the true reward provided by the reward function and the quality metric provided by the critic network creates a gradient to be used for back propagation while training the critic.

With an accurate critic network, it is possible to train the actor network to produce actions which maximize the output from the reward function. Figure 12.18 demonstrates the state-action-reward loop for the notching problem with the DDPG agent.

The DDPG is allowed to train on this environment for one million episodes, which produces the learning curve shown in Figure 12.19.

This figure shows that the algorithm is able to learn the environment relatively quickly. The improvement in performance is smooth, and appears to have reached its maximum value, since improvement is minimal at later training episodes. The trained DDPG algorithm is able to produce phase-coded waveforms with deep spectral notches, reaching 80 dB of attenuation as shown in Figure 12.20.

This notch is desirable in that it has relatively constant values for each region of the spectrum. It is important to note that this waveform has not been optimized for anything other than spectral notching, but it demonstrates the use of DDPG for similar applications.

12.3.8 Cautionary topics

In RL, the mechanism for incentivizing improvement is the reward function. A reward function is a scoring metric that helps the algorithm to improve by incentivizing

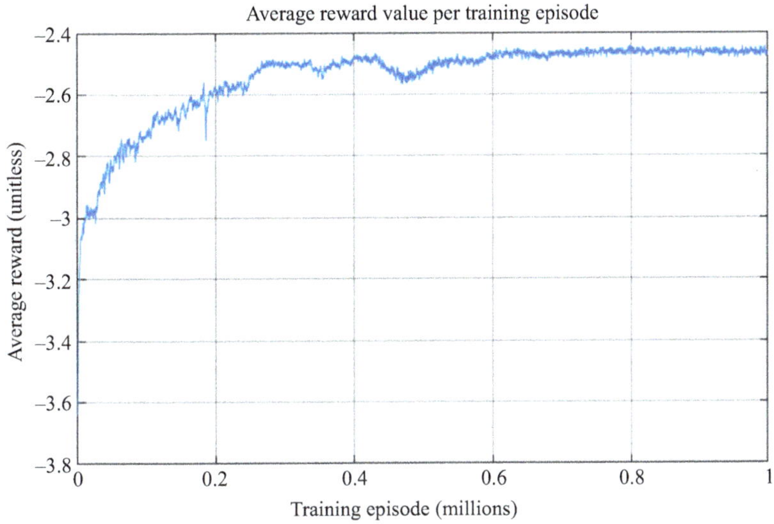

Figure 12.19 Training curve

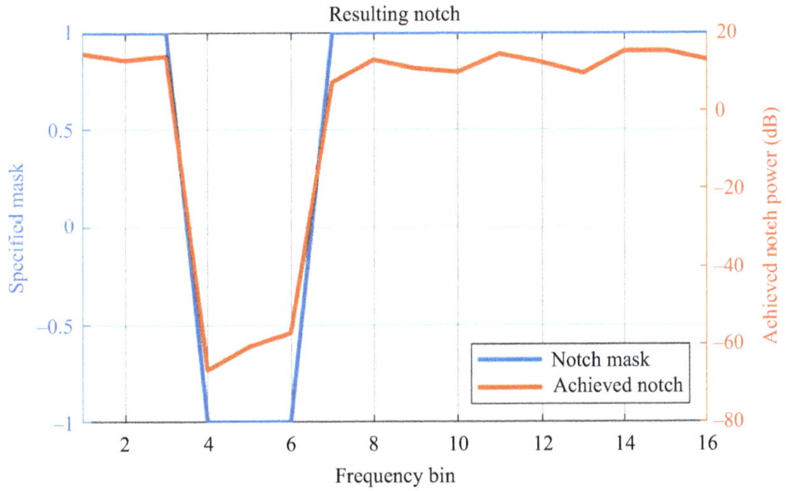

Figure 12.20 Best notch

high-quality solutions over low-quality solutions. The DDPG algorithm attempts to maximize the reward function by training an actor network to take actions in the given environment, while the critic network attempts to learn the reward structure. In model-free RL, like DDPG, the agent is free to explore the environment, but does not have access to the reward function directly. Instead, it explores the environment through experience, and learns the behavior that maximizes the reward it receives.

For this reason, the reward function is one of the most important factors in successfully implementing a DDPG algorithm. The critic network's entire purpose is to learn how to approximate the quality of state-action pairs as closely to the reward function as possible. It is the very nature of this feature that allows users to train neural networks to solve complex problems with high accuracy by attempting to achieve the same success metrics specified by the designer. The reward functions can be simple, or complex, and can accommodate human-readable metrics for success. It is also worth mentioning the concept of inverse reinforcement learning (IRL), which attempts to learn the reward function from observed optimal behavior [70].

In [64], several reward functions were observed to quantify their effects on performance. Specifically, the following reward function was evaluated:

$$R_1 = \frac{P[f]}{P[\bar{f}]} \tag{12.10}$$

where R_1 is the reward, P is the power spectrum of the phase-coded signal, and f and $\bar{f}$ are the transmit and notching frequencies, respectively.

Here, a new reward function is evaluated. Instead of taking the ratio of the transmit power to the notch power, we create a reward that incentivizes conforming to idealized notch power levels. The idealized notch power for each bin is calculated by considering that energy is conserved, meaning energy taken away from the notch region is placed in other frequency bins. By setting an optimistic target for notch depth and distributing the remainder of the power evenly amongst the transmit regions, an idealized power value for each frequency bin can be determined.

Now with the desired power for each frequency bin known, the reward function simply penalizes the solutions for deviating from these values. Equations (12.11)–(12.13) combine to produce a reward function that leverages the knowledge of a desired power level for each frequency bin:

$$D_{Notch} = \sum \left|\Delta(P_{NotchAchieved}, P_{NotchTarget})\right| \tag{12.11}$$

$$D_{Transmit} = \sum \left|\Delta(P_{TxAchieved}, P_{TxTarget})\right| \tag{12.12}$$

$$R_2 = -(D_{Notch} + D_{Transmit}) \tag{12.13}$$

Note the negative sign in the reward function, since we want to minimize deviation, and the reward function is to be maximized.

After training the DDPG with the reward function outlined in Equation (12.10), some interesting behavior is observed. This reward function seems useful, as it rewards the ratio of transmit power to notch power for being large, and a notch power of near zero would rightfully produce an extremely high reward value. However, there exists a "cheat" case in which the DDPG algorithm learns to place all the power in one of the transmit bins, usually by setting all the phases to zero. The trivial case of $\phi_n \approx c$ for all n, and where c is any constant number, achieves an extremely high reward, while simultaneously producing a very poor notch.

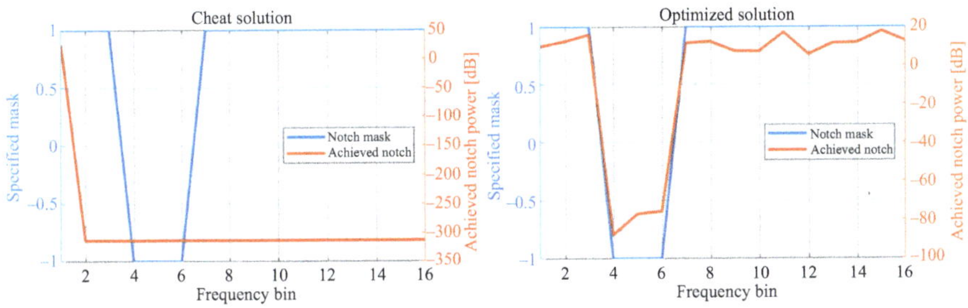

Figure 12.21 Cheat solution vs. intended solution

Figure 12.21 shows a comparison of two solutions to the notching problem. On the left is an example of a PS resulting from a cheat solution that produces an extremely high reward. On the right is an example of a notch that more accurately reflects the goals of the designer but scores a lower reward, which is problematic.

This example highlights the importance of being thorough when designing the reward function so that unintended solutions are not learned. One of the main benefits of using a reward function that maximizes conformity to an ideal, is the very low risk for "cheat" solutions, and is desirable for this study.

12.3.9 Angular action spaces

When using gradient-based optimization algorithms or neural networks, the algorithms are asked to solve a minimization or maximization problem on an equation. In these types of problems, the starting values of the independent variables are typically randomly initialized, after which they follow the gradient of the function until the minimum or maximum output value is reached. As an example, let us examine a minimization problem on a simple function with the independent variable being an angle, and bounded between $[-180°, 180°]$:

$$V_1 = f(\theta) \tag{12.14}$$

where V_1 is the output value of the function, and θ is the lone input variable. The algorithm will seek to find the value for θ which minimizes the resulting output value, V_1. Figure 12.22 shows two different initialization points for θ before following the gradient of the curve to minimize V_1:

Notice that the two different starting points yield different solutions, since Initialization Point 2 hits the upper bound of the input variable, though the bound itself is arbitrary and has no physical meaning. This demonstrates the effect that initialization can have on finding a high-quality solution to the problem. In most design problem domains, the independent variables are naturally bounded, but the cyclical nature of angles pose a unique challenge. When training a neural network, the function has thousands or millions of independent variables, which exacerbate the problem. One option for resolving this issue is to break the phase values down into two components

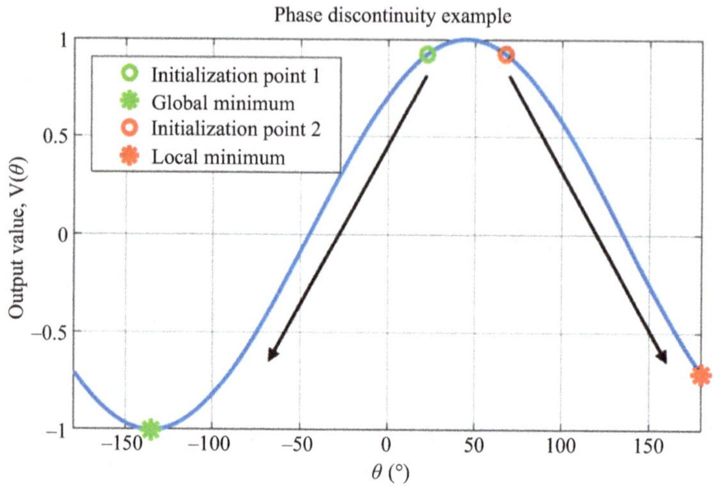

Figure 12.22 Minimization outcome highly dependent on starting location

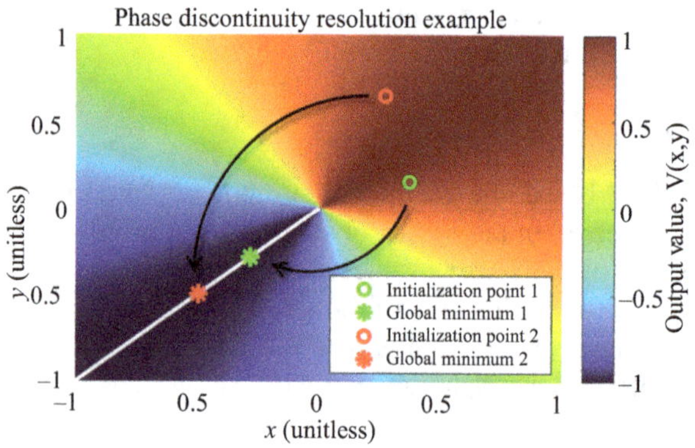

Figure 12.23 Minimization outcome independent of starting location

and use the arctangent function to produce angles from the components, as shown in (12.15):

$$V_2 = f(x, y) \tag{12.15}$$

While this increases the dimensionality of the problem, it removes the discontinuity in angle values at their limits. Figure 12.23 depicts the same problem depicted in Figure 12.22, but with the phase component method.

Notice that the starting point can no longer preclude the algorithm from determining the optimal solution, as it is always possible to reach it by following the gradient of the surface. Also notice that there are many optimal solutions to this problem, as

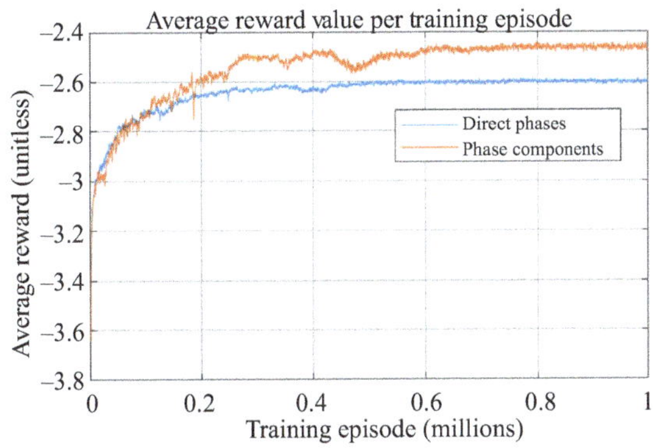

Figure 12.24 Learning curve comparison between direct phase and phase component environments

denoted by the white line. Any point on the line will produce an identical result, due to the nature of the arctangent function discussed earlier.

The result of this configuration is to allow for more reliable minimization, which results in higher quality outputs, as well as a more stable neural network training process which produces better outputs at completion.

In order to compare the two action spaces posed here, the DDPG algorithm is run with both the direct phase environment and the phase component environment. Figure 12.24 shows the learning curves for both environments trained for 1 million episodes.

It is clear from these results that the phase component environment trains to a higher quality than the direct phase environment, which can also be seen in the resulting notches.

Figure 12.25 shows notching results for the two fully trained DDPG algorithms at the conclusion of the 1 million episode training process.

There is a clear improvement in performance when using the phase component environment which achieves notch depths of 80 dB as opposed to the 60 dB notches achieved with the direct phase environment. It is thus concluded that the phase component environment is superior and will be used exclusively for this effort moving forward.

12.3.10 Accuracy of environment during training

As previously discussed, RL algorithms learn by exploring their environments and adapting their behavior in order to maximize performance. Consequently, it is important to be aware of how well the training process will translate to the final implementation. For games and simulations, the translation can be nearly perfect, but for real-world applications this is not the case. Additionally, real-world applications will suffer from deviations between true states and the observations made by the

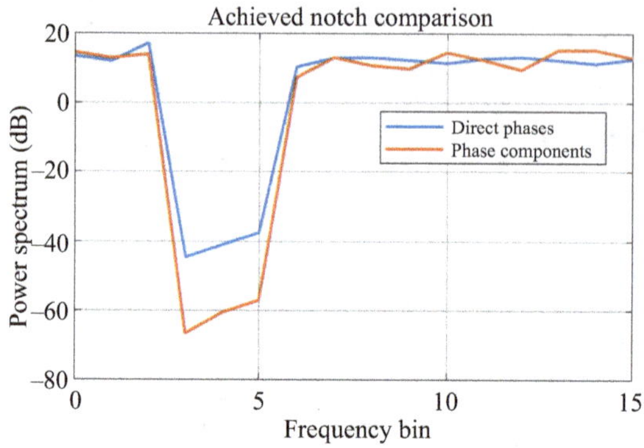

Figure 12.25 Achieved notch comparison between direct phase and phase component environments

system ($o_t \neq s_t$), adding another level of complexity. In order to train the RL agent to perform well in these environments, it is important to address these issues before training the algorithm. When using simulations for training, the agent will only be as good as the simulation it is trained on. While this limitation is not unique to RL solutions, it is likely the greatest hurdle for implementing high-performing AI in the field of cognitive radar.

12.4 End-to-end learning for jointly optimizing data to decision pipeline

A fundamental requirement for the implementation of a perception–action cycle is the addition of feedback in the transceiver, which requires real-time signal processing and ML in the decision process. This has led to a new area of research relating to end-to-end learning which aims to jointly optimize the data to decision pipeline.

A radar system observes and tries to understand the world around it through the data it acquired. As in many sensing systems, radar employs an acquisition, processing to decision pipeline to extract actionable information about the environment as illustrated in Figure 12.26. In this conventional approach first, the radar sensor acquires measurements from the environment. Next, the acquired data is processed to generate the signal representation that is suitable for further inference or decision. At this stage multiple processing steps can be applied and various types of products such as range-Doppler, range-Angle maps, spectrograms or synthetic aperture radar (SAR) images can be constructed. The final goal for many radar applications is not to reconstruct these various processing products but rather apply an inference or decision task such as detection of a target, estimation of a specific variable or classification.

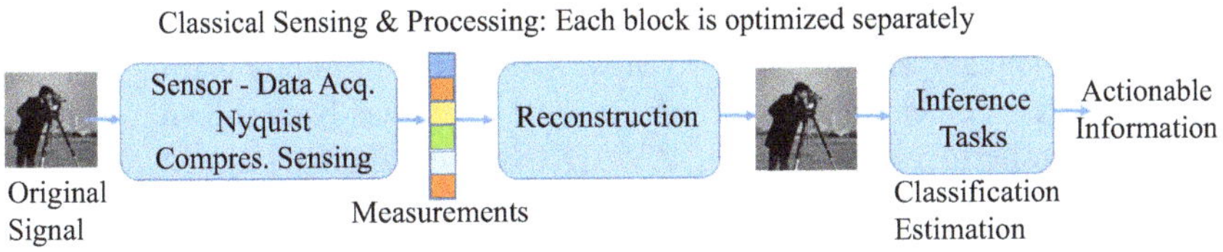

Figure 12.26 Classical sensing and data processing pipeline

Hence, the third stage in the classical data to decision pipeline is inference stage that is applied to the output of processing stage.

In conventional processing, each stage in the pipeline is independent of one another and optimized separately. The data acquisition is generally implemented as conventional Nyquist rate uniform sampling. Another approach is to do compressive sensing (CS) through less number of random linear measurements if the underlying signal can be represented sparsely in some domain. However, for both data acquisition, the main criteria is to be able to reconstruct the original signal. The data acquisition or specifically the measurements are not designed considering the final inference goal.

The second stage in conventional processing is the signal reconstruction stage. While in some cases this might be a linear transformation operation such as 2D Fourier transforms as in range-Doppler processing, in many radar applications such as SAR, it is a multi-stage non-linear reconstruction. Specifically for CS-based reconstruction, regularized optimization problems are solved to achieve final reconstruction. Signal reconstruction, specifically for CS, is a computationally expensive operation. While reconstruction could be the final goal in some cases, i.e., radar imaging, in many cases, a further inference is desired on the reconstructed signal. Detection of a target in a range-Doppler map, or estimation of the speed of a target, or classification of a target in a SAR image are some examples of varying inference applications on reconstructed radar signals. However, reconstruction approaches are generally designed to optimize only related metrics to its own goal such as maximizing signal to noise ratio (SNR) or minimizing mean squared error (MSE). They do not take into account in which inference task that reconstructed output will be used.

The final inference task is applied to the reconstruction stage outputs. This multi-stage approach although allowing already developed inference techniques to be easily utilized on varying signal domains, it is sub-optimal since the whole data to decision pipeline is not designed for the specific task-dependent goal. Hence, approaches that will allow inference directly from the acquired measurements in an end-to-end manner are required. For this goal, ML or specifically DL-based approaches offer potential solutions. ML or statistical signal processing and pattern recognition techniques have been historically applied only for the final inference stage of the conventional processing pipeline. However, in recent years, several studies have shown successful results on utilizing DL-based architectures to reconstruct signals from their compressed measurements [71–76]. These approaches utilize a variety of DNN structures including autoencoders, CNNs, or generative adversarial networks (GANs) to learn the mapping from low dimensional data space to the original signal space for the given signal type in a data-driven way. In addition, jointly learning optimal measurements together with the reconstruction process has been shown to enhance reconstruction performance [72–75]. Also DNN-based approaches generally show enhanced reconstruction performance compared to traditional CS based reconstruction techniques [77,78] while also providing much lower reconstruction times given a trained model. These recent results pave the way for an end-to-end learning framework where a joint data-to-decision pipeline combining acquisition, reconstruction, and inference stages can be learned.

12.4.1 End-to-end learning architecture

In this section, we will show an example end-to-end learning architecture that will combine acquisition, reconstruction and inference (i.e., classification) stages within a DNN to jointly learn the optimal measurements and the direct inference from these measurements for the task of classification. An illustration of the end-to-end learning framework is shown in Figure 12.27. For simplicity, the framework is shown for general images but the developed concept is also valid for radar data with additional modifications. It can be seen that the general structure of the end-to-end architecture include a combination of three distinct but jointly learned stages: sensing, reconstruction, and classification.

We model the sensing process as a general linear data acquisition as done in CS. In CS, a fixed measurement matrix (MM) with entries selected randomly from a given distribution to generate a small number of linear measurements are used. The measurements y from the original signal x can be modeled through a general MM $\mathbf{\Phi}$

$$y = \mathbf{\Phi}x. \tag{12.16}$$

The goal of the sensing part is to learn the MM hence the measurements for the original signal x. In order to learn the MM, the linear data acquisition in (12.16) is modeled using a fully connected (FC) layer with linear activation functions. In this

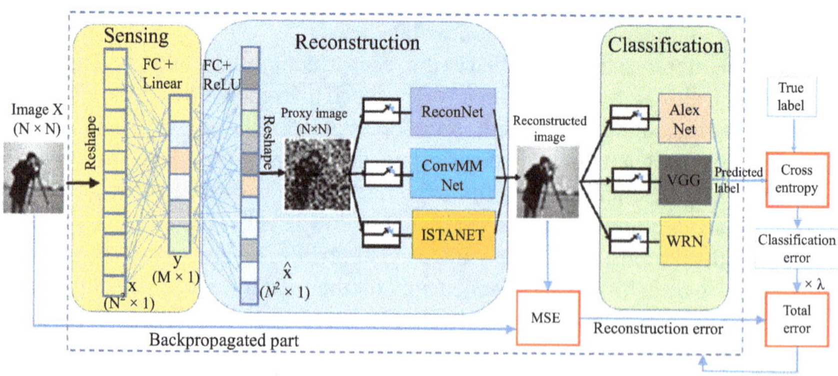

Figure 12.27 An example end-to-end learning framework

example framework, we assume that the input signal to the DNN are images hence $\mathbf{X} \in R^{N \times N}$. First the input image is vectorized via reshaping, i.e., $\mathbf{X} \rightarrow \mathbf{x} : R^{N \times N} \rightarrow R^{N^2 \times 1}$ and fed into FC layer to obtain compressed linear measurements $\mathbf{y} \in R^{M \times 1}$. Since only linear activations are used in this layer, the output of the FC layer models (12.16) exactly, where entries of $\boldsymbol{\Phi}$ are the weights used in the FC layer. Once the DNN is trained, this sensing stage and the remaining part of the DNN can be detached and the parameters of FC can be used to acquire the optimal measurements, while the remaining network can accomplish the direct inference from these compressed measurements.

The second part of the proposed end-to-end learning framework is the reconstruction network that takes the compressed measurements $\mathbf{y}$ from the sensing stage and reconstructs a signal estimate $\widehat{\mathbf{X}}$ that will be input to the classification network following this stage. In [79,80], a single FC layer with ReLu activation is used to imitate the adjoint operator by learning to create pseudo images. However, in recent years, DL-based signal recovery from compressed measurements led to several successful DNN structures that show enhanced signal recovery performance for the class of signals as they are trained on with much less computational complexity compared to classical CS recovery approaches. Enhanced DL-based reconstruction networks that map the compressed measurements to the original signal domain as opposed to a single FC layer can be used as the reconstruction stage of the overall framework. For this goal, we specifically focus on the recent and comparably successful reconstruction networks such as ReconNet [71,72], IstaNet [76], or ConvMMNet [75]. Each of these reconstruction networks differs by their DNN architectures on how they perform reconstruction from the given compressed measurements as explained in more detail in their respective publications. Such reconstruction networks are trained by minimizing a type of reconstruction-specific loss as the average squared reconstruction error in (12.17)

$$\mathscr{L}_R(\boldsymbol{\Theta}) = \frac{1}{T} \sum_{i=1}^{T} \| f(y_i, \boldsymbol{\Theta}) - \mathbf{X}_{\mathrm{i}} \|_F. \tag{12.17}$$

where T is the total number of training samples, and $f(y_i, \boldsymbol{\Theta})$ is the reconstruction network model with parameters, $\boldsymbol{\Theta}$ and the input compressed measurement samples y_i. In the end-to-end learning framework instead of a specific loss for only reconstruction or inference stages, a joint loss function that will derive joint learning of the end-to-end network architecture is utilized.

The final stage of the end-to-end learning framework is the inference network. Inference is a task-dependent goal and for the example architecture presented here a classification task is utilized. There have been a variety of successful DL-based classification architectures that have been applied to computer vision tasks such as AlexNet [81], VGG [82], and wide residual network (WRN) [83]. While a classification network can be chosen or designed to be used in this stage, for our example framework the WRN architecture is utilized. WRN is an extension of residual network

(RESNet) [84] utilizing skip connections and residual blocks. If just the classification task is considered, the networks are trained by minimizing the CEL that is defined as

$$\mathscr{L}_C(\widehat{\ell_i}, \ell_i) = -\sum_{i=1}^{T}\sum_{c=1}^{C} \ell_{i,c} log S(\hat{\ell}_{i,c}). \quad (12.18)$$

where $S(\hat{\ell}_{i,c})$ is the soft-max layer output that gives the probability that sample i belongs to class c.

12.4.2 Loss function of the end-to-end architecture

In the end-to-end framework, we jointly learn a MM that maps an original signal x_i to compressed measurements $y_i = \mathbf{\Phi}_S x_i$ and an inference network mapping y_i to a class label $\boldsymbol{\ell}_i$ over a training set of T samples. Learning the parameters of both sensing and inference networks can be done jointly through solving an optimization problem minimizing a defined loss. Minimizing the CEL in (12.18) is a natural selection for the whole end-to-end architecture since the final goal of DNN is to obtain the best classification performance. However, we show in our simulation results that it is not the case and a hybrid loss that incorporates a weighted combination of the reconstruction and classification losses increases the final classification performance. The main reason for this is that minimizing only the cross-entropy cost does not directly force the reconstruction DNN architecture to generate better reconstruction outputs as inputs to the classification part of the end-to-end network. Hence, we utilize a hybrid loss that incorporates a weighted combination of the reconstruction and classification losses. The goal by injecting the reconstruction loss into the total loss is to force the reconstruction network to generate better image estimates that will lead to enhanced classification performance. Thus, the proposed architecture is trained by minimizing the following total loss:

$$\mathscr{L}_{\mathscr{T}} = \mathscr{L}_{\mathscr{R}}(f_{\Theta_R}(y_i), x_i) + \lambda \mathscr{L}_C(f_{\Theta_C}(f_{\Theta_R}(\mathbf{\Phi}_s x_i)), \ell_i) \quad (12.19)$$

where $\mathscr{L}_R$ is the mean-squared reconstruction loss defined in (12.17), λ is the hyper-parameter that defines the ratio between $\mathscr{L}_C$ and $\mathscr{L}_R$ losses and it can be selected over a validation set.

12.4.3 Simulation results

CIFAR-10 [81] dataset is used for simulation and quantitative analysis of end-to-end learning. CIFAR-10 dataset consists of 60,000 color images of size 32 × 32 in 10 classes, with 6,000 images per class. The object classes are airplane, automobile, bird, cat, deer, dog, frog, horse, ship, and truck. We used gray-scale versions of all images in the dataset. Out of 60,000 total images, 50,000 images are used for training, and validation, while the remaining 10,000 are used for testing. Out of the training set of images, 80% of them are used for training, and 20% are used for validation purpose in a random manner. We present our results using two evaluation metrics. The classification results are presented using accuracy, while reconstruction performance is measured using peak-to-signal noise ratio (PSNR)[85]. The classification accuracy is defined as the ratio of the total number of correctly predicted labels to the total number of ground truth labels.

We compute the backpropagation of the compared DNNs by using mini-batch gradient descent routine. We have selected a batch size of 32 and epoch size of 500 with the help of ADAM optimization for a varying learning rate from 0.1 to 0.0001 to determine the network parameters. We have used Tensorflow [86], the open source DL framework, for training, validation, and testing purposes. All simulations are run on a DL machine with 3 NVIDIA Titan RTX GPUs to carry out the training, validation, and testing tasks.

In order to provide a baseline performance, we first present the classification results on the original signal domain. The original images in CIFAR-10 dataset is used without any compression. Three different classification networks, being AlexNet, VGG-3, and WRN, are trained and their performances are compared. All classification network parameters are initialized with random weights. The obtained accuracy over the test dataset is reported in Table 12.2.

WRN is the best-performing network over the original image domain among the compared techniques with 97% accuracy level. Nevertheless, we utilized both WRN and VGG-3 networks in the joint framework and provide results for both networks since they provide the two best accuracy results on the original signal domain.

Next we present the performance of the end-to-end deep joint learning (DJL) framework. We trained and tested for a set of compressed measurement numbers varying from $M = 64$ to $M = 768$. Since the images are 32×32, the dimension of original signal domain is $N = 1,024$ and utilized measurement rates correspond to M/N ratios of 0.0625 to 0.75. The proposed DJL framework is implemented with a variety of choices of reconstruction and classification networks. While ReconNet, ConvMMNet, or ISTANET$^+$ are the reconstruction network choices, VGG-3 or WRN are used as classification networks. Each combination case is trained and tested over the grayscale CIFAR-10 dataset using the same set of measurement numbers. We trained each DJL framework case with two different loss functions; either only the CEL defined in (12.18) or the proposed weighted total loss defined in (12.19). For all scenarios, all the network parameters are randomly initialized before training. Obtained accuracy results over the test datasets are shown in Table 12.3 for DJL.

There are several important conclusions that can be observed from the results presented in Table 12.3. First, if the end-to-end network is trained with the proposed weighted loss (WL) that combines both cross entropy loss (CEL) and reconstruction

Table 12.2 Classification accuracy on original images

Classification network	Accuracy
Simple Deep CNN/AlexNet	79%
VGG-3	88%
Wide residual network (WRN)	**97%**

Table 12.3 Classification accuracy for DJL framework with all tested cases

Acc.		**Cross-entropy loss**		**Weighted loss**	
Reconst.	**M**	**VGG-3**	**WRN**	**VGG-3**	**WRN**
ReconNet	64	36%	42%	59%	66%
ConvMMNet		41%	49%	65%	73%
ISTANET$^+$		**43%**	**50%**	**66%**	**74%**
ReconNet	128	43%	50%	65%	72%
ConvMMNet		48%	55%	69%	79%
ISTANET$^+$		**50%**	**57%**	**70%**	**80%**
ReconNet	256	50%	57%	71%	80%
ConvMMNet		55%	61%	76%	83%
ISTANET$^+$		**57%**	**63%**	**79%**	**85%**
ReconNet	512	59%	65%	80%	86%
ConvMMNet		63%	69%	85%	86%
ISTANET$^+$		**63%**	**69%**	**86%**	**90%**
ReconNet	768	63%	72%	85%	91%
ConvMMNet		66%	74%	87%	94%
ISTANET$^+$		**67%**	**75%**	**88%**	**96%**

loss (RL), the achieved accuracy levels are much higher than utilizing only the CEL. In addition, DJL framework allows using a choice of reconstruction and classification networks. It can be seen that from three possible reconstruction and two classification network combinations ISTANET$^+$ and WRN combination generally provides the best accuracy levels for both loss function cases and at all tested number of measurements. Another important observation is that achieved accuracy levels with the proposed networks and training with WL achieves similar accuracy levels obtained over original image domain for a higher number of measurements. This is because the proposed structure jointly reconstructs and classifies with a loss function that combines both reconstruction and classification errors in a weighted manner.

In order to understand the effect of weighting between CEL and RL, a simulation study is performed. The total loss is defined in (12.19) and the parameter λ controls how much CEL is added. If $\lambda = 0$, the total loss is the same as only RL while for very high λ total loss is dominated by only CEL. For the case of using ISTANET$^+$ and WRN combination in DJCL framework the achieved validation accuracy levels for a set of λ values is shown in Figure 12.28. The reconstruction network generates an image to be the input for the classification network as a midproduct of DJCL and the PSNR of that image is also shown in Figure 12.28. It can be seen that for smaller λ, the network focuses more on reconstruction and generates a high PSNR midproduct image but final accuracy levels are low. Increasing λ up to a level increases the achieved accuracy while sacrificing from the PSNR of the midproduct image. Although increasing λ more means for the network to pay much more importance to CEL, the accuracy levels decrease since the network cannot generate higher PSNR

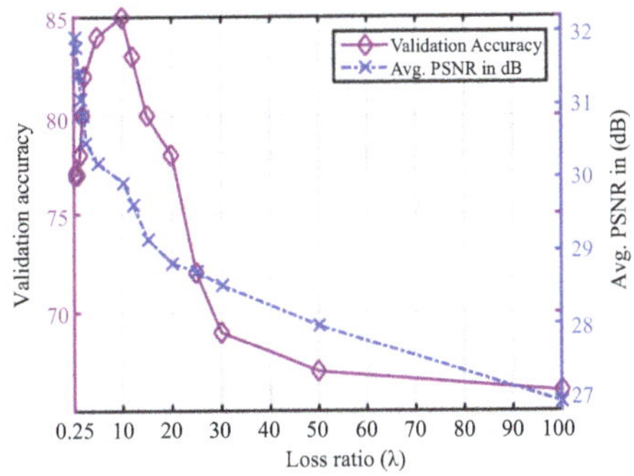

Figure 12.28 *Effect of loss ratio parameter λ on validation accuracy and PSNR of reconstruction network output for $M = 256$*

images that will be input to the classification network. Using such analysis an optimal λ parameter can be selected for the WL using the validation set and the performance of the selected parameter is tested with the independent test dataset.

12.5 Conclusion

This chapter has presented an in-depth overview and discussion of ways in which researchers are currently investigating to exploitation of DL in cognitive radar design. As cognitive radar architectures are inspired by human cognition, a principle area in which DL can play a role is in cognitive process modeling, including decision mechanisms that implement the perception–action cycle. This includes considerations of not just memory and attention, but the representation of knowledge and more sophisticated means for representing different levels of information resulting in more sophisticated actions. This chapter also focused specifically on certain timely topics critical to cognitive radar development, which have been topics of recent focus. This includes challenges to real-time, end-to-end DL, which is essential to being able to "close-the-loop" by reducing computational latency, RL, and the integration of knowledge via physics-aware DL, enabling trade-offs between data-driven design and computational complexity. Indeed, DL is a dynamic and rapidly advancing field, where there is still much opportunity towards investigating in the context of cognitive radar.

Acknowledgments

The authors would like to acknowledge the support of the National Science Foundation Awards #1932547, #1931861, and #2047771 towards generated the results shared in this work.

References

[1] Gurbuz SZ, Griffiths HD, Charlish A, *et al.* An overview of cognitive radar: past, present, and future. *IEEE Aerospace and Electronic Systems Magazine*. 2019;34(12):6–18.

[2] Fuster JM. *Cortex and Mind: Unifying Cognition*. Oxford, UK: Oxford University Press; 2003.

[3] Anderson JR. Cognitive psychology. *Artificial Intelligence*. 1984;23(1):1–11.

[4] Binet A and Simon T. Le développement de l'intelligence chez les enfants. *L'année psychologique*. 1907;14(1):1–94.

[5] Stern W. Die psychologischen Methoden der Intelligenzprüfung; 1912.

[6] Wechsler D. Die Messung der Intelligenz Erwachsener. Textband zum Hamburg-Wechsler-Intelligenztest für Erwachsene (HAWIE); Deutsche Bearbeitung Anne von Hardesty, und Hans Lauber. 1956.

[7] Sternberg RJ and Salter W. Conceptions of intelligence. In RJ Sternberg (ed.), *Handbook of Human Intelligence* (pp. 3–28). New York, NY: Cambridge University Press; 1982.

[8] Legg S and Hutter M. Universal intelligence: a definition of machine intelligence. *Minds and Machines*. 2007;17(4):391–444.

[9] Newel A and Simon HA. Computer science as empirical inquiry: symbols and search. *Communications of the ACM*. 1976;19(3):113–126.

[10] Haugeland J. *Artificial Intelligence: The Very Idea*. Cambridge, MA: MIT Press; 1989.

[11] Vidulich M, Dominguez C, Vogel E, *et al. Situation Awareness: Papers and Annotated Bibliography*. Armstrong Lab Wright-Patterson AFB; 1994.

[12] Sarter NB and Woods DD. How in the world did we ever get into that mode? Mode error and awareness in supervisory control. *Human Factors*. 1995;37(1):5–19.

[13] Endsley MR. Situation awareness global assessment technique (SAGAT). In: *Proceedings of the IEEE 1988 National Aerospace and Electronics Conference*. IEEE; 1988. p. 789–795.

[14] Wickens CD. *Engineering Psychology and Human Performance* (2nd edn). HarperCollins Publishers, 1992

[15] Miller GA. The magical number seven, plus or minus two: some limits on our capacity for processing information. *Psychological Review*. 1956;63(2):81.

[16] Zadeh LA. Fuzzy sets. In: *Fuzzy Sets, Fuzzy Logic, and Fuzzy Systems: Selected Papers by Lotfi A Zadeh*. Singapore: World Scientific; 1996. p. 394–432.

[17] McCulloch WS and Pitts W. A logical calculus of the ideas immanent in nervous activity. *The Bulletin of Mathematical Biophysics*. 1943;5(4):115–133.

[18] Ram SS, Gurbuz SZ, and Chen V. Modeling and simulation of human motions for micro-Doppler signatures. In: *Radar for In-Door Monitoring: Detection, Localization, and Assessment*. Boca Raton, FL: CRC Press; 2017.

[19] Greco MS and Watts S. *Radar Clutter Modeling and Analysis*. Academic Press Library in Signal Processing, Amsterdam: Elsevier; 2014.

[20] Rahman S and Robertson DA. Radar micro-Doppler signatures of drones and birds at K-band and W-band. *Scientific Reports*. 2018;8.

[21] Willard JD, Jia X, Xu S, Steinbach MS, and Kumar, V. Integrating scientific knowledge with machine learning for engineering and environmental systems. *ACM Computing Surveys*. 2020;55:1–37.

[22] Gurbuz SZ, Sun S, and Tahmoush D. Radar systems, signals, and phenomenology. In: Gurbuz SZ, editor, *Deep Neural Network Design for Radar Applications*. IET; 2020.

[23] Gurbuz SZ and Amin MG. Deep neural networks for radar micro-Doppler signature classification. In: *Micro-Doppler Radar and Its Applications*. London: IET; 2020.

[24] Gurbuz SZ, Gurbuz AC, Crawford C, *et al.* Radar-based methods and apparatus for communication and interpretation of sign languages. In: U.S. Patent App. No. US2020/0334452 (Invention Disclosure filed Feb. 2018; Prov. Patent App. filed Apr. 2019); 2020.

[25] Boulic R, Thalmann NM, and Thalmann D. A global human walking model with real-time kinematic personification. *The Visual Computer*. 1990;6(6):344–358.

[26] Bengio Y and Lecun Y. In: Bottou L, Chapelle O, Decoste D, *et al.*, editors, *Scaling Learning Algorithms towards AI*. Cambridge, MA: MIT Press; 2007.

[27] Pan SJ and Yang Q. A survey on transfer learning. *IEEE Transactions on Knowledge and Data Engineering*. 2010;22:1345–1359.

[28] Erhan D, Bengio Y, Courville A, *et al.* Why does unsupervised pre-training help deep learning? *Journal of Machine Learning Research*. 2010;11(19):625–660. Available from: http://jmlr.org/papers/v11/erhan10a.html.

[29] Seyfioğlu MS and Gürbüz SZ. Deep neural network initialization methods for micro-Doppler classification with low training sample support. *IEEE Geoscience and Remote Sensing Letters*. 2017;14(12):2462–2466.

[30] Seyfioğlu MS, Özbayoğlu AM, and Gürbüz SZ. Deep convolutional autoencoder for radar-based classification of similar aided and unaided human activities. *IEEE Transactions on Aerospace and Electronic Systems*. 2018;54(4):1709–1723.

[31] Seyfioglu MS, Erol B, Gurbuz SZ, *et al.* DNN transfer learning from diversified micro-Doppler for motion classification. *IEEE Transactions on Aerospace and Electronic Systems*. 2019;55(5):2164–2180.

[32] Gurbuz SZ and Amin MG. Radar-based human-motion recognition with deep learning: promising applications for indoor monitoring. *IEEE Signal Processing Magazine*. 2019;36(4):16–28.

[33] Shrivastava A, Pfister T, Tuzel O, *et al.* Learning from simulated and unsupervised images through adversarial training. In: *2017 IEEE Conference on Computer Vision and Pattern Recognition (CVPR)*; 2017. p. 2242–2251.

[34] Wang M and Deng W. Deep visual domain adaptation: a survey. *Neurocomputing*. 2018;312:135–153.

[35] Gurbuz S, Rahman M, Kurtoglu E, *et al.* Cross-frequency training with adversarial learning for radar micro-Doppler signature classification. *Proceedings of SPIE*. 2020;11408:1–11.

[36] Gurbuz SZ, Mahbubur Rahman M, Kurtoglu E, *et al.* Multi-frequency RF sensor fusion for word-level fluent ASL recognition. *IEEE Sensors Journal*. 2021;1–1.

[37] Isola P, Zhu JY, Zhou T, *et al.* Image-to-image translation with conditional adversarial networks. In: *2017 IEEE Conference on Computer Vision and Pattern Recognition (CVPR)*; 2017. p. 5967–5976.

[38] Zhu JY, Park T, Isola P, *et al.* Unpaired image-to-image translation using cycle-consistent adversarial networks. In: *2017 IEEE International Conference on Computer Vision (ICCV)*; 2017. p. 2242–2251.

[39] Ronneberger O, Fischer P, and Brox U-Net: Convolutional networks for biomedical image segmentation. In: Navab N, Hornegger J, Wells W, and Frangi A. (eds), *Medical Image Computing and Computer-Assisted Intervention – MICCAI 2015*. Lecture Notes in Computer Science, vol. 9351. Cham: Springer; 2015.

[40] Li C and Wand M. Precomputed real-time texture synthesis with Markovian generative adversarial networks. *CoRR*. 2016;abs/1604.04382.

[41] Ye F, Luo W, Dong M, *et al.* SAR image retrieval based on unsupervised domain adaptation and clustering. *IEEE Geoscience and Remote Sensing Letters*. 2019;16:1482–1486.

[42] Jennison A, Lewis B, DeLuna A, *et al.* Convolutional and generative pairing for SAR cross-target transfer learning. In: Zelnio E, Garber FD, editors, *Algorithms for Synthetic Aperture Radar Imagery XXVIII*. vol. 11728. International Society for Optics and Photonics. SPIE; 2021. p. 13–19. Available from: https://doi.org/10.1117/12.2585898.

[43] Mahbubur Rahman M and Gurbuz SZ. Multi-frequency RF sensor data adaptation for motion recognition with multi-modal deep learning. In: *2021 IEEE Radar Conference (RadarConf21)*; 2021. p. 1–6.

[44] Mahbubur Rahman M, Malaia E, Gurbuz AC, *et al.* Effect of kinematics and fluency in adversarial synthetic data generation for ASL recognition with RF sensors. *IEEE Transactions on Aerospace and Electronics Systems*. 2021;1–1.

[45] Goodfellow I, Pouget-Abadie J, Mirza M, *et al.* Generative adversarial nets. In Ghahramani Z, Welling M, Cortes C, Lawrence ND, and Weinberger KQ (eds), *Advances in Neural Information Processing Systems*. Boston: Curran Associates, Inc., 2014, pp. 2672–2680.

[46] Arjovsky M, Chintala S, and Bottou L. Wasserstein GAN. ArXiv. 2017;abs/1701.07875.

[47] Erol B, Gurbuz SZ, and Amin MG. Motion classification using kinematically shifted ACGAN-synthesized radar micro-Doppler signatures. *IEEE Transactions on Aerospace and Electronic Systems*. 2020;56(4):3197–3213.

[48] Forssell U and Lindskog P. Combining semi-physical and neural network modeling: an example of its usefulness. *IFAC Proceedings Volumes*. 1997;30(11):767–770. IFAC Symposium on System Identification (SYSID'97), Kitakyushu, Fukuoka, Japan, 8–11 July 1997. Available from: https://www.sciencedirect.com/science/article/pii/S1474667017429387.

[49] Erol B, Gurbuz SZ, and Amin MG. Synthesis of micro-Doppler signatures for abnormal gait using multi-branch discriminator with embedded kinematics. In: *2020 IEEE International Radar Conference (RADAR)*; 2020. p. 175–179.

[50] Rahman MM, Gurbuz SZ, and Amin MG. Physics-aware design of multi-branch GAN for human RF micro-Doppler signature synthesis. In: *2021 IEEE Radar Conference (RadarConf21)*; 2021. p. 1–6.

[51] Amin MG, Zeng Z, and Shan T. Arm motion classification using curve matching of maximum instantaneous Doppler frequency signatures. In: *2020 IEEE International Radar Conference (RADAR)*; 2020. p. 303–308.

[52] Efrat A, Fan Q, and Venkatasubramanian S. Curve matching, time warping, and light fields: new algorithms for computing similarity between curves. *Journal of Mathematical Imaging and Vision*. 2007;27:203–216.

[53] Wang S, Song J, Lien J, *et al.* Interacting with soli: exploring fine-grained dynamic gesture recognition in the radio-frequency spectrum. In: *Proceedings of the 29th Annual Symposium on User Interface Software and Technology*. New York, NY: ACM; 2016. p. 851–860.

[54] Zhang Z, Tian Z, and Zhou M. Latern: dynamic continuous hand gesture recognition using FMCW radar sensor. *IEEE Sensors Journal*. 2018;18(8): 3278–3289.

[55] Wang M, Cui G, Yang X, *et al.* Human body and limb motion recognition via stacked gated recurrent units network. *IET Radar, Sonar Navigation*. 2018;12(9):1046–1051.

[56] Li H, Shrestha A, Heidari H, *et al.* Activities recognition and fall detection in continuous data streams using radar sensor. In: *2019 IEEE MTT-S International Microwave Biomedical Conference (IMBioC)*; 2019. p. 1–4.

[57] Kurtoglu E, Gurbuz AC, Malaia E, *et al.* ASL trigger recognition in mixed activity/signing sequences for RF sensor-based user interfaces. *IEEE Transactions on Human-Machine Systems*. 2021;52:699–712.

[58] Yoon J, Jarrett D, and van der Schaar M. Time-series generative adversarial networks. In: Wallach H, Larochelle H, Beygelzimer A, *et al.*, editors, *Advances in Neural Information Processing Systems*, vol. 32. Red Hook, NY: Curran Associates, Inc.; 2019. Available from: https://proceedings.neurips.cc/paper/2019/file/c9efe5f26cd17ba6216bbe2a7d26d490-Paper.pdf.

[59] Silver D, Huang A, Maddison CJ, *et al.* Mastering the game of Go with deep neural networks and tree search. *Nature*. 2016;529. Available from: https://www.nature.com/articles/nature16961.

[60] Artificial Intelligence: Go Master Lee Se-dol Wins Against AlphaGo Program; 2016. Available from: https://www.bbc.com/news/technology-35797102.

[61] NVIDIA DLSS 2.0: A Big Leap In AI Rendering; 2020. Available from: https://www.nvidia.com/en-us/geforce/news/nvidia-dlss-2-0-a-big-leap-in-ai-rendering/.

[62] Selvi E, Buehrer RM, Martone A, *et al.* Reinforcement learning for adaptable bandwidth tracking radars. *IEEE Transactions on Aerospace and Electronic Systems*. 2020;56:3904–3921.

[63] Thornton CE, Kozy MA, Buehrer RM, *et al.* Deep reinforcement learning control for radar detection and tracking in congested spectral environments. *IEEE Transactions on Cognitive Communications and Networking*. 2020;6(4):1335–1349.

[64] Smith GE and Reininger TJ. Reinforcement learning for waveform design. In: *Proceedings of the IEEE Radar Conference 2021*; 2021. p. 1–6.

[65] Mnih V, Kavukcuoglu K, Silver D, *et al.* Human-level control through deep reinforcement learning. *Nature*. 2015;518(7540):529–533.

[66] Nunn C and Moyer LR. Spectrally-compliant waveforms for wideband radar. *IEEE Aerospace and Electronic Systems Magazine*. 2012;27(8):11–15.

[67] Horne CP, Jones AM, Smith GE, *et al.* Fast fully adaptive signalling for target matching. *IEEE Aerospace and Electronic Systems Magazine*. 2020;35(6): 46–62.

[68] Kulpa JS, Krawczyk G, and Kurowska A. Pseudonoise waveform design for spectrum sharing systems. *IEEE Aerospace and Electronic Systems Magazine*. 2020;35(10):30–39.

[69] Ravenscroft B, Owen JW, Jakabosky J, *et al.* Experimental demonstration and analysis of cognitive spectrum sensing and notching for radar. *IET Radar, Sonar Navigation*. 2018;12(12):1466–1475.

[70] Ng AY and Russell SJ. Algorithms for inverse reinforcement learning. In Langley P. (ed), *ICML*. Morgan Kaufmann; 2000, pp. 663–670.

[71] Kulkarni K, Lohit S, Turaga P, *et al.* Reconnet: non-iterative reconstruction of images from compressively sensed measurements. In: *Proceedings of the IEEE Conference on Computer Vision and Pattern Recognition*; 2016. p. 449–458.

[72] Lohit S, Kulkarni K, Kerviche R, *et al.* Convolutional neural networks for noniterative reconstruction of compressively sensed images. *IEEE Transactions on Computational Imaging*. 2018;4(3):326–340.

[73] MdRafi R and Gurbuz AC. Learning to sense and reconstruct a class of signals. In: *2019 IEEE Radar Conference (RadarConf)*; 2019. p. 1–5.

[74] MdRafi R and Gurbuz AC. Data driven measurement matrix learning for sparse reconstruction. In: *2019 IEEE Data Science Workshop (DSW)*; 2019. p. 253–257.

[75] Mdrafi R and Gurbuz AC. Joint learning of measurement matrix and signal reconstruction via deep learning. *IEEE Transactions on Computational Imaging*. 2020;6:818–829.

[76] Zhang J and Ghanem B. ISTA-Net: iterative shrinkage-thresholding algorithm inspired deep network for image compressive sensing. *CoRR*. 2017;abs/1706.07929. Available from: http://arxiv.org/abs/1706.07929.

[77] Candès EJ and Romberg J. Magic: Recovery of Sparse Signals via Convex Programming; 2005. http://users.ece.gatech.edu/justin/l1magic/downloads/l1magic.pdf.

[78] Donoho DL, Maleki A, and Montanari A. Message-passing algorithms for compressed sensing. *Proceedings of the National Academy of Sciences*. 2009;106(45):18914–18919. Available from: https://www.pnas.org/content/106/45/18914.

[79] Adler A, Elad M, and Zibulevsky M. Compressed learning: a deep neural network approach. arXiv preprint arXiv:161009615; 2016.

[80] Zisselman E, Adler A, and Elad M. Compressed learning for image classification: a deep neural network approach. In: *Handbook of Numerical Analysis*, vol. 19. Amsterdam: Elsevier; 2018. p. 3–17.

[81] Krizhevsky A, Sutskever I, and Hinton GE. Imagenet classification with deep convolutional neural networks. In: *Advances in Neural Information Processing Systems*; 2012. p. 1097–1105.

[82] Simonyan K and Zisserman A. Very deep convolutional networks for large-scale image recognition. arXiv preprint arXiv:14091556; 2014.

[83] Zagoruyko S and Komodakis N. Wide residual networks. arXiv preprint arXiv:160507146; 2016.

[84] He K, Zhang X, Ren S, *et al.* Deep residual learning for image recognition. In: *Proceedings of the IEEE Conference on Computer Vision and Pattern Recognition*; 2016. p. 770–778.

[85] Gonzalez RC and Woods RE. *Digital Image Processing*. Upper Saddle River, NJ: Prentice Hall; 2008. Available from: http://www.amazon.com/Digital-Image-Processing-3rd-Edition/dp/013168728X.

[86] Abadi M, Agarwal A, Barham P, *et al.* TensorFlow: Large-Scale Machine Learning on Heterogeneous Systems; 2015. Software available from tensorflow.org. Available from: https://www.tensorflow.org/.

Part III

Beyond cognitive radar—from theory to practice

Chapter 13
One-bit cognitive radar

Arindam Bose[1], Jian Li[2] and Mojtaba Soltanalian[1]

Target parameter estimation in active sensing, and particularly radar signal processing, is a long-standing problem that has been studied extensively. In this chapter, a novel approach for target parameter estimation is discussed for cases where one-bit analog-to-digital-converters (ADCs), also known as signal comparators with time-varying thresholds, are employed to sample the received radar signal instead of high-resolution ADCs. The considered problem has potential applications in the design of inexpensive radar and sensing devices in civilian applications and paves the way for future radar systems employing low-resolution ADCs for faster sampling and high-resolution target determination. The target estimation has been formulated as a multivariate weighted least-squares optimization problem that can be solved in a cyclic manner.

Additionally, an important problem in cognitive radar is to enhance the estimation performance of the system by a joint design of its probing signal and receive filter using the a priori information on interference. In such cases, the knowledge of interference statistics (particularly the covariance) plays a vital role in an effective design of the radar waveforms. In most practical scenarios, however, the received signal and interference statistics are available subject to some uncertainty. Particularly, an extreme manifestation of this practical observation occurs for radars employing one-bit receivers, where only a normalized version of the interference covariance matrix can be obtained. This chapter also formulates a waveform optimization problem and discusses an algorithm to design the transmit waveform and the receive filter of one-bit radars given such uncertainties in acquired interference statistics. The effectiveness of the proposed algorithms is corroborated through numerical analysis.

13.1 Introduction

The problem of target parameter estimation permeates the field of active sensing and radar signal processing. Active sensing systems aim to uncover the location and other useful properties such as velocity information and reflectance properties of a target of interest by dispatching a transmit waveform toward the target and studying

[1]Department of Electrical and Computer Engineering, University of Illinois Chicago, Chicago, IL, USA
[2]Department of Electrical and Computer Engineering, University of Florida, Gainesville, FL, USA

the received echo reflected by it. For example, the complete dynamics of a moving vehicle including its location and velocity with respect to the observer can easily be found by simply measuring the difference between the transmitted and received electromagnetic waves in the time and frequency domain. Further analysis of the received signal can reveal more information about the target vehicle of interest.

Since the two world wars, radar systems have been developed, improved, and have made their way into diverse applications such as meteorology [1,2], air traffic control [3,4], structural health monitoring [5,6], synthetic aperture imaging [7,8], and underwater sensing [9,10], among others. The unwanted echoes of the transmit signal that are received as delayed and frequency-shifted versions of the transmitted signal and are correlated with the main backscattered signal from the target of interest are generally referred to as clutter. Furthermore, noise is the term usually used for signal-independent interference such as effects of adverse jamming signals [11] as well as thermal noise and atmospheric disturbances. Note that both clutter and noise degrade the accuracy of target parameter estimation; thus, making the receive filter heavily dependent not only on the transmit signal but interference as well. Since the concept of cognitive radar was first introduced by [12] in 2006, it has gained significant interest among researchers for its superior performance in resolving target parameters. According to [12], a cognitive radar *"continuously learns about the environment through experience gained from interactions with the environment, the transmitter adjusts its illumination of the environment in an intelligent manner, the whole radar system constitutes a dynamic closed feedback loop encompassing the transmitter, environment, and receiver."* A judicious design of cognitive radar where both the transmit signal and the receive filter are optimized in a joint manner can consequently lead to a more accurate estimation of the target parameters and a more tractable computational cost for the radar system.

One immediate and well-known choice for the receive filter would be the matched filter (MF) that maximizes the signal-to-noise ratio (SNR) in the presence of additive white noise. The MF multiplies the received signal with a mirrored and delayed image of the transmitted signal [11]. By locating the peak of the output signal, MF discovers the time delay of the received signal, which facilitates the estimation of the distance of the target from the radar, otherwise known as the range. On the other hand, a relative difference in motion between the target and the radar results in a Doppler frequency shift in the received signal spectrum. In the case of a perceivable Doppler shift in the received signal, a bank of MFs is adopted to estimate the Doppler shift, where the critical frequency of each MF is tuned to a different Doppler frequency [13]. However, MF performs poorly in the presence of interference with arbitrary covariance in the received signal [14]. Such limitations of MF have led researchers to search for new avenues to suppress interference/clutter. For instance, a simple solution can be to minimize the autocorrelation sidelobes of the transmit signal [15–17] along with designing the corresponding MF for the receiver. Another line of research can be found in [18–21] where the effects of the clutter are mitigated by minimizing the sidelobes of an ambiguity function (AF). In addition, the negative effects of interference, especially due to jamming, can be avoided by putting little energy of the transmit signal into the frequency bands where the presence of jamming is significant. Furthermore, different hardware constraints such as maximum clipping of power amplifiers and analog-to-digital converters (ADC) decrease the performance of MF estimation.

For a more efficient estimation of the target parameters, one can aim to maximize the signal-to-clutter-plus-interference ratio (SCIR) in lieu of SNR. Such a scenario arises when the target detection performance of the radar is deteriorated by clutter or jamming. In such cases, one can use a mismatched filter (MMF) instead of an MF [22]. An MMF is basically a pulse compression approach where the filter is not perfectly *matched* to the expected received signal but rather is *matched* to an altered time function [23]. In comparison with the MF, an MMF allows more degrees of freedom by introducing a receive filter that performs as well as or possibly better than the MF in most clutter conditions. Furthermore, the formulation of MMF is not subject to different practical power constraints of the transmit signals such as constant-modulus (CM) constraint, similarity constraint (SC), and low peak-to-average power ratio (PAPR). Hence, a joint design of the transmit signal and the MMF receive filter can offer a more robust parameter estimation framework.

The quest for jointly designing radar sequences and corresponding receive filters for clutter/interference rejection has been longstanding among researchers [14,22,24–30]. In [24], the authors presented a joint design scheme of the transmit waveform and receive filter by minimizing the mean-square error (MSE) of the estimate of a target's scattering coefficient in the presence of clutter and interference subject to some of the practical constraints mentioned earlier. In particular, they presented three flavors of their algorithm: Cognitive REceiver and Waveform design (CREW); namely, CREW (gra), CREW (fre), and CREW (mat) based on how they seek to optimize their problem objective. The CREW (gra) algorithm uses a gradient-based approach to minimize the MSE that stems from using the optimum MMF. On the other hand, the CREW (fre) transforms the objective into the frequency domain and focuses on obtaining an optimum power spectrum for the transmit sequence. They further showed that the CREW (fre) algorithm can be specialized to the MF case and, thus, provided a new algorithm called CREW (mat) which can be viewed as an extension of the cyclic algorithms presented in [15]. Another variation of CREW, namely, CREW (cyclic), can be found in [14], where the authors formulated a cyclic approach to jointly design the transmit waveform and receive filter coefficients. For more on this topic, see [14] and the references within.

It is important to note that sampling and quantization of the signal of interest is the first step in digital signal processing. The analog-to-digital conversion ideally requires an infinite number of bits to identically represent the original analog signal, which is not feasible in practice. In fact, the aforementioned techniques assume that the received signal is available in full precision. The resulting error of quantization can then be modeled as additive noise that usually has little to no impact on algorithms that assume the infinite precision case, provided that the sampling resolution is high enough [31]. The signals of interest in many modern applications, albeit, are extremely wideband and may pass through several RF chains that require multitudinous uses of ADCs. Such modern applications include spectral sensing for cognitive radio [32,33], cognitive radars [33], radio astronomy [34], automotive short-range radars [35], and driver assistant systems [36], to name a few.

The assumption of high-precision data is, however, not appropriate when the measurements are extremely quantized to very low bit rates. Note that the cost and the power consumption of ADCs grow exponentially with their number of quantization bits and sampling rate [37]. Such issues can be mitigated by a reduction in the number of quantization bits. In the most extreme case, the sampling process is carried out by utilizing only *one bit per sample*. This can be achieved by repeatedly comparing the signal of interest with a time-varying threshold (reference) level. On the plus side, one-bit comparators can provide an extremely high sampling rate and are very cheap and easy to manufacture [37]. Moreover, the one-bit ADCs operate on very low power, and they can significantly reduce the data flow in the system, which further reduces the overall energy consumption. One-bit sampling has been studied from a classical statistical signal processing viewpoint in [38–45], a compressive sensing viewpoint in [46–50], and a sampling and reconstruction viewpoint in [51,52]. It has been shown in [46–49] that under certain assumptions, with enough one-bit samples, one can recover the full-precision data with bounded error. Further, note that in many recent works, one-bit ADCs on the receiver side are shown to be quite efficient in resolving target parameters when accompanied by suitable one-bit digital-to-analog converters (DACs) on the transmit side [53].

In this chapter, we focus on the radar processing and parameter estimation schemes for both stationary and moving targets using one-bit samplers with time-varying thresholds. For both cases of stationary and moving targets, we provide an approach that is formulated as minimization of a multivariate weighted least-squares objective with linear constraints that can be solved in an iterative manner [54]. As stated earlier, the mentioned approach is cost-effective and computationally efficient. Moreover, one-bit radar covariance estimation will be discussed.

The rest of this chapter is organized as follows. In Section 13.2, we discuss and formulate the estimation problem in the case of a stationary target. Section 13.3 describes a state-of-the-art approach to recover target parameters based on the *Bussgang theorem*. The radar algorithm to estimate the aforementioned parameters is presented in Section 13.4 for a stationary target. In Section 13.5, we extend the problem formulation, as well as the estimation algorithm, for parameter estimation in moving target scenarios. We further extend the parameter estimation formulations for a stationary target scenario to more advanced setups in Section 13.6. Numerical results that verify the validity of claims and examine the performance of the proposed algorithms are presented in Section 13.7. One-bit radar waveform and receive filter design is discussed in Section 13.8, followed by relevant numerical investigations in Section 13.9. Finally, Section 13.10 concludes the chapter.

Notation: We use bold lowercase letters for vectors and bold uppercase letters for matrices. x_i denotes the ith component of the vector $\mathbf{x}$. $(\cdot)^T$ and $(\cdot)^H$ denote the transpose and the conjugate transpose of the vector or matrix argument, respectively. $(\cdot)^*$ denotes the complex conjugate of a complex matrix, vector, or scalar. $\|\cdot\|$ denotes the l_2 norm of a vector, while $\|\cdot\|_F$ denotes the Frobenius norm of a matrix. $\Re(\cdot)$ and $\Im(\cdot)$ are the real and imaginary parts of a complex vector or scalar, respectively. Furthermore, the sets of real, complex, and natural numbers are denoted by $\mathbb{R}$, $\mathbb{C}$, and $\mathbb{N}$, respectively. $\operatorname{sgn}(\cdot)$ is the element-wise sign operator with an output of $+1$

for non-negative numbers and -1 otherwise. Moreover, tr $(\cdot)$ and $\mathcal{N}(\cdot)$ represent the trace and the normalization operator on a matrix argument. $\mathbf{I}$ is the identity matrix. In addition, diag $(\cdot)$ and Diag $(\cdot)$ represent the diagonal vector of its argument matrix and the diagonal matrix made with its argument vector, respectively. $\mathbb{E}\{\cdot\}$ and Cov$(\cdot)$ denote the expectation and the covariance operator, respectively. Finally, the symbol $\odot$ represents the Hadamard product of matrices.

13.2 System model

Let

$$\mathbf{s} = \begin{bmatrix} s_1 & s_2 & \dots & s_N \end{bmatrix}^T \in \mathbb{C}^N \tag{13.1}$$

denotes the complex-valued radar transmit sequence of length N that will be used to modulate a train of subpulses [13]. The energy of $\{s_k\}_{k=1}^N$ is constrained to be N:

$$\|\mathbf{s}\|^2 = N \tag{13.2}$$

without any loss of generality. We shall first adopt the discrete data model described in [24,55] in order to lay out the problem formulation for the simpler case of non-moving targets. Under the assumptions of negligible intrapulse Doppler shift and that the sampling is synchronized to the pulse rate, the received discrete-time baseband signal after pulse compression and proper alignment to the range cell of interest will satisfy the following [22,55]:

$$\begin{aligned}\mathbf{y} = \alpha_0 \begin{bmatrix} s_1 \\ \vdots \\ s_{N-1} \\ s_N \end{bmatrix} + \alpha_1 \begin{bmatrix} 0 \\ s_1 \\ \vdots \\ s_{N-1} \end{bmatrix} + \dots + \alpha_{N-1} \begin{bmatrix} 0 \\ \vdots \\ 0 \\ s_1 \end{bmatrix} \\ + \alpha_{-1} \begin{bmatrix} s_2 \\ \vdots \\ s_N \\ 0 \end{bmatrix} + \dots + \alpha_{-N+1} \begin{bmatrix} s_N \\ 0 \\ \vdots \\ 0 \end{bmatrix} + \boldsymbol{\varepsilon}\end{aligned} \tag{13.3}$$

where $\alpha_0 \in \mathbb{C}$ is the scattering coefficient of the current range cell, $\{\alpha_k\}_{k \neq 0}$ are the scattering coefficients of the adjacent range cells that contribute to the clutter in $\mathbf{y}$, and $\boldsymbol{\varepsilon}$ is the signal-independent noise which comprises of measurement noise as well as other disturbances such as jamming. By collecting all the delayed samples of the transmitted signal into a matrix, the data model in (13.3) can be simplified as

$$\mathbf{y} = \mathbf{A}^H \boldsymbol{\alpha} + \boldsymbol{\varepsilon} \tag{13.4}$$

where

$$\mathbf{A}^H = \begin{bmatrix} s_1 & 0 & \dots & 0 & s_N & s_{N-1} & \dots & s_2 \\ s_2 & s_1 & \dots & \vdots & 0 & s_N & \dots & \vdots \\ \vdots & \vdots & \ddots & 0 & \vdots & \vdots & \ddots & s_N \\ s_N & s_{N-1} & \dots & s_1 & 0 & 0 & \dots & 0 \end{bmatrix}, \tag{13.5}$$

and

$$\boldsymbol{\alpha} = \left[\alpha_0 \; \alpha_1 \; \dots \; \alpha_{N-1} \; \alpha_{-(N-1)} \; \dots \; \alpha_{-1}\right]^T \tag{13.6}$$

is the corresponding scattering coefficient vector. In (13.5), the first column of $\mathbf{A}^H$ represents the principal reflection from the target after range cell alignment, and the columns from the second to the last of $\mathbf{A}^H$ are, in fact, the different delayed echos of the transmit signal $\mathbf{s}$ (see [55] for more details). Furthermore, if the Doppler shifts are not negligible due to the relative difference in motion between the target and the radar system, the data model in (13.4) needs to be modified to accommodate the same and has been discussed in Section 13.5.

By applying one-bit comparators at the receiver, the sampled baseband signal can be written as:

$$\begin{aligned} \boldsymbol{\gamma}_r &= \operatorname{sgn}\left(\Re\{\mathbf{A}^H\boldsymbol{\alpha} + \boldsymbol{\varepsilon} - \boldsymbol{\lambda}\}\right), \\ \boldsymbol{\gamma}_i &= \operatorname{sgn}\left(\Im\{\mathbf{A}^H\boldsymbol{\alpha} + \boldsymbol{\varepsilon} - \boldsymbol{\lambda}\}\right), \\ \boldsymbol{\gamma} &= \tfrac{1}{\sqrt{2}}(\boldsymbol{\gamma}_r + j\boldsymbol{\gamma}_i), \end{aligned} \tag{13.7}$$

where $\boldsymbol{\lambda}$ is the tunable complex-valued threshold level vector at the comparators, whose design is discussed in Section 13.4. Note that in (13.7), we sample both real and imaginary parts of the received signal in order to preserve the phase information. We further assume that the clutter coefficients $\{\alpha_k\}_{k\neq 0}$ are zero-mean and their variance,

$$\beta \triangleq \mathbb{E}\{|\alpha_k|^2\}, \qquad k \neq 0, \tag{13.8}$$

and the covariance matrix of $\boldsymbol{\varepsilon}$,

$$\boldsymbol{\Gamma} \triangleq \mathbb{E}\{\boldsymbol{\varepsilon}\boldsymbol{\varepsilon}^H\}, \tag{13.9}$$

are known quantities. We further assume that $\{\alpha_k\}_{k\neq 0}$ are independent of each other and of $\boldsymbol{\varepsilon}$ as well. Note that, in radar applications, both $\boldsymbol{\Gamma}$ and β can be acquired using various preprocessing techniques, *e.g.*, by employing pre-scans, and are typically assumed to be known *a priori* [24]. A detailed discussion of the pre-scanning process can be found in [12].

As mentioned earlier, once the received signal $\mathbf{y}$ is available, one can estimate the target backscattering coefficient α_0 by exploiting the signal model in (13.4) using an MMF. The MMF estimate of α_0 has the following linear form in $\mathbf{y}$ [14]:

$$\hat{\alpha}_0 = \frac{\mathbf{w}^H\mathbf{y}}{\mathbf{w}^H\mathbf{s}} \tag{13.10}$$

where $\mathbf{w} \in \mathbb{C}^N$ is the MMF vector of the receive filter. The MSE of the mentioned estimate can be derived as

$$\text{MSE}(\hat{\alpha}_0) = \mathbb{E}\left\{\left|\frac{\mathbf{w}^H\mathbf{y}}{\mathbf{w}^H\mathbf{s}} - \alpha_0\right|^2\right\} = \frac{\mathbf{w}^H\mathbf{R}\mathbf{w}}{\left|\mathbf{w}^H\mathbf{s}\right|^2} \tag{13.11}$$

where

$$\mathbf{R} = \beta \sum_{0<|k|\leq(N-1)} \mathbf{J}_k\mathbf{s}\mathbf{s}^H\mathbf{J}_k^H + \mathbf{\Gamma}, \tag{13.12}$$

is the covariance matrix of the interference terms in (13.4), and $\{\mathbf{J}_k\}$ are the shift matrices formulated as

$$\mathbf{J}_k = \mathbf{J}_{-k}^H = \begin{bmatrix} 0 & \dots & 0 & 1 & \dots & 0 \\ \vdots & & & & \ddots & \\ & & & & & 1 \\ \underbrace{0 & \dots & 0}_{k} & \dots & & \end{bmatrix}_{N\times N}^H, \quad k = 0, 1, \cdots, N-1. \tag{13.13}$$

Note that the denominator of the MSE in (13.11) is the power of the signal at the receiver, and its numerator is the power of the interferences. Therefore, minimizing the MSE is identical to maximizing the SCIR.

Note that one can exploit the relationship between the covariance matrices of the received signals before and after the non-linear transformation of one-bit sampling in order to estimate the target parameter α_0. This relationship is provided by the Bussgang theorem in a *normalized* sense where the interference is assumed to be Gaussian in nature [56]. In the following section, we briefly discuss a state-of-the-art Bussgang-theorem-aided procedure to estimate α_0. Afterward, in Section 13.4, we present an algorithm that, through minimizing the MSE, jointly recovers the scattering coefficient of the current range cell, α_0, and the received signal, $\mathbf{y}$, from the one-bit sampled received data $\boldsymbol{\gamma}$, as introduced in (13.7).

13.3 Bussgang-theorem-aided estimation

In this section, we describe a state-of-the-art Bussgang-theorem-aided approach to estimate the target parameters [56]. Let $Y(t)$ be a real-valued, scalar, and stationary Gaussian process that undergoes the one-bit sampling process $Z(t) = \text{sgn}(Y(t))$. The autocorrelation function of the process $Z(t)$, denoted by $R_Z(\tau)$, is given by

$$R_Z(\tau) = \mathbb{E}\{Z(t+\tau)Z(t)\} = \frac{2}{\pi}\sin^{-1}\left(R_Y(\tau)\right) \tag{13.14}$$

where $\bar{R}_Y(\tau) = R_Y(\tau)/R_Y(0)$ is the normalized autocorrelation function of the process $Y(t)$ [57]. The good news is that the Bussgang theorem [56] states that the cross-correlation function of the processes $Y(t)$ and $Z(t)$ is proportional to the autocorrelation function of $Y(t)$, *i.e.*, $R_{ZY}(\tau) = \mu R_Y(\tau)$, where the factor μ depends on the power of the process $Y(t)$.

The case of complex-valued vector processes, which was studied in [58], can be extended in a similar manner. Let $\mathbf{y}$ be the complex-valued vector whose one-bit samples are given by $\boldsymbol{\gamma} = \frac{1}{\sqrt{2}}(\mathrm{sgn}(\Re(\mathbf{y})) + j\mathrm{sgn}(\Im(\mathbf{y})))$, as in (13.7). Then the normalized autocorrelation of the vector $\mathbf{y}$ is given by

$$\bar{\mathbf{R}}_{\mathrm{y}} = \mathcal{N}(\mathbf{R}_{\mathrm{y}}) \triangleq \mathbf{D}^{-1/2}\mathbf{R}_{\mathrm{y}}\mathbf{D}^{-1/2} \tag{13.15}$$

where $\mathbf{D} = \mathbf{R}_{\mathrm{y}} \odot \mathbf{I}$ is a diagonal matrix containing only the diagonal entries of $\mathbf{R}_{\mathrm{y}}$. It has been shown in [58] that the following covariance equality holds:

$$\bar{\mathbf{R}}_{\mathrm{y}} = \sin\left(\frac{\pi}{2}\mathbf{R}_{\gamma}\right), \tag{13.16}$$

where $\mathbf{R}_{\gamma}$ is the autocorrelation matrix of the one-bit sampled data, $\boldsymbol{\gamma}$.

In order to apply the above results to the one-bit radar processing problem using a threshold level vector $\boldsymbol{\lambda} \in \mathbb{C}^N$, we can derive the covariance matrix of the difference between the received signal and the time-varying threshold, viz.

$$\mathbf{R}_{\mathrm{y}-\boldsymbol{\lambda}} = |\alpha_0|^2\mathbf{s}\mathbf{s}^H + \boldsymbol{\lambda}\boldsymbol{\lambda}^H + \mathbf{R} - 2\Re(\alpha_0\mathbf{s}\boldsymbol{\lambda}^H). \tag{13.17}$$

Therefore, one can compute the scattering coefficient α_0 by solving the following non-convex optimization problem:

$$\min_{\alpha_0} \quad \left\|\bar{\mathbf{R}}_{\mathrm{y}-\boldsymbol{\lambda}} - \mathcal{N}(|\alpha_0|^2\mathbf{s}\mathbf{s}^H + \boldsymbol{\lambda}\boldsymbol{\lambda}^H + \mathbf{R} - 2\Re(\alpha_0\mathbf{s}\boldsymbol{\lambda}^H))\right\|_F \tag{13.18}$$

in which $\bar{\mathbf{R}}_{\mathrm{y}-\boldsymbol{\lambda}}$ is obtained via (13.16), and using only one observation or *snapshot* of $\boldsymbol{\gamma}$.

13.4 Radar processing for stationary targets

In this section, we discuss the proposed approach to recover both the received signal $\mathbf{y}$ and the scattering coefficient α_0 from the one-bit sampled received signal $\boldsymbol{\gamma}$ for a stationary target by minimizing the aforementioned MSE in (13.11).

For a given transmit sequence $\mathbf{s}$, the optimum receive filter $\mathbf{w}$ can be simply given as closed-form solution [14,24]:

$$\mathbf{w} = \mathbf{R}^{-1}\mathbf{s} \tag{13.19}$$

up to a multiplicative constant. Nevertheless, the MMF approach to recover α_0, discussed in (13.10), requires the availability of the un-quantized (or high-resolution quantized) received signal $\mathbf{y}$, which is unfortunately not available directly due to the one-bit sampling of the received signal. Therefore, we shall resort to an alternative optimization approach that utilizes the one-bit sampled data $\boldsymbol{\gamma}$ in lieu of $\mathbf{y}$ in order to estimate the target parameter. In pursuance of radar parameter recovery using one-bit sampled data with time-varying thresholds, we analyze two matters of major significance:

1. The recovery of $\mathbf{y}$ and estimation of α_0 by employing the one-bit data procured at the receiver.
2. The design of the next set of thresholds to be used at the one-bit ADCs.

13.4.1 Estimation of stationary target parameters

In order to efficiently estimate the received signal $\mathbf{y}$ and target parameter α_0, we consider minimizing the following weighted-least-squares (WLS) objective in a more generalized sense:

$$Q(\mathbf{y}, \alpha_0) \triangleq (\mathbf{y} - \alpha_0\mathbf{s})^H \mathbf{R}^{-1}(\mathbf{y} - \alpha_0\mathbf{s}). \tag{13.20}$$

It should be noted that the usage of the above criterion has the following advantages:

1. Unlike the MMF in (13.10), Q does not require a knowledge of $\mathbf{y}$.
2. It is a function of both $\mathbf{y}$ and α_0, laying the ground for their respective joint recovery.
3. It can easily be observed that, for any given $\mathbf{y}$, the optimum α_0 in (13.20) is identical to that of MMF in (13.10) with the use of (13.19)—thus, making it a natural choice for parameter recovery.
4. In effect, the minimization of (13.20) enforces the system model introduced in (13.3). Note that the model mismatch can be written as

   $$\mathbf{y} - \alpha_0\mathbf{s} = \tilde{\mathbf{A}}^H\tilde{\boldsymbol{\alpha}} + \boldsymbol{\varepsilon} \tag{13.21}$$

   where $\tilde{\mathbf{A}}^H$ and $\tilde{\boldsymbol{\alpha}}$ are derived from $\mathbf{A}^H$ and $\boldsymbol{\alpha}$ with their first column and first entry dropped, respectively. It can easily be established that the objective function in (13.20) penalizes the model mismatch based on the second-order mismatch statistics derived as

   $$\begin{aligned}&\mathbb{E}\left\{\left(\tilde{\mathbf{A}}^H\tilde{\boldsymbol{\alpha}} + \boldsymbol{\varepsilon}\right)\left(\tilde{\mathbf{A}}^H\tilde{\boldsymbol{\alpha}} + \boldsymbol{\varepsilon}\right)^H\right\}\\ &= \mathbb{E}\left\{\tilde{\mathbf{A}}^H\tilde{\boldsymbol{\alpha}}\tilde{\boldsymbol{\alpha}}^H\tilde{\mathbf{A}}\right\} + \mathbb{E}\left\{\boldsymbol{\varepsilon}\boldsymbol{\varepsilon}^H\right\}\\ &= \mathbf{R}.\end{aligned} \tag{13.22}$$

Hence, based on the property 3, and by substituting the optimum α_0, the objective function (13.20) can be reformulated as

$$\begin{aligned}Q(\mathbf{y}, \hat{\alpha}_0) &\triangleq Q(\mathbf{y})\\ &= \mathbf{y}^H\left(\mathbf{I} - \frac{\mathbf{s}\mathbf{w}^H}{\mathbf{w}^H\mathbf{s}}\right)^H \mathbf{R}^{-1}\left(\mathbf{I} - \frac{\mathbf{s}\mathbf{w}^H}{\mathbf{w}^H\mathbf{s}}\right)\mathbf{y}.\end{aligned} \tag{13.23}$$

Hence, the problem of jointly estimating α_0 and $\mathbf{y}$ boils down to:

$$\begin{aligned}\min_{\mathbf{y}}\quad & \mathbf{y}^H\left[\left(\mathbf{I} - \frac{\mathbf{s}\mathbf{w}^H}{\mathbf{w}^H\mathbf{s}}\right)^H \mathbf{R}^{-1}\left(\mathbf{I} - \frac{\mathbf{s}\mathbf{w}^H}{\mathbf{w}^H\mathbf{s}}\right)\right]\mathbf{y}\\ \text{s.t.}\quad & \boldsymbol{\Omega}_r\left(\mathbf{y}_r - \boldsymbol{\lambda}_r\right) \geq \mathbf{0},\\ & \boldsymbol{\Omega}_i\left(\mathbf{y}_i - \boldsymbol{\lambda}_i\right) \geq \mathbf{0},\end{aligned} \tag{13.24}$$

where $(\mathbf{y}_r, \mathbf{y}_i)$ and $(\boldsymbol{\lambda}_r, \boldsymbol{\lambda}_i)$ denote the real and imaginary parts of $\mathbf{y}$ and $\boldsymbol{\lambda}$, respectively, and $\boldsymbol{\Omega}_r \triangleq \text{Diag}\left(\boldsymbol{\gamma}_r\right)$, $\boldsymbol{\Omega}_i \triangleq \text{Diag}\left(\boldsymbol{\gamma}_i\right)$. One can easily verify that the optimization problem in (13.24) is a convex quadratic program with linear constraints that can be

Algorithm 1: One-bit radar parameter estimation for stationary targets

Ensure: The transmit sequence $\mathbf{s}$ and set the threshold vector $\boldsymbol{\lambda}$ arbitrarily, or generate according to (13.25) and (13.26).

1: Compute the optimal MMF vector $\mathbf{w}$ using (13.19).
2: Compute the optimal vector $\mathbf{y}$ by solving (13.24).
3: Estimate the target scattering coefficient α_0 using the MMF estimator in (13.10).
4: In case of tracking, set $\boldsymbol{\lambda}$ according to (13.25) and (13.26) and go to Step 2.

solved efficiently. Upon finding the optimal $\mathbf{y}$, the optimal α_0 can be calculated using the MMF estimate in (13.10).

It is worth noting that the optimization problem in (13.24) can easily be translated into a non-negative least-squares (NNLS) optimization problem. This can be achieved by changing variables such that $\tilde{\mathbf{y}}_r \triangleq \boldsymbol{\Omega}_r(\mathbf{y}_r - \boldsymbol{\lambda}_r)$ and $\tilde{\mathbf{y}}_i \triangleq \boldsymbol{\Omega}_i(\mathbf{y}_i - \boldsymbol{\lambda}_i)$, adding up to $\tilde{\mathbf{y}} \triangleq \tilde{\mathbf{y}}_r + j\tilde{\mathbf{y}}_i$. As a result, fast NNLS approaches can be exploited to expedite the recovery process [59].

13.4.2 Time-varying threshold design

In the following, we briefly discuss how one can choose the threshold vector $\boldsymbol{\lambda}$ depending on the number of one-bit ADCs being used on the receiver side.

13.4.2.1 Sampling with a single one-bit ADC

From an information-theoretic viewpoint, in order to collect the *most* information on $\mathbf{y}$, one could expect $\boldsymbol{\lambda}$ to be set in such a way that by considering the *a priori* information, observing any of the two outcomes in the set $\{-1, +1\}$ at the output of the one-bit sampler for a single sample has the same likelihood. In a general case, $\boldsymbol{\lambda}$ is expected to partition the set of likely events into subsets with similar *cardinality*. When the probability density function (pdf) of the received signal follows a Gaussian distribution, this goal is achieved by setting $\boldsymbol{\lambda}$ as close as possible to the expected value of $\mathbf{y}$. More precisely, we choose:

$$\boldsymbol{\lambda} = \mathbb{E}\{\alpha_0\}\,\mathbf{s}. \tag{13.25}$$

In other words, the choice of $\boldsymbol{\lambda}$ will be governed by our *future* expectation of the value of α_0. This is particularly pertinent to target-tracking scenarios.

13.4.2.2 Sampling with multiple one-bit ADCs

We note that our estimation method can easily be extended to cases where the signal is sampled by several ADCs in parallel. This only leads to extra linear constraints in (13.24). Assuming that K number of ADCs is used, and the thresholds are set *a priori*, in the single sample case, the thresholds are optimal if they partition the set of likely events into $K+1$ subsets with similar *cardinality*. The determination of the thresholds will be even more difficult when the number of samples or the number of ADCs grows large. However, a close *approximation* of the optimal threshold vectors

$\{\boldsymbol{\lambda}_k\}_{k=1}^K$ can be obtained by assuming $\{\boldsymbol{\lambda}_k\}_{k=1}^K$ to be random variables [37]. In other words, a good set of random sampling threshold vectors $\{\boldsymbol{\lambda}_k\}_{k=1}^K$ should mimic the behavior of $\mathbf{y}$. In particular, we generate $\{\boldsymbol{\lambda}_k\}_{k=1}^K$ as a set of random vectors similar to $\mathbf{y}$ that has the same (Gaussian) distribution:

$$\begin{aligned} \mathbb{E}\{\boldsymbol{\lambda}\} &= \mathbb{E}\{\alpha_0\}\,\mathbf{s}, \\ \operatorname{Cov}(\boldsymbol{\lambda}) &= \mathbb{E}\left\{|\alpha_0|^2\right\}\,\mathbf{s}\mathbf{s}^H + \mathbf{R}. \end{aligned} \tag{13.26}$$

The steps of the proposed approach are summarized in Algorithm 1 for reader's convenience.

13.5 Radar processing for moving targets

In this section, we consider the moving targets scenario where the Doppler effect can no longer be neglected. In order to perform a recovery of radar parameters, *i.e.*, the backscattering coefficient and the Doppler shift of the target, we first update the system model of (13.3) and then modify the proposed approach discussed in Section 13.4 to recover the normalized Doppler shift as well.

13.5.1 Problem formulation for moving targets

Let $\mathbf{s} \in \mathbb{C}^N$ denote the discrete-time transmit sequence of a digital system, as in (13.1). After alignment to the range-azimuth cell of interest, the new discrete-time complex-valued received baseband data vector, which is backscattered from the moving target in the corresponding range-azimuth cell, can be formulated as (see [30,60–62])

$$\mathbf{y} = \alpha_0(\mathbf{s} \odot \mathbf{p}(\nu)) + \mathbf{c} + \mathbf{n}, \tag{13.27}$$

where α_0 is the complex backscattering coefficient of the target in the current range-azimuth cell and $\mathbf{p}(\nu) = \left[e^{j2\pi(0)\nu},\ e^{j2\pi(1)\nu},\ \ldots,\ e^{j2\pi(N-1)\nu}\right]^T$ is the propagation effect vector with $\nu \in [-0.5, 0.5)$ being the normalized Doppler shift of the target. The N-dimensional vectors $\mathbf{c}$ and $\mathbf{n}$ denote the signal-dependent clutter and signal-independent noise, respectively. The clutter vector $\mathbf{c}$ is comprised of returned echos from uncorrelated scatterers at different range-azimuth cells [62] (as depicted in Figure 13.1), which are spread in Doppler frequency due to the possible clutter motion and can be formulated as

$$\mathbf{c} = \sum_{k=0}^{N_c\ 1}\sum_{l=0}^{L\ 1} \alpha_{(k,l)}\mathbf{J}_k\left[\mathbf{s} \odot \mathbf{p}(\nu_{d_{(k,l)}})\right], \tag{13.28}$$

where $N_c \leq N$ is the number of range-rings, L is the number of various azimuth sectors, and $\alpha_{(k,l)}$ and $\nu_{d_{(k,l)}}$ are the scattering coefficient and normalized Doppler shift of the (k,l)-th range-azimuth cell, respectively, interfering with the range-azimuth cell of interest. The matrix $\mathbf{J}_k$ is defined in the same way as in (13.13).

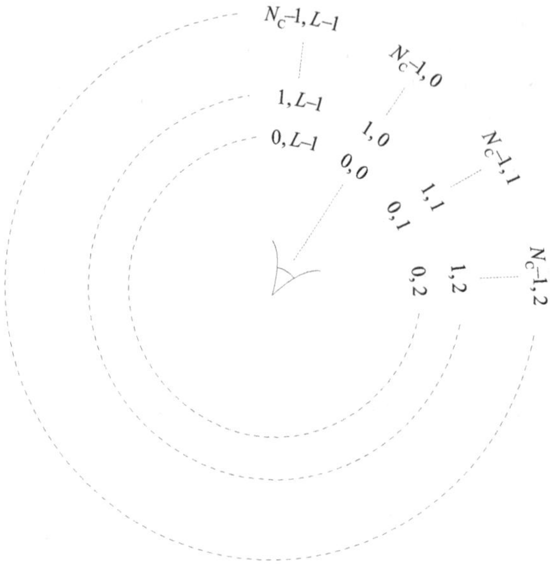

Figure 13.1 The setting for different range-azimuth cells. All the cell numbers are shown in (range, azimuth) pairs.

The covariance matrix of the clutter vector $\mathbf{c}$ can be written as

$$\mathbf{\Sigma_c} = \sum_{k=0}^{N_c-1}\sum_{l=0}^{L-1} \sigma^2_{(k,l)} \mathbf{J}_k \mathbf{\Phi}\left(s,(k,l)\right)\mathbf{J}_k^T \tag{13.29}$$

with $\sigma^2_{(k,l)}$ being the average scattering power of the scatterer in (k,l)-th range-azimuth cell. Furthermore, $\mathbf{\Phi}$ in (13.29) can be expressed as

$$\mathbf{\Phi}\left(s,(k,l)\right) = \text{Diag}\left(\mathbf{s}\right)\mathbf{C}_\nu(k,l)\text{Diag}\left(\mathbf{s}\right)^H,$$

where $\mathbf{C}_\nu(k,l)$ is the covariance matrix of the propagation effect vector of the (k,l)th cell [62], defined as

$$\mathbf{C}_\nu(k,l) = \begin{cases} 1 & k = l \\ e^{j(k-l)\bar{\nu}_{d(k,l)}} \dfrac{\sin\left(\frac{k-l}{2}\varepsilon_{d(k,l)}\right)}{\sin\left(\frac{k-l}{2}\varepsilon_{d(k,l)}\right)} & k \neq l. \end{cases} \tag{13.30}$$

Similar to (13.9), we denote the covariance matrix of the signal-independent interference by $\mathbf{\Gamma}$, and re-define the covariance matrix of the interference as follows:

$$\mathbf{R} = \text{Cov}(\mathbf{c}+\mathbf{n}) = \mathbf{\Sigma_c} + \mathbf{\Gamma}. \tag{13.31}$$

Moreover, in (13.29), the clutter patches in each range-azimuth cell are assumed to have uniform Doppler shifts in the interval $\Omega_c = \left(\bar{\nu}_{d(k,l)} - \frac{\varepsilon_{d(k,l)}}{2}, \bar{\nu}_{d(k,l)} + \frac{\varepsilon_{d(k,l)}}{2}\right)$ [61]. Note that the assumption of having uniform Doppler shifts in each range-azimuth cell

Algorithm 2: One-bit radar parameter estimation for moving targets

Ensure: The transmit sequence $\mathbf{s}$ and set the threshold vector $\boldsymbol{\lambda}$ arbitrarily or generate according to (13.40).

1: For fixed $\mathbf{y}$, α_0, and ν, compute the optimal MMF vector $\mathbf{w}$ according to (13.33).
2: For fixed $\mathbf{w}$ and ν, compute the optimal vector $\mathbf{y}$ by solving the criterion in (13.35) with respect to $\mathbf{y}$.
3: For fixed $\mathbf{y}$ and $\mathbf{w}$, compute the optimal target normalized Doppler shift ν by minimizing the criterion in (13.37).
4: If convergence is reached, go to Step 5; otherwise, go to Step 1.
5: For fixed $\mathbf{w}$, $\mathbf{y}$, and ν, estimate the target backscattering coefficient α_0 using (13.32).
6: In case of tracking, set $\boldsymbol{\lambda}$ according to (13.40) and go to Step 1.

results from the fact that the clutter patches in these cells can be any object in our environment, with some of them moving. If such objects are moving, even slightly, the echoes reflected from the corresponding cells will have a Doppler shift associated with that movement. Examples of such objects include vehicles, ocean waves, and trees with moving leaves due to the wind [63,64]. These contribute to a small Doppler frequency shift in clutter input which is assumed to be distributed uniformly.

13.5.2 Estimation of moving target parameters

When the received signal is available, and the Doppler shift is known, an estimation of the backscattering coefficient α_0 with minimal MSE can be achieved by using an MMF, in a similar manner as in (13.10). The estimate of the target backscattering coefficient given by MMF is

$$\hat{\alpha}_0 = \frac{\mathbf{w}^H \mathbf{y}}{\mathbf{w}^H (\mathbf{s} \odot \mathbf{p}(\nu))}. \tag{13.32}$$

Additionally, it can be verified that the optimal $\mathbf{w}$ that minimizes the MSE criterion is given by

$$\mathbf{w} = \mathbf{R}^{-1} (\mathbf{s} \odot \mathbf{p}(\nu)). \tag{13.33}$$

up to a multiplicative constant.

Note once again that, due to using one-bit ADCs, the access to the received signal $\mathbf{y}$ is restricted to only its one-bit samples, given by (13.7). In order to tackle the problem of estimating the backscattering coefficient α_0 and the normalized Doppler shift ν, we form a modified version of the weighted-least-squares objective function in (13.20), in compliance with the system model defined in (13.27):

$$\tilde{Q}(\mathbf{y}, \alpha_0, \nu) \triangleq [\mathbf{y} - \alpha_0(\mathbf{s} \odot \mathbf{p}(\nu))]^H \mathbf{R}^{-1} [\mathbf{y} - \alpha_0(\mathbf{s} \odot \mathbf{p}(\nu))] \tag{13.34}$$

with $\mathbf{R}$ being the covariance matrix of the interference defined in (13.31).

Similar to the stationary target case, the aforementioned objective function is chosen to have the following properties:

1. It does not rely on the knowledge of the received signal $\mathbf{y}$ and yet enforces the system model in (13.27).
2. For given $\mathbf{y}$ and ν, the optimal α_0 of (13.34) is identical to the MMF estimate of α_0 in (13.32).
3. It is a function of $\mathbf{y}$, α_0, and ν, which permits their joint estimation, and last but not least.
4. The recovery of $\mathbf{y}$ using $\tilde{Q}$ paves the way for the usage of other classical signal processing methods that rely on the knowledge of $\mathbf{y}$.

The problem of jointly estimating $\mathbf{y}$, α_0, and ν for moving target determination thus becomes

$$\begin{aligned} \min_{\mathbf{y},\alpha_0,\nu} \quad & [\mathbf{y} - \alpha_0(\mathbf{s} \odot \mathbf{p}(\nu))]^H \mathbf{R}^{-1} [\mathbf{y} - \alpha_0(\mathbf{s} \odot \mathbf{p}(\nu))] \\ \text{s.t.} \quad & \boldsymbol{\Omega}_r (\mathbf{y}_r - \boldsymbol{\lambda}_r) \geq \mathbf{0}, \\ & \boldsymbol{\Omega}_i (\mathbf{y}_i - \boldsymbol{\lambda}_i) \geq \mathbf{0}. \end{aligned} \tag{13.35}$$

However, by substituting the optimal α_0 in (13.32) into the objective function of (13.35), we achieve a more simplified optimization problem:

$$\begin{aligned} \min_{\mathbf{w},\mathbf{y},\nu} \quad & \left\| \mathbf{R}^{-1/2} \left(\mathbf{I} - \frac{[\mathbf{s} \odot \mathbf{p}(\nu)]\mathbf{w}^H}{\mathbf{w}^H[\mathbf{s} \odot \mathbf{p}(\nu)]} \right) \mathbf{y} \right\|_2^2 \\ \text{s.t.} \quad & \boldsymbol{\Omega}_r (\mathbf{y}_r - \boldsymbol{\lambda}_r) \geq \mathbf{0}, \\ & \boldsymbol{\Omega}_i (\mathbf{y}_i - \boldsymbol{\lambda}_i) \geq \mathbf{0}. \end{aligned} \tag{13.36}$$

In order to solve the above minimization problem, we resort to cyclic optimization over $\mathbf{w}$, $\mathbf{y}$, and ν, until convergence. The optimal $\mathbf{w}$ for fixed $\mathbf{y}$ and ν can be obtained using (13.33). Next, for fixed $\mathbf{w}$ and ν, it is easy to see that the above optimization problem is a convex linearly-constrained quadratic program with respect to $\mathbf{y}$, which can be efficiently solved. Lastly, in order to find the optimal normalized Doppler shift ν when $\mathbf{w}$ and $\mathbf{y}$ are fixed, we can rewrite the optimization problem (13.35) with respect to ν as:

$$\min_{\nu} g(\nu) \tag{13.37}$$

$$\text{s.t. } \mathbf{p}(\nu) = \left[e^{j2\pi(0)\nu} \;\; e^{j2\pi(1)\nu} \;\; \ldots \;\; e^{j2\pi(N-1)\nu} \right]^T$$

where

$$g(\nu) \triangleq \begin{bmatrix} 1 \\ \mathbf{p}(\nu) \end{bmatrix}^H \begin{bmatrix} 0 & -(\hat{\alpha}_0\mathbf{s})^T \odot (\mathbf{y}^H\mathbf{R}^{-1}) \\ -(\hat{\alpha}_0\mathbf{s})^* \odot (\mathbf{R}^{-1}\mathbf{y}) & |\hat{\alpha}_0|^2\mathbf{R}^{-1} \odot (\mathbf{s}\mathbf{s}^H)^* \end{bmatrix} \begin{bmatrix} 1 \\ \mathbf{p}(\nu) \end{bmatrix} \tag{13.38}$$

and where $\hat{\alpha}_0$ is calculated using (13.32).

The optimization problem in (13.37) resembles that of estimating the direction-of-arrival (DOA) in uniform linear arrays (ULAs) and can be dealt with using one of the many algorithms for estimating the DOA—see [65] for details. We repeat the cyclic optimization procedure until a pre-defined convergence criterion is satisfied. Once the $\mathbf{w}$, $\mathbf{y}$, and ν are estimated, the backscattering coefficient α_0 can be easily retrieved via (13.32).

As to the design of the threshold vector $\boldsymbol{\lambda}$, the same arguments discussed in Section 13.4.2 hold. However, the statistics of the (Gaussian) randomly generated threshold vector $\boldsymbol{\lambda}$ change as follows:

$$\begin{aligned}\mathbb{E}\{\boldsymbol{\lambda}\} &= \mathbb{E}\{\mathbf{y}\} = \mathbb{E}\{\alpha_0\}(\mathbf{s} \odot \mathbb{E}\{\mathbf{p}(\nu)\}),\\ \mathrm{Cov}(\boldsymbol{\lambda}) &= \mathrm{Cov}(\mathbf{y}) \\ &= \mathbb{E}\left\{|\alpha_0|^2\right\}(\mathbf{s}\mathbf{s}^H) \odot \mathbb{E}\{\mathbf{p}(\nu)\mathbf{p}^H(\nu)\} + \mathbf{R}.\end{aligned} \tag{13.39}$$

For readers' convenience, the steps of the proposed approach for moving target radar parameter estimation are summarized in Algorithm 2.

13.6 Other low-resolution sampling scenarios

In this section, we study the extensions of the proposed method discussed in Section 13.4 to different cases for the stationary target scenario. We further note that the same extensions can be applied to the case of the moving target as well.

13.6.1 *Extension to parallel one-bit comparators*

It can be noted that the problem formulation in (13.24) can be extended to the implementation of an array of K number of one-bit comparators in parallel with different time-varying thresholds, denoted by $\boldsymbol{\lambda}^{(k)}$, $k = 1, \cdots, K$. In this way, the optimization problem requires the recovered signal to comply with all the comparison information that is produced by the one-bit comparators. Thus, the constraints in (13.24) can be updated as

$$\begin{aligned}\boldsymbol{\Omega}_r^{(k)}\left(\mathbf{y}_r - \boldsymbol{\lambda}_r^{(k)}\right) &\geq \mathbf{0}, \quad \forall\, k \in \{1, \cdots, K\},\\ \boldsymbol{\Omega}_i^{(k)}\left(\mathbf{y}_i - \boldsymbol{\lambda}_i^{(k)}\right) &\geq \mathbf{0}, \quad \forall\, k \in \{1, \cdots, K\},\end{aligned} \tag{13.40}$$

where $\boldsymbol{\Omega}_r^{(k)} = \mathrm{Diag}\left(\boldsymbol{\lambda}_r^{(k)}\right)$ and $\boldsymbol{\Omega}_i^{(k)} = \mathrm{Diag}\left(\boldsymbol{\lambda}_i^{(k)}\right)$.

13.6.2 *Extension to p-bit ADCs*

Another alternative way to glean more information from the received signal $\mathbf{y}$ is to use multi-bit ADCs. For a generic p-bit ADC, we have $(2^p - 1) + 2$ thresholds such that $\lambda_0 < \lambda_1 < \lambda_2 < \cdots < \lambda_{2^p-1} < \lambda_{2^p}$, where we define $\lambda_0 \triangleq -\infty$ and $\lambda_{2^p} \triangleq +\infty$ for the ease of notation. Thus, each sample of the input signal can fall into any of the 2^p quantization regions, which further indicates that each input sampled data has to lie in an interval $[\lambda_k, \lambda_{k+1}]$, for some $0 \leq k \leq (2^p - 1), k \in \mathbb{N} \cup \{0\}$. Thus, if q number of p-bit ADCs are used instead of one-bit comparators, each of the ADCs will have $(2^p - 1) + 2$ thresholds—leading to a total number of $q(2^p + 1)$ thresholds.

Observe that the optimization problem in this case requires enforcing the following constraints:

$$[\mathbf{y}_r]_n \in \left[[\boldsymbol{\lambda}_r]_n^{(k_n)}, [\boldsymbol{\lambda}_r]_n^{(k_n+1)}\right],$$
$$[\mathbf{y}_i]_m \in \left[[\boldsymbol{\lambda}_i]_m^{(k_m)}, [\boldsymbol{\lambda}_i]_m^{(k_m+1)}\right],$$

for all $n, m \in \{1, \cdots, N\}$ and for integers k_n and k_m provided by the p-bit ADCs, where $[\boldsymbol{\lambda}_r]_n^{(k_n)}$ and $[\boldsymbol{\lambda}_i]_m^{(k_m)}$ denote the k_nth and k_mth components of $[\boldsymbol{\lambda}_r]_n$ and $[\boldsymbol{\lambda}_i]_m$, respectively. Let,

$$\boldsymbol{\lambda}_r^{lower} \triangleq \left[[\boldsymbol{\lambda}_r]_1^{(k_1)}, \cdots, [\boldsymbol{\lambda}_r]_N^{(k_N)}\right]^T,$$
$$\boldsymbol{\lambda}_r^{upper} \triangleq \left[[\boldsymbol{\lambda}_r]_1^{(k_1+1)}, \cdots, [\boldsymbol{\lambda}_r]_N^{(k_N+1)}\right]^T, \qquad (13.41)$$

and define $\boldsymbol{\lambda}_i^{lower}$ and $\boldsymbol{\lambda}_i^{upper}$ in a similar manner. Then, the constraints of the optimization problem in (13.24) can be updated, in this case, as

$$+1 \cdot (\mathbf{y}_r - \boldsymbol{\lambda}_r^{lower}) \geq \mathbf{0},$$
$$+1 \cdot (\mathbf{y}_i - \boldsymbol{\lambda}_i^{lower}) \geq \mathbf{0},$$
$$-1 \cdot (\mathbf{y}_r - \boldsymbol{\lambda}_r^{upper}) \geq \mathbf{0},$$
$$-1 \cdot (\mathbf{y}_i - \boldsymbol{\lambda}_i^{upper}) \geq \mathbf{0}. \qquad (13.42)$$

13.7 Numerical analysis for one-bit radar signal processing

In this section, we delve into examining the performance of the proposed target parameter estimation methods. The estimation error of our proposed approaches is compared with that of the Bussgang-aided approach of Section 13.3, and estimation using un-quantized received signal, denoted by ∞-precision. We first consider the case of stationary targets and employ the approach discussed in Section 13.4 and then move on to the case of moving targets discussed in Section 13.5.

13.7.1 Stationary targets

For the simulations, we assume that the noise is additive, white, and Gaussian with a variance of 0.1, the average clutter power β is 0.1, and that the transmit sequence is generated using the method in [14] with a PAPR of 1. The results in all cases are averaged over 100 runs of the algorithms unless mentioned otherwise.

Let $\hat{\alpha}_0$ denote the estimate of α_0, and further define the normalized estimation error as $|\alpha_0 - \hat{\alpha}_0|/|\alpha_0|$. In Figure 13.2, the normalized estimation error obtained via a Monte-Carlo method with randomly generated ground truths for α_0 is plotted against the transmit sequence length N for the proposed algorithm, the Bussgang-aided approach of Section 13.3, and the $\infty - precision$ case. In addition, for visualization, Figure 13.3 shows the results of the estimations, in the Monte-Carlo method for $\alpha_0 = (0.5 + j0.5)$ for $N \in \{50, 100, 1,000\}$ on the complex plane. It can be seen from

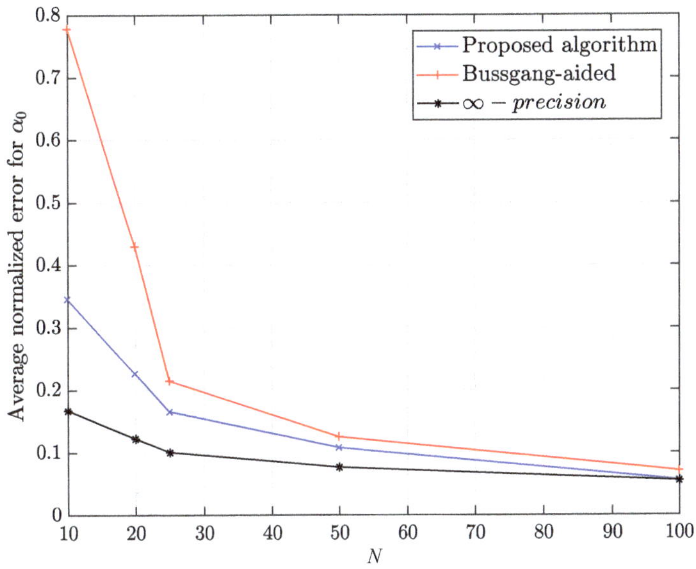

Figure 13.2 Average normalized estimation error of stationary target scattering coefficient α_0, defined by the ratio $|\alpha_0 - \hat{\alpha}_0|/|\alpha_0|$, for different transmit sequence lengths $N \in \{10, 25, 50, 100\}$

both figures that the estimate of the proposed algorithm approaches that of the ∞-precision as N grows large. This is expected because when N grows large, the number of comparisons grows large at the same rate revealing the true nature of the unquantized data. From an information-theoretic point of view, this translates to more available information on the received signal through its one-bit samples that contribute to the amelioration of the scattering coefficient recovery. Consequently, the estimation performances of all three approaches are enhanced with an increase of N, as is apparent in both figures.

Figure 13.4 shows the performance of the proposed algorithm in the presence of different noise power levels for the stationary target and compares it with that of $\infty - precision$ case. For this experiment, we keep $N = 25$, and again assume that the noise is additive, white, and Gaussian with variance $\sigma^2 \in \{10^{-5}, 10^{-4}, 10^{-3}, 10^{-2}, 10^{-1}, 1, 10\}$. As it can be seen from Figure 13.4, the average normalized error of estimating α_0 remains very low for $\sigma^2 < 0.1$, however, the performance of the algorithm decreases rapidly after $\sigma^2 = 1$.

13.7.2 Moving targets

Herein we present the simulation results for radar parameter estimation in the case of moving targets. Similar to the stationary target scenario, we assume that the noise is additive, white, and Gaussian with a variance of 0.1, and that the transmit sequence is

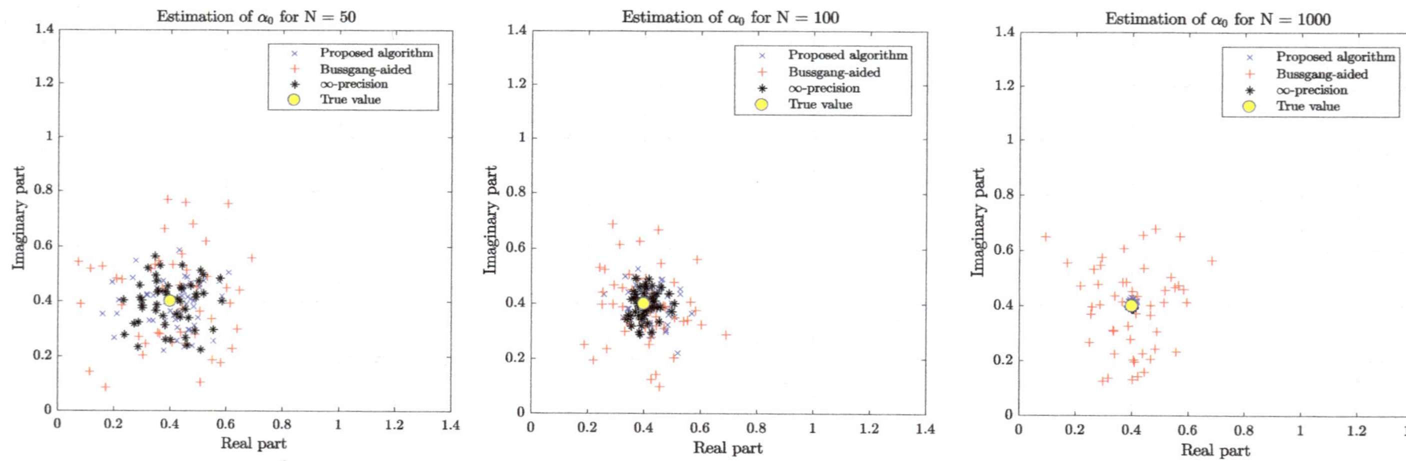

Figure 13.3 *Comparison of stationary target scattering coefficient (α_0) estimation performances for $N \in \{50, 100, 1,000\}$. The results of estimation using the proposed algorithm, the Bussgang-aided approach, and the $\infty-$precision case are shown on complex plane along with the true value of $\alpha_0 = (0.5 + j0.5)$*

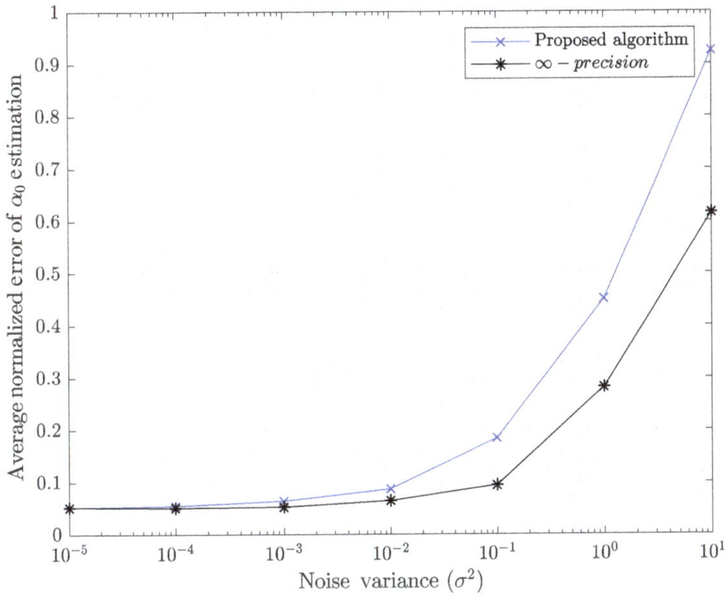

Figure 13.4 Average normalized estimation error of stationary target scattering coefficient α_0, defined by the ratio $|\alpha_0 - \hat{\alpha}_0|/|\alpha_0|$, for different noise variances $\sigma^2 \in \{10^{-5}, 10^{-4}, 10^{-3}, 10^{-2}, 10^{-1}, 1, 10\}$

generated using the method in [14] with a PAPR of 1. The number of the interfering range rings N_c and the number of azimuth sectors L are set to 2 and 10, respectively. Additionally, the normalized Doppler shifts of the adjacent range-azimuth cells are assumed to be uniformly distributed over the interval $\Omega_c = [-0.1, 0.1]$; see [66] for further details.

The estimation performance of different approaches in moving target scenarios is examined via a Monte-Carlo method with randomly generated ground truths for target parameters and presented in Figure 13.5. More precisely, the normalized estimation error of the backscattering coefficient α_0 of a moving target for the case of the proposed algorithm, as well as the Bussgang-aided approach (modified for moving targets) along with the $\infty - $*precision* case are shown in Figure 13.5(a). The estimation errors for the normalized Doppler shift ν are depicted in Figure 13.5(b) for the same scenarios. Furthermore, as in the case of a stationary target, Figure 13.6 plots the radar parameter estimates for the case of a moving target through the Monte Carlo method. The upper plots in Figure 13.6 show the results of estimating the backscattering coefficient α_0, along with its true value, for $N \in \{50, 100, 1,000\}$ on the complex plane. On the other hand, the lower plots in Figure 13.6 show the estimates of the normalized Doppler shift on the polar plane. The result of estimation for different approaches is shown on circles with slightly different radii for the sake of clarity.

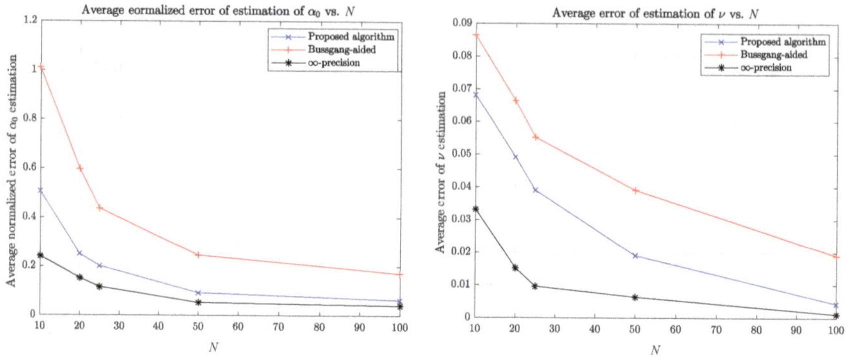

(a) (b)

Figure 13.5 Performance comparison of moving target parameter estimation using the proposed algorithm, the Bussgang-aided approach, and the ∞ − precision case: (a) average normalized error of estimating the backscattering coefficient, ($|\alpha_0 - \hat{\alpha}_0|/|\alpha_0|$) and (b) average error of estimating the normalized Doppler shift ν, for different transmit sequence lengths $N \in \{10, 20, 25, 50, 100\}$

As in the case of stationary targets, it can be observed from Figure 13.5 that estimates α_0 and ν become more precise as N grows larger. In fact, as N grows large, the performance of our proposed estimator improves and gets closer to the estimate made by the ∞-precision case. Further note that in order to attain the same performance in the estimation of the parameters of a moving target, the proposed algorithm requires more samples than the stationary target case, as can be seen through Figures 13.2 and 13.5.

Finally, Figure 13.7 demonstrates the performance of the proposed algorithm in the presence of different noise power levels for the moving targets and compares it with that of ∞-precision case. For this experiment, we again use the same settings as used for the stationary case, *i.e.*, $N = 25$, and the noise is additive, white, and Gaussian with variance $\sigma^2 \in \{10^{-5}, 10^{-4}, 10^{-3}, 10^{-2}, 10^{-1}, 1, 10\}$. The average normalized error of estimating α_0 is shown in Figure 13.7(a), while the errors for estimating the normalized Doppler shift ν are depicted in Figure 13.7(b). Similar to the case of stationary targets, it can be seen from Figure 13.7 that the average error of estimating both α_0 and ν stay very low for $\sigma^2 < 0.1$, however, their performance degrades strongly after $\sigma^2 = 1$, even for ∞-precision case.

13.8 One-bit radar waveform design under uncertain statistics

In a one-bit cognitive radar system, it is often well-known that full-scale signals can indeed be recovered with high accuracy from one-bit measurements, however, at an

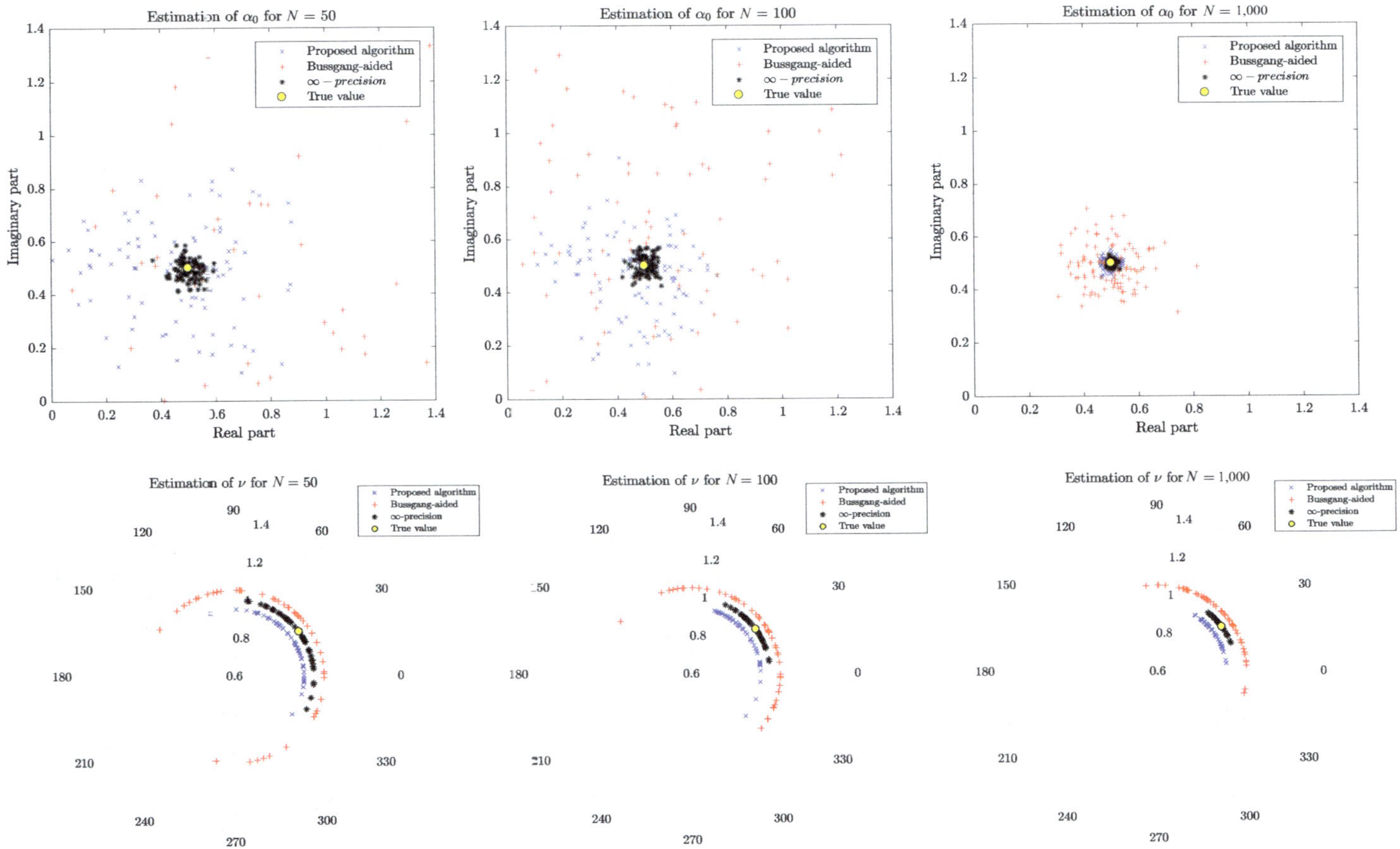

Figure 13.6 *Performance comparison of moving target parameter estimation using the proposed approach, the Bussgang-aided approach, and the ∞ − precision case for $N \in \{50, 100, 1,000\}$. The upper plots show the results of estimating α_0 on the complex plane, while the lower plots show the results of estimating ν on the polar plane, where different radii are used for different approaches for visual clarity.*

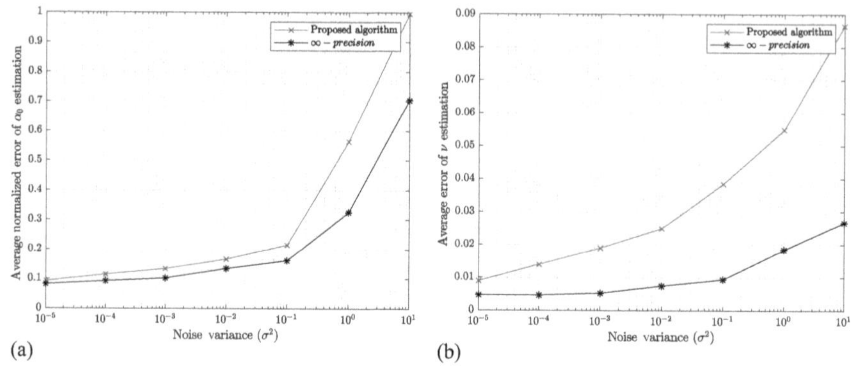

Figure 13.7 Performance comparison of moving target parameter estimation using the proposed algorithm, and the ∞*-precision case: (a) average normalized error of estimating the backscattering coefficient,* $(|\alpha_0 - \hat{\alpha}_0|/|\alpha_0|)$*; (b) average error of estimating the normalized Doppler shift* ν*, for different noise variances* $\sigma^2 \in \{10^{-5}, 10^{-4}, 10^{-3}, 10^{-2}, 10^{-1}, 1, 10\}$

increased computational cost [67]. This increased cost mainly incurs from the fact that by using a one-bit receiver, the knowledge of interference statistics is available only in a *normalized* sense and such uncertainties prohibit one from using any traditional tools to jointly design transmit sequences, as well as the receive filter coefficients [37]. As mentioned earlier, one such traditional tool is CREW (cyclic) [14] for which the authors seek to minimize the MSE of the estimate of a target's scattering coefficient in the presence of clutter and interference and present an alternating minimization algorithm. Although this algorithm is computationally efficient, it requires full knowledge of the interference covariance matrix. In the following, we introduce a specialized variation of CREW (cyclic) to tackle the problem of jointly designing the transmit signals and the receive filter coefficients in the presence of uncertainty in interference statistics [68].

13.8.1 Problem formulation for waveform design

In the case of receivers with one-bit ADC, the quantizer is nothing but a simple sign comparator and each measurement is represented using only one bit, *i.e.*, $+1$ or -1, and the auto-correlation of the received signal is only obtainable in a *normalized* sense [54]. In particular, the matrix $\mathbf{R}$ in (13.12) is obtainable only in a normalized sense, *i.e.*, one only has access to

$$\bar{\mathbf{R}} = \boldsymbol{D}^{-\frac{1}{2}}\mathbf{R}\boldsymbol{D}^{-\frac{1}{2}}, \tag{13.43}$$

where $\boldsymbol{D} = \mathbf{R} \odot \boldsymbol{I}$. Then, the problem of interest is to design the transmit waveform $\mathbf{s}$ and the receive filter $\mathbf{w}$ given the normalized interference statistics $\bar{\mathbf{R}}$. In the following, we denote $\boldsymbol{d} = \operatorname{diag}\left(\boldsymbol{D}^{\frac{1}{2}}\right)$.

In such a case, a meaningful approach to the aforementioned design problem is to consider minimization of the expected MSE criterion:

$$\min_{\mathbf{w},\mathbf{s}} \quad \mathbb{E}\left\{\frac{\mathbf{w}^H \boldsymbol{D}^{\frac{1}{2}} \bar{\mathbf{R}} \boldsymbol{D}^{\frac{1}{2}} \mathbf{w}}{|\mathbf{w}^H \mathbf{s}|^2}\right\}, \tag{13.44}$$

under some practical signal power constraints. Note that the expectation is taken over $\boldsymbol{D}$. The above problem is clearly non-convex. In the following, we handle the non-convexity of the optimization objective in (13.44) with respect to (w.r.t.) the probing sequence $\mathbf{s}$ and the receive filter $\mathbf{w}$ using an alternating approach and propose a specialized flavor of CREW (cyclic), named as CREW (one-bit).

13.8.2 Joint design method: CREW (one-bit)

13.8.2.1 Optimization of the radar waveform

The numerator of the MSE in (13.11) can be rearranged, for a fixed receive filter $\mathbf{w}$, as

$$\begin{aligned} \mathbf{w}^H \mathbf{R} \mathbf{w} &= \mathbf{w}^H \left(\beta \sum_{\substack{k=-N+1 \\ k \neq 0}}^{N-1} \boldsymbol{J}_k \mathbf{s} \mathbf{s}^H \boldsymbol{J}_k^H + \boldsymbol{\Gamma} \right) \mathbf{w} \\ &= \mathbf{s}^H \left(\underbrace{\beta \sum_{\substack{k=-N+1 \\ k \neq 0}}^{N-1} \boldsymbol{J}_k \mathbf{w} \mathbf{w}^H \boldsymbol{J}_k^H}_{\chi} \right) \mathbf{s} + \mathbf{w}^H \boldsymbol{\Gamma} \mathbf{w}. \end{aligned} \tag{13.45}$$

Thus, the criterion in (13.11) can be reformulated as,

$$\frac{\mathrm{MSE}(\hat{\alpha}_0)}{\beta} = \frac{\mathbf{s}^H \chi \mathbf{s} + \mu}{\mathbf{s}^H \mathbf{W} \mathbf{s}}, \tag{13.46}$$

where $\mu = (\mathbf{w}^H \boldsymbol{\Gamma} \mathbf{w})/\beta$ and $\mathbf{W} = \mathbf{w}\mathbf{w}^H$. It is interesting to note that μ is unknown; however, independent of $\mathbf{s}$, and thus merely a constant scalar w.r.t. $\mathbf{s}$. To deal with the optimization problem of (13.46), we follow the identical framework as [14] that exploits the idea of fractional programming [69].

Let $a(\mathbf{s}) = \mathbf{s}^H \chi \mathbf{s} + \mu$, and $b(\mathbf{s}) = \mathbf{s}^H \mathbf{W} \mathbf{s} > 0$ (MSE needs to be finite). Further, let $f(\mathbf{s}) = a(\mathbf{s})/b(\mathbf{s})$ and $\mathbf{s}_*$ denote the current value of $\mathbf{s}$. We define $g(\mathbf{s}) \triangleq a(\mathbf{s}) - f(\mathbf{s}_*) b(\mathbf{s})$, and $\mathbf{s}_\dagger \triangleq \arg\min_{\mathbf{s}} g(\mathbf{s})$. It can be easily verified that $g(\mathbf{s}_\dagger) \leq g(\mathbf{s}_*) = 0$. As a result, we have that $g(\mathbf{s}_\dagger) = a(\mathbf{s}_\dagger) - f(\mathbf{s}_*) b(\mathbf{s}_\dagger) \leq 0$ which indicates to $f(\mathbf{s}_\dagger) \leq f(\mathbf{s}_*)$ as $b(\mathbf{s}_\dagger) > 0$. Therefore, $\mathbf{s}_\dagger$ can be considered as a new vector $\mathbf{s}$ that monotonically decreases $f(\mathbf{s})$. Note that $\mathbf{s}_\dagger$ does not necessarily have to be a minimizer of $g(\mathbf{s})$; instead, it is enough if $g(\mathbf{s}_\dagger) \leq g(\mathbf{s}_*)$.

Under the assumption that $\|\mathbf{s}\|_2^2 = N$, for a fixed $\mathbf{w}$ and any arbitrary $\mathbf{s}_*$ of the minimizer $\mathbf{s}$ of (13.46), we have:

$$g(\mathbf{s}) = \mathbf{s}^H(\chi - f(\mathbf{s}_*)\mathbf{W})\mathbf{s} + \mu = \mathbf{s}^H \boldsymbol{T}\mathbf{s} + \mu, \tag{13.47}$$

where $\boldsymbol{T} \triangleq \chi - f(\mathbf{s}_*)\mathbf{W}$. Then the problem of (13.46) w.r.t. unimodular $\mathbf{s}$ can be recast as the following unimodular quadratic program (UQP) [70]:

$$\max_{\mathbf{s}} \ \mathbf{s}^H \tilde{\boldsymbol{T}} \mathbf{s} \qquad \text{s.t. } |s_k| = 1, \qquad 1 \leq k \leq N, \tag{13.48}$$

where $\tilde{\boldsymbol{T}} \triangleq \lambda \boldsymbol{I} - \boldsymbol{T}$ is a positive definite matrix and λ is a real scalar greater than the maximum eigenvalue of $\boldsymbol{T}$. Note that (13.48) is NP-hard in general, and a sub-optimal solution can be sought by semi-definite relaxation (SDR). To tackle this problem efficiently, in [70] a set of *power method-like* iterations was suggested that can be used to monotonically increase the criterion in (13.48); namely, the vector $\mathbf{s}$ is updated in each iteration n using the nearest-vector problem:

$$\begin{aligned} &\min_{\mathbf{s}^{(n+1)}} \quad \left\| \mathbf{s}^{(n+1)} - \tilde{\boldsymbol{T}}\mathbf{s}^{(n)} \right\|_2 \\ &\text{s.t.} \quad \left| s_k^{(n+1)} \right| = 1, \qquad 1 \leq k \leq N. \end{aligned} \tag{13.49}$$

Fortunately, the solution to (13.49) is simply given analytically by $\mathbf{s}^{(n+1)} = e^{j\arg(\tilde{\boldsymbol{T}}\mathbf{s}^{(n)})}$. A proof of monotonically increasing behavior of the UQP objective in (13.48) can be found in [14].

13.8.2.2 Optimization of the receive filter

For a fixed $\mathbf{s}$, the objective of (13.44) can be further simplified as,

$$\begin{aligned} \mathbb{E}\left\{ \frac{\mathbf{w}^H \boldsymbol{D}^{\frac{1}{2}} \bar{\mathbf{R}} \boldsymbol{D}^{\frac{1}{2}} \mathbf{w}}{|\mathbf{w}^H \mathbf{s}|^2} \right\} &= \frac{\mathbb{E}\left\{ \operatorname{tr}\left(\mathbf{w}\mathbf{w}^H \boldsymbol{D}^{\frac{1}{2}} \bar{\mathbf{R}} \boldsymbol{D}^{\frac{1}{2}} \right) \right\}}{|\mathbf{w}^H \mathbf{s}|^2} \\ &= \frac{\mathbb{E}\left\{ \boldsymbol{d}^H \left(\mathbf{w}\mathbf{w}^H \odot \bar{\mathbf{R}}^H \right) \boldsymbol{d} \right\}}{|\mathbf{w}^H \mathbf{s}|^2} \\ &= \frac{\operatorname{tr}\left(\left(\mathbf{w}\mathbf{w}^H \odot \bar{\mathbf{R}}^H \right) \mathbb{E}\left\{ \boldsymbol{d}\boldsymbol{d}^H \right\} \right)}{|\mathbf{w}^H \mathbf{s}|^2}. \end{aligned} \tag{13.50}$$

It is evident that the knowledge of $\boldsymbol{d}$ indirectly demands more information on β and $\boldsymbol{\Gamma}$. However, assuming the statistics of the noise is unchanging, one can estimate $\boldsymbol{\Gamma}$ in a normalized sense by just listening to the environment while not transmitting any waveform. As a result, from the one-bit receiver, the normalized interference covariance matrix $\bar{\boldsymbol{\Gamma}}$ can be obtained in a similar fashion as, $\bar{\boldsymbol{\Gamma}} \triangleq \boldsymbol{A}^{-\frac{1}{2}} \boldsymbol{\Gamma} \boldsymbol{A}^{-\frac{1}{2}}$, where $\boldsymbol{A} = \boldsymbol{\Gamma} \odot \boldsymbol{I}$. Thus, the interference covariance matrix $\mathbf{R}$ in (13.12) can be reformulated as,

$$\mathbf{R} = \boldsymbol{D}^{\frac{1}{2}} \bar{\mathbf{R}} \boldsymbol{D}^{\frac{1}{2}} = \beta \mathbf{S} + \boldsymbol{A}^{\frac{1}{2}} \bar{\boldsymbol{\Gamma}} \boldsymbol{A}^{\frac{1}{2}}, \tag{13.51}$$

where $\mathbf{S} = \sum_{k \neq 0} \boldsymbol{J}_k \mathbf{s}\mathbf{s}^H \boldsymbol{J}_k^H$ is constant for a known $\mathbf{s}$. Hence, a judicious approach is to solve the following problem in order to optimize $\boldsymbol{d}, \boldsymbol{a}$, and β in a joint manner:

$$\{\hat{\boldsymbol{d}}, \hat{\boldsymbol{a}}, \hat{\beta}\} = \arg \min_{\boldsymbol{d}, \boldsymbol{a}, \beta} \quad \left\| \text{Diag}\,(\boldsymbol{d})^{\frac{1}{2}} \; \bar{\mathbf{R}} \; \text{Diag}\,(\boldsymbol{d})^{\frac{1}{2}} - \beta \mathbf{S} + \text{Diag}\,(\boldsymbol{a})^{\frac{1}{2}} \; \bar{\boldsymbol{\Gamma}} \; \text{Diag}\,(\boldsymbol{a})^{\frac{1}{2}} \right\|_F^2,$$
$$\text{s.t.} \quad \boldsymbol{d} > 0, \boldsymbol{a} > 0, \beta > 0. \tag{13.52}$$

The above minimization problem is non-convex, and, hence, in order to efficiently solve it, we resort to an alternating approach; *i.e.*, solving for each variable while keeping the other two variables as constant. By doing so, w.r.t. each variable, the problem becomes convex and can be solved using a number of available numerical solvers, such as the "fmincon" function in MATLAB® that implements the Broyden–Fletcher–Goldfarb–Shanno (BFGS) algorithm. Note that by solving (13.52), one can obtain β and $\boldsymbol{d}$ in an average sense, which in other words justifies the usage of expectation in the formulation of (13.50).

With this information in mind, let $\sum_{k=1}^{N} \nu_k \boldsymbol{u}_k \boldsymbol{u}_k^H$ represent the eigenvalue decomposition (EVD) of $\mathbb{E}\left\{\boldsymbol{d}\boldsymbol{d}^H\right\}$, where $\{\nu_k\}$ and $\{\boldsymbol{u}_k\}$ represent the eigenvalues and the eigenvectors, respectively. As a result, the numerator of (13.50) can be further simplified as,

$$\begin{aligned} \text{tr}\left(\left(\mathbf{w}\mathbf{w}^H \odot \bar{\mathbf{R}}^H\right) \sum_{k=1}^{N} \nu_k \boldsymbol{u}_k \boldsymbol{u}_k^H \right) &= \sum_{k=1}^{N} \nu_k \boldsymbol{u}_k^H \left(\mathbf{w}\mathbf{w}^H \odot \bar{\mathbf{R}}^H\right) \boldsymbol{u}_k \\ &= \left(\mathbf{w}\mathbf{w}^H\right) \sum_{k=1}^{N} \nu_k \; \text{Diag}\,(\boldsymbol{u}_k) \; \bar{\mathbf{R}} \; \text{Diag}\left(\boldsymbol{u}_k^H\right) \\ &= \mathbf{w}^H \boldsymbol{Q} \mathbf{w}, \end{aligned} \tag{13.53}$$

where

$$\boldsymbol{Q} = \sum_{k=1}^{N} \nu_k \; \text{Diag}\,(\boldsymbol{u}_k) \; \bar{\mathbf{R}} \; \text{Diag}\left(\boldsymbol{u}_k^H\right). \tag{13.54}$$

It is interesting to note that $\boldsymbol{Q}$ can be viewed as the expected value of $\mathbf{R}$. By using $\boldsymbol{D}^{\frac{1}{2}} = \text{Diag}\,(\boldsymbol{d})$, the following can be deduced:

$$\mathbb{E}\{\mathbf{R}\} = \mathbb{E}\left\{\boldsymbol{D}^{\frac{1}{2}} \bar{\mathbf{R}} \boldsymbol{D}^{\frac{1}{2}}\right\} = \mathbb{E}\left\{\boldsymbol{d}\boldsymbol{d}^H\right\} \odot \bar{\mathbf{R}}. \tag{13.55}$$

Assuming $\mathbb{E}\left\{\boldsymbol{d}\boldsymbol{d}^H\right\} = \eta\eta^H + \boldsymbol{\Sigma}$, (13.55) can be reformulated as

$$\begin{aligned} \mathbb{E}\{\mathbf{R}\} &= \left(\eta\eta^H + \boldsymbol{\Sigma}\right) \odot \bar{\mathbf{R}} \\ &= \sum_{k=1}^{N} \nu_k \boldsymbol{u}_k \boldsymbol{u}_k^H \odot \bar{\mathbf{R}} \\ &= \sum_{k=1}^{N} \nu_k \; \text{Diag}\,(\boldsymbol{u}_k) \; \bar{\mathbf{R}} \; \text{Diag}\left(\boldsymbol{u}_k^H\right), \end{aligned} \tag{13.56}$$

which proves the claim.

Algorithm 3: CREW (ONE-BIT)

Ensure: $\mathbf{s}^{(0)} \leftarrow$ unimodular (or low PAPR) vector in $\mathbb{C}^N$, $\mathbf{w}^{(0)} \leftarrow$ random vector in $\mathbb{C}^N$, the outer loop index $t \leftarrow 1$.

1: **repeat**

2: **For fixed w**,

i) Compute $\chi, \mathbf{W}$ using (13.45), and thus, in turn find $\tilde{\boldsymbol{T}}$.

ii) Solve the power method like iterations discussed in (13.49), and calculate $\mathbf{s}^{(t)}$ in each iteration until convergence.

3: Measure $\bar{\mathbf{\Gamma}}$ at the output of the one-bit receiver and compute $\bar{\mathbf{R}}$ using $\mathbf{s}^{(t)}$.

4: **For fixed s**,

i) Solve (13.52) to obtain $\boldsymbol{d}$ and β in average sense.

ii) Compute the EVD of $\mathbb{E}\left\{\boldsymbol{d}\boldsymbol{d}^H\right\}$, and form $\boldsymbol{Q}$.

iii) Update $\mathbf{w}^{(t)}$ as $\boldsymbol{Q}^{-1}\mathbf{s}^{(t)}$.

5: **until** convergence, *e.g.*, $\left|\text{MSE}^{(t+1)} - \text{MSE}^{(t)}\right| < \varepsilon$ for some given $\varepsilon > 0$.

In light of the above, the receive filter optimization problem translates to:

$$\min_{\mathbf{w}} \frac{\mathbf{w}^H \boldsymbol{Q} \mathbf{w}}{|\mathbf{w}^H \mathbf{s}|^2}. \tag{13.57}$$

Hence, for a given $\mathbf{s}$, the optimization problem in (13.57) w.r.t. $\mathbf{w}$ results in the closed-form solution, $\mathbf{w} = \boldsymbol{Q}^{-1}\mathbf{s}$, within a multiplicative constant. Finally, the algorithm CREW (one-bit) is summarized in Algorithm 3 in a concise manner.

13.9 Waveform design examples

In this section, we evaluate the performance of CREW (one-bit) and compare it with three state-of-the-art methods, namely CAN-MMF, CREW (fre), and CREW (cyclic). The CAN-MMF method employs the CAN algorithm in [15] to simply design a transmit waveform with good correlation properties and independent of the receive filter. Note that no prior knowledge of interference is used in the waveform design of CAN-MMF.

We adopt the same simulation setups as in [14]. Especially, for the interference covariance matrix, we consider the following:

$$\mathbf{\Gamma} = \sigma_J^2 \mathbf{\Gamma}_J + \sigma^2 \boldsymbol{I},$$

where $\sigma_J^2 = 100$, and $\sigma^2 = 0.1$ are the jamming and noise powers, respectively. Furthermore, the jamming covariance matrix $\mathbf{\Gamma}_J$ is given by $[\mathbf{\Gamma}_J]_{k,l} = \gamma_{k-l}$ where $[\gamma_0, \gamma_1, \cdots, \gamma_{N-1}, \gamma_{-(N-1)}, \cdots, \gamma_{-1}]^T$ can be obtained by an inverse FFT (IFFT) of the jamming power spectrum $\{\eta_p\}$ at frequencies $(p-1)/(2N-1)$, $p = 1, \cdots, 2N-1$. For CREW(fre) and CREW(cyclic), we fix the average clutter power to $\beta = 1$. Finally,

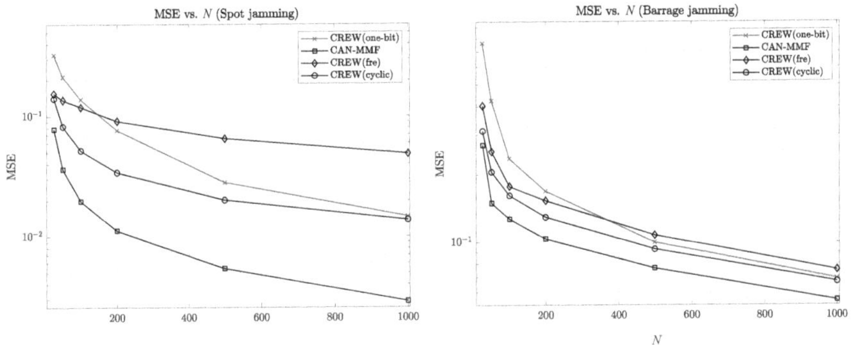

(a) (b)

Figure 13.8 MSE values obtained by the different design algorithms for (a) spot jamming with normalized frequency $f_0 = 0.2$*, and (b) barrage jamming in the normalized frequency interval* $[f_1, f_2] = [0.2, 0.3]$ *for the unimodularity constraint on the transmit sequence*

we use the Golomb sequence in order to initialize the transmit waveform $\mathbf{s}$ for all algorithms.

We consider two modes of jamming: spot and barrage. Spot jamming is concentrated power directed toward one channel or frequency. In our example, we use a spot jamming located at a normalized frequency $f_0 = 0.2$. On the other hand, barrage jamming is power spread over several frequencies or channels at the same time. We consider a barrage jamming located in the normalized frequency bands $[f_1, f_2] = [0.2, 0.3]$.

Figure 13.8(a) and (b) depict the MSE values for spot and barrage jamming, respectively, corresponding to CAN-MMF, CREW(fre), and CREW(cyclic), under the unimodularity constraint, for various sequence lengths. It is evident from the figures that when the sequence length N is small, the MSE is higher for CREW (one-bit) compared to other algorithms. However, as N increases, CREW (one-bit) shows similar performance as CREW (cyclic), and eventually, they coincide with one another for higher values of N. Consequently, it is implied that higher signal length introduces more degrees of freedom in designing transmit waveform and thus, compensates for the uncertainties in interference statistics. It is further important to observe that the knowledge of the one-bit measurements impacts the design of the receive filter, and alternatively, the design of the receive filter coefficients impacts the design of transmit waveform, which justifies the role of a cognitive radar.

13.10 Concluding remarks

High-resolution sampling with conventional analog-to-digital converters (ADCs) can be very costly and energy-consuming for many modern applications. This is further accentuated as recent applications, including those in sensing and radar signal processing, show a growing appetite in even larger than usual sampling rates—thus making the mainstream ADCs a rather unsuitable choice. To overcome these shortcomings, it was shown that in lieu of using the conventional ADCs in radar parameters estimation, one can use inexpensive comparators with time-varying thresholds and solve an optimization problem to recover the target parameters with satisfactory performance. This is very beneficial at high frequencies as it is both practical and economical, while it can also pave the way for future applications to sample at much higher rates. Simulation results were presented that verify the efficiency of one-bit target parameter estimation for both stationary and moving targets, especially as the length of the transmit sequence N grows large. Finally, radar waveform and receive filter design were studied under uncertain statistics that are common due to low-resolution sampling.

References

[1] Stepanenko VD. *Radar in Meteorology*. Wright-Patterson AFB, OH: Foreign Technology Div; 1975.

[2] Browning K. Uses of radar in meteorology. *Contemporary Physics*. 1986;27(6): 499–517. Available from: https://doi.org/10.1080/00107518608211028.

[3] Bussolari S, Bernays J. Mode S data link applications for general aviation. In: *Proceedings of 14th Digital Avionics Systems Conference*; 1995. p. 199–206.

[4] Nolan M. Fundamentals of air traffic control. In: *Cengage Learning*. Boston, MA: Cengage; 2010.

[5] Ihn JB and Chang FK. Pitch-catch active sensing methods in structural health monitoring for aircraft structures. *Structural Health Monitoring*. 2008;7(1): 5–19.

[6] Lynch JP, Sundararajan A, Law KH, *et al.* Design of a wireless active sensing unit for structural health monitoring. In: *Health Monitoring and Smart Nondestructive Evaluation of Structural and Biological Systems III*, vol. 5394. International Society for Optics and Photonics; 2004. p. 157–169.

[7] Soumekh M. *Synthetic Aperture Radar Signal Processing*, vol. 7. New York: Wiley; 1999.

[8] Curlander JC and McDonough RN. *Synthetic Aperture Radar*, vol. 396. New York, NY: John Wiley & Sons; 1991.

[9] Heidemann J, Stojanovic M, and Zorzi M. Underwater sensor networks: applications, advances and challenges. *Philosophical Transactions of the Royal Society A*. 2012;370(1958):158–175.

[10] Farr N, Bowen A, Ware J, *et al.* An integrated, underwater optical/acoustic communications system. In: *OCEANS*. IEEE; 2010. p. 1–6.

[11] He H, Li J, and Stoica P. *Waveform Design for Active Sensing Systems: A Computational Approach*. Cambridge: Cambridge University Press; 2012.

[12] Haykin S. Cognitive radar: a way of the future. *IEEE Signal Processing Magazine*. 2006;23(1):30–40.

[13] Levanon N and Mozeson E. *Radar Signals*. New York, NY: John Wiley & Sons; 2004.

[14] Soltanalian M, Tang B, Li J, *et al*. Joint design of the receive filter and transmit sequence for active sensing. *IEEE Signal Processing Letters*. 2013;20(5): 423–426.

[15] Stoica P, He H, and Li J. New algorithms for designing unimodular sequences with good correlation properties. *IEEE Transactions on Signal Processing*. 2009;57(4):1415–1425.

[16] Soltanalian M and Stoica P. Computational design of sequences with good correlation properties. *IEEE Transactions on Signal Processing*. 2012;60(5):2180–2193.

[17] Bose A and Soltanalian M. Constructing binary sequences with good correlation properties: an efficient analytical-computational interplay. *IEEE Transactions on Signal Processing*. 2018;66(11):2998–3007.

[18] Woodward PM. *Probability and Information Theory, with Applications to Radar: International Series of Monographs on Electronics and Instrumentation*, vol. 3. Amsterdam: Elsevier; 2014.

[19] Sussman S. Least-square synthesis of radar ambiguity functions. *IRE Transactions on Information Theory*. 1962;8(3):246–254.

[20] Wolf JD, Lee GM, and Suyo CE. Radar waveform synthesis by mean square optimization techniques. *IEEE Transactions on Aerospace and Electronic Systems*. 1969;AES-5(4):611–619.

[21] Wu L, Babu P, and Palomar DP. Cognitive radar-based sequence design via SINR maximization. *IEEE Transactions on Signal Processing*. 2017;65(3):779–793.

[22] Stoica P, Li J, and Xue M. Transmit codes and receive filters for radar. *IEEE Signal Processing Magazine*. 2008;25(6):94–109.

[23] Spafford L. Optimum radar signal processing in clutter. *IEEE Transactions on Information Theory*. 1968;14(5):734–743. Available from: https://doi.org/10.1109/TIT.1968.1054205.

[24] Stoica P, He H, and Li J. Optimization of the receive filter and transmit sequence for active sensing. *IEEE Transactions on Signal Processing*. 2012;60(4): 1730–1740.

[25] Rummler WD. A technique for improving the clutter performance of coherent pulse train signals. *IEEE Transactions on Aerospace and Electronic Systems*. 1967;AES-3:898–906.

[26] Pillai SU, Youla DC, Oh HS, *et al*. Optimum transmit-receiver design in the presence of signal-dependent interference and channel noise. In: *Conference Record of the 33rd Asilomar Conference on Signals, Systems, and Computers, 1999*, vol. 2. Piscataway, NJ: IEEE; 1999. p. 870–875.

[27] DeLong D and Hofstetter E. On the design of optimum radar waveforms for clutter rejection. *IEEE Transactions on Information Theory*. 1967;13(3): 454–463.

[28] Bell MR. Information theory and radar waveform design. *IEEE Transactions on Information Theory*. 1993;39(5):1578–1597.

[29] Kay S. Optimal signal design for detection of Gaussian point targets in stationary clutter/reverberation. *IEEE Journal of Selected Topics in Signal Processing*. 2007;1(1):31–41.

[30] Naghsh MM, Soltanalian M, Stoica P, *et al.* A Doppler robust design of transmit sequence and receive filter in the presence of signal-dependent interference. *IEEE Transactions on Signal Processing*. 2014;62(4):772–785.

[31] Gianelli C, Xu L, Li J, *et al.* One-bit compressive sampling with time-varying thresholds for sparse parameter estimation. In: *Sensor Array and Multichannel Signal Processing Workshop (SAM), 2016 IEEE*. Piscataway, NJ: IEEE; 2016. p. 1–5.

[32] Sun H, Nallanathan A, Wang CX, *et al.* Wideband spectrum sensing for cognitive radio networks: a survey. *IEEE Wireless Communications*. 2013;20(2):74–81.

[33] Lunden J, Koivunen V, and Poor HV. Spectrum exploration and exploitation for cognitive radio: Recent advances. *IEEE Signal Processing Magazine*. 2015;32(3):123–140.

[34] Burke BF and Graham-Smith F. *An Introduction to Radio Astronomy*. Cambridge: Cambridge University Press; 2009.

[35] Strohm KM, Bloecher HL, Schneider R, *et al.* Development of future short range radar technology. In: *Radar Conference, 2005. EURAD 2005. European*. Piscataway, NJ: IEEE; 2005. p. 165–168.

[36] Hasch J, Topak E, Schnabel R, *et al.* Millimeter-wave technology for automotive radar sensors in the 77 GHz frequency band. *IEEE Transactions on Microwave Theory and Techniques*. 2012;60(3):845–860.

[37] Khobahi S and Soltanalian M. Signal recovery from 1-bit quantized noisy samples via adaptive thresholding. In: *2018 52nd Asilomar Conference on Signals, Systems, and Computers*; 2018. p. 1757–1761.

[38] Ribeiro A and Giannakis GB. Bandwidth-constrained distributed estimation for wireless sensor networks—Part I: Gaussian case. *IEEE Transactions on Signal Processing*. 2006;54(3):1131–1143.

[39] Ribeiro A and Giannakis GB. Bandwidth-constrained distributed estimation for wireless sensor networks—Part II: unknown probability density function. *IEEE Transactions on Signal Processing*. 2006;54(7):2784–2796.

[40] Host-Madsen A and Handel P. Effects of sampling and quantization on single-tone frequency estimation. *IEEE Transactions on Signal Processing*. 2000;48(3):650–662.

[41] Bar-Shalom O and Weiss AJ. DOA estimation using one-bit quantized measurements. *IEEE Transactions on Aerospace and Electronic Systems*. 2002;38(3):868–884.

[42] Dabeer O and Karnik A. Signal parameter estimation using 1-bit dithered quantization. *IEEE Transactions on Information Theory*. 2006;52(12): 5389–5405.

[43] Dabeer O and Masry E. Multivariate signal parameter estimation under dependent noise from 1-bit dithered quantized data. *IEEE Transactions on Information Theory*. 2008;54(4):1637–1654.

[44] Khobahi S, Naimipour N, Soltanalian M, *et al.* Deep signal recovery with one-bit quantization. In: *2019 IEEE International Conference on Acoustics, Speech and Signal Processing (ICASSP)*; 2019. p. 2987–2991.

[45] Hu H, Soltanalian M, Stoica P, *et al.* Locating the few: sparsity-aware waveform design for active radar. *IEEE Transactions on Signal Processing*. 2017;65(3):651–662.

[46] Boufounos P and Baraniuk R. 1-Bit compressive sensing. In: *42nd Annual Conference on Information Sciences and Systems*; 2008. p. 16–21.

[47] Knudson K, Saab R, and Ward R. One-bit compressive sensing with norm estimation. *IEEE Transactions on Information Theory*. 2016;62(5):2748–2758.

[48] Zymnis A, Boyd S, and Candes E. Compressed sensing with quantized measurements. *IEEE Signal Processing Letters*. 2010;17(2):149–152.

[49] Plan Y and Vershynin R. Robust 1-bit compressed sensing and sparse logistic regression: a convex programming approach. *IEEE Transactions on Information Theory*. 2013;59(1):482–494.

[50] Dong X and Zhang Y. A MAP approach for 1-bit compressive sensing in synthetic aperture radar imaging. *IEEE Geoscience and Remote Sensing Letters*. 2015;12(6):1237–1241.

[51] Masry E. *The Reconstruction of Analog Signals from the Sign of Their Noisy Samples*. California University of San Diego, La Jolla, Department of Electrical Engineering and Computer Sciences; 1980.

[52] Cvetkovic Z and Daubechies I. Single-bit oversampled A/D conversion with exponential accuracy in the bit-rate. In: *Data Compression Conference, 2000. Proceedings. DCC 2000*. Piscataway, NJ: IEEE; 2000. p. 343–352.

[53] Cheng Z, Liao B, He Z, *et al.* Transmit signal design for large-scale MIMO system with 1-bit DACs. *IEEE Transactions on Wireless Communications*. 2019;18(9):4466–4478.

[54] Ameri A, Bose A, Li J, *et al.* One-bit radar processing with time-varying sampling thresholds. *IEEE Transactions on Signal Processing*. 2019;67(20): 5297–5308.

[55] Blunt S and Gerlach K. Adaptive pulse compression via MMSE estimation. *IEEE Transactions on Aerospace and Electronic Systems*. 2006;42(2):572–584.

[56] Bussgang JJ. *Cross-correlation Functions of Amplitude-Distorted Gaussian Signals*. Cambridge, MA: MIT Res. Lab.; 1952.

[57] Van Vleck JH and Middleton D. The spectrum of clipped noise. *Proceedings of the IEEE*. 1966;54(1):2–19.

[58] Liu CL and Vaidyanathan PP. One-bit sparse array DOA estimation. In: *2017 IEEE International Conference on Acoustics, Speech and Signal Processing (ICASSP)*. Piscataway, NJ: IEEE; 2017. p. 3126–3130.

[59] Bro R and De Jong S. A fast non-negativity-constrained least squares algorithm. *Journal of Chemometrics*. 1997;11(5):393–401.

[60] Aubry A, Maio AD, Piezzo M, *et al.* Cognitive design of the receive filter and transmitted phase code in reverberating environment. *IET Radar, Sonar Navigation*. 2012;6(9):822–833.

[61] Gini F, De Maio A, and Patton L. *Waveform Design and Diversity for Advanced Radar Systems*. London: Institution of Engineering and Technology; 2012.

[62] Aubry A, Maio AD, Farina A, *et al*. Knowledge-aided (potentially cognitive) transmit signal and receive filter design in signal-dependent clutter. *IEEE Transactions on Aerospace and Electronic Systems*. 2013;49(1):93–117.

[63] Wang CJ, Wen BY, Ma ZG, *et al*. Measurement of river surface currents with UHF FMCW radar systems. *Journal of Electromagnetic Waves and Applications*. 2007;21(3):375–386. Available from: https://doi.org/10.1163/156939307779367350.

[64] Kondo M, Kawai K, Hirano H, *et al*. Ocean wave observation by CW mm-wave radar with narrow beam. In: *Proceedings of the 2001 IEEE Radar Conference*; 2001. p. 398–403.

[65] Stoica P and Moses RL. *Spectral Analysis of Signals*. Englewood Cliffs, NJ: Pearson/Prentice Hall; 2005.

[66] Skolnik MI. *Radar Handbook*. New York, NY: McGraw-Hill; 1970.

[67] Laska JN, Wen Z, Yin W, *et al*. Trust, but verify: fast and accurate signal recovery from 1-bit compressive measurements. *IEEE Transactions on Signal Processing*. 2011;59(11):5289–5301.

[68] Bose A, Ameri A, and Soltanalian M. Waveform design for one-bit radar systems under uncertain interference statistics. In: *2019 53rd Asilomar Conference on Signals, Systems, and Computers*; 2019. p. 1167–1171.

[69] Dinkelbach W. On nonlinear fractional programming. *Management Science*. 1967;13(7):492–498. Available from: http://www.jstor.org/stable/2627691.

[70] Soltanalian M and Stoica P. Designing unimodular codes via quadratic optimization. *IEEE Transactions on Signal Processing*. 2014;62(5):1221–1234.

Chapter 14
Cognitive radar and spectrum sharing

Hugh Griffiths[1] *and Matthew Ritchie*[1]

14.1 The spectrum problem

14.1.1 Introduction

The RF and microwave electromagnetic spectrum, extending from perhaps 100 kHz to 100 GHz, is a precious resource. It is shared by many types of users, for a diverse range of applications including broadcasting, communications, radionavigation and sensing. All users need greater bandwidth, yet the resource is strictly finite. Therefore, there is a need to develop more efficient means of spectral coexistence. This need has increased significantly in recent years and the resulting demand for spectrum is resulting in a wide range of innovative concepts which aim to address the challenge.

The purpose of this chapter is to describe and investigate one particular set of techniques based on adaptive spectrum allocation achieved by means of intelligent, cognitive processing. We first describe the nature of the spectrum problem and of spectrum allocation, and introduce the concept of cognitive radar. This is followed by a description of techniques to generate waveforms that fulfill both communications and radar functions, and then of intelligent adaptive waveform design. Finally, the ideas and results are summarised and some conclusions are drawn.

14.1.2 Spectrum and spectrum allocation

At an early stage, it was realised that mutually agreed allocation of different frequency bands for different purposes was necessary to avoid interference. One of the first attempts to do this was at the International Radiotelegraph Convention held in Washington DC in 1927 [1], allocating frequencies of 56–60 MHz for 'amateurs and experiments' and above 60 MHz as 'not reserved'. By 1937 [2], the maximum frequency covered by the allocations had risen to 300 MHz 'for experimental purposes'.

As radar was developed, the radar frequency bands were to a large extent dictated by the applications, but also by the dimensions of the waveguides employed – so,

[1]Electronic and Electrical Engineering Department, University College London, UK

for example, the WG16 waveguide has internal dimensions of 0.9″ by 0.4″, and is still widely used at X-band. These radar bands were designated by letters, partly for convenience and partly to avoid having to reveal exact radar frequencies, and that convention has stood ever since [3].

The process of allocation of spectrum is regulated by the International Telecommunications Union (ITU) and is reviewed at an international level by the World Radiocommunication Conference (WRC), held every 3–4 years. A chart which shows the frequency allocations for the USA, based on this plan, can be found on a US Government website [4]. This is detailed and complex, and is worth careful study. It shows the different services (Aeronautical Mobile, Aeronautical Mobile Satellite, Aeronautical Radionavigation, Amateur, Space Research, Standard Frequency and Time Signal, Standard Frequency and Time Signal Satellite), all colour coded. Some are allocated on an exclusive basis, and some are shared. The chart also shows which are Government Exclusive, which are Non-Government Exclusive, and which are Government/Non-Government Shared. The ITU has completed frequency allocations up until 275 GHz, discussions are currently underway on the spectra beyond this into the sub-THz region.

The last decade of the twentieth century and the first decades of the twenty-first century have witnessed phenomenal growth in the use of the RF and microwave spectrum for radio and TV broadcasting, mobile phones, broadband Wi-Fi and radionavigation. Whereas previously a small number of terrestrial UHF TV channels was considered adequate, now there is demand for hundreds of high-definition channels, streamable to mobile devices. It is these factors that have led to the problem that is described as the 'spectrum crunch'. An illustration of this is the prices paid by telecommunications companies in different countries for spectrum for mobile phones: in the 5G spectrum auction in 2021 in the United Kingdom, the price was approximately €4.8m per MHz [5]. At the same time, radars require ever more spectrum, particularly for high-resolution imaging applications. It may be expected that the problem will only ever get worse.

A 2015 publication [6] considered this spectrum problem from a radar point of view, in both technical and regulatory terms. There are several key issues:

- Signals of all kinds are spectrally impure, and radiate energy to a greater or lesser extent outside of their nominal bandwidth, which limits the degree to which channels may be closely spaced. A somewhat extreme example is given in Figure 14.1, which shows a measurement of the spectrum of the emission from an X-band magnetron transmitter from a marine navigation radar. The pulse length is 100 ns, giving a nominal bandwidth of 10 MHz, but the emitted energy extends well beyond this bandwidth, and in fact there is radiation over the whole of X-band. It follows that designing and realising signals with clean spectrum characteristics is key, and this is an active area of research.
- Linearisation techniques may be used to optimise power amplifier responses and hence to reduce the level of out of band spectral products. Such techniques may be open-loop or closed-loop, the choice representing a trade-off between performance (including degradation with time and with temperature)

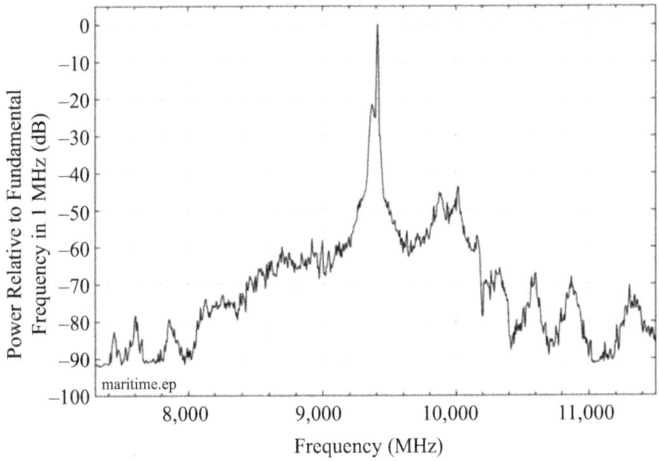

Figure 14.1 Power spectral density of the emission from a radar using a pulsed magnetron (Furono Model 1953C X-band maritime surface radar), with a pulse length of 100 ns giving a nominal bandwidth of 10 MHz. However, the radiated signal extends over the whole of X-band. Reproduced, with permission, from [6].

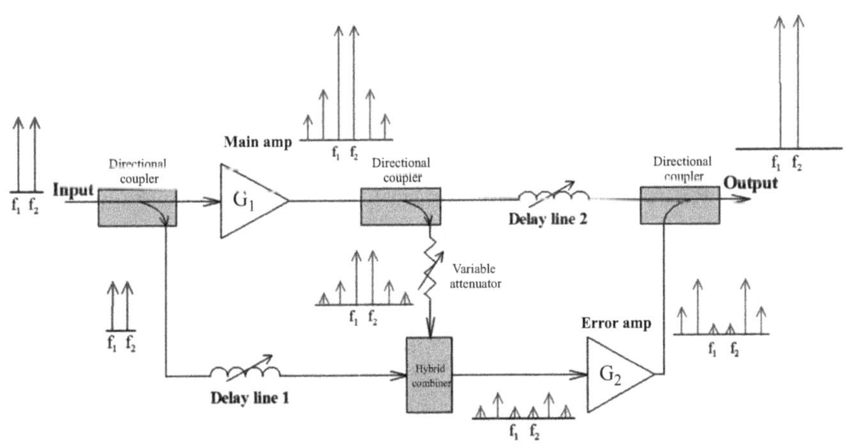

Figure 14.2 Open-loop linearisation technique [13]

and complexity or cost. An example of an open-loop linearisation technique is shown in Figure 14.2.

- Modern digital signal processing (DSP) allows precise, broadband radar signals to be generated and adaptively varied, potentially on a pulse-by-pulse basis. It is

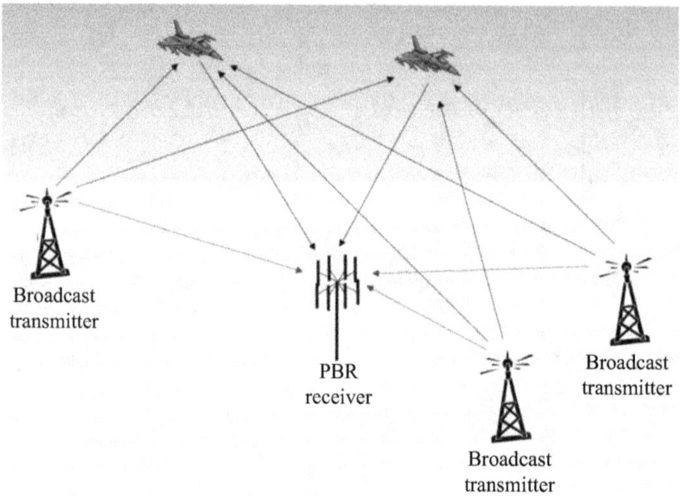

Figure 14.3 Passive Radar. Reproduced, with permission, from [6].

also possible to control the spectral characteristics of the signal, reducing the level of out-of-band emissions. This set of techniques has become known as 'waveform diversity' and the general subject as 'spectrum engineering' [7–12].

- Passive radar, using broadcast, communications or radionavigation signals as illuminators of opportunity (Figure 14.3), has obvious attractions since it requires no additional spectrum of its own, and as such it has been termed 'green radar'. There are many other advantages too, including the potential of improved detection performance against low observable targets, and that it allows the use of bands (particularly at VHF) that are not normally available for radar use. However, the waveforms are not necessarily ideal for radar purposes, and significant processing effort is necessary to suppress the direct signal at the passive radar receiver. Passive radar has come of age in the past decade, with practical systems built and demonstrated in several countries [7]. Further, it should be possible to design the signals and the coverage of broadcast, communications or navigation transmissions so that they not only fulfill their primary purpose but are also in some sense optimised as passive radar illuminators. This has been termed 'commensal radar'.
- The regulations that govern spectrum allocation should be framed taking into account the effect of interference of one service on another as a function of signal level, which may be serious (e.g., for communications associated with emergency services) or less serious (e.g., for everyday phone, text or e-mail communications), rather than taking the view that nothing should ever interfere with anything else. In many cases, a small degree of interference can actually be tolerated. This will require careful measurement and analysis.

- Another approach may be described as dynamic spectrum sharing – in other words to dynamically dispose signals in time, frequency, direction, polarisation, … so as to minimise co-interference. In spite of the multitude of signals, the actual spectrum occupancy at a given point, as a function of frequency, time, direction and polarisation may actually be quite low. This can be visualised in terms of what has been called the 'Radio Frequency Transmission Hypercube'. With suitable real-time sensing of the spectrum occupancy and control of the emitted signals, it should be possible for multiple signals to co-exist, dynamically minimising their mutual interference by disposing their energy in the available domains. This is part of cognitive radar, which is described in the next sub-section.

14.1.3 Cognitive radar definition

Adaptive signal processing has been incorporated in radars for many decades, for example, in the form of adaptive array processing on receive to suppress jamming, or adaptive setting of detection thresholds in CFAR processing. In a multifunction phased array radar, the dwell time and update time on a particular target can be varied adaptively as part of the tracking process, according to the target behaviour [15]. This principle can be taken further to give what has become known as cognitive radar, in which not only the receive processing but also the form of the transmitted signal are varied adaptively and dynamically, in response to a changing target scene. This will be especially true in the case of military radar, where targets may have low RCS and the radars have to operate in an environment of jamming and other countermeasures, which are constantly changing.

The concept of cognitive radar was introduced and developed by Professor Simon Haykin of McMaster University, Canada [14,16,17], and has been the subject of much interest and research in the past decade. It is fair to say that there is not universal agreement on the definition of cognitive radar. There are some who prefer the term 'Fully Adaptive Radar' [18], arguing that Cognitive Radar promises an intelligent radar that thinks for itself and has vastly superior performance.

The IEEE P-686 radar terminology standard [19] defines cognitive radar as:

> *A radar system that in some sense displays intelligence, adapting its operation and its processing in response to a changing environment and target scene. In comparison to adaptive radar, cognitive radar learns to adapt operating parameters as well as processing parameters and may do so over extended time periods.*

Inherent in this definition is the fact that a cognitive radar learns – in other words, if it is presented with a dynamically changing target scene, and then at some later time is presented with the same dynamically changing target scene, then it will perform better the second time because of what it has learned the first time. This may be considered to mimic the behaviour of natural systems such as bats, which vary the form of their (acoustic) transmitted signal in the course of navigation and of the detection and interception of prey [20] in an intelligent and sophisticated way. This is an example

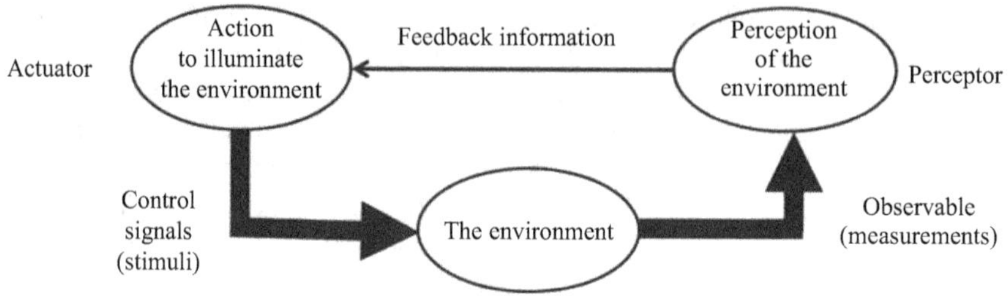

Figure 14.4 The perception–action cycle in cognitive radar [14]

of biologically inspired signal processing [21]. Modern radars are still far from this level of biologically capable adaptation, but are moving increasingly towards demonstrated elements of each of the requirements needed to be deemed to be cognitive. This learning might be rapid, or might take place over much longer timescales.

Haykin's concept also emphasises the perception–action cycle (Figure 14.4), in which the radar dynamically alters the form of its transmitted signal in response to the changing nature of the target scene and the information that the radar wants to acquire.

In many cases, cognitive radar will involve dynamically changing the form of the transmitted signal in response to the changing spectrum environment (including other signals of all kinds, and interference and jamming). Later within this chapter, examples of cognitive spectral management methods are described as well as examples of adaptive and/or cognitive methods using real experimental systems are discussed in Section 14.2.5.

The comparison of levels of cognition is challenging when describing either and simulation-based scenario or/and experimental system. As seen in the AI community, it may be sensible to consider cognition in terms of a graded level design rather than a binary state that a system is either cognitive or not. A paper by Horne in 2018 aimed to define an ontology for cognitive radars, to enable a comparison between them and a framework to measure their level of cognition [22]. The dimensions by which cognition of a given system can be judged are shown in Figure 14.5. These are the degrees of sophistication of (a) decision making, (b) memory, and (c) planning. The space is shown to range from a minimally Fully Adaptive Radar (FAR) system which uses rule-based decision making, fixed memory, and very little capability to 'plan', all the way to a fully cognitive system that excels in all of these areas.

14.1.4 Target-matched illumination

Another intuitively appealing example of waveform diversity, dating back to the 1980s, is target-matched illumination, devised by the Norwegian Dag Gjessing and described by him in his book [23]. This exploits the idea that a given target, viewed from a particular incidence angle, will have a particular impulse response (range profile) made up of the various scattering centres that make up the target, at delays appropriate to the incidence angle (Figure 14.6). It follows that there will be an optimum

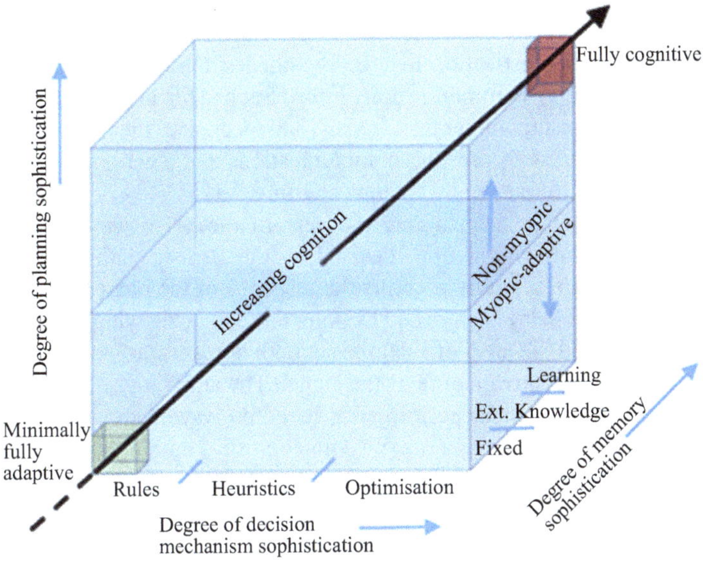

Figure 14.5 3-D space of radar cognition characteristics [22]

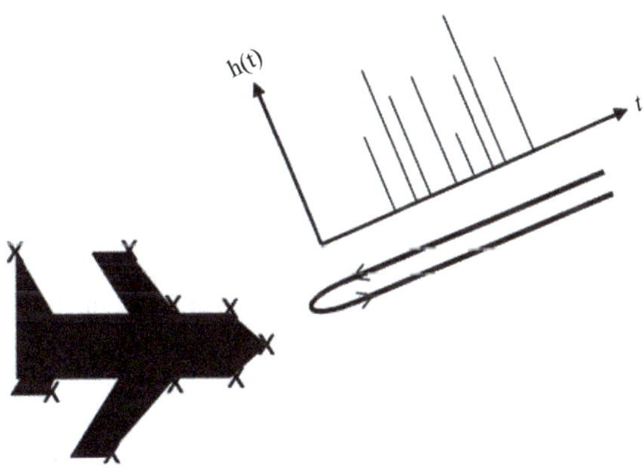

Figure 14.6 Impulse response of a target viewed from a given incidence angle. Each scattering centre (marked as a cross) contributes to the overall target impulse response h(t).

waveform, matched to that impulse response, to detect that target against a uniform noise background.

Guerci and Pillai developed the theory to show that there should also be an optimum waveform to maximise the probability of correctly distinguishing between two given target types [24–26]. Furthermore, Guerci describes in [24, Chapter 5] some of the practical considerations for implementing a cognitive radar.

14.1.5 Embedded communications

A further degree of sophistication may be provided if the transmitted radar signal is also arranged to carry communications information. This aligns with the idea of using the term 'RF sensing' rather than 'radar', and recognises the trend towards multifunction RF systems, with sensing, communications and electronic warfare (EW) functions carried out from a single antenna aperture.

An important tool in determining the performance of a waveform for radar purposes is the ambiguity function, devised by the British mathematician Philip Woodward [27]. This essentially presents the response of the radar to a single point target, as a function of delay (range) and Doppler (velocity), showing the resolution in range and velocity and the level of sidelobes in both domains, as well as ambiguities that result from the repetitive nature of the signal. The ambiguity function is widely used to analyse expected radar performance based on waveform parameters. However, it should be noted that the ambiguity function says nothing about the spectral confinement of a radar signal, so does not tell the whole story.

Approaches to embedded communications include:

- Seeking communications modulation formats (usually digital) that also have favourable ambiguity functions.
- Generating radar signals with spectral nulls, in which a carrier can be inserted, modulated with communications information. Further, the frequency of the spectral null and the carrier can be varied in some prescribed manner [28,29].

This research domain is reviewed further in Section 14.2.

14.1.6 Low probability of intercept (LPI)

Yet a further consideration is the ease (or difficulty) of interception of our radar signals by an adversary. In some cases, it is desirable to make a signal difficult to intercept, by spreading the energy in time in a long pulse (and hence reducing the peak transmitted power) and/or coding the waveform in a way that is not known by the intercepting receiver [30]. This is a standard technique in EW. It should be noted that to an intercept receiver, the radar signal power varies with range as $1/R^2$, while the radar echo varies as $1/R^4$. This gives the intercept receiver an advantage, which means that an LPI radar needs to have substantial processing gain, by virtue of the time-bandwidth product of its waveform, to offset the difference in range dependence, if the radar is to give good detection performance at the same time as low detectability. Example LPI radar and communications concepts are discussed in Section 14.2.4.

14.1.7 Summary

This section has introduced the spectrum congestion problem and some of the issues that surround it. The RF and microwave spectrum is a strictly finite resource, and all users want more bandwidth. It appears that the problem is only ever going to get worse. However, there are some grounds for optimism, since there are means of dynamically using spectrum in an intelligent, adaptive way. One important set of techniques is based on cognitive processing, in which the radar 'learns' about the disposition of targets, clutter, and other signals in the environment, and arranges the form of its transmitted signal accordingly. The process is dynamic, so as the environment changes, then so does the form of the transmitted signal. In the following sections, concepts relating the spectrum sharing, joint radar and communications concepts and example of cognitive radar will be discussed in further detail, referring to cutting-edge research output examples.

14.2 Joint radar and communications research

The fields of radar and communications have remained close partners over many years of development, jointly benefiting from improvements in DSP and Radio Frequency (RF) hardware upgrades. They have mostly remained separate in their application areas with relatively little crossover. In recent years, however, there has been significant interest in the research area which aims to merge the capabilities into joint systems that perform both radar and communications functions. This research area attracts a wide range of implementations which look to design waveforms which are suitable for both tasks [31], to inject communications signals within radar bands [32], or to perform sensing using purely communications-based waveforms [33]. As flexible programmable advanced FPGA and RF devices increase in both their accessibility and capabilities, the future of these developments is likely to move more and more towards real-world implementations and in-service systems. In some prior literature, the term DFRC (Dual Function Radar Communications) is used to describe a system which fulfils both functions. As part of this, cognition will play a vital element in the future of both communication and sensing. The ability of a system to have intelligence and adapt to an environment will become fundamental to its ability to perform the complex roles that it undertakes.

Within this section, types of radar and communication systems and concepts will be reviewed, focusing on either jointly designed waveform research or coexisting radar and communication systems. In recent years, a number of review papers have effectively summarised the research area; these include [34–36]. The principle of a DFRC can be seen in Figure 14.7. In this image, a single DFRC can be seen simultaneously communicating with communications nodes as well as sensing airborne targets using energy directed from the antenna array.

In defence and military applications, there are often scenarios where a user needs both to sense the environment for potential threats and to transmit information in a covert or secure manner. Achieving these goals within a congested and contested spectrum environment is challenging and requires advanced hardware and algorithms working together to ensure military dominance in the EW battlespace in this way. Traditional methodologies usually treat radar sensing and communication as totally separate tasks which are unrelated, and which use separate hardware and antennas. This results in inefficient use of energy and spectrum resources. There are several

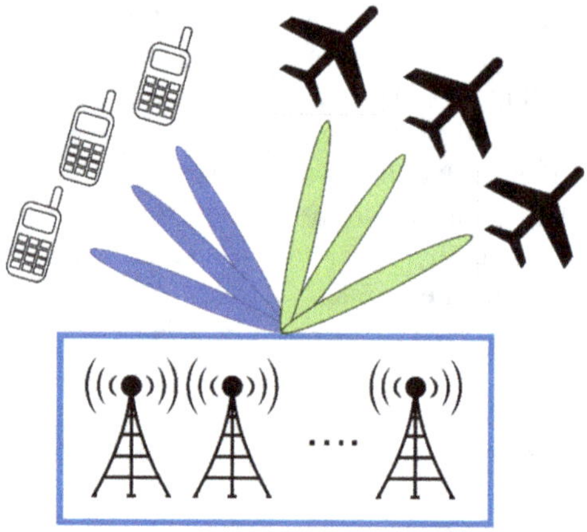

Figure 14.7 Example radar and communication co-existence configurations

reasons why the idea of being able to fulfill both the RF sensing role of a radar and the communications using a single system. A non-exhaustive list of these is as follows:

1. Using a single piece of hardware for both roles smaller compact platforms that have strict size, weight and power (SWAP) issues.
2. A reduction in spectrum requirement if both tasks use the same RF bandwidth.
3. Reduction in costs by having one device that is performing multiple roles.
4. Reduction of platform RCS by reducing the number of antennas required.
5. Deconflicting EM compatibility issues between competing RF devices by unifying them into a single transmitter/receiver system.
6. The sharing of both sensing and communications information for passive bistatic nodes.
7. Improved fusion of radar and communication which will be vital for future networked multistatic radar systems.

The history of DFRC systems is believed to have started in 1963 when the world's first system was proposed in which digital communications bits were modulated on the radar pulse interval [34]. A NASA project in 1978 demonstrated a Ku-band radar and communications system, as a niche real-world implementation of a DFRC system [37]. There was then a long period of inactivity in open research until the early 1990s when the first patent for multiple-input-multiple-output (MIMO) radar was filed (1994) and then the USA Office of Naval Research (ONR) started a programmed named Advanced Multifunction RF Concept (AMRFC) in 1996 [38]. These were very important milestones, as many DFRC systems rely on MIMO waveform design in order to perform both tasks and the ONR program was fundamental in starting the concept that a single, RF device can perform multiple roles rather than having discrete devices per task on a platform. In more recent years, DARPA has launched a challenge on spectrum utilisation which looks to push research towards finding viable solutions for future DFRC systems. This challenge was named the Shared Spectrum Access for Radar and Communications (SS-PARC) project [39]. The focus of this was on the spectrum around S-band (2–4 GHz) as this is often the most heavily competed section of the RF spectrum where naval radar systems as well as both 3G and 4G networks are in direct competition.

A key aspect of the co-existing of both radar and communications is the ability to be spectrally aware. This is the ability to have up-to-date knowledge of the level of activity of both co-operative and non-co-operative RF systems activity within the possible operational bands that a system could utilise. An ontology of the levels of spectral awareness was proposed within [40]. This describes the various approaches that can be applied to provide the required knowledge of the RF spectrum. These were broken down into the areas of signal sensing, geo-location or beacon sharing. For the signal sensing route, a variety of methods can be applied to provide info on the current presence of RF signals. These included:

- Covariance methods – RF signals can be detected by comparison of correlation between the received signal (Rx) and a non-zero lag version with Rx and a zero-lag version of itself.
- Energy detection – This technique evaluates the energy received in the specific frequency band defined against a threshold. This type of method struggles with lower SNR signals but requires no knowledge of the signals structure.
- Matched filtering – This method would use a correlation of known signals against the observed spectrum. *A priori* knowledge of the signals of interest is required for this to be effective.
- Cyclostationary processing – This is a type of correlation processing which compared (Rx) against a series of modulation formats (carrier tones, pulses, or cyclic prefixes). A spectral correlation process is then applied to the outputs in order to detect the presence of signals.

Alternative methods to achieve spectral awareness are either based around geo-location use agreed information of proposed RF system operation space to deconflict interference scenarios. Of course, the effectiveness of this is limited to how well RF systems adhere to these geo-fences and it is ineffective in a non-cooperative situation. Or the use of Beacon Sharing which uses mutual communication between primary and secondary uses of the spectrum via beacon signals in order to enable joint access without interference. Again, this methodology is limited to co-operative parties.

Broadly, the area of joint radar and communication waveforms can be broken into three key sub-categories. These are (i) concepts that aim to provide solutions which enable the co-existence of radar and communication, (ii) joint radar and communication systems (often referred to as DFRCs), and (iii) a specific sub-category where the objective is to implement waveforms that have LPI properties. These areas of research are now summarised in the sections below.

14.2.1 *Applications of joint radar and communication*

There is a broad range of applications for a system that can jointly perform radar sensing and communication, and a discussion on these applications can be found in [41]. In the civilian domain, automotive sensing is a significantly growing area where radars are used extensively. Self-driving vehicles inherently need to both sense and communicate, it would be desirable to achieve both these functions in a short time as vehicles pass each other. Each vehicle could share the knowledge it had learned about the road behind it via the same sensing signal it needed to generate to avoid collisions and steer itself to the end location. This form of Vehicle-to-Vehicle (V2V) communication is an active area of research and will continue to be so in the next few years.

A growing area of application for joint radar and communication is as part of unmanned air vehicles (UAVs) which are rapidly increasing in their usage. These platforms are required to sense their environment to avoid collisions and safely navigate but actively use communication both for their telemetry commands and also for data streaming of video or other information. As they are often small compact devices, the requirements on Size, Weight and Power (SWAP) are very strict. This lends itself well to the idea of having a single RF device that can provide both the sensing and communication capabilities. The benefits of a joint radar and communication system on a network of UAVs are simulated within [42].

Research has been undertaken in the area of using devices such as RFID tags as part of a sensing/communication network. RFID tags have been in use for a long time as ways of passively reacting to an imparted signal. This can be configured as a means of encoding data in a reflection back to a radar system by modulating the radar signal that is incident on the tag in the first place. Research in [43–46] shows a wide range of implementations in which tag devices are used to communicate information with backscattering techniques. This type of joint radar and communication is more directional and less capable of high data rates, but is very useful for niche application areas that may require this. Examples include Internet of Things (IoT)-based sensing and communicating concepts, as described within [47].

All modern military scenarios require resilient sensing and communication to ensure missions can be completed. An example may be the deployment of a missile system and how that is able to both sense the target as well as communicate. An example case is the use of a stand-off platform with an active radar to provide the illumination on the target while encoding bearing and position information to a missile seeker or forward deployed UAV platform which is operating passively.

14.2.2 *Co-existence radar and communication research*

Without strong coordination between radar systems and communication systems, it is possible to produce scenarios which result in high amounts of mutual interference where communication signal affects the performance of the radars and vice versa. By centralising and coordinating the way that they perform their tasks, it is feasible to reduce the negative impacts of a congested and contested RF environment as well as improve efficient use of the RF spectrum. Figure 14.8 shows a range of different ways

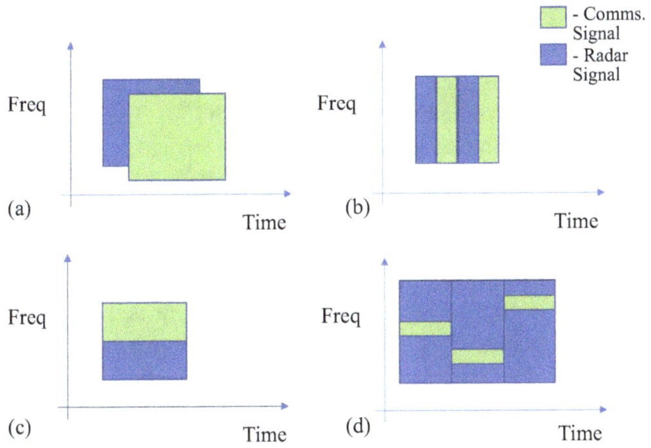

Figure 14.8 *Example radar and communication co-existence configurations: (a) congested overlap. (b) Time division multiplexing. (c) Frequency division multiplexing. (d) Agile embedding of communications signals within radar signals.*

that radar and communication systems can overlap and therefore co-exist. The first example Figure 14.8(a) is when no co-ordination is attempted, and signals therefore overlap and interfere with each other. Figure 14.8(b) and (c) shows when signals are either time or frequency division multiplexed in order to provide separation between them. The last example Figure 14.8(d) shows a more elaborate methodology which allows for communications signals to be embedded within radar signals in an agile manner using a notching process. The frequency and time domain embedding of signals is discussed further within [48], which is related to OFMDA signal encoding as it allows for both time and frequency distinction of signal components.

The time and frequency multiplexing concepts are well understood and are the simplest to implement. They bring the advantage of having a closely integrated signal RF aperture and DSP backend performing both tasks but may disadvantage one mode over the other by either narrowing its bandwidth or increasing time between transmits for a given mode. In the example case in which a required Pulse Repetition Frequency (PRF) needs to be met to sense a target unambiguously in Doppler, then this compromised may not be acceptable. For the last example, Figure 14.8(d), where the communications signals are embedded within the radar signals, an adaptive and very agile processing methodology is required to effectively achieve this. The benefit of doing this is that it is possible to tune the size of the notch made in the radar waveform depending on the priority of the communications vs. the radar signal, on a pulse-by-pulse basis.

Key publications in the area of adaptive waveform notching include [32,49]. This research describes how a real-time demonstrator was created which was capable of designing a notched FM waveform that was coexisting with changing in-band communication signals. The justification for this work was clearly made by describing the demand for the spectrum usage (see Section 14.1.1). The method used was described as "sense-predict-and-notch (SPAN)" which used a feedback loop that is similar to the perception-action cycle (Figure 14.3) showing how intelligent joint radar and communication is a wider part of cognitive radar research [50]. The experimental results show how the system was able to retrain predicted performance, despite the existence of the interference signals. The frameworks created allow for a more efficient use of the spectrum as well as mitigating for errors introduced the notched waveform.

The co-existence of both radar and communications can also be enabled via the management of authorisation of the bands which are used. As described in Section 14.1.2, the spectrum is an allocated resource where there is a primary user within that band. This inflexible legacy system will likely result in an increasingly difficult situation to meet the demand for RF as a limited resource. Proposals have been made to move away from this fixed solution to a more dynamic setup which allows for primary and secondary user access to the spectrum [51]. The primary users are labelled the Incumbent Users (IUs) who have legacy rights to access that particular frequency band, while the new users named Secondary Users (SUs) can also access this frequency band but must give way to the IUs. This type of solution is labelled a Dynamic-Shared Spectrum solution and is attracting a lot of attention currently as it may help solve some of the issues many countries are facing.

The co-existence of heterogenous communications protocols in the same band is commonplace, for example, Bluetooth and Wi-Fi. But in the case of radar and communications, additional challenges exist including the high power of transmitted signals and more significant repercussions for military systems if signals are missed. A real-world example of this type of solution being implemented was the usage of "white space" in TV bands. In the UK in 2015, the regulator Ofcom [52] allowed SUs to access the Digital Television bands with a frequency range 470–790 MHz via dynamic spectrum sharing. This successful trial demonstrated that this is very much the direction of how spectrum can be managed in the future. It is likely that tiers of access to given sections of the spectrum will be seen more to try and balance the increasing demand for spectrum.

Further cases where radar and communication systems are configured to jointly share spectrum are reviewed within [40]. This includes the description of how cognitive radars, communications, and joint systems could be deployed. A cognitive radar as described in [38] is deemed to be aware of the RF environment and can dynamically adapt to avoid interference. The same concept of adapting to the RF environment is used to describe the cognitive communications system. Finally, the joint cognitive system would collaborate to avoid creating harmful interference to each other. This description of cognition is very limited and does not include the wider concepts of memory and learning which are widely quoted to be the key attributes of a cognitive radar over and above an adaptive radar [22].

Efforts in the joint radar waveform DFRC research are now increasingly looking to include the practical challenges to its implementation. It is a domain that is clearly well suited to the advanced processing and significant potential benefits that cognitive sensing promises. As the TRL of the concept increases, the fidelity of the modelling is moving away from demonstrating simply what the theoretic benefits are and moving on to including real-world limitations of equipment to ensure that these benefits are still present within the constraints of a realistic scenario. This includes recent analysis focusing on constraining the complexity of the RF hardware by bounding the number of channels or using hybrid pre-coding to reduce the required DSP [53]. By limiting the number of required channels feeding a large-scale antenna area in order to maximise energy efficiency with reduced hardware complexity while still delivering the DFRC capability.

14.2.3 Single waveform tasked with both radar and communication

A different approach to undertaking the task of jointly sensing and communicating is using the same waveform to perform both tasks. This is quite challenging compared to separate co-existing signals as the optimisation needs to trade-off between communication and sensing performance. This section focuses on a couple of examples of applied methods in this area. A full review of single waveform sensing and communication can be found within [36].

A form of waveform adaptation, where a single signal is used for both radar and comms., was demonstrated by sequencing of the transmission of the pulses using different degrees of freedom. This communication method is called Index Modulation and a key publication in the area of index modulation is the MaJoRCOM method described in [54]. In this chapter, the transmission of radar signals is indexed on a set of variables, which in this case are which antenna is transmitting (N out of M) and the central frequency of the waveform (f_0 to f_n). The data is encoded into these parameters such that a communications node within the scene receives the sensing signal. The example shown uses four transmitting antennas and two central frequencies, but these can be arbitrarily scaled (Figure 14.9). As the information itself is encoded in the parameters used for each pulse transmitted the data rate is tied directly to the PRF of the radar. Hence, to double the data rate, it would be necessary to double the PRF, this may introduce some challenges as the PRF utilised by a radar is often constrained by other factors. The unambiguous range and Doppler of a series of pulses are defined by the PRF and hence there may be a conflict between the desired data rate and the requirements for unambiguously sensing targets at given ranges and speeds.

One of the key advantages of the MaJoRCOM method is that it is relatively simple to implement and that the modulation has no direct effect on the radar-sensing capabilities, other than the previously mentioned ambiguity affects. You can still design your idealised radar waveform and simply modulate a set of parameters in its transmission which will have no effect on the range resolution, sidelobes, SNR, etc.

The MaJoRCOM index modulation method itself is not cognitive and is focused on enabling spectrum sharing for a DFRC system. But, the optimisation of the potential trade-offs between radar and communication depending on the scenario and environment could be a cognitive decision-making process. For example, if bandwidth was a waveform parameter that was used in the index modulation, the system would have to decide to use codebook waveforms with lower bandwidths to increase the possible bits that can be sent. This would reduce the radar performance as the range resolution would be affected when using smaller bandwidth signals.

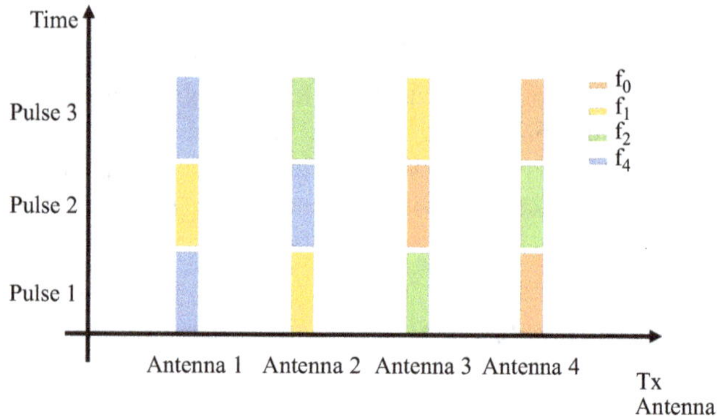

Figure 14.9 Transmit example of MaJoRCOM radar and communication index modulation method [54]

When designing DFRC waveforms, research has been undertaken into using information theory to describe fundamental performance limits of a joint system [55], funded via the DARPA SS-PARC program. This work shows that radar signals can be combined with communication signals in order to complete both tasks using only a single antenna solution. Upon reception, two processing chains perform either the traditional radar signal processing this is completed after the communication contributions have been decoded and removed from the received signal. The Fisher information was then applied to evaluate the optimal joint receiver format when a water-filling communications configuration was modelled.

Multiple antenna DFRC systems are also well suited to being able to communication and sensing through their beam forming capabilities. MIMO radar and Active Electronically Scanned Arrays (AESAs) are both technologies that are enablers for DFRC systems. This form of spatial processing enables beams to be created where energy is directed either towards a target for sensing purposes or towards a communications node to send data. MIMO systems are used extensively in the communications domain to enable higher throughput and better resilience within a communications signal. Radar systems also leverage MIMO techniques and can benefit significantly from the beamforming capabilities it enables. An example MIMO DFRC outputs can be seen in [56]. This work shows how using the precoding method, it is possible to create a joint radar waveform which includes communication symbols that are designed to be optimised for the performance radar transmit beamforming as well as being able to meet SINR constraints for the communication users. The simulation-based result within [56] demonstrated that it was possible to achieve angular estimation radar performance that is comparable to a non-joint DFRC waveform. Some of the challenges in this space include the optimisation step of waveform design which often requires a large processing capacity which may not be feasible in a real-time scenario. In order to mitigate the processing requirements, various methods have been used to reduce its complexity without compromising on performance.

14.2.4 LPI radar and communication waveforms

Radar waveforms are often designed to satisfy the requirement of Low Probability of Intercept (LPI). This means that the radar waveform is designed such that an Electronic Surveillance (ES) receiver cannot easily detect its presence while the radar can still perform its RF sensing objectives. This differs from Low Probability of Exploit (LPE) where the signal may be detectable by adversarial receiver systems, but it cannot be used easily to gain information on the radar system. In future, cognitive radar systems will need to take account of this requirement while adaptive and learning the RF environment they have to operate within. When designing joint radar and communications waveforms some literature has also tried to apply these requirements to the waveform design to enable an LPI radar and communication waveform satisfying a broad range of requirements.

From a radar perspective, a commonly used approach for LPI is FMCW (Frequency Modulated Continuous Wave) signals. These are low peak power transmitted signals, in comparison to pulsed-Doppler mode radars, and are constantly changing their central frequency which would reduce integration gain within an ES receiver. Many other LPI waveform designs also exist for radar including noise waveforms, phase-coded signals or producing an array with very narrow beam patterns which, therefore, reduce the probability of sidelobes being detected.

With regard to the communications signals, covert transmission methods traditionally relied on signals which were designed to rapidly hop in frequency randomly or the use of noise-like waveforms [57]. Obviously, encryption is a part of stopping the information being exploited but before that steps can be taken to reduce it being even intercepted in the first place. By creating a noise like signal or one that is very agile in frequency over a broad bandwidth, it is very challenging to eavesdrop on the data within the comms signal. Some recent work [58] shows how eavesdroppers can be modelled within a DFRC scenario with the objective of minimising the potential information they can obtain.

An intuitively attractive concept in the area of LPI radar and communications was introduced by Blunt in [28,29], involving intra-pulse radar-embedded communications. The concept was to create communications signals by re-radiating radar signals that are already in the Electromagnetic Environment (EME) with additional phase modulation components. This means that eavesdroppers are likely to not be able to differentiate the radar signal from the radar and communications signal and, in some cases, the pure radar signal may dominate for the ES receiver. This type of communication may be achieved passively using RFID tag devices or actively using Digital RF Memory (DRFM) devices. A key challenge for the active re-transmission is to have sufficient speed to capture, modify and re-transmit the signal with characteristics that match the original transmission.

Future systems may look to apply quantum-sensing methodologies, which have gathered significant interest in recent years. Quantum radar [59] and quantum communication could also be exploited to provide both LPI sensing, with embedded encrypted communications that are impossible to decipher with non-quantum-based methodologies. The promise of this type of methodology is leading to significant investment in these capabilities. Quantum communication has been shown to be experimentally implemented while quantum radar currently has a low Technology Readiness Level (TRL), mostly at the technical white paper study status.

14.2.5 Adaptive/cognitive radar concepts and examples

A key aspect of intelligent spectrum usage and sharing is the ability to adapt to any given EM environment. Cognitive or Fully Adaptive (FAR) radar systems are designed with the ability to react to every changing EM environment that they will have to operate within. A few experimental radar systems have been developed which aim to demonstrate cognition as part of their capabilities. This is difficult to do, since it means presenting a dynamically changing target scene to the radar and observing how it adapts and 'learns'. However, this adaptation and cognition do not necessarily have to be done in real time, and the system may step through a scenario in slow time. This section reviews example cases of where such systems have been used to demonstrate these capabilities or novel concepts that define how such adaptions could take place.

A key example system that has been developed in recent years is the CREW radar (Cognitive Radar Experimental Workspace) [60]. This system combined four transmitter and receiver nodes operating at W-band (92–96 GHz) with 4 GHz of total instantaneous bandwidth and PRFs up to 30 kHz. The system was developed at the Ohio State University, USA, and has been the focus of several experimental validations of cognitive radar concepts. The parameters under control in some of the experiments include the waveform PRF, transmitted power, the waveform itself and the number of pulses integrated as part of the processing chain. By varying these parameters in a dynamic way, real targets have been sensed with the system which adapts as a scenario is played out.

A cognitive radar publication that demonstrates dynamic spectrum Radio Frequency Interference (RFI) avoidance can be found in [61]. The hardware used to demonstrate this was a USRP device which is a commonly used Software Defined Receiver (SDR) that lends itself well to the task of sensing in a reconfigurable manner. The scenario is based around the concept that a radar sensor can perform its normal task of sensing the environment for targets as well as interleaving this with RF sensing for active interference. By interleaving this task, the system can actively sense potential sources of interference (LTE signals were used in this example case) and adapt the radar waveform to avoid these. The results show order of magnitude improvement in false alarm rates due to this adaption compared to the non-adapting baseline case.

Clearly spectrum sensing and adaption bring significant advantages for a system that must operate in a congested environment. But the challenge for radar systems is that moving a narrow band signal to a non-congested portion of the spectra may be sufficient in some cases, but this may not be feasible for wideband signals. For example, Synthetic Aperture Radar (SAR) systems require very significant bandwidths in order for them to perform their imaging tasks successfully. In those cases, it will be necessary to adapt by notching the waveform and using interpolation and sparse

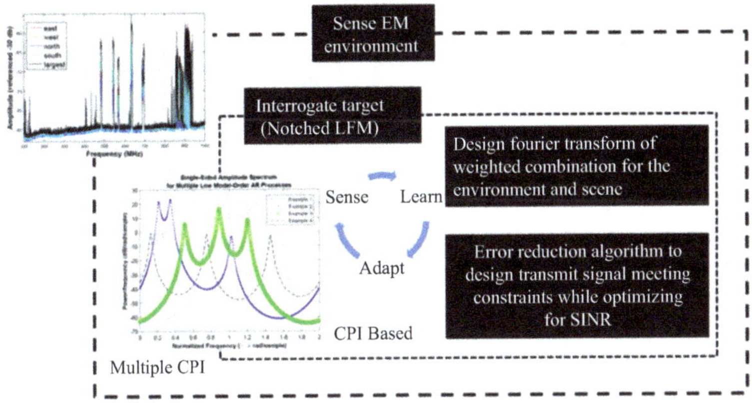

Figure 14.10 Proposed adaptive/autonomous signalling strategy hierarchy for interference avoidance and target-matched illumination [62]

sampling methods to partially reconstruct the performance back to the result that a full uninterrupted bandwidth would produce.

An example method of waveform adaptivity that has been demonstrated experimentally is the notching of waveforms to enable communication signals to be embedded within the full bandwidth as a sub-band. The radar signals themselves can use interpolation to recover from the introduced notched section in the bands in order to mitigate for their negative impact. This was demonstrated both via simulation and experiment within [62]. The concept design can be seen in Figure 14.10, showing how Haykin's perception–action cycle forms a key element of the concept. The EM environment is sensed, a target is detected using a Linear Frequency Modulated (LFM) chirp which has been designed to have notches placed in the areas where RFI exists within the spectrum and this waveform design is then iterated based on both target and environment changes.

A different example research prototype cognitive radar system named CODIR is described in [63,64]. This system was designed to operate in the S- and X-frequency bands in comparison with the much higher frequency range of the CREW radar. Within [65], it is shown how the system can be used to adapt to a congested spectrum. The examples used include a range of jamming scenarios from no interference, moderate jamming (−135 dB/Hz) to strong jamming (−123 dB/Hz) and finally maximum jamming (−113 dB/Hz). It was found that using special gapped LFM waveforms to reduce overlap with the noise jamming that target SINR was increased significantly. In a situation with no interference, the normal LFM waveforms were shown be preferable but upon the introduction of strong noise jamming, the gapped waveforms outperformed the non-gapped LFM signals. Therefore, a practical system must be spectrally aware to understand what signals are currently out there, and respond in such a way as to ensure it is affected in a reduced way.

Although several example cases of experimental systems and research concepts have been described in this section, adaptive waveform design and cognitive radar are very much a rapidly changing field. For future systems to benefit from these concepts, the TRL of these ideas will need to be significantly increased and a leap across the "valley of death" of innovation is needed to show real-world deployment. A key issue in the delivery of a real-world radar system, particularly in the military case, is delivering to a given specified performance. If adaptive systems have non-predictable or repeatable decision-making elements, then guaranteeing their performance in a range of scenarios is challenging to prove. For cognitive radars to be fully taken up by end users, the challenge of how to prove they work to a given specification will need to be addressed.

14.3 Summary and conclusions

This chapter has provided a summary of the key aspects of cognitive radar with a focus on the vital challenge of spectrum sharing. Clearly this is a broad domain which has a range of applications and challenges which have been partially addressed via the recent developments of both algorithms and adaptive hardware. Cognitive radar is a relatively recent concept that shows significant potential but presents many challenges in implementation as trust. Due to both the demand for spectrum as well as the requirement for more agile, dynamic and multi-role RF sensors cognitive radar is a concept that is here to stay and is expected to significantly increase in its TRL in the coming decade. The Holy Grail of cognitive radars that can perform to the same level as bio-inspired bat or dolphin biosensors is still some way off. But early experimental research in this field has already demonstrated how cognition can be applied to real world examples.

With respect to trust, radar systems are often used in vital roles which cannot afford to miss a detection, therefore, acceptance of delivery of a system is an important aspect of real-world systems. Cognitive radar, due to its non-static and learning abilities, may react in a way that is not 100% predictable and, therefore, may not be accepted.

The future of intelligent RF systems clearly requires significant changes from legacy static single-tasked systems. In order for both communications and radar to co-exist within a common spectrum, that is in demand more than ever, then they will have to share the space or be able to jointly perform both tasks with single signals. Concepts that enable either a single waveform that can perform both tasks or the use of spectrum sharing via notched waveform design or dynamic access have been summarised within the chapter.

Acknowledgments

The authors of this chapter would like to thank Shannon Blunt, Graeme Smith, Chris Baker, Christos Masouros and Colin Horne for their contribution to the research domain and their partnership in the research outputs that have been described. Their inputs have been vital in the creation of this chapter and their support is very much appreciated.

We would also like to acknowledge the funding provided by the Engineering Physical Sciences Research Council (EPSRC), the IET, Thales UK, the Office of Naval Research Global (ONR-G) and the Defence Science and Technology Laboratory (Dstl).

References

[1] ITU. 'General regulations annexed to the international radiotelegraph convention'. 1927. Accessed 10 October 2021. Available from: http://search.itu.int/history/HistoryDigitalCollectionDocLibrary/1.4.48.en.100.pdf.

[2] ITU. 'Documents of the Moscow Telecommunications Conference'. 1946. Accessed 10 October 2021. Available from: http://search.itu.int/history/HistoryDigitalCollectionDocLibrary/4.6.51.en.100.pdf.

[3] IEEE Std 521-2002, IEEE Standard Letter Designations for Radar-Frequency Bands. 2002.

[4] NTIA. 'United States Frequency Allocations'. 2011. Accessed 10 October 2021. Available from: https://www.ntia.doc.gov/files/ntia/publications/spectrum_wall_chart_aug2011.pdf.

[5] Barbara Veronese. '5G spectrum: the varying price of a key element of the 5G revolution'. 2021. Accessed 10 October 2021. Available from: https://www.oxera.com/insights/agenda/articles/5g-spectrum-the-varying-price-of-a-key-element-of-the-5g-revolution/.

[6] Griffiths, H.D., Cohen, L., Watts, S., *et al.* 'Radar spectrum engineering and management: technical and regulatory approaches', *IEEE Proceedings*, 2015, **103**, (1), pp. 85–102.

[7] Griffiths, H.D., Blunt, S.D., Cohen, L.C., and Savy, L. 'Challenge problems in spectrum engineering and waveform diversity'. *IEEE Radar Conference 2013*.

[8] Griffiths, H.D. 'The challenge of spectrum engineering'. *EuRAD Conference 2014*, 2014.

[9] Blunt, S.D. and Perrins, E.S. *Radar & Communication Spectrum Sharing*. Stevenage: IET, 2018.

[10] Griffiths, H.D. 'The spectrum crunch – a radar perspective'. In: Blunt, S.D. and Perrins, E.S., editors. *Radar & Communication Spectrum Sharing*. Stevenage: IET, 2018. pp. 266–290.

[11] Griffiths, H.D. and Baker, C.J., 'Passive bistatic radar for spectrum sharing'. In: Blunt, S.D. and Perrins, E.S., editors. *Radar & Communication Spectrum Sharing*. Stevenage: IET, 2018. pp. 266–290.

[12] Griffiths, H.D. and Baker, C.J. *An Introduction to Passive Radar*. Norwood, MA: Artech House, 2017.

[13] Cripps, S.C. *Advanced Techniques in RF Power Amplifier Design*. Norwood, MA: Artech House, 2002.

[14] Haykin, S. *Cognitive Dynamic Systems: Perception-Action Cycle, Radar and Radio*. Cambridge: Cambridge University Press, 2012.

[15] van Keuk, G. and Blackman, S.S. 'On phased-array radar tracking and parameter control', *IEEE Trans Aerospace and Electronic Systems*, 1993, **29**, (1), pp. 186–194.
[16] Haykin, S. 'Cognitive radar: a way of the future', *IEEE Signal Processing Magazine*, 2006, **23**, (1), pp. 30–40.
[17] Haykin, S., Xue, Y., and Davidson, T.N. 'Optimal waveform design for cognitive radar'. *42nd Asilomar Conference on Signals, System and Computers*, 2008.
[18] Guerci, J.R. *Cognitive Radar: the Knowledge-Aided Fully Adaptive Approach*. Norwood, MA: Artech House, 2010.
[19] IEEE radar terminology standard. IEEE Std 686-2017.
[20] Baker, C.J., Smith, G.E., Balleri, A., Holdereid, M.W., and Griffiths, H.D. 'Biomimetic echolocation with application to radar and sonar sensing', *IEEE Proceedings*, 2014, **102**, (4), pp. 447–458.
[21] Balleri, A., Griffiths, H.D., and Baker, C.J., editors. *Cognitive Radar: the Knowledge-Aided Fully Adaptive Approach*. Stevenage: IET, 2017.
[22] Horne, C., Ritchie, M., and Griffiths, H.D. 'Proposed ontology for cognitive radar systems', *IET Radar, Sonar and Navigation*, 2018, **12**, (12), pp. 1363–1370.
[23] Gjessing, D. *Target Adaptive Matched Illumination Radar*. Stevenage: IEE, 1986.
[24] Guerci, J.R. 'Knowledge-aided adaptive waveform design'. *Waveform Diversity and Design Workshop*, 2004.
[25] Guerci, J.R. and Pillai, S.U. 'Theory and application of optimum transmit-receive radar'. *IEEE International Radar Conference*, 2000.
[26] Pillai, S.U., Guerci, J.R., and Pillai, S.R. 'Joint optimal Tx-Rx design for multiple target identification problem'. *IEEE Sensor Array and Multichannel Signal Processing Workshop*, 2002.
[27] Woodward, P.M., *Probability and Information Theory, with Applications to Radar*. Stevenage: Pergamon Press, 1953.
[28] Blunt, S.D., Yatham, P. and Stiles, J. 'Intrapulse radar-embedded communications', *IEEE Trans Aerospace and Electronic Systems*, 2010, **46**, (3), pp. 1185–1200.
[29] Blunt, S.D., Metcalf, J.G., Biggs, C.R. and Perrins, E. 'Performance characteristics and metrics for intra-pulse radar-embedded communication', *IEEE Journal on Selected Areas in Communications*, 2011, **29**, (10), pp. 2057–2066.
[30] Pace, P.E., *Detecting and Classifying Low Probability of Intercept Radar*, 2nd ed. Norwood, MA: Artech House, 2009.
[31] Hassanien, A., Amin, M.G., Zhang, Y.D., and Ahmad, F. 'Dual-function radar-communications: information embedding using sidelobe control and waveform diversity', *IEEE Transactions on Signal Processing*, 2016, **64**, (8), pp. 2168–2181.
[32] Kovarskiy, J.A., Owen, J.W., Narayanan, R.M., Blunt, S.D., Martone, A.F., and Sherbondy, K.D. 'Spectral prediction and notching of RF emitters for

cognitive radar coexistence'. *IEEE International Radar Conference*, 2020. pp. 61–66.

[33] Palmer, J.E., Harms, H.A., Searle, S.J., and Davis, L.M. 'DVB-T passive radar signal processing', *IEEE Transactions on Signal Processing*, 2013, **61**, (8), pp. 2116–2126.

[34] Liu, F., Masouros, C., Petropulu, A.P., Griffiths, H.D., and Hanzo, L. 'Joint radar and communication design: applications, state-of-the-art, and the road ahead', *IEEE Transactions on Communications*, 2020, **68**, (6), pp. 3834–3862.

[35] Luong, N.C., Lu, X., Hoang, D.T., Niyato, D., and Kim, D.I. 'Radio resource management in joint radar and communication: a comprehensive survey', *IEEE Communications Surveys Tutorials*, 2021, **23**, (2), pp. 780–814.

[36] Mishra, K.V., Bhavani, Shankar, M.R., Koivunen, V., Ottersten, B., and Vorobyov, S.A. 'Toward millimeter-wave joint radar communications: a signal processing perspective', *IEEE Signal Processing Magazine*, 2019, **36**, (5), pp. 100–114.

[37] Cager, R., LaFlame, D., and Parode, L. 'Orbiter Ku-band integrated radar and communications subsystem', *IEEE Transactions on Communications*, 1978, **11**, (26), pp. 1604–1619.

[38] Hughes, P.K. and Choe, J.Y. 'Overview of advanced multifunction RF system (AMRFS)'. *Proceedings of the IEEE International Conference on Phased Array Systems and Technology*, 2002. pp. 21–24.

[39] DARPA. 'Shared Spectrum Access for Radar and Communications'. 2013. Accessed 10 October 2021. Available from: https://www.darpa.mil/program/shared-spectrum-access-for-radar-and-communications.

[40] Labib, M., Marojevic, V., Martone, A.F., Reed, J.H., and Zaghloul, A.I. 'Coexistence between communication and radar systems – a survey'. 2017. Accessed 10 October 2021. Available from: http://arxiv.org/abs/1709.08573.

[41] Paul, B., Chiriyath, A.R., and Bliss, D.W. 'Survey of RF communications and sensing convergence research', *IEEE Access*, 2017, **5**, pp. 252–270.

[42] Chen, X., Wei, Z., Fang, Z., Ma, H., Feng, Z., and Wu, H. 'Performance of joint radar-communication enabled cooperative UAV network'. *ICSIDP – IEEE International Conference on Signal, Information and Data Processing*, 2019.

[43] Weinstein, R. 'RFID: a technical overview and its application to the enterprise', *IT Professional*, 2005, **7**, (3), pp. 27–33.

[44] Yen, C.C., Gutierrez, A.E., Veeramani, D., and van der Weide, D. 'Radar cross-section analysis of backscattering RFID tags', *IEEE Antennas and Wireless Propagation Letters*, 2007, **6**, pp. 279–281.

[45] Farris, J., Iera, A., Militano, L., and Spinella, S.C. 'Performance evaluation of RFID tag-based "virtual" communication channels'. *IEEE International Conference on Communications*, ICC. 1, 2014. p. 2897–2902.

[46] Decarli, N., Guidi, F., and Dardari, D. 'A novel joint RFID and radar sensor network for passive localization: design and performance bounds', *IEEE Journal on Selected Topics in Signal Processing*, 2014, **8**, (1), pp. 80–95.

[47] Akan, O.B. and Arik, M. 'Internet of radars: Sensing versus sending with joint radar-communications', *IEEE Communications Magazine*, 2020, **58**, (9), pp. 13–19.

[48] Dokhanchi, S.H., Mysore, B.S., Mishra, K.V., and Ottersten, B. 'A mmwave automotive joint radar-communications system', *IEEE Transactions on Aerospace and Electronic Systems*, 2019, **55**, (3), pp. 1241–1260.

[49] Owen, J.W., Ravenscroft, B., Kirk, B.H., *et al.* 'Experimental demonstration of cognitive spectrum sensing & notching for radar'. *IEEE Radar Conference*, 2018. pp. 0957–0962.

[50] Haykin, S. and Fuster, J.M. 'On cognitive dynamic systems: cognitive neuroscience and engineering learning from each other', *Proceedings of the IEEE*, 2014, **4**, (102), pp. 608–628.

[51] Bhattarai, S., Park, J.M., Gao, B., Bian, K., and Lehr, W. 'An overview of dynamic spectrum sharing: ongoing initiatives, challenges, and a roadmap for future research', *IEEE Transactions on Cognitive Communications and Networking*, 2016, **2**, (2), pp. 110–128.

[52] Ofcom. *Implementing TV White Spaces*, 2015.

[53] Kaushik, A., Masouros, C., and Liu, F. 'Hardware efficient joint radar-communications with hybrid precoding and rf chain optimization'. *IEEE International Conference on Communications (ICC)*, 2021.

[54] Huang, T., Shlezinger, N., Xu, X., Liu, Y., and Eldar, Y.C. 'Majorcom: a dual-function radar communication system using index modulation', *IEEE Transactions on Signal Processing*, 2020, **68**, pp. 3423–3438.

[55] Chiriyath, A.R., Paul, B., Jacyna, G.M., and Bliss, D.W. 'Inner bounds on performance of radar and communications co-existence', *IEEE Transactions on Signal Processing*, 2016, **64**, (2), pp. 464–474.

[56] Liu, X., Huang, T., Shlezinger, N., Liu, Y., Zhou, J., and Eldar, Y.C. 'Joint transmit beamforming for multiuser MIMO communications and MIMO radar', *IEEE Transactions on Signal Processing*, 2020, **68**, pp. 3929–3944.

[57] Nicholson, D.L. *Spread Spectrum Signal Design: Low Probability of Exploitation and Anti-Jam Systems*. San Francisco, CA: Freeman, 1988.

[58] Su, N., Liu, F., and Masouros, C. 'Secure radar-communication systems with malicious targets: integrating radar, communications and jamming functionalities', *IEEE Transactions on Wireless Communications*, 2021, **20**, (1), pp. 83–95.

[59] Lanzagorta, M. *Quantum Radar*. San Rafael, CA: Morgan & Claypool Publishers, 2021.

[60] Smith, G.E., Cammenga, Z., Mitchell, A., *et al.* 'Experiments with cognitive radar'. *IEEE 6th International Workshop on Computational Advances in Multi-Sensor Adaptive Processing (CAMSAP)*, 2015. pp. 293–296.

[61] Kirk, B.H., Kozy, M.A., Gallagher, K.A., *et al.* 'Cognitive software-defined radar: evaluation of target detection with RFI avoidance'. *IEEE Radar Conference*, 2019.

[62] Jones, A.M., Horne, C.P., Griffiths, H.D., Smith, G., Mitchell, A., and John Baptiste, P. 'Experimentation of an adaptive and autonomous RF signalling strategy for detection'. *IEEE Radar Conference*, 2018. pp. 1213–1218.

[63] Oechslin, R., Wieland, S., Hinrichsen, S., Aulenbacher, U., and Wellig, P. 'A cognitive radar testbed for outdoor experiments', *IEEE Aerospace and Electronic Systems Magazine*, 2019, **34**, (12), pp. 40–48.

[64] Oechslin, R., Aulenbacher, U., Rech, K., Hinrichsen, S., Wieland, S., and Wellig, P. 'Cognitive radar experiments with CODIR'. *International Conference on Radar Systems*, 2017. pp. 1–6.

[65] Oechslin, R., Wellig, P., Hinrichsen, S., Wieland, S., Aulenbacher, U., and Rech, K. 'Cognitive radar parameter optimization in a congested spectrum environment'. *IEEE Radar Conference*, 2018. pp. 0218–0223.

Chapter 15
Cognition in automotive radars

Sian Jin[1], Xiangyu Gao[1] and Sumit Roy[1]

15.1 Introduction

The Society of Automotive Engineers (SAE) has defined six levels of driving automation [1,2] starting from Level-0 (no automation), Level-1 (low-level driver assistance), Level-2 (advanced driver assisted systems or ADAS), Level-3 (conditional driving automation with human-in-loop override), Level-4 (high automation in most conditions), and ending with Level-5 (full automation in all conditions). Each level of progression is critically dependent on the increasing integration of vehicular sensors – cameras, radars, ultrasonic sensors, and various inertial navigation units – and their rapidly evolving capabilities. Our focus in this chapter is on automotive radars that are the backbone of Level-2 (ADAS) automation today.

Automotive radars are typically used for desired target (object) detection followed by extraction of object features; with the advent of high-resolution radars, these now include imaging of targets and the background [4]. Recent advances in post-processing intelligence added to traditional radar signal processing have further improved radar functionality. For example, in complex environments with radar interference, adaptive signal processing techniques allow each radar to adapt its waveforms to reduce the impact of co-channel interference. New machine learning methods potentially enable enhanced classification of targets (vehicles, pedestrians, etc.) and predict the complex behavior of objects (lane changes, braking, etc.) from radar signal [5]. These advances in signal processing highlight a path towards realization of cognition in automotive radar.

An early proponent of cognitive radar was Haykin [6]. In contrast to traditional radar that senses the environment without any transmitter-side adaptation based on initial returns from prospective targets and/or environment being mapped, a cognitive radar can iteratively adapt illuminations based on prior information, using a broad class of decision-estimation driven learning algorithms (as will be shown in Figure 15.4) [6]. The primary goals of cognitive automotive radar can be classified into two objectives [7–9] for our purposes: (1) mitigating multi-radar interference in dense

[1]Department of Electrical and Computer Engineering, University of Washington, Seattle, WA, USA

vehicular radar scenarios; (2) improving imaging and classification performance of desired targets. In this chapter, we first briefly review the basic concepts of traditional automotive radar and the need for greater operational intelligence under dense vehicular radar cases. We then introduce emerging themes involving intelligent signal processing and machine intelligence, and explore their impact on radar imaging and object recognition.

15.2 Review of automotive radar

15.2.1 Automotive radar

Automotive radars are designed to extract information (range, velocity, angle, etc.) about targets of interest [4] in time and space. Such information constitutes key elements of driver-assistive automotive functions such as adaptive cruise control, cross-traffic alert, parking assistance, rear collision warning [4], etc. To achieve higher levels of driving automation, further improvements to sensors (e.g., notably cameras, radar/lidar, etc.) robustness and functionality are required. The grand challenge for driverless automotive operation is the realization of ultra-reliable sensor-based navigation in dense urban environments – i.e., inclusive of the ability to map a complex navigational scenario, comprising "desired" and "undesired" objects, in a scene with background clutter. In such scenarios, the multi-radar interference problem has garnered very little attention to date, wherein automotive radars must continue to detect and estimate true target information (e.g., other vehicles, pedestrians, etc.) with high accuracy and avoid false targets. The advent of new techniques, such as multi-input multi-output (MIMO) radar adapted to the vehicular environment for high-resolution imaging, offer ingredients for reliable autonomous driving that are yet to be fully exploited.

Given the need for large radar bandwidths to support high-resolution operation, millimeter-wave (mm-wave) bands, e.g., 76–81 GHz, have been permitted for automotive radar operations by FCC. This enables radars with small-aperture antennas to achieve desired antenna gain with low-cost antenna-on-chip and radar-on-chip systems [11]. Presently, frequency-modulated continuous wave (FMCW) modulation is widely used for automotive radars due to its low cost and good range–velocity estimation performance. Further, MIMO radar principles are also being adopted to achieve greater angle resolution as well as robustness in varying operational conditions resulting from mobility. Hence, FMCW-MIMO radar is considered to be a desired benchmark in the near future for autonomous operation; we next review the principles and properties of FMCW and MIMO radar operation.

15.2.2 FMCW radar

The FMCW radar transmits periodic wideband linear frequency-modulated signal (chirps) as shown in Figures 15.1 and 15.2. The transmit (TX) signal is reflected from targets (vehicles, pedestrians, etc.) and received at the radar receiver. The FMCW radar can detect targets' range and velocity from the receive (RX) signal using the

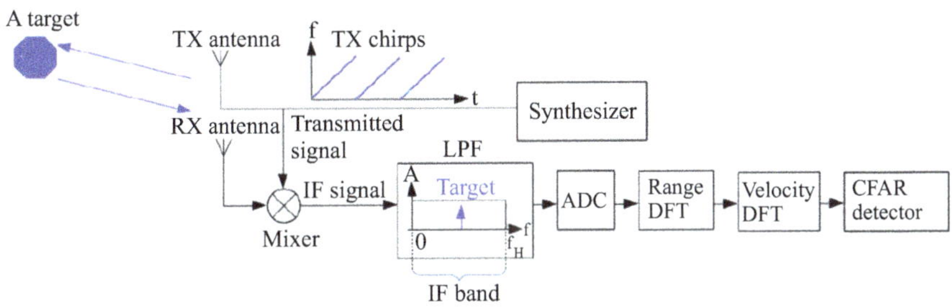

Figure 15.1 Basic FMCW radar block diagram

stretch processing structure in Figure 15.1. In this structure, a mixer at the receiver multiplies the RX signal with the TX signal to produce an intermediate frequency (IF) signal. Since the RX signal and the TX signal are linear frequency-modulated signals that have constant frequency difference, the IF signal is a single tone signal. For target at range d, the IF signal frequency is $f_I = \frac{2d}{c}h$ is the time delay $\frac{2d}{c}$ times the chirp slope h (see Figure 15.2), where c denotes the speed of propagation. Thus, detecting the frequency of the IF signal f_I can give the target range d. The IF signal is then passed into an anti-aliasing low-pass filter (LPF) and an analog-to-digital converter (ADC) for digital signal processing (DSP). The discrete Fourier transform (DFT) is a widely adopted DSP technique that can estimate f_I to infer d, and hence the range DFT.

The maximum detectable range and the range resolution (the minimum range difference that can be resolved) are two key performance metrics relating to range estimation. The maximum detectable range or the cutoff frequency of LPF shown in Figure 15.1 depends on the IF bandwidth f_H via $d_{max} = \frac{f_H c}{2h}$, obtained by solving $f_I = f_H$. The range resolution depends on the duration of the TX chirp T_c. The minimum frequency difference that can be resolved by range DFT is $1/T_c$ and thus the range DFT partitions the IF band $[0, f_H]$ into $N = \frac{f_H}{1/T_c}$ resolvable bins [12]. This means that the range resolution is $d_{res} = \frac{d_{max}}{N} = \frac{c}{2B_c}$, where $B_c = T_c h$ is the chirp's RF bandwidth. From this, we see that the range resolution d_{res} benefits from the high RF bandwidth B_c. The processing structure in Figure 15.1 supports high RF bandwidth B_c (of the order of GHz) while keeping the IF bandwidth f_H small (of the order of MHz). This avoids using a high-bandwidth matched filter and a high-speed ADC, making it cost-effective. Thus, the low-cost and high-range resolution properties render FMCW-based radar designs that are very competitive for automotive applications [10].

FMCW radar transmits chirps periodically with inter-chirp duration T_g. The movement of the object or radar makes the IF frequency difference between the two consecutive chirp $\Delta f_I = \frac{2vT_g}{c}h$ is the time delay difference $\frac{2vT_g}{c}$ multiplies chirp slope h, where v is the relative velocity of the object. Typically, the frequency change $\Delta f_I \ll f_I$ and f_I can be regarded as nearly constant over a time duration called the coherent processing interval (CPI). In practice, we choose the CPI to be the length of P chirp transmission cycles, i.e., PT_g. Within a CPI, although the IF frequency

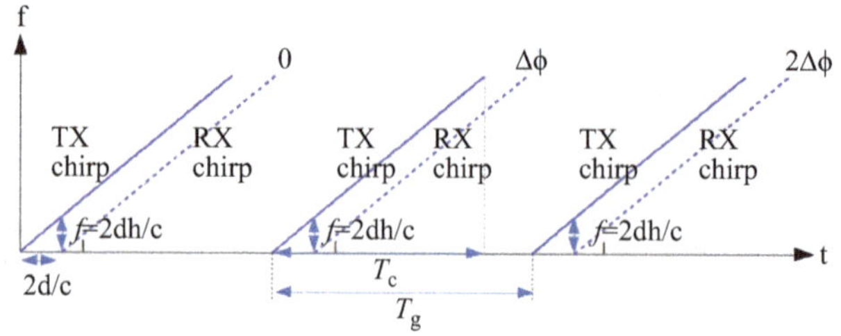

Figure 15.2 FMCW radar TX chirps and RX chirps

remains almost unchanged, the relative motion vT_g in a T_g causes a significant phase difference between two consecutive chirps. Such phase difference $\Delta\Phi = 2\pi f_c \frac{2vT_g}{c}$ is the angular frequency $2\pi f_c$ multiplied by the time-delay difference $\frac{2vT_g}{c}$, where f_c is the radio frequency of the chirp. Thus, detecting the phase change $\Delta\Phi$ can solve the target velocity v. As the phase linearly increases across the P chirps in a CPI (see Figure 15.2), a DFT over the P chirps is widely adopted to estimate $\Delta\Phi$ that is used to infer v, and hence such DFT is called the velocity/Doppler DFT.

The maximum detectable velocity and the velocity resolution are two performance metrics of velocity estimation. The maximum detectable velocity depends on the inter-chirp duration T_g, and is shown to be $v_{max} = \frac{c}{4f_c T_g} = \frac{\lambda}{4T_g}$ [13], where λ is the RF wavelength of the transmitted chirp. The velocity DFT partitions the detectable velocity range $[-v_{max}, v_{max}]$ into P velocity bins. Thus, the velocity resolution – the minimum velocity difference that can be resolved – is $v_{res} = \frac{2v_{max}}{P}$.

Following the range and velocity DFT operations, the signals are input into the constant-false-alarm rate (CFAR) detector. The output of the CFAR detector is an N-by-P two-dimensional (2D) range–velocity spectrum, where N is the number of range and P the number of velocity bins, respectively. The CFAR detector detects and localizes the target by detecting peaks in the 2D range–velocity spectrum.

15.2.3 MIMO radar and angle estimation

FMCW radar can estimate the range–velocity of targets with a *single* RX antenna as shown in Figure 15.1. For estimating the angle or direction of arrival (DoA) of targets, an FMCW radar requires multiple RX antennas (a RX antenna array) [10]. A phased array radar with single TX and N_r RX antennas results in single-input-multiple-output (SIMO) radar configuration [14], depicted in Figure 15.3(a) with $N_r = 8$ RX antennas uniformly distributed on a line with any two consecutive antennas separated by distance L. For a far-field target located at angle θ with respect to antenna broadside, the difference of radio propagation distance between two consecutive RX antenna elements is $L\sin(\theta)$, leading to the phase difference $\omega = 2\pi f_c \frac{L\sin(\theta)}{c}$ between the RX signals on two consecutive RX antennas. As the phase linearly increases across the N_r RX antennas (see Figure 15.3(a)), DFT over N_r RX antennas is widely adopted to estimate ω to infer θ, and hence such DFT is called the angle DFT. Increasing

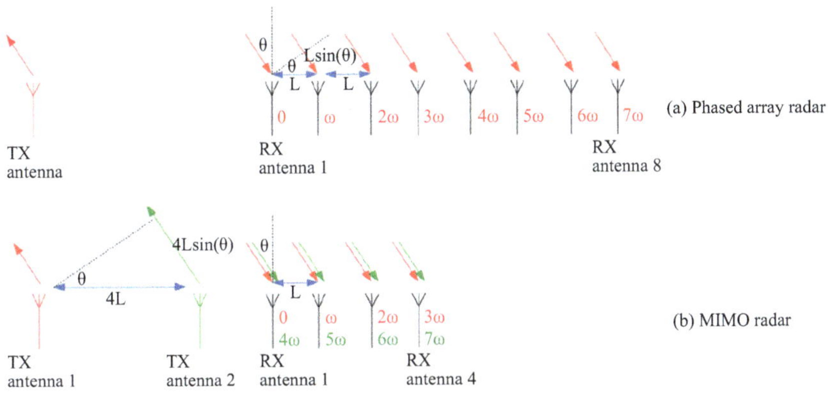

Figure 15.3 Principle of phased array radar and MIMO radar

Table 15.1 Summary of range, velocity, and angle estimation performance

Metric	Formula[a]	Metric	Formula
Range resolution	$d_{res} = \frac{c}{2B_c}$	Max operating range	$R_{\max} = \frac{f_{IF}c}{2h}$
Velocity resolution	$v_{\text{res}} = \frac{\lambda}{2PT_g}$	Max operating velocity	$v_{\max} = \frac{\lambda}{4T_g}$
Angle resolution	$\theta_{\text{res}} = \frac{\lambda}{N_{\text{Rx}} L \cos\theta}$	Max operating angle	$\theta_{\max} = \sin^{-1}\left(\frac{\lambda}{2L}\right)$

[a] c is the speed of propagation, B_c is the chirp bandwidth, λ is the wavelength of the transmitted signal, T_g is the inter-chirp duration, and P is the number of chirps per CPI. $N_{\text{Rx}} = N_r$ for SIMO radar, $N_{\text{Rx}} = N_r \times N_t$ for MIMO radar, N_r is the number of RX antenna, N_t is the number of TX antenna, L is the separation between receive antenna pairs.

the number of antennas results in an angle DFT with a higher resolution (sharper peak) [14]. For detecting a target at angle θ, the angle resolution – the minimum angle difference that can be resolved – is given by $\theta_{res} = \frac{\lambda}{N_r L \cos\theta}$. We summarize the range, velocity and angle estimation performance in Table 15.1.

Note that for achieving Level-4 and Level-5 autonomous driving, the angular resolution achieved needs to be $\leq 1°$ [10]. To achieve this, it can be shown that approximately 100 RX ULA-phased array elements with inter-antenna spacing $L = \frac{\lambda}{2}$ are needed [10] with angle DFT processing. However, each RX antenna element in Figure 15.3(a) requires a full RX chain including a mixer, low pass filter, and ADC shown in Figure 15.1 [14], incurring significant hardware cost. To mitigate this, a widely adopted approach in the automotive industry is to resort to MIMO radar [10,14,15].

A MIMO radar uses N_t TX antennas along with N_r receive antennas with suitably designed TX waveforms sent simultaneously, for achieving small angle resolution. Figure 15.3(b) shows a MIMO radar that has $N_t = 2$ TX antennas separated by distance $4L$ and $N_r = 4$ RX antennas with any two consecutive RX antennas separated by distance L. Similar to the phased array radar, the TX signal from the first TX antenna (TX antenna 1) is received at the 4 RX antennas with phase $0, \omega, 2\omega, 3\omega$ (the phase of RX antenna 1 is 0 as a reference) [14]. Because TX antenna 2 is placed $4L$ from TX antenna 2, as shown in Figure 15.3(b), the TX signal from TX antenna 2 propagates an additional $4L\sin(\theta)$ distance than the signal from TX antenna 1. Thus, at the 4 RX antennas, the RX signal phases from the TX antenna 2 are $4\omega, 5\omega, 6\omega, 7\omega$. This makes the MIMO system equivalent to the phased array system in Figure 15.3(a), but with a much smaller number of RX chains. In general, the MIMO radar can achieve an equivalent virtual phased array with N_tN_r elements using only N_t TX antennas and N_r RX antennas [10]. This significantly reduces the size and the cost of the radar, making it suitable for automotive applications.

From the above angle estimation principle, MIMO radar operation requires separating the received waveforms originating from different transmit antennas. Thus, MIMO radar waveforms on different TX antennas should be orthogonal or near-orthogonal [4,11]; such orthogonality can be achieved in frequency, time, or code domain [4,14,16]. In practice, ideal orthogonality is hard to achieve, especially over a range of Doppler shifts [17]. Different waveform separation schemes can lead to a different performance in terms of resolution, sidelobes, range–velocity–angle coupling. A review of different waveform separation schemes for MIMO radars can be found in [16].

15.3 Cognitive radar

According to the IEEE standard [18], cognitive radar is "a radar system that in some sense displays intelligence, adapting its operation and its processing in response to a changing environment and target scene." In contrast to conventional adaptive radars that restrict their adaptation to the receiver only, cognitive radars adapt both the transmitter and the receiver for optimal performance [19]. The transmitter-side adaptation of cognitive radar includes waveform adaptation [20] and antenna beam pattern reconfiguration [21]. These adaptations require reconfigurable components such as adaptable waveform generator and antenna arrays [21–23]. The transmitter adapts waveform and antenna beam pattern according to the feedback from the receiver, based upon intelligence about the system and environment state [24]. Such intelligence is gained through observations of results from interactions with the environment [20] (e.g., stored perception results in similar environments [21]) processed via learning algorithms and using other domain knowledge. The operational principles of cognitive radar can be summarized in the perception–action cycle (PAC) that will be detailed in Section 15.3.1. The PAC includes the perception, learning, and action parts that are explained in detail in the subsequent subsections.

15.3.1 Perception–action cycle

Cognitive radar relies on observations of dynamic interaction [6] in the form of a cycle – known as the PAC [21]. This is fundamental to realizing cognition in radar [25]

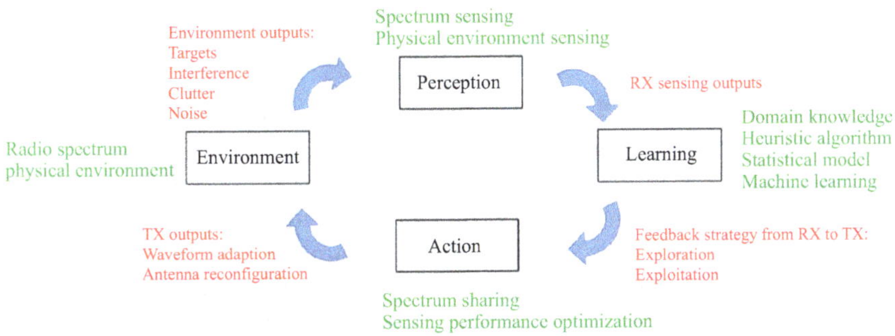

Figure 15.4 Block diagram of PAC [7,26,27]

and improving physical environment perception (detection, tracking, imaging, classification, etc.) performance from one cycle to the next [20]. Specifically, the RX processes the received signal, obtains the information about the environment, and feeds back such information to the TX. The TX receives the feedback from the RX, conducts actions in an intelligent manner, and adaptively changes its emission to improve radar performance.

A typical PAC shown in Figure 15.4 consists of four blocks: environment, perception, learning, and action. The abstract block "environment" refers to the radio spectrum or the physical environment (cars, pedestrians, road, buildings, etc.). Accordingly, the perception block refers to sensing the occupation of the radio spectrum and extraction of physical environment features via detection and classification of objects in the scene [28]. Subsequently, the learning block represents the RX action that selects a suitable decision rule towards a certain goal (e.g., improving detection or imaging performance) using spectrum sensing and target detection/imaging/classification results. The learning component interprets the results via the use of domain knowledge, learning algorithms, and observations stored in memory [28]. For example, in the interference scenario where different types of interference (in and out of network) are potentially experienced, the learning part classifies the interference using learning algorithms and stored data, and then relies on signal processing techniques related to domain knowledge for interference mitigation. The learning outcome is fed back to the TX for taking actions – such as radar waveform adaptation [20] or beam pattern revision [21] – for improving spectrum-sharing efficiency or optimizing detection/tracking performance. Through the PAC, the cognitive radar ties together radar sensing and control to improve its sensing performance [20].

15.3.2 Perception

In a dynamic environment, the cognitive radar system should be able to perceive the environment, explore environmental information and possibly change to rules or parameters accordingly [29]. Hence, the perception of the environment is an important part for supporting the subsequent adaptive actions. As shown in Figure 15.4, the perception block of cognitive radar includes physical environment sensing and radio spectrum sensing.

To sense the physical environment, automotive radars are required to accurately estimate the range, angle, and velocity of objects, as introduced in Section 15.2. Advanced physical environment perception can be achieved by combining low-level signal processing algorithms – radar imaging [30] – and high-level machine learning approaches for object recognition [31]. We introduce advanced environment perception algorithms for automotive radar in Section 15.4.

In general, determining which parts of the radio spectrum are currently available for use is another important aspect of cognitive radar perception. Such determination is based on spectrum sensing, i.e., using an RX to estimate the energy at the input in any desired sub-band and comparing it to a pre-determined threshold [25]. The challenge for spectrum sensing is that a large bandwidth needs to be sensed [7]. If the received signal on a large sub-band is sampled at the Nyquist rate, an ADC with high sampling rates is required. This increases the costs and is impractical [7] for portable scenarios with size, weight, and power (SWAP) constraints. Mitigation approaches such as fast spectrum sensing (FSS) [32] reduce the sampling rate in estimating the power spectrum [25], which would typically lead to loss of spectrum resolution. However of late, the theory and practice of compressive sensing (CS) [19,21,33] have matured to enable high-resolution sensing with reduced rate sampling. In CS-based spectrum sensing, the spectrum is sensed using sub-Nyquist sampled ADCs based on the assumption of spectrum sparsity and suitable algorithms used to estimate instantaneous signal amplitude in each sub-band [21]. In Section 15.5.2, we will introduce an efficient spectrum sensing method suitable for automotive FMCW radar that is different from the above surveyed spectrum sensing schemes.

15.3.3 Learning

The dynamic environment (moving targets, time-varying interference, etc.) has a large impact on radar performance. Learning is a process that uses the immediate or past RX perception information to learn the dynamic environment or update the control policy in the dynamic environment [20,34]. The learning algorithms are broadly based on knowledge-based approach, Bayesian approach, and machine learning approaches [6, 34]. The knowledge-based approach uses prior knowledge of the environment, domain knowledge, and heuristic algorithms to learn the dynamic environment and update control policy [6]. The Bayesian approach uses prior knowledge and a statistical model that is updated using immediate or past perception information [6,34]. The machine-learning approach is based only on the perception information and knowledge of actions [34].

For complex environments, reinforcement learning (RL) is especially suitable for cognitive radar and is recommended by Haykin [6,35]. The block diagram of RL for cognitive radar as shown in Figure 15.5 has a one-to-one correspondence to the block diagram of PAC in Figure 15.4; the environment and radar perception parts in Figure 15.5 already exist in Figure 15.4. The policy is a map of actions (output)

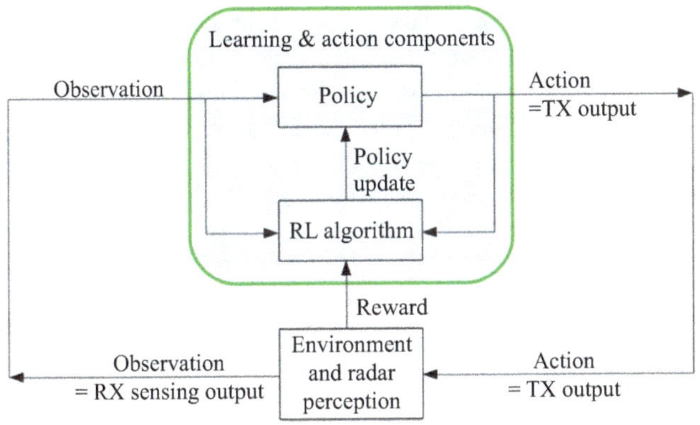

Figure 15.5 Block diagram of RL for cognitive radar [35,36]

based on the RX perception information (input) [36]. The RL algorithm continuously updates the policy based on the actions, RX perception information, and reward from the environment [36]. The reward from the environment is typically a scalar performance metric for a specific task [37]. For example, if the task is improving detection performance, the reward can be the signal-to-interference-plus-noise ratio (SINR) at the radar receiver. The goal of the learning algorithm is to maximize the average cumulative reward by finding an optimal policy [36,37].

15.3.4 Action

As mentioned earlier, cognitive radars continuously learn about the environment through experience gained from interactions with the environment, and continually updates the receiver with relevant information on the environment. The action block makes adjustments to the transmitted waveforms and beampattern in order to maximize its performance of multiple radar functions (e.g., imaging, detection, tracking, and classification) over time [7,21,34,38].

A RL algorithm can be used to adapt chirp parameters [8,39]. For example, in [8], the radars learn to adapt bandwidth and center frequency of the chirp to mitigate interference. However, as indicated in Section 15.2.1, changing bandwidth impacts range resolution. A good spectrum-sharing scheme needs to achieve adaptability to improve SINR while minimizing impact on other radar performance metrics (e.g., range resolution) [7]. Beam-pattern optimization [40,41] is a promising way to achieve this goal via reducing the overall spectrum congestion in the environment [7] in the spatial domain.

The adaptive transmit waveforms are used in other applications besides interference mitigation. Tan *et al.* [42] proposed to adapt waveforms for enhancing target recognition. Specifically, when the cognitive radar measures a micro-Doppler spectrogram from a target class that is not represented in its pre-stored training set, an anomaly detection algorithm will provide a trigger to the cognitive radar scheduler to collect micro-Doppler spectrograms from this unknown target.

The adaptation of cognitive radars to scenes can be summarized from two perspectives. In good detection conditions, sensor resources are saved by minimizing the time, bandwidth [29], or antenna samples [43]. In less favorable conditions (large target ranges, small SNR, moving clutter, small target velocity), appropriate radar resources need to be allocated to the sensor in terms of time or bandwidth, in order to improve environment perception performance [44].

15.4 Physical environment perception for FMCW automotive radars

An important aspect of cognitive radar PAC is physical environment perception. This section illustrates advanced physical environment perception of the cognitive radar including imaging and object recognition. The basic radar imaging algorithms in this section include range–velocity imaging, micro-Doppler imaging, and range–angle imaging. For the purpose of improving angle resolution, we introduce the multiple signal classification (MUSIC) high-resolution DoA estimation algorithm [10] and synthetic aperture radar (SAR) [45]. In the end, we briefly review automotive radar object recognition based on radar images.

15.4.1 Range–velocity imaging

As illustrated in Section 15.2.1, FMCW radars transmit a linear frequency-modulated signal; the received signal reflected from a target is mixed with the transmitted signal to obtain the beat frequency, which is a function of the round-trip delay and, therefore, can be mapped directly to the range.* Similarly, transmitting a train of periodic FMCW chirps allows velocity/Doppler estimation for targets that undergo relative radial motion. Such radial motion induces a phase shift over the chirps in a range resolution cell, which is used to compute the radial velocity.

For FMCW radars, it is common to perform the traditional DFT approach in the fast time (samples) and slow time (chirps) of IF signal for real-time range–velocity estimation, named the range DFT and velocity DFT [46]. The obtained results of two DFT operations are a 2D spectrum, named range–velocity (or range-Doppler) image. In Figure 15.6(a), we present one example of a range–velocity image for a pedestrian moving away from radar in the parking lot (see Figure 15.7(a)). Note that with huge available frequency bandwidth (up to 4 GHz) and a sufficient number of chirp observations, FMCW radars can provide highly accurate, fine-resolution range and velocity estimation results, as shown in Table 15.1.

*The beat frequency is also a function of the Doppler frequency due to range-Doppler coupling. However, under the low-speed scenario, the round-trip delay is more dominant, and hence we can ignore the impact of the Doppler in range estimation.

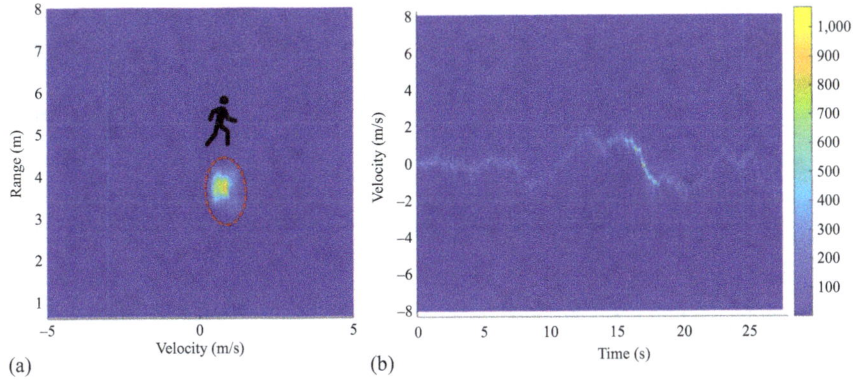

Figure 15.6 (a) The range–velocity image of a moving pedestrian and (b) the micro-Doppler image obtained by STFT [46] (© 2019 IEEE)

15.4.2 Micro-Doppler imaging

The range–velocity map does not show how the velocity spectrum changes with time. A technique to visualize the velocity spectrum as a function of time is the micro-Doppler map using the short-time Fourier transform (STFT). Compared to the usual velocity DFT measurement mentioned above, STFT divides a longer train of FMCW chirps into shorter segments of equal length and computes the Fourier transform separately on each shorter segment. For the same pedestrian, we plot the micro-Doppler map in Figure 15.6(b), which can be used to track object velocity and display specific micro-Doppler movement signatures among different objects [31,46].

15.4.3 Range–angle imaging

For automotive MIMO radars with a virtual array, DoA angle finding can be done with digital beamforming by performing DFTs (named angle DFT) on snapshots across the array elements. If we perform the angle DFT on the range–velocity images on different array elements, we would obtain a 3D spectrum cube, named range–velocity–angle (RVA) image. If we perform the range DFT and angle DFT on the sampled signal of a single chirp, we would obtain a 2D range–angle spectrum, named range–angle (RA) image. One example of the RA image for a pedestrian is displayed in Figure 15.7(b).

Angle DFT can be implemented efficiently in an embedded DSP to save computation time. However, angle DFT suffers from low resolution. Super-resolution DoA estimation can be achieved with subspace-based methods such as MUSIC, CS-based methods, the iterative adaptive approach, etc.

MUSIC is one of the most popular super-resolution DoA estimation algorithms. MUSIC performs eigen-decomposition on the covariance matrix of received signals, and then constructs the K-dimension signal subspace and $(N-K)$-dimension noise

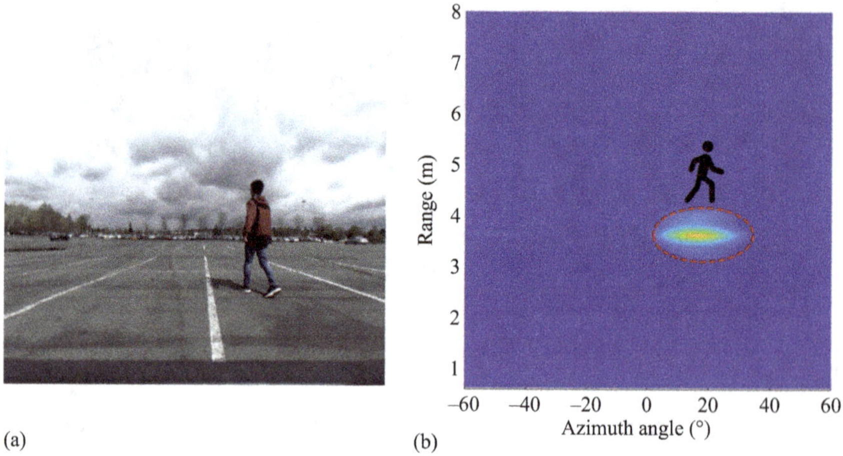

Figure 15.7 (a) A pedestrian moving away from an empty parking lot and (b) the range–angle map obtained by range DFT and angle DFT [46]

subspace U_n, where K denotes the number of targets and N denotes the number of array elements. The MUSIC estimates K DoAs $\{\theta_k\}_{k=1}^{K}$ are given by the locations of K maxima of the pseudospectrum [47]:

$$P_{\text{MUSIC}}(\theta) = \frac{1}{a^*(\theta)U_n U_n^* a(\theta)} \tag{15.1}$$

where $a(\theta)$ denotes the steering vector corresponding to the search direction θ.

We compare the performance of DFT-based and MUSIC-based DoA estimation algorithms via an experiment with two corner reflectors. Two corner reflectors are placed on the table with 20° angle difference. From Figure 15.8(a), DFT-based DoA estimation generated a single peak for two corner reflectors, while the MUSIC-based DoA estimation could separate two peaks which represent the angles of two corner reflectors as shown in Figure 15.8(b).

It is worth noting that the high-resolution of MUSIC algorithm comes at the cost of accurately estimating the covariance matrix of the received signal and the number of targets, which usually takes multiple snapshots and increases computation load [10].

15.4.4 Synthetic aperture radar imaging

SAR is a well-known imaging technique that overcomes the limits of small physical aperture on the angular resolution by coherently processing the returns from a series of transmitted pulses to create a large *synthetic* aperture [45,48]. While SAR was originally created for airborne radar remote sensing and mapping [49], it can also be applied in side-looking automotive radars by exploiting the vehicle motion [45,50], for supporting some ADAS applications such as lane changing or keeping assist, blind-spot detection, and rear cross-traffic alert.

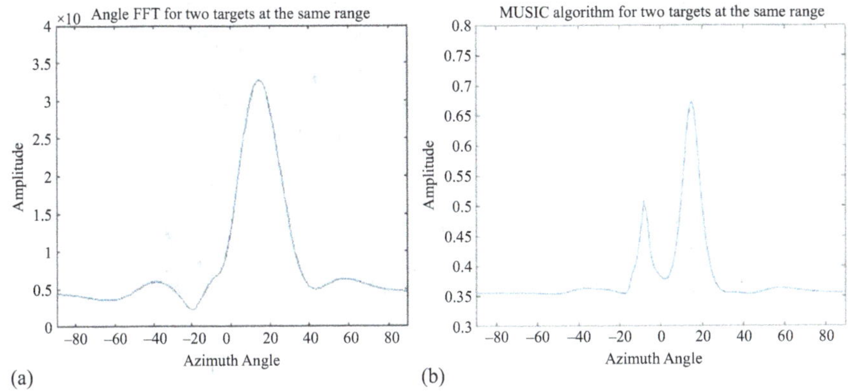

Figure 15.8 (a) DFT-based DoA estimation result for two corner reflectors with 20° azimuth difference and (b) MUSIC-based DoA estimation result for the same scenario as (a) [46] (© 2019 IEEE)

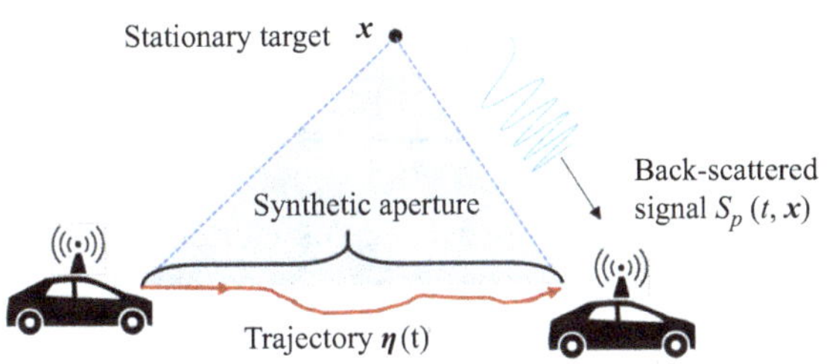

Figure 15.9 Automotive SAR system model

As shown in Figure 15.9, assuming the radar is moving past a stationary point target at $\mathbf{x} \in \mathbb{R}^2$, the back-scattered signal is recorded at each measurement position, and the sensor trajectory is denoted by $\eta(t) \in \mathbb{R}^2$. A general back-scattered signal model is given by

$$S_{\mathrm{p}}(t, \mathbf{x}) = \underbrace{\exp\left(j\phi\left(\|\eta(t) - \mathbf{x}\|_2\right)\right)}_{\text{phase history}} \cdot \mathrm{A}\left(\|\eta(t) - \mathbf{x}\|_2\right) \tag{15.2}$$

where $\phi(\cdot)$ is the phase history that describes the phase change of the back-scattered signal along the sensor trajectory as a function of the two-way traveled target distance [51], $A(\cdot)$ is the amplitude that is a function of target distance as well.

The synthetic aperture is formed by the in-phase addition of this coherently measured phase history. The in-phase addition is described mathematically as a cross-correlation (matched filter) between the input signal and a reference signal along

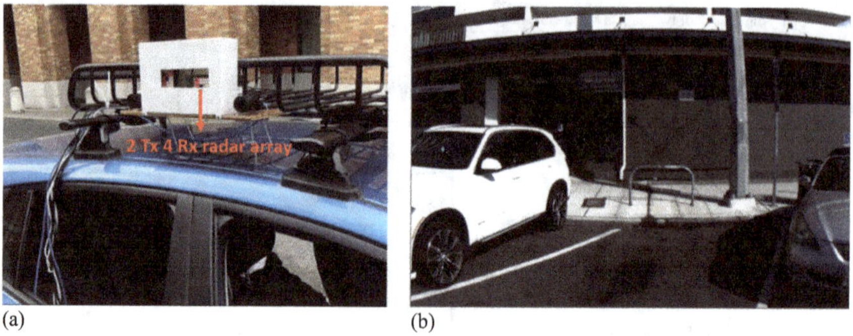

Figure 15.10 *(a) Integrated radar-camera data capture platform mounted on one side of vehicle and (b) two inclinedly parked cars at the road side [45] (© 2021 IEEE)*

the trajectory [45,51,52], which is also called azimuth compression in many SAR algorithms and is given by

$$I(\mathbf{x}) = \sum_{t} S_p(t, \mathbf{x}) \underbrace{\exp\left(2\pi j f_c \frac{\|\eta(t) - \mathbf{x}\|_2}{c}\right)}_{\text{reference phase}} \tag{15.3}$$

where $I(\mathbf{x})$ describes the synthetic signal along $\eta(t)$ for pixel $\mathbf{x}$.

Prior research has explored extensive SAR algorithms applied for automotive radar applications. References [51,53,54] perform range migration, backprojection, and range-Doppler algorithms, respectively, for imaging vehicles in the parking lot. To fulfill the need for real-time capability for automotive radars, there is research [45,52,55,56] focusing on speeding up post-processing via techniques to reduce the algorithmic run-time complexity. For example, references [52,56] achieve parallel implementation utilizing GPUs for backprojection algorithm. Fembacher *et al.* [55] used approximate SAR techniques to reduce computation resources at the cost of image quality degradation.

Gao *et al.* [45] proposed a hierarchical MIMO-SAR algorithm that applies coherent SAR principles to vehicular MIMO radar to improve the side-view (angular) resolution. The proposed 2-stage hierarchical processing workflow drastically reduces the computation load by selecting the "region of interest" in the first stage. We show one experiment example from [45] in Figure 15.11, where the radar data of road-side scene is collected via a vehicle-mounted radar platform as shown in Figure 15.10. The MIMO-SAR algorithm reduces computational time to 1.17 s per frame [45] and the obtained imaging result (presented in Figure 15.11(a)) shows an enormous performance gain provided by large synthetic, compared to its single-shot range–angle imaging counterpart (presented in Figure 15.11(b)).

As discussed earlier, coherent SAR processing requires accurate vehicle trajectory estimation to achieve coherent in-phase processing. The traditional ego-motion

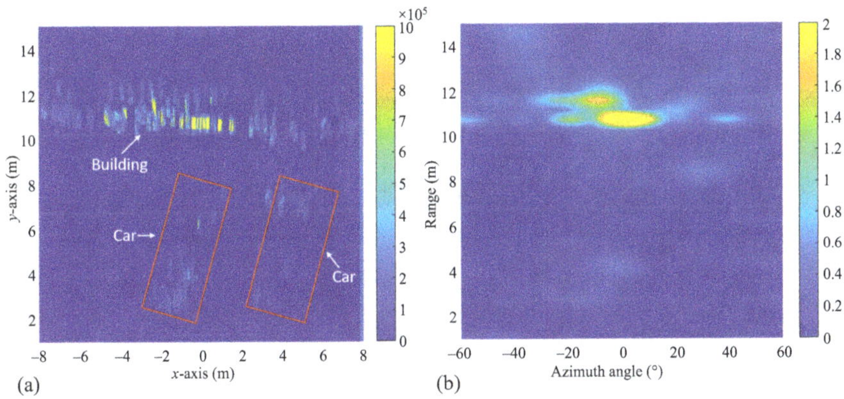

Figure 15.11 (a) The MIMO-SAR imaging result (magnitude I of (15.3)*) where we use two rectangles to mark the parked cars and (b) range–angle map imaging for single-frame radar data [45] (© 2021 IEEE)*

estimation solutions include combining the measurement from inertial measurement units (IMU), wheel speed sensors, and global positioning systems (GPS) [57].

Recently the need for self-contained alternatives has been inspired to reduce the cost of the high-precision expensive inertial navigation systems. One promising approach is to perform the radar odometry algorithm [58] to determine the velocity and direction of radar motion [45,51], without relying on any other external sensors.

15.4.5 Radar object recognition based on radar image

Pedestrians, cyclists, and other road users are especially vulnerable in road accidents and, therefore, it is essential to identify them in a timely manner to foresee dangerous situations [59]. Radar sensors are excellent candidates for this task since they are able to simultaneously measure range, radial velocity, and angle while remaining robust in adverse weather and lighting conditions (e.g., rain, snow, darkness).

Prior works have attempted radar object recognition for three main classes of objects (pedestrian, cyclist, and car) with various radar data input formats, e.g., RVA image [31], range-Doppler image [60], range–angle image [61], micro-Doppler image [46], etc. Based on that, many deep learning-based radar object recognition algorithms have been proposed to enhance the system performance. Next, we introduce a state-of-the-art radar recognition algorithm RAMP-CNN [31], which processes the RVA image sequences with the fusion of three encoder–decoder structures.

The RAMP-CNN model is designed to relieve the computation burden of processing 4D RVA image sequences (range, velocity, angle, and time/frames), while keeping all necessary information. The 4D data is assumed to embed all information we need: temporal information behind the chirps in one frame, as well as the change of spatial information across frames [31]. As shown in Figure 15.12, the

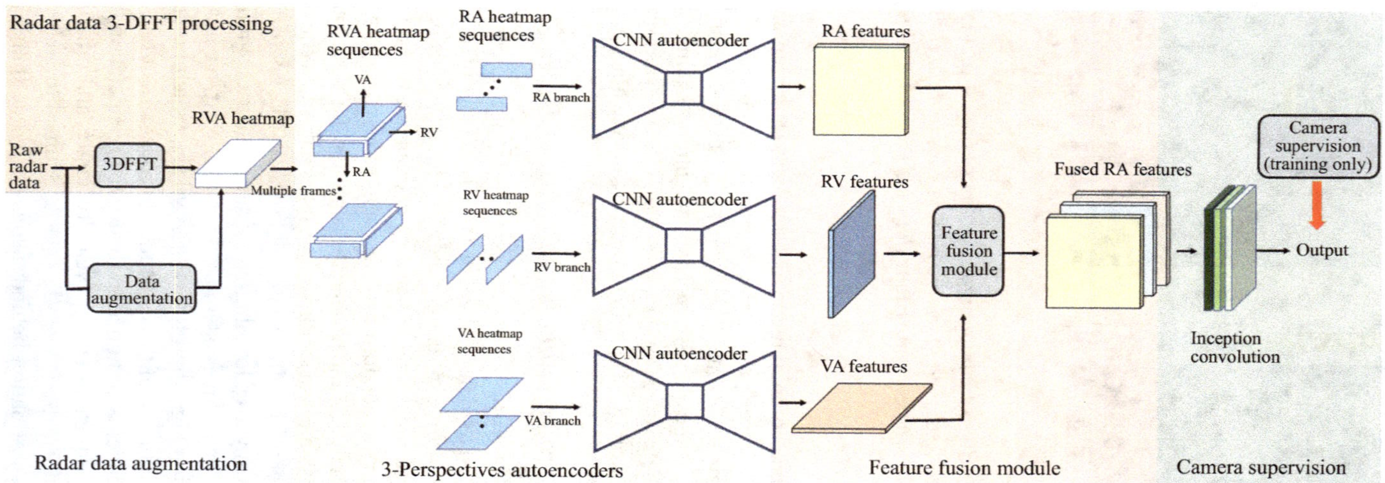

Figure 15.12 The architecture of RAMP-CNN model [31] for road user detection (© 2020 IEEE)

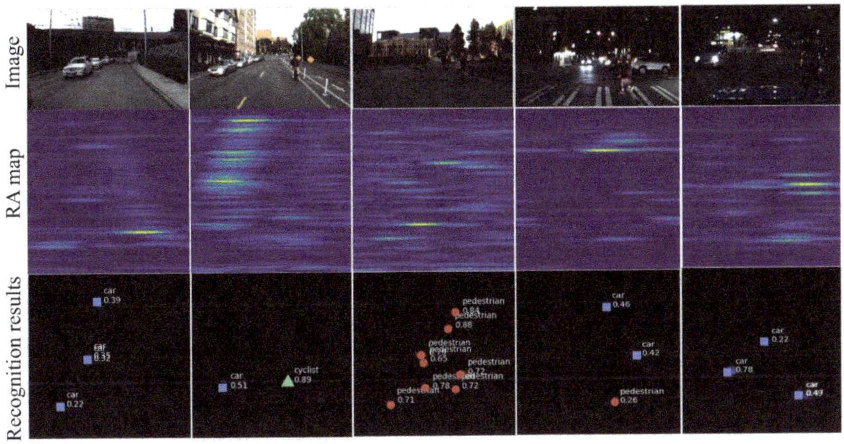

Figure 15.13 Five test examples from the day-time on-road and night-time on-road scenario. The top row shows the synchronized camera image only for visualization, the second row shows the corresponding radar range–angle (RA) image, and the bottom row shows the visualization of the RAMP-CNN [31] model results (© 2020 IEEE).

RAMP-CNN model slices each RVA image into 2D maps from 3 perspectives, that is, range–angle (RA) image, range–velocity (RV) image, and velocity–angle (VA) image. Then the RA, RV, and VA image sequences are processed by three parallel encoder–decoder structures to generate feature bases, which are then fused to support the object recognition decision.

Gao *et al.* train and test the RAMP-CNN model on the UWCR dataset [31], which includes the raw radar data (I–Q samples post-demodulated at the receiver) and synchronized camera images for various scenarios – daytime/nighttime, parking lot, curbside, campus road, city road, freeway, etc. We show several running results (test examples) of RAMP-CNN in Figure 15.13.

15.5 Cognitive spectrum sharing in automotive radar network

An important application of cognitive radar is cognitive interference avoidance. In a high-density automotive radar scenario where multiple radars operate closely and simultaneously in the same band [62], mutual interference can become a serious issue [63]. This issue can be partially addressed by signal processing at the receiver, e.g., receiver-side interference excision [64]. While removing interference signals, these receiver-side interference mitigations may distort useful target signals and may introduce artifacts [64]. In contrast, cognitive radar mitigates mutual radar interference via avoidance instead of post-treatment [65]. Such cognitive interference avoidance can be achieved by spectrum sensing, and then adaptively changing waveform or transmit beampattern [21–23]. These avoidance operations mitigate interference and almost do not introduce distortion on the target signals. Thus, the cognitive interference avoidance represents a promising path for automotive radar to improve the interference situation [65]. In this chapter, we first illustrate the spectrum congestion issues in a vehicular FMCW radar network. Then, we use two cognitive radar examples to show how cognition can reduce the impact of interference. Detailed PAC principles including spectrum perception, learning, and action algorithms will be illustrated and performance evaluation is provided.

15.5.1 Spectrum congestion, interference issue, and MAC schemes

Short-range automotive radars are enabled for operation in the 77−81 GHz band [64,65] due to the significant bandwidth available. With the increasing number of vehicular FMCW radars in this band, mitigating the mutual interference between radars in urban congestion scenarios is a potential future issue. When different FMCW radars have distinct chirp slopes, the dechirped interference at the FMCW radar IF band appears like noise-floor elevation that reduces target detectability. With the same chirp slope, the dechirped interference at FMCW radar IF band leads to ghosts (fake targets), causing false alarms. In Figure 15.14, we show the range DFT result, range–velocity DFT result (RV image) and CFAR detection result at a victim radar receiver when there is another interfering radar with an identical chirp slope and a target. We can observe from Figure 15.14 that the interference leads to high-power ghost and the ghost-introduced side-lobe at the victim radar's receiver. The ghost as well as the ghost-introduced side-lobe leads to false alarms after CFAR detection.

A general approach to limit mutual-radar interference is sharing the spectrum using media access control (MAC) schemes in time, frequency, and spatial domains. The state-of-the-art radar MAC schemes can be classified as centralized MAC, non-adaptive distributed MAC, and cognitive distributed MAC schemes. Centralized MAC schemes in [66–68] require a centralized controller to orthogonally allocate time, frequency, and spatial resources. Non-cognitive distributed MAC schemes for avoiding radar interference include random frequency hopping [13,69] and phase coding [70–72]. Existing centralized MAC schemes and non-cognitive distributed MAC schemes do not require feedback from radar RX to radar TX to optimize transmit waveforms and are non-cognitive. Cognitive distributed MAC schemes adapt timing, center frequency, and bandwidth of the transmit waveforms [7,8,25,39,73] or adapt transmit beam pattern [40] according to the feedback from the RX. This feedback comes from RX-side spectrum sensing, which aims at recognizing time, frequency, and spatial resources used by other radars in real-time [21,26].

In this section, we focus on illustrating the concept of cognitive radar in the context of spectrum sensing and spectrum sharing. The spectrum sensing corresponds to the perception block in PAC, the spectrum-sharing logic corresponds to the learning block in PAC, and the TX-side adaptation corresponds to the action block in PAC (see Figure 15.4). We consider the automotive FMCW radar network where all radars adopt the same chirp slope and transmit period. A distributed MAC scheme – carrier sensing multiple access (CSMA) [73], and its cognitive enhancement are introduced as two examples of cognitive automotive radar.

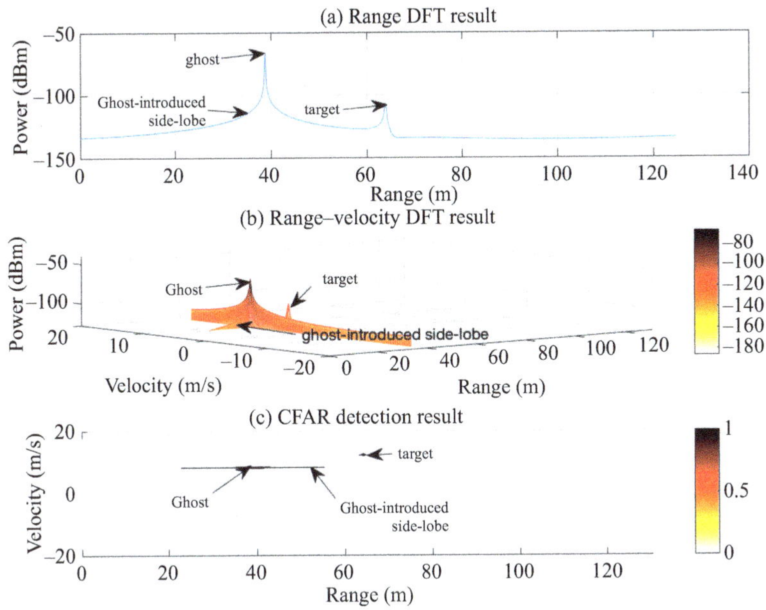

Figure 15.14 Ghost introduced by interference

15.5.2 FMCW-CSMA-based spectrum sharing

Applying CSMA in FMCW radar for spectrum sharing is a good illustration of the cognition concept for vehicular radar. This idea was first explored in [73], leading to FMCW-CSMA radar. The FMCW-CSMA radar shown in Figure 15.15 differs from the original FMCW radar in Figure 15.1 due to additional feedback from RX to TX. Such feedback is based on a carrier sensing block that senses interference from the IF band at the RX and outputs timing control to the chirp synthesizer. The chirp synthesizer then reconfigures the chirp transmit starting instant according to the timing control for avoiding interference.

Carrier sensing, corresponding to perception block in PAC, is key for avoiding interference. The principle of carrier sensing for interference avoidance is shown in Figure 15.16. Suppose there is an interference source (radar I) periodically transmitting chirps (the solid chirps in Figure 15.16) at chirp starting instant $t_I, t_I + T_p, t_I + 2T_p, \ldots$ Another radar (radar 0) starts transmitting at $t_0, t_0 + T_p, t_0 + 2T_p, \ldots$ receives the interference if [74]

$$t_0 \leq t_I + \frac{d_I}{c} \leq t_0 + f_H/h, \tag{15.4}$$

where d_I is the distance from the interference source to this radar. To check whether the condition in (15.4) happens, radar 0 first uses carrier sensing, or listen-before-talk. That is, radar 0 stops transmitting but dechirps the received signal on the whole

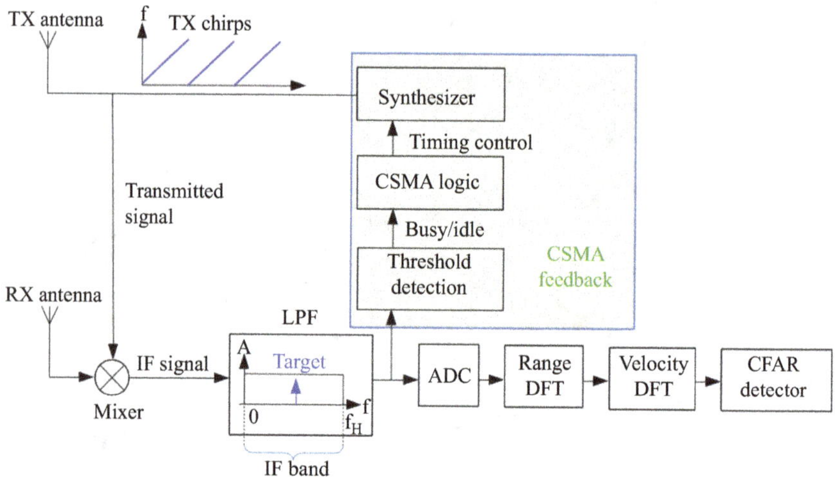

Figure 15.15 Block diagram of FMCW-CSMA radar

transmitting band at the first chirp cycle starting from instant t_0. If there is an interference being dechirped, a ghost will appear in radar 0's IF band (see Figure 15.16) – a clear peak much higher than the noise floor.† A simple threshold detector then reliably declares the band at the chirp start to be busy.

As the interference is periodic, from next chirp cycle, radar 0 backs-off its chirp starting instant to $t_0' + T_p, t_0' + 2T_p, t_0' + 3T_p, \ldots$ to avoid interference, where t_0' is chosen according to a random backoff strategy [73]. After the radar chooses new chirp starting instants $t_0' + T_p, t_0' + 2T_p, t_0' + 3T_p, \ldots$, it senses the transmitting band again at instant $t_0' + T_p$. If in the next chirp cycle, there is no ghost detected in carrier sensing (as indicated in the second chirp cycle in Figure 15.16), the band at that chirp starting instant is sensed to be idle. Then, radar 0 starts transmitting at $t_0' + 2T_p, t_0' + 3T_p, t_0' + 4T_p, \ldots$ from the next chirp cycle. Otherwise, the radar backs off and re-senses the band again.

Compared to radars that use energy detection (ED) [25] for sensing interference, the carrier sensing structure in Figure 15.16 uses the output post de-chirping of the interference source and results in more reliable interference detection. Furthermore, such detection is done in the analog domain without the need for ADC. Thus, it can sense a large bandwidth of the band with low cost and low computational complexity. This is different from the existing FSS [32] technique and CS-based spectrum sharing [19,33] that require ADC. In particular, FSS reduces the sampling rate for

†When the slopes of the interference radars are different from the slope of the victim radar, the interference can raise the noise floor at the victim radar. In this case, we need to have more complicated signal and interference models to quantify the impact of the interference. Here, we consider a simpler scenario where all radars adopt the same slope and assume ghosts lead to false peaks for ease of illustrating the advantage of the cognitive radar.

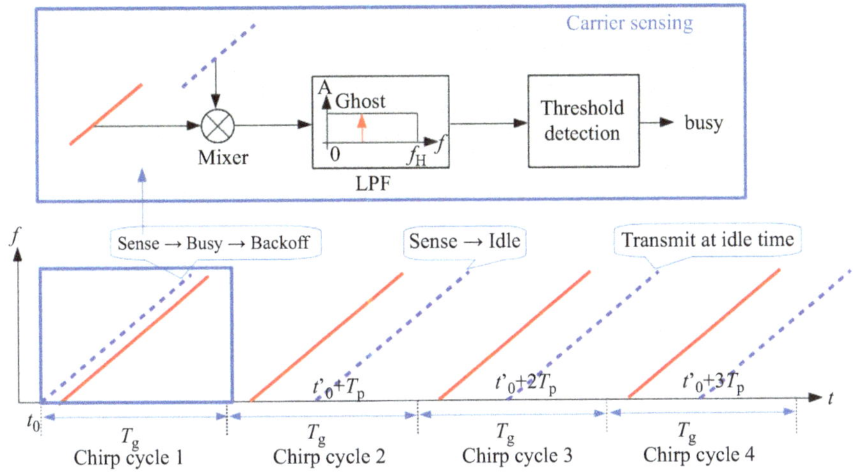

Figure 15.16 Carrier sensing for FMCW radar. The solid chirps are the existing transmitting chirps from an interfering radar, and the dash chirps are the victim radar's chirps.

increasing spectrum sensing speed, but the resulting spectrum sensing resolution is reduced. CS-based spectrum sharing uses compressed sensing algorithms while only requiring sub-sampled ADC, but the computational complexity is high.

We now discuss the CSMA logic that corresponds to the learning block in PAC. Each radar first enters the initialization step in Figure 15.17 where it randomly accesses the channel with probability p (or access the channel following Bernoulli distribution of mean p). If the radar gains access, it transmits in the current chirp cycle. Otherwise, it senses the channel at a random chirp starting instant before its transmission. The goal of using this initial access probability p is to achieve the desired trade-off between the access ratio and collision probability. The access ratio equals to the number of radars accessing the spectrum after CSMA contention divided by the total number of radars. When the number of chirp cycles L_{sense} for CSMA contention is fixed, larger initial access probability p yields a larger access ratio after L_{sense} chirp cycles. The collision probability refers to the number of radars detecting ghosts divided by the total number of radars. The larger the initial access probability p, the larger collision probability. Hence, as p increases, both the access ratio and the collision probability increase. Thus, appropriately choosing p in the initialization step can increase the average number of successfully scheduled radars (radars without observing ghosts) by avoiding a small access ratio or a large collision probability. In FMCW-CSMA radar, our simulation shows that $p = 0.1$ is an appropriate constant value to achieve a large number of successfully scheduled radars.

If a radar does not gain access initially, it enters the CSMA loop shown in Figure 15.17. The first step is carrier sensing at a random chirp starting instant. If the channel is sensed to be busy, then the radar backs off its chirp starting instant and

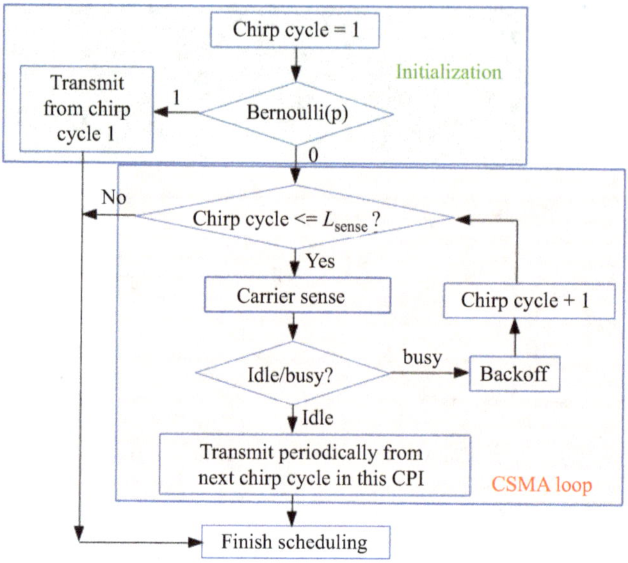

Figure 15.17 CSMA logic for FMCW radar

re-senses again in the next chirp cycle. This is indicated in the chirp cycles 1 and 2 in Figure 15.16. If the channel is sensed to be idle, then the radar starts transmitting from the next chirp cycle. This is indicated in the chirp cycles 2, 3, and 4 in Figure 15.16. The chirp transmission and re-sense are conducted in the next chirp cycle because the carrier sensing requires a chirp's duration. If the radar is not scheduled after L_{sense} chirp cycles, the radar avoids transmitting at the current CPI, and uses the CSMA logic again from the next CPI. We will show the performance of FMCW-CSMA in the next subsection.

FMCW-CSMA radar has a strong relation to PAC introduced in Section 15.3.1. First, FMCW-CSMA radar contains a spectrum perception (carrier sensing) block which allows it to perceive the environment. Second, FMCW-CSMA radar uses the CSMA logic in Figure 15.17 to learn the environment (interferers' transmit start instants) and provides feedback to the TX, corresponding to the learning block in Figure 15.4. The TX reconfigures the transmit start instants and impacts the environment via the action part in Figure 15.4. Thus, FMCW-CSMA radar can be regarded as an exemplar cognitive radar. In Section 15.5.3, we will introduce a new cognitive component to improve FMCW-CSMA radar performance.

15.5.3 FMCW-cognitive-CSMA-based spectrum sharing

In this part, we improve FMCW-CSMA radar by adding additional cognitive feedback to improve spectrum-sharing performance, yielding an FMCW-cognitive-CSMA radar. The block diagram of the FMCW-cognitive-CSMA radar is shown in

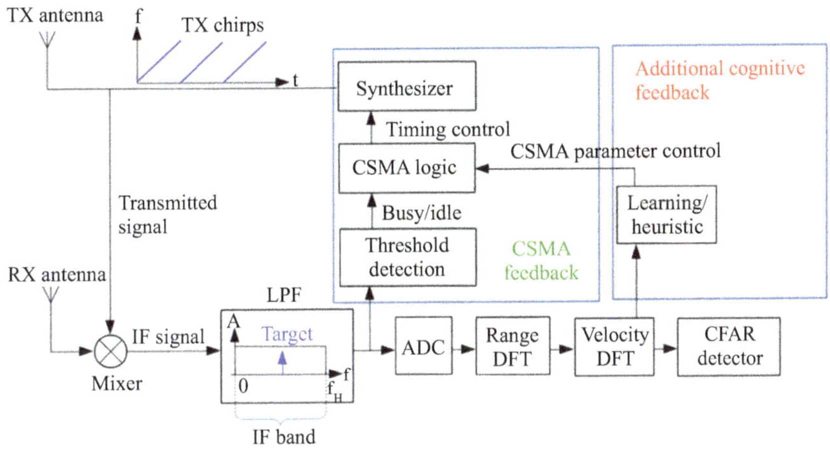

Figure 15.18 Block diagram of FMCW-cognitive-CSMA radar

Figure 15.18. Compare to the FMCW-CSMA radar in Figure 15.15, the FMCW-cognitive-CSMA radar has additional cognitive feedback to control the initial access probability p in CSMA. As mentioned in Section 15.5.2, the initial access probability p balances the trade-off between the access ratio and collision probability. However, the access ratio and collision probability also depend on the number of co-channel radars K in the system. When the number of co-channel radars K is small, the collision probability is small, and we hope to increase the access ratio by increasing the initial access probability p, and vice versa. Suppose the number of co-channel radars K is known, the following heuristic control of p can improve the number of successfully scheduled radars:

$$p = \gamma + \beta \exp(-\alpha K), \tag{15.5}$$

where $\alpha, \beta, \gamma > 0$ and $\gamma + \beta \leq 1$. In (15.5), the initial access probability p decreases with the number of co-channel radars K to avoid collisions. α, β, γ are tuning parameters for achieving a large number of successfully scheduled radars.

The number of co-channel radars K can be inferred from learning block in Figure 15.18. The learning block contains the heuristic control in (15.5) and the estimator on the number of co-channel radars. The estimator uses the range–velocity (RV) image to classify co-channel vehicles and other objects using machine learning algorithms as mentioned in Section 15.4.5. Based on the classification results, the number of co-channel radars K can be estimated.

The FMCW-cognitive-CSMA radar in Figure 15.18 has a stronger sense of cognition compared to FMCW-CSMA radar in Figure 15.15. While the FMCW-CSMA radar only perceives the spectrum using carrier sensing, the FMCW-cognitive-CSMA also perceives the physical environment using the RV image. In the learning part, the FMCW-cognitive-CSMA learns the number of co-channel radars K and uses the

Table 15.2 Simulation setup for different spectrum-sharing schemes

Parameter	Value
IF bandwidth	$f_H = 10$ MHz
Chirp slope	$h = 12$ MHz/μs
Inter-chirp duration	$T_g = 25$ μs
Road length	$1,000$ m
Max number of CSMA cycles	$L_{sense} = 10$

heuristic in (15.5) to realize adaptive control of the CSMA parameter p at the TX. The adaptive control of the CSMA parameter p changes the action (transmit start instants) of the FMCW radar TX. This shows another example of applying the concept of PAC in automotive radar.

We consider the setup in Table 15.2 for simulating the spectrum-sharing performance. We assume K co-channel FMCW radars with identical chirp setup randomly distributed on a road of length $1,000$ m. CSMA adopts a random backoff scheme with backoff time 0.5 us $\times$ randi([1, 20]), where randi([1, 20]) outputs a uniform random integer in [1, 20]. For the FMCW-CSMA radar, the initial access probability $p = 0.1$, while in the FMCW-cognitive-CSMA radar, the initial access probability p in (15.5) with $\alpha = 0.01, \beta = 0.9, \gamma = 0.1$, and the number of co-channel radars K is perfectly estimated. We simulate the performance of three schemes: (1) FMCW-random access – a pure random access scheme – using the structure in Figure 15.1; (2) FMCW-CSMA as in Figure 15.15; (3) the FMCW-cognitive-CSMA in Figure 15.18. Figure 15.19 shows the spectrum-sharing performance of FMCW-random access, FMCW-CSMA, and FMCW-cognitive-CSMA. The number of successfully scheduled radars first increases with the number of co-channel FMCW radars in the network when the spectrum is not congested. However, with the increase of the collision, the number of successfully scheduled radars decreases. Over all regions, we observe that FMCW-cognitive-CSMA achieves the largest number of successfully scheduled radars, FMCW-random access the smallest, and FMCW-CSMA performs in between. This indicates that, with the increasing level of cognition, the spectrum-sharing performance improves.

15.5.4 Comments on spectrum sharing for cognitive radar

The FMCW-cognitive-CSMA radar in Figure 15.18 uses the co-channel radar-aware heuristic in (15.5) to improve the spectrum-sharing performance. Such heuristic, although effective and simple, is not optimal in scheduling. More advanced cognitive schemes in automotive FMCW radar network can adopt RL introduced in Section 15.3.3 for optimizing the parameters in the CSMA logic.

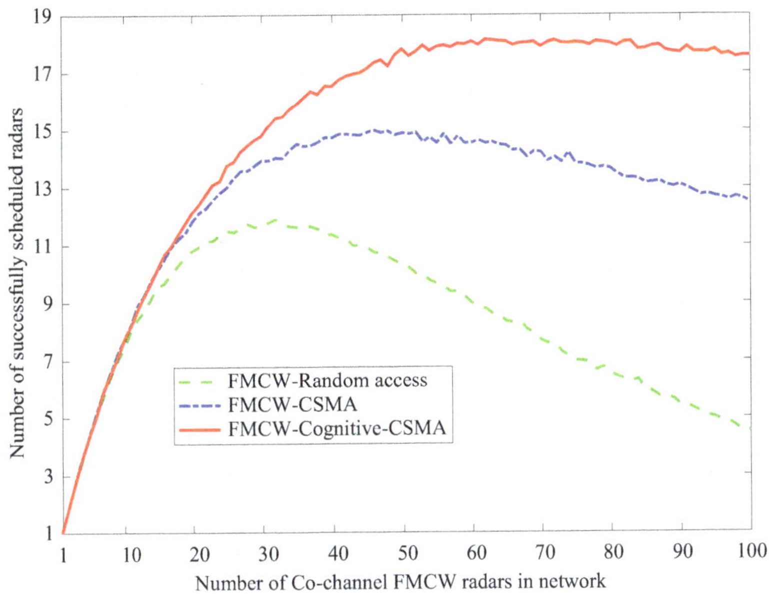

Figure 15.19 Spectrum-sharing performance of FMCW-random access, FMCW-CSMA, and FMCW-cognitive-CSMA

The reaction time of spectrum sensing and spectrum sharing is another important aspect. In Sections 15.5.2 and 15.5.3, when all FMCW radars adopt the same chirp slope and chirp duration, the spectrum environment changes slowly. Thus, the scheduling in the next chirp cycle is based on the information in the current chirp cycle, as indicated in Figure 15.16. In more complex scenarios where different FMCW radars adopt different chirp slopes and chirp durations, the spectrum environment changes within each chirp cycle. In such a case, the chirp waveform needs to be quickly adapted within each chirp cycle to avoid interference [75]. This quick adaptation requires reducing the spectrum sensing time, and the spectrum sensing time reduction can possibly lead to poorer accuracy in spectrum sensing, causing mismanagement in waveform adaptation [75]. In these complex scenarios, implementing FSS and spectrum sharing without degrading performance greatly is an open problem and requires future research.

15.6 Concluding remarks

Automotive radars require efficient and low-cost solutions for improving radar functionality in the congested spectrum and complex physical environments. Cognitive radar is a promising concept to improve automotive radars via the feedback from the radar receiver to the radar transmitter. This can be easily realized in automotive radars as each automotive radar has co-located receiver and transmitter. We reviewed advanced environment perceptions (imaging and object recognition) and discussed the promising application of cognitive radar in automotive radar imaging and object recognition. The concept of cognitive spectrum sharing was introduced using two cognitive radar systems – FMCW-CSMA and FMCW-cognitive-CSMA. We showed that the spectrum-sharing gain increases with the increasing level of cognition in FMCW radar.

There are several other new research directions in cognitive radar for automotive applications. The first direction is bandwidth sharing within a vehicular system. In the near future, automotive radar may share bandwidth with vehicular communication systems for achieving higher spectrum utilization [62]. Existing cognitive co-existence between the radar and communication systems can be found in [21] and references therein. The second direction is applying networking concepts to vehicular cognitive radars. The cognitive radar reviewed in this chapter only considers cognition in a single automotive radar. The so-called Internet of Radars (IoR) can merge the observations of other radars into a radar's cognitive learning process via the support of communication networks [76]. In such an IoR system, the cognitive feedback includes information sharing between different radars, leading to notions of distributed cognition. The radar perception (detection, imaging, and object recognition) performance is expected to improve with feedback from multiple radar sources.

References

[1] Synopsys. The 6 Levels of Vehicle Autonomy Explained; 2021. Available from: https://www.synopsys.com/automotive/autonomous-driving-levels.html.

[2] NHTSA. Automated Vehicles for Safety; 2021. Available from: https://www.nhtsa.gov/technology-innovation/automated-vehicles-safety.

[3] Lambert F. Tesla is adding a new '4D' radar with twice the range for self-driving; 2020. Available from: https://electrek.co/2020/10/22/tesla-4d-radar-twice-range-self-driving/.

[4] Patole SM, Torlak M, Wang D, *et al.* Automotive radars: a review of signal processing techniques. *IEEE Signal Processing Magazine*. 2017;34(2):22–35.

[5] Li Y and Ibanez-Guzman J. Lidar for autonomous driving: the principles, challenges, and trends for automotive lidar and perception systems. *IEEE Signal Processing Magazine*. 2020;37(4):50–61.

[6] Haykin S. Cognitive radar: a way of the future. *IEEE Signal Processing Magazine*. 2006;23(1):30–40.

[7] Hakobyan G, Armanious K, and Yang B. Interference-aware cognitive radar: a remedy to the automotive interference problem. *IEEE Transactions on Aerospace and Electronic Systems*. 2020;56(3):2326–2339.

[8] Thornton CE, Kozy MA, Buehrer RM, *et al.* Deep reinforcement learning control for radar detection and tracking in congested spectral environments. *IEEE Transactions on Cognitive Communications and Networking*. 2020;6(4):1335–1349.

[9] Selvi E, Buehrer RM, Martone A, *et al.* Reinforcement learning for adaptable bandwidth tracking radars. *IEEE Transactions on Aerospace and Electronic Systems*. 2020;56(5):3904–3921.

[10] Sun S, Petropulu AP, and Poor HV. MIMO radar for advanced driver-assistance systems and autonomous driving: advantages and challenges. *IEEE Signal Processing Magazine*. 2020;37(4):98–117.

[11] Bilik I, Longman O, Villeval S, *et al.* The rise of radar for autonomous vehicles: signal processing solutions and future research directions. *IEEE Signal Processing Magazine*. 2019;36(5):20–31.

[12] Rao S. Introduction to mmWave sensing: FMCW radars. In: *Texas Instruments (TI) mmWave Training Series*; 2017.

[13] Roos F, Bechter J, Knill C, *et al.* Radar sensors for autonomous driving: modulation schemes and interference mitigation. *IEEE Microwave Magazine*. 2019;20(9):58–72.

[14] MIMO Radar [TI Application Report]; 2017.

[15] Li J and Stoica P. MIMO radar with colocated antennas. *IEEE Signal Processing Magazine*. 2007;24(5):106–114.

[16] Sun H, Brigui F, and Lesturgie M. Analysis and comparison of MIMO radar waveforms. In: *2014 International Radar Conference*; 2014. p. 1–6.

[17] Wang P, Boufounos P, Mansour H, *et al.* Slow-time MIMO-FMCW automotive radar detection with imperfect waveform separation. In: *ICASSP 2020 – 2020 IEEE International Conference on Acoustics, Speech and Signal Processing (ICASSP)*; 2020. p. 8634–8638.

[18] IEEE Standard for Radar Definitions. IEEE Std 686-2017 (Revision of IEEE Std 686-2008). 2017;p. 1–54.

[19] Mishra KV, Eldar YC, Shoshan E, *et al.* A cognitive sub-Nyquist MIMO radar prototype. *IEEE Transactions on Aerospace and Electronic Systems*. 2020;56(2):937–955.

[20] Haykin S, Xue Y, and Setoodeh P. Cognitive radar: step toward bridging the gap between neuroscience and engineering. *Proceedings of the IEEE*. 2012;100(11):3102–3130.

[21] Greco MS, Gini F, Stinco P, *et al.* Cognitive radars: on the road to reality: progress thus far and possibilities for the future. *IEEE Signal Processing Magazine*. 2018;35(4):112–125.

[22] Baylis C, Fellows M, Cohen L, *et al.* Solving the spectrum crisis: intelligent, reconfigurable microwave transmitter amplifiers for cognitive radar. *IEEE Microwave Magazine*. 2014;15(5):94–107.

[23] Elbir AM, Mishra KV, and Eldar YC. Cognitive radar antenna selection via deep learning. *IET Radar, Sonar and Navigation*. 2019;13(6):871–880.

[24] Huizing A, Heiligers M, Dekker B, *et al.* Deep learning for classification of mini-UAVs using micro-Doppler spectrograms in cognitive radar. *IEEE Aerospace and Electronic Systems Magazine*. 2019;34(11):46–56.

[25] Kirk BH, Narayanan RM, Gallagher KA, *et al.* Avoidance of time-varying radio frequency interference with software-defined cognitive radar. *IEEE Transactions on Aerospace and Electronic Systems*. 2019;55(3):1090–1107.

[26] Gurbuz SZ, Griffiths HD, Charlish A, *et al.* An overview of cognitive radar: past, present, and future. *IEEE Aerospace and Electronic Systems Magazine*. 2019;34(12):6–18.

[27] Martone AF, Sherbondy KD, Kovarskiy JA, *et al.* Metacognition for radar coexistence. In: *2020 IEEE International Radar Conference (RADAR)*; 2020. p. 55–60.

[28] Haykin S. Cognitive dynamic systems: radar, control, and radio [point of view]. *Proceedings of the IEEE*. 2012;100(7):2095–2103.

[29] Giusti E, Saverino AL, Martorella M, *et al.* A rule-based cognitive radar design for target detection and imaging. *IEEE Aerospace and Electronic Systems Magazine*. 2020;35(6):34–44.

[30] Gao X, Roy S, Xing G, *et al.* Perception through 2D-MIMO FMCW automotive radar under adverse weather. In: *2021 IEEE International Conference on Autonomous Systems (ICAS)*; 2021. p. 1–5.

[31] Gao X, Xing G, Roy S, *et al.* RAMP-CNN: a novel neural network for enhanced automotive radar object recognition. *IEEE Sensors Journal*. 2021;21(4): 5119–5132.

[32] Martone AF, Ranney KI, Sherbondy K, *et al.* Spectrum allocation for noncooperative radar coexistence. *IEEE Transactions on Aerospace and Electronic Systems*. 2018;54(1):90–105.

[33] Cohen D, Mishra KV, and Eldar YC. Spectrum sharing radar: coexistence via Xampling. *IEEE Transactions on Aerospace and Electronic Systems*. 2018;54(3):1279–1296.

[34] Greco MS, Gini F, and Stinco P. Cognitive radars: some applications. In: *2016 IEEE Global Conference on Signal and Information Processing (GlobalSIP)*; 2016. p. 1077–1082.

[35] Smith GE, Gurbuz SZ, Brüggenwirth S, *et al.* Neural networks & machine learning in cognitive radar. In: *2020 IEEE Radar Conference (RadarConf20)*; 2020. p. 1–6.

[36] MathWorks. Reinforcement Learning Agents; 2021.

[37] Arulkumaran K, Deisenroth MP, Brundage M, *et al.* Deep reinforcement learning: a brief survey. *IEEE Signal Processing Magazine*. 2017;34(6):26–38.

[38] Haykin S. Cognitive radar: a way of the future. *IEEE Signal Processing Magazine*. 2006;23(1):30–40.

[39] Liu P, Liu Y, Huang T, *et al.* Decentralized automotive radar spectrum allocation to avoid mutual interference using reinforcement learning. *IEEE Transactions on Aerospace and Electronic Systems*. 2021;57(1):190–205.

[40] Bianco S, Napoletano P, Raimondi A, *et al.* AESA adaptive beamforming using deep learning. In: *2020 IEEE Radar Conference (RadarConf20)*; 2020. p. 1–6.

[41] Gurbuz AC, Mdrafi R, and Cetiner BA. Cognitive radar target detection and tracking with multifunctional reconfigurable antennas. *IEEE Aerospace and Electronic Systems Magazine*. 2020;35(6):64–76.

[42] Tan QJO, Romero RA, and Jenn DC. Target recognition with adaptive waveforms in cognitive radar using practical target RCS responses. In: *2018 IEEE Radar Conference (RadarConf18)*; 2018. p. 0606–0611.

[43] Elbir AM, Mulleti S, Cohen R, *et al.* Deep-sparse array cognitive radar. In: *2019 13th International conference on Sampling Theory and Applications (SampTA)*; 2019. p. 1–4.

[44] Oechslin R, Aulenbacher U, Rech K, *et al.* Cognitive radar experiments with CODIR. In: *International Conference on Radar Systems (Radar 2017)*; 2017. p. 1–6.

[45] Gao X, Roy S, and Xing G. MIMO-SAR: a hierarchical high-resolution imaging algorithm for mmWave FMCW radar in autonomous driving. *IEEE Transactions on Vehicular Technology*. 2021;70(8):7322–7334.

[46] Gao X, Xing G, Roy S, *et al.* Experiments with mmWave automotive radar test-bed. In: *2019 53rd Asilomar Conference on Signals, Systems, and Computers*; 2019. p. 1–6.

[47] Schmidt R. Multiple emitter location and signal parameter estimation. *IEEE Transactions on Antennas and Propagation*. 1986;34(3):276–280.

[48] Richards MA, editor. Principles of modern radar: basic principles. In: *Radar, Sonar and amp; Navigation*. Institution of Engineering and Technology; 2010. Available from: https://digital-library.theiet.org/content/books/ra/sbra021e.

[49] Krieger G. MIMO-SAR: opportunities and pitfalls. *IEEE Transactions on Geoscience and Remote Sensing*. 2014;52(5):2628–2645.

[50] Kan T, Xin G, Xiaowei L, *et al.* Implementation of real-time automotive SAR imaging. In: *2020 IEEE 11th Sensor Array and Multichannel Signal Processing Workshop (SAM)*; 2020. p. 1–4.

[51] Gisder T, Meinecke MM, and Biebl E. Synthetic aperture radar towards automotive applications. In: *2019 20th International Radar Symposium (IRS)*; 2019. p. 1–10.

[52] Harrer F, Pfeiffer F, Löffler A, *et al.* Synthetic aperture radar algorithm for a global amplitude map. In: *2017 14th Workshop on Positioning, Navigation and Communications (WPNC)*; 2017. p. 1–6.

[53] Wu H and Zwick T. Automotive SAR for parking lot detection. In: *2009 German Microwave Conference*; 2009. p. 1–8.

[54] Mure-Dubois J, Vincent F, and Bonacci D. Sonar and radar SAR processing for parking lot detection. In: *2011 12th International Radar Symposium (IRS)*; 2011. p. 471–476.

[55] Fembacher F, Khalid FB, Balazs G, *et al.* Real-time synthetic aperture radar for automotive embedded systems. In: *2018 15th European Radar Conference (EuRAD)*; 2018. p. 517–520.

[56] Gisder T, Harrer F, and Biebl E. Application of a stream-based SAR-backprojection approach for automotive environment perception. In: *2018 19th International Radar Symposium (IRS)*; 2018. p. 1–10.

[57] Kong X. Inertial navigation system algorithms for low cost IMU [dissertation]. The University of Sydney; 2000.

[58] Kellner D, Barjenbruch M, Klappstein J, *et al.* Instantaneous ego-motion estimation using Doppler radar. In: *16th International IEEE Conference on Intelligent Transportation Systems (ITSC 2013)*; 2013. p. 869–874.

[59] Pérez R, Schubert F, Rasshofer R, *et al.* Single-frame vulnerable road users classification with a 77 GHz FMCW radar sensor and a convolutional neural network. In: *2018 19th International Radar Symposium (IRS)*; 2018. p. 1–10.

[60] Zhang G, Li H, and Wenger F. Object detection and 3d estimation via an FMCW radar using a fully convolutional network. In: *ICASSP 2020 – 2020 IEEE International Conference on Acoustics, Speech and Signal Processing (ICASSP)*; 2020. p. 4487–4491.

[61] Wang Y, Jiang Z, Gao X, *et al.* RODNet: radar object detection using cross-modal supervision. In: *2021 IEEE Winter Conference on Applications of Computer Vision (WACV)*; 2021. p. 504–513.

[62] Saponara S, Greco MS, and Gini F. Radar-on-chip/in-package in autonomous driving vehicles and intelligent transport systems: opportunities and challenges. *IEEE Signal Processing Magazine*. 2019;36(5):71–84.

[63] Jin S and Roy S. FMCW radar network: multiple access and interference mitigation. *IEEE Journal of Selected Topics in Signal Processing*. 2021;15(4):968–979.

[64] Alland S, Stark W, Ali M, *et al.* Interference in automotive radar systems: characteristics, mitigation techniques, and current and future research. *IEEE Signal Processing Magazine*. 2019;36(5):45–59.

[65] Hakobyan G and Yang B. High-performance automotive radar: a review of signal processing algorithms and modulation schemes. *IEEE Signal Processing Magazine*. 2019;36(5):32–44.

[66] Aydogdu C, Keskin MF, Garcia N, *et al.* RadChat: spectrum sharing for automotive radar interference mitigation. *IEEE Transactions on Intelligent Transportation Systems*. 2021;22(1):416–429.

[67] Khoury J, Ramanathan R, McCloskey D, *et al.* RadarMAC: mitigating radar interference in self-driving cars. In: *2016 13th Annual IEEE International Conference on Sensing, Communication, and Networking (SECON)*; 2016. p. 1–9.

[68] Mazher KU, Heath RW, Gulati K, *et al.* Automotive radar interference characterization and reduction by partial coordination. In: *2020 IEEE Radar Conference (RadarConf20)*; 2020. p. 1–6.

[69] Luo TN, Wu CHE, and Chen YJE. A 77-GHz CMOS automotive radar transceiver with anti-interference function. *IEEE Transactions on Circuits and Systems I: Regular Papers*. 2013;60(12):3247–3255.

[70] Tang B, Huang W, and Li J. Slow-time coding for mutual interference mitigation. In: *2018 IEEE International Conference on Acoustics, Speech and Signal Processing (ICASSP)*; 2018. p. 6508–6512.

[71] TI. Interference Mitigation For AWR/IWR Devices; 2020.

[72] Rao S and Mani AV. Interference characterization in FMCW radars. In: *2020 IEEE Radar Conference (RadarConf20)*; 2020. p. 1–6.

[73] Ishikawa S, Kurosawa M, Umehira M, *et al.* Packet-based FMCW radar using CSMA technique to avoid narrowband interference. In: *2019 International Radar Conference (RADAR)*; 2019. p. 1–5.

[74] Jin S and Roy S. Cross-layer interference modeling and performance analysis in FMCW radar multiple access network. In: *2020 IEEE 92nd Vehicular Technology Conference (VTC2020-Fall)*; 2020. p. 1–6.

[75] Martone AF, Sherbondy KD, Kovarskiy JA, *et al.* Practical aspects of cognitive radar. In: *2020 IEEE Radar Conference (RadarConf20)*; 2020. p. 1–6.

[76] Akan OB and Arik M. Internet of radars: sensing versus sending with joint radar-communications. *IEEE Communications Magazine*. 2020;58(9):13–19.

Chapter 16
A canonical cognitive radar architecture

Joseph R. Guerci[1], *Sandeep Gogineni*[1], *Hoan K. Nguyen*[1], *Jameson S. Bergin*[1] *and Muralidhar Rangaswamy*[2]

Various cognitive radar (CR) definitions or architectures have been described in recent years [1–9]. The emphasis of these approaches varies from "learning-on-the-fly," "intelligent" behavior via knowledge-aided (KA) and/or other artificial intelligence (AI) approaches, adaptive spatio-temporal waveforms, and full adaptivity (transmit and receive). The impetuses for CR also vary from advanced military applications in contested environments, civilian applications in highly congested electromagnetic spectrum operations, to advanced autonomous vehicle applications [10]. In this chapter, we provide a generalized canonical CR architecture that can accommodate all the known CR elements currently described, and how it can be implemented using existing and emerging embedded computing architectures.

We begin with a canonical cognitive fully adaptive (CoFAR) high-level architecture description, then drill down on each of the key constituent elements in terms of both potential functionality and implementation using existing or soon to be available technologies (e.g., neuromorphic computing).

16.1 A canonical CR architecture

Figure 16.1 depicts the canonical cognitive fully adaptive radar (CoFAR) architecture that is the focus of this chapter. The major salient and enabling functionalities include:

- **Intelligent resource allocation and radar scheduler**. A CR, like any radar, has finite resources—timeline, transmit energy, compute resources, etc. To fully realize the performance gains of a CR radar, it must optimize the utilization of its limited resources—and then "layout" the spatio-temporal timeline on both transmit and receive.
- **Mission computer and interface**. All advanced radar systems, particularly those for military applications, are a part of a much larger system-of-systems. It is, therefore, necessary to have a mission computer and/or interface to higher level

[1]Information Systems Laboratories, Inc., San Diego, CA, USA
[2]Air Force Research Laboratory, Wright Patterson Air Force Base, Dayton, OH, USA

functionality. This is a means whereby high-level mission goals and rules of engagement (e.g.) can be translated into operating requirements for the CoFAR system. In some systems, this mission-level computer/interface can be combined with the above radar resource allocation and scheduler function. Later in this chapter, we will adopt this approach. Lastly, for older systems, a human operator may perform some of these functions.

- **A real-time channel estimation (RTCE) function**. It is true to say: "as goes channel knowledge, so goes performance." In general, the radar channel consists of open-air propagation properties, ground clutter (land/sea), targets (including background targets), electromagnetic interference (EMI), and/or electronic warfare (EW) effects. Additionally, it includes all internal transmit–receive channel effects such as channel mismatch, and antenna array manifold (e.g., mutual coupling) effects [11]. The only way to approach transmit–receive optimality in practice is to have an accurate channel model (see below and [2]). As will be discussed later, exploiting CR's flexibility and power to enhance channel knowledge is key.
- **Full transmit and receive adaptivity**. In addition to full receive spatio-temporal (and possibly polarimetric) adaptivity [11], full adaptivity on transmit is assumed. This is no mean feat. It requires very high-speed embedded computing supporting a fully digital arbitrary waveform generator (DAWG) and transmit amplifier and antenna capabilities [2]. This can be a major discriminating factor for CR compared to conventional adaptive radar. In addition to optimizing full transmit–receive operation for traditional radar metrics (detection, track accuracy, etc.), these degrees-of-freedom (DoFs) can also be exploited to enhance channel knowledge, particularly signal-dependent sources (targets, clutter, etc.).
- **Distributed high-performance embedded computing (HPEC)**. It should be no surprise that CR can require substantial real-time computer resources and that those resources are distributed throughout the CR architecture. The exact nature of the computing architectures (e.g., FPGA, GPU, neuromorphic, etc.) are highly dependent on the specific functions and methods. Note that it is assumed that the architecture includes front-end HPEC (dubbed CoFAR co-processor in Figure 16.1) since some adaptivity requires extremely low latency (e.g., intrapulse response).

The above operate in a real-time closed-loop fashion to achieve maximum theoretical performance, i.e., the optimum transmit–receive configuration to maximize (e.g.) probability of detection (for a prescribed false alarm rate), and maximum information extraction overall (e.g., tracking performance, target ID, etc.). What precisely is meant by "optimal," and how a CR can approach this level of performance is discussed in the remaining sections—along with practical implementation considerations. Before leaving this section, we present an abstracted version of the above diagram that magnifies the more salient major elements of the CR "OODA loop" (observe–orient–decide–act), see Figure 16.2. Given requisite adaptive DoFs (ADoFs), arguably the most important process is the scheduler and resource allocation. Note also that the "learning" function in the sense–learn–adapt (SLA) decision

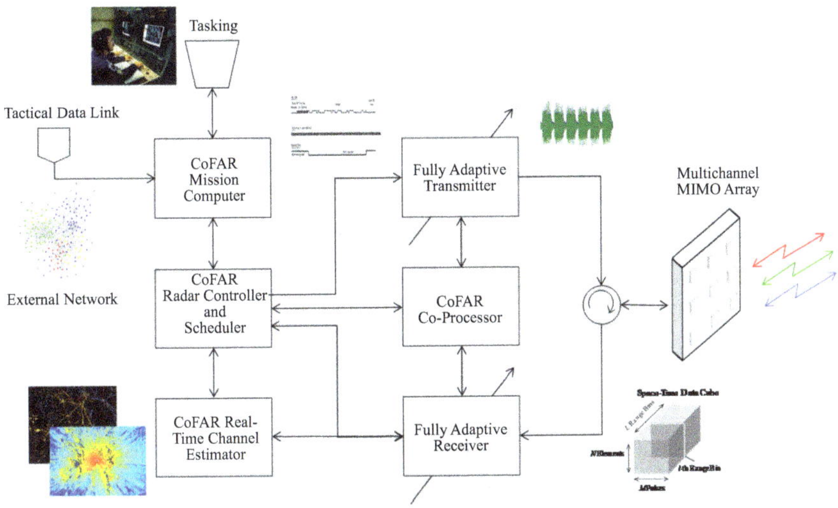

Figure 16.1 A canonical cognitive fully adaptive radar (CoFAR) architecture [12]

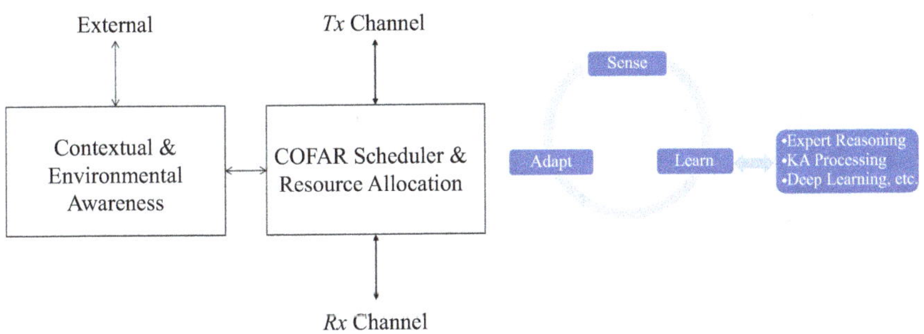

Figure 16.2 Simplified CoFAR OODA and SLA cycles emphasizing the essential role of the scheduler and resource allocation functions

cycle is supervised in the sense that prior contextual and knowledge-aided (KA) methods are assumed. This is modeled after a human "subject matter expert" (SME) who brings vast experience and "decisioning" tools to bear on new observations, as well as an ability to solve problems "on-the-fly."

16.2 Full transmit–receive adaptivity

A key attribute of the CoFAR architecture is full adaptivity on both transmit and receive. This is in contrast to "traditional" adaptive radar wherein adaptivity was mostly confined to the receive channel, for example, space–time adaptive processing

(STAP), and KA STAP [11]. Moreover, in a CoFAR architecture, the level of sophistication employed in the adaptive process is far greater than traditional methods such as "sample covariance" approaches [11]. This additional sophistication can entail KA methods, active probing of the channel, and other machine learning/AI methods.

Fundamentally, both the transmit and receive adaptivity functions are based on a real-time channel estimator (RTCE) that is continually updated via both onboard (organic) sensing and computing, as well as from offboard (exogenous) data via external networking. How the RTCE informs transmit–receive adaptivity is described later. Implementation issues associated with the RTCE will be discussed later in this chapter. It is important to re-emphasize: "*As goes channel knowledge, so goes performance!*"

16.2.1 Full transmit adaptivity

Full transmit adaptivity entails spatio-temporal optimization of all transmit degrees-of-freedom (DoFs) such as fast-time (spectral), slow-time (e.g., Doppler), scan-time, 3D spatial (antenna), and possibly polarimetric [2]. We will derive the optimum transmit function first assuming knowledge of the requisite channel (targets, clutter, interference, etc.). How this channel model can be estimated in practice is discussed later in the chapter.

The signal-dependent components of the channel are targets and clutter (and possibly responsive jamming, which will not be addressed here). We will further focus our efforts on pulsed Doppler radars, though the results obtained are readily extensible to continuous wave (CW) radars. Lastly, we will assume that we can ignore intra-pulse (within a single pulse) time dilation effects due to Doppler. This again is not very restrictive for most practical radars. Again, the results we obtain can be extended via subbanding (e.g.), and/or other approaches to the more extreme operating conditions (e.g., wideband).

Formally, the far-field vector space Green's function representations for a target and clutter channel are given by [13]

$$\boldsymbol{y}_T(t) = \int \boldsymbol{H}'_T(\tau)\,\boldsymbol{s}(t-\tau)\,d\tau \tag{16.1}$$

$$\mathbf{y}_C(t) = \int \mathbf{H}'_C(\tau)\,\boldsymbol{s}(t-\tau)\,d\tau \tag{16.2}$$

where without loss of generality, we have assumed that the number of transmit and receive DoFs are equal to N, such that $\mathbf{y}_T(t), \mathbf{y}_C(t), \boldsymbol{s}(t) \in \mathbb{C}^N$, and thus $\mathbf{H}_T(t), \mathbf{H}_C(t) \in \mathbb{C}^{N\times N}$. $\boldsymbol{s}(t)$ is the total transmit space–time waveform, $\mathbf{H}_T(t), \mathbf{H}_C(t)$ denote the target and clutter Green's functions respectively, and $\mathbf{y}_T(t), \mathbf{y}_C(t)$ denote the target and clutter received signals, respectively.

Since all signals and systems in practice are finite bandwidth, we will further assume a fully discrete model of the form

$$\mathbf{y}_T = \mathbf{H}_T\boldsymbol{s}, \tag{16.3}$$

$$\mathbf{y}_C = \mathbf{H}_C\boldsymbol{s}. \tag{16.4}$$

For example, if $\mathbf{s}_1, \mathbf{s}_2, \ldots, \mathbf{s}_M \in \mathbb{C}^N$ denote the transmit spatial steering vectors for M-pulses [11], then the full transmit vector $\boldsymbol{s} \in \mathbb{C}^{NM}$ is given by

$$\boldsymbol{s} = \begin{bmatrix} \mathbf{s}_1 \\ \vdots \\ \mathbf{s}_M \end{bmatrix} \tag{16.5}$$

and thus $\mathbf{H}_T, \mathbf{H}_C \in \mathbb{C}^{NM \times NM}$. Although theoretically the target and clutter Green's functions are deterministic, it will be of practical benefit to assume they are at least partially stochastic, due to the unavoidable presence of modeling errors, etc., [2]. A tilde notation will be employed to denote a random variable, i.e., $\widetilde{\mathbf{H}}_T, \widetilde{\mathbf{H}}_C \in \mathbb{C}^{NM \times NM}$.

Even at this stage, despite not having detailed the operation of a CR, it is possible to determine the absolute performance bounds on any radar, cognitive or not, with respect to basic operations such as detection and ID. This is possible because we have utilized a channel Green's function approach for targets and/or clutter. Important note: At the time of this writing, there remains a lot of confusion about just what benefits a CR can afford. What is made clear here, is that the best performance any radar can achieve is computable from the physics of the problem.

Consider Figure 16.3, which depicts the classic input–output configuration for a known target in noise. Our goal will be to jointly optimize both the receiver function $\boldsymbol{w} \in \mathbb{C}^{NM}$ and the transmit waveform $\boldsymbol{s} \in \mathbb{C}^{NM}$ to maximize, for example, output signal-to-noise ratio (SNR) for a finite energy, finite duration transmit signal [2].

As detailed in [2], the joint optimization for maximizing the output SINR from the receiver is obtained by "working backwards" from the receiver. For the known

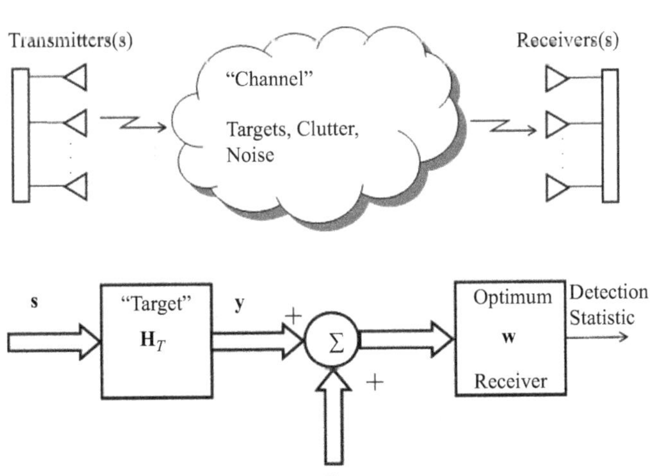

Figure 16.3 Basic target-in-noise problem for joint transmit–receive optimization [2]

signal in noise (generally colored noise), the optimum receiver is a whitening matched filter [14], i.e.,

$$\boldsymbol{w} = \boldsymbol{R}^{-1}\boldsymbol{y}, \tag{16.6}$$

where $\boldsymbol{w} \in \mathbb{C}^{NM}$ is the optimum receiver (weight vector), $\boldsymbol{R} \in \mathbb{C}^{NM \times NM}$ is the total interference plus receiver noise covariance matrix [11], $\boldsymbol{y} \in \mathbb{C}^{NM}$ is the target echo (note in general $\boldsymbol{y} \neq \boldsymbol{s}$, since $\boldsymbol{y} = \boldsymbol{H}_T\boldsymbol{s}$), $H_T \in \mathbb{C}^{NM \times NM}$ is the target Green's function (generally stochastic), and $\boldsymbol{s} \in \mathbb{C}^{NM}$ is the transmit space–time waveform. As further shown in [2], the optimum transmit waveform is that eigenfunction associated with maximum eigenvalue for the following eigensystem:

$$\left(\overline{H_T' H_T}\right) \boldsymbol{s} = \lambda \boldsymbol{s}. \tag{16.7}$$

The overbar notation denotes the expectation operator, and λ is a scalar eigenvalue.

The above has been extended to the case of signal-dependent clutter interference, in which case the optimum transmit waveform is given by

$$\left(\overline{\boldsymbol{H}_C' \boldsymbol{H}_C} + \sigma^2 \boldsymbol{I}\right)^{-1} \left(\overline{\boldsymbol{H}_T' \boldsymbol{H}_T}\right) \boldsymbol{s} = \lambda \boldsymbol{s}, \tag{16.8}$$

where $\boldsymbol{H}_C \in \mathbb{C}^{NM \times NM}$ denotes the clutter channel Green's function (generally stochastic), and $\sigma^2 \boldsymbol{I} \in \mathbb{C}^{NM \times NM}$ is the receiver (white) noise component.

The above hold for all finite energy signals. Additional waveform constraints (such as constant modulus) generally lead to nonlinear (and generally non-convex) optimization problems. Again the reader is referred to [2] for further details.

16.2.2 Full receive adaptivity

Arguably, "full" receive adaptivity has existed in practice for some time (see e.g., [2]). Additionally, the early "ancestor" of CR is KA radar that greatly increased the sophistication of the adaptive filtering process [7]. What CR brings to the table is ever more sophisticated channel knowledge that, in turn, can be used by the receiver to improve performance. Traditionally, the major functional adaptivity in the receiver is STAP, that is, multidimensional space–time adaptive filtering [2]. The essence of STAP can be distilled into the familiar Wiener–Hopf equation:

$$\boldsymbol{w} = \boldsymbol{R}^{-1}\boldsymbol{s}, \tag{16.9}$$

where $\boldsymbol{w} \in \mathbb{C}^{NM}$ are the receiver filter weights that maximize output SINR, $\boldsymbol{s} \in \mathbb{C}^{NM}$ is the desired target steering vector, and $\boldsymbol{R} \in \mathbb{C}^{NM \times NM}$ is the total interference covariance matrix (guaranteed to be positive definite due to the presence of receiver thermal noise). While knowledge of both $\boldsymbol{s}$ and $\boldsymbol{R}$ are required, it is $\boldsymbol{R}$ that is the major driver of real-world performance (knowledge of $\boldsymbol{s}$ is primarily a calibration issue). Until KA STAP was developed in the late 1990s to 2000s, the only techniques available were variants on the maximum-likelihood (ML) sample covariance approach [15]. However, as the number of adaptive DoFs increased, so too did the requirements for i.i.d. training data (target free)—no mean feat in practice. While reduced-rank STAP methods helped alleviate this problem for some applications, it was not a universal solution.

KA-STAP can be viewed as a means for augmenting the sample data training methods with exogenous information derived from a plurality of possible sources [7]. The Bayesian framework [14] provides a compact (and rigorous) framework for incorporating priors into the estimation framework [7]. For example, for a variety of radar operating frequencies, it is possible to estimate the requisite clutter covariance matrix by performing real-time ray tracing using digital terrain databases or other reflectivity data [7]. Wishart [16] established that the elements of a sample covariance matrix $\left[L\hat{\mathbf{R}}\right]_{i,j}$ formed from an outer product sum of L Gaussian i.i.d. samples, i.e.,

$$\hat{\mathbf{R}} = \frac{1}{L}\sum_{i=1}^{L}\mathbf{x}_i\mathbf{x}_i' \tag{16.10}$$

obey a Wishart distribution (actually a complex Wishart for the case at hand [17]) of degree L, i.e., $\hat{\mathbf{R}}\ W\left(L\hat{\mathbf{R}}, L\right)$. (1.10) also corresponds to the ML estimate of the underlying covariance matrix $\boldsymbol{R}$ [17,18].

If a prior estimate of the covariance matrix exists, say $\hat{\mathbf{R}}_0$, it is not unreasonable to assume it too is similarly Wishart distributed—particularly if it was formed over the same geographical region. Note: The author has found through empirical analysis of real-world clutter that "de-trending" of the sample data may need to be performed to preserve Gaussianity [7]. The simplest example is the detection and removal of large clutter discretes. The corresponding Bayesian (maximum a posteriori) estimate that combines $\hat{\mathbf{R}}$ and $\hat{\mathbf{R}}_0$ is easily derived: let L_1 and L_0 denote the degrees of $\hat{\mathbf{R}}$ and $\hat{\mathbf{R}}_0$, respectively, which are further assumed to be i.i.d. and complex Wishart [17]. Then $\hat{\mathbf{R}}$ and $\hat{\mathbf{R}}_0$ are collectively sufficient statistics for $L_0 + L_1$ i.i.d. samples $\{\mathbf{x}_i : i = 1, \ldots L_0 + L_1\}$. Thus, the maximum a posteriori solution of $\hat{\mathbf{R}}$ given prior $\hat{\mathbf{R}}_0$ is equivalent to the ML solution based on all of the data $\{\mathbf{x}_i : i = 1, \ldots L_0 + L_1\}$, [18], i.e.,

$$\hat{\mathbf{R}} = \max_{\boldsymbol{R}} f\left(\boldsymbol{x}_i : i = 1, \ldots, L_0 | \boldsymbol{R}\right) f_0\left(\boldsymbol{R}\right) \tag{16.11}$$

$$= \max_{\boldsymbol{R}} f\left(\boldsymbol{x}_i : i = 1, \ldots, L_0 + L_1 | \boldsymbol{R}\right) \tag{16.12}$$

$$= \frac{1}{L_0 + L_1}\left(L_0\hat{\mathbf{R}}_0 + L_1\hat{\mathbf{R}}_1\right), \tag{16.13}$$

where $f_0\left(\mathbf{R}\right)$ denotes the prior pdf associated with the prior covariance estimate $\hat{\mathbf{R}}_0$ based on L_0 samples—and thus is $W\left(L_0\hat{\mathbf{R}}_0, L_0\right)$; and $\hat{\mathbf{R}}_1$ denotes the ML estimate based on L_1 samples. Equation (15.11) has an obvious intuitive appeal: the a posteriori covariance estimate is formed as a weighted sum of the prior and current estimates with weighting factors proportional to the amount of data used in the formation of the respective sample covariances.

An obvious yet useful generalization of (15.11) is

$$\hat{\mathbf{R}} = \alpha\hat{\mathbf{R}}_0 + \beta\hat{\mathbf{R}}_1, \tag{16.14}$$

$$\alpha + \beta = 1, \tag{16.15}$$

which is the familiar "colored loading" or "blending" approach of Hiemstra [19] and Bergin *et al.* [20], respectively. The practical advantages of (15.14) versus (15.11) are many. For example, the data used to form the prior covariance might lose its relevance with time—the so-called stale weights problem [21]. In that case, even though $\hat{\mathbf{R}}_0$ might have been formed from L_0 samples, it "effectively" has less information and should be commensurately "de-weighted." A common method for accomplishing this, borrowed from adaptive Kalman filtering, is the fading memory approach in which case α in (15.14) is replaced with

$$\alpha = e^{-\gamma t} L_0, \tag{16.16}$$

where t is the time elapsed since the prior covariance estimate was formed, and the positive scalar γ is the fading memory constant [22]. β is chosen to satisfy the weighted sum constraint.

In a more general setting, the blending parameters (α, β) could be chosen based on the relative "confidence" in the estimates. For example, $\hat{\mathbf{R}}_0$ could be derived from a physical scattering model of the terrain. In which case it is also typically of the form (15.10) with the distinction that the outer products represent clutter patch steering vectors weighted by the estimated reflectivity [23,24], i.e.,

$$\hat{\mathbf{R}}_0 = \frac{1}{N_c} \sum_{i=1}^{N_c} G_i v_i v_i', \tag{16.17}$$

where N_c clutter "patches" have been utilized in the formation of $\hat{\mathbf{R}}_0$ (typically corresponding to a particular iso-range ring [23,24]) where $v_i \in C^{NM}$ is the space–time (angel-Doppler) steering vector corresponding to the ith clutter patch and G_i its corresponding power [23]. Such information could be available *a priori* from SAR imagery [7,25] (essentially a high-resolution clutter reflectivity map) or physics-based models [26].

Though the "confidence" metric to apply, in the form of the weighting parameter α, is difficult to ascribe in practice since the quality of the *a priori* estimate is vulnerable to a number of error sources, a straightforward remedy is to choose α adaptively so as to maximally "whiten" the observed interference data [27]. For example:

$$\min_\alpha Z_L(\alpha), \tag{16.18}$$

where

$$Z_L(\alpha) = \| \sum_i y_i y_i' - \boldsymbol{I} \|, \tag{16.19}$$

and

$$\mathbf{y}_i = \left(\alpha \hat{\mathbf{R}}_0 + \beta \hat{\mathbf{R}}_1\right)^{-\frac{1}{2}} \mathbf{x}_i \tag{16.20}$$

$\mathbf{x}_i$ is the received space–time snapshot measurement vector for the ith range bin, $\left(\alpha \hat{\mathbf{R}}_0 + \beta \hat{\mathbf{R}}_1\right)^{-\frac{1}{2}}$ is the whitening matrix corresponding to a particular α, $\mathbf{y}_i$ is the vector residue with $\dim(y_i) = \dim(x_i)$, and the summation is performed over a suitable subset of the radar observations for which $\hat{\mathbf{R}}_0$ is believed valid.

The above adaptive α approach is but a special case of an entire class of direct filtering methods incorporating prior information, viz., data pre-whitening (or simply data de-trending). In a more general setting, the space–time vector residues, $\mathbf{y}_i$, can be viewed as a "de-trended" vector time series using prior knowledge in the form of a covariance-based whitening filter. The major potential advantage of this is to remove (or attenuate) the major quasi-deterministic trends in the data (e.g., clutter discretes, mountains, etc.) so that the resulting residue vector time series is less non-stationary or inhomogeneous (see [2] for further details).

CR, with its ability to achieve sophisticated channel knowledge (see below), has opened new possibilities for enhanced performance. For example, if the clutter channel Green's function is known, the data "pre-whitening" step in the colored noise matched filter can be achieved via coherent subtraction, i.e.,

$$z = y - H_C s_T, \tag{16.21}$$

where $y \in \mathbb{C}^{NM}$ is the total received space–time signal, $H_C \in \mathbb{C}^{NM \times NM}$ is the clutter Green's function, and $\mathbf{s}_T \in \mathbb{C}^{NM}$ is the transmit steering vector. Comparison of (15.21) with (15.9) shows the fundamental difference in the approaches. Interestingly, they are not mutually exclusive. For example, a Green's function pre-whitening approach could be followed by a sample covariance whitening approach to address the inevitable imperfections in estimating H_C. Lastly, the clutter covariance can also be deduced from the clutter Green's function as follows:

$$R = \left(\overline{(H_C \mathbf{s}_T)\,(H_C \mathbf{s}_T)'} \right). \tag{16.22}$$

The purpose of this brief subsection was to make clear that an optimal radar entails both joint transmit-receive optimization. If CR is to be of any use, it must be compared to the optimum—not simply to other suboptimal, albeit more traditional adaptive radar approaches. Hence, the real essence of CR must be to obtain the best channel knowledge possible given all possible tools that can be implemented in a modern radar architecture including KA processing, proactive transmit probing, and/or other advanced AI methods (see below). "*As goes channel knowledge, so goes performance.*"

16.3 CR real-time channel estimation (RTCE)

Though not often explicitly stated, all adaptive radar schemes require channel estimation whether explicit or implicit. In CR, channel estimation is a corner stone of operation and performance. Since targets, clutter (all types), and some responsive jamming are all signal (transmit) dependent sources, and moreover these sources are the most important in practice, significant attention must be paid to the design of RTCE of signal-dependent sources.

In traditional adaptive radar, the channel estimation was done indirectly and "passively" by using variants of the aforementioned sample covariance (ML approach). An evolution towards CR was KA radar that brought a number of AI approaches (e.g., expert systems, etc.) to aid the covariance estimation process (see above and [7]).

With the introduction of transmit diversity/adaptivity, an entirely new era in channel estimation has emerged. The first book on this subject which also introduced an engineering CR architecture was introduced in 2010 [2,5].

In fully adaptive CR, the transmit function can play an integral part of the channel estimation process for signal (transmit) dependent targets, clutter, and interference. Additive receiver noise and transmit independent EMI are not dependent on the transmit function in general—though they will depend on operating frequency.

A particularly powerful and general technique is multi-input, multi-output (MIMO) probing [2]. For example, if there are strong clutter discretes present in a field of view they can cause false alarms and desensitization due to covariance contamination [28]. It would thus be desirable to estimate the spatial Green's functions associated with them $\left\{\hat{H}_{D_i}\right\}$, $i = 1, \ldots, N_D$, and apply a multidimensional pre-filtering stage

$$z = \boldsymbol{y} - \sum_{i=1}^{N_D} \hat{H}_{D_i}\mathbf{s}, \tag{16.23}$$

where $\boldsymbol{y}$ is the measured (received) space–time snapshot vector, $\boldsymbol{s}$ is the transmit steering vector, and z is the pre-whitened residual. The term pre-whitened is used since this step only removes the clutter discretes, and not the clutter in general.

In [28], simultaneous orthogonal waveforms were transmitted from each transmit subarray to rapidly estimate the clutter discrete Green's functions. For some applications, this "interruption" in the normal transmit timeline may be acceptable in exchange for enhanced clutter discrete suppression. However, it has recently been shown that it is possible to extract the Green's functions using non-orthogonal, but linearly independent waveforms [2,29].

Consider a conventional (non-orthogonal) spatial transmit steering vector $\boldsymbol{s}_i$. The corresponding Green's function response for a signal-dependent channel source is given by

$$\boldsymbol{y}_i = \boldsymbol{H}\boldsymbol{s}_i. \tag{16.24}$$

Clearly this one input–output response is insufficient to solve for $\boldsymbol{H}$. Assuming $\boldsymbol{H} \in \mathbb{C}^{N\times N}$, we thus examine the case where N generally linearly independent transmit steering vectors are utilized. This is represented by

$$\boldsymbol{y} = \boldsymbol{S}\boldsymbol{h}, \tag{16.25}$$

$$\begin{bmatrix} \boldsymbol{y}_1 \\ \vdots \\ \boldsymbol{y}_N \end{bmatrix} = \begin{bmatrix} \boldsymbol{S}_1 \\ \vdots \\ \boldsymbol{S}_N \end{bmatrix} \begin{bmatrix} \boldsymbol{h}_1 \\ \vdots \\ \boldsymbol{h}_N \end{bmatrix}, \tag{16.26}$$

where $\boldsymbol{h}_i$ is the column vector ($N \times 1$) of the ith row of $\boldsymbol{H}$, and the ($N \times N^2$) matrix $\boldsymbol{S}_i$ is given by

$$\boldsymbol{S}_i = \begin{bmatrix} \boldsymbol{s}_i^T & \boldsymbol{0} & \cdots & \boldsymbol{0} \\ \boldsymbol{0} & \boldsymbol{s}_i^T & & \boldsymbol{0} \\ \vdots & & \ddots & \vdots \\ \boldsymbol{0} & \boldsymbol{0} & \cdots & \boldsymbol{s}_i^T \end{bmatrix} \tag{16.27}$$

where $\boldsymbol{s}_i^T$ ($1 \times N$) denotes the transpose (without conjugation) of the ith steering vector. Note that $\boldsymbol{h}$, ($N^2 \times 1$) which is the vector concatenation of the rows of $\boldsymbol{H}$, represents the N^2 unknown elements of $\boldsymbol{H}$. One can readily see that a necessary (not sufficient) condition is that at least N different transmit steering vectors must be sent. This ensures that $\boldsymbol{S}$ in (15.25) is at least ($N^2 \times N^2$), and thus potentially invertible. Sufficiency is achieved if every row of $\boldsymbol{S}$ is linearly independent (not necessarily orthogonal). This leads to the following final set of necessary and sufficient conditions:

$$\text{rank}(\boldsymbol{S}_i) = N, \forall i = 1, \ldots, N, \tag{16.28}$$

$$\|\boldsymbol{s}_i \boldsymbol{s}_j\|^2 < \|\boldsymbol{s}_i\| \|\boldsymbol{s}_j\|, \forall i,j : i \neq j, \tag{16.29}$$

$$\|\boldsymbol{s}_i\| \neq 0, \forall i = 1, \ldots, N, \tag{16.30}$$

The non-zero norm requirement on $\boldsymbol{s}_i$ ensures that the rank of $\boldsymbol{S}_i$ is N, while Schwarz's inequality ensures that the steering vectors are not co-linear (and thus linearly independent). Note that these conditions were developed without invoking any symmetry or reciprocity properties of H. In the absence of transmit–receive manifold and RF pathway errors, $\boldsymbol{H}$ can indeed be symmetric due to the electromagnetic reciprocity theorem [30]. In this case, the number of unknowns is almost cut in half.

The conditions of (15.28)–(15.30), while necessary and sufficient, to ensure mathematical solubility, do not account for ever-present additive noise and interference. Thus, an extremely important area for future investigation is the selection of an optimal set of linearly independent steering vectors as a function of H and the channel interference. For example, choosing the transmit steering vectors according to the maximum a posteriori criterion [14]

$$max_{\{s_1,\ldots,s_N\}} f\left(\boldsymbol{H} | \boldsymbol{y}_1, \ldots, \boldsymbol{y}_N\right), \tag{16.31}$$

where $f(\cdot)$ denotes the posterior conditional probability density function of $\boldsymbol{H}$ conditioned on the observed measurements. When more than N transmit steering vectors are used, (15.25) becomes an overdetermined set of linear equations. This provides additional opportunities for filtering in the presence of noise.

For the additive Gaussian white nose case (AGWN), the least squares pseudo-inverse is optimum in an ML sense, i.e.,

$$\hat{\boldsymbol{h}} = \left(\boldsymbol{S}'\boldsymbol{S}\right)^{-1} \boldsymbol{S}'\boldsymbol{y}. \tag{16.32}$$

Using recursive least squares (RLS) methods, this estimate can be updated recursively and allows for fading memory to address the dynamically varying channel case [31].

An interesting special case that satisfies the observability conditions is when $\boldsymbol{s}_i = \alpha_i \boldsymbol{e}_i$, where $\boldsymbol{e}_i$ is the Euclidean basis vector (α_i an arbitrary non-zero constant)

$$\boldsymbol{e}_i = \left[0 \ \cdots \ 0 \ 1 \ 0 \ \cdots \ 0\right]^T, \tag{16.33}$$

where 1 is the *i*th entry. This corresponds to the time division multiple access (TDMA) case in which only one transmit DoF is utilized at any given time. It is also a useful technique for calibrating the transmit array manifold for unknown errors [5]. Once the transmit array manifold is calibrated, other channel probing techniques could be employed that have less system regret than TDMA that can have a deleterious impact on radar timeline.

In practice, physical insights into the problem at hand can significantly guide the selection of transmit steering vectors. For example, if we wish to characterize the spatial clutter channel that consists of a finite set of distributed large clutter discretes, "it makes sense" to choose a linear set of spatial transmit steering vectors that scan the relevant field-of-regard. This ensures that each clutter discrete has a good chance of receiving a reasonable amount of illumination energy, thus aiding the estimate of its Green's parameters in the presence of additive interference. This is the approach we will follow.

In [28], a new MIMO "probing" technique was introduced to assist in identifying large clutter discretes so that they could be "pre-nulled" via transmit adaptivity—either virtually via the transmit virtual array in the receiver, or with the actual physical transmit antenna pattern. Pre-nulling large discretes, even if only partially effective, reduces the requirements on the receive-only adaptivity for STAP [11,15]. However, to achieve the requisite clutter discrete channel model, orthogonal transmit waveforms were utilized. Moreover, the technique in [28] did not explicitly estimate the composite Green's function H. Guided by the discussion above, we will choose a set of non-orthogonal (but linearly independent) steering vectors that scan a field-of-regard for the purpose of estimating the MIMO Green's function response to a finite set of strong clutter discretes (strong point scatterers). This probing modality is also consistent with a wide area surveillance (WAS) mode that most surveillance radars naturally perform. In this way, the regret of radar resources of MIMO probing of the channel is minimized or eliminated altogether. The theoretical spatial Green's function response to a point target or clutter discrete, assuming a uniform linear array (ULA), with half-wavelength element spacing, under a narrowband signal model, is given by [5]

$$[H_c]_{m,n} = \alpha_c e^{j2\pi(m-n)\overline{\theta}_c}, \tag{16.34}$$

where $[H_c]_{m,n}$ denotes the $(m,n)^{th}$ element of the spatial Green's function $\boldsymbol{H}_c$, α_c, $\overline{\theta}_c$ denote the clutter discrete's amplitude and normalized angle-of-arrival (AoA), respectively [5].

When N_c spatially distinct point scatterers are present, the total MIMO Green's function is given by

$$[H_c]_{m,n} = \sum_{q=1} N_c \alpha_{c_q} e^{j2\pi(m-n)\overline{\theta}_{cq}}. \tag{16.35}$$

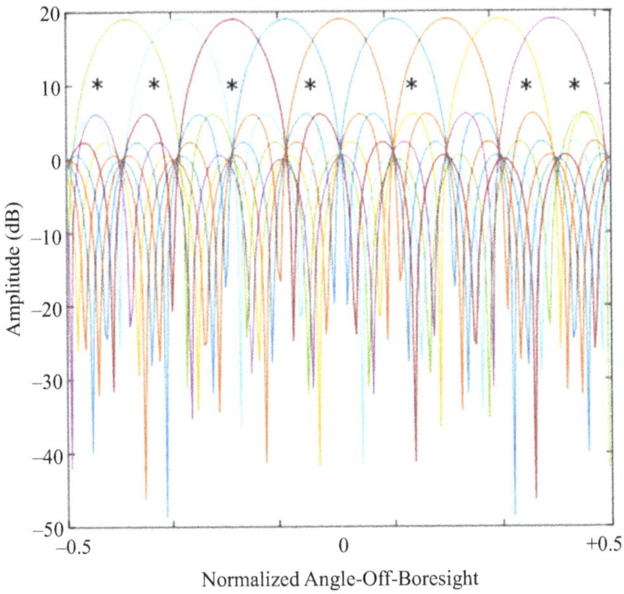

Figure 16.4 Illustration of the non-orthogonal MIMO probing beams to measure the clutter discrete Green's function

In the following example, we will assume a finite set of clutter discretes are randomly distributed uniformly across the field-of-regard and are not known *a priori*. Figure 16.4 shows both the distribution of random clutter discretes and the selection of non-orthogonal probing beams. Again, the selection of the beams was to ensure that each region that may contain a clutter discrete receives enough transmit energy to enable an effective estimate of the channel Green's function in the presence of additive noise/interference.

Figure 16.5 shows the *L*2 matrix norm error between the estimated and actual channel Green's function versus clutter-to-noise ratio (CNR) for a single discrete. We have assumed (for convenience) that each clutter scatterer is the same strength. As expected, performance improves with increasing CNR.

The optimum transmit steering vector satisfies the generalized eigenvector system given by

$$\lambda \left(E\left\{\boldsymbol{H}_C^H \boldsymbol{H}_C\right\} + \sigma^2 \boldsymbol{I}\right) \boldsymbol{s} = E\left\{\boldsymbol{H}_T^H \boldsymbol{H}_T\right\} \boldsymbol{s}, \tag{16.36}$$

$$\lambda \boldsymbol{s} = \left(E\left\{\boldsymbol{H}_C^H \boldsymbol{H}_C\right\} + \sigma^2 \boldsymbol{I}\right)^{-1} E\left\{\boldsymbol{H}_T^H \boldsymbol{H}_T\right\} \boldsymbol{s}, \tag{16.37}$$

where $E(\cdot)$ denotes the usual expectation operator, $\sigma^2 \boldsymbol{I}$ denotes the covariance matrix of additive receiver noise. $\boldsymbol{H}_T$ is the desired target Green's function. For example at hand, we will assume that it is a boresight-aligned target.

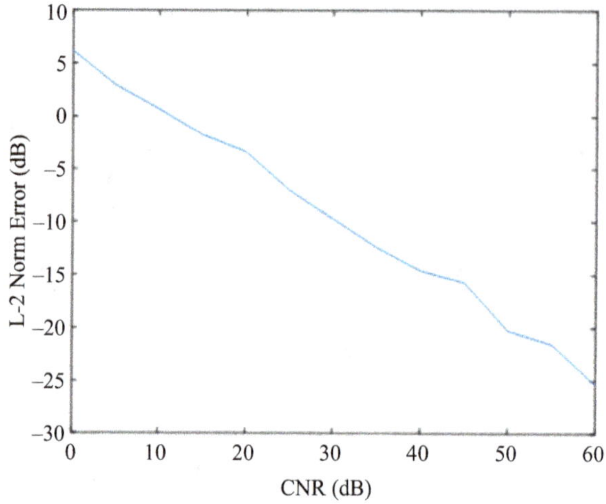

Figure 16.5 Estimation error versus CNR of the clutter discretes

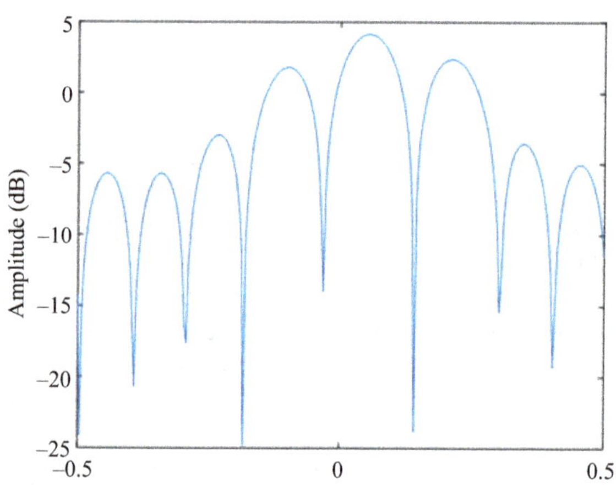

Figure 16.6 Optimum transmit spatial waveform response after estimating clutter discrete locations (indirectly through the Green's function)

Figure 16.6 shows that adapted optimum transmit pattern (7 nulls). Note that transmit nulls are placed in the direction of the discretes. Also note the squinting of the mainbeam off boresight due to the mainbeam nulling of one of the close-in discretes.

There are of course many possibilities for enhancing the RTCE function. Some incorporate externally networked information sources in a KA architecture, prior physics-based and/or database information, and other active probing in a SLA CR framework. Advanced AI methods can also be incorporated to further enhance the process [2]. This is clearly a new area of research and development for many years to come.

16.4 CR radar scheduler

In real-world challenging environments, a radar will be confronted with many competing requirements not all of which can be satisfied with limited (finite) resources. For example, in the ground surveillance radar case, there will arise conflicts between allocating valuable radar resources and timeline to both wide-area search (WAS) and high-value target (HVT) tracking [2]. Indeed it is this fundamental conflict that in no small part led to the development and proliferation of active electronically scanned antennas (AESAs) that have the ability to rapidly and electronically switch transmit/receive beams, thus minimizing "regret" when conducting simultaneous track and search (see [32] for an excellent introduction to AESAs).

The CoFAR radar scheduler performs several major functions: Creates a RTCE model derived from all the aforementioned mechanisms, for example, active MIMO probing and KA channel estimation. Performs a real-time resource allocation optimization that attempts to divvy up the finite resources (energy, time, computation, etc.) to best meet both mission level requirements and evolving tactical requirements (e.g., electronic protection, emissions control (EMCON), etc.) Finally, a transmit timeline is created down to each transmit pulse (and possibly intra-pulse) the time-horizon of which can vary widely from a fraction of a second to many minutes. Figure 16.7 shows a high-level "flow-down" from the main aircraft mission computer (in this case, ISR Mission Computer), to the CoFAR Radar Scheduler. The scheduler, in turn, performs all functions necessary to populate the complete radar timeline—down to potentially sub-PRI timeframes.

Mathematically, the scheduler receives real-time inputs from the mission computer and crafts an objective function J that both quantifies overall performance relative to mission objectives, and applies hard and soft constraints. As an example, consider the following example ISR objective function:

$$J(\boldsymbol{x}) = J_{\mathrm{WAS}}(\boldsymbol{x}) + J_{\mathrm{HVT}}(\boldsymbol{x}) + J_{\mathrm{SAR}}(\boldsymbol{x}) + J_{\mathrm{CR}}(\boldsymbol{x}), \tag{16.38}$$

where, for the ground surveillance example, $J_{\mathrm{WAS}}(\boldsymbol{x})$ represents the WAS metric, $J_{\mathrm{HVT}}(\boldsymbol{x})$ represents the HVT track metric, $J_{\mathrm{SAR}}(\boldsymbol{x})$ denotes the synthetic aperture radar metric, and $J_{\mathrm{CR}}(\boldsymbol{x})$ denotes CR functions (e.g., MIMO probing).

The above would, of course, be subject to a multitude of constraints on a variety of factors including for example transmit duty factor, peak power, spatial and/or frequency keep-out zones, etc. To date, for practical applications, (15.31) is in general a non-convex, nonlinear programming problem for which only approximate optimization techniques are available [33,34].

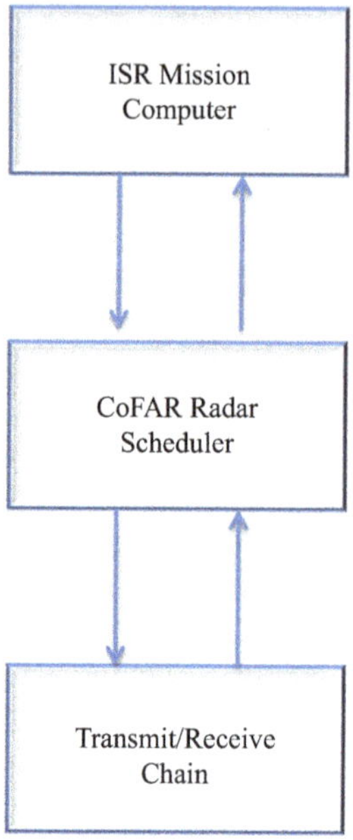

Figure 16.7 High-level control flow-down (and up) illustration of the CR scheduler

16.5 Cognitive radar and artificial intelligence

Currently there remains a degree of ambiguity in the terms "cognitive systems" and AI and/or machine intelligence/learning. To be clear, CR is a subset of AI systems. Machine learning more recently referred to as "deep learning," is also a subset of AI, albeit a more recent addition. A CR can be succinctly described as an automated system that has sophisticated real-time environmental/contextual knowledge that can be used to effect advanced real-time adaptivity of both transmit and receive functions. Note the emphasis on "sophisticated" and "advanced." An amoeba has rudimentary environmental awareness, but you would not use it as a meaningful exemplar of a cognitive system. A *ca.* 1940s analog CFAR circuit is an example of rudimentary real-time adaptivity, but clearly not a good example of "advanced" adaptivity. The degree of sophistication in environmental awareness and real-time adaptivity is what warrants the new moniker of "cognitive." The original example of the bat proposed by Haykin remains a good original biological analog for cognitive active sensing [6]. A useful mapping from psychological/biological to engineering terminology is provided in Figure 16.8.

It is absolutely possible to build and deploy a fully CR as described herein without ever using any of the so-called modern machine learning techniques such as deep learning (e.g., convolutional neural networks [35]). Of course, it is also possible to develop a CR that fully leverages the latest advances in AI. Figure 16.9 depicts the

Cognitive Property	Cognitive Radar Equivalent
Perceiving	Sensing
Thinking, Reasoning, Judging, Problem Solving	Expert Systems, Rule-Based Reasoning, Adaptive Algorithms & Computation, AI, Machine Learning
Remembering	Memory, Environmental Database

Figure 16.8 Mapping of biological cognition properties to a cognitive radar (see [2])

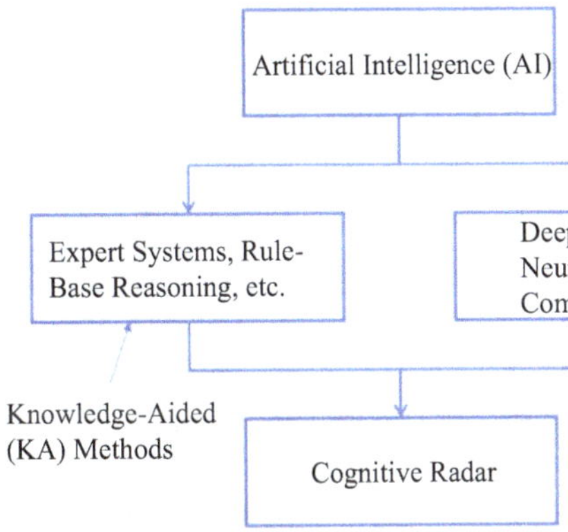

Figure 16.9 Relationship between cognitive radar and AI. Note that CR is a consumer of AI.

basic relationship between AI and CR. Note that knowledge-aided (KA) methods, as described earlier, draw on the more traditional AI methods (e.g., expert systems).

16.6 Implementation considerations

Some of the author's careers span over 30 years. Needless to say, the state-of-the-art in enabling RF subsystems and high-performance embedded computing (HPEC) have evolved many folds in performance C-SWAP (cost, size, weight, and power) over that period. Many seminal radar textbooks were written prior to this evolution, and, in many cases, reflect the restrictions imposed by available implementation hardware and software. For example, transmit waveforms were restricted to special classes of constant modulus signals such as linear frequency modulation (LFM) to achieve maximum transmit effective radiated power (ERP) and real-time radar signal processing such as pulse compression and Doppler filtering [36]. Real-time digital signal processing requirements, especially for multichannel systems with relatively large bandwidths, also greatly restricted the types of algorithms that could be implemented [37].

With the development and maturation of fully digital front-ends including DAWGs, digital receiver exciters (DREX) [38], and solid-state amplifiers [39], it is now possible to transmit a much broader class of waveforms. Supported by advanced HPEC, adaptive transmit waveforms are also now possible. Thus, the technical (and cost) excuse for avoiding arbitrary/adaptive transmit waveforms has all but been eliminated.

Advances in field programmable gate arrays (FPGAs) have enabled processing speeds and costs that approach that of traditional application-specific integrated circuits (ASICs) [40]. General purpose graphical processing units (GP GPUs) have an internal parallel processing architecture particularly amenable to ray tracing calculations [41]. It is thus now possible to implement high performance parallel processing and systolic array processing systems using entirely commercial chips. Lastly, neuromorphic computing chips are becoming available that will allow for the real-time implementation of advanced deep learning (neural network) solutions [42].

16.7 Advanced modeling and simulation to support cognitive radar

The raison d'être for CR is operation in extremely complex environments. Consequently, to assess and quantify the value of CR either elaborate and expensive field testing must be conducted, or high-fidelity modeling and simulation (M&S) must be used—the latter being the obvious choice for most radar researchers for obvious reasons.

Fortunately, with investments from private companies, and funding institutions such as the Defense Advanced Research Projects Agency (DARPA) and the Air Force Research Laboratory (AFRL), a sophisticated radar M&S tool is now commercially available. RFView® is a physics-based, high fidelity M&S tool that covers operating frequencies from VHF to X-band [43]. Arguably the most powerful feature of RFView is the ability to model site-specific terrain clutter. This is accomplished by importing digital terrain and cover maps, and then employing an appropriate physics solver for the clutter reflections (including multipath).

Now, we present an example to demonstrate the potential theoretical gains that can be achieved by a CR system as the radar platform is flying along a path. Figure 16.10 shows an airborne X-band radar surveillance scenario of a northbound offshore aircraft flying off the coast of southern California. The path flown by the radar-carrying aircraft is also marked in the figure. The region that is illuminated by the radar is also shown. The region has significant heterogeneous terrain features and thus presents an interesting real-world clutter challenge. In the presence of flat terrain with no clutter discretes, the spectrum will be flat and as a result there would not be any advantage as a result of adapting the transmit waveform. The heterogeneous terrain in this example ensures strong CoFAR performance potential. This example was generated using high-fidelity physics-based M&S tool RFView.

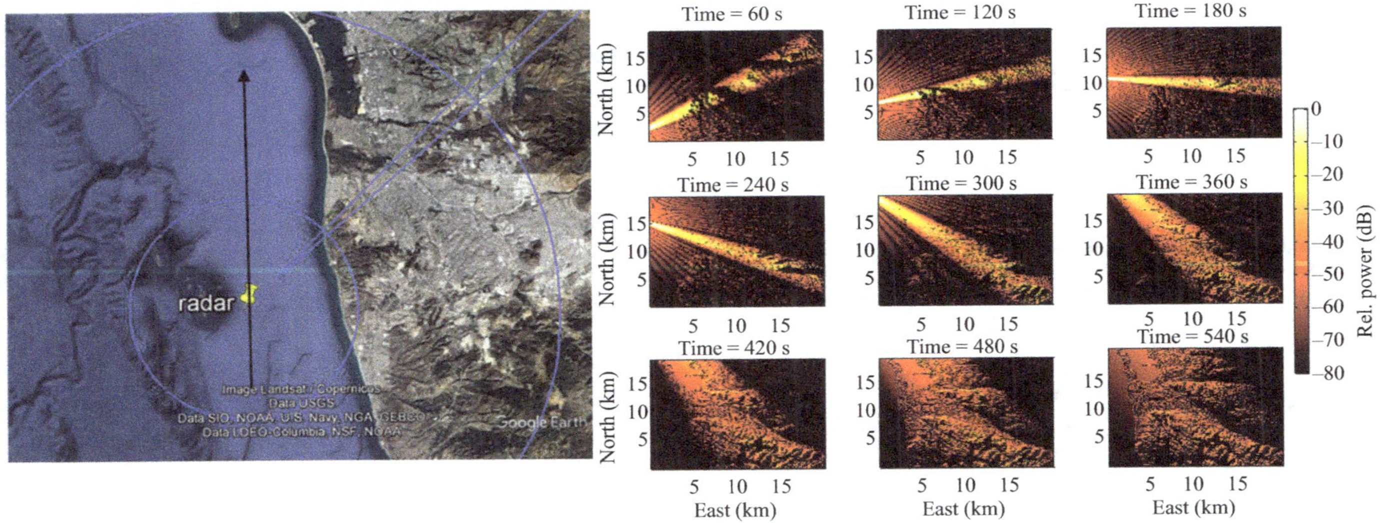

Figure 16.10 X-band site-specific airborne GMTI radar scenario off the coast of southern California. Left: Scenario location and geometry. Right: Radar beam pointing positions at different portions of the flight.

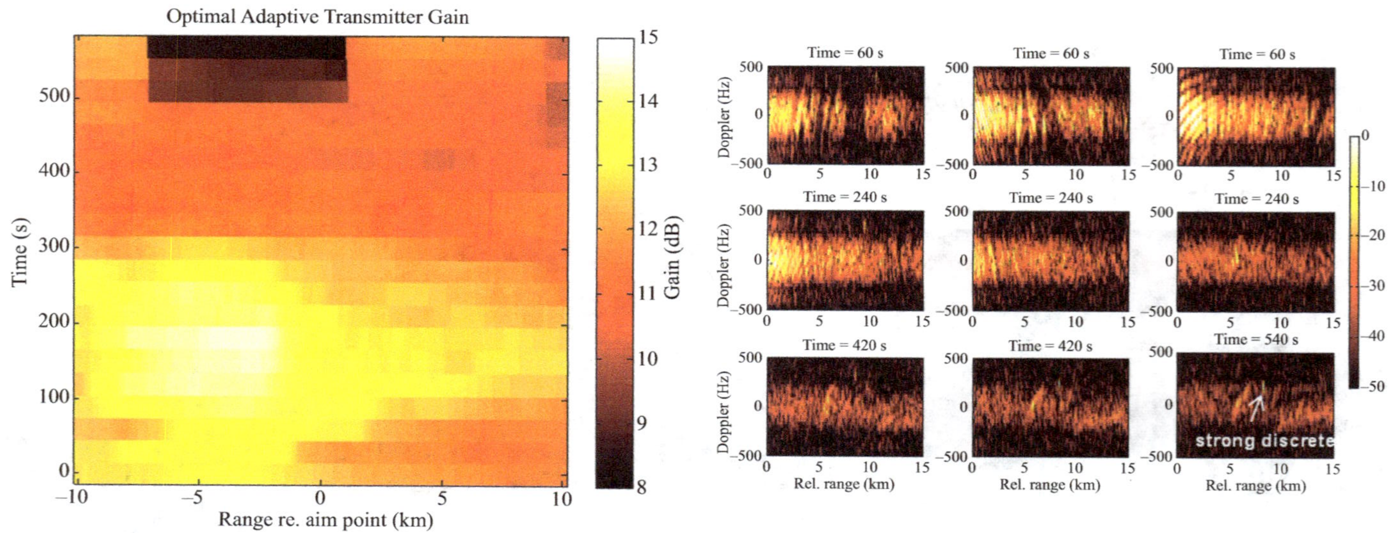

Figure 16.11 *Left: Optimal maximum gain (dB) using adaptive waveforms as a function of time and range of interest. Note in general maximum gain is achieved in regions with the strongest heterogeneous clutter (as expected). Right: Corresponding range-Doppler plots. Note that there is no gain when the clutter is weak with the presence of a single discrete (again to be expected).*

Shown in Figure 16.11 is the theoretical performance gain (tight bound) using the optimal transmit pulse shape. As expected, the maximum potential gain is achieved in those regions with the strongest heterogeneous clutter since this produces significant eigenvalue spread. Also as expected, there is no gain in areas of weak clutter where only a single large discrete (impulse) is present since this yields a relatively flat eigenspectrum. Note however that these results assume the optimizer has access to the true channel transfer functions. However, in reality, these channel impulse responses and their corresponding transfer functions are not known ahead of time and they need to be estimated from the measured data. This may involve sending out strong probing signals to enhance the accuracy of channel estimation.

16.8 Remaining challenges and areas for future research

In [44], a number of potential pitfalls of CR are raised. A careful reading and reflection reveals that these concerns are the very same confronting all modern AI systems, particularly deep learning—not just CR. Many of the potential pitfalls are related to "how" the CR radar was "trained." It goes without saying that if you did not train for an unforeseen event, one cannot expect good performance. So, "How does one train CR so that it is robust to any and all physical realizable possible future events?" Mathematically, this is of course ill-posed unless you can somehow characterize all possible future events (deterministically or statistically). One practical remedy is to use high-fidelity M&S to generate copious amounts of synthetic training data. This is precisely what was discussed in Section 16.7. To make this approach even more powerful, one could use AI methods to develop ever more challenging synthetic environments. In this way, human bias (and of course ignorance) could potentially be removed. Indeed, this approach was very successful in the training of DeepMind AlphaGo to play the Chinese game of Go.

With the steady advances in HPEC, and key enabling RF technologies as discussed in Section 16.6, one needs to continually assess "what is implementable." At the time of this writing, neuromorphic integrated circuits (ICs) are just becoming commercially available. These are not traditional digital ICs mimicking neural networks. They are in fact hardware-implemented neural networks. Consequently, these chips can potentially perform functions in real-time that would require many (e.g., multi-core) FPGAs, GPUs, or ASICs to achieve similar performance. The more CR can leverage these types of advances, the more CR performance can commensurately advance.

Lastly, the question of the similarities/differences has been raised between the CR architecture described herein (and in the prior references) with Haykin's original CR concept [6]. The simple answer is the CoFAR architecture herein is a particular engineering implementation of the original biological concept presented in [6]. To transform from biological to human-engineered systems, a functional translation must be performed. The first direct translation was presented in 2010 [5] and can also be seen in Figure 16.8. It shows the translation from Biological Cognitive Property to Cognitive Radar Equivalent. Since the CR equivalent functions typically entail advanced engineered systems, their actual capabilities are highly dependent on the current state-of-the-art.

References

[1] A. Farina, S. Haykin, and A. D. Maio, *The Impact of Cognition on Radar Technology*. London: Institution of Engineering & Technology, 2017.
[2] J. R. Guerci, *Cognitive Radar: The Knowledge-Aided Fully Adaptive Approach*, 2nd ed. Norwood, MA: Artech House, 2020.
[3] K. L. Bell, C. J. Baker, G. E. Smith, J. T. Johnson, and M. Rangaswamy, "Fully adaptive radar for target tracking part i: Single target tracking," In: *IEEE Radar Conference*, 2014, pp. 0303–0308.
[4] J. R. Guerci, R. M. Guerci, M. Ranagaswamy, J. S. Bergin, and M. C. Wicks, "CoFAR: cognitive fully adaptive radar," In: *IEEE Radar Conference*, 2014, pp. 0984–0989.
[5] J. R. Guerci, *Cognitive Radar: The Knowledge-Aided Fully Adaptive Approach*, Norwood, MA: Artech House, 2010.
[6] S. Haykin, "Cognitive radar: a way of the future," *IEEE Signal Processing Magazine*, vol. 23, pp. 30–40, 2006.
[7] J. R. Guerci and E. J. Baranoski, "Knowledge-aided adaptive radar at DARPA: an overview," *IEEE Signal Processing Magazine*, vol. 23, pp. 41–50, 2006.
[8] A. F. Martone, K. D. Sherbondy, J. A. Kovarskiy, *et al.*, "Practical aspects of cognitive radar," In: *IEEE Radar Conference*, 2020, pp. 1–6.
[9] J. W. Owen, C. A. Mohr, B. H. Kirk, S. D. Blunt, A. F. Martone, and K. D. Sherbondy, "Demonstration of real-time cognitive radar using spectrally-notched random FM waveforms," In: *IEEE International Radar Conference*, 2020, pp. 123–128.
[10] M. Harris. Echodyne Shows Off Its Cognitive Radar for Self-Driving Cars. [Online]. Available: https://spectrum.ieee.org/cars-that-think/transportation/self-driving/echodyne-cognitive-radar-self-driving-cars.
[11] J. R. Guerci, *Space-Time Adaptive Processing for Radar* 2nd ed. Norwood, MA: Artech House, 2014.
[12] S. Gogineni, J. R. Guerci, H. K. Nguyen, J. S. Bergin, B. C. Watson, and M. Rangaswamy, "Modeling and simulation of cognitive radar," in *IEEE Radar Conference*, Florence, Italy, Sep. 2020.
[13] M. D. Greenberg, *Applications of Green's Functions in Science and Engineering*, Mineola, NY: Courier Dover Publications, 2015.
[14] H. L. V. Trees, *Detection, Estimation and Modulation Theory. Part I*, New York: Wiley, 1968.
[15] J. Ward, "Space-time adaptive processing for airborne radar (ref. no. 1998/241)," In: *IEE Colloquium on Space-Time Adaptive Processing*, 1998.
[16] J. Wishart, The generalized product moment distribution in samples from a normal multivariate population. *Biometrika*, vol. 20A, pp. 32–52, 1928.

[17] S. Pillai and C. Burns, *Array Signal Processing*, New York, NY: Springer-Verlag, 1989.

[18] T. W. Anderson, *An Introduction to Multivariate Statistical Analysis*, 3rd ed. Hoboken, NJ: Wiley-Interscience, 2003.

[19] J. Hiemstra and C. SAIC, "Colored diagonal loading," In: *Proceedings of the IEEE Radar Conference*, 2002, pp. 386–390.

[20] J. S. Bergin, C. M. Teixeira, P. M. Techau, and J. R. Guerci, "STAP with knowledge-aided data pre-whitening," In: *Proceedings of the IEEE Radar Conference*, pp. 289–294, 2004.

[21] J. R. Guerci, "Theory and application of covariance matrix tapers for robust-adaptive beamforming," *IEEE Transactions on Signal Processing*, vol. 47, pp. 977–985, 1999.

[22] A. Gelb, *Applied Optimal Estimation*, Cambridge, MA: MIT Press, 2002.

[23] J. R. Guerci, *Space-Time Adaptive Processing for Radar (Artech House Radar Library)*, Norwood, MA: Artech House, 2003.

[24] P. M. Techau, J. R. Guerci, T. H. Slocumb, and L. J. Griffiths, "Performance bounds for hot and cold clutter mitigation," *IEEE Transactions on Aerospace and Electronic Systems*, vol. 35, pp. 1253–1265, 1999.

[25] J. R. Guerci, "Knowledge-aided sensor signal processing and expert reasoning (KASSPER)," In: *Proceedings of 1st Annual DARPA KASSPER Workshop*, Apr. 2002.

[26] G. R. Legters and J. R. Guerci, "Physics-based airborne GMTI radar signal processing," In: *IEEE Radar Conference*, pp. 283–288, 2004.

[27] P. Stoica, J. Li, X. Zhu, and J. Guerci, "On using a priori knowledge in space–time adaptive processing," *IEEE Transactions on Signal Processing*, vol. 56, pp. 2598–2602, 2008.

[28] J. S. Bergin, J. R. Guerci, R. M. Guerci, and M. Rangaswamy, "MIMO clutter discrete probing for cognitive radar," *IEEE International Radar Conference*, pp. 1666–1670, 2015.

[29] J. R. Guerci, J. S. Bergin, S. Gogineni, and M. Rangaswamy, "Non-orthogonal radar probing for MIMO channel estimation," In: *Proceedings of the IEEE Radar Conference*, Boston, MA, Apr. 2019.

[30] L. Rayleigh, "On the law of reciprocity in diffuse reflection," *Philosophical Magazine*, vol. 49, pp. 324–325, 1900.

[31] D. Manolakis, F. Ling, and J. Proakis, "Efficient time-recursive least-squares algorithms for finite-memory adaptive filtering," *IEEE Transactions on Circuits and Systems*, vol. 34, pp. 400–408, 1987.

[32] R. Sturdivant, C. Quan, and E. Chang, *Systems Engineering of Phased Arrays (Radar and Electronic Warfare)*, Norwood, MA: Artech House, 2018.

[33] M. S. Bazaraa, H. D. Sherali, and C. M. Shetty, *Nonlinear Programming: Theory and Algorithms*, New York: John Wiley & Sons, 2013.

[34] L. Weinberg, "Scheduling multifunction radar systems," *Imaginative Engineering through Education and Experience*, pp. 10-4A-10-4J, 1977.

[35] A. Krizhevsky, I. Sutskever, and G. E. Hinton, "Imagenet classification with deep convolutional neural networks," *Advances in Neural Information Processing Systems*, pp. 1097–1105, 2012.

[36] M. I. Skolnik, *Radar Handbook*, New York, NY: McGraw-Hill Education, 2008.

[37] A. Farina, *Antenna-Based Signal Processing for Radar Systems*, Norwood, MA: Artech House, 1992.

[38] J. S. Herd and M. D. Conway, "The evolution to modern phased array architectures," *Proceedings of the IEEE*, vol. 104, pp. 519–529, 2015.

[39] A. Biondi, S. D'Angelo, F. Scappaviva, D. Resca, and V. A. Monaco, "Compact GaN MMIC T/R module front-end for x-band pulsed radar," In: *11th European Microwave Integrated Circuits Conference (EuMIC)*, pp. 297–300, 2016.

[40] T. Stratoudakis, *Introduction to LabVIEW FPGA for RF, Radar, and Electronic Warfare Applications*, Norwood, MA: Artech House, 2021.

[41] M. Larsen, J. S. Meredith, P. A. Navratil, and H. Childs, "Ray tracing within a data parallel framework," In: *IEEE Pacific Visualization Symposium (Pacificvis)*, pp. 279–286, 2015.

[42] S. Greengard, "Neuromorphic chips take shape," *Communications of the ACM*, vol. 63, pp. 9–11, 2020.

[43] Available: https://rfview.islinc.com/RFView/login.jsp.

[44] M. Greenspan, "Potential pitfalls of cognitive radars," In: *IEEE Radar Conference*, Sep. 2014, pp. 1288–1290.

Chapter 17
Advances in cognitive radar experiments

Graeme E. Smith[1,], Jonas Myhre Christiansen[2] and Roland Oechslin[3]*

17.1 The need for cognitive radar experiments

The emergence of full digital arrays [1, and refs. therein] and software-defined radar [2–6, among many], created radars with expansive parameter spaces that required managing; this led to the emergence of cognitive radar [7–11] as a method to managing these parameter spaces that, in turn, required a new class of testbeds for research experiments.

The relationship between the cognitive radar and the sensed environment is highly dynamic due to the tight feedback loops required to implement the principles of cognition as set out by Haykin and Fuster, based on Fuster's neuropsychology research [9–11]. Essentially, every pulse that the radar receives provides it with more knowledge about the state of the local environment and this knowledge can be used to update the radar's operating parameters to maximize performance. It should be noted that the feedback loop could also operate at slower speeds, such as coherent processing interval (CPI)-to-CPI, dwell-to-dwell, or target-update-to-target-update. Whatever the time frame, the principles of cognition connect the cognitive process directly to the local environment through the probing radar transmissions and the reception of the scattered waveforms. With such tight coupling, it becomes apparent that testbeds are needed early in the research process to ensure sufficient fidelity in the received signal to allow proper evaluation of the cognitive algorithm.

17.1.1 Cognition for radar sensing

To understand how cognitive concepts can be applied to radar sensing, we first need to understand something of cognition itself.

[1]Johns Hopkins Applied Physics Laboratory, Laurel, MD, USA
[2]Norwegian Defence Research Establishment (FFI), Kjeller, Norway
[3]armasuisse Science and Technology, Bern, Switzerland
*This chapter is dedicated to Graeme Smith, who died on September 6th, 2022 after contributing a major part of this book chapter.

Fuster's cognitive framework [12] includes five components that are exhibited by the cognitive structure of the cortical networks in the brain. He calls the cognitive structures *cognits* and defines them as

> *...a generic term for any representation of knowledge in the cerebral cortex. A cognit is an item of knowledge about the world, the self, the relations between them. Its network structure is made up of elementary representations of perception or action that have been associated with one and other by learning or past experience. These small units of representation constitute the nodes of the network, which themselves may also have a network structure at a more simple level (thus cognits within cognits).*

The five components that Fuster says all cognitive structures exhibit are defined as follows:

Perception–action cycle: The perception–action cycle is an endless feedback loop between a cognitive organism's perception of the environment and the actions it undertakes in response to that perception. Since actions change the relationship between the organism and the environment, with each action the perception changes requiring new actions to be selected. A view of the perception–action cycle as it applies to radar is provided in Figure 17.1.

Memory: The memory comprises both short-term and long-term aspects and can be thought of as containing anything from raw sensory inputs, to previous action selections, to the state of the cognit at some prior time.

Attention: A cognitive organism must be able to pay attention to the relevant parts of its sensory inputs, memories, and perception to avoid becoming overwhelmed as it selects actions.

Intelligence: Selection of actions, deciding what things to pay attention to, and what to place in memory require intelligence. Fuster does not define what intelligence is, he only states that it is required.

Language: Language provides a structure for information within the cognit. In higher-level organisms, such as humans, this language can extend to a spoken component for transfer of information, although this is not required for an organism to be cognitive.

Fuster asserts that each cognit has all five of these properties, just on different scales. Low-level cognits, which are physically smaller, e.g. contain fewer neurons, control low-level aspect of the organism, such as fine motor control of the fingers. High-level cognits, which are large and contain many, many neurons, control high-level aspects of the organism behavior, such as abstract reasoning. Critically, the cognits are hierarchical—*cognits within cognits*—so even high-level cognits are ultimately connected to sensory inputs and action outputs. While it may not seem the case, Fuster would argue that the radar engineer sitting stationary at their desk, eyes closed, reflecting on an advanced space–time adaptive processing paper they have just read is very much involved in a sensory-in, action-out process that drives their cognition.

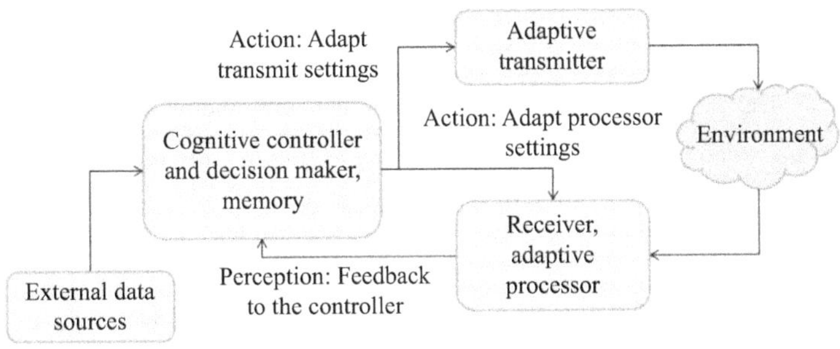

Figure 17.1 The perception–action cycle for cognitive radar processing

Within the majority of cognitive radar research, the principal focus has been on the perception–action cycle since it provides a convenient framework to describe the feedback loop that can exist between the transmit and receive chains of a radar. A notable exception is the work of Guerci [7, and references therein] who takes a top–down approach to developing a cognitive radar by looking for ways to model all the various components of the cognitive process.

The perception–action cycle is shown in Figure 17.1 as a feedback loop between perception and action. The perception of the environment is provided by the radar receiver and processor, which converts the received radar data to an input information to the controller. The latter decides on the next action taken by the transmitter and processor. The perception will be updated solely by actions (working in conjunction with the other aspects of cognition listed above) and not directly from the radar sensor. Similarly, the actions should not be thought of as emanating from the action block: instead they should be thought of as arising from the entire perception–action cycle process.

The cognitive radar testbeds that will be discussed in this chapter have primarily been designed to implement the perception–action cycle as a foundational unit of cognition. Fuster's other components of cognition are present in the experiments the testbeds have been used in, but in more abstract ways. For example, the Kalman tracking process or a neural network inherently includes the concept of memory. Meanwhile, a detection process can be viewed as a form of attention since it selects samples of incoming data containing target information, which should be focused on, versus samples containing noise that should be ignored. As such, the testbeds presented, and the associated experiments highlighted, constitute some of the first attempts to actually build cognitive radars.

17.1.2 Chapter overview

This chapter will present three representative cognitive radar testbeds: the cognitive radar engineering workspace (CREW) developed at The Ohio State University, cognitive detection, identification and ranging (CODIR) developed at armasuisse Science and Technology, and a universal software radio peripheral (USRP) system developed at the Norwegian Defence Research Establishment (FFI). The discussion of the CODIR

Table 17.1 Overview of the testbeds discussed in the chapter

Testbed	Typical experiment setup	Enabling components for CR experiments
CREW	Mono- and multistatic laboratory setup with up to 4 TX and 4 RX channels in W-Band	• Programmable AWG Keysight M8190A • PAC implemented in MATLAB®
CODIR	Short-range outdoor setup with up to 2 TX and 4 RX channels in X-Band	• LFM signal generation on a custom board with an AD9914 DDS and a Raspberry Pi controller • PAC implemented in MATLAB
USRP-based	Mid-range outdoor setup with 2 RX channels in S-Band	• Signal generation with an Ettus X310 USRP • PAC implemented in C++
mini-CODIR	Short-range distributed sensing using up to 4 nodes with 2 RX channel each in X-Band	• Signal generation with a Zynq 7030 SoC • PAC implemented in MATLAB

system will be broken into two parts, the first discussing the original CODIR and the second the newer miniature cognitive detection, identification and ranging (mini-CODIR) that is a distributed sensor. The key figures of the discussed testbeds above are summarized in Table 17.1. In a further section, other hardware realizations of cognitive radar reported in the literature are discussed. The lessons learned from the architectures, design considerations, and experimental achievements of those systems are similar to the three testbeds discussed in Sections 17.2 to 17.5. As such, we believe the content of this chapter may be considered general.

To close the chapter, we will consider the need for future cognitive testbeds. The research into cognitive radar continues to advance with new ideas and concepts emerging all the time. It naturally follows that the testbeds needed to address these new research goals will be an advance on the current state of the art.

17.2 The CREW test bed

17.2.1 The CREW design

A block diagram of the CREW is provided in Figure 17.2 where each of the major components is shown with an indication of how they interact. In the CREW, there are four IF upconverters, IF downconverters, Tx heads, and Rx heads to achieve simultaneous 4 Tx-by-4 Rx operation. Only one of each was shown in the diagram for clarity. All components, other than the Tx and Rx heads and cables that connect them, were mounted in a standard rack. The two local oscillators (LOs) were bench-top signal generators that have very low-phase noise. The arbitrary waveform generators (AWGs) and analog-to-digital converters (ADCs) were connected to the personal

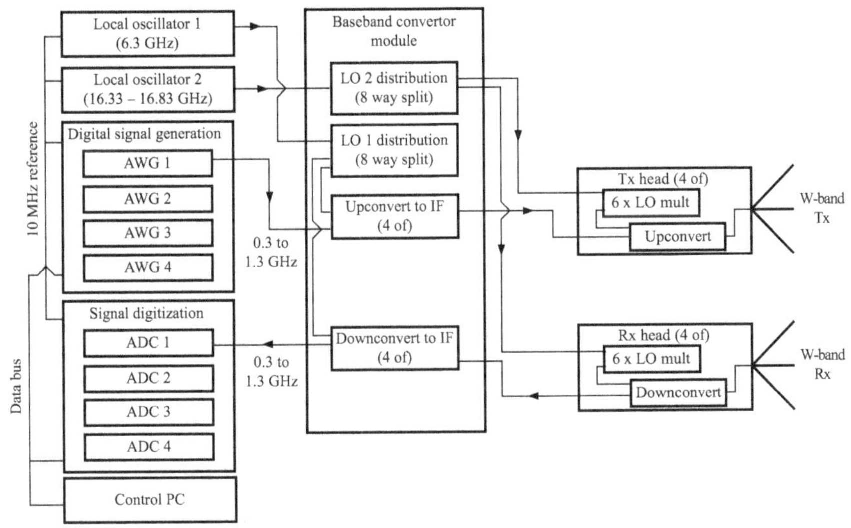

Figure 17.2 A block diagram of the principal components of the CREW. Only a single intermediate frequency (IF) upconverter, IF downconverter, downconverter, transmit (Tx) head and receive (Rx) head are shown. The actual system includes four of each of these. Figure taken from [13]. ©2016 IEEE.

computer (PC) using a PXIe bus extension that permits high-speed transfer of data. The reference signal was provided by one of the bench-top signal generators as a standard 10 MHz clock signal used to phase lock radio frequency (RF) equipment. The digital components operated at a sampling frequency of 3 GHz. The baseband complex-sample bandwidth was 1 GHz. Digital upconversion and downconversion were employed such that the outputs of the digital-to-analog converters (DACs) and inputs to the ADCs were at an IF bandwidth between 0.3 and 1.3 GHz. The baseband converter module* used the 6.3 GHz LO signal to convert between the baseband and IF frequencies. The IF was centered on 5.5 GHz. The converter module also distributed the $\approx$16 GHz second LO signal. Each of the Tx and Rx heads received the distributed LO signal that was six times the frequency multiplied to permit the conversion between IF and the final RF frequency, centered at 94 GHz (W-band). The second LO frequency could be adjusted to allow the 1 GHz of instantaneous bandwidth to be stepped over a 4 GHz operational bandwidth. Table 17.2 lists the principal operating and hardware parameters. Polarization diversity was via rotation of the RF heads and hence could only be achieved on an experiment-to-experiment basis.

*Strictly, the signals being processed through the baseband convertor module are at the 0.3 to 1.3 GHz IF. However, the module has been referred to as the "baseband converter" since the system was delivered and we continue to use that term here to be consistent with prior publications.

Table 17.2 Summary of the CREW principal operating parameters. Table taken from [13]. ©2016 IEEE.

Parameter	Value
Max Tx power	25 dBm
Frequency range	92 GHz to 96 GHz
Pulse width	1 ns to 100 μs
pulse repetition frequency (PRF)	Up to 30 kHz
Waveform	Programmable pulse-to-pulse
Number of transmitters	4
Antenna gain (Tx and Rx)	33 dB
Antenna azimuth beamwidth	9°
Antenna elevation beamwidth	11°
Number of receivers	4
ADC Keysight M9703A	Four 3.2 GSa/s 12-bit channels
AWG Keysight M8190A	14-bit resolution at 8 GSa/s or 12-bit resolution at 12 GSa/s and spurious free dynamic range of 90 dBc
Baseband converter	Input freq: 0.3–1.3 GHz Output freq: 5–6 GHz Tx gain: 20 dB with 19 dB noise figure Rx gain: 12 dB with 5 dB noise figure

The AWGs and ADCs used in the CREW enable significant levels of waveform adaption. The ADC card has 16 GB of onboard memory that can be used to record bursts of pulses ahead of downloading to the PC. Strictly, there are two AWG cards arranged in a primary–secondary configuration to behave as if there are four individual AWGs. Each card can store up to 2048 Giga Samples (GSa) of waveform data. The primary AWG card provides a trigger that initiates both the transmit and receive processes. A delay can be placed between the ADCs receiving the trigger and when they start to record data. Further, the number of samples collected by the ADCs can be limited such that only those relating to the range swath of interest are recorded, rather than having to digitize the entire pulse repetition interval (PRI). Operation in this manner limits the amount of data that must be transferred to the PC and processed each CPI.

For waveform adaption experiments, there are two modes of operation available. In the first instance, individual pulse operation is possible. A waveform is sent to each AWG, these are transmitted, the echoes digitized and the samples sent to the PC for analysis. Based on the analysis, new waveforms are designed and sent to the AWG. This process then continues for the duration of the experiment. Operation in this mode allows maximum flexibility, but it must be noted the PRFs that may be used must be less than approximately 500 Hz due to the data transfer overhead. Alternatively, a better use of the system, is to capture an entire CPI at a time on the ADC card and preload a library of waveform blocks onto the AWGs that can be selected from dynamically. This mode is known as dynamic linking. In this mode of operation, a new waveform block sequence is calculated for each CPI and sent to the AWG, while the ADC is primed to store an appropriate number of samples. Operating in this manner allows PRFs that exceed 30 kHz, although the precise PRF limit depends on a number of parameters. While the dynamic linking method does limit the waveforms that can be used in cognitive processing, this constraint is not significant. The shortest waveform that can be stored is 320 samples long, meaning that 64×10^9 different waveforms can be stored. The dynamic linking allows longer waveforms to be constructed from these short waveforms. With so many available, the number of permutations is sufficient that the finite AWG memory should not restrict research.

A photograph of the CREW hardware is provided in Figure 17.3. The digital back end, LOs, and baseband converter modules are shown on the right. A single pair of the Tx and Rx heads is shown on the left. It is clear that the equipment can easily be moved around the laboratory allowing a variety of multistatic radar-target geometries to be investigated.

As described so far, the CREW seems of fairly standard design that is, conceptually, similar to that used by many current radar systems. However, feeding the received echoes to the PC allows for real-time signal analysis, interpretation, and selection of future waveforms and system parameters. It was this facility that enables cognitive concepts to be supported and explored. In addition, the PC was able to store memories,

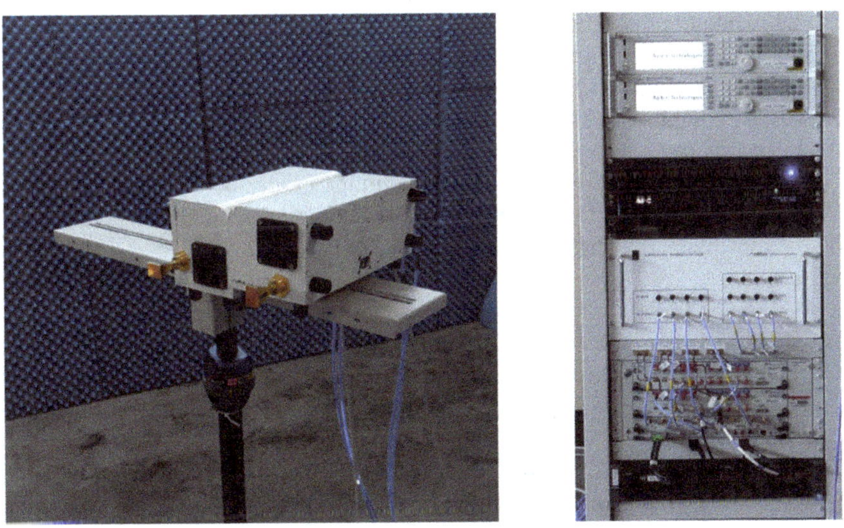

Figure 17.3 The CREW equipment. The radar RF module comprising a single transmitter and receiver is shown on the left-hand side. The waveform generators, digitizers, based band converter and local oscillator are mounted in a standard rack, shown on the right-hand side. Figure taken from [13]. ©2016 IEEE.

created by the radar or imported from external databases, and can exploit these memories to help it interpret current received echoes and set future transmit parameters. Equally importantly the PC can be programmed directly in MATLAB®. This allows algorithms to be developed off-line using simulations and then ported to the radar by uploading the script to the PC. As a result, as new cognitive concepts were developed, they could be very quickly tested and evaluated allowing rapid development.

17.2.2 CREW demonstration experiments

The CREW testbed is controlled by a MATLAB software suite known as Fully-Adaptive Radar Modelling and Simulation Environment (FARMS) that implements the perception–action cycle on which the fully-adaptive radar (FAR) concepts outlined in [14–17] are based. FARMS was developed to allow the FAR engine to be connected to either the CREW hardware or a modeling and simulation environment that uses the same software interface as the hardware. The software allows the radar engineer to seamlessly switch from modeling and simulation to experimental testing. This approach has allowed the FAR framework approach to cognitive radar to be validated using both simulation and experiments. Given the criticality of the FAR framework to the operation of the CREW, we shall summarize it before looking at some example experiments.

A simplified block diagram of the FAR framework is shown in Figure 17.4. It is an attempt to model the perception–action cycle from Fuster's description of cognition, [9,12], for a radar sensor. The framework comprises two key modules. First, a *perceptual processor* that outputs a perception of the local environment based on measurements received from the hardware, prior knowledge of the environment, knowledge of the actions undertaken, and a prior prediction of what the perception will be, conditioned on the actions taken. Second, an *executive processor* that outputs actions to be undertaken by the hardware based upon a forward prediction of the perception, conditioned on possible actions, a consideration for the cost of actions, and required performance objectives for the radar. Frequently, the framework is implemented using a combination of Bayesian estimation and optimization methods.

Essentially, the FAR framework selects actions by answering the following question, [13,14]: *What are the lowest-cost actions that can be undertaken and still obtain a useful perception?* In the early examples of FAR work an adaptive single-target, range-Doppler tracking example was employed. The actions were waveform parameters, specifically, PRF and pulse width (PW), and the perception was the target state estimate and the associated covariance matrix. The cost of the actions was based on the radar timeline, with parameter combinations that resulted in short-duration CPIs that used less of the radar timeline thus having lower cost. The quality of the perception was based on the state covariance matrix with a useful perception defined as one in which the range and range-rate variances were below goal values specified by the operator. The actions—waveform parameters—were selected by minimizing the action cost, constrained by the predicted perception quality, target state and covariance, conditioned on the hypothesized action selection. The results of [13,14] demonstrated this

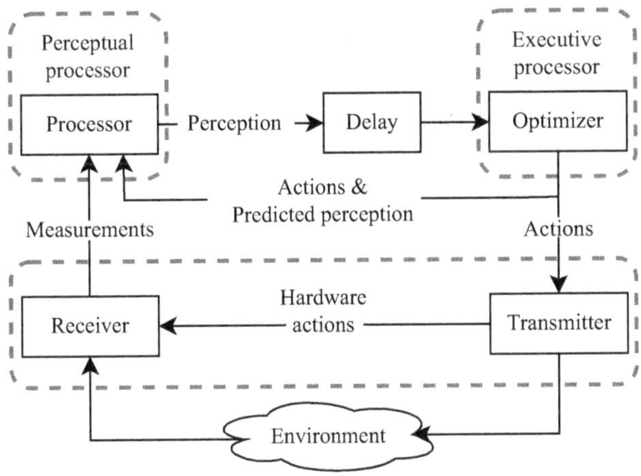

Figure 17.4 Simplified block diagram of the FAR framework

approach to provide an optimized track using a perception–action cycle approach and to be implementable on the CREW hardware.

Below, we shall highlight key experimental results demonstrated using CREW. Detailed analysis will be limited and references will be provided to the papers where the underlying theory and results are fully discussed.

17.2.2.1 Cartesian tracking using hierarchical, fully-adaptive radar (HFAR)

In [16], the original FAR concept from [14] was extended to allow multiple perception–action cycles to operate both in hierarchical and in parallel structures to facilitate more advanced radar operations. This new form of FAR was named hierarchical, fully-adaptive radar (HFAR) and included a refinement of the FAR block diagram, generalized to better facilitate hierarchical operation; see [16, Figures 1 and 2]. The CREW was used to demonstrate the HFAR concept with a Cartesian tracking experiment in which updates from two monostatic radars, each capable of range-Doppler tracking but with no angular information, were combined.

To implement hierarchy, Mitchell introduces the idea of information passing between the different perception–action cycles, each implemented using a restructured FAR framework [16]. While Mitchell's formulation does not require a specific form for the information, it is most likely to represent some combination of details about selected actions, performance goals, and measurements. This is the case in example presented here.

The experimental setup, implemented in an indoor laboratory at The Ohio State University's ElecroScience Laboratory, for demonstrating the HFAR concept is shown in Figure 17.5. Two monostatic radars with overlapping fields of view are hypothesized, each capable of range-Doppler tracking, but unable to make angular measurements. While each radar is unable to perform Cartesian tracking alone, if their positions are known and their measurements are passed to a fusion center to be

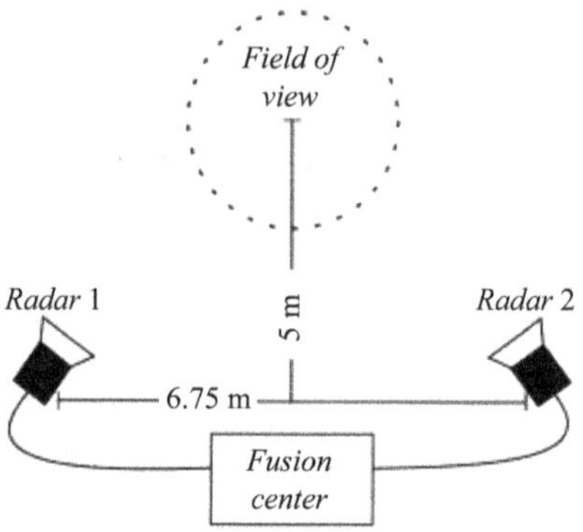

Figure 17.5 The radar concept for demonstrating HFAR Cartesian tracking that was implemented using the CREW testbed. Figure taken from [16]. ©2018 IET.

combined, then a Cartesian track may be established for targets within the common field of view.

The setup from Figure 17.5 can be achieved using the CREW. Two pairs of transmit and receive heads are used: one pair to implement radar 1 and another pair to implement radar 2. Since all received data for the CREW is processed by the computer in the main rack fusion can readily be achieved.

For a target, a person walking in an inward spiral was used. The motion pattern was selected as it presented a wide variety of ranges and velocity projections to each radar. To ensure the person walked the same path each time, the spiral pattern was marked on the floor with tape. While the person's speed did vary between experiments, after several practice trials, the time it took them to walk the spiral only varied by a few seconds out of ≈25 s, implying they were walking with the comparable velocity in each trial. Given the dynamic nature of the cognitive algorithms being implemented, this was deemed to be an acceptable level of variation.

The cognitive algorithm implemented for the experiments consisted of a two-tier hierarchy, with two parallel perception–action cycles operating in the lower tier. In accordance with the conventions set out in [16], the lower tier is referred to as tier 1 and the higher tier as tier 2. Each perception–action cycle was implemented using the FAR framework.

The tier 1 FAR implementations operated on each of the radar nodes shown in Figure 17.5 to perform adaptive single-target range-Doppler tracking in a manner similar to [13,14]. Each perception–action cycle was responsible for selecting the radar waveform parameters to optimize the quality of the range-Doppler track estimate obtained for each CPI. The full description of this process is given in [14,16]. The CREW waveform parameters are listed in Table 17.3. Note that for this experiment, the PRF has a fixed value of 4 kHz. This was necessary to ensure the measurements at radar 1 and 2 were time-synchronized. The dimensionality of the action space was

Table 17.3 Operating parameters for the CREW during the HFAR experiments. Table modified from [16]. ©2018 IET.

Parameter	Parameter bounds	# of Values	Goal value	Weight
Center frequency	92.5 GHz	N/A	N/A	N/A
PRF	4 kHz	N/A	N/A	N/A
Number of pulses	[64, 512]	10	64	10
Pulse width	[0.3 μs, 1.2 μs]	10	0.3 μs	10
Bandwidth	[100 MHz, 1 GHz]	10	100 MHz	1
Transmit power	[−1 dBm, 15 dBm]	10	−1 dBm	1

increased from the initial experiments, [13], to include four waveform parameters, as listed in the table. The parameter goal values were selected to minimize the number of samples collected per CPI and minimize the bandwidth and transmit power. The action cost was then a function of how far the selected values deviated from these goals. The quality of the perception was evaluated by setting goal values for the range and range-rate covariance.

The tier 2 FAR implementation was responsible for optimizing a Cartesian track of the target. Its perception was also the target state estimate and associated covariance, although this time the state included position and velocity in both x and y directions. This resulted in a state space with four components: x position, y position, x-rate, and y-rate. Its measurements were the raw range and range–rate measurements from the two radar nodes that were passed up the hierarchy by the tier 1 perception–action cycles. Since the tier 2 perception–action cycle is not connected directly to the radar hardware, it cannot directly control the radar. Instead, its actions took the form of additional performance constraints for the tier 1 perception–action cycles. The flow of measurement information up the hierarchy and constraints down the hierarchy is the information flow that Mitchell described in [16].

The experimental study compared the impact of having *hierarchical feedback* in the HFAR structure and not having it. In [16], the hierarchical feedback consists of actions output by the tier 2 perception–action cycle, i.e., the placing of additional constraints onto the tier 1 perception goals.

Figure 17.6 shows a comparison of state estimates from the tier 2 Cartesian tracker with and without hierarchical feedback. In the figure, the red lines with circle markers represent the case without feedback and the blue lines with “x” markers the case with feedback. In both cases, the tracker is able to track the target—a person walking along an inward spiral—successfully. However, it is clear that the track quality is lower for the case without hierarchical feedback: the curves for position and rate are smoother when feedback is employed.

The improved performance of the hierarchical feedback case is made clear in Figure 17.7. The figure shows the standard deviation for each of the four elements of the target state vector, which is the metric that indicates the quality of the track and hence tier 2 perception quality. Again, the red line with circle markers is the

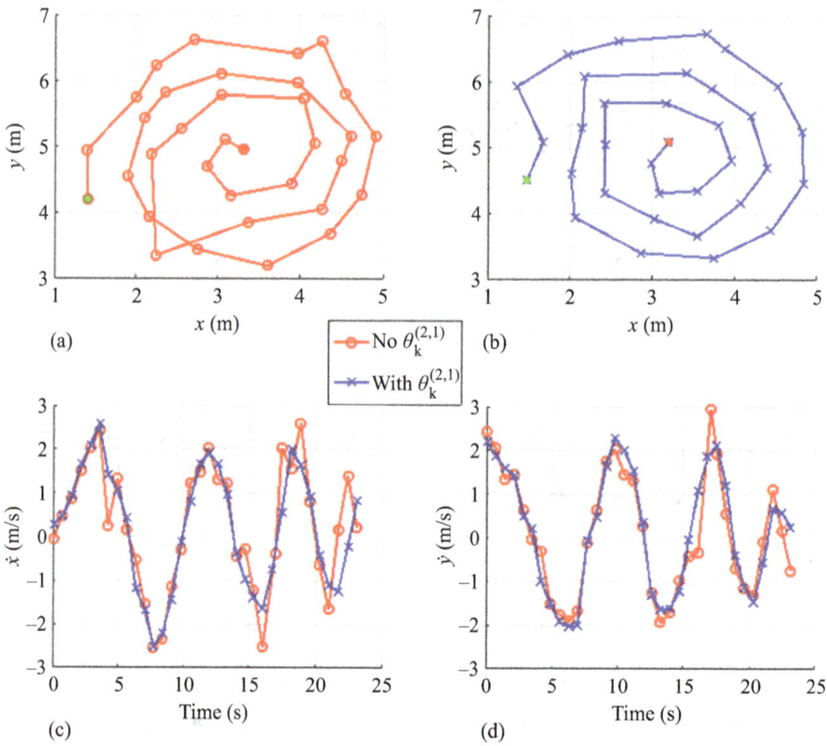

Figure 17.6 Cartesian state estimates from the tier 2 perception–action cycle that performs fusion, with and without hierarchical feedback. Subplots (a) and (b) show the target position estimate, in x and y, without and with hierarchical feedback, respectively. Subplots (c) and (d) contrast the x-rate and y-rate without and with hierarchical feedback cases. In the legend, the symbol $\theta_k^{(2,1)}$ represents the hierarchical feedback and the meaning of the symbol is clearly defined in [16]. Figure taken from [16]. ©2018 IET.

case without hierarchical feedback and the blue line with "x" markers is the case with feedback. The performance goals set by the operator are shown with dashed black lines. It is apparent from the figure that when there is no hierarchical feedback, the tier 2 perception quality goals are not met, despite a target track being achieved. Conversely, when feedback is incorporated, the perception quality quickly converges to a level that exceeds the goal.

The final result, shown in Figure 17.8, shows the quality of perception for the two tier 1 perception–action cycles when hierarchical feedback is in effect. The two tier 1 perception–action cycles control the range-Doppler track on each of the two radar nodes, and the perception is again the target state, with quality evaluated against the state covariance. Subplots (a) and (c) show radar node 1's range and range–rate

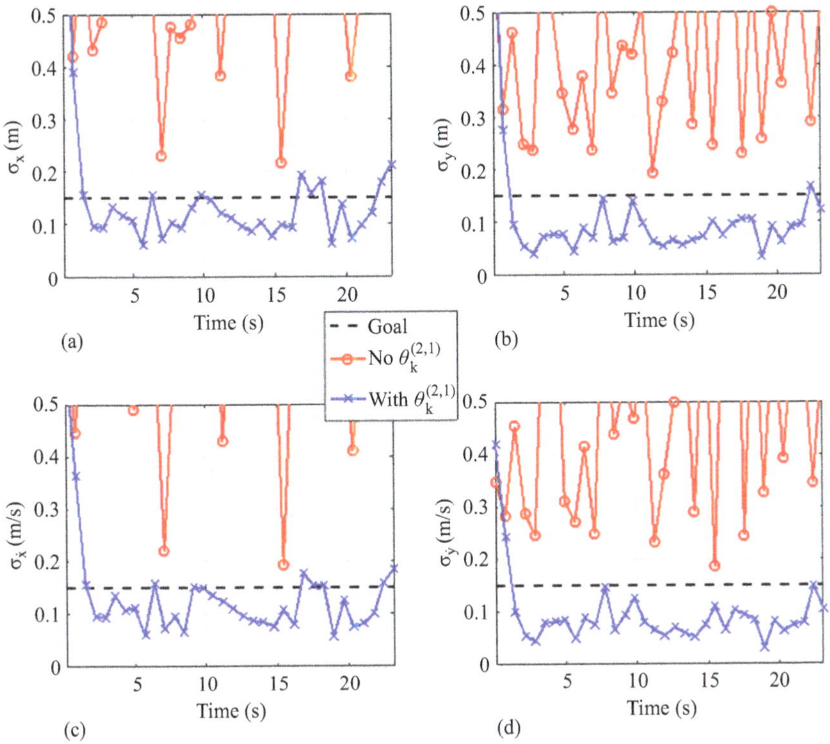

Figure 17.7 Cartesian track standard deviations for each element of the target state vector with and without hierarchical feedback: (a) x position; (b) y position; (c) x-rate; and (d) y-rate. Figure taken from [16]. ©2018 IET.

covariance respectively, while subplots (b) and (d) show the same information for radar 2. The dashed black line again shows the goal value for perception quality. In this case, the quality goal varies as a function of time because the tier 2 actions, that provide the feedback, continuously provide changing goals that are tighter than the default values for the tier 1 perception performance goals. We see that the achieved performance is consistently better than the requested goal. This is why the tier 2 perception quality is better when feedback is in effect: as tier 2 tightens the tier 1 performance goals it receives more accurate measurements from tier 1 and is able to establish a better track.

The results presented here demonstrate that multiple instances of the FAR framework can be connected in a hierarchy to undertake more advanced adaptive radar processing than the original framework permitted. The multiple levels of the hierarchy provide a consistent means of handling problems of varying scales. The HFAR approach to cognitive radar is in keeping with both Fuster's theories from neuropsychology, [12], and existing research into cognitive radar [11,14, for example].

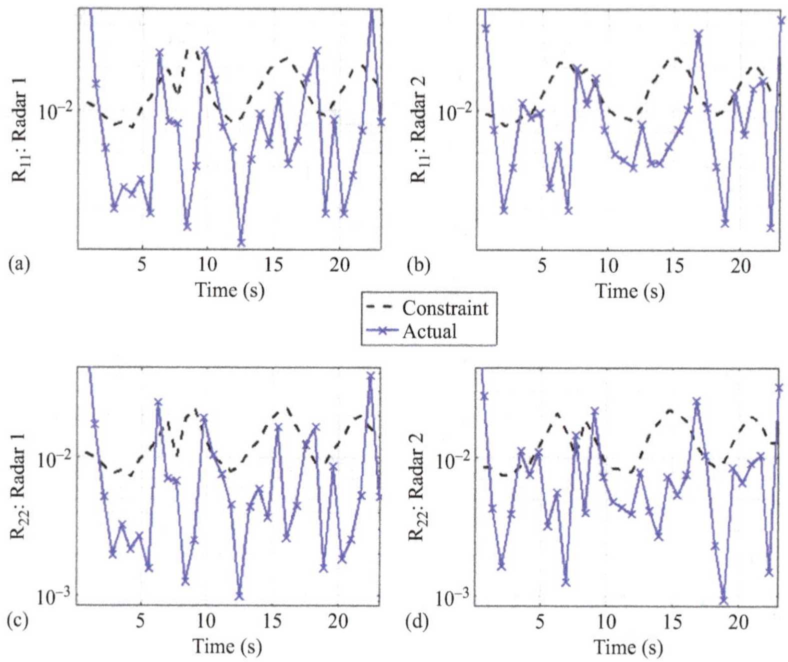

Figure 17.8 The diagonal elements of each tier 1 perception–action cycle measurement covariance matrix compared to the constraints set by the tier 2 actions: (a) range covariance for radar 1; (b) range covariance for radar 2; (c) range–rate covariance for radar 1; and (c) range–rate covariance for radar 2. Figure taken from [16]. ©2018 IET.

A critical component of the results presented is the need for hierarchical feedback of information. It was only by allowing the tiers of the hierarchy that are detached from the sensing hardware to pass information down to the lower levels that desired performance levels could be achieved.

17.2.2.2 Neural network-based fully–adaptive radar (FAR)

The CREW has been used to demonstrate that neural networks can be utilized within a perception–action cycle to select actions, as reported in [18]. Here we shall provide a summary of that research and its experimental results.

The majority of the work with the CREW uses a commercial off-the-shelf solver to provide the optimization module of the executive processor, Figure 17.4 top right. Selection of the best possible actions, constrained by perception quality, typically takes the form of a constrained, non-linear optimization. In [15], a generalized form of the action cost function is presented, and it is shown that it can work for a variety of radar operations. However, there remain two distinct problems with this approach.

First, despite work to provide generalized cost functions, the FAR approach to cognitive radar still requires a skilled engineer to have significant input to creating the function. While this has been demonstrated possible for small-scale problems, such as adaptive single-target range-Doppler tracking, the design task will become intractably complicated for larger-scale problems, e.g., advanced radar resource management where there are many competing objectives and radar tasks to be included [19, for example].

Second, solvers for optimization problems have undesirable properties for cognitive radar. The time it takes the optimizer to converge on a solution can be substantial, potentially longer than the desirable update rate of the radar. Although outside the scope of this chapter, this challenge had to be addressed in the HFAR experiments and was a guiding influence for the selection of the waveform parameter goal values, see [16]. Perhaps worse is that the time it takes the solver to find the solution is not constant. The time to convergence for the solver often depends on the initialization value and this can vary between cognitive radar measurements.

To overcome these two challenges, it would be desirable to use a learning method to select actions based on the predicted perception of the local environment. The desire arises because, for many learning algorithms, after an initial training stage, the online operation is fast and execution time is fixed. Such an approach was explored using the CREW—a neural network was trained to replicate the behavior of the optimization process in single-target range-Doppler tracking example from [13]. Essentially, the optimizer block of the FAR framework was replaced with a neural network, see [18, Figure 2].

To research neural networks, an adaptive single-target range-Doppler tracking problem was used. The radar was responsible for tracking a target, a person jogging back and forth in front of the radar antenna and adapting the waveform parameters for each CPI to optimize track performance. This situation is described in detail in [13,14] and is comparable to the operation of the tier 1 perception–action cycles from the HFAR experiments described earlier. A single perception–action cycle, implemented using the FAR framework, was used to control the radar. The actions were the waveform parameters, the perception was the target state and its covariance, and the radar was created using one pair of transmit and receive heads from the CREW. The experiment environment was set up at The Ohio State University's ElectroScience Laboratory. A schematic diagram of the experiment is shown in Figure 17.9. The CREW operating parameters and the adaption ranges are shown in Table 17.4.

The neural network used in the experiments was a feed-forward neural network trained using the Levenberg–Marquardt training algorithm paired with a generalized regression neural network structure. Full details of the network are provided in [18]. Here we shall simply note that the network is modest in size, comprising five neurons in the input layer, approximately 240 neurons in the hidden layer, and 5 neurons in the output layer. The number of neurons in the hidden layer is approximate as it varied between 231 and 243 depending on the number of waveform parameters being adapted.

A critical component of this research was that the neural network was trained on simulated data. The FARMS software that controls the CREW can be operated in a

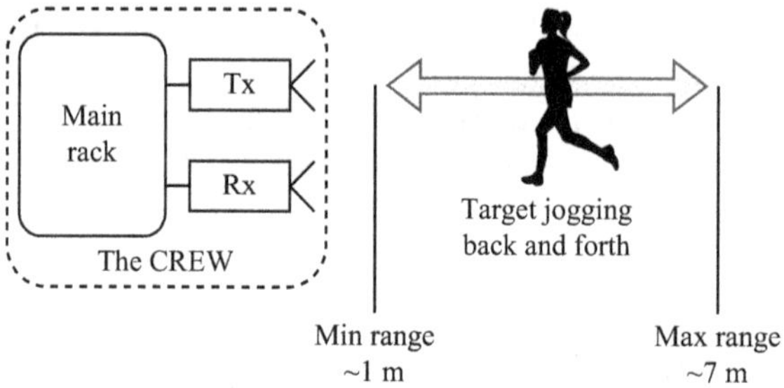

Figure 17.9 Setup for the adaptive single-target range-Doppler experiment used to demonstrate neural network use in the FAR framework using the CREW testbed. Figure modified from [18]. ©2022 IEEE.

Table 17.4 Operating parameters for the CREW during the neural network experiments. Table modified from [18]. ©2022 IEEE.

Parameter	Parameter bounds	# of Values	Goal value	Weight
PRF	[2 kHz, 12 kHz]	10	12 kHz	20
Number of pulses	[64, 1024]	10	64	10
Pulse width	[0.15 μs, 1.2 μs]	10	0.15 μs	1
Bandwidth	[100 MHz, 1 GHz]	10	100 MHz	1
Transmit power	−13 dBm	N/A	N/A	N/A

simulation mode. One of the provided simulation environments is for a single-target range-Doppler tracking case. This simulation has been tuned so that the target radar cross section (RCS) fluctuations and variations in return signal-to-noise ratio (SNR) are consistent with the setup shown in Figure 17.9. A training database for the neural network was created using this simulation environment.

The perception–action cycle process selects a set of waveform parameters (the action) in response to the target state estimate (perception) for each CPI. This happens regardless of whether FARMS is running a simulation or experiment. By saving the perceptions and selected actions (target states and waveform parameters), a training database for the neural network was created. Once trained, the neural network could be used in the FAR process instead of the solver. It is important to note that the neural network was trained to approximate the operation of the solver undertaking a specific constrained non-linear optimization and that this is different to training the neural network to approximate a generalized solver. Each time the form of the constrained non-linear optimization is changed, e.g., changing the number of radar parameters adapted in the actions, the neural network must be retrained.

Cases were studied in which the number of parameters to be adapted ranged from one to four. For each case, the neural network was retrained using the simulator with the solver-based optimization problem set to work with the appropriate number of waveform parameters.

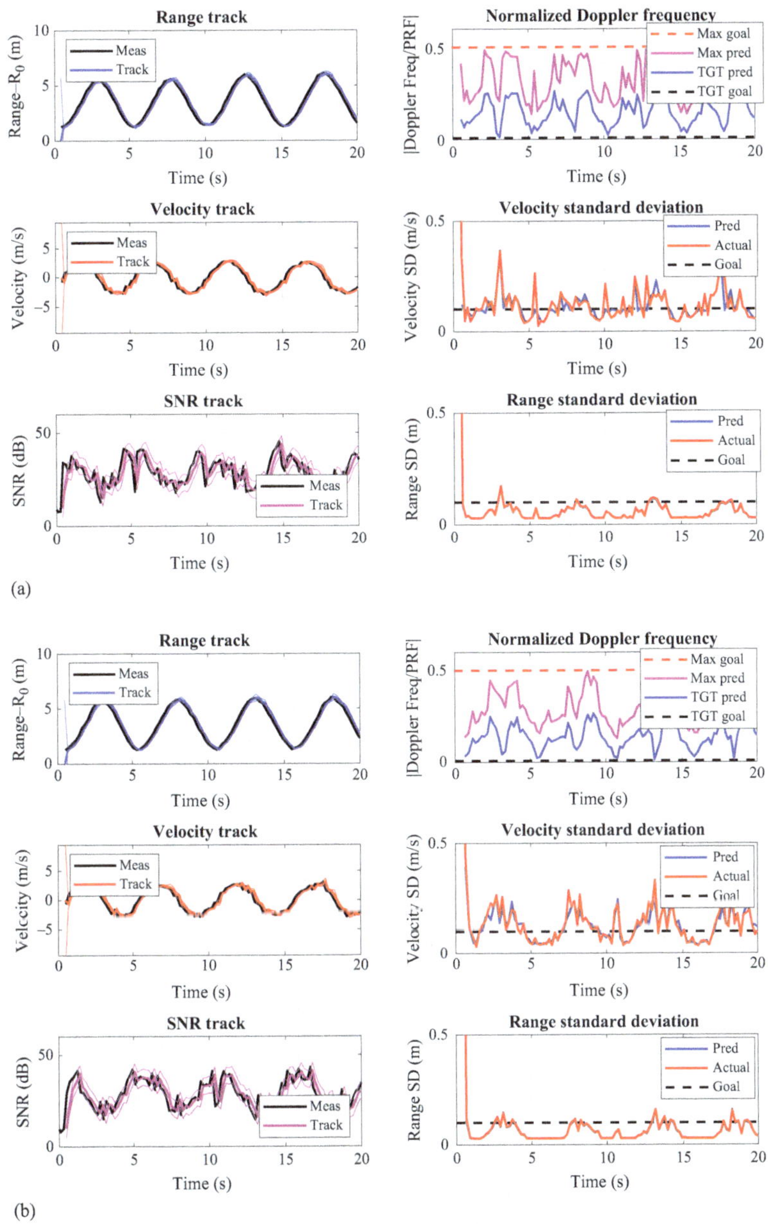

Figure 17.10 Comparison of the perceptions obtained for adapting a single waveform parameter (PRF) using (a) a numerical solver and (b) a neural network. Figure taken from [18]. ©2022 IEEE.

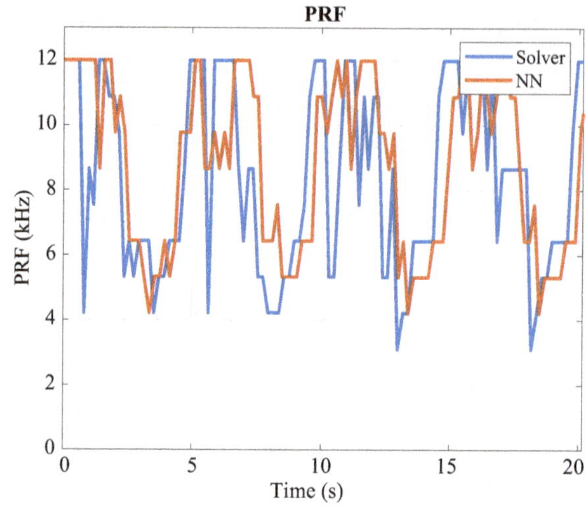

Figure 17.11 Comparison of the action selections (chosen PRFs) for the solver-based and neural-network-based FAR implementations. Figure taken from [18]. ©2022 IEEE.

The results for the adaption of a single parameter, the PRF, are shown in Figures 17.10 and 17.11. The first figure shows the perception progressing over time for (a) the numerical solver and (b) the neural network. The second figure compares the actions taken over time—PRF selections—for the two methods. In Figure 17.10 the left of the plot shows the progression over time of components of the target state vectors and the associated radar measurements; the right of the plot shows the predicted normalized Doppler of the target (top), the standard deviation of the target velocity compared to the goal value (middle), and the range standard deviation compared to the goal value (bottom).

What is striking about Figures 17.10 and 17.11 is how closely the two approaches match. We can see from the range and velocity tracks that the motion of the target, the jogging person, is very similar in both experiments. We also see that the obtained perceptions and the action selections match reasonably well. A more detailed analysis of the similarity is provided in [18].

To compare the solver and neural network approaches, John-Baptiste calculates the optimizer cost function values for the neural network solutions and then compares these values with those obtained during the optimizer-based experiments. Comparison is performed by histograming the cost function values obtained during the experiment and then using Student's *t*-test to compare the histogram distributions. In [18], it is shown that, to a 5% statistical significance, both the optimizer and neural network solutions are the same for all considered parameter adaption cases.

A major significance of the neural network approach is its faster execution time. [18, Table VI] reports the meantime for the optimizer to run and the neural network to execute for the different parameter adaption cases run. As the number of parameters adapted increases from one to four, the mean optimizer time increases from 0.010 s to 0.015 s. The execution times for the neural network are an order of magnitude faster, with an average time of 0.004 s and a stand deviation of 0.0003 s across the various cases. The low standard deviation shows that the neural network execution time is not affected by the number of parameters adapted. The improved execution time means that the neural network approach should be able to obtain more measurement updates per unit time and this would be expected to result in better tracking performance.

Overall, these experiments with the CREW demonstrate that a neural network can be used to approximate the optimization process inherent to the FAR approach to implementing the perception–action cycle from cognition. It was notable that the neural network approach had a faster execution time, which has the potential to lead to increased measurement rates and hence improved performance (although that was not explicitly shown). More excitingly, the introduction of neural networks to cognitive radar opens the door to using learning techniques to replace the cost functions at the core of the FAR framework that require engineer input to fine-tune.

17.3 The cognitive detection, identification, and ranging testbed

The development of the cognitive detection, identification and ranging (CODIR) testbed was initiated in 2015 with the start of the NATO SET-227 research task group [20]. It was decided to modify an existing short-range frequency-modulated continuous-wave (FMCW) radar demonstrator, operating in X-band, to perform cognitive radar experimentation.

17.3.1 Development considerations

At the beginning of the development, the following key requirements were placed on the system:

- The waveform generator shall be software programmable and able to adapt the waveform within a CPI.
- The radar signal processor shall be able to process the radar data including single target detection and tracking in (near) real-time.
- A software module (optimizer) shall be able to choose the optimal waveform and processing parameters for the next track update, based on the recent radar processing output.

In the first stage, the waveform generator and radar signal processor have been developed and tested independently of the optimizer development. The modules were first tested similarly to a traditional, non-cognitive radar sensor. The waveform adaptivity was initially tested with an external trigger. In parallel, the optimizer has been developed and tested first with simulated in-phase and quadrature (IQ) data from a digital twin representing the actual sensor. In the second phase, sensor data with a reproducible target environment and with different waveform and processing parameters have been recorded as an input to the optimizer for further testing and optimization. The development concluded with the integration of sensor and optimizer for real-time testing of the system.

17.3.2 The CODIR design

The final CODIR system consists of an adaptive radar sensor that perceives the environment using optimized radar parameter settings and a controller that tracks the target and selects the optimal radar parameters for every new measurement. The sensor is composed of an adaptive waveform generator, an RF front end, an ADC converter, and a real-time signal processor and display. The signal processor consists of a range-Doppler processing step and a maximum search to detect the target. The controller consists of a Kalman filter tracker and the optimizer that selects both optimal waveform and processing parameters in real-time, based on the latest measurement (detection). More details on the CODIR system are available in [21]. A functional block diagram is provided in Figure 17.12 and a photo of the CODIR sensor is shown in Figure 17.13.

Both the controller and the processor are coded in MATLAB® and run on an Ubuntu Linux PC. The control of the waveform generator and the transfer of the raw data from the sensor backend to the processor are managed using software written in C/C++. The operational parameters of the sensor are summarized in Table 17.5.

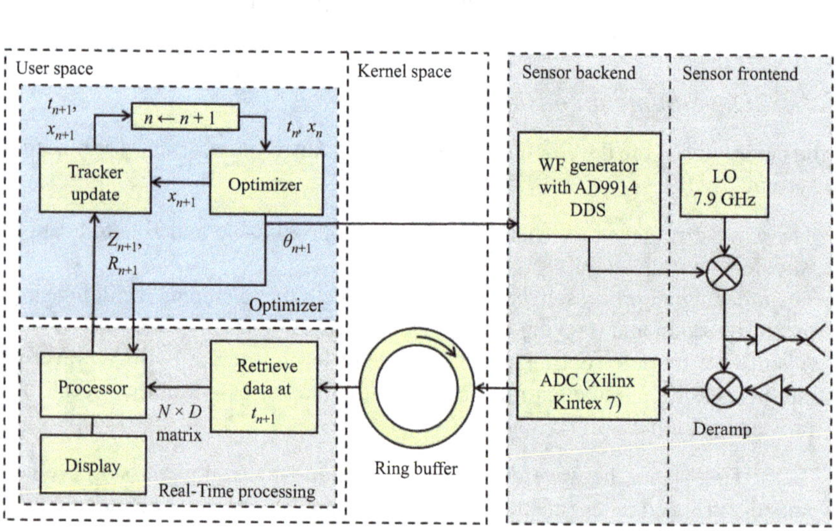

Figure 17.12 Cognitive detection, identification and ranging (CODIR) functional block diagram. The black arrows show the data flow and indicate the perception–action cycle feedback loop. Figure taken from [21]. ©2019 IEEE.

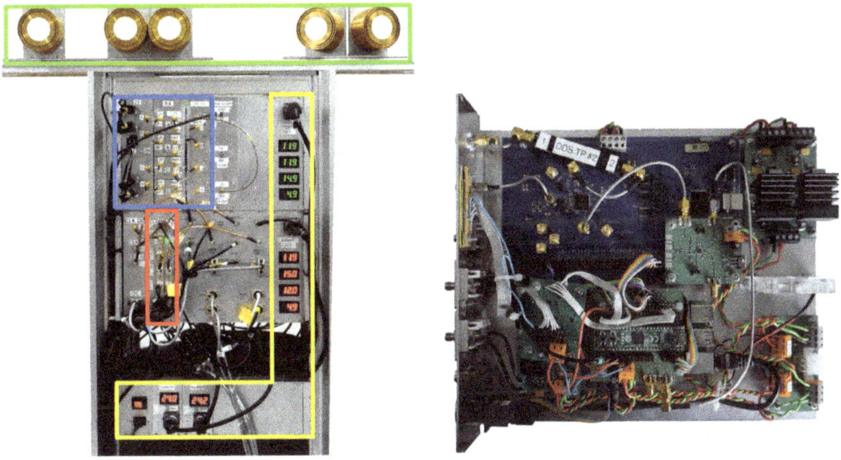

Figure 17.13 The CODIR sensor. A picture of the sensor with antennae (green box), frontend RF modules (blue box), signal generator board (red box), and power supply (yellow box) is shown on the left. The signal generation module with the DDS board is shown on the right.

Table 17.5 CODIR operating parameters

Parameter	Value
Tunable frequency range	8.1–9.3 GHz
Bandwidth	up to 1.2 GHz
TX power	up to 33 dBm
IF band frequency range	3–23 MHz
PRF	up to 83 kHz
Waveform	LFM generated with a DDS
Number of TX channels	2
Number of RX channels	4

The following cognitive radar-specific components have been developed:

- Waveform generator: The waveform generator module consists of an AD9914 DDS evaluation board generating LFM pulses, a small field programmable gate array (FPGA) that triggers the DDS with a given PRF and a Raspberry Pi that controls and adapts both components. The waveform is changed by a request from the controller via a local area network interface. Through this interface, the chirp bandwidth, the chirp length, and the PRF can be adapted within microseconds.
- Optimizer: The optimizer is a software module within the controller and implements a perception–action cycle using the formulation proposed in [22] to choose

the optimal waveform and processing parameters for the next track update. Selection is based on the latest track state vector, the track covariance and the sensor performance (measurement covariance, SNR) estimated using all waveform and processing parameter settings considered for optimization. The final parameter selection is done by minimizing a scalar cost function that balances the cost to assure a good tracking performance with the cost for the sensor to do the measurement. More details on the optimization method and the choice of cost function are given in [21].

17.3.3 Experimental work with CODIR

Several experiments in different environments and with different optimization goals have been performed with the CODIR testbed.

17.3.3.1 Cognitive radar performance analysis with different types of targets

In [23], the performance of the optimization using a perception–action cycle was examined by considering a variety of targets (motor car, bicycle, and radio controlled buggy) in an outdoor environment. The radar parameters were real-time adapted by minimizing a scalar cost function to meet the three (partly conflicting) objectives "best track accuracy," "minimal bandwidth usage," and "minimal radar usage" with different prioritization. As an example, Figure 17.14 presents five measurements with different cost functions and shows how the target (measured and tracked) range accuracy changes when the optimization priority shifts from "best track accuracy" to "minimal bandwidth usage." This work shows that the choice of cost function is crucial and influences the radar performance decisively. The design of the cost function is also discussed in [15].

17.3.3.2 Waveform adaptation in a jammed and congested spectrum environment

In this experiment, a congested and noise-diode-jammed spectral environment were considered. The environment was defined by an allowed frequency band of 8.5–9.3 GHz and a jammed part of the band between 8.7 and 9.1 GHz. The jammer signal was realized with a power-adjustable noise diode, band-filtered, and injected into the receive channel directly after the antenna. Details to the experiment are available in [24]. The catalog of possible waveforms (WFs) consisted of LFM WFs with different chirp bandwidths and of gapped LFM WFs with a spectral notch at the jammed part of the allowed band. The jammer interference with both a standard LFM waveform and a gapped LFM WFs are shown in Figure 17.15.

The gapped WFs were synthesized by assembling LFM subchirps with different start and stop frequencies and start phases adjusted for an overall phase-continuity of the gapped waveform. The task of the optimizer was to select an optimal waveform from the catalog to optimize the target SNR and to decide between an LFM waveform with large bandwidth but potential interference from the jammer and a gapped LFM waveform with less bandwidth but less interference potential. Several experiments with different jammer power levels were conducted and demonstrated

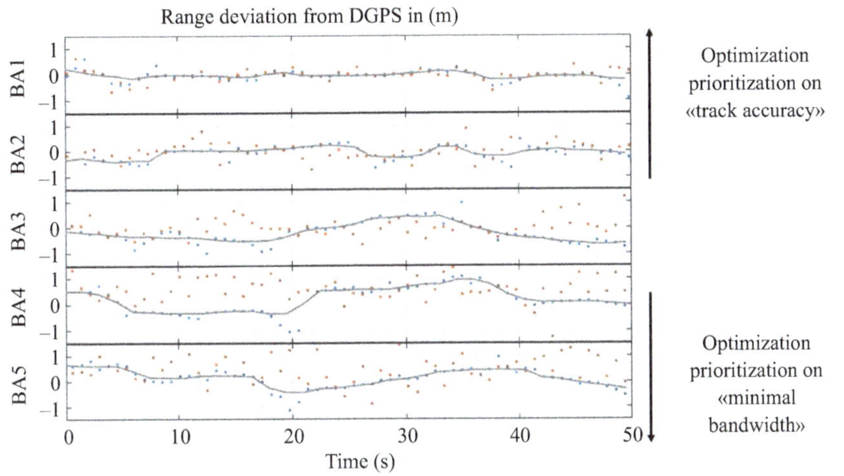

Figure 17.14 Deviation of measured (red) and tracked (blue) target range from the ground truth value (differential GPS). Series of five measurements (BA1–BA5) with optimization priorities varying from "best track accuracy" (BA1) to "minimal bandwidth" (BA5). The track updates are smoothed with a sliding window (grey line) for an indication of the track accuracy over time. Figure adapted from [23]. ©2019 IEEE.

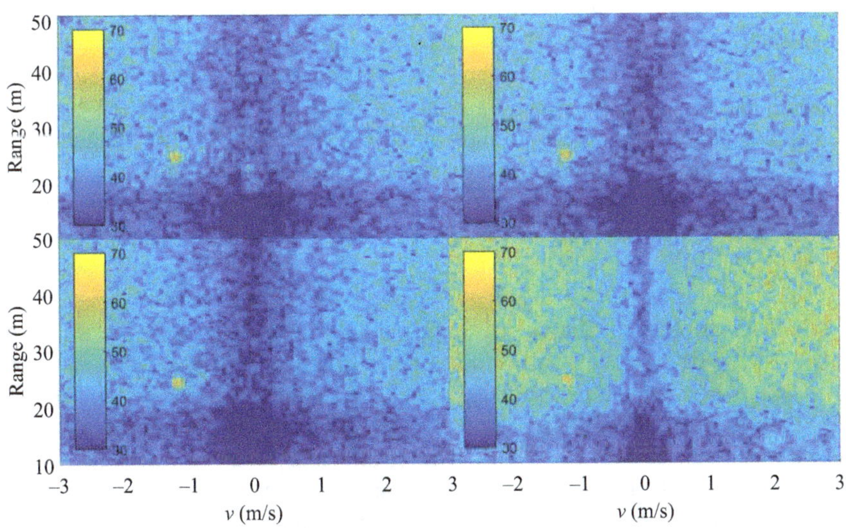

Figure 17.15 Impact of waveform selection and jamming on the range-Doppler map. Panels from top left to bottom right: (a) gapped LFM, no jamming; (b) gapped LFM, moderate jamming; (c) standard LFM, no jamming; (d) standard LFM, moderate jamming. Figure taken from [24]. ©2018 IEEE.

the FAR optimizer to adapt the radar to an increasing interference by switching from large bandwidth LFM waveform to gapped LFM waveform.

17.4 Universal software radio peripheral-based cognitive radar testbed

17.4.1 USRP testbed design

A USRP-based cognitive radar testbed was developed by the Norwegian Defence Research Establishment (FFI) to support a PhD program at The Ohio State University researching how cognitive methods could be implemented on low-cost radars [25]. A full system description is provided in [25,26]. A X310 USRP was selected as the digital backend of the platform because it provided a high digitized-signal quality with real-time software control at a modest price point. The high signal quality facilitates high-sensitivity radar applications, such as unmanned aerial vehicle (UAV) detection and tracking. The response time of the software control facilitated real-time radar control of waveforms and other parameters required for cognitive methods.

The testbed uses a X310 USRP for waveform generation and digital sampling. The X310 device can operate at many frequency ranges depending on the RF downconvertor used with it. In USRP parlance, the downconvertors are referred to as "daughter boards" that connect in a plug-and-play fashion to the main USRP device. The specification of the UBX-160 cards used in the testbed is 160 MHz instantaneous bandwidth with complex samples and a tuning bandwidth of 10 MHz to 6 GHz RF frequency. The X310 has two channels, each of which connects to its own UBX-160 daughterboard. Utilizing the dual channels of the USRP allows for phase comparison monopulse, dual polarization or other dual channel applications that depend on phase coherency. The USRP is connected to a workstation computer via a dual 10 Gbps ethernet link. The waveform to be transmitted is sent over the link to the USRP and the sampled data is sent back to the computer for processing. The RF frontend consists of filters, amplifiers, and antennas optimized for operation in a frequency range of 3–3.3 GHz, part of S-band. A full parts list is given in [26]. A block diagram of the system is shown in Figure 17.16 and a picture of the RF frontend is shown in Figure 17.17.

The testbed is capable of adapting the radar waveform and operating parameters on a dwell-to-dwell basis. During operation, there is a delay when sending waveforms to the USRP, a delay when transferring samples from the USRP to the workstation, and a delay when calculating the radar parameters for the upcoming dwell. The total delay is in the order of milliseconds, which would limit adaptation on a pulse-by-pulse basis to unacceptably low PRFs. The waveform generator is a true AWG, and all signals that can be represented digitally in an IQ format may be transmitted.

The signal processing software developed for this testbed is published on Github in its entirety,† and is based on the C++ and CUDA programming languages to

†https://github.com/jonasmc83/USRP_Software_defined_radar

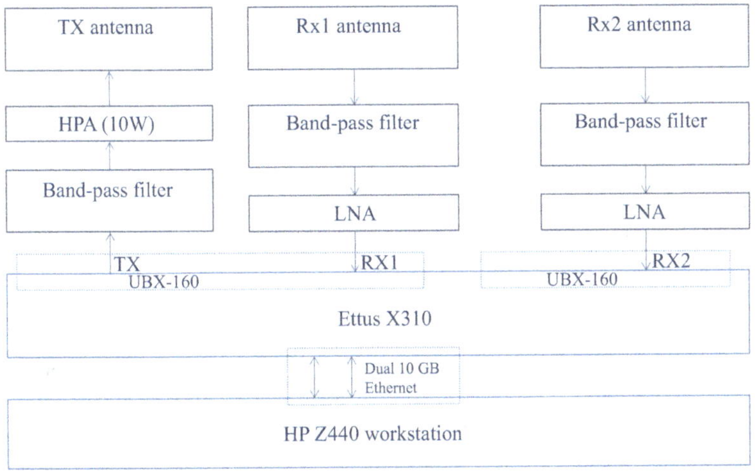

Figure 17.16 Block diagram of the USRP-based cognitive radar testbed. Figure taken from [26].

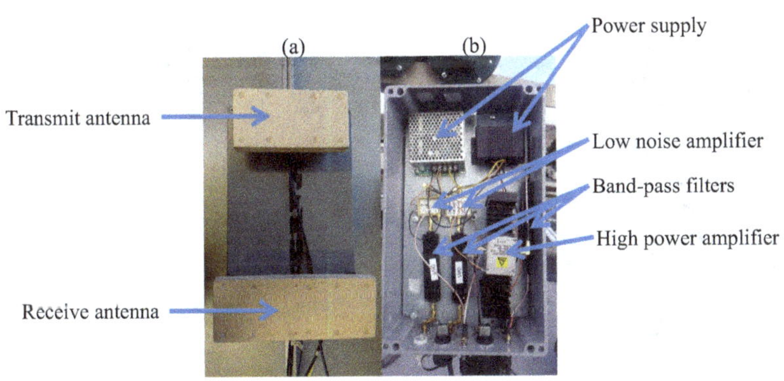

Figure 17.17 Pictures of the different components of the USRP-based cognitive radar: (a) shows the antenna setup with the transmit antenna on top and receive antennas at the bottom; and (b) shows the RF front-end electronics with filters and amplifiers. Figure adapted from [26].

achieve real-time operation. The real-time signal processing is a requirement for testing adaptive and cognitive radar algorithms because the system adaption must happen at a rate corresponding to the changes in the sensed environment. An example of such adaptive algorithm implemented on this testbed is the adaptive track update interval algorithm [27], where the update interval for the next track update is calculated using the FAR framework [14] to minimize the track covariance. An experimental demonstration of this algorithm is described in the next section using a UAV as a target.

17.4.2 USRP testbed demonstration experiments

17.4.2.1 Detection range

The system utilizes a 10-W power amplifier on transmit and initial predictions of the detection performance [28] showed a detection range greater than 5 km for airliner size targets and a maximum detection range of 3 km for light aircraft. Figure 17.18(a) shows a detection of a small aircraft at 2.8 km distance, and Figure 17.18(b) shows a detection of a large airliner aircraft detected at 5.5 km. These results confirm the radar range equation performance estimates using real-world target of opportunity.

17.4.2.2 Tracking small targets

The USRP radar is able to track small targets at short range. Using a short pulse allows it to monitor a surveillance volume close to the radar. An experiment with a small UAV—a DJI Mavic—was performed. The pulse length was set to 0.8 μs, which gives a minimum range of 120 m. The PRF was set to 100 kHz, which gives

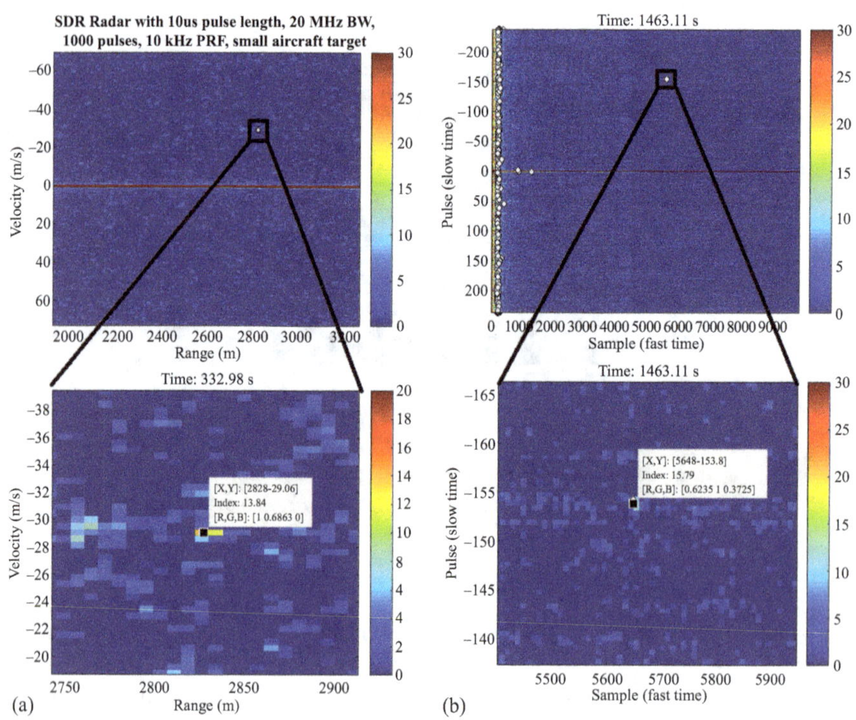

Figure 17.18 Validation of predicted detection ranges of the USRP based radar testbed. Figure taken from [26].

an unambiguous range of 1.5 km. The experiment is shown in Figure 17.19(a), which shows a picture of the UAV, a satellite photo of the test range and the position of the radar, and the track of the UAV. The UAV path in the overlay is the radar track of the UAV and goes someway to illustrating the angular accuracy of the testbed's angle of arrival (AoA) estimation.

Figure 17.19(b) shows plots of the range, Doppler velocity, and SNR of the UAV as it first flies away from, and then back towards the radar. The black dots are the raw radar detections and green squares are the tracker output. The tracker maintains track until 350 m. It loses track when the target stops as it prepares to change direction. The lost track happens because the track merges into the stationary clutter and is not detected for several dwells. The operator initiates a new track as soon as the target resumes motion and becomes separate from the clutter. It is clear from the figure that the UAV is successfully tracked in both directions.

The results suggest that the maximum detection range for the UAV is ≈600 m. In Figure 17.19(a), we see that the UAV is at a range of 300 m at times ≈18 s and ≈45 s. From Figure 17.19(c), we see that the SNR at these times is ≈25 dB. Doubling the range to a target reduces the SNR by 12 dB, so for a UAV at 600 m we would expect the SNR to drop to 13 dB which is just sufficient for detection.

17.4.2.3 Adaptive update interval method using the FAR framework

The paper by Christiansen [29] presents a method of calculating a track update based on a cost function that optimizes a measure of resource usage weighted against the

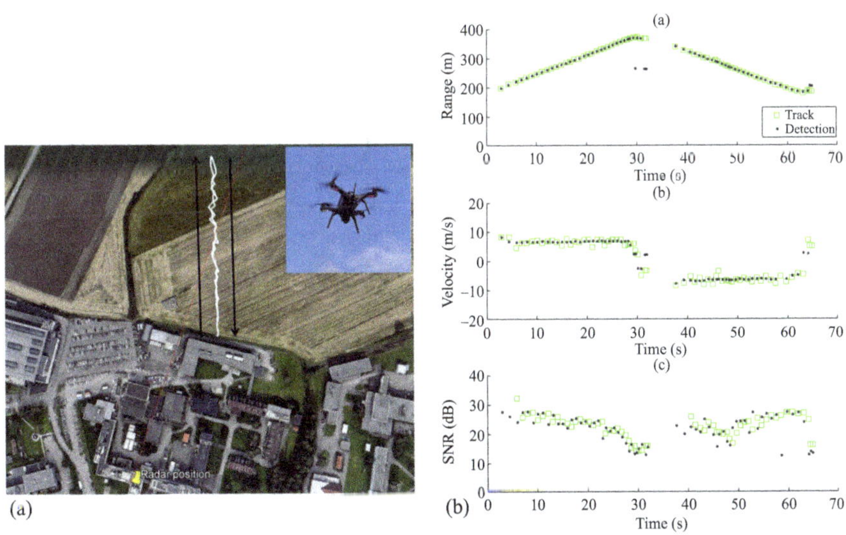

Figure 17.19 Experiment showing tracking of small UAV target out to 350 m. Figure taken from [26]. (a) Experimental setup of UAV flying outwards, turning, and flying back. (b) Range, range-rate (Doppler velocity), and SNR plot respectively, from top to bottom.

global mean square error of the tracker [14]. The article shows the method applied to a simulation study. The same approach was implemented on the USRP radar and the required interval until the next track update was dynamically adapted based on the predicted track error.

The USRP cognitive radar testbed, running software that implemented the cognitive adaptive update interval tracker from [29] was used to track the DJI Mavic UAV flying a course comparable to the track shown in Figure 17.19(a).

Figure 17.20(a) shows the tracker parameters of range uncertainty, velocity uncertainty, and selected update interval respectively as a function of experiment time. The tracker uncertainty is approximately constant since the optimization method decreases the update interval as the target moves further from the radar. The measurement uncertainty increases when the target is further from the radar due to reduced SNR. The optimizer selects shorter update intervals since the tracker has lower state uncertainty with shorter update intervals and hence the global track error remains approximately constant despite the SNR drop. Figure 17.20(b) compares two flight trials: the first, the red line is for a case where the update interval is fixed at 1 s; the second, the blue line is for the adaptive update interval case. The value of the cost function at the core of the adaptive approach can still be calculated for the fixed interval case. As such, it is a convenient metric for comparison of the two trials. The cost function values,

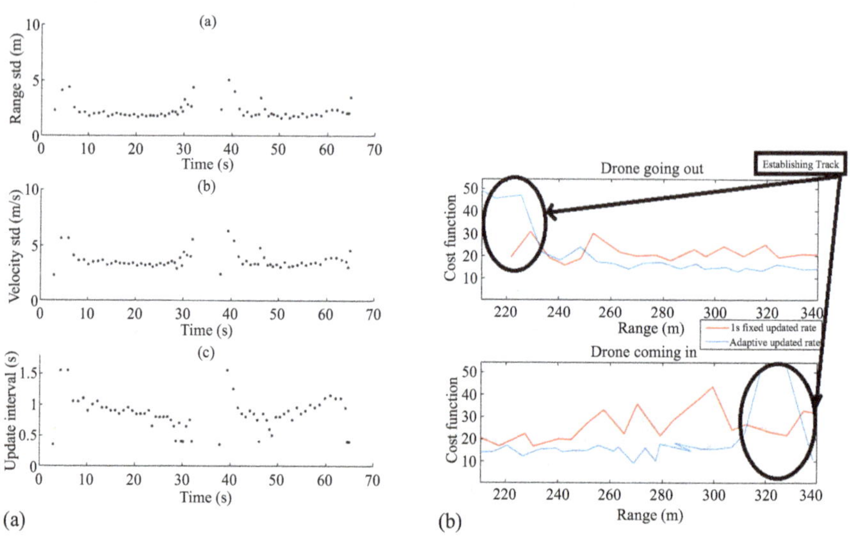

Figure 17.20 Experiment showing tracking of small UAV target, with showing track parameters and cost function values. Figure taken from [26]. (a) Tracker parameters, with plots of range uncertainty, velocity uncertainty, and the selected next update interval, respectively. (b) Comparison of the cost function values for two trials. The first, red line has a fixed update interval of 1 s. The second, blue line uses the adaptive update interval algorithm.

based on the true posterior information, were smaller when using the adaptive update interval method. This result was expected since the adaptive algorithm was designed to minimize the cost function based on predicted posterior information.

17.5 The miniature cognitive detection, identification, and ranging testbed

While the CODIR testbed, presented in Section 17.3, was a good starting system for cognitive radar experiments, it had some limitations in adaptability and mobility. For the development of a successor system an existing X-band FMCW radar based on a Xilinx Zynq 7030 system on a chip (SoC) as a backend was the starting point. A new RF frontend with improved sensitivity and suppression of unwanted mixing products in the high frequency (HF) and the IF band was developed. A further requirement for the development was the real-time operation with AWG waveforms and netted operation using different radar sensors.

17.5.1 The miniCODIR design

The miniature cognitive detection, identification and ranging (miniCODIR) testbed consists of four monostatic X-band radar sensors linked to a central processor. Each sensor consists of a backend and an RF-frontend. The frontend designed is similar to a classical (non-cognitive) radar and includes a coherent, double superheterodyne mixer board for up- and down-conversion of the transmit and receive signals, a common LO and clock generation board, a power amplifier (PA) driver on the transmit path and low noise amplifier (LNA) boards close to the receive antennae. The LO frequency for the second mixing stage defines the center frequency of the transmit signal and is software adjustable. The backend consists of an Ethernet I/O interface, a GPS receiver for time stamping, the ADCs and DACs modules, and a Xilinx Zynq 7030 SoC. The latter generates the Tx waveform and the pre-processes the sampled IQ data stream from the receive channels. A simplified sensor block diagram is shown in Figure 17.21 and a photo of the sensor is provided in Figure 17.22. Typical operational parameter ranges are summarized in Table 17.6.

The central processor contains the signal processor for each sensor, the controller, and the operator GUI. These processing modules run in parallel as separate MATLAB instances with a common shared memory space for data exchange. Each signal processor is assigned to one of the sensors and takes the decimated IQ data from the corresponding sensor as input for range-Doppler processing. A constant false alarm rate (CFAR) detector extracts a list of measurements (detections) and stores them in the share memory space as input for the controller module. This module fuses the detections from all sensors to a target track using a centralized extended information filter (EIF) tracker and decides on the optimal waveform or set of radar parameters for the next track update based on the predicted track performance and external objectives, such as a minimal spectral footprint. The waveform optimization is designed similar to the optimization module of the CODIR testbed (see Section 17.3) but has

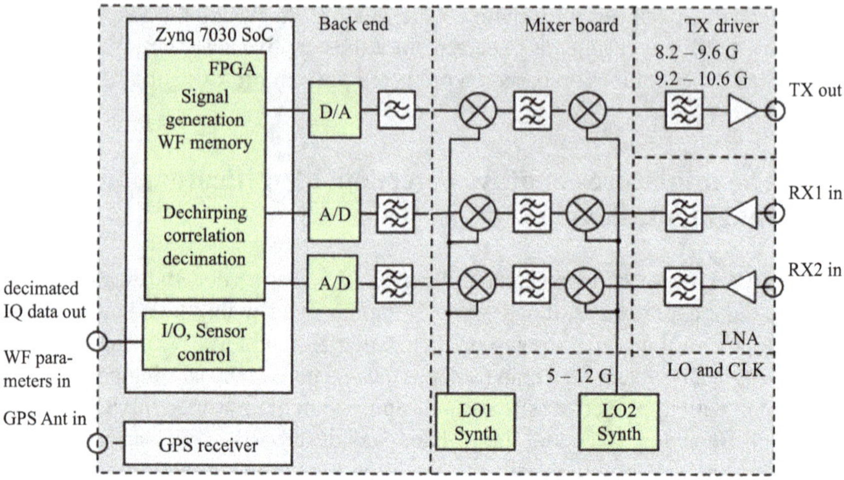

Figure 17.21 Simplified block diagram of one of the miniCODIR sensor nodes

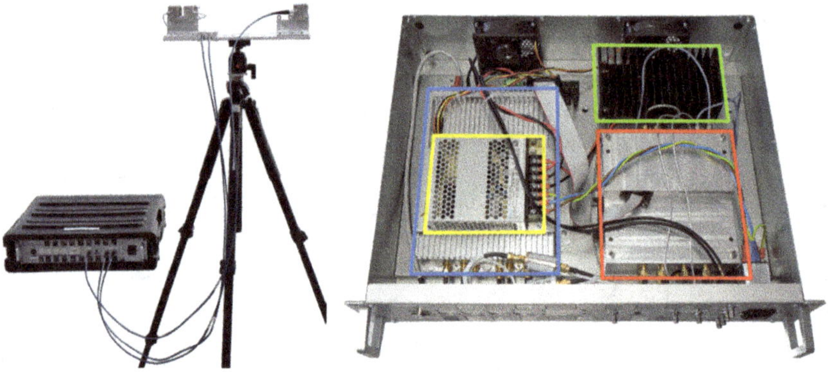

Figure 17.22 The miniCODIR testbed. A picture of the deployed sensor node with antennae and sensor rack is shown on the left. The view into the sensor rack (right panel) shows the backend (blue box) and the power supply (yellow box) on the left side of the rack while on the right half, the frontend modules (red box), and the power amplifier (green box) are visible.

been extended to optimize the parameters of multiple sensor nodes. It can be switched off for classical (non-cognitive) radar operation. More details on the processing chain are given in [30]. A simplified diagram of the miniCODIR central processor is shown in Figure 17.23.

The sensors are connected to the central computer via a Gigabit Ethernet connection, to transfer raw sensor data to the central processor and control statements to

Table 17.6 Typical miniCODIR operation parameter ranges use during experiments and tests

Parameter	Value
Tunable frequency range	8.2–10.6 GHz
Instantaneous Bandwidth	up to 80 MHz
TX Power	up to 33 dBm
Waveform	LFM or AWG
PRF	1.7–13.6 kHz
processed pulse length	80%–90% of the pulse repetition interval PRI = 1/PRF
Number of TX channels per node	1
Number of RX channels per node	2

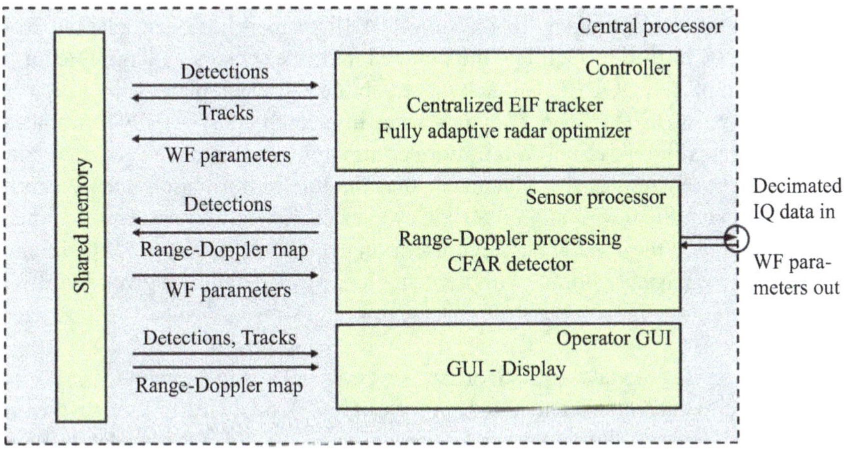

Figure 17.23 miniCODIR central processor functional block diagram. For simplicity, only one sensor processor is shown. The arrows indicate the data flows between the different processing modules.

the sensors. To control the sensor from the central computer, a MATLAB interface based on the `libiio` library from Analog Devices has been developed.

The miniCODIR system has the following key characteristics for cognitive radar experimentation:

- Signal generation: A LFM mode and an AWG mode are available. For the LFM waveform, the baseband signal is generated with a DDS by defining the corresponding LFM waveform parameters (start and stop frequency, pulse duration). To change the waveform, the corresponding waveform parameters need to be changed by the central processor. To use the AWG waveform mode, the waveform as a list of complex values that need to be transferred to the FPGA memory. In operation, the stored waveform is sent repetitively to the transmit chain. To change the waveform, the central processor needs to send a new waveform to the FPGA memory.

- Received signal processing: The received baseband signal is ADC converted with a sampling rate of 250 MHz and sent to the FPGA. In the LFM mode, the received chirp r_n is digitally dechirped by multiplying r_n with a complex conjugate copy of the transmit chirp s_n. The dechirped signal $y_n = r_n s_n^*$ is decimated using a cascaded-Integrator-Comb (CIC) and inverse sinc finite impulse response (FIR) filter and transferred as an IQ data stream to the central computer for further processing. In the AWG mode, the received signal r_n is correlated with the transmit signal s_n using a Fourier transform-based correlation. The correlation product in frequency space $\hat{y}_n = \hat{r}_n \hat{s}_n$ is decimated and streamed to the central computer for further processing. Here, the hat accent represents the Fourier transform of the signal.
- Controller: The optimizer for the miniCODIR testbed has been adapted from the CODIR testbed to facilitate the planned multiple sensors and multiple target operation of miniCODIR. To track targets in the Cartesian plane with a network of sensors, an EIF tracker [31] has been implemented. The EIF formulation is algebraically equivalent to the more commonly used extended Kalman filter (EKF) tracker but has the advantage, that the track information update can be written using the information matrix (inverse covariance matrix). This is similar to the notation used in the perception–action cycle formulation of [22], which is used for the current optimizer implementation. More details are given in [30].

The processing modules (sensor processors, controller, and operator GUI) are not synchronized to each other. The sensor processors work on a CPI-to-CPI base while the controller updates the target tracks according to the current track update rate (typically 0.5–1 s). The tracker takes the latest detections from each sensor to update the current target tracks. When the optimizer decides to change the radar parameters or WF, the new waveform parameters are stored in the share memory. At the next CPI, the sensor processors pass this information to the sensor backend which updates the waveform immediately ready for the next pulse transmission. The change of DDS parameter in the LFM mode and the transfer of a new waveform to the FPGA memory in the AWG mode takes a few milliseconds. The change of center frequency takes 100–200 ms, because the parameters for the phase locked loop (PLL) circuitry to synthesize the LO signal for the second mixing state need to be recomputed. The complete waveform change process including waveform computation, transmission, reception, and processing of the data with the new parameters takes roughly 300 ms which is enough for a FAR adaptation a rate commensurate with the track update interval. Within the current architecture, a code optimization for a CPI-to-CPI adaptation is achievable. However, for pulse-to-pulse adaption, crucial parts of the processing and optimization need to be moved to the FPGA in the sensor backend. Such a pulse-to-pulse waveform adaptation using a waveform generation in the FPGA has been demonstrated in [32].

17.5.2 miniCODIR experiments

The new optimization degree of freedom—*sensor diversity*—with multiple sensors was explored in a series of outdoor experiments in an urban environment with limited sensor coverage due to building obstruction [30]. The focus of the experiment was to minimize the sensor resources used (number of sensors used and/or bandwidth usage) for a given tracking scenario. For the following experiments, an early version of the miniCODIR frontend with limited waveform adaptation capabilities (only LFM with a maximal chirp bandwidth of $B = 200$ MHz) was used.

17.5.2.1 Resource optimization in radar networks

In the first experiment, the sensor waveform parameters were kept constant (non-adaptive with a chirp bandwidth of 200 MHz) and the optimization possibilities were limited to the sensor usage (transmit on or off). The task of the optimizer was both to minimize the total sensor usage (objective 1) and to optimize the track accuracy (objective 2). A single target was present in the scene. The track and the optimization cost function results are shown in Figure 17.24(a) and (b), respectively. In line with the optimization goal, only one sensor is operating most of the time. In areas with a single-sensor coverage, near sensors S3 and S4, only these sensors are operating. In an area with a multi-sensor coverage, near sensor S1 visible also from S2 and S4, an alternate illumination and detection by sensors S1 and S2 address both optimization objectives. Operation with one sensor at a time minimizes the sensor usage (objective 1). Conversely, illumination by two sensors from different aspect angles that have complementary measurement covariance minimizes the tracking covariance (objective 2).

In a second experiment, the setup of the first experiment was modified by allowing the sensors to change their chirp bandwidth. Each sensor had four possible transmit states for optimization: transmit on/off as one state; and bandwidth choices of either 50 MHz, 100 MHz or 200 MHz as the other three states. Objective 2—*track accuracy*—was adopted from the first experiment while objective 1 was to minimize the total amount of used bandwidth. With the extended optimization space and the possibility of using a smaller bandwidth, with a corresponding reduction in sensor cost, the controller decided to use all sensors with visibility to the target, S1, S2, and S4, in parallel but with the smallest possible bandwidth of 50 MHz. The corresponding total amount of bandwidth is 150 MHz which is less than the 200 MHz used in the first experiment. Thus, the controller has been able to improve on objective 1, without neglecting objective 2.

17.6 Other cognitive radar testbeds

In the following section, three other testbeds to explore cognitive radar (CR) strategies are briefly discussed. First, two laboratory testbeds with real-time radar processing

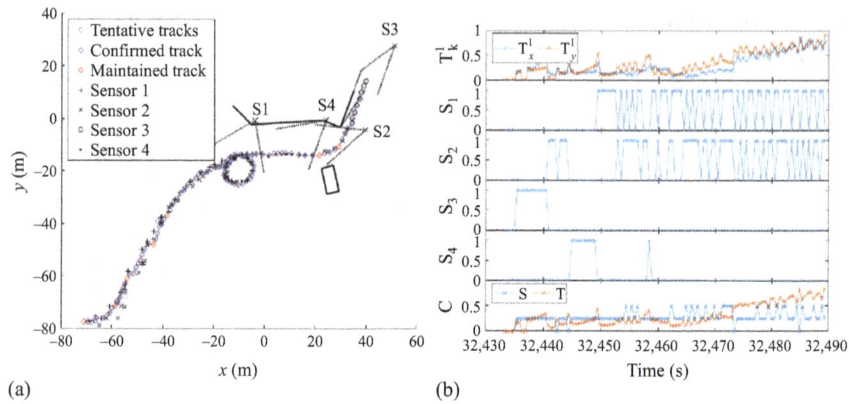

Figure 17.24 (a) Single target track updates and sensor detections. (b) Track accuracy and sensor costs. From top to bottom, the panels show the track accuracy cost terms related to the x and y coordinate components (T_x and T_y), the cost to operation (S_i for sensor i) and the combined costs for sensor operation (S) and track accuracy (T) related costs. Figure taken from [30]. ©2020 IEEE.

functionality and cognitive adaptation of the transmit waveform are presented. Here, the transmit signal is delayed and Doppler shifted and mixed with an externally generated radio frequency interference (RFI) signal. The combined signal is then injected into the receiver chain to emulate a (set of) radar targets with additional interference. Finally, this section is concluded with the introduction of an outdoor testbed with a data recording functionality to explore cognitive strategies in post-processing.

17.6.1 SDRadar: cognitive radar for spectrum sharing

The SDRadar testbed was developed jointly by the Pennsylvania State University, the University of Kansas and the Army Research Laboratory [33] is devoted to mitigate RFI using a PAC to optimization of transmit waveform. The cognitive radar-enabling component of the testbed is an X310 USRP which adapts, transmits, receives, and processes pulses in real time. The PAC consists of a spectrum-sensing function to locate the RFI in the operational band and a waveform adaptation function which either avoids or notches out the frequencies affected by the RFI. These functionalities work on a pulse-to-pulse basis and the PAC achieves sub-millisecond response times, if implemented completely on the FPGA of the SDR [34]. This fast PAC on signal level is complementary to the PAC discussed in Sections 17.2 to 17.5 which consist of high-level functions such as target detection and tracking and work on a CPI-to-CPI basis. Together, the PAC on different adaptation time scales may contribute to a CR model with a hierarchy of PAC loops.

With this testbed, laboratory experiments with different waveform adaptation strategies for RFI mitigation were tested in real-time. The first approach is

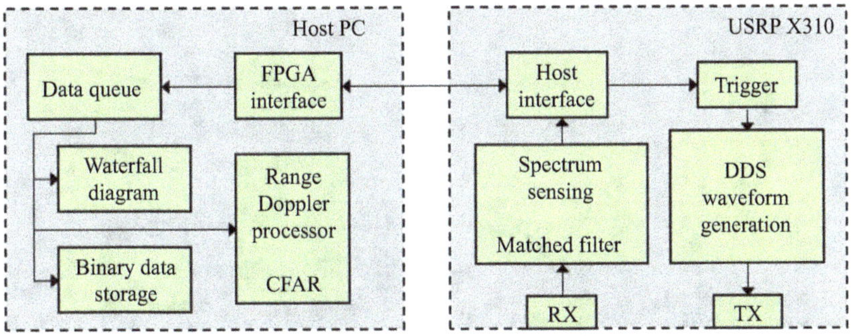

Figure 17.25 Simplified block diagram of the SDRadar testbed with the control PC (left) and the USRP (right). The PAC is implemented in the USRP and consists of a spectrum sensing functionality and a DDS waveform generator. Picture redrawn based on [35].

RFI-avoidance, where the transmit WF is restricted to the largest contiguous, non-RFI-affected band interval [4]. The advantages of this approach are a simplicity of waveform generation with a DDS. The other approach is to mitigate the RFI by generating a transmit WF on the fly with a frequency notch in the RFI-affected band [32]. The advantage of this approach is that the full operational band except for the notched RFI-affected interval can be used.

17.6.2 Spectral coexistence via xampling (SpeCX)

The SpeCX prototype was developed at the Technion in Haifa [36] and combines a sub-Nyquist radar prototype [37] with a cognitive radio receiver for spectrum sensing. The key feature of this testbed is the use of compressed sensing techniques to process the incoming data at the RX chain. Such techniques sample and process only small narrow sub-bands of the original signal and achieve a target resolution similar to classical Nyquist processing. As a consequence, the transmit signal can be restricted to the sub-bands processed by the compressed sensing techniques. The non-used sub-bands can be used for other users in the spectrum.

In the demonstration in [36], the SpeCX testbed implements a PAC feedback loop that combines a spectrum sensing algorithm, a sub-band selection method to maximize SINR, and a compressed sensing radar processor for target position estimation. Based on the decision of the PAC, an optimal pre-recording waveform is sent out by the signal generator and injected into the RX chain. A photo of the system is shown in Figure 17.26.

17.6.3 Anticipation in NetRad

NetRad, shown in Figure 17.27, is an S-band multi-static radar developed jointly by University College London (UCL) and the University of Cape Town. The system

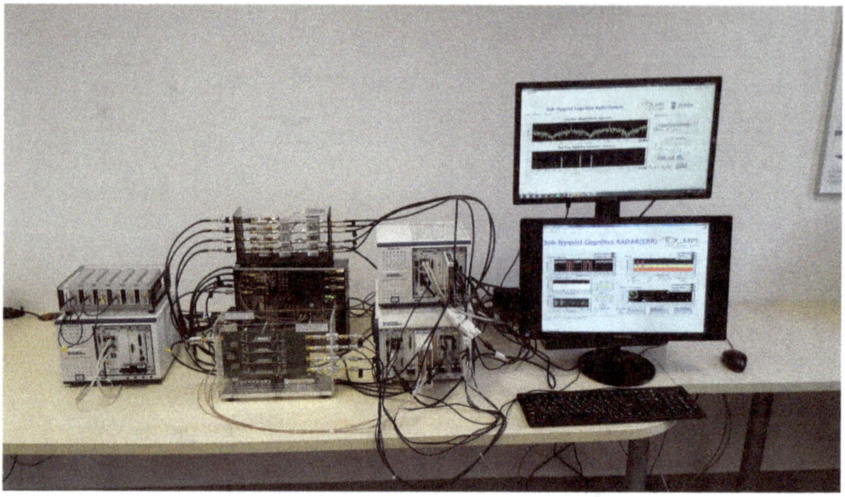

Figure 17.26 Photo of the SpeCX prototype. The system is composed of a signal generator (left), a cognitive radio receiver for spectrum sensing (middle, top), and a cognitive radar receiver for radar processing (middle, bottom). The PAC has been implemented on an external laptop, not visible in this picture. Photo courtesy of Dr Kumar Vijay Mishra.

consists of several nodes which are synchronized in time and frequency using GPS disciplined oscillators developed at the University of Cape Town. The system has been used in several experiments throughout the years. The system is inherently not adaptive and, hence, the experiment and collection need to be planned in order to get the data necessary to test cognitive/adaptive radar algorithms.

The system was used to evaluate an algorithm for anticipation in a cognitive radar [38]. The radar collected data, which in post-processing, utilized a POMDP-based algorithm which may look ahead and estimate obstructions. The estimate allowed the radar to perform more frequent track updates in advance of the obstruction, illustrating the property of anticipation. The track updates with and without anticipation is illustrated in Figure 17.28.

17.7 Future cognitive radar testbed considerations

We posit that the requirements for future cognitive radar testbeds and experimental studies will be driven by three key research objectives: a shift from individual sensors to fully distributed sensor systems; inclusion of the remaining building block of cognition, as defined by Fuster [12], beyond the perception–action cycle; and the introduction of *learning*, another critical aspect of cognition in humans and mammals. This view is based on trends in the open literature and visibility of activities such as the

Figure 17.27 Photo of the NetRad system. Photo courtesy of Dr Colin Horne of UCL.

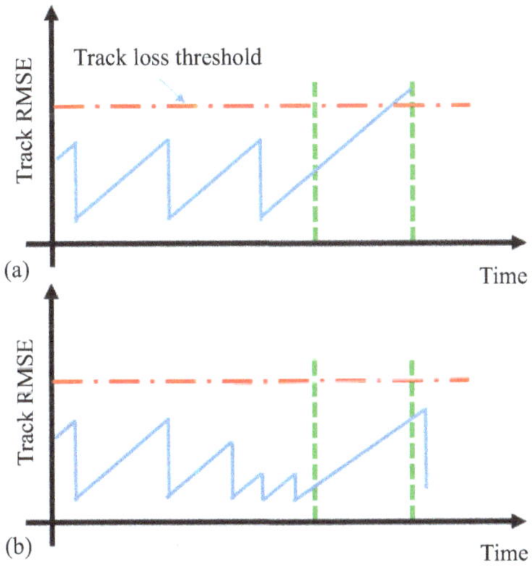

Figure 17.28 Results of anticipation experiment utilizing the NetRad system. The figure illustrates how the track update will adapt in beforehand to the obstacle, showing the property of anticipation in the algorithm. Photo courtesy of Dr Colin Horne of UCL [38].

NATO research task group SET-302, initiated in 2021 and titled *Cognitive Radar*, that has specific objectives in its technical activity proposal to conduct a "…demonstration of distributed cognitive radar…" and to undertake a "Conceptual demonstration of a cognitive radar that can learn…"

17.7.1 Distributed cognitive radar systems

The first of these objectives, research into distributed cognitive radar systems, is the most straightforward to consider. Systems such as miniCODIR and the CREW are able to simulate a degree of distributed operation as their transmit and receive nodes can be spatially separated. However, the nodes are cabled together, limiting baselines, and there is a dependency on a central processor. Future cognitive radar testbeds will have to comprise independent nodes, each capable of operating as a fully functioning cognitive radar and with sufficient processing power to both execute their own cognitive algorithms and to contribute to the overall, distributed cognitive algorithm. In addition to processing power, the testbeds will need radios for communication over extended baselines, the ability to determine their location, knowledge of the common time, and, for spatially coherent operation in which multiple apertures form composite beams, a distributed shared phase reference. From a research perspective, many of these requirements can be met with currently available commercial off-the-shelf components, such as the use of the cellular phone data network and global positioning system equipment. Conversely, future testbed designers should remember that one of the principal uses of radar is for defense and that in real-world situations, these research conveniences cannot always be relied upon.

17.7.2 Machine learning techniques

The latter two objectives are likely to be met through the use of artificial intelligence (AI) and machine learning (ML), and there is already active research into the application of these techniques [39–45]. One of the more common approaches to implement AI/ML techniques is to use (deep) neural networks. The ability of neural networks to learn, recognize patterns, and approximate functions means that they can stand in for various aspects of AI/ML algorithms that would otherwise prove extremely difficult to implement.

While a neural network can be executed on a conventional central processing unit (CPU), the nature of the network means that alternative computing architectures are more appropriate. The most well-known alternative to the CPU is the graphics processing unit (GPU) that has a highly parallelized architecture and is, therefore, able to execute parallelizable algorithms extremely quickly. The tensor mathematics that underpins neural networks is highly parallelizable and the use of GPU for its execution is, therefore, preferable [46]. Recently, hardware manufacturers have started to move past the GPU to develop tensor processing units (TPUs) [47–50]. These are dedicated chips specifically optimized to undertake the tensor mathematics of the neural network and thus allow highly efficient AI/ML operation.

It follows that as cognitive radar research begins to include the concepts of attention, memory, intelligence, and learning more directly, the experimental testbeds will require GPU and TPU for rapid processing. Systems such as the USRP-based cognitive testbeds described in Sections 17.4 and 17.6.1 do already utilize a GPU.

17.7.3 Confluence of algorithms—metacognition

A future CR must cope with changes to the local environment that occur on very different timescales. It needs, for example, quickly respond to a nearby RFI while, at the same time, new targets need to be tracked and the mission goals could change. Therefore, more complex learning and optimization strategies are needed. A CR testbed should be prepared for a hierarchy of multiple PAC running in parallel working at different time scales in combination with more complex high-level ML-based learning and optimization schemes. The integration of multiple parallel CR strategies on a single platform will be possible in the near future. For the management and regulation of such a set of parallel CR strategies, an additional metacognitive feedback loop is proposed in [51,52]. Such a cycle acts as a top-level control process that schedules the various CR strategies and decides which algorithm shall be applied in a given environment and target. A first proof-of-concept simulation with a metacognitive regulation cycle that manages a set of CR algorithms for RFI mitigation is presented in [51].

Considering the increasing hardware resources available for real-time computing we posit, that the implementation of a metacognitive feedback cycle together with a set of parallel cognitive algorithms poses no conceptual problem to a future real-time CR testbed. This conclusion is supported by the fact that the requirements of the RF frontend electronics do not change as a result of the evolution to a metacognitive system. However, the synchronization and temporal coordination of such a set of CR algorithms in real-time, running in parallel on potentially distributed hardware units each with different latency behaviors (FPGA, local control PC, centralized server at a data fusion engine) may become a major challenge.

17.8 Summary

This book chapter has presented the CREW, CODIR, miniCODIR, and a USRP-based cognitive radar testbeds and highlighted some experimental demonstrations conducted with each.

The CREW testbed is a multistatic, laboratory-scale system operating in W-band with a 1 GHz instantaneous bandwidth and capable of arbitrary waveform generation. It has successfully demonstrated waveform adaption in a Cartesian tracking experiment in which data from two adaptive nodes was fused in a central processor. The underlying cognitive algorithm used a hierarchical implementation of the FAR framework to both adapt the waveform parameters at the two sensing nodes and to continuously update the sensor goal values to minimize the covariance of the Cartesian track. Another set of experiments demonstrated that a neural network could be used to replace the optimization process used in FAR-based cognitive algorithms. Critically, the neural network was trained on simulated data and is still able to operate successfully in a real-world range-Doppler tracking experiment.

The CODIR and miniCODIR cognitive testbeds are X-band FMCW systems. CODIR is a monostatic system with two transmit channels, four receive channels, an instantaneous bandwidth of up to 1.2 GHz, and LFM waveforms. The miniCODIR testbed is a multistatic system with four nodes connected to a central processor. Each node has one transmit channel, two receive channels, up to 80 MHz of instantaneous bandwidth, and arbitrary waveform generation capability. Two sets of CODIR experiments were presented. The first demonstrated how cognitive radars can be used to balance different priorities that may arise, such as balancing track accuracy against radar resource usage. The second demonstrated a cognitive radar algorithm that selected "gapped" or "notched" LFM chirps from a waveform catalog to successfully mitigate a noise jammer operating within the radar's instantaneous bandwidth. miniCODIR has been used to demonstrate the management of resources within a network of radars. In an initial experiment, the waveforms of the radars in the network were fixed and the cognitive algorithm minimized the number of active sensors in the network while simultaneously minimizing the track covariance. In a second experiment, the radar nodes were able to change the bandwidth of their waveform and the cognitive algorithm was shown capable of minimizing the total bandwidth used while simultaneously minimizing the track covariance.

The USRP-based cognitive radar testbed is a demonstration of how low-cost digital components can be used to develop cognitive radar capabilities. The total system cost was under \$20 k for a S-band radar with arbitrary waveform generation and phase-comparison monopulse for target bearing estimation. The system demonstrated capable of detecting a Boeing 737 at a range of 5.5 km. A DJI Mavic UAV was detected at a range of $\approx$350 m. Based on SNR analysis, the maximum detection range for the UAV was extrapolated to be $\approx$600 m.

A FAR-based adaptive update interval tracking algorithm was demonstrated on the USRP-based testbed. The DJI Mavic UAV was used as the target and the algorithm minimizing the tracking covariance as the target flew back and forth in front of the radar. The covariance minimization was achieved by adapting the track update interval to compensate for the increased measurement error, due to the SNR roll-off, when the target was at long-ranges.

The chapter closed with a short discussion on other systems to test CR strategies and some considerations of the requirements for future cognitive radar testbeds. Based on their experience developing the highlighted testbeds and the trends in the open research literature, the authors believe that the next generation of cognitive radar testbed will need to be fully distributed in nature, i.e., not dependent on a central processor, and be equipped with GPU and TPU to enable the use of neural-network-based cognitive radar algorithms that utilize the latest AI/ML advances.

Acknowledgments

Dr. Smith thanks the Johns Hopkins University Applied Physics Laboratory for their generous support in writing this chapter. Dr. Smith also thanks Dr. Adam Mitchell, Dr. Peter John-Baptiste, and Dr. Kristine L. Bell for their research contributions and numerous stimulating conversations on cognitive radar.

Dr. Christiansen thanks The Norwegian Defence Research Establishment (FFI) for their support in writing this chapter. Dr. Christiansen also thanks PhD advisors

Dr. Smith and Dr. Baker for their guidance, and Dr. Torvik and Dr. Olsen for their fruitful conversations.

Dr. Oechslin thanks Armasuisse Science and Technology for funding the research activities related to the CODIR and miniCODIR testbeds. Dr. Oechslin also thanks Dr. Uwe Aulenbacher and Andreas Zutter for their extensive contributions and fruitful discussions.

References

[1] S. H. Talisa, K. W. O'Haver, T. M. Comberiate, M. D. Sharp, and O. F. Somerlock, "Benefits of digital phased array radars," *Proceedings of the IEEE*, vol. 104, no. 3, pp. 530–543, 2016.

[2] S. Prager, G. Sexstone, D. McGrath, J. Fulton, and M. Moghaddam, "Snow depth retrieval with an autonomous UAV-mounted software-defined radar," *IEEE Transactions on Geoscience and Remote Sensing*, vol. 60, pp. 1–16, 2021.

[3] S. Prager, T. Thrivikraman, M. S. Haynes, J. Stang, D. Hawkins, and M. Moghaddam, "Ultrawideband synthesis for high-range-resolution software-defined radar," *IEEE Transactions on Instrumentation and Measurement*, vol. 69, no. 6, pp. 3789–3803, 2020.

[4] B. H. Kirk, R. M. Narayanan, K. A. Gallagher, A. F. Martone, and K. D. Sherbondy, "Avoidance of time-varying radio frequency interference with software-defined cognitive radar," *IEEE Transactions on Aerospace and Electronic Systems*, vol. 55, no. 3, pp. 1090–1107, 2019.

[5] P. Liu, J. Mendoza, H. Hu, P. G. Burkett, J. V. Urbina, S. Anandakrishnan, and S. G. Bilén, "Software-defined radar systems for polar ice-sheet research," *IEEE Journal of Selected Topics in Applied Earth Observations and Remote Sensing*, vol. 12, no. 3, pp. 803–820, 2019.

[6] K. El-Darymli, N. Hansen, B. Dawe, E. W. Gill, and W. Huang, "Design and implementation of a high-frequency software-defined radar for coastal ocean applications," *IEEE Aerospace and Electronic Systems Magazine*, vol. 33, no. 3, pp. 14–21, 2018.

[7] J. R. Guerci, *Cognitive Radar: The Knowledge-Aided Fully Adaptive Approach*, 2nd ed. Boston, MA: Artech House, 2020.

[8] C. Horne, M. Ritchie, and H. D. Griffiths, "Proposed ontology for cognitive radar systems," *IET Radar, Sonar & Navigation*, vol. 12, no. 12, pp. 1363–1370, 2018.

[9] S. Haykin and J. M. Fuster, "On cognitive dynamic systems: Cognitive neuroscience and engineering learning from each other," *Proceedings of the IEEE*, vol. 102, no. 4, pp. 608–628, 2014.

[10] S. Haykin, Y. Xue, and P. Setoodeh, "Cognitive radar: Step toward bridging the gap between neuroscience and engineering," *Proceedings of the IEEE*, vol. 100, no. 11, pp. 3102–3130, 2012.

[11] S. Haykin, *Cognitive Dynamic Systems Perception-Action Cycle, Radar & Radio*. Cambridge: Cambridge University Press, 2012.

[12] J. M. Fuster, *Cortex and Mind Unify Cognition.* New York, NY: Oxford University Press, 2003.

[13] G. E. Smith, Z. Cammenga, A. Mitchell, K. L. Bell, J. Johnson, M. Rangaswamy, and C. Baker, "Experiments with cognitive radar," *IEEE Aerospace and Electronic Systems Magazine*, vol. 31, no. 12, pp. 34–46, 2016.

[14] K. L. Bell, C. J. Baker, G. E. Smith, J. T. Johnson, and M. Rangaswamy, "Cognitive radar framework for target detection and tracking," *IEEE Journal of Selected Topics in Signal Processing*, vol. 4553, pp. 1–1, 2015. [Online]. Available: http://ieeexplore.ieee.org/lpdocs/epic03/wrapper.htm?arnumber=7181639.

[15] A. E. Mitchell, G. E. Smith, K. L. Bell, A. J. Duly, and M. Rangaswamy, "Cost function design for the fully adaptive radar framework," *IET Radar, Sonar & Navigation*, vol. 12, no. 12, pp. 1380–1389, 2018.

[16] A. E. Mitchell, G. E. Smith, K. L. Bell, and M. Rangaswamy, "Hierarchical fully adaptive radar," *IET Radar, Sonar & Navigation*, vol. 12, no. 12, pp. 1371–1379, 2018.

[17] P. John-Baptiste, K. L. Bell, J. T. Johnson, and G. E. Smith, "Fully adaptive radar for multiple target tracking," *IEEE Transactions on Aerospace and Electronic Systems*, vol. 58, no. 6, pp. 5749–5765, 2022.

[18] P. John-Baptiste, J. T. Johnson, and G. E. Smith, "Neural network-based control of an adaptive radar," *IEEE Transactions on Aerospace and Electronic Systems*, vol. 58, no. 1, pp. 168–179, 2022.

[19] P. W. Moo and Z. Ding, *Adaptive Radar Resource Management.* New York, NY: Elsevier, 2015.

[20] "Cognitive radar (radar cognitif) – final report of task group set-227," NATO Science and Technology Organisation, Tech. Rep., October 2020.

[21] R. Oechslin, S. Wieland, S. Hinrichsen, U. Aulenbacher, and P. Wellig, "A cognitive radar testbed for outdoor experiments," *IEEE Aerospace and Electronic Systems Magazine*, vol. 34, no. 12, pp. 40–48, 2019.

[22] K. L. Bell, C. J. Baker, G. E. Smith, J. T. Johnson, and M. Rangaswamy, "Fully adaptive radar for target tracking part i: Single target tracking," in *2014 IEEE Radar Conference*, 2014, pp. 0303–0308.

[23] R. Oechslin, P. Wellig, U. Aulenbacher, S. Wieland, and S. Hinrichsen, "Cognitive radar performance analysis with different types of targets," in *2019 IEEE Radar Conference (RadarConf)*, April 2019, pp. 1–6.

[24] R. Oechslin, P. Wellig, S. Hinrichsen, S. Wieland, U. Aulenbacher, and K. Rech, "Cognitive radar parameter optimization in a congested spectrum environment," in *2018 IEEE Radar Conference (RadarConf18)*, April 2018, pp. 0218–0223.

[25] J. M. Christiansen, "Fully adaptive radar for detection and tracking," 2020. Available: http://rave.ohiolink.edu/etdc/view?acc_num=osu1587093543249087.

[26] J. M. Christiansen and G. E. Smith, "Development and calibration of a low-cost radar testbed based on the universal software radio peripheral," *IEEE Aerospace and Electronic Systems Magazine*, vol. 34, no. 12, pp. 50–60, 2019.

[27] J. M. Christiansen and G. E. Smith, "Parameter selection in a fully adaptive tracking radar," in *2019 International Radar Conference (RADAR)*, 2019, pp. 1–6.
[28] J. M. Christiansen, G. E. Smith, and K. E. Olsen, "Usrp based cognitive radar testbed," in *2017 IEEE Radar Conference (RadarConf)*, May 2017, pp. 1115–1118.
[29] J. M. Christiansen, K. E. Olsen, and G. E. Smith, "Fully adaptive radar for track update-interval control," in *2018 IEEE Radar Conference (RadarConf18)*, April 2018, pp. 0400–0404.
[30] R. Oechslin, S. Wieland, A. Zutter, U. Aulenbacher, and P. Wellig, "Fully adaptive resource management in radar networks," in *2020 IEEE Radar Conference (RadarConf20)*, September 2020, pp. 1–6.
[31] S. Thrun, W. Burgard, and D. Fox, *Probabilistic robotics*. Cambridge, MA: MIT Press, 2005.
[32] J. W. Owen, C. A. Mohr, B. H. Kirk, S. D. Blunt, A. F. Martone, and K. D. Sherbondy, "Demonstration of real-time cognitive radar using spectrally-notched random fm waveforms," in *2020 IEEE International Radar Conference (RADAR)*, 2020, pp. 123–128.
[33] B. H. Kirk, J. W. Owen, R. M. Narayanan, S. D. Blunt, A. F. Martone, and K. D. Sherbondy, "Cognitive software defined radar: waveform design for clutter and interference suppression," in *Radar Sensor Technology XXI*, K. I. Ranney and A. Doerry, Eds., vol. 10188, International Society for Optics and Photonics. SPIE, 2017, pp. 446– 461. Available: https://doi.org/10.1117/12.2262305.
[34] B. H. Kirk, A. F. Martone, K. D. Sherbondy, and R. M. Narayanan, "Performance analysis of pulse-agile SDRadar with hardware accelerated processing," in *2020 IEEE International Radar Conference (RADAR)*, 2020, pp. 117–122.
[35] A. F. Martone, K. D. Sherbondy, J. A. Kovarskiy, *et al.*, "Closing the loop on cognitive radar for spectrum sharing," *IEEE Aerospace and Electronic Systems Magazine*, vol. 36, no. 9, pp. 44–55, 2021.
[36] D. Cohen, K. V. Mishra, and Y. C. Eldar, "Spectrum sharing radar: Coexistence via xampling," *IEEE Transactions on Aerospace and Electronic Systems*, vol. 54, no. 3, pp. 1279–1296, 2018.
[37] E. Baransky, G. Itzhak, N. Wagner, I. Shmuel, E. Shoshan, and Y. Eldar, "Sub-Nyquist radar prototype: hardware and algorithm," *IEEE Transactions on Aerospace and Electronic Systems*, vol. 50, no. 2, pp. 809–822, 2014.
[38] C. P. Horne, M. Ritchie, H. D. Griffiths, F. Hoffmann, and A. Charlish, "Experimental validation of cognitive radar anticipation using stochastic control," in *Proceedings of Asilomar Conference on Signals, Systems, and Computers*, 2016, pp. 1–5.
[39] Y. Shi, B. Jiu, J. Yan, and H. Liu, "Data-driven radar selection and power allocation method for target tracking in multiple radar system," *IEEE Sensors Journal*, vol. 21, no. 17, pp. 19296–19306, 2021.

[40] R. J. G. van Sloun, R. Cohen, and Y. C. Eldar, "Deep learning in ultrasound imaging," *Proceedings of the IEEE*, vol. 108, no. 1, pp. 11–29, Jan 2020.

[41] C. E. Thornton, M. A. Kozy, R. M. Buehrer, A. F. Martone, and K. D. Sherbondy, "Deep reinforcement learning control for radar detection and tracking in congested spectral environments," *IEEE Transactions on Cognitive Communications and Networking*, vol. 6, no. 4, pp. 1335–1349, 2020.

[42] J. Wang, S. Guan, C. Jiang, D. Alanis, Y. Ren, and L. Hanzo, "Network association in machine-learning aided cognitive radar and communication co-design," *IEEE Journal on Selected Areas in Communications*, vol. 37, no. 10, pp. 2322–2336, 2019.

[43] G. E. Smith, S. Z. Gurbuz, S. Brüggenwirth, and P. John-Baptiste, "Neural networks & machine learning in cognitive radar," in *2020 IEEE Radar Conference (RadarConf20)*, September 2020, pp. 1–6.

[44] E. Selvi, R. M. Buehrer, A. Martone, and K. Sherbondy, "Reinforcement learning for adaptable bandwidth tracking radars," *IEEE Transactions on Aerospace and Electronic Systems*, vol. 56, no. 5, pp. 3904–3921, 2020.

[45] G. E. Smith and T. J. Reininger, "Reinforcement learning for waveform design," in *2021 IEEE Radar Conference (RadarConf21)*, May 2021, pp. 1–6.

[46] C. A. Navarro, R. Carrasco, R. J. Barrientos, J. A. Riquelme, and R. Vega, "Gpu tensor cores for fast arithmetic reductions," *IEEE Transactions on Parallel and Distributed Systems*, vol. 32, no. 1, pp. 72–84, 2021.

[47] M. Ibrahim. What is a tensor processing unit (tpu) and how does it work?, 2021. Available: https://towardsdatascience.com/what-is-a-tensor-processing-unit-tpu-and-how-does-it-work-dbbe6ecbd8ad.

[48] Google. Cloud tensor processing units (tpus), 2021. Available: https://cloud.google.com/tpu/docs/tpus.

[49] Y. Wang, G.-Y. Wei, and D. M. Brooks, "Benchmarking tpu, gpu, and cpu platforms for deep learning," *ArXiv*, vol. abs/1907.10701, 2019.

[50] N. Jouppi, C. Young, N. Patil, and D. Patterson, "Motivation for and evaluation of the first tensor processing unit," *IEEE Micro*, vol. 38, no. 3, pp. 10–19, 2018.

[51] A. F. Martone, K. D. Sherbondy, J. A. Kovarskiy, *et al.*, "Metacognition for radar coexistence," in *2020 IEEE International Radar Conference (RADAR)*, April 2020, pp. 55–60.

[52] K. V. Mishra, M. R. B. Shankar, and B. Ottersten, "Toward metacognitive radars: Concept and applications," in *2020 IEEE International Radar Conference (RADAR)*, 2020, pp. 77–82.

Chapter 18

Quantum radar and cognition: looking for a potential cross fertilization

Alfonso Farina[1], Marco Frasca[2] and Bhashyam Balaji[3]

"What a piece of work is a man! How noble in reason, how infinite in faculty! In form and moving how express and admirable! In action how like an angel, in apprehension how like a god! The beauty of the world. The paragon of animals. And yet, to me, what is this quintessence of dust?" (Hamlet, William Shakespeare)

18.1 Introduction

The purpose of this chapter is to try to establish a connection between cognitive radar (being the focus of this entire book)—which is the most recent evolution of the state-of-the-art radar—with quantum physics, quantum technology, and quantum sensing when it will be practically feasible.

This chapter develops along a narration moving from cognitive radar to quantum radar up to a vision of a potential cognitive-quantum radar.

The "fil rouge" we are trying to weave is to bridge cognitive radar and quantum physics by:

- learning from nature and mimicking nature: neurons and neural networks, Hebb's rule, mirror neurons, perception–action cycle (PAC), and
- inferring new ideas and insights from the major scientists of modern physics: microtubules: a quantum interpretation of brain, and a model of microtubule based learning for perception–action behavior control.

Mimicking nature motivates the onset of cognitive radar, as a "Brain Inspired Radar Design" and provides the neuroscience base of adaptivity and cognitivity in radar signal processing.

Cognitivity was preceded by adaptivity, waveform diversity and design, and knowledge-based systems (KBS) [1,2].

[1]Selex-ES (retired), Private Consultant, Rome, Italy
[2]MBDA Italia S.p.A., Rome, Italy
[3]Department of Systems and Computer Engineering, Canada

Neurons and neural networks can be related to adaptive signal processing in radar and sonar. In a general sense, the D. Hebb's rule* to update the synapse has been mimicked by Howells–Applebaum [3] and Brennan—Reed algorithm in [4]. From the point of view of artificial neurons and artificial neural networks, Hebb's principle can be described as a method of determining how to alter the weights between model neurons. The weight between two neurons increases if the two neurons activate simultaneously, and reduces if they activate separately. Nodes that tend to be either both positive or both negative at the same time have strong positive weights, while those that tend to be opposite have strong negative weights (excerpt from reference in footnote ∗).

The discovering of mirror-neurons effect is one of the most exciting events in neuroscience.† Mirror-neurons are related to empathy, imitation, the chameleon effect, and probably language (sing, calls, etc.) development. Researchers in cognitive neuroscience and cognitive psychology consider that this system provides the physiological mechanism for the PAC. The mirror-neurons may be important for understanding the actions and intentions of other people, and for learning new skills by imitation. It is also suggested that mirror-neuron systems may simulate observed actions, and thus contribute to the theory of mind skills. It is felt that mirror neurons are the neural basis of the human capacity to feel empathy, and namely to resonate with another's emotional states. Thanks to the visual/audio-motor coupling mediated by the mirror system, some processes, such as understanding others' motor goals and intentions, are faster compared to systems based on mere cognitive, inferential processes. In the 1980s and 1990s, Prof. Giacomo Rizzolatti [5] was working with G. Di Pellegrino, L. Fadiga, L. Fogassi, and V. Gallese at the University of Parma, Italy, and discovered this phenomenon [6,7].

That being said, just now it does not seem possible to suggest how this neuroscience view can be embedded in the math and engineering of cognitive radar. This could be the avenue for future research (see Chapter 19 "Metacognitive radars" in this book). Getting radar experts to collaborate with neuroscientists would, in our opinion, be extremely welcome.

Hebbian learning accounts of mirror neurons. Hebbian learning and spike-timing-dependent plasticity have been used in an influential theory of how mirror neurons emerge. Mirror neurons are neurons that fire both when an individual performs an action and when the individual sees or hears another performing a similar action. The discovery of these neurons has been very influential in explaining how individuals make sense of the actions of others, by showing that, when a person perceives the actions of others, the person activates the motor programs which they would use to perform similar actions. The activation of these motor programs then adds information to the perception and helps predict what the person will do next based on the perceiver's own motor program. A challenge has been to explain how individuals come to have neurons that respond both while performing an action and

*https://en.wikipedia.org/wiki/Hebbian_theory

†Mirror neuron. [Online]. Available: https://en.wikipedia.org/wiki/Mirror_neuron.

while hearing or seeing another perform similar actions (excerpt from reference of footnote ∗).

Mirror neurons and the perception–action link.[‡] Social life depends in large part on the capacity to understand the intentional behavior of others. What are the origins of this capacity? The classic cognitive view claims that intentional understanding can be explained only in terms of the ability to intellectually read the mind of others. Over the past few years, this view has been challenged by several neuroscientific findings regarding social cognition. In particular, the functional properties of mirror neurons and their direct matching mechanism indicate that intentional understanding is based primarily on the motor cognition that underpins one's own potentiality to act (from the abstract of reference in the footnote ‡).

PAC was introduced by Prof. Simon Haykin as the basis of cognitive radar for maximizing information gain. Further steps to pave the way to radar cognition are: memory for predicting the consequences of actions, prioritizing the allocation of available resources, and intelligence for decision-making. These steps constitute Haykin's paradigm.

The mirror neurons as the underlying neuroscience mechanism of the PAC for radar was hypothesized and described in [8] and in [1] on pp. 256–258 and even more recently in [9].

18.2 Cognitive radar

In describing the current status of cognitive radar, we follow [10] for an introductory survey. Cognitive dynamic systems have been inspired by the unique neural computational capability of the brain and the viewpoint that cognition (in particular the human one) is a supreme form of computation [11–14]. Some exemplifications within this new class of systems, which is undoubtedly among the hallmarks of the 21st century, are cognitive radar, control, radio, and some other engineering dynamic architectures. S. Haykin published two pioneering articles in the context of the cognitive radar [14,15], a recent review is given in [16]. The key idea behind this new paradigm is to mimic the human brain as well as that of other mammals with echolocation capabilities (bats, dolphins, whales, etc.). They continuously learn and react to stimulations from the surrounding environment according to four basic processes: PAC, memory, attention, and intelligence. This last observation highlights the importance of specifying which are the "equivalents" of the aforementioned activities in a cognitive radar. We have real-time, tactical and strategic (for instance, at command and control and intelligence levels) reactions. Thus, cognitivity will hopefully apply at these increasing scales of time. In this chapter, we mainly focus on cognitivity for the

[‡]https://www.oxfordhandbooks.com/view/10.1093/oxfordhb/9780199988709.001.0001/oxfordhb-9780199988709-e-016, V. Gallese, "Mirror Neurons and the Perception-Action Link", Edited by Kevin N. Ochsner and Stephen Kosslyn, Print Publication Date: Dec 2013, DOI:10.1093/oxfordhb/9780199988709.013.0016

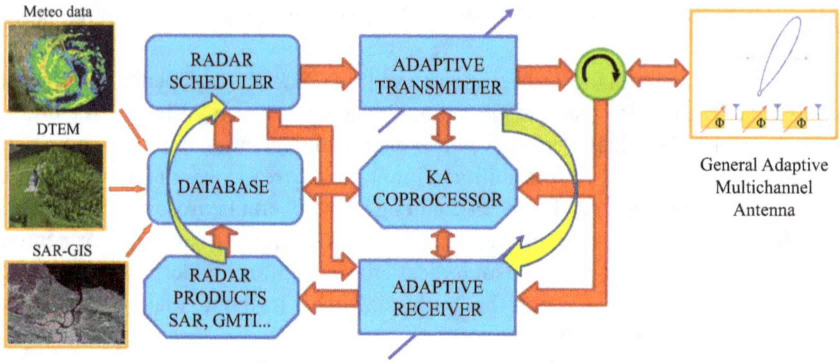

Figure 18.1 Notional scheme of a cognitive radar highlighting the PAC (KA: knowledge aided)

first time scale. In the other time scales, the PAC scheme will be more articulated and will "breadth" at increasing time scales. This is thoroughly discussed in [1,13,17–20].

The perception–action process, schematized as in Figure 18.1, has the fundamental task of probing the environment. A remarkable feature of cognitive radar, which complements the adaptivity in reception—a landmark of adaptive radar—is the ability to be adaptive also in the transmitted beam pattern, with easily arguable operational advantages. In particular, knowing *a priori* the angular directions along which to put nulls in the transmitted beam, these can be formed by exploiting a technique of small perturbation of the array antenna illumination function, as described in [21]. Otherwise, this prior knowledge can be gained by exploiting either an ESM (Electronic Support Measurement) or a cognitive MIMO (Multiple Input Multiple Output) as described in Section 9.2 "Cognitive MIMO radar beampattern shaping" of [1]. A further step towards adaptive spatial nulling in transmission is detailed in [22] with experimental results.

Algorithmic challenges are also present to exploit as efficiently as possible both the feedback information and the *a priori* knowledge. Interesting to read Dr J. Guerci recent book [13] and the comments by A. Farina.[§]

18.2.1 Cognitive radar scheduler

The topic of radar scheduler, also referred to as radar system manager, has been discussed historically and from an industrial point of view in the Chapters 1 and 10 of [1] and [23].

[§]Insider blog: https://blog.artechhouse.com/2020/09/21/1311/ J. Guerci, "Cognitive Radar: The Knowledge-Aided Fully Adaptive Approach", 2nd ed., a Review by A. Farina by Artech House on Monday, September 21, 2020.

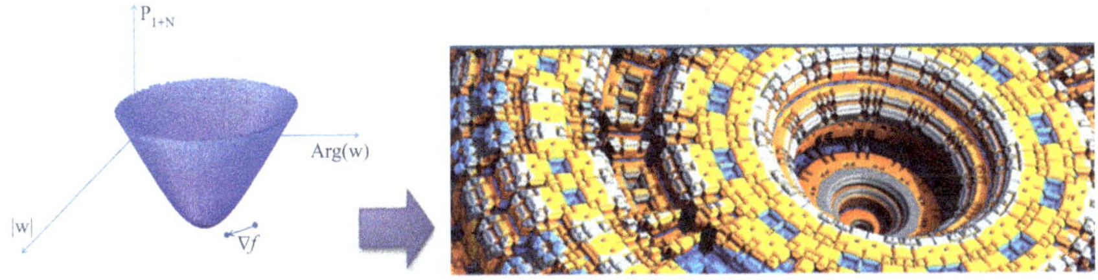

Figure 18.2 *The bowl on the left shows the power of the interference at the output (vertical axis) of a two-element array antenna ($N = 2$) with the weight of the first element set to $w_1 = 1$. The amplitude and phase of the weight w_2 of the second array element are shown on the x and y coordinates. The figure on the right is an imaginative view of a multidimensional—angle, frequency, waveform, tasks—objective function to estimate in time and space and minimize in a cognitive radar.*

The key difference between adaptive and cognitive radar is the following: adaptive radar knows the objective function to minimize (e.g. the interference power with respect to the useful target signal), while cognitive radar needs to estimate the objective function in an opaque,‖ deceiving, and threatening environment and then minimize it. An illustration of the concept is in Figure 18.2.

The key question is "how can the radar perform this operation in an opaque, deceiving and threatening environment?" The answer may be found in control theory. What follows is a quick ride—though not exhaustive—across the history of control theory. Our goal is to guide the reader to the breakthrough of Chapter 6 of [1].

The control theory deals with the control of continuously operating dynamical systems in engineered processes and machines. The objective is to develop a control model for such systems using a control action in an optimum manner without delay or overshoot and ensuring control stability. Control theory is subfield of mathematics, computer science, and control engineering.¶ Control theory dates from the 19th century, when James Clerk Maxwell first described the theoretical basis for the operation of governors. WWII and the space race gave rise to open-loop and closed-loop (feedback) control** [24].

‖A. Farina, "Causal Inference in Statistics. An attempt at some reflection." *ISIF Perspectives*, vol. 3, May 2020.

¶Most recently, control theory has been proposed to mitigate the effects of COVID-19. https://spectrum.ieee.org/biomedical/diagnostics/how-control-theory-can-help-control-covid19 "How Control Theory Can Help Us Control COVID-19," By Greg Stewart, Klaske van Heusden and Guy A. Dumont, Posted 17 Apr 2020 | 11:00 GMT, IEEE Spectrum. Using feedback, a standard tool in control engineering, we can manage our response to the novel coronavirus pandemic for maximum survival while containing the damage to our economies.

**https://en.wikipedia.org/wiki/Control_theory

The book by Aström and Murray [24] is really an excellent reference for many fields of the knowledge, including biology. Biological systems provide perhaps the richest source of feedback and control examples. The basic problem of homeostasis, in which a quantity such as temperature or blood sugar level is regulated to a fixed value, is but one of the many types of complex feedback interactions that can occur in molecular machines, cells, organisms, and ecosystems. The Example 2.13 (pp. 58–60) describes in some detail the "Transcriptional regulation," the process by which messenger RNA (mRNA) is generated from a segment of DNA. Example 2.14 "Wave propagation in neuronal networks" (pp. 60–61) describes accurately the control system of a simple biological system well described by the Hodgkin–Huxley equations. The dynamics of the membrane potential in a cell are a fundamental mechanism in understanding signaling in cells, particularly in neurons and muscle cells. The Hodgkin–Huxley equations give a simple model for studying propagation waves in networks of neurons. The Hodgkin–Huxley model was originally developed as a means to predict the quantitative behavior of the squid giant axon. Hodgkin and Huxley shared the 1963 Nobel Prize in Physiology (along with J. C. Eccles) for analysis of the electrical and chemical events in nerve cell discharges. Adaptive control uses on-line identification of the process parameters, or modification of controller gains, thereby obtaining strong robustness properties. Adaptive controls were applied for the first time in the aerospace industry in the 1950s, and have found particular success in that field.[**] Optimal control is a particular control technique in which the control signal optimizes a certain "cost index."[**] Stochastic control deals with control design with uncertainty in the model. In typical stochastic control problems, it is assumed that there exist random noise and disturbances in the model and the controller, and the control design must take into account these random deviations.[**] LQG, Linear (in the state and measurement equations) Quadratic (in the cost function) Gaussian (the probability density functions of the forcing and measurement noises) control, is characterized by the separation/dual principle (simply stated, the optimal control law is proportional to the state estimate of the dynamic system to control) and it is the optimal stochastic control.[††]

In the more general case not covered by LQG, the Bellman equation, named after Richard E. Bellman, is a necessary condition for optimality associated with the mathematical optimization method known as dynamic programming. It writes the "value" of a decision problem at a certain point in time in terms of the payoff from some initial choices and the "value" of the remaining decision problem that results from those initial choices. This breaks a dynamic optimization problem into a sequence of simpler sub-problems, as Bellman's "principle of optimality" prescribes. The Bellman equation was first put in engineering control theory and to other topics in applied mathematics, and subsequently became an important tool in economic theory; though the basic concepts of dynamic programming are prefigured in John von Neumann and Oskar Morgenstern's theory of games and economic behavior [25] and Abraham Wald's sequential analysis [26]. Almost any problem that can be solved using optimal

[††]https://en.wikipedia.org/wiki/Linear%E2%80%93quadratic%E2%80%93Gaussian_control and https://en.wikipedia.org/wiki/Stochastic_control

control theory can also be solved by analyzing the appropriate Bellman equation. However, the term "Bellman equation" usually refers to the dynamic programming equation associated with discrete-time optimization problems.[‡‡] The idea of solving a control problem by applying Bellman's principle of optimality and then working out backwards in time an optimizing strategy can be generalized to stochastic control problems, a practical example is described in [27]. In continuous-time optimization problems, the analogous equation is a partial differential equation that is usually called the Hamilton–Jacobi–Bellman (HJB) equation.[§§]

Pontryagin's maximum principle is a necessary but not sufficient condition for optimum, by maximizing a Hamiltonian, but this has the advantage over HJB of only needing to be satisfied over the single trajectory being considered.[‖‖]

Apart from the linear-quadratic case, the case of incomplete state information is not well-known. A *Partially Observable Markov Decision Process (POMDP)* is a generalization of a Markov decision process (MDP).[¶¶] A POMDP models an agent decision process in which it is assumed that the system dynamics are determined by an MDP, but the agent cannot directly observe the underlying state. Instead, it must maintain a probability distribution over the set of possible states, based on a set of observations and observation probabilities, and the underlying MDP. The POMDP framework is general enough to model a variety of real-world sequential decision processes. Applications include robot navigation problems, machine maintenance, and *planning under uncertainty in general*. The general framework of Markov decision processes with incomplete information was described by Karl Johan Aström in 1965 [28] in the case of a discrete state space, and it was further studied in the operations research community where the acronym POMDP was coined. It was later adapted for problems in artificial intelligence and automated planning by Leslie P. Kaelbling and Michael L. Littman. An exact solution to a POMDP yields the optimal action for each possible belief over the world states. The optimal action maximizes (or minimizes) the expected reward (or cost) of the agent over a possibly infinite horizon. The sequence of optimal actions is known as the optimal policy of the agent for interacting with its environment.[¶¶] Monte Carlo POMDP (MC-POMDP) is the particle filter version for the POMDP algorithm. In MC-POMDP, particles filters are used to update and approximate the beliefs, and the algorithm is applicable to continuous valued states, actions, and measurements.[***] Today POMDP has started to be adopted for cognitive radar functions including the cognitive radar scheduler [15,17].

A breaking point of view is represented by the cognitive stochastic control applied also to cognitive radar scheduler (M. Fatemi, S. Haykin, "Cognitive Control Theory with an Application," Chapter 6 of [1]). Prof. S. Haykin is the leader of this further advancement of control theory.

[‡‡]https://en.wikipedia.org/wiki/Bellman_equation
[§§]https://en.wikipedia.org/wiki/Hamilton%E2%80%93Jacobi%E2%80%93Bellman_equation
[‖‖]https://en.wikipedia.org/wiki/Pontryagin%27s_maximum_principle
[¶¶]https://en.wikipedia.org/wiki/Partially_observable_Markov_decision_process
[***]https://en.wikipedia.org/wiki/Monte_Carlo_POMDP

Speaking of cognitive control, we naturally think of cognitive control in the brain: please remember also Examples 2.13 and 2.14 of [24]. Cognitive control resides in the executive part of the brain, reciprocally coupled to its perceptual part via the working memory. The net result of this three-fold combination is the perception–action cycle that embodies the environment, thereby constituting a closed-loop feedback system of a global kind. The PAC awareness prompted the engineering need for bringing cognitive control into the specific formalism of cognitive dynamic systems and cognitive radar specifically. Two underpinnings of cognitive control are learning and planning, each of which is based on two notions: (i) The two-state model, which embodies target state of the environment and entropic state of the perceptor. (ii) The cyclic directed information flow, which follows from the global perception–action cycle: the first principle of cognition. Next, mathematical formalism of the learning process in cognitive control results in a state-free cognitive control learning algorithm, where computational complexity follows the linear law. One-step further is the cognitive control learning algorithm which is shown to be a special case of the celebrated Bellman's dynamic programming; hence, convergence and optimality of the new algorithm. The structural composition of the cognitive controller is addressed in Section 6.7 of [1]. Then, Section 6.8 of [1] validates an engineering application of the cognitive controller by presenting a computational experiment involving a cognitive tracking radar. Specifically, the new cognitive controller has been compared against two other different sub-optimal cognitive controllers: one controller involves dynamic optimization that is computationally expensive; the other controller involves the use of traditional Q-learning[†††] that is computationally tractable, but inefficient in performance.

18.2.2 Within the cognitive radar

The radar transmitter, through waveform emission, stimulates the background with the goal to obtain a response (i.e., a radar echo) from it. The mentioned response is perceived by the radar receiver which plays the equivalent role of the human senses. Actually, human response develops on a spectrum of times. It may be almost immediate and also taking more time implying intimate reflections. [‡‡‡]

Memory is the process through which information is registered, stored, and retrieved; new perceptions determine new memories. Tasks such as target recognition and identification are strongly informed by memories.

Attention requires processing the perceptor output to extract information and to selectively concentrate on some discrete aspect of information. It can require system actions for prioritizing the allocation of available resources (the radar time, power, etc. [8,16,29,30]) in accordance with their importance (for instance, a detection in a given range–azimuth-Doppler bin usually involves a confirmation process which calls for a specific radar waveform optimized to the actual interference/clutter conditions and the Doppler bin under test [31,32]). As to intelligence, among the aforementioned four

[†††] https://en.wikipedia.org/wiki/Q-learning

[‡‡‡] Cognition correctly involves the sense of time as one of the reviewers correctly argues and suggests. These authors thank this referee for bringing the concept to their attention.

functions, it is by far the most difficult to describe. While intelligence functionalities are based on the perception–action cycle, memory, and attention, it is the presence of a feedback at multiple levels that allows the system to take intelligent decisions in the face of inevitable uncertainties in the environment. The presence of such a closed-loop feedback between the actuator (transmitter) and the perceptor (receiver) represents the main ingredient which makes the cognitive radar unique and clearly distinguishes it from the classic adaptive architecture. In this last case, adaptivity is mainly confined to the receiver branch except for some static forms of transmit diversity [13,33], usually implemented in terms of mode selection (i.e. long-range versus short-range, search versus tracking). The information sharing involved in the feedback process, as already said, is complemented by the use of a memory, which in the radar case is constituted by a dynamic database. It contains knowledge sources about the operating context such as:

- geographic features of the illuminated area [2]: type of terrain, presence of clutter discretes, terrain elevation profiles (for instance gathered through Geographic Information Systems (GISs) or Digital Terrain Elevation Models (DTEMs));
- electromagnetic characteristics of the overlaid radiators: operating frequency, modulation and policy, activity profiles, location of transmitters. This information can be obtained through Radio Environment Maps (REMs) [34] which localize surrounding emissions in time, frequency, and space (spectrum sensing modules which continuously sound the environment and acquire fresh information on the external electromagnetic interference can be possibly used to update the content of the REMs);
- data from other sensors [35] (synthetic aperture radars (SARs), infrared devices, meteorological measurements, etc.).

The overall information flow coordinates and triggers the actions of the system. For example, with reference to the search process, it is exploited to devise the new transmit waveform [33,34,36], to select the training data for receiver adaptation [2,35], to censor data containing clutter discretes, to choose the most suitable detector within a bank/cluster available at the receiver. This implies a continuous adaptation of the pair perceptor–actuator ruled by the available information flow and usually coordinated by a system manager [23]. The quoted diversity in the transmit–receive chain is actually already present in nature making the cognitive radar a bio-inspired concept. Many mammals with echolocation capabilities, in particular the bats ([31], Chapter 6), in their natural behavioral phases, change the waveform in a spontaneous and systematic way producing through tongue clicking a variety of modulated sonar signals. A nice example is the Eptesicus Nilssonii bat ([31], Chapter 6). While attempting to feed on prey, it changes the Pulse Repetition Time (PRT) and the waveform shape between the approach phase and the terminal phase. In fact, studying the wideband ambiguity function during the search phase, researchers have understood that the bat is capable of resolving both in range and Doppler, then during the terminal phase, it improves the range resolution but the signal becomes quite Doppler tolerant [37].

18.2.3 Verification and validation

One relevant point to ponder is the "Challenges with Verification and Validation." Here is an excerpt from [38][§§§] reported in the summary. "The topic of modern advanced radar systems are complex and specifying and measuring their dynamic performance in practical trials' conditions is not straightforward. Computer modelling and simulation will thus play greater roles. The authors...concentrate on the problems of specifying and measuring the detection performance of modern adaptive radars. They stress the pivotal customer supplier relationship, and emphasize the distinction between the end user's interest in capability and what can be measured to provide a basis for acceptance. Drawing on the lessons from past projects, the problems of adaptive radars can only be contained by close cooperation between all involved in the procurement process, flexibility in recognizing and defining the necessary dependence on modelling and simulation, and the continuing need for research to lend substance to their scope and validity."

This challenge has been focused on cognitive radar and described in [30]. The definition of a true cognitive system is that performance for a specific task will improve on repetition; however, it is this adaptive behavior which makes verification challenging. Verification can be applied to the system at a particular point, but not the learned or improved system. Discussed in the paper are examples of potential ways to overcome this challenge, especially for use in defense applications: machine learning in the design phase, system level optimization, and apprentice learning.

In the chapter, particular care has been dedicated to challenges with verification and validation. In this area, it has been demonstrated that, by applying machine learning in conjunction with human decisions, the algorithm should not learn a decision that a human would not make. Apprentice learning aims to mimic the operator's response to a particular task; therefore, in the future, the same task could be completed autonomously with confidence. At the end of the day, the authors of [30] feel that important future research directions are the development of radar equations for cognitive radar, the evaluation of the impact of errors in the databases, the quantification of training time for system Operators, the verification and validation procedures, and the establishment of a technology readiness level achievable for cognitive radar. All these efforts aim to establish the gains of cognitive radar versus the resulting engineering and financial effort, and the attractiveness for customers.

18.3 Quantum mechanics in a nutshell

We give here a cursory look at quantum mechanics. The reason to present such a more formal section arises from a possible need for the reader to learn some rudiments of quantum mechanics to go deeper into the current literature on quantum sensors that we are going to cite. For a more extended treatment, we refer to the books given in [39–42]. We will take some concepts like Hilbert space (complex vector space

[§§§]Thanks to Dr S. Watts for providing the reference.

with inner product), operator algebra and some linear algebra for granted. These are normally apprehended in standard undergraduate courses. For notation purposes, in agreement with common usage, vectors in a finite or infinite dimensional space are indicated with bra ($\langle\ldots|$) and ket ($|\ldots\rangle$) so that, touching each other makes a scalar product. Similarly, uppercase letters ($A,\ B,\ C,\ldots$) are operators acting on them that could be represented by matrices whose elements are given by touching with bra on the left and ket on the right of them.

A quantum system is generally described by a vector $|\psi\rangle$ belonging to a Hilbert space $\mathscr{H}$, representing its state. A state of a quantum system is characterized by a set of observables, $\{A, B, C, \ldots\}$, that are represented by operators acting on the given Hilbert space. These operators have the following properties:

1. They are Hermitian self-adjoint, $A = A^\dagger,\ B = B^\dagger,\ C = C^\dagger,\ \ldots$ and so on, their eigenvalues are real.
2. Their eigenvalues represent the set of all possible results of measurements of an observable.
3. The eigenvectors of the Hermitian operators form a complete orthonormal basis.
4. They are simultaneously measurable if the operators are commuting with each other that is, $[A,B] = [A,C] = [B,C] = \ldots = 0$, being the commutator $[A,B] = AB - BA$ and so on.
5. The operators acting on the Hilbert space $\mathscr{H}$ form an algebra. As, in this algebra, there is an involutive operation, self-adjointness, then it is said a *-algebra.[‖‖] This generalizes the idea of complex conjugation from numbers to operators.

In order to describe the time evolution of a quantum system, we need to introduce the Hamiltonian H that characterizes the physics of the quantum system. This operator, belonging to the *-algebra, represents the energy of the system which does not depend explicitly on time. We give below some examples of Hamiltonian operators:

1. Free particle of mass m and momentum $\mathbf{p} = (p_x, p_y, p_z)$

$$H = \frac{\mathbf{p}^2}{2m} = \frac{p_x^2}{2m} + \frac{p_y^2}{2m} + \frac{p_z^2}{2m}. \tag{18.1}$$

   Here $p_x,\ p_y,\ p_z$ are self-adjoint operators belonging to the *-algebra, so is their square, thus granting the same property for H. We remember that the velocity operator can be obtained by $\mathbf{v} = \mathbf{p}/m$.
2. Particle of mass m in a field force with potential $V(\mathbf{x})$

$$H = \frac{\mathbf{p}^2}{2m} + V(\mathbf{x}). \tag{18.2}$$

   The coordinates $\mathbf{x} = (x, y, z)$ belong to the *-algebra and so does the potential V.
3. Particle of mass m and charge $-e$ in an electromagnetic field with potentials $\mathbf{A}(\mathbf{x}, t)$ and $\phi = \phi(\mathbf{x}, t)$:

$$H = \frac{1}{2m}\left(\mathbf{p} + \frac{e}{c}\mathbf{A}(\mathbf{x}, t)\right)^2 - e\phi(\mathbf{x}, t). \tag{18.3}$$

[‖‖] https://en.wikipedia.org/wiki/*-algebra.

Here c is the speed of light. This Hamiltonian depends explicitly on time and so does not represent the energy of the system being not conserved.

From Heisenberg commutation relations, which are the quantum analogs of the classical Poisson brackets, we know that

$$[x_i, x_j] = 0, \quad [p_i, p_j] = 0, \quad [x_i, p_j] = i\hbar\delta_{ij}, \tag{18.4}$$

where $\hbar = h/2\pi$ with h the Planck's constant and $i,j = x,\ y,\ z$. This implies that, generally, the potential V and the kinetic energy, proportional to $\mathbf{p}^2$, do not commute. This forces us to choose two possible representations for the state vector of the system, in coordinate $\mathbf{x}$ space or in momenta $\mathbf{p}$ space, and the choice of the former excludes the latter. Both are acceptable spaces. A representation can be built by using the eigenvectors of the chosen one. For instance, we consider the eigenvectors of the coordinate representation given by (we introduce a hat just to distinguish an operator from its eigenvalues)

$$\hat{x}_i|\mathbf{x}\rangle = x_i|\mathbf{x}\rangle, \tag{18.5}$$

and project the state vector on them giving the wave function $\psi(\mathbf{x}) = \langle\mathbf{x}|\psi\rangle$.

The evolution in time of a state vector is given by the Schrödinger equation

$$H|\psi\rangle = i\hbar\frac{\partial}{\partial t}|\psi\rangle. \tag{18.6}$$

When we assume that we are studying a quantum system by the time evolution of its state vector we are working in the Schrödinger representation and, in this case, the operators do not evolve in time. Otherwise, when we work with the time evolution of operators, we are working in the Heisenberg representation. The time evolution of an operator A will be given by the Heisenberg equation

$$\frac{dA}{dt} = \frac{\partial A}{\partial t} + \frac{i}{\hbar}[H, A]. \tag{18.7}$$

In this case, the states do not evolve in time. Whatever representation we choose, the dynamics is always originated from the Hamiltonian H that, in this way, assumes a key role in quantum mechanics.

From the Heisenberg equation of motion (18.7), it is easy to see that, when a given operator A does not depend explicitly on time and commutes with the Hamiltonian, $[A, H] = 0$, then, it represents a conserved quantity. Hence, for the conservation of energy, it is sufficient that the Hamiltonian does not explicitly depend on time as it trivially commutes with itself.

For each operator belonging to the *-algebra, we have a complete set of orthonormal eigenvectors with a corresponding set of eigenvalues, the spectrum that here we assume to be discrete to avoid cluttering formulas. So, let us assume for the moment, for the self-adjoint operator A,

$$A|a_n\rangle = a_n|a_n\rangle, \tag{18.8}$$

where $\{|a_n\rangle\}$ is the set of eigenvectors and $\{a_n\}$ is the corresponding set of eigenvalues. We exclude degeneracy (i.e., some subset of identical eigenvalues for different eigenvectors) for the sake of simplicity. This implies that

$$\langle a_m|a_n\rangle = \delta_{nm}. \tag{18.9}$$

This means that any arbitrary state vector belonging to the Hilbert space $\mathscr{H}$ can be expressed through this set of eigenvectors. In particular, we can write

$$|\psi\rangle = \sum_n p_n|a_n\rangle \tag{18.10}$$

where the coefficients are given by $p_n = \langle a_n|\psi\rangle$. We call these complex numbers probability amplitudes. Then, whenever we will do an experiment to measure the observable A, we will get the value a_n with a probability $|p_n|^2$. This is the so-called Born's rule and is at the foundation of the probabilistic interpretation of quantum mechanics. Generally speaking, the state vector is a probability amplitude and its squared modulus a probability. This implies that a state vector is always normalized to unity

$$\langle\psi|\psi\rangle = 1, \tag{18.11}$$

therefore, the following identity holds for probability amplitudes:

$$\sum_n |p_n|^2 = 1. \tag{18.12}$$

The following completeness relation also holds for the identity operator I:

$$I = \sum_n |a_n\rangle\langle a_n| \tag{18.13}$$

and is a consequence of the completeness of the set of eigenvectors in the Hilbert space $\mathscr{H}$. A consequence of this probabilistic interpretation is that, when a pair of operators A and B do not commute, the following theorem, due to Weyl and embodying the Heisenberg uncertainty principle, holds

$$\Delta A \Delta B \geq \frac{1}{2}\langle\psi|[A,B]|\psi\rangle, \tag{18.14}$$

where

$$\begin{aligned}\Delta A &= \sqrt{\langle\psi|A^2|\psi\rangle - (\langle\psi|A|\psi\rangle)^2},\\ \Delta B &= \sqrt{\langle\psi|B^2|\psi\rangle - (\langle\psi|B|\psi\rangle)^2}.\end{aligned} \tag{18.15}$$

When applied to (18.4), one recovers the standard formulation of the Heisenberg uncertainty principle for positions and momenta ($i = x,\ y,\ z$)

$$\Delta x_i \Delta p_i \geq \frac{\hbar}{2}. \tag{18.16}$$

This implies that the more precise a measurement of position is the less precise the measurement of momentum is and vice versa. So, these kinds of observables are said to be complementary. When the observables commute they are said to be compatible. As stated at the start of this section, a quantum system is completely characterized by a complete set of compatible observables. Anyway, a measurement of a complementary observable is always possible and so, we can always expand our state vector with the eigenvectors of the given observable and compute the corresponding probabilities for its eigenvalues.

Finally, let us consider a free particle and use the coordinate representation. We just note that the coordinate operator is rather singular. It has a continuous spectrum of values and the following relations hold:

$$\langle x_i|x_j'\rangle = \delta_{ij}\delta(x_i - x_i'), \tag{18.17}$$

$$\int d^3x|\mathbf{x}\rangle\langle\mathbf{x}| = I.$$

In this representation, one has the wave function $\psi(\mathbf{x}) = \langle\mathbf{x}|\psi\rangle$. This means that the equation

$$[x_i, p_j]\psi(\mathbf{x}) = i\hbar\delta_{ij}\psi(\mathbf{x}) \tag{18.18}$$

has the solution

$$\mathbf{p} = -i\hbar\left(\frac{\partial}{\partial x}, \frac{\partial}{\partial y}, \frac{\partial}{\partial z}\right) = -i\hbar\nabla \tag{18.19}$$

yielding the momentum operator in differential form. In this solution, we have assumed a possible additive function to be zero to keep translation invariance while a constant would be inessential. We can substitute this solution into (18.1) providing the Hamiltonian for the free particle in differential form:

$$H = -\frac{\hbar^2}{2m}\left(\frac{\partial^2}{\partial x^2} + \frac{\partial^2}{\partial y^2} + \frac{\partial^2}{\partial z^2}\right) = -\frac{\hbar^2}{2m}\Delta_2. \tag{18.20}$$

So, the Schrödinger equation takes the form, of a state evolving in time,

$$-\frac{\hbar^2}{2m}\Delta_2\psi(\mathbf{x}, t) = i\hbar\frac{\partial\psi(\mathbf{x}, t)}{\partial t}. \tag{18.21}$$

This extends quite simply to several other cases.

18.4 Quantum harmonic oscillator

In the following, a quantum system will be essential to understand the main points of the operation of a quantum radar. The system is a simple harmonic oscillator (SHO). In classical mechanics, the SHO is widely known as it represents the behavior of a small elastic deformation by the effect of Hooke's law. One has a restoring force going like $F = -kx$ with spring constant k being a constant specific to the system and x being the corresponding deformation. Then, one gets the Newton equation

$$\frac{d^2x(t)}{dt^2} = -\frac{k}{m}x(t) \tag{18.22}$$

that yields simple oscillations, given the proper initial conditions. This force admits a potential given by

$$V(x) = \frac{1}{2}kx^2. \tag{18.23}$$

Therefore, referring to (18.2) for the Hamiltonian, in the coordinate representation, we get the following Schrödinger equation:

$$-\frac{\hbar^2}{2m}\frac{\partial^2\psi}{\partial x^2} + \frac{1}{2}kx^2\psi = i\hbar\frac{\partial\psi}{\partial t}. \tag{18.24}$$

We just note that the Hamiltonian does not depend on time and so, energy is conserved as also happens for the classical system. Indeed, this property of classical systems is preserved when quantized. This means that we can look for a solution of the form

$$\psi(x,t) = e^{-\frac{i}{\hbar}Et}\varphi(x) \tag{18.25}$$

and we have to solve the time-independent Schrödinger equation

$$-\frac{\hbar^2}{2m}\frac{d^2\varphi}{dx^2} + \frac{1}{2}kx^2\varphi = E\varphi. \tag{18.26}$$

We recognize here a typical eigenvalue problem in ordinary differential equations.

In order to solve this kind of eigenvalue problem, we consider an algebraic approach due to Dirac [39] that will be helpful in the following. First of all, we make (18.26) dimensionless. We introduce the new variables

$$y = \sqrt{\frac{m\omega}{\hbar}}x \tag{18.27}$$

$$\varepsilon = \frac{2E}{\hbar\omega}.$$

We have set $\omega^2 = k/m$. We are left with the equation

$$-\frac{d^2\varphi}{dy^2} + y^2\varphi = \varepsilon\varphi. \tag{18.28}$$

We can now introduce the creation and annihilation operators as

$$a^\dagger = \frac{1}{\sqrt{2}}\left(-\frac{d}{dy} + y\right) \tag{18.29}$$

$$a = \frac{1}{\sqrt{2}}\left(\frac{d}{dy} + y\right).$$

It is easy to see that

$$[a, a^\dagger] = \frac{1}{2}\left[\frac{d}{dy} + y, -\frac{d}{dy} + y\right] = aa^\dagger - a^\dagger a = I. \tag{18.30}$$

This means that (18.28) can be rewritten using the creation and annihilation operators as

$$\left(a^\dagger a + \frac{1}{2}\right)\varphi = \frac{\varepsilon}{2}\varphi. \tag{18.31}$$

Now, we can change representation and move away from the x-coordinate. This can be down by introducing the number operator [39]

$$N = a^\dagger a \tag{18.32}$$

that solves the following eigenvalue problem:

$$N|n\rangle = n|n\rangle \tag{18.33}$$

where $n \in \mathbb{N}_0$. The integer number n is called an occupation number. Then, we get the nth energy eigenvalue for E given by

$$E_n = \left(n + \frac{1}{2}\right)\hbar\omega. \tag{18.34}$$

Furthermore, one has

$$\begin{aligned} a^\dagger|n\rangle &= \sqrt{n+1}|n+1\rangle \\ a|n\rangle &= \sqrt{n}|n-1\rangle. \end{aligned} \tag{18.35}$$

This is the reason for the name of these operators. The creation operator increases the occupation number by one and the annihilation operator decreases the occupation number by one. If n would be the number of photons, as this is also the quantum representation of the electromagnetic field, these operators just create or annihilate a photon in the field.

The quantum harmonic oscillator has a ground state defined by

$$a|0\rangle = 0 \tag{18.36}$$

with energy $E_0 = \hbar\omega/2$. For an electromagnetic field, this means the absence of photons but, anyway, this state has an energy different from zero. Then, returning to the coordinate representation, the above equation becomes

$$\frac{1}{\sqrt{2}}\left(\frac{d}{dy} + y\right)\varphi_0(y) = 0. \tag{18.37}$$

This can be solved immediately to give

$$\varphi_0(x) = \left(\frac{m\omega}{\pi\hbar}\right)^{\frac{1}{4}} e^{-\frac{m\omega}{2\hbar}x^2}. \tag{18.38}$$

We have undone the scaling of the y variable and normalized it to unity. This is the wave function of the ground state of the quantum harmonic oscillator and is a coherent state. A coherent state is a state where the uncertainty inequality becomes an equality, that is

$$\Delta x \Delta p = \frac{\hbar}{2}. \tag{18.39}$$

So, a coherent state minimizes the uncertainty product between two complementary variables. Finally, any other eigenfunction of the harmonic oscillator can be obtained by the equation

$$|n\rangle = (a^\dagger)^n|0\rangle \tag{18.40}$$

that in coordinate representation becomes

$$\varphi_n(x) = \left[\frac{1}{\sqrt{2}}\left(-\frac{d}{dy}+y\right)\right]^n \varphi_0(x). \tag{18.41}$$

18.5 Quantum electromagnetic field

In Section 18.4, we discussed the quantum SHO. In this section, we shall see that the quantum SHO plays a fundamental role in the description of the electromagnetic field.

Maxwell's equations describe the electromagnetic field. The quantum mechanical description of the electromagnetic field was initiated in the 1920s and the foundation laid by Dirac. Quantum electrodynamics (QED) was ultimately described by Feynman, Schwinger and Tomanaga in the 1940s. It is now one of the most successful theories in physics.

The QED description of the electromagnetic field treats the electromagnetic field as an infinite collection of independent travelling modes, where each mode is described by a unique frequency (and hence wavelength) and all travelling at the speed of light with the energy of the field quantized with energy $E = hf$, where h is Planck's constant. Maxwell's equations are still valid in quantum physics, but the electromagnetic field is no longer described by numbers. The quantization of the EM field leads to the photon description of the EM field and the electric field is described by a complex operator function.

18.5.1 Single mode

Specifically, recall that in classical electromagnetism, the electric field is given by

$$E(t,f) = \frac{E_0}{\sqrt{2}}\left(a^*(t,f)e^{i(2\pi ft+\phi)} + a(t,f)e^{-i(2\pi ft+\phi)}\right), \tag{18.42}$$

where $a(t,f), a^*(t,f)$ are complex numbers. In QED, the electric field operator is

$$\hat{E}(t,f) = \frac{E_0}{\sqrt{2}}\left(\hat{a}^\dagger(t,f)e^{i(2\pi ft+\phi)} + \hat{a}(t,f)e^{-i(2\pi ft+\phi)}\right), \tag{18.43}$$

where $\hat{a}(t,f), \hat{a}^\dagger(t,f)$ are complex linear annihilation and creation operators that satisfy the quantum SHO commutation relations

$$[\hat{a}(t,f), \hat{a}^\dagger(t',f)] = \hat{a}(t,f), \hat{a}^\dagger(t',f) - \hat{a}^\dagger(t,f), \hat{a}(t',f) = \delta(t-t'). \tag{18.44}$$

From our discussion in Section 18.4, the single-mode electromagnetic field states are number states $|n\rangle_f, n = 0, 1, \ldots$, where the subscript f indicates the frequency f corresponding to the mode. Therefore, it follows that

$$\begin{aligned} \hat{a}(t,f)|N\rangle_f &= \sqrt{N}|N-1\rangle_f, \\ \hat{a}^\dagger(t,f)|N\rangle_f &= \sqrt{N+1}|N+1\rangle_f, \\ \hat{a}^\dagger(t,f)a(t,f)|N\rangle_f &= N|N\rangle_f. \end{aligned} \tag{18.45}$$

Another representation of the electric field is in terms of in-phase and quadrature voltages

$$E(t,f) = E_0(I(t,f)\cos(2\pi ft) + Q(t,f)\sin(2\pi ft), \tag{18.46}$$

where

$$\begin{aligned} I(t,f) &= X(t,f)\cos\phi + P(t,f)\sin\phi, \\ Q(t,f) &= -X(t,f)\sin\phi + P(t,f)\cos\phi, \end{aligned} \tag{18.47}$$

and the "position" and "momentum" variables, $X(t,f$ and $P(t,f)$, are

$$\begin{aligned} X(t,f) &= \frac{1}{\sqrt{2}}(a^*(t,f) + a(t,f)), \\ P(t,f) &= \frac{i}{\sqrt{2}}(a^*(t,f) - a(t,f)). \end{aligned} \tag{18.48}$$

In QED, the electric field operator can be written in terms of the in-phase and quadrature operators

$$\hat{E}(t,f) = E_0(\hat{I}(t,f)\cos(2\pi ft) + \hat{Q}(t,f)\sin(2\pi ft), \tag{18.49}$$

where the electric field quadrature operators, namely the amplitude quadrature operator

$$\hat{X}(t,f) = \frac{1}{\sqrt{2}}(\hat{a}^\dagger(t,f) + \hat{a}(t,f)), \tag{18.50}$$

and the phase quadrature operator

$$\hat{P}(t,f) = \frac{j}{\sqrt{2}}(\hat{a}^\dagger(t,f) - \hat{a}(t,f)). \tag{18.51}$$

The in-phase and quadrature operators are given by

$$\begin{aligned} \hat{I}(t,f) &= \hat{X}(t,f)\cos\phi + \hat{P}(t,f)\sin\phi, \\ \hat{Q}(t,f) &= -\hat{X}(t,f)\sin\phi + \hat{P}(t,f)\cos\phi. \end{aligned} \tag{18.52}$$

From the commutation relations (18.44), it follows that the field quadratures do not commute, i.e., in units where $\hbar = 1$,

$$[\hat{I}(t,f), \hat{Q}(t,f)] = [\hat{X}(t,f), \hat{P}(t,f)] = i. \tag{18.53}$$

For an arbitrary state $|\psi\rangle$, the expectation value and the variance of the electric field are defined as

$$\begin{aligned} \langle\psi|\hat{E}(t,f)|\psi\rangle &\equiv \langle\hat{E}(t,f)\rangle_\psi, \\ \mathrm{var}[\hat{E}(t,f)]_\psi &= \langle\hat{E}^2(t,f)\rangle_\psi - \langle\hat{E}(t,f)\rangle_\psi^2. \end{aligned} \tag{18.54}$$

It can be shown that for any state $|\psi\rangle$

$$\mathrm{var}[\hat{I}(t,f)]_\psi \mathrm{var}[\hat{Q}(t,f)]_\psi \geq \frac{1}{4}. \tag{18.55}$$

One special case is the vacuum state $|\psi\rangle = |0\rangle$ for which

$$\text{var}[\hat{I}(t,f)]_{\text{vac}} = \text{var}[\hat{Q}(t,f)]_{\text{vac}} = \frac{1}{2}, \tag{18.56}$$

so that

$$\text{var}[\hat{I}(t,f)]_{\text{vac}}\text{var}[\hat{Q}(t,f)]_{\text{vac}} = \frac{1}{4}. \tag{18.57}$$

Another state is the coherent state $|\psi\rangle = |\alpha\rangle$ where

$$|\alpha\rangle = e^{-\frac{1}{2}|\alpha|^2} \sum_{n=0}^{\infty} \frac{\alpha^n}{\sqrt{n!}} |n\rangle, \tag{18.58}$$

so that the average number of photons is $N = \langle a^\dagger a\rangle_\alpha = |\alpha|^2$. It can be shown that the photon number in this state is Poisson distributed. The expectation value of the electric field for a coherent state is

$$\langle \hat{E}(t,f)\rangle = E_0\alpha(e^{i\phi}\cos(2\pi ft) + e^{-i\phi}\sin(2\pi ft)), \tag{18.59}$$

which is similar to that of a classical field of amplitude $E_0\alpha$ and phase ϕ. Since the quadrature variances are

$$\text{var}[\hat{I}]_\alpha = \text{var}[\hat{Q}]_\alpha = \frac{1}{2}, \tag{18.60}$$

a coherent state has a well-defined amplitude and phase (unlike the vacuum state) and has the least amount of noise allowed by quantum physics (ideal noiseless classical field analog in quantum theory). In the optical regime, such states are created by lasers.

Another example of a classical state is the thermal state, which represents the thermal radiation from a black body at temperature T. At f, the thermal state has

$$\langle \hat{a}\rangle_T = \langle \hat{a}^\dagger\rangle = 0, \tag{18.61}$$

and

$$\langle \hat{a}^\dagger \hat{a}\rangle_T = N_T, \quad N_T = \frac{1}{e^{\frac{hf}{k_B T}} - 1}, \tag{18.62}$$

where k_B is the Boltzmann constant. The thermal field quadratures can be shown to have

$$\langle \hat{I}(t)\rangle_T = \langle \hat{Q}(t)\rangle_T = 0 \tag{18.63}$$

and

$$\text{var}[\hat{I}(t)]_T = \text{var}[\hat{Q}(t)]_T = N_t + \frac{1}{2}, \tag{18.64}$$

with the zero-temperature state reducing to the vacuum state.

18.5.2 Multiple modes

The extension to the multiple modes follows from the single-mode discussion with some extensions. In particular, the photon number states are

$$|n_1, n_2, \ldots, n_k\rangle_{f_1, f_2, \ldots, f_k} \equiv |n_1\rangle_{f_1} |n_2\rangle_{f_2} \cdots |n_k\rangle_{f_k}. \tag{18.65}$$

The multi-mode extension of the single-mode states is possible to define. For instance, the k-mode coherent state is

$$|\alpha_1, \ldots, \alpha_k\rangle = e^{-\frac{1}{2}|\sum_{i=1}^{k} \alpha_i|^2} \sum_{n_1=0}^{\infty} \frac{\alpha_1^n}{\sqrt{n_1!}} \cdots \sum_{n_k=0}^{\infty} \frac{\alpha_k^n}{\sqrt{n_k!}} |n_1, \ldots, n_k\rangle. \tag{18.66}$$

The k-mode thermal state can be similarly defined. The vacuum state, coherent state, and thermal states are all Gaussian states.

A Gaussian state is specified by a Wigner covariance matrix, and the resulting Gaussian distribution with the Wigner covariance matrix is termed the Wigner function. The Wigner covariance matrix is a physical covariance matrix and differs from covariance matrices that are traditionally studied in radar signal processing. For one, the Wigner covariance matrices are real, not complex. Furthermore, they are normalized to the single-photon level, and hence their estimation requires very careful system calibration to the single-photon level [43].

For instance, in the single-mode case, the Wigner covariance matrix is given by

$$\begin{bmatrix} \cosh 2r + \cos\theta \sinh 2r & \sin\theta \sinh 2r \\ \sin\theta \sinh 2r & \cosh 2r - \cos\theta \sinh 2r \end{bmatrix} \tag{18.67}$$

Two special cases are the vacuum state and the squeezed vacuum state with Wigner matrices

$$V = \begin{bmatrix} 1 & 0 \\ 0 & 1 \end{bmatrix} \quad \text{(Vacuum)}, \quad V = \begin{bmatrix} e^{2r} & 0 \\ 0 & e^{-2r} \end{bmatrix} \quad \text{(Squeezed Vacuum)} \tag{18.68}$$

Here $V = E[xx^T]$ where $x = \begin{bmatrix} I & Q \end{bmatrix}$. The squeezing is observed in the case of the squeezed vacuum state wherein the variance along one direction is smaller than the variance along the other direction. The Wigner function in the one-mode Gaussian state case can be visualized in a contour plot as tilted ellipses in general, with the vacuum state corresponding to a circle in the origin and the squeezed vacuum state as in ellipse with the lengths of the major (minor) axis of e^{2r} (e^{-2r}). Apart from being positive-definite and real, the Wigner covariance matrix elements also satisfy the Heisenberg uncertainty principle which constrains the variances

$$V_{11} V_{22} \geq 1. \tag{18.69}$$

Since for a pure Gaussian state $\det V = 1$

$$V_{11} V_{22} = 1 + V_{12}^2 = 1 + \sin\theta \cosh^2 2r. \tag{18.70}$$

There is no notion of Heisenberg uncertainty principle in covariance matrices studied in radar signal processing. Wigner covariance matrices are also used to study N-mode case.

18.6 Quantum illumination

Quantum illumination is a quantum effect, due to entangled states of photons, in a noisy and lossy environment, that yields an advantage for the observation of a target with respect to an ordinary incoherent or coherent source. As the name suggests, a quantum illumination radar is based on novel types of transmitters that are fundamentally quantum mechanical.

In order to explain the concept of quantum illumination, it is essential to define properly an entangled state. Let us consider a set of photon states $\{|n\rangle,\ n \in \mathbb{N}\}$, where n is the number of photons in a single state, belonging to a Hilbert space $\mathscr{H}$. We will have the product states

$$|n_1, n_2, \ldots\rangle = |n_1\rangle|n_2\rangle \ldots \tag{18.71}$$

that belongs to a Fock space $\mathscr{F}_+(\mathscr{H})$. This space generalizes the Hilbert space of states for a single particle to a space for states of an arbitrary number of particles. The states in (18.71) form a basis for this space. A generic state given by the sum of product states that cannot be reduced to a product state is an entangled state. We can write this as

$$|\psi_E\rangle = \sum_{n_1,n_2,\ldots} a_{n_1,n_2,\ldots}|n_1, n_2, \ldots\rangle. \tag{18.72}$$

This state is normalized in such a way that

$$\sum_{n_1,n_2,\ldots} |a_{n_1,n_2,\ldots}|^2 = 1. \tag{18.73}$$

An example of entangled state can be given by considering a state with a pair of photons, that we labels 1 and 2, with a left-handed polarization L and a right-handed polarization R. Therefore,

$$|L, R\rangle_E = a|1, L\rangle_1|2, R\rangle_2 + b|1, R\rangle|2, L\rangle \tag{18.74}$$

with the condition $|a|^2 + |b|^2 = 1$. The only way to get a non-trivial product state is by having one of two probability amplitudes, a or b, equal to zero. In this way, we realize immediately the main property of an entangled state. The effect of a measurement on the first photon of the pair will give polarization L with probability $|a|^2$ and polarization R with probability $|b|^2$. But, as one of these outcomes is obtained, immediately the state of the second photon is determined and known to the experimenter, being R or L, respectively. The unobserved photon could be very far displaced in space and time. This is an example of quantum non-locality. It does not imply any causality violation as the information is retained by the experimenter doing the measurement and, transmitting it to a second experimenter implies using ordinary communication channels.

The polarization entanglement is an example of discrete variable entanglement. It is also possible to have entanglement in quadrature voltages. The quadrature entanglement is an example of a continuous variable entanglement. The simplest possibility of continuous variable entanglement is an entanglement using two modes. This particular example of quantum illumination was first investigated theoretically [44]. The Wigner covariance matrix for the two-mode case $x = [I_1, Q_1, I_2, Q_2]$ is given by

$$E[xx^T] = \begin{bmatrix} \sigma_1^2 \mathbf{1}_2 & \rho\sigma_1\sigma_2\mathbf{R}(\phi) \\ \rho\sigma_1\sigma_2\mathbf{R}(\phi) & \sigma_2^2\mathbf{1}_2 \end{bmatrix}, \tag{18.75}$$

where ρ is the Pearson correlation coefficient, σ_1^2 and σ_2^2 are the measured signal powers, ϕ is the phase shift between signals, $\mathbf{1}_2$ is the 2×2 identity matrix and $\mathbf{R}(\phi)$ is the reflection matrix

$$\mathbf{R}(\phi) = \begin{bmatrix} \cos\phi & \sin\phi \\ \sin\phi & -\cos\phi \end{bmatrix}. \tag{18.76}$$

Noise radars also have a traditional covariance matrix but with a rotation matrix instead of reflection matrix.

Apart from the positive definiteness and reality property, two-mode Gaussian states also satisfy the Heisenberg inequality as it is a physical covariance matrix. In addition, there is now a possibility of entanglement among the quadratures. Specifically, it can be shown that when $\sigma_1 = \sigma_2 = \sigma$ and $\phi = 0$, then the two-mode Gaussian state is entangled if

$$\rho > 1 - \frac{1}{2\sigma^2}. \tag{18.77}$$

The classical analog of the two-mode squeezed state is the two-mode noise state where the above inequality is not respected. Entangled states, such as two-mode squeezed states, cannot be generated using technologies currently used to build radar transmitters.

Some of the multi-mode Gaussian states are also entangled states, i.e., such states can be explained classically, or be created using technologies currently used to build radars.

There are infinitely many other possibilities possible from quantum physics, and Gaussian states are a subset of all possible states. This richness of possibilities gives novel possibilities for quantum radars.

18.7 An experimental demonstration

In summary, quantum physics expands the possibilities of radar transmitters and receivers. While there are many theoretical possibilities, these ideas would be uninteresting if they were not practically realizable. Interestingly, it is indeed possible to explore such concepts in laboratory settings.[¶¶¶] We describe one such example in microwave regime, namely the quantum two-mode squeezed (QTMS) radar. This

[¶¶¶] There are two other chapters in this book on cognitive radar hardware/prototypes to refer to: "A Canonical Cognitive Radar Architecture" by Joe Guerci *et al.* and "Advances in Cognitive Radar Experiments" by Graeme Smith *et al.*

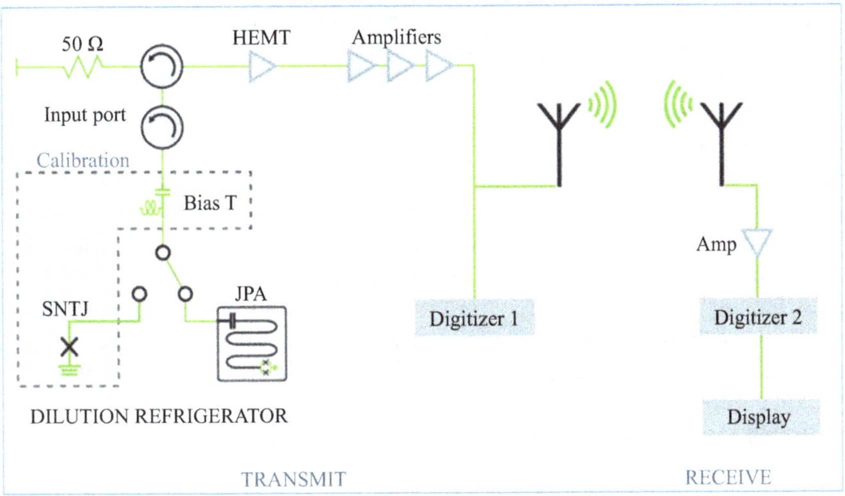

Figure 18.3 Schematic of a quantum two-mode squeezed radar

kind of quantum radar exploits a quantum enhancement rather than quantum illumination in its full glory. Anyway, experimental results prove unequivocally the gain with respect to an equivalent classical noise radar.

One approach is to use Josephson Parametric Amplifier (JPA). JPAs are quantum limited amplifiers, i.e., they add the least amount of noise allowed by quantum mechanics. JPAs are built using Josephson junctions, which are nonlinear circuit elements that are non-dissipative. Nonlinearity is key to the generation of entangled signals and the nonlinearity that is feasible using this approach in microwave regime is considerably larger than is possible in the optical regime. In fact, JPAs have been shown to exhibit SPDC (Spontaneous Parametric Down Conversion) and are among the brightest sources of entangled fields across the electromagnetic spectrum [45,46]. Figure 18.3 is an example realization of a microwave quantum radar, where the experiment was carried out by Chris Wilson's group at the Institute of Quantum Computing, University of Waterloo [47]. The JPA is the key ingredient of the radar. The superconducting noise tunnel junction (SNTJ) is there to validate that a microwave-entangled signal was indeed generated. The two modes are then amplified using a low-noise amplifier (a high-electron-mobility transistor (HEMT o HFET) amplifier in this case) and other amplifiers outside the dilution refrigerator. One of the modes is then digitized and stored and the other mode transmitted. The received signal is then digitized and then the stored mode signal is used as the matched filter.

In order to generate the signals, dilution refrigerators are needed. One reason is that the superconducting quantum interference device (SQUID) needs to be superconducting. A more important reason is that vacuum needs to be generated to generate the microwave vacuum that is needed to generate the two-mode squeezed vacuum (TMSV) source [48].

The QTMS radar was analyzed using conventional signal processing techniques and it was shown that the data fit a noise radar model with the difference being that the entanglement resulted in a higher correlation for the same transmit power [47,49]. This also means that classical noise radar signal processing techniques could be applied to QTMS radar thus enabling the leverage of decades of classical radar expertise [50]. This also motivates a different perspective for understanding and analyzing classical noise radars [51].

From the experiments at Waterloo and Wien (see [52]), there are many challenges and open questions on a practical system. The entangled signals are of very low output power (femto-Watt regime) and require a very controlled and extremely cold (around 10–30 mK) environment. The entangled signals have to be somehow brought to room temperature. An important step would be to transmit and detect entangled microwave signal in the open atmosphere.

An objection could be made that the signals emerging from the dilution refrigerator are not entangled. This is because symmetric amplification does break entanglement (as noted in [45,47] and also experimentally validated in [53]). Quantum mechanically, when one amplifies the signal, the noise is also amplified. It is seen that when the amplified noise exceeds the noise of the classical source, there is no quantum entanglement or practical advantage.

Nevertheless, the fact is that the two modes were entangled at the JPA source. This leaves open the objection that our results could be reproduced simply by building a better classical system—in principle. This was indeed confirmed experimentally recently [53]. Such objections can be unambiguously countered by removing the amplifier chain and extracting the entangled quantum signal directly. If successful, the resulting transmitted signal would be theoretically impossible to reproduce without an entangled source. Furthermore, asymmetric amplification may provide a way to preserve entanglement during amplification, as has been demonstrated in the optical regime [54].

18.8 Hybridization of cognitive and quantum radar: what recent research in neuroscience can tell about

In the quest for new inspiration aimed at an hybridization of cognitive and quantum radar, properly combining the above presentations, we look to recent researches in neuroscience from top physicists and from a Nobel laureate too.

Kurzweil [55] suggests that the brain contains a hierarchy of pattern recognizers. These hierarchical pattern recognizers are used not just for sensing the world, but also for nearly all aspects of thought. He says the neocortex contains 300 million very general pattern recognition circuits and argues that they are responsible for most aspects of human thought. It would employ techniques such as hidden Markov models and genetic algorithms. Challenge: exploiting these concepts in the radar signal processing context.

A connectome**** is a comprehensive map of neural connections in the brain and may be thought of as its "wiring diagram." More broadly, a connectome would include the mapping of all neural connections within an organism's nervous system. Challenge: Connectome can inspire distributed radar signal processing.

The relevance of nanometer-scale electronic synaptic components is reported in the DARPA Project: Systems of Neuromorphic Adaptive Plastic Scalable Electronics (SyNAPSE).††††

The brain does not store information about simple tasks or other knowledge all by itself. Information is stored within the context of events and ideas. Also, a person can apply the knowledge learned about one situation to another. This type of learning is called generalization. Generalization [56]: experiences obtained in one situation are applicable to other situations. Challenge: exploiting the "generalization" concept to radar as suggested in [57]. An example of generalization is illustrated in Figure 18.4.

Moving to the investigations on neuroscientists, an insightful book on brain and mind is the best-seller authored by the famous Michio Kaku [58]. Quite interesting is also the review published in *The New York Times* by Prof. Adam Frank [59].

The role of quantum physics in neuroscience has been hypothesized by Prof. Roger Penrose and colleagues.

Orchestrated objective reduction (Orch OR)‡‡‡‡ is a controversial hypothesis that postulates that consciousness originates at the quantum level inside neurons, rather than the conventional view that it is a product of connections between neurons. The mechanism is held to be a quantum process called objective reduction that is orchestrated by cellular structures called microtubules. It is proposed that the theory may answer the hard problem of consciousness and provide a mechanism for free will. The hypothesis was first put forward in the early 1990s by Nobel laureate for physics, Roger Penrose, and anesthesiologist and psychologist Stuart Hameroff. The hypothesis combines approaches from molecular biology, neuroscience, pharmacology, philosophy, quantum information theory, and quantum gravity. While mainstream theories assert that consciousness emerges as the complexity of the computations performed by cerebral neurons increases, Orch OR posits that consciousness is based on non-computable quantum processing performed by qubits formed collectively on cellular microtubules, a process significantly amplified in the neurons. The qubits are based on oscillating dipoles forming superposed resonance rings in helical pathways throughout lattices of microtubules. The oscillations are either electric, due to charge separation from London forces, or magnetic, due to electron spin—and possibly also due to nuclear spins (that can remain isolated for longer periods) that occur in gigahertz, megahertz, and kilohertz frequency ranges. Orchestration refers to the hypothetical process by which connective proteins, such as microtubule-associated proteins (MAPs), influence or orchestrate qubit state reduction by modifying the space–time-separation of their superimposed states. The latter is based on Penrose's objective-collapse theory for interpreting quantum mechanics, which postulates the

****https://en.wikipedia.org/wiki/Connectome

††††https://www.darpa.mil/program/systems-of-neuromorphic-adaptive-plastic-scalable-electronics

‡‡‡‡https://en.wikipedia.org/wiki/Orchestrated_objective_reduction

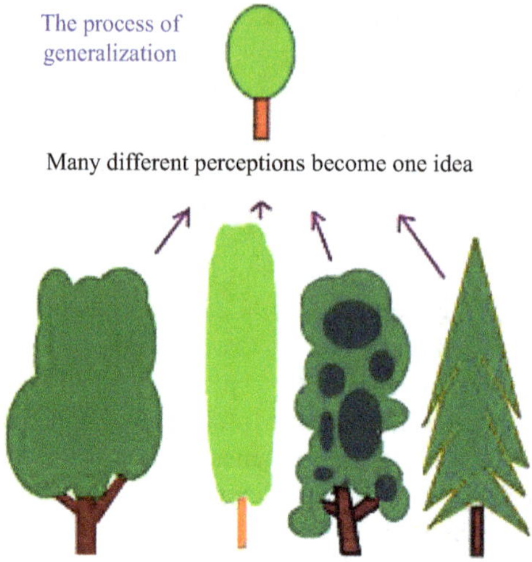

Figure 18.4 The concept of generalization

existence of an objective threshold governing the collapse of quantum states, related to the difference of the spacetime curvature of these states in the universe's fine-scale structure. Orchestrated objective reduction has been criticized from its inception by mathematicians, philosophers, and scientists. The criticism concentrated on three issues: Penrose's interpretation of Gödel's theorem; Penrose's abductive reasoning linking non-computability to quantum events; and the brain's unsuitability to host the quantum phenomena required by the theory, since it is considered too "warm, wet and noisy" to avoid decoherence. In 2014, Penrose and Hameroff published lengthy responses to these criticisms and revisions to many of the theory's peripheral assumptions, while retaining the core hypothesis.

The main message is that the brain, through thin filaments present in all cells (including neurons), called microtubules, which determine the structural integrity of the cell and play a role in cell division, these microtubules have a role in the processing of information inside the cell. Roger Penrose, and later Stuart Hameroff, hypothesized and supported quantum processing capabilities linked to microtubules.

The subsequent references have further discussed and investigated the role of quantum physics in neuroscience [60–63].

This last reference [64] is a doctoral thesis which inspires to bring new lymph to connect cognitive processing (i.e., PAC) to quantum processing (quantum PAC) and eventually move from cognitive radar to Cognitive-Quantum Radar (CQR). This is the challenge that we aim to face and seek an initial answer.

18.9 Quantum and cognitive radar

The schematic of Figure 18.5 illustrates a vision concept of a hybridization of cognitive and quantum radar, that we call as, CQR. The scheme foresees the exploitation of a

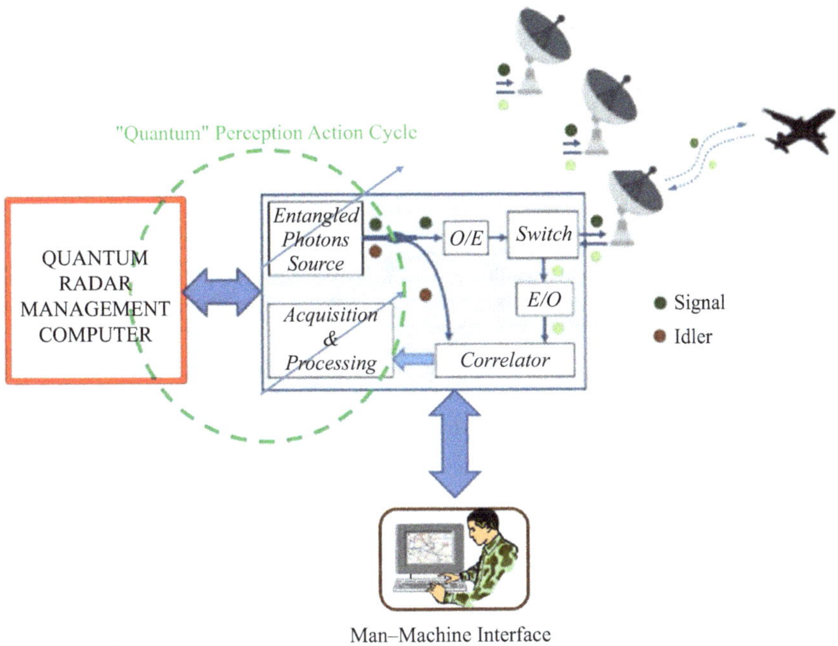

Figure 18.5 Schematic representation of a possible cognitive-quantum radar

quantum PAC, an adaptive transmission of entangled photons, an adaptive reception of the reflected photons, and a quantum radar management computer. The antenna may be an array of antennas in a suitable lattice to achieve accurate radar target measurements.

Actually, we have tried to merge cognitivity and quantum concepts. A new "technology animal" could be envisaged to come out in the hopefully not so distant future. This new system should enjoy the reinforcement of cognition capabilities due to the argued quantum base of cognition.

18.10 Conclusions

The purpose of this chapter has been to capture, from the technical literature, points of tangency between the concepts of cognition and quantum physics. This might open the way to a potential hybridization between cognitive radar and quantum radar, a CQR. Being an initial vision, a lot of work is expected to be done to give more scientific fundamentals to this vision. We feel that the quantum remote sensing will be deployed on a longer time scale. We are in the stage of proving the concept and realizing even more powerful test beds. Open literature is witnessing this continuous upgrade. We believe that a cross-disciplinary approach between experts in different fields would be rewarding and would be essential to make progress in the field.

Acknowledgments

The authors would like to thank Prof. Antonio De Maio and Prof. Simon Haykin for their contributions to the section on the cognitive radar of this chapter.

References

[1] A. Farina, A. De Maio, and S. Haykin, editors. *The Impact of Cognition on Radar Technology*. Radar, Sonar & Navigation. Stevenage: SciTech: An Imprint of IET, 2017.

[2] C.T. Capraro, G. T. Capraro, A. De Maio, A. Farina, and M. Wicks. Demonstration of knowledge-aided space–time adaptive processing using measured airborne data. *IEE Proceedings – Radar, Sonar and Navigation*, 153(6):487–494(7), 2006.

[3] A. Farina. *Antenna Based Signal Processing Techniques for Radar Systems*. Boston, MA: Artech House, 1992.

[4] L.E. Brennan and L.S. Reed. Theory of adaptive radar. *IEEE Transactions on Aerospace and Electronic Systems*, AES-9(2):237–252, 1973.

[5] N. Canessa, M. Motterlini, C. Di Dio, *et al*. Understanding others' regret: A fMRI study. *PLoS One*, 4(10):1–10, 2009.

[6] G. Di Pellegrino, L. Fadiga, L. Fogassi, V. Gallese, and G. Rizzolatti. Understanding motor events: a neurophysiological study. *Experimental Brain Research*, 91:176–180, 1992.

[7] G. Rizzolatti, L. Fadiga, V. Gallese, and L. Fogassi. Premotor cortex and the recognition of motor actions. *Cognitive Brain Research*, 3(2):131–141, 1996.

[8] A. De Maio, A. Farina (Keynote Speaker), A. Aubry, and V. Carotenuto. Cognitive radar, inspiring principles, architecture, signal processing and challenging signal processing applications. In *2015 IET International Radar Conference*, 14–16 October 2015, Hangzhou, China, Oct. 2015.

[9] A. Farina, *Causal Inference in Statistics. An Attempt at Some Reflection*, Book Review Reflection, ISIF Perspectives On Information Fusion, vol. 3, 2020, pp. 36–39.

[10] A. De Maio and A. Farina. The role of cognition in radar sensing. In *2020 IEEE Radar Conference (RadarConf20)*, pp. 1–6, Sep. 2020.

[11] S. Haykin. Cognitive dynamic systems. *Proceedings of the IEEE*, 94(11): 1910–1911, 2006.

[12] S. Haykin. *Cognitive Dynamic Systems: Perception–Action Cycle, Radar and Radio*. Cambridge: Cambridge University Press, 2012.

[13] J. Guerci. *Cognitive Radar: The Knowledge-Aided Fully Adaptive Approach*. Norwood, MA: Artech House, 2010.

[14] S. Haykin. Cognitive radar: a way of the future. *IEEE Signal Processing Magazine*, 23(1):30–40, 2006.

[15] S. Haykin. Cognitive dynamic systems: Radar, control, and radio [point of view]. *Proceedings of the IEEE*, 100(7):2095–2103, 2012.

[16] S. Z. Gurbuz, H. D. Griffiths, A. Charlish, M. Rangaswamy, M. S. Greco, and K. Bell. An overview of cognitive radar: past, present, and future. *IEEE Aerospace and Electronic Systems Magazine*, 34(12):6–18, 2019.

[17] J. Ender and S. Brüggenwirth. Cognitive radar – enabling techniques for next generation radar systems. In *2015 16th International Radar Symposium (IRS)*, pp. 3–12, Jun. 2015.

[18] M. Wicks. Spectrum crowding and cognitive radar. In *2010 2nd International Workshop on Cognitive Information Processing*, pp. 452–457, Jun. 2010.

[19] F. Smits, A. Huizing, W. van Rossum, and P. Hiemstra. A cognitive radar network: architecture and application to multiplatform radar management. In *2008 European Radar Conference*, pp. 312–315, Oct. 2008.

[20] S. Brüggenwirth, A. Huizing, and A. Charlish. Cognitive radar special issue—Part 1. *IEEE Aerospace and Electronic Systems Magazine*, 34(12):4–5, 2019.

[21] H. Steyskal, R. Shore, and R. Haupt. Methods for null control and their effects on the radiation pattern. *IEEE Transactions on Antennas and Propagation*, 34(3):404–409, 1986.

[22] T. Webster, T. Higgins, A. K. Shackelford, J. Jakabosky, and P. McCormick. Phase-only adaptive spatial transmit nulling. In *2015 IEEE Radar Conference (RadarCon)*, pp. 0931–0936, 2015.

[23] S. Sabatini and M. Tarantino. *Multifunction Array Radar-System Design and Analysis*. Norwood, MA: Artech House, 1994.

[24] K.J. Aström and R.M. Murray. *Feedback Systems. An Introduction for Scientists and Engineers*. Norwood, MA: Artech House, Norwood, MA, Sep. 2012. http://www.cds.caltech.edu/~murray/books/AM08/pdf/am08complete_28Sep12.pdf.

[25] J. von Neumann and O. Morgenstern. *Theory of Games and Economic Behavior*. Princeton, NJ: Princeton University Press, 1944.

[26] A. Wald. *Sequential Analysis*. New York: John Wiley & Sons, 1947.

[27] A. Benavoli, A. Balleri, and A. Farina. Joint waveform and guidance control optimization for target rendezvous. *IEEE Transactions on Signal Processing*, 67(16):4357–4369, 2019.

[28] K.J. Aström. Optimal control of Markov processes with incomplete state information. *Journal of Mathematical Analysis And Applications*, 10:174–205, 1965. Submitted by Richard Bellman.

[29] A. Farina and F. A. Studer. *Radar Data Processing: Introduction and Tracking*, vol. 1. Research Studies Press Ltd, 1985.

[30] A. Farina, L. Timmoneri, K. Hocking, C. Tierney, and D. Greig. Thoughts on cognitive systems: applications, benefits and challenges for modern radar. *Polaris Innovation Journal*, (40):8–11, 2019. Special Issue: Artificial Intelligence Innovations, developments and capabilities, first paper in the journal.

[31] F. Gini, A. De Maio, and L. K. Patton. *Waveform Design and Diversity for Advanced Radar Systems*. IET Series 22. London: Institution of Engineering and Technology, 2012.

[32] A. De Maio, S. De Nicola, Y. Huang, S. Zhang, and A. Farina. Code design to optimize radar detection performance under accuracy and similarity constraints. *IEEE Transactions on Signal Processing*, 56(11):5618–5629, 2008.

[33] W.L. Melvin and J.A. Scheer, editors. *Principles of Modern Radar: Volume 3: Radar Applications*. Radar, Sonar & Navigation. London: Institution of Engineering and Technology, 2013.

[34] A. Aubry, A. De Maio, M. Piezzo, and A. Farina. Radar waveform design in a spectrally crowded environment via nonconvex quadratic optimization. *IEEE Transactions on Aerospace and Electronic Systems*, 50(2):1138–1152, 2014.

[35] F. Gini and M. Rangaswamy. *Knowledge Based Radar Detection, Tracking and Classification*, vol. 52. New York: John Wiley & Sons, 2008.

[36] A. Aubry, A. DeMaio, A. Farina, and M. Wicks. Knowledge-aided (potentially cognitive) transmit signal and receive filter design in signal-dependent clutter. *IEEE Transactions on Aerospace and Electronic Systems*, 49(1):93–117, 2013.

[37] A. Balleri and A. Farina. Ambiguity function and accuracy of the hyperbolic chirp: comparison with the linear chirp. *IET Radar, Sonar & Navigation*, 11(1):142–153, 2017.

[38] S. Watts, H.D. Griffiths, A.M. Kinghorn, *et al.* The specification and measurement of radar performance – future research challenges. *Journal of Defence Science*, 8:83–91, 2003.

[39] P. A. M. Dirac. *The Principles of Quantum Mechanics*. Oxford: Oxford University Press, 1988.

[40] J. J. Sakurai and J. Napolitano. *Modern Quantum Mechanics*. Cambridge: Cambridge University Press, 2015.

[41] S. Weinberg. *Lectures in Quantum Mechanics*. Cambridge: Cambridge University Press, 2015.

[42] L. Susskind and A. Friedman. *Quantum Mechanics: The Theoretical Minimum*. New York: Basic Books, 2014.

[43] C.W. Sandbo Chang, M. Simoen, J. Aumentado, *et al.* Generating multimode entangled microwaves with a superconducting parametric cavity. *Physics Review Applied*, 10:044019, 2018.

[44] Si-Hui Tan, B.I. Erkmen, V. Giovannetti, *et al.* Quantum illumination with Gaussian states. *Physics Review Letters*, 101:253601, 2008.

[45] J. Bourassa and C.M. Wilson. Progress toward an all-microwave quantum illumination radar. *IEEE Aerospace and Electronic Systems Magazine*, 35(11):58–69, 2020.

[46] R. Simon. Peres-Horodecki separability criterion for continuous variable systems. *Physics Review Letters*, 84:2726–2729, 2000.

[47] D. Luong, C.W. Sandbo Chang, A.M. Vadiraj, A. Damini, C.M. Wilson, and B. Balaji. Receiver operating characteristics for a prototype quantum two-mode squeezing radar. *IEEE Transactions on Aerospace and Electronic Systems*, 56(3):2041–2060, 2020.

[48] M. Sanz, K.G. Fedorov, F. Deppe, and E. Solano. Challenges in open-air microwave quantum communication and sensing. In *2018 IEEE Conference on Antenna Measurements Applications (CAMA)*, pp. 1–4, Sep. 2018.

[49] David Luong, Bhashyam Balaji, and Sreeraman Rajan. Quantum two-mode squeezing radar and noise radar: Correlation coefficient and integration time. *IEEE Access*, 8:185544–185547, 2020.

[50] D. Luong, S. Rajan, and B. Balaji. Entanglement-based quantum radar: From myth to reality. *IEEE Aerospace and Electronic Systems Magazine*, 35(4): 22–35, 2020.

[51] D. Luong, S. Rajan, and B. Balaji. Quantum two-mode squeezing radar and noise radar: correlation coefficients for target detection. *IEEE Sensors Journal*, 20(10):5221–5228, 2020.

[52] S. Barzanjeh, S. Pirandola, D. Vitali, and J. M. Fink. Microwave quantum illumination using a digital receiver. *Science Advances*, 6(19):eabb0451, 2020.

[53] N. Messaoudi, C. W. Chang Sandbo, A. M. Vadiraj, C. M. Wilson, J. Bourassa, and B. Balaji. Practical advantage in microwave quantum illumination. In *2020 IEEE Radar Conference (RadarConf20)*, pp. 1–5, Sep. 2020.

[54] G.S. Agarwal and S. Chaturvedi. How much quantum noise of amplifiers is detrimental to entanglement. *Optics Communications*, 283(5):839–842, 2010. Quo vadis Quantum Optics?

[55] R. Kurzweil. *How to Create a Mind: The Secret of Human Thought Revealed.* Viking, Nov. 2012. https://en.wikipedia.org/wiki/How_to_Create_a_Mind.

[56] C. H. Judd. Generalized experience. In: C. H. Judd, editor, *Psychology of Secondary Education*. Oxford: Ginn & Company, 1927, pp. 414–441.

[57] M. C. Wicks. Radar the next generation – sensors as robots. In *2003 Proceedings of the International Conference on Radar (IEEE Cat. No.03EX695)*, pp. 8–14, 2003.

[58] M. Kaku. *The Future of the Mind: The Scientific Quest to Understand, Enhance, and Empower the Mind.* Jan. 2015. https://shp.shoppipubherenow.icu/0307473341.

[59] A. Frank. Dreaming in code. *The New York Times*, Mar. 2014. https://www.nytimes.com/2014/03/09/books/review/michio-kakus-future-of-the-mind.html.

[60] J. M. Schwartz, H. P. Stapp, and M. Beauregard. Quantum physics in neuroscience and psychology: a neurophysical model of mind-brain interaction. *Philosophical Transactions of the Royal Society B*, 360:1309–1327, 2005. https://www.nytimes.com/2014/03/09/books/review/michio-kakus-future-of-the-mind.html.

[61] M. P. A. Fisher. Quantum cognition: The possibility of processing with nuclear spins in the brain. *Annals of Physics*, 362:593–602, 2015.

[62] J. Ouellette. A new spin on the quantum brain. *Quanta Magazine*, Nov. 2016. https://www.quantamagazine.org/a-new-spin-on-the-quantum-brain-20161102/.

[63] B. Adams and F. Petruccione. Quantum effects in the brain: A review. *AVS Quantum Science*, 2(2):022901, 2020.

[64] J. O. Pfaffmann. *A Model of Microtubule Based Learning For Perception-Action Behavior Control*. PhD thesis, Wayne State University, 2003.

Chapter 19
Metacognitive radar

Kumar Vijay Mishra[1], Bhavani Shankar M.R.[2] and Björn Ottersten[2]

The key strength of cognitive radars is their ability to learn the channel or target environment and then adapt both the transmitter and the receiver to provide an enhanced performance [1,2]. On the other hand, a conventional radar optimizes only receive processing in response to changes in the target environment. During the past two decades, cognition in radar has matured from a conceptual stage [3–5] to the implementation in hardware [6–8]. Several diverse applications of cognitive radar have been suggested such as spectrum sharing [9], adaptable beampattern design [10], enhanced tracking [11,12], and resource allocation [13].

In general, a cognitive radar is designed to apply a single specific framework or algorithm to achieve its desired performance based on pre-determined criteria. Since radars perform a variety of tasks such as detection, estimation, and tracking, in practice, a single cognitive radar framework is insufficient to address changes in the system hardware and channel environment over long periods of time. For example, a radar that cognitively selects the beam direction to avoid jamming [14] may need a different cognitive strategy when the jammer is co-located with the target. Further, as the complexity of the radar system increases (e.g. use of multiple antennas and waveforms) and the channel conditions worsen (low signal-to-noise-ratio/SNR and presence of clutter), a single cognitive algorithm is unable to address the changing performance requirements. In such cases, a strategy to combine various cognitive frameworks in a *metacognitive radar* is highly desirable.

In its original concept, the cognitive radar drew upon the definition of cognition from neurobiology, wherein it is a process through which humans and animals sense and interact with their environment [1]. Similarly, metacognition is a well-studied concept in both neurobiology and educational psychology, often associated with the definition proposed by John Flavell [15]. It is formally defined as higher-order thinking that actively controls the cognitive processes engaged in learning [16]. This definition often summarizes metacognition as *learning about learning* or *knowing about knowing* [17,18]. At the heart of the metacognitive system lie four components:

[1]United States DEVCOM Army Research Laboratory, Adelphi, MD, USA
[2]Interdisciplinary Centre for Security, Reliability and Trust, University of Luxembourg, Luxembourg

acquiring *knowledge* about the environment; *monitoring* different cognition methods; a *strategy* to use the information obtained; and *transfer* the learned strategy to a new environment [16].

In wireless communications, some recent works have discussed applications of metacognition. In [19], a metacognitive radio was proposed in the context of efficient spectrum utilization by a cognitive communications system that constantly monitors and acquires the channel state information (CSI). Then, an appropriate cognitive method is selected from a suite of strategies such as genetic algorithms, reinforcement learning, or artificial neural networks. An example of this strategy is the precoder–decoder design for interference management which coordinates co-existing multiple transmitters such that their mutual interference aligns at the receivers and occupies only a portion of the signal space. In this context, the CSI is routinely sensed and estimated. Later, in [20], this concept of metacognitive radio was expanded to a general metacognitive engine which included addressing multiple applications. However, these works only incorporate the first three components of metacognition: acquisition, monitoring, and strategy selection. The transfer of learned knowledge to new scenarios, which is essential to demonstrate metacognition, was ignored in these studies.

So far, metacognition in radar remains relatively unexamined. In this work, we introduce the concept of metacognitive radar and explain its key features. Unlike the aforementioned previous works on metacognition in wireless communications, we include all four components of metacognition in our radar formulation. We illustrate this concept through some examples of resource selection and sharing. In our proposed configuration, learning-based methods are critical in enabling metacognition. In fact, the use of techniques such as deep learning to empower cognition is synonymous with the original definition of metacognition as *learning about learning*. Another key technique in this context is the application of control- and game-theoretic methods which are capable of modeling decision-making in an environment of conflict and cooperation between rational players [21]. For example, in [21], various games are modeled depending on the information available on the radar and jammer about each other. A metacognitive radar could model various performance objectives as different games and then select the most appropriate one for the situation. A metacognitive radar holds the promise of making cognitive radars more realistic and efficient by expanding their original sensing cycles.

19.1 Metacognitive concepts in radar

A conventional cognitive radar [22] may be viewed as a dynamic closed-loop system employing three steps (Figure 19.1). In the *sense* or *observe* stage, the radar gathers all the information from the target environment. Then, it *decides* or *learns* by applying some degree of intelligence which includes learning, planning, and decision-making methods. Finally, the radar *adapts* to the change in the target channel by reconfiguring the transmitter and receiver in order to be as flexible as possible to enhance performance. This constitutes a typical *cognitive cycle* and is also a common feature in cognitive radio systems. All three steps are performed cyclically. In a two-step model of cognitive radar, the sense and decide stages are grouped under *perception* and the adapt step as an *action*.

19.1.1 Metacognitive cycle

It is pertinent to remark on the importance of metacognition as separate from *plain* cognition. Since the cognitive cycle is a closed-loop system, there is no provision for altering any of the steps once the cycle has kicked in an operational system. This leads to an inherent inflexibility of the system to adapt to drastic changes in the channel conditions, changes in engineering modules, the operating objective, or all of these. Hence, the radar must include multiple strategies with their own cognitive cycles. The selection of the appropriate strategy is handled by metacognition.

Motivated by its psychological definitions [16], we define a metacognitive cycle with four key steps Figure 19.1. The *knowledge* acquisition step collates the assumptions inherent to all individual cognitive cycles to generate their decisions. For example, in a radar-communications spectrum-sharing scenario, the three cycles

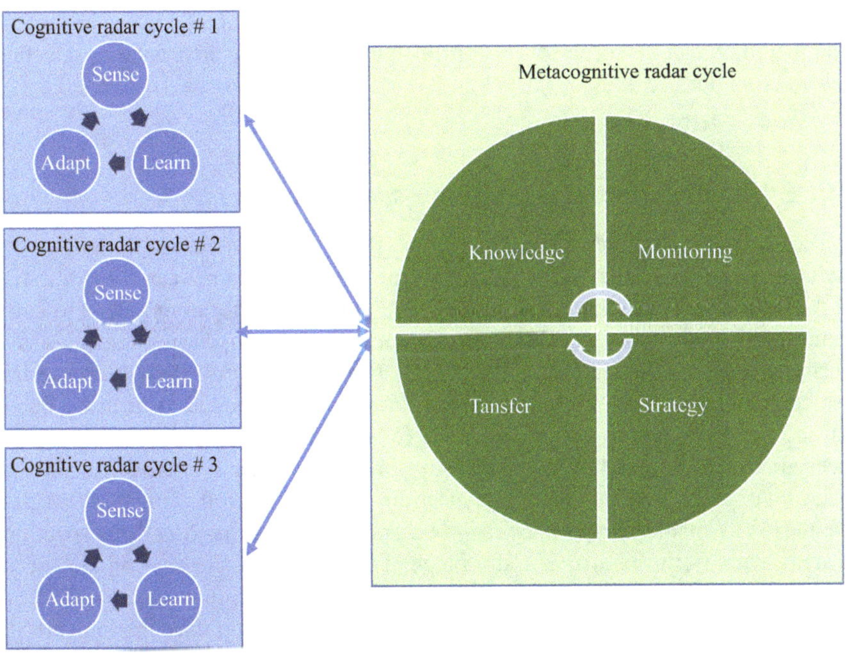

Figure 19.1 Metacognitive radar cycle consists of knowledge acquisition, monitoring, strategy or control, and transfer stages. A metacognitive engine monitors various cognitive strategies employed by the radar. Each of the cognitive strategies has its conventional sense–learn–adapt cycle with perception and action stages.

may each be tailored to the three different waveforms which only the metacognitive knowledge keeps a record of. In the *monitoring* step, the metacognitive radar evaluates the strengths and drawbacks of choosing a particular cycle. For example, choosing a reinforcement learning for spectrum sharing would require devising a policy and reward while a deep learning strategy would operate only with a large training set.

The monitoring step is viewed as information flowing from the cognitive cycle to the metacognitive cycle [23]. The reverse of this flow is the *strategy/control* stage of metacognition, wherein the radar applies a learning tool to select a specific cognitive cycle. For example, a deep learning engine may decide to choose between various strategies for spectrum sharing. As discussed in [20], the addition of this secondary, metacognitive cycle permits control and independent judgment of primary cognition behavior. Metacognition increases the confidence of the radar in its cognitive judgments. When the radar is deployed in slightly different environments that none of the cognitive cycles were prepared for, metacognition imparts the ability to apply knowledge accumulated in the first three steps to the new domain. This final stage of knowledge *transfer* completes the metacognitive cycle.

We now elaborate on some sample applications of a metacognitive radar. Our first application of spectrum sharing follows from prior works on metacognitive radio [19]. The second example is concerned with the allocation of resources such as power in a radar-communications scenario. The third example of metacognitive antenna selection follows from our previous works [13,24,25].

19.1.2 Applications: metacognitive spectrum sharing

As seen in metacognitive radio applications, the concept of metacognition is very useful for addressing the spectrum-sharing problem. Spectrum scarcity is a major current concern for the radio community and, in particular, for both radar and communications applications [26,27]. As a consequence, spectrum-sharing strategies have recently gained considerable attention [28,29]. Current standardization effort in wireless communications for finding new spectrum opportunities aims at using bands that are not dedicated to communications for communications use. The European Licence Shared Access (LSA) effort [30] aims at allowing a licensee, to use the spectrum of an incumbent who has a spectrum access right, following certain rules. The idea is to enable mobile telephony (4G and 5G) operators to use this spectrum in areas apart from the airport radio range. The LSA relies for that on the use of geo-location databases (GLDB) that are consulted before the connection of the LSA users. Licensed Shared Access is a big step forward to save cost once implemented in the operators' network and solve spectrum scarcity in the lower bands below 6 GHz. On the other hand, the US Citizen Band Radio System (CBRS) [26] initiative has been launched as a way to compensate for US delay compared to the LSA initiative. Then in order to bring some added value compared to LSA, US standardization effort consisted of adding an extended version of LSA to be used in the 3.5 GHz band (3,550–3,700 MHz).

A metacognitive spectrum-sharing radar in such a situation should operate as follows. There are various learning and non-learning cognitive approaches to spectrum sharing depending on the application, e.g., reinforcement learning at cognitive spectrum sharing for autonomous driving [31] and Xampling for surveillance cognitive radar-radio applications [9]. A metacognitive engine could choose between the two methods depending on the deployment of the radar in those situations. However, when the system is deployed in an application that none of the two techniques are developed for, the metacognitive radar exploits its knowledge and adapts the system to the new application.

19.1.3 Applications: metacognitive power allocation

Similar to spectrum sharing, power allocation between a radar and communications systems can also be modeled as a metacognitive radar problem [32]. For instance, let the radar and communications transmit power be P_R and P_C, respectively. Define the complex-Gaussian-distributed gains for the various discrete-time channel impulse responses with zero mean and variances σ_t^2, σ_i^2, σ_c^2, σ_f^2, and σ_r^2, respectively, as follows: $h_T \sim \mathscr{CN}(0, \sigma_t^2)$: radar transmitter to the target and back to the radar receiver; $h_I \sim \mathscr{CN}(0, \sigma_i^2)$: radar transmitter to clutter and back to the radar receiver; $h_C \sim \mathscr{CN}(0, \sigma_c^2)$: radar transmitter to the communications receiver; $h_F \sim \mathscr{CN}(0, \sigma_f^2)$: communications transmitter to the communications receiver; $h_R \sim \mathscr{CN}(0, \sigma_r^2)$: radar transmitter to the target and clutter and then to the communications receiver; and $w[n] \sim \mathscr{CN}(0, \sigma_w^2)$ and $v[n] \sim \mathscr{CN}(0, \sigma_v^2)$ denote the noise trails at the radar and communications receivers, respectively.

At the radar receiver, the signal-to-interference-plus-clutter-plus-noise-ratio (SICNR) is

$$\mathrm{SICNR}_R = \frac{\sigma_t^2 P_R}{\sigma_c^2 P_C + \sigma_i^2 P_R + \sigma_w^2} \tag{19.1}$$

The radar may also have a maximum power, maximum interference, and minimum SICNR constraints so that

$$0 \leq P_R \leq P_{R,\max}, \tag{19.2}$$

$$\sigma_c^2 P_C \leq T_{C,\max}, \tag{19.3}$$

$$\text{and} \quad \mathrm{SICNR}_R \geq \mathrm{SICNR}_{R,\min}, \tag{19.4}$$

where $P_{R,\max}$, $T_{C,\max}$, and $\mathrm{SICNR}_{R,\min}$ are pre-defined constants.

At the communications receiver, the signal-to-interference-plus-noise-ratio (SINR) is

$$\mathrm{SINR}_C = \frac{\sigma_f^2 P_C}{\sigma_r^2 P_R + \sigma_v^2} \tag{19.5}$$

The communications receiver may have the maximum power, maximum interference, and minimum SINR constraints as

$$0 \leq P_C \leq P_{C,\max}, \tag{19.6}$$

$$\sigma_r^2 P_R \leq T_{R,\max}, \tag{19.7}$$

$$\text{and} \quad \text{SINR}_C \geq \text{SINR}_{C,\min}, \tag{19.8}$$

where $P_{C,\max}$, $T_{R,\max}$, and $\text{SINR}_{C,\min}$ are pre-defined constants.

Let the game be the triplet $\mathscr{G} = \langle \mathscr{K}, \mathscr{S}, \mathscr{U} \rangle$ where $\mathscr{K}$ is the set of players with cardinality $|\mathscr{K}| = K$, $\mathscr{S} = S_1 \times \cdots \times S_K$ is the space comprising of strategies $\{S_i\}_{i=1}^K$ of all players, and $\mathscr{U} = \{u_1, \cdots, u_K\}$ is the set of utility functions of each player which maps their strategies to a real line, i.e., $u_i : S_i \rightarrow \mathbb{R}$, $i = 1, \cdots, K$. In our spectral coexistence problem, $|\mathscr{K}| = 2$ and $K = 1$ and 2 corresponds to radar and communications, respectively. Further $S_1 = [0, P_{R,\max}]$ and $S_2 = [0, P_{C,\max}]$.

The utility functions are given by the difference in payoff (maximization of SICNR) and the cost functions (minimization of power). As an example, the respective utility functions could be

$$u_1 = \ln(\text{SICNR}_R - \text{SICNR}_{R,\min}) - (\mu_1 \sigma_t^2 P_R + \gamma_1 \sigma_r^2 P_R) \tag{19.9}$$

$$\text{and } u_2 = \ln(\text{SINR}_C - \text{SINR}_{C,\min}) - (\mu_2 \sigma_f^2 P_C + \gamma_2 \sigma_c^2 P_C), \tag{19.10}$$

where μ_i and γ_i, $i = 1, 2$ are to be determined. The power allocation is determined by solving for the values of these utility functions. The metacognitive system models the interaction between the radar and communications as a cooperative or non-cooperative game depending on the information available and solves the resulting optimization problem for the desired resource allocation.

19.1.4 Applications: Metacognitive antenna selection

Sparse array selection is one of the common tasks performed by cognitive radars [13]. Larger arrays have a high associated cost, area, and computational load. To address this problem, a cognitive radar deploys a full array and then selects an optimal (sparse) subarray to transmit and receive the signals in response to changes in the target environment. This task is achieved through a variety of techniques such as optimization, greedy search, random array selection, and deep learning [13,33].

Briefly, the cognitive cycle in this problem operates as follows. For a fixed array geometry, assume a phased array employed by a radar that performs angle estimation using sparse recovery methods. During the very first scan, full array is active and the received signal from this scan is fed to the network. The cognitive radar goal is to find an optimal antenna array for the next few scans in which fewer antennas than the full array will be used. For deep learning, the radar employs a trained network that chooses an optimal array using the covariance of the received signal. The optimal array provides the lowest estimation error in the direction-of-arrival. The same criterion is used by the optimization and greedy search methods For random array selection, no such criterion is used. All four methods differ in their computation speeds. Thus, a metacognitive radar chooses between these different strategies depending on the available computing resources. When the array geometry is changed, a metacognitive radar will apply a technique for knowledge transfer. In the following section, we explain this last stage with a concrete example. For all other stages, we refer the reader to our prior works [13,33]. The example is based on our recent work on sparse array selection for arbitrary geometries [25].

19.2 Cognition masking

Recent cognition literature [34,35] envisages situations when the target itself may become cognitive. In this *inverse cognition* scenario, a target may be equipped with cognitive abilities that predict the actions of a cognitive radar trying to detect the target and guard against it. In general, this can be extended to any attacker–defender system. Another example is an interactive learning model, where the instructor aims to estimate the learner's knowledge and absorption of the material provided by the instructor.

In inverse cognition, it is imperative to first identify if the adversary is cognitive. To this end, the authors [34] developed stochastic revealed preferences-based algorithms to ascertain if the adversary optimizes a utility function and, if so, estimate that function. This framework may also be viewed as a generalization of inverse reinforcement learning (IRL), where the reward function associated with optimal behavior is learned passively [36]. The inverse cognition system, on the other hand, actively probes its adversarial system [34]. After detecting the cognitive system, the defender desires to estimate the information learned by the attacker. In [35], this problem was modeled as an inverse Bayesian filtering problem. A Bayesian filter provides a posterior distribution for an underlying state given its noisy observations. Its inverse filter reconstructs this posterior distribution given the actual state and noisy measurements of the posterior [35]. The model in [35] was a linear Gaussian state-space model where the attacker employed a Kalman filter (KF) to estimate the defender's state while the latter estimated the *former's estimate of the defender* using inverse Kalman filter (IKF). In practice, counter-adversarial systems are highly non-linear. Very recently, inverse cognition has been extended to non-linear system dynamics through several inverse systems including inverse extended Kalman filter (IEKF) [37,38], inverse unscented Kalman filter (IUKF) [39], and inverse cubature/quadrature filter (ICKF/IQKF) [40].

When the target is equipped with such cognitive abilities, the cognitive radar must simultaneously consider two actions: mask its cognitive abilities and continue to estimate the target parameters. The former objective has been considered metacognition [41] while the latter may be more appropriately termed *inverse–inverse cognition*. This brings us to the inevitable question of precisely defining a metacognitive radar. Unfortunately, no such definition exists. In the classical metacognitive application, several subsystems may have individual cognition cycles that are managed by a metacognitive engine. When the same cognitive task is evaluated via different performance metrics, then techniques such as multi-objective optimization may be used. When an *invasive* target is actively trying to determine the cognitive abilities of a radar (e.g. knowing which sector the radar is next scheduling its beam), then a metacognitive

Table 19.1 Summary of metacognitive radar concepts

Problem scenario	Architecture/strategy	Sample applications
Multiple cognitive cycles	Metacognitive engine	Spectrum sharing
Multiple performance metrics	Multi-objective optimization	Power allocation
Invasive target	Cognition masking	Beam scheduling
Cognitive target	Inverse–inverse cognition	Target tracking

approach could be to hide the radar's cognition. In an inversely cognitive target, the radar desires to update its inverse–inverse filters accordingly. Table 19.1 summarizes some of these metacognitive concepts and applications.

19.3 Example: antenna selection across geometries

Consider an M-element antenna array receiving a signal $s(t_i)$ from the direction $\Theta = (\theta, \phi)$ where θ and ϕ are the elevation and the azimuth angles of the source with respect to the antenna array. We assume that the received signal is narrowband and the source is in the far-field of the antenna array. Then, the output of the antenna array is given by

$$\mathbf{y}(t_i) = \mathbf{a}(\Theta)s(t_i) + \mathbf{n}(t_i), \quad 1 \leq i \leq T, \tag{19.11}$$

where T is the number of snapshots, $\mathbf{y}(t_i) = [y_1(t_i), \ldots, y_M(t_i)]^T$ and $y_m(t_i)$ denotes the output of the mth antenna for the ith snapshot. $\mathbf{n}(t_i) = [n_1(t_i), \ldots, n_M(t_i)]^T$ is the noise vector and $n_m(t_i)$ is additive white Gaussian noise (AWGN) with variance σ_n^2. $\mathbf{a}(\Theta) = [a_1(\Theta), \ldots, a_M(\Theta)]^T$ is the steering vector and $a_m(\Theta) = \exp\{-j\frac{2\pi}{\lambda}\mathbf{p}_m^T\mathbf{r}(\Theta)\}$ where $\mathbf{p}_m = [x_m, y_m, z_m]^T$ is the position of the mth antenna in a Cartesian coordinate system, $\mathbf{r}(\Theta) = [\cos(\phi)\sin(\theta), \sin(\phi)\sin(\theta), \cos(\theta)]^T$ and λ is the wavelength of the baseband signal.

19.3.1 Cognitive cycle

In the sparse array selection problem, we consider an M-element antenna array where K out of M are to be selected in the sense that the selected subarray provides the "best" DOA estimation performance. Then, we have $C = \binom{M}{K}$ possible subarray choices. Hence, it can be considered as a classification problem with C classes. We formulate the problem statement as: predicting the class which corresponds to the "best" subarray when the array output is given.

We consider the antenna selection problem in a deep learning context [13] and design a deep CNN to classify the input data of the network (the antenna array output) to select the "best" subarray for DOA estimation. We transfer the knowledge in the training data to the *target* domain for antenna selection with another array geometry. In the sequel, we first discuss the generation of the training data.

19.3.2 Knowledge transfer across different array geometries

We treat the antenna selection problem as a classification problem with C classes (i.e., C many subarray configurations). Each class is labeled with the positions of the antennas of the subarray corresponding to that class. Let $P_k^c = \{p_{x_k}^c, p_{y_k}^c, p_{z_k}^c\}$ be the set of antenna coordinates in the cth subarray for $k = 1, \ldots, K$. The positions of the antennas in the cth subarray constitute the set $\mathscr{S}_c = \{P_1^c, P_2^c, \ldots, P_K^c\}$, and the set of all classes are given as $\mathscr{S} = \{\mathscr{S}_1, \mathscr{S}_2, \ldots, \mathscr{S}_C\}$, which contains all the subarray configurations. In order to select the best subarrays in $\mathscr{S}$, we compute the CRB $\forall \mathscr{S}_c$. We define the absolute CRB for the direction Θ and $\mathscr{S}_c$ as

$$\kappa(\Theta, \mathscr{S}_c) = 1/2(\kappa(\theta, \mathscr{S}_c)^2 + \kappa(\phi, \mathscr{S}_c)^2)^2, \tag{19.12}$$

where $\kappa(\theta, \mathscr{S}_c)$ and $\kappa(\phi, \mathscr{S}_c)$ are the CRB terms for elevation and azimuth angles respectively [13,42]. The CRB for θ in a single source scenario is computed as follows [42]:

$$\kappa(\theta, \mathscr{S}_c) = \frac{\sigma_n^2}{2T\,\mathbb{R}\mathrm{e}\{(\dot{\mathbf{a}}_{c_\theta}^H[\mathbf{I}_K - \mathbf{a}_c\mathbf{a}_c^H/K]\dot{\mathbf{a}}_{c_\phi}) \odot (\sigma_s^4\mathbf{a}_c^H\mathbf{R}_c^{-1}\mathbf{a}_c)\}}, \tag{19.13}$$

and, for the azimuth angle, $\kappa(\phi, \mathscr{S}_c)$ as

$$\kappa(\phi, \mathscr{S}_c) = \frac{\sigma_n^2}{2T\,\mathbb{R}\mathrm{e}\{(\dot{\mathbf{a}}_{c_\phi}^H[\mathbf{I}_K - \mathbf{a}_c\mathbf{a}_c^H/K]\dot{\mathbf{a}}_{c_\theta}) \odot (\sigma_s^4\mathbf{a}_c^H\mathbf{R}_c^{-1}\mathbf{a}_c)\}}, \tag{19.14}$$

where $\odot$ is the dot product and $\mathbf{a}_c \in \mathbb{C}^K$ denotes the steering vector corresponding to the subarray with positions $\mathscr{S}_c$. $\mathbf{R}_c = 1/T\sum_{i=1}^{T}\mathbf{y}_c(t_i)\mathbf{y}_c^H(t_i)$ is the sample covariance matrix for the subarray output $\mathbf{y}_c(t_i) = \mathbf{a}_c s(t_i) + \mathbf{n}_c(t_i)$ and $\mathbf{I}_K$ is a $K \times K$ unit matrix. $\dot{\mathbf{a}}_{c_\theta} = \frac{\partial \mathbf{a}_c}{\partial \theta}$ and $\dot{\mathbf{a}}_{c_\phi} = \frac{\partial \mathbf{a}_c}{\partial \phi}$ are the partial derivatives of $\mathbf{a}_c$ with respect to θ and ϕ, respectively. σ_s^2 and σ_n^2 are the signal and noise variances. For simplicity, we select $\sigma_s^2 = 1$ and define the signal-to-noise ratio in the training data as $\mathrm{SNR}_{\mathrm{TRAIN}} = 10\log_{10}(\sigma_s^2/\sigma_n^2)$.

Among C subarray configurations, we found that the number of subarrays that provide the "best" DOA estimation performance (i.e., the lowest CRB) is much less than C. Hence, we labeled them in the set $\mathscr{B} = \{b_1, b_2, \ldots, b_{\bar{C}}\}$ where

$$b_{\bar{c}} = \arg\min_{c=1,\ldots,C} \kappa(\Theta, \mathscr{S}_c), \tag{19.15}$$

which implies that there are $\bar{C}$ subarrays among C that provides the lowest CRB. This is due to the configuration of the array structure [13].

The training dataset for the CNN structure is $\mathscr{D} = (\mathbf{X}, z)$, where $z \in \mathscr{B}$ denotes the label that is the best subarray index. $\mathbf{X} = \{\mathbf{X}_h\}_{h=1}^3$ is the input which is a three-channel data. The first channel, $\mathbf{X}_1 \in \mathbb{R}^{M \times M}$, contains the angle information of the array covariance matrix as $[\mathbf{X}_1]_{i,j} = \angle\{[\mathbf{R}]_{i,j}\}$ for the (i,j)th entry. The second and the third channels are the real and imaginary parts of the covariance matrix, respectively. In particular, $[\mathbf{X}_2]_{i,j} = \mathbb{R}\mathrm{e}\{[\mathbf{R}]_{i,j}\}$ and $[\mathbf{X}_3]_{i,j} = \mathbb{I}\mathrm{m}\{[\mathbf{R}]_{i,j}\}$.

We design a deep CNN structure which is composed of 14 layers. The first layer is the input layer which accepts the input of size $M \times M \times 3$. The $\{2, 4, 6, 8\}$th layers are convolutional layers each of which has 64 filter with the size of 2×2. In the

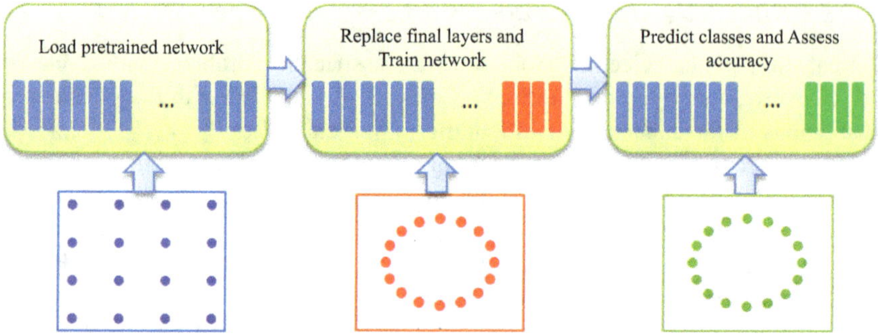

Figure 19.2 Transfer learning framework for knowledge transfer across URA and UCA geometries

10th and 12th layers, there are fully connected layers with 1,024 units whose 50% is randomly selected to avoid overfitting. After each convolutional and fully connected layers (in the $\{3, 5, 7, 9, 11, 13\}$th layers), there is a rectified linear unit (ReLU) layer where $\text{ReLU}(x) = \max(x, 0)$. In the last layer, there is a classification layer with $\bar{C}$ units where a softmax function is used to obtain the probability distribution of the classes. In order to train the network, the training data is collected for P directions and L realizations. Then, the network is realized in MATLAB® on a PC with 768-core GPU. A stochastic gradient decent algorithm is used to update the network parameters with the learning rate 0.01 and mini-batch size 500 for 50 epochs.

Once the training data is in the *source* domain, $\mathscr{D}_S$ is generated with the *source* array geometry. Then, it is used to train the network and we obtain CNN_S where the subscript stands for the *source* domain. The pre-trained network (i.e., CNN_S) is modified by replacing the classification layer with the one that is appropriate with *target* data labels. The weights in all of the convolutional layers are kept fixed to preserve the learned features in the *source* domain. We call the resulting network CNN_{TR} which is trained with the *target* training set, $\mathscr{D}_T$, generated with the *target* array geometry.

19.4 Numerical simulations

We first present the performance of the proposed CNN approach for the *source* domain case where different array geometries such as URA and UCA are considered with different settings. We collected data for $P_{\text{TRAIN}} = 100$ equally spaced directions in the interval $[0°, 359°]$ azimuth plane and for $L_{\text{TRAIN}} = 100$ realizations. $\text{SNR}_{\text{TRAIN}} = 20$ dB and $T_{\text{TRAIN}} = 100$ is selected. The network is tested for different SNR_{TEST} levels for $J_T = 100$ Monte-Carlo trials. We obtain above 90% validation accuracy for the training data in all cases as in [13]. This means that the proposed CNN structure accurately selects the antenna subarray in the "best" sense. In order to present the

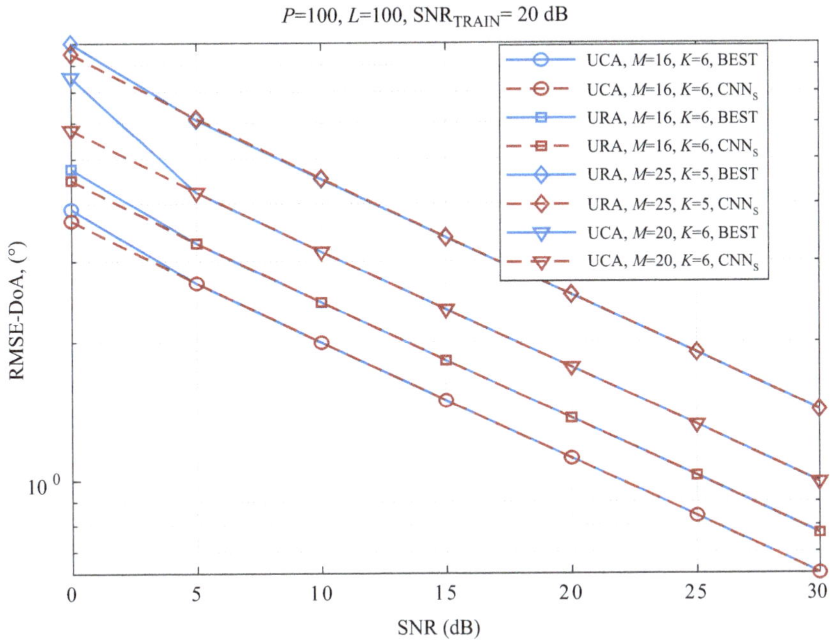

Figure 19.3 DOA estimation performance for the CNN_S for different array settings [25]

DOA estimation performance of the selected subarray, the MUSIC algorithm [43] is employed. In this case, we prepared a test data which is separately generated from the training data with different DOA angles which are selected uniformly randomly. In Figure 19.3, the DOA estimation performance for different arrays are given and compared with the "best" subarray performance which refers to the subarray that provides the lowest CRB value. As it is seen from the figure, our CNN approach performs well and it closely follows the best subarray performance.

To evaluate the performance of our CNN approach for transfer learning, *target* domain data is generated for UCA geometry with $M = 16, K = 6$ and element spacing is $\lambda/2$. We consider 1D scenario where the azimuth space is sampled with $P_T = 10$ directions and $L_T = 10$ for the *target* dataset. The *source* domain data is generated for URA geometry with $M = 16, K = 6$ with $\lambda/2$ element spacing. The transfer learning performance is investigated as the size of the *source* dataset varies. In Figure 19.4, P_S, the number of directions in the *source* dataset, varies as the other parameters are kept fixed as $L_S = 100$ and $\text{SNR}_{\text{TRAIN}} = 15$ dB. For each P_S, CNN_{TR} is generated and trained with *target* dataset. As seen from Figure 19.4, CNN_{TR} performs much better than CNN_{T} even if they are trained with the same data (i.e., *target* data). The performance of CNN_{TR} is attributed to the use of features from the *source* domain that are not available in the *target* domain. We obtain approximately 20% increase in the

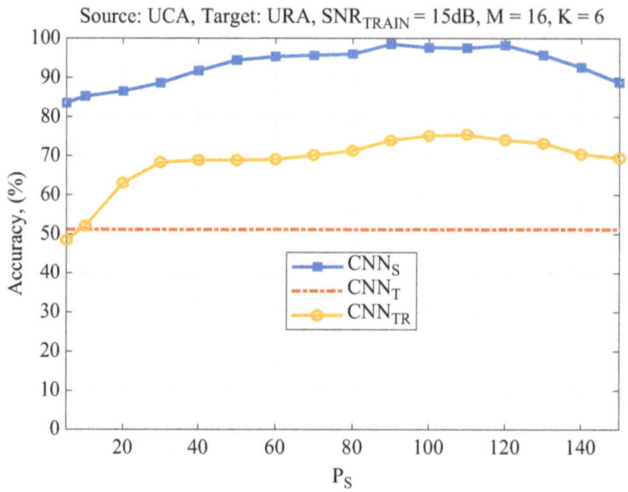

Figure 19.4 Performance of CNN_S, CNN_T, and CNN_{TR} vs P_S. $L_S = 100$ and $SNR_{TRAIN} = 15dB$ [25].

classification accuracy. We observe that transfer learning works if the *source* dataset is large so that more features can be transferred to *target* domain. While the accuracy of CNN_S is similar for different P_S values, it significantly affects the performance of CNN_{TR}. From Figure19.4, we see that $P_S = 100$ is a good choice for transfer learning where the *source* data is 100 times larger than *target* data. The performance of CNN_{TR} decreases after a certain P_S value since collecting more dense direction data leads to ambiguities in labeling.

19.5 Summary

We introduced the concept of metacognitive radar whose aim is to impart additional confidence to the decisions of a cognitive radar. This is achieved by collating several cognitive cycles in the system for various scenarios and selecting the most appropriate one for a given situation. We discussed key elements of a metacognitive radar and its possible applications. Finally, we illustrated a cognitive radar antenna selection by including the knowledge transfer step.

References

[1] Haykin S. Cognitive radar: A way of the future. *IEEE Signal Processing Magazine*. 2006;23(1):30–40.

[2] Guerci JR. *Cognitive Radar: The Knowledge-Aided Fully Adaptive Approach*. Boston, MA: Artech House; 2010.

[3] Mishra KV and Eldar YC. Performance of time delay estimation in a cognitive radar. In: *IEEE International Conference on Acoustics, Speech and Signal Processing*; 2017. p. 3141–3145.

[4] Mishra KV and Eldar YC. Sub-Nyquist radar: principles and prototypes. In: Maio AD, Eldar YC, Haimovich A, editors, *Compressed Sensing in Radar Signal Processing*. Cambridge: Cambridge University Press; 2019. p. 1–48.

[5] Martone AF, Ranney KI, Sherbondy K, *et al.* Spectrum allocation for noncooperative radar coexistence. *IEEE Transactions on Aerospace and Electronic Systems*. 2017;54(1):90–105.

[6] Slavik Z and Mishra KV. Cognitive interference mitigation in automotive radars. In: *IEEE Radar Conference*; 2019.

[7] Owen JW, Ravenscroft B, Kirk BH, *et al.* Experimental demonstration of cognitive spectrum sensing & notching for radar. In: *IEEE Radar Conference*; 2018. p. 0957–0962.

[8] Ravenscroft B, Owen JW, Jakabosky J, *et al.* Experimental demonstration and analysis of cognitive spectrum sensing and notching for radar. *IET Radar, Sonar & Navigation*. 2018;12(12):1466–1475.

[9] Cohen D, Mishra KV, and Eldar YC. Spectrum sharing radar: coexistence via Xampling. *IEEE Transactions on Aerospace Electronic Systems*. 2018 3;29:1279–1296.

[10] Alaee-Kerahroodi M, Mishra KV, Shankar MRB, *et al.* Discrete phase sequence design for coexistence of MIMO radar and MIMO communications. In: *IEEE International Workshop on Signal Processing Advances in Wireless Communications*; 2019. p. 1–5.

[11] Sharaga N, Tabrikian J, and Messer H. Optimal cognitive beamforming for target tracking in MIMO radar/sonar. *IEEE Journal of Selected Topics in Signal Processing*. 2015;9(8):1440–1450.

[12] Bell KL, Baker CJ, Smith GE, *et al.* Cognitive radar framework for target detection and tracking. *IEEE Journal of Selected Topics in Signal Processing*. 2015;9(8):1427–1439.

[13] Elbir AM, Mishra KV, and Eldar YC. Cognitive radar antenna selection via deep learning. *IET Radar, Sonar & Navigation*. 2019;13(6):871–880.

[14] Jiang X, Zhou F, Jian Y, *et al.* An optimal POMDP-based anti-jamming policy for cognitive radar. In: *IEEE Conference on Automation Science and Engineering*; 2017. p. 938–943.

[15] Flavell JH. Metacognition and cognitive monitoring: a new area of cognitive-developmental inquiry. *American Psychologist*. 1979;34(10):906.

[16] Livingston JA. *Metacognition: An Overview*. Educational Resources Information Center, U. S. Department of Education; 2003. TM034808.

[17] Metcalfe J and Shimamura AP. *Metacognition: Knowing About Knowing*. Cambridge, MA: MIT Press; 1994.

[18] Dunlosky J and Metcalfe J. *Metacognition*. Thousand Oaks, CA: Sage Publications; 2008.

[19] Asadi H, Volos H, Marefat MM, *et al.* Metacognitive radio engine design and standardization. *IEEE Journal on Selected Areas in Communications*. 2015;33(4):711–724.

[20] Asadi H, Volos H, Marefat MM, *et al.* Metacognition and the next generation of cognitive radio engines. *IEEE Communications Magazine*. 2016;54(1):76–82.

[21] Song X, Willett P, Zhou S, *et al.* The MIMO radar and jammer games. *IEEE Transactions on Signal Processing*. 2012;60(2):687–699.

[22] Mitola J. *Cognitive Radio: An Integrated Agent Architecture for Software Defined Radio*. Stockholm: Royal Institute of Technology; 2000.

[23] Achtziger A, Martiny SE, Oettingen G, *et al.* Metacognitive processes in the self-regulation of goal pursuit. *Social Metacognition*. 2012;p. 121–139.

[24] Elbir AM and Mishra KV. Joint Antenna Selection and Hybrid Beamformer Design using Unquantized and Quantized Deep Learning Networks. arXiv e-prints. 2019.

[25] Elbir AM and Mishra KV. Sparse array selection across arbitrary sensor geometries with deep transfer learning. *IEEE Transactions on Cognitive Communications and Networking*. 2020;7(1):255–264.

[26] Mishra KV, Bhavani Shankar MR, Koivunen V, *et al.* Toward millimeter wave joint radar communications: a signal processing perspective. *IEEE Signal Processing Magazine*. 2019;36:100–114.

[27] Ayyar A and Mishra KV. Robust communications-centric coexistence for turbo-coded OFDM with non-traditional radar interference models. In: *IEEE Radar Conference*; 2019. p. 1–6.

[28] Dokhanchi SH, Mysore BS, Mishra KV, *et al.* A mmWave automotive joint radar-communications system. *IEEE Transactions on Aerospace and Electronic Systems*. 2019;55:1241–1260.

[29] Duggal G, Vishwakarma S, Mishra KV, *et al.* Doppler-resilient 802.11ad-based ultra-short range automotive radar. arXiv preprint arXiv:190201306. 2019.

[30] Palola M, Matinmikko M, Prokkola J, *et al.* Live field trial of Licensed Shared Access (LSA) concept using LTE network in 2.3 GHz band. In: *IEEE International Symposium on Dynamic Spectrum Access Networks*; 2014. p. 38–47.

[31] Liu P, Liu Y, Huang T, *et al.* Cognitive Radar Using Reinforcement Learning in Automotive Applications. arXiv preprint arXiv:190410739. 2019.

[32] Mishra KV, Martone A, and Zaghloul AI. Power allocation games for overlaid radar and communications. In: *URSI Asia-Pacific Radio Science Conference (AP-RASC)*; 2019. p. 1–4.

[33] Elbir AM and Mishra KV. Joint Antenna Selection and Hybrid Beamformer Design using Unquantized and Quantized Deep Learning Networks. arXiv preprint arXiv:190503107. 2019.

[34] Krishnamurthy V, Angley D, Evans R, *et al.* Identifying cognitive radars – inverse reinforcement learning using revealed preferences. *IEEE Transactions on Signal Processing*. 2020;68:4529–4542.

[35] Krishnamurthy V and Rangaswamy M. How to calibrate your adversary's capabilities? Inverse filtering for counter-autonomous systems. *IEEE Transactions on Signal Processing*. 2019;67(24):6511–6525.

[36] Ng AY and Russell SJ. Algorithms for inverse reinforcement learning. In: *International Conference on Machine Learning*. vol. 1; 2000. p. 2.
[37] Singh H, Chattopadhyay A, and Mishra KV. Inverse extended Kalman filter—Part I: fundamentals. *IEEE Transactions on Signal Processing*. 2023;71:2936–2951.
[38] Singh H, Chattopadhyay A, and Mishra KV. Inverse extended Kalman filter—Part II: highly non-linear and uncertain systems. *IEEE Transactions on Signal Processing*. 2023;71:2936–2951.
[39] Singh H, Mishra KV, and Chattopadhyay A. Inverse Unscented Kalman Filter. arXiv preprint arXiv:230401698. 2023.
[40] Singh H, Mishra KV, and Chattopadhyay A. Inverse Cubature and Quadrature Kalman Filters. arXiv preprint arXiv:230310322. 2023.
[41] Pattanayak K, Krishnamurthy V, and Berry C. How can a cognitive radar mask its cognition? In: *IEEE International Conference on Acoustics, Speech and Signal Processing*; 2022. p. 5897–5901.
[42] Stoica P and Nehorai A. MUSIC, maximum likelihood, and Cramér–Rao bound: further results and comparisons. *IEEE Transactions on Acoustics, Speech, and Signal Processing*. 1990;38(12):2140–2150.
[43] Schmidt RO. Multiple emitter location and signal parameter estimation. *IEEE Transactions on Antennas and Propagation*. 1986;34(3):276–280.

Epilogue

With each year that passes, the current trends show a continual growth in the field of cognitive radar theory, modeling, processing, and applications. The US Department of Defense, NATO, and many industry units are actively engaged in developing cognitive radars while also fulfilling the requirements of the field-worthiness of such systems. As evidenced from the breadth and depth of each chapter, significant progress has been made in the past 5 years. However, many gaps still exist in our understanding of (a) methods to raise the radar's cognitive capabilities beyond Level 3 of Bloom's taxonomy, (b) effect of multiple cognitive cycles simultaneously operating in different subsystems, (c) theoretical guarantees of radar's cognitive behavior, (d) real-time deployability of existing solutions for cognition at and below Level 2, and (e) impart cognition to legacy field-tested systems, where modifying the older hardware or even software is impractical or expensive. In this context, new tools and techniques mentioned in this book offer insights into several of the features listed above. A substantial leap from theory to practice is still required to achieve completely cognitive radar system. Fortunately, steps are being taken in this direction leading cognitive radars out of their infancy.

With such significant advances, cognitive radars also bring bigger security concerns. Independent of any technical and algorithmic challenges, a cognitive radar by its very nature has remained limited in its applicability to many critical arenas. Since cognitive radars adapt and change their behavior based on observations of external environment, they are prone to manipulation by hostile entities either through cyber-hacking or intentional jamming. As a result, conventional radars are still preferred to operate deterministically because they offer reasonable resistance to electronic countermeasures. These are the key drivers for research to look beyond cognitive radar through novel concepts of cognition masking, inverse cognition, and metacognition.

Concomitant to cognition is the concept of autonomy. Unsurprisingly, recent civilian applications of cognitive radars are automotive sensors in self-driving cars, unmanned aerial vehicles, and multi-agent systems. These applications not only take cognitive radars closer to the Norbert Weiner's definition of "cybernetics" but also allow the application of stochastic control and group decision making to radars. As has been the case with other radar applications, deep learning has emerged as the preferred enabling technology for cognitive radars. The emerging area of integrated sensing and communications (ISAC) is also one of the most researched problems for cognitive wireless systems. Taken together, progress in all these areas offers a promising future to the cognitive radar community.

Today, it is very apparent that advances in cognitive radar are no longer limited to a conceptual level or to a small set of applications. Signal processing applied to cognitive radars has triggered new methods for waveform and filter design, signal retrieval, learning models, and beamforming algorithms. As many chapters from this book illustrate, there are also serious ongoing efforts to theoretically guarantee cognitive performance.

In essence, this book encapsulates the continuously and quickly evolving landscape of modern cognitive radars, including deployment in diverse scenarios; synergistic evolution of several new theories on cognitive sensing; and with applications spanning the from classical to quantum domains. We believe that these trends strongly point toward productive and useful future research in our cognitive radar community.

Kumar Vijay Mishra
Adelphi, MD, USA

Bhavani Shankar M.R.
University of Luxembourg

Muralidhar Rangaswamy
Wright-Patterson Air Force Base, OH, USA

Index

www.ingramcontent.com/pod-product-compliance
Lightning Source LLC
LaVergne TN
LVHW011537200926
848484LV00001B/274
* 9 7 8 1 8 3 9 5 3 4 7 4 4 *